Randomness and the Riemann Hypothesis

Probability, Statistics, and the Primes

Randomness and the Riemann Hypothesis

Probability, Statistics, and the Primes

Andrey Feuerverger
Greg Martin

AMERICAN MATHEMATICAL SOCIETY
Providence, Rhode Island

2020 *Mathematics Subject Classification.* Primary 11-01, 11-02; Secondary 11K65, 11M99, 11N05, 11N37, 11N99, 60B20, 62F99, 65C20.

Cover image of G. F. B. Riemann is a public domain image: `https://commons.wikimedia.org/wiki/File:Georg_Friedrich_Bernhard_Riemann.jpeg`.

For additional information and updates on this book, visit
www.ams.org/bookpages/mbk-156

Library of Congress Cataloging-in-Publication Data

Names: Feuerverger, Andrey author | Martin, Greg, 1970- author
Title: Randomness and the Riemann hypothesis : probability, statistics, and the primes / Andrey Feuerverger, Greg Martin.
Description: Providence, Rhode Island : American Mathematical Society, [2026] | Includes bibliographical references and index.
Identifiers: LCCN 2026007740 | ISBN 9781470483807 paperback | ISBN 9781470485726 ebook
Subjects: LCSH: Riemann hypothesis | Number theory | Probabilities | Mathematical statistics | AMS: Number theory – Instructional exposition (textbooks, tutorial papers, etc.) | Number theory – Research exposition (monographs, survey articles) | Number theory – Probabilistic theory: distribution modulo 1; metric theory of algorithms – Arithmetic functions | Number theory – Zeta and L-functions: analytic theory – None of the above, but in this section | Number theory – Multiplicative number theory – Distribution of primes | Number theory – Multiplicative number theory – Asymptotic results on arithmetic functions | Number theory – Multiplicative number theory – None of the above, but in this section | Probability theory and stochastic processes – Probability theory on algebraic and topological structures – Random matrices (probabilistic aspects) | Statistics – Parametric inference – None of the above, but in this section | Numerical analysis – Probabilistic methods, simulation and stochastic differential equations – Models, numerical methods
Classification: LCC QA246 .F484 2026
LC record available at https://lccn.loc.gov/2026007740

∞ The paper used in this book is acid-free and falls within the guidelines established to ensure permanence and durability.
Visit the AMS home page at `https://www.ams.org/`

10 9 8 7 6 5 4 3 2 1 31 30 29 28 27 26

For Grace

from whom I learned of angels

–A. F.

Contents

Preface

Only once before did I stand in a comparable awe before the mysterious harmony of thought and nature. That was when together with Erdős in the days of our youth we had found that primes play a game of chance.

Marek (Mark) Kac (1979, [**614**])

We have written this book with three closely related aims in mind.

First and foremost, we wanted to retell one of the most remarkable mathematical stories ever told. But we wanted to tell it through what has proven, time and again, to be a singularly fruitful lens, one characterized by notions of randomness and by probabilistic and statistical reasoning. This is a viewpoint that has led to enormous progress with spectacular successes along the way, as well as occasional failures which have in equal parts been as surprising as they are enlightening. It is a viewpoint that has served as an ofttimes reluctant backbone for the subject over hundreds of years and yet has guided and challenged the intuitions of what have unquestionably been some of the most able explorers who ever lived.

Secondly, we wanted to tell this story in a radically different way, and for it to be as self-contained and as accessible as possible to a broad class of readers who have some mathematical background. In particular, our hope is that such readers will find it more rewarding to follow a treatment such as this rather than one intended for a lay audience. But at the same time, we wanted to tell the story without pulling any of its punches; to tell it in a way that is authentic and, in certain respects, complete. Nevertheless, this book has not been written for experts (much as we would be gratified if some were to find it useful) and neither was it meant to be a textbook. As *first* envisioned, our expectation of the reader was to have just a little more than the maturity gained from a *rigorous* course in calculus—one that included the rudiments of real analysis—together with an adventurous spirit and a desire to expand horizons. However, as our project grew it became evident that more than this—in particular, having encountered some rigorous probability as well as some complex analysis—would be desirable as well. But since—initially, at least—we had never meant to *demand* this of the reader, we have made amends. These appear in the form of self-contained and carefully devised appendices that cover these topics (among others) in lively and meticulous detail. The appendices offer opportunities for extending and solidifying one's understanding of the material they cover, and constitute an integral part of this book. To be sure, readers with less background will encounter passages they may find challenging. But we encourage them to acknowledge and set such difficulties aside, and skip to other sections and other chapters. Such passages can be returned to over time after more background and experience have been acquired.

Our third aim was to pay homage to the central construct that has driven this subject for nearly two hundred years, namely the Riemann zeta function, and the seemingly impenetrable mysteries concerning its zeros. Therefore, as much as our perspective was meant to centre on notions of randomness and probability, we have nevertheless paid considerable attention to the zeta function throughout. Needless to say, however, the probabilistic aspects of this function have not been ignored.

This book is the result of what was at first meant to be a much smaller endeavour, a labour of love attempting to capture the grandeur of its subject matter. For one author this project was undertaken as a way to learn this remarkable subject and to faithfully record that journey; for the other it was a way to share some beautiful things acquired over many years (whilst holding the other author in check). But as this project progressed it grew, and it became clear to us that it could serve a broader purpose than originally intended. We realized that it could, on the one hand, become a work that a mathematically seasoned reader would enjoy reading over an extended period of time, purely for pleasure. But we realized too that it might serve as a flexible and uninhibiting resource for certain kinds of courses, one that would allow an adventurous instructor to plot their own flight path, augmenting the material with some of their own favourite topics, while offering students what we hope is a book they will *want* to read.

But there is more. Our book has turned out to include a considerable amount of detail, with many of its topics covered rather intensively. As a consequence, this book can also serve as a handy reference for a broad range of material, although more casual readers are, of course, free to decide which topics they wish to pursue in depth, and which to read more lightly. Furthermore, in our efforts to keep this book as self-contained as possible, we have included the previously mentioned carefully prepared appendices, most of which can—and should—be read attentively. These can be used for study, for review, for envisioning the big pictures, or to be savoured solely for the pleasure of their content. At the same time, the appendices also include useful material that, in many cases, is widely strewn across the literature. More about what this book contains and how it is organized is spelled out in Chapter 1; in particular, Section 1.2 provides guidance on the best ways to navigate the material.

We have avoided the temptation to compete with any of the splendid accounts that are now available of number theory, and of analytic or probabilistic number theory in particular. We have therefore refrained from attempting to reprove every last result from among those widely available elsewhere, although references for proofs that have been omitted are liberally provided. Nevertheless, for readers who are game, we have retained a judicious selection of proofs, particularly those that highlight the statistical and probabilistic aspects of the subject. Otherwise, we have aimed to be resolutely rigorous (but without being unduly pedantic) and to quote best, or close-to-best, results whenever possible, mindful of the value this can have when serving as a reference. We have selected our coverage to stay as close as possible to our main themes, namely probabilistic and statistical thinking about the primes, about the Riemann zeta function, and about other closely related aspects of the theory. Above all, we did aim "to write an interesting book, and one unlike other books", one that will be worthwhile as well as fun to read [**499**, Preface]. And if we may further quote Hardy & Wright—with no presumption intended!—"We may have succeeded at the price of too much eccentricity, or we may have failed; but we can hardly have failed completely, the subject matter being so attractive that only extravagant incompetence could make it dull" [**499**, Preface]. Needless to say, however, while we did intend for our scope to be broad, it could not be unlimited, and not every important topic could be covered.

And now, we invite the reader: *read on, and enjoy*!

CHAPTER 1

Introduction

With caution judge of probability. Things deemed unlikely, e'en impossible, experience oft hath proved to be true.

William Shakespeare [**294**, p. 242][1]

As a first and admittedly rough approximation, one might say that this book is an exploration of the connections between two topics: on the one hand, probability, statistics, and randomness; and on the other, analytic number theory—particularly the primes and the Riemann zeta function. Our focus on these connections (as well as on the two topics themselves) is the lens through which our subject will be viewed. Our intent has been to write a highly accessible book, one that will be of interest to a wide technical audience, and particularly to non-specialists who have a background in mathematics and who are keen to expand their horizons.

While the primes have, quite literally, been an object of study by the purest of mathematicians for millennia, number theory has, in more recent years, been vastly expanded in scope owing to its applications in the computer sciences, and particularly to cryptography and the security of networks and communications. Speaking of his work in this field, G. H. Hardy is famously quoted as having once claimed that he has never done anything 'useful': "No discovery of mine has made, or is likely to make, directly or indirectly, for good or ill, the least difference to the amenity of the world." An avowed pacifist, he added: "No one has yet discovered any warlike purpose to be served by the theory of numbers" (Hardy, 1967, [**491**, pp. 44, 49]). Ironically, cryptographic methods are now widely classified as munitions—with the security of nations sometimes at stake—and government agencies routinely demand that 'back doors' be installed in encryption devices. Behind many such applications lies the fact that primes are plentiful and can be produced efficiently, yet when large primes (of, say, several hundred decimal digits) are multiplied together, no practical algorithm to factor their product has ever been found.[2]

In focusing on connections between probability, statistics, and analytic number theory, one is especially interested in windows onto randomness (or more precisely, *pseudo*-randomness) which these connections may give rise to; indeed, the seemingly unpredictable behaviours which occur can potentially serve to deepen our understanding of the nature and the limitations of randomness itself. In this story,

[1] As commonly attributed, but did Shakespeare really say this? See the analysis in Section 1.7.

[2] Indeed, RSA Security LLC (currently owned by Dell Technologies) once offered a $200,000 USD prize to anyone who could factor a certain 2048 bit (617 digit) number. See, for example, Davenport (2008, [**263**, Chapter VIII]), Fine & Rosenberger (2016, [**381**, Chapter 5]), or Guy (1975, [**473**]). Wagstaff (2013, [**1095**]) provides an introduction to the general theory.

a starring role is played by the *zeta function*

$$\zeta(s) = \sum_{n=1}^{\infty} \frac{1}{n^s} \tag{1.1}$$

on account of its connection to the primes, as evidenced by the remarkable formula

$$\zeta(s) = \sum_{n=1}^{\infty} \frac{1}{n^s} = \prod_{\text{primes } p} \left(1 + \frac{1}{p^s} + \frac{1}{p^{2s}} + \cdots\right) = \prod_{p} \left(1 - \frac{1}{p^s}\right)^{-1}, \tag{1.2}$$

which is valid for all real-valued $s > 1$. This *product representation*, due to Euler, holds essentially because when the product over the primes is multiplied out, and the terms suitably rearranged, it reproduces $1/n^s$ exactly once for every integer n. That this is so is a straightforward consequence of the (not entirely trivial) theorem concerning the unique factorization of integers into products of primes, to which (1.2) is essentially equivalent.[3] The function $\zeta(s)$ thus encodes, in some sense, *all* of the available information about the primes. At the same time, however, the *statistical* behaviour of $\zeta(s)$ itself introduces new and remarkable forms of (pseudo-) randomness, above and beyond the apparent randomness which dominates the sequence of primes.

The literature on this subject is vast, and unparalleled for the depth and reach of its ideas. Most of its available results have been hard won, often through subtle, ingenious, and intricate analysis, with wide gaps of time having elapsed between its significant milestones. To be utterly convinced of this, one needs only to read the elegant and humbling introduction of Iwaniec and Kowalski (2004, [**594**]) who make reference to "great ideas ... camouflaged ... in arguments of enormous complexity." Hardy (1920, [**486**]) referred to it as "the most difficult of all branches of a difficult science" even though many of its "actual results are such as can be stated in a simple and striking form." In forging a path through such material we had to be selective, although we have tried hard to provide a reasonably complete picture at the cost of sometimes omitting details whose interest is primarily technical. In particular, when stating results, every effort has been made to state them in their best, or at least in their close-to-best possible forms, and to provide extensive and up-to-date references throughout. Even so, however, we have taken great pains to try to keep the exposition as accessible as possible to readers whose background in mathematics is not quite complete, but who have at least had some exposure to complex analysis and rigorous probability. We hope that readers having only minimal background (but other readers as well!) will find the carefully prepared appendices that have been included to be worthwhile and illuminating. *The appendices are meant to be an integral part of this book.*

Some remarks are in order concerning the degree of mathematical rigour adopted. Our intention has been to be fully rigorous, particularly in stating results, but at the same time to avoid burdening the text with the kind of technical detail that impedes the flow of the discussion. One such example has already occurred: In (1.2), the legitimacy of the manipulations involving the infinite product and rearranged summation has been glossed over. In this instance, it is not too difficult to supply the details, and a fully rigorous proof is given in Section 1.6. Another example occurs in Chapter 2 at one of the proofs of the infinitude of primes, where

[3] Of course, the geometric series result $1/(1-a) = 1 + a + a^2 + \cdots$ for $|a| < 1$ was used in (1.2) as well.

it is simply glossed over that π^2 is irrational (Legendre was the first to prove it). And a still further example—which arises in Appendix A—concerns the precise definitions for the interior and exterior of a closed curve in the complex plane. For details such as these, as well as for any proofs that have been omitted, the reader will be referred to sources that have been provided throughout.

Our primary concern will be not with the vast scope of number theory *per se*, but rather with the statistical and probabilistic properties appearing to arise within it. To bound the discussions, however, certain topics have had to be omitted. These include sieve theory; exponential (or trigonometric) sums and the circle method; as well as modular forms and elliptic curves. Dirichlet characters, L-functions, and primes in arithmetic progression (which also possess interesting 'random' behaviours) are doubtless the most lamentable among the many omissions, but a line did have to be drawn somewhere. Nevertheless, our bibliography is extensive, and as mentioned, citations have been liberally scattered throughout.

1.1. Summary of the chapters and appendices

Including this introduction, this book consists of twelve chapters and six appendices.

Chapter 2—our longest chapter—is devoted to the primes. We discuss what is known about their basic as well as many of their deeper properties. The Prime Number Theorem (PNT) and related results such as the Mertens theorems are treated in some detail, with emphasis on critically important asymptotic properties and their associated error terms. Among other topics, the Poisson character of the primes, as well as the nature of the gaps between them (both large and small), are also discussed along with a variety of open problems. Some historical notes are provided as well. In this chapter, as in the others, proofs are placed mainly at the end so as to not impede the flow of the discussion. For proofs that have not been included, references are provided as to where they may be found.

Chapter 3 takes time out to introduce arithmetic functions (also known as arithmetical functions) and to explore their general properties. Such functions, particularly those which are 'additive' or 'multiplicative', play essential roles in the theory. Many well-known arithmetic functions are then defined, and their key properties laid out in some detail. These discussions again emphasize the critically important asymptotic characteristics of the functions, as well as of their various averages. It is also highlighted that equivalents to the Prime Number Theorem, as well as to the Riemann hypothesis, can be expressed in terms of many of the functions described. This rather intensive chapter can serve as a reference that one can turn to whenever background is needed on any of the functions discussed there.

The material of Chapters 2 and 3 sets the stage for the discussions in Chapter 4 of statistical models for the primes, in particular the heuristic model introduced by Harald Cramér in the 1930s, as well as a model proposed by David Hawkins in the 1950s. Such models have furnished indispensable frameworks for formulating questions and conjectures about the behaviour of the sequence of primes. In reasoning from such models, and in developing their consequences, many standard (as well as not-so-standard) results of probability theory play an essential role. Among other topics, the surprising result of Helmut Maier which demonstrates the limitations of such models is discussed in detail. Nevertheless, in spite of such shortcomings, statistical models continue to this day to influence and to provide direction for explorations into properties of the primes and the quantities related to them.

The material of Chapters 5, 6, and 7 includes much that is classical, constituting the fundamental threads that hold our story together. Chapter 5 presents the theory of the Riemann zeta function—viewed as an analytic function over the complex plane—and discusses its connections to the primes. Coverage includes the analytic continuation and functional equation, as well as the general behaviour of this function, including its symmetries, together with some initial comments about its mysterious zeros. In Chapter 6 these zeros and their statistical properties are examined in considerably greater detail. Topics covered include spacings of the zeros, as well as discussions of zero-density results and zero-free regions. The pair correlation conjecture of Montgomery then leads into a discussion of random matrix ensembles (where some basic linear algebra is assumed); all available evidence suggests that the resulting eigenvalue distributions belong to a previously unrecognized 'universality class' to which the zeros of the zeta function also belong. Chapters 5 and 6 each end with a series of remarks on closely related topics, followed by selected proofs. Finally, in Chapter 7 we discuss the Lindelöf hypothesis, as well as the statistical and mean value properties of the zeta function on the so-called 'critical line'.

Chapter 8 is devoted to the celebrated Riemann hypothesis (RH). Many statements equivalent to RH are presented, and reasons for believing in RH, as well as for not believing in RH, are laid out in some detail. A few other difficult open problems are noted as well.

Chapter 9 is primarily concerned with what is commonly referred to as *probabilistic number theory*. There we discuss the Turán–Kubilius inequalities, and the theorems of Hardy and Ramanujan, Erdős and Kac, Delange, and Erdős and Wintner. A number of other topics, including Landau's Poisson distribution result for prime factor counts, the arcsine law for the divisor function, the Dickman–de Bruijn distribution, and the asymptotic distribution of prime factors are also considered briefly. Proofs for some of the key results are given as well.

Chapter 10 is closely aligned with classical statistical theory, and focuses on the natural exponential family of infinitely divisible distributions determined by the zeta function. Among other aspects, it deals (from a purely theoretical standpoint) with maximum likelihood estimation and likelihood ratio testing for this family. The associated generalized linear model together with its novel 'natural variance function' are developed as well. A connection between the Riemann hypothesis and the infinite divisibility of the zeta distributions is also addressed. Some knowledge of statistical theory is needed for approaching this material; however, this chapter can be read selectively by readers without this background as there is still much there that can be learned.

Finally, Chapters 11 and 12—our last two chapters—deal with an assortment of topics associated with number-theoretic probabilistic models. In particular, Chapter 11 deals with the properties of the random multiplicative functions known as the Radamacher and Steinhaus chaos models, while Chapter 12 deals with Kac's randomness model for additive arithmetic functions, a certain randomness model for the zeta function, and a randomized version of the Mertens function. The purpose of all of these models (at least originally) had been to shed light on, and to help build intuition for, understanding the Riemann and Lindelöf hypotheses. In Chapter 12, the so-called Chowla conjectures—which concern the correlational and other 'time series' properties of certain arithmetic function sequences—are discussed as well.

It should be mentioned that a limited number of repetitions have been allowed to occur throughout. These are intended as a convenience for the reader, to reduce the need to keep looking back, and to allow the discussions to be more fully self-contained. Two notable instances of this are associated with the content of Chapter 3. In our treatment of the various arithmetic functions in Chapter 3, equivalences to the Prime Number Theorem or to the Riemann hypothesis (which can be expressed in terms of those functions) are noted in each case. However, these PNT equivalences are repeated in a summary in Section 3.10, and the RH equivalences reappear in the discussions of Chapter 8.

A substantial number of illuminating and judiciously selected proofs are provided throughout and, as already mentioned, have mainly been placed at the end of each chapter. Due to the scope of our coverage, and to keep this book to a manageable size, not every assertion could be accompanied by a proof, and proofs that are unusually lengthy or exceedingly technical have been omitted. Nevertheless, references are invariably given to where such proofs may be found. (More casual readers may bypass proofs without much loss of continuity.) Simpler assertions that can sometimes be recognized as being straightforward exercises do occasionally occur. Yet this is not a textbook, and explicit exercises are few in number. Those that appear are usually fairly simple (though sometimes only after some reflection), and are intended only to keep the reader engaged with the material. Traditional exercise sets can be found in many standard textbooks.

Six carefully designed appendices have also been included, which the authors hope readers at *all* levels will find to be interesting as well as worthwhile. *In spite of being labelled 'appendices' they are nonetheless intended as an integral part of this book and are meant to hold equal status to the chapters*! Each of these appendices has been crafted to provide not only a convenient review of its subject matter for anyone who may have encountered it previously, but also an accessible and essentially self-contained overview for material that may be only partially familiar to the reader. The appendices also provide further details and reference material on many technical points relevant for the discussions. We believe that it will be found very worthwhile to move freely between the chapters of the book and the various appendices, as needs and interests arise.

Appendix A provides a detailed account of the ideas and main results from the elegant theory of complex functions of a complex variable. This is relevant, for example, to the discussion of analytic continuation in Chapter 5, and is occasionally alluded to elsewhere in the book. Appendix B describes the properties of some basic functions of a complex variable, with particular emphasis on the Gamma function which happens to be linked with the zeta function in rather surprising ways. Appendix C is devoted to properties of Dirichlet series, of which the zeta function is the most important example. Lambert series and Mellin transforms are considered there as well. Appendix D provides a comprehensive overview of the probability background needed to fully appreciate the key statistical limit laws. It includes not only the various Laws of Large Numbers and Central Limit Theorems, but also the zero-one laws and Laws of the Iterated Logarithm, as well as their generalizations and certain related topics. Probabilistic notions arise frequently when reasoning with heuristic models and constitute an undercurrent that runs throughout this book. Appendix E provides a detailed and handy compilation of results for sums and products involving primes and natural numbers, as well as their powers and their logarithms. Such sums and products are not always trivial, and yet arise frequently in both heuristic as well as in rigorous reasoning. Finally, Appendix F contains a summary of some standard and frequently used mathematical results. Readers who have had only modest exposure to the material of Appendix A and Appendix D are encouraged to read those two appendices conscientiously; they will be a revelation.

1.2. How to read this book

The unconventional arrangement of the contents of this book merits comment. We have chosen to fuse together themes that are interrelated so that each chapter of this book focuses primarily on one concrete aspect of the subject. This differs from more typical treatments in which the presentation is arranged in a sequential theorem–proof manner. Our rationale was to try to hold key topics together in coherent arrangements that were more aligned with our aims. In traditional expositions, developments are frequently interrupted in order to lay out the supplementary materials needed before one can proceed. One practical (if admittedly inadvertent) advantage of the approach we have adopted is that it may lend itself more conveniently for use as a reference. In fact, many of the topics we discuss are treated broadly and with considerable detail. Nevertheless, all needed material does get covered appropriately at some point in the book.

One way to read this book—given how it has been designed—is to treat it as an *immersion experience* to be savoured over an extended period of time. It is not a textbook, and we trust that no harm will befall a reader who chooses to gloss over the occasional segment whose details lie afield from their range of interests. The chapters *can be*, but they do not *have to be*, read sequentially. The reader is encouraged to skip ahead of as well as to look back from their then-current chapter whenever additional background is desired. Thus—as but one instance of where this might prove useful—passing references are made to both the zeta function and to the Riemann hypothesis in Chapters 2 and 3 (and elsewhere as well) before these topics are actually covered in Chapters 5 and 6. Likewise, the reading of any chapter may be interrupted whenever the need or interest arises to consult any of the appendices. Thus when reading Chapter 4 on heuristic models, for example, it may make sense to pause to review and read up on the elegant theory of probability—the subject of Appendix D. Likewise, when reading Chapter 5 on the zeta function, one may wish to pause to review the theory of functions of a complex variable—the subject of Appendix A. We believe that reading in this spirited way, as well as consulting external sources *à volonté*, will translate into an enlightening as well as an enjoyable experience. One alternate possibility—if we may dare to suggest it!—might be to read this book twice, with the second pass focusing on points that seemed obscure on the first. But whatever approach one adopts, our single most important goal has been to offer the reader an opportunity to reflect on the grandeur and the mysteries of the story we are striving to share.

On the understanding that it is to be taken together with its appendices we have sought to make this book substantially self-contained. For the diffident reader who prefers right at the get-go to have everything laid out in strict and orderly succession we offer the following advice: conscientiously read Appendices A, B, C, and D first, then read Chapters 5 and 6, and finally read the remaining chapters in sequence. But we do also urge any such reader to enjoy the forest, and not only its trees.

1.3. Further reading

Here are a few remarks about references and external sources. Owing to their fundamental and alluring nature, the subjects we cover have been the object of study for hundreds of years and it would be difficult to overstate the number of papers, books, and monographs devoted to them. These number in the tens if not hundreds of thousands of publications. Even our own (very limited) bibliography contains over one thousand entries, and yet hardly includes every item that is relevant or important (although a link to a more complete list of references is provided). Some of our references date back some one hundred years and more, but one must appreciate that in a subject as challenging as this a century is not an unduly long time. We encourage the interested reader to seek out and read as many external sources as they may wish.

A number of books and monographs may be regarded as being especially important, with many of these also being works of singular elegance. Among the historically important sources available in the English language, the versions of Hardy & Wright (2008, [**499**]) and Titchmarsh (1986, [**1059**])—both of which were updated by Roger Heath-Brown—are indispensable classics, while Ingham (1932, [**580**]), and Davenport (2000, [**262**]) must surely be counted amongst them. Among accessible treatments currently available, those by Apostol (1976, [**25**]), Ayoub (1963, [**46**]), Bach & Shallit (1996, [**54**]), Baker (1984, [**66**]), (2012, [**67**]), Bateman & Diamond (2004, [**82**]), Chandrasekharan (1968, [**188**]), de Koninck & Luca (2012, [**650**]), Mazur & Stein (2016, [**745**]), Miller & Takloo-Bighash (2006, [**756**]), Overholt (2015, [**828**]), Rose (1994, [**914**]), Stopple (2003, [**1034**]), and Tenenbaum & Mendès-France (2000, [**1055**])—a list that is *very far* from complete—can all be recommended. Somewhat more advanced are the texts by Edwards (1974, [**310**]), Ivić (1985, [**586**]), Patterson (1995, [**836**]), and Tenenbaum (2015, [**1053**]), while advanced volumes—each a veritable *tour de force*—are those of Iwaniec & Kowalski (2004, [**594**]), Karatsuba & Voronin (1992, [**622**]), and Montgomery & Vaughan (2006, [**774**]), to which we would also add Narkiewicz (2000, [**797**]) and Ivić (2013, [**590**]), the first on account of its extraordinary historical and mathematical detail, and the second on account of its sheer elegance. Among more recent publications, the volumes by Ellison & Mendès-France (1975, [**325**]), Andrew Granville (2025, [**457**]), Dimitris Koukoulopoulos (2019, [**658**]), and Emmanuel Kowalski (2021, [**661**]) are noteworthy. Titles of many other worthwhile books may be found in our bibliography and many of them will be referred to in our discussions.

The reader will find that we have made many references to papers appearing in academic journals. This is intended solely as a convenience, their primary purpose being to document assertions, and provide entry points to the literature for topics one might wish to pursue. It is regrettable that some of the papers are trapped behind paywalls their authors would never have purposely wished for. However, in many cases alternate versions can be found via the `arXiv`, `Google`, or other websites, while access to such papers—and especially to the `MathSciNet` resource—can often also be arranged through appropriate libraries. In the case of more recent publications, copies can sometimes be obtained from their authors directly. Let us also call attention the many entries relating to mathematical subjects on `Wikipedia`, and which, in recent years, have undergone steady development.

Finally, we would be remiss not to mention Edmund Landau's landmark *Handbuch* (1909, [**676**]), a historically important work which provided the first systematic treatment of the considerable progress made up till the time of its publication. The volume by Prachar (1957, [**866**]) must be mentioned here as well. Neither of these has been translated into English, but Landau's *Handbuch* has been reprinted in 1953 and 1974 in editions containing detailed appendices by P. T. Bateman in English providing highly informative summaries on the progress made in the intervening periods since 1909.

1.4. Notation, abbreviations, and limits

We provide now a few remarks concerning notation. The real and complex numbers will be denoted by $\mathbb{R}$ and $\mathbb{C}$, respectively, and the natural numbers by $\mathbb{N}$. A time-honoured tradition which dates to Riemann's 1859 memoir, and one we will follow religiously (other than in Appendix A), is to denote complex variables as $s = \sigma + it$ instead of the more familiar $z = x + iy$. A statement such as $\sigma > 1$ should then be interpreted to mean that $s \in \mathbb{C}$, so that it is complex-valued, but with real part $\Re(s) > 1$, while a statement such as $s > 1$ is meant to also imply that $s \in \mathbb{R}$ (or equivalently has imaginary part $\Im s = 0$) meaning that it is real-valued.

The letters n and p, throughout, with or without subscripts, will *invariably* refer to *natural numbers* $n = 1, 2, 3, \ldots$, and to *primes*, $p_1 = 2, p_2 = 3, p_3 = 5, \ldots$, respectively, particularly when they appear as arguments in sums or products. We will also use $\equiv$ to signify equalities that represent definitions, or that involve no more than some trivial rearrangement.

The definitions of standard arithmetic functions such as μ, φ, τ, σ, ω, Ω, λ, Λ, and so on are dealt with collectively in Chapter 3; other more localized notations will be introduced when needed. We refrain from the annoying practice of obliging readers to hunt down terms, acronyms and abbreviations that have appeared only many pages earlier. A few exceptions are: the abbreviation PNT which stands for the Prime Number Theorem, RH and LH which stand for the Riemann hypothesis and the Lindelöf hypothesis, respectively, and RMT which refers to random matrix theory. Aside from these, 'non-local' acronyms will usually be avoided.

Certain notation concerning limits will arise constantly throughout this book. Suppose that $f(x)$ is a real-valued function of a real-valued variable $x \in \mathbb{R}$. We expect that the reader will have encountered the notions of the *inferior* and *superior limits* of $f(x)$ when (for example) $x \to \infty$. These are defined, respectively, as

$$\liminf_{x\to\infty} f(x) \equiv \lim_{x\to\infty} \inf_{y>x} f(y) \quad \text{and} \quad \limsup_{x\to\infty} f(x) \equiv \lim_{x\to\infty} \sup_{y>x} f(y),$$

and represent the 'asymptotic' greatest lower bound and least upper bound of $f(x)$. If a and A represent the lim inf and lim sup values, respectively, and $\epsilon > 0$ is arbitrary, then $f(x)$ eventually stays above $a - \epsilon$ but keeps falling below $a + \epsilon$, and eventually stays below $A + \epsilon$ but keeps rising above $A - \epsilon$, as x increases. These values allow us to gauge the range of $f(x)$ in the $x \to \infty$ limit. For example, the sequence $(-1)^n\left(1+\frac{1}{n}\right)$, with $n = 1, 2, \ldots$, has $\limsup = 1$ and $\liminf = -1$ as $n \to \infty$. When $\liminf f(x)$ and $\limsup f(x)$ are equal, their common value is called simply the *limit* of $f(x)$ and is denoted by $\lim_{x\to\infty} f(x)$. Also prevalent in our subject is the notation $f(x) \sim g(x)$, typically used when $f(x)$ and $g(x)$ both increase without bound; it means that $\lim_{x\to\infty} f(x)/g(x) = 1$. The definitions given above all extend naturally to cases where x approaches a value other than ∞. More discussion concerning limits may be found in Rudin (1976, [**924**]), Stein & Shakarchi (2005, [**1023**]), (2003, [**1021**]), and indeed in many other textbooks.

1.5. Orders of magnitude

As will quickly become apparent, the accuracy of approximations is of vital importance throughout our subject and the so-called "little o" and "big O" notations will be used extensively. Let $f(x)$ and $g(x)$ be real functions of a real variable with $g(x) > 0$. Then "little o", as in $f(x) = o(g(x))$, means that $f(x)/g(x) \to 0$, while "big O", as in $f(x) = O(g(x))$, means that there is a constant $C > 0$ such that, eventually, $|f(x)| \le Cg(x)$. In these cases x is approaching some limit, either specified or understood, although most commonly $x \to \infty$ (or occasionally $x \to 0$). The "big O" concept can be expressed in different ways and the following three notations (which are used widely in the literature) are *equivalent* to one another:

$$f(x) = O(g(x)), \quad f(x) \ll g(x), \quad g(x) \gg f(x).$$

For example, smaller powers of common functions have smaller orders of magnitude, implying statements such as $x^{1/3} = o(x^{1/2})$ and $(\log x)^2 = o((\log x)^3)$ as $x \to \infty$. It is immediate from the definitions that $f(x) = o(g(x))$ implies $f(x) = O(g(x))$, so that $x^{1/3} = O(x^{1/2})$ and $(\log x)^2 = O((\log x)^3)$ are also true; it is also clear that $f(x) \le g(x)$ implies $f(x) = O(g(x))$ when f is a nonnegative function (although not necessarily if f can take negative values). Another important fact is that any (positive) power of x grows more quickly than any power of $\log x$. We state this more precisely as

$$(\log x)^C = o(x^\epsilon), \ \textit{as } x \to \infty,$$

for any fixed positive real numbers C (no matter how large) and ϵ (no matter how small); a proof of this fact has been included in Section 1.6.

There are also negated versions of the "little o" notation: we write $f(x) = \Omega(g(x))$ to mean that $f(x) \ne o(g(x))$. This will be the case if at least one of

$$\limsup_{x\to\infty} \frac{f(x)}{g(x)} > 0 \quad \text{and} \quad \liminf_{x\to\infty} \frac{f(x)}{g(x)} < 0$$

were to hold. One may represent these two conditions as $f(x) = \Omega_+(g(x))$ and $f(x) = \Omega_-(g(x))$, respectively, and use the notation $f(x) = \Omega_\pm(g(x))$ to mean that both must hold. For example, when $x \to \infty$, we have $x \sin x = \Omega_\pm(x)$, and $x(1+\sin x) = \Omega_+(x)$ (and thus $= \Omega(x)$ as well), but it is not $= \Omega_-(x)$.

Finally, the notation $f(x) \asymp g(x)$ means that $g(x) \ll f(x) \ll g(x)$, so that $f(x) \ll g(x)$ and $g(x) \ll f(x)$ both hold, or equivalently that $f(x)/g(x)$ is eventually bounded away from both 0 and ∞. One can express this as $cg(x) \le |f(x)| \le Cg(x)$ for some constants $0 < c \le C$ when x large enough, or as

$$0 < \liminf \frac{f(x)}{g(x)} \le \limsup \frac{f(x)}{g(x)} < \infty.$$

When $f(x) \asymp g(x)$ one says that f and g have the same *asymptotic order* (or that they are asymptotically of the same size).

Notations such as these are attributed mainly to Bachman, Hardy, Landau, and Vinogradov. However, we caution that in the literature these notations are sometimes given alternate definitions which are not all compatible with one another.

Results in our subject are often stated with complicated-looking compositions of exponential, power, and logarithm functions (such as those in (1.3)). To a newcomer, it can be hard at first to decide which such expressions grow faster or slower. To this end there is a **crucial lemma** well worth mastering:

If $f(x)$ and $g(x)$ both tend to ∞, and $\log f(x) = o(\log g(x))$, then $f(x) = o(g(x))$.

We prove this crucial lemma in Section 1.6, where we also provide solutions to the exercises below. They are meant to help readers calibrate themselves to the definitions and to the various helpful facts that have already been mentioned. (They can also be returned to later for further review.) In all parts of these exercises, assume that both $\lim_{x\to\infty} f(x) = \infty$ and $\lim_{x\to\infty} g(x) = \infty$.

EXERCISES 1.1.

(a) Show that $\exp\big(\sqrt{\log x}\,\big) = o(x)$, and that $x^{-2+\sin x} = \Omega(1/x)$, as $x \to \infty$.

(b) For any real numbers $a < b$, show that $f(x)^a = o\big(f(x)^b\big)$, as $x \to \infty$.

(c) Show that the converse of the above crucial lemma is false, by giving examples of functions $f(x)$ and $g(x)$, both tending to infinity, such that $f(x) = o(g(x))$, but $\log f(x) \neq o(\log g(x))$.

(d) Show that the big O version of the crucial lemma is also false, by giving examples of functions $f(x)$ and $g(x)$, both tending to infinity, such that $\log f(x) = O(\log g(x))$, but $f(x) \neq O(g(x))$.

(e) On the other hand, show that $\log f(x) = o(\log g(x))$ does imply that $f(x) = O(g(x))$. (This is an effective strategy in practice for proving $f(x) \ll g(x)$ when the functions involved have nested logarithms and exponentials.)

The tremendous importance attached to the accuracy of some of the approximations to be discussed can be exemplified by an admittedly crude analogy. To express this analogy we need to borrow some terms and concepts from Appendix D, but if these are unfamiliar to you this paragraph can just be skimmed on a first reading. Suppose then that $S_n \equiv \sum_{j=1}^{n} X_j$ is a sum of n independent and identically distributed random variables X_j with mean 0 and finite variance. Then when trying to understand the limiting behaviour of S_n as $n \to \infty$, it matters greatly if S_n is divided by n, by $\sqrt{n}$, or by $\sqrt{n \log\log n}$, to indicate but three possibilities. These choices correspond, respectively, to three foundational results in the theory of probability, namely the Law of Large Numbers, the Central Limit Theorem, and the Law of the Iterated Logarithm. (Details are spelled out in Appendix D.)

By the same token, if the function $\mathrm{li}(x)$ is meant to approximate the function $\pi(x)$ in a statement such as $\pi(x) = \mathrm{li}(x) + \mathrm{Error}$ (the functions $\pi(x)$ and $\mathrm{li}(x)$ will be defined in Chapter 2), then it turns out to matter greatly if the "Error" in this approximation is

$$o\left(\frac{x}{\log x}\right),\quad O\left(xe^{-C\sqrt{\log x}}\right),\quad O\left(xe^{-c(\log x)^{3/5}/(\log\log x)^{1/5}}\right),$$
$$O\left(x^{1/2+\epsilon}\right),\quad \text{or } O\left(\sqrt{x}\log x\right). \tag{1.3}$$

In expressions such as these, C and c invariably represent unspecified positive constants, and it is understood that ϵ may take on any positive value, however small.

EXERCISES 1.2 (Hint: use the crucial lemma!).

(a) Verify that the order terms shown in equation (1.3) represent successively stronger assertions (as $x \to \infty$).

(b) Show that $e^{C\sqrt{\log x}}$ approaches ∞ more slowly than any power of x (as $x \to \infty$), but approaches ∞ more quickly than any power of $\log x$. Furthermore, show that the same is also true of $e^{c(\log x)^{3/5}/(\log\log x)^{1/5}}$.

1.6. Selected proofs

PROOF OF THE EULER PRODUCT FORMULA (1.1). Here, we give a fully rigorous proof that the formula

$$\zeta(s) = \sum_{n=1}^{\infty}\frac{1}{n^s} = \prod_p\left(1+\frac{1}{p^s}+\frac{1}{p^{2s}}+\cdots\right) = \prod_p\left(1-p^{-s}\right)^{-1}$$

is valid not only for all real $s > 1$, but also for all complex s with $\Re s > 1$.

We first show that the series for $\zeta(s)$ converges in these cases. Represent the complex number $s \in \mathbb{C}$ as $s = \sigma + it$, where $\sigma, t \in \mathbb{R}$ are both real-valued, and observe that

$$\left|n^{-s}\right| = \left|e^{-(\sigma+it)\log n}\right| = e^{-\sigma\log n} = n^{-\sigma}.$$

We therefore have

$$\left|\sum_{n=1}^{\infty}\frac{1}{n^s}\right| \le \sum_{n=1}^{\infty}\left|\frac{1}{n^s}\right| = \sum_{n=1}^{\infty}\frac{1}{n^\sigma} < 1 + \int_1^{\infty}\frac{dx}{x^\sigma} = 1 + \frac{1}{\sigma-1} < \infty \quad \text{for } \sigma > 1.$$

In particular, if $\sigma > 1$, we will have $\sum_{n>k}|1/n^s| \to 0$ as $k \to \infty$; we will use this fact below.

Next, since all primes p exceed 1, the formula for geometric series gives

$$1\Big/\left(1-\frac{1}{p^s}\right) = 1 + \frac{1}{p^s} + \frac{1}{p^{2s}} + \frac{1}{p^{3s}} + \cdots.$$

We may then consider the partial products

$$\prod_{p\le k}\left(1-p^{-s}\right)^{-1} = \prod_{p\le k}\left(1+\frac{1}{p^s}+\frac{1}{p^{2s}}+\cdots\right) = \sum_{p|n \implies p\le k}\frac{1}{n^s} = \sum_{n=1}^{k}\frac{1}{n^s} + R. \tag{1.4}$$

Note that the middle equality holds because each n being summed over has a unique factorization into powers of the primes $p \le k$ and thus arises from choosing the appropriate terms from each of the factors in the second product. The remainder R here is a sum of the form $\sum 1/n^s$ in which all $n > k$ (although it excludes those n having prime factors exceeding k); hence $|R|$ is bounded by $\sum_{n>k}|1/n^s|$. The result follows on taking the $k \to \infty$ limit on both sides of (1.4). □

Many proofs of the Euler product formula can be found in the literature. The one in Stein & Shakarchi (2003, [**1021**, pp. 168–169, 182–184]) is particularly elegant.

PROOF THAT $(\log x)^C = o(x^\epsilon)$ AS $x \to \infty$. What we need to prove here is that $\lim_{x\to\infty}(\log x)^C/x^\epsilon = 0$ for all positive real numbers C and ϵ. First, the special case $C = \epsilon = 1$ can be handled using l'Hôpital's rule (which is valid since we have an ∞/∞ indeterminate form):

$$\lim_{x\to\infty} \frac{\log x}{x} = \lim_{x\to\infty} \frac{\frac{d}{dx}\log x}{\frac{d}{dx}x} = \lim_{x\to\infty} \frac{1/x}{1} = 0.$$

It turns out that this special case is all we need if we use a little trickery. First, for general ϵ, write $(\log x)/x^\epsilon = \frac{1}{\epsilon}(\log x^\epsilon)/x^\epsilon$; then since $u = x^\epsilon$ tends to infinity as $x \to \infty$, the limit $\lim_{x\to\infty}(\log x^\epsilon)/x^\epsilon$ equals $\lim_{u\to\infty}(\log u)/u = 0$. Finally, for general C, write $(\log x)^C/x^\epsilon = \big((\log x)/x^\delta\big)^C$ with $\delta = \epsilon/C$; but we have just shown that $(\log x)/x^\delta \to 0$ regardless of the value of δ, so the same is true of its Cth power. □

PROOF OF THE CRUCIAL LEMMA. Let $f(x)$ and $g(x)$ be functions that both tend to infinity as $x \to \infty$, and assume that $\log f(x) = o(\log g(x))$, so that $\lim_{x\to\infty}(\log f(x))/(\log g(x)) = 0$. Then in particular, $(\log f(x))/(\log g(x)) \le \frac{1}{2}$ when x is sufficiently large. Next, exponentiating both sides of $\log f(x) \le \frac{1}{2}\log g(x)$ shows that $f(x) \le g(x)^{1/2}$, and so $f(x)/g(x) \le 1/g(x)^{1/2}$. Since $g(x) \to \infty$, we know that $1/g(x)^{1/2} \to 0$, and therefore $f(x)/g(x) \to 0$ which proves that $f(x) = o(g(x))$. □

SOLUTIONS TO EXERCISES 1.1 AND 1.2.

(a) Write $f(x) = \exp\big(\sqrt{\log x}\big)$ and $g(x) = x$; these both tend to infinity as $x \to \infty$. We have $\log f(x) = \sqrt{\log x}$ and $\log g(x) = \log x$, so that $(\log f(x))/(\log g(x)) = 1/\sqrt{\log x}$ which tends to 0 as $x \to \infty$. Therefore $\log f(x) = o(\log g(x))$, and therefore $f(x) = o(g(x))$ by the crucial lemma. As for the second example, we have $x^{-2+\sin x}/\frac{1}{x} = x^{-1+\sin x}$, which equals 1 whenever $\sin x = 1$, that is, whenever x is of the form $(2k+\frac{1}{2})\pi$. Since there are numbers of this form larger than any bound, the definition of lim sup implies that $\limsup_{x\to\infty} x^{-2+\sin x}/\frac{1}{x} \ge 1$, which confirms the assertion $x^{-2+\sin x} = \Omega(\frac{1}{x})$.

(b) Let $f(x) \to \infty$ as $x \to \infty$, and let $a < b$ be real numbers. Then $f(x)^a/f(x)^b = 1/f(x)^{b-a}$ where $b - a$ is positive; thus $f(x)^{b-a} \to \infty$ still and so $1/f(x)^{b-a} \to 0$, confirming the assertion $f(x)^a = o\big(f(x)^b\big)$.

(c) If we set $f(x) = x$ and $g(x) = x^2$, then certainly $f(x) = o(g(x))$; however, after taking logarithms, $(\log f(x))/(\log g(x)) = (\log x)/(\log x^2) = \frac{1}{2}$ and this certainly does not tend to 0, so $\log f(x) \ne o(\log g(x))$.

(d) Similar to the previous exercise, take $f(x) = x^2$ and $g(x) = x$. Then $\log f(x) = 2\log x = O(\log x) = O(\log g(x))$, but $f(x) = x^2 \ne o(x) = o(g(x))$.

(e) This is just a combination of two implications: $\log f(x) = o(\log g(x))$ implies $f(x) = o(g(x))$ by the crucial lemma, while $f(x) = o(g(x))$ implies $f(x) = O(g(x))$ directly from their definitions. □

1.7. William Shakespeare and probability ???

The quote that heads this chapter is frequently attributed to Shakespeare and reads like one he could well have written. It appears (verbatim) in the 2016 book "4044 William Shakespeare Quotes" by A. A. Douglas. Travelling back in time leads to versions that appeared in two books by E.D.E.N. Southworth: *The Hidden Hand* (1859), and *India: The Pearl of Pearl River* (1856). Southworth may have obtained the quote from the remarkable volume *Aphorisms from Shakespeare Arranged According to the Plays...* by Capel Lofft (1812). It appears there—in the section on *The Taming of the Shrew*—in the form:

> With caution judge of Possibility,
> Things *thought* unlikely, e'en impossible,
> Experience often shews us to be true.'

However, in that play by Shakespeare there is no more than:

> TRANIO: I pray, sir, tell me, is it possible
> That love should of a sudden take such hold?
> LUCENTIO: O Tranio, till I found it to be true,
> I never thought it possible or likely. [**980**]

(Nothing remotely comparable to this appears in any of Shakespeare's other known writings.)

Beyond this, it is not clear from where Lofft got his quote. In a footnote to it, he refers to Aristotle and Boileau. Searching Nicolas Boileau's *L'art Poétique* (1740) one finds (in the third of its four 'chants') "Le vrai peut quelquefois n'être pas vraisemblable [The truth may sometimes not be probable]." As for Aristotle, in §XVIII of his *Poetics*, he quotes Agathon (whose works have been lost) saying: "Such an event is probable in Agathon's sense of the word: 'it is probable,' he says, 'that many things should happen contrary to probability.'"

Conclusion: While he certainly *knew* it—and had even *penned* words to the same effect—the quote is not in the *exact* form that Shakespeare had written.

Note: The ??? in the heading for this segment on Shakespeare was shamelessly adapted from a wonderful article by Perci Diaconis (2002, [**278**]) titled *G. H. Hardy and probability ???*.

CHAPTER 2

The Prime Numbers

I have never stopped thinking about the theory of numbers. I will give you some idea of my preoccupation with this area of research by confessing to you that even without any hope of success, I still prefer it to other work which might interest me while I think about it, and which is sure to yield results.

Sophie Germain [**412**][1]

All of the statistical behaviours we will encounter in this book stem—whether directly or indirectly—from the seemingly random nature of the primes. Upon examining them it is hard not to get "the feeling of being in the presence of one of the inexplicable secrets of creation."[2] We therefore begin our explorations with an in-depth look at the prime numbers and the properties that can be expressed explicitly in terms of them. The material in this chapter is arranged in topic order—not in historical or mathematical sequence. But although we have chosen to keep like results together, some historical notes (as well as a selection of proofs) will be found towards the end of the chapter.

There will, however, be two conspicuous abuses to what has just been stated. The first of these is that on a few occasions we will make reference to the Riemann zeta function as well as to its so-called 'nontrivial zeros', even though this material will not be treated until Chapters 5 and 6. There are several ways to deal with this, and one option is to simply skim over such references and return to them after having read those other chapters. We are, after all, dealing with an immense subject, and there is much to be said for savouring its components, at times separately, and at times all together.

A second abuse that occurs—although mainly in just this chapter—is the approach we have adopted for dealing with error terms. As had been pointed out in Chapter 1, the quality of certain approximations—and in particular the error terms corresponding to them—lie at the heart of the subject and turn out to be connected with questions that are as intriguing as they are difficult. To take just one key example, one is concerned with approximating—as accurately as possible—the number of primes up to some value, say, x; and for any such approximation, one wants to know just how accurate it is in the $x \to \infty$ limit. It turns out that for all available approximations the exact order of this error is not known; it depends on the location of the zeros of the zeta function for which only limited information is available. As more is discovered about its zeros, more can be said about the error of these approximations. Because it simplifies the discussion to not constantly base

[1] Translation courtesy of David Pengelley. For more on Sophie Germain, see footnote 19 in Section 2.7, and for an account of her work in number theory, see [**680**].

[2] Don Zagier (1977, [**1122**, p. 8]).

it on the best available error terms, statements of best known results will typically be postponed and then collected for reference in Section 2.11.

With these preliminaries out of the way, let us begin.

2.1. The distribution of the primes

The *prime numbers* are those integers $p > 1$ which have no positive divisors other than 1 and p. They consist of $2, 3, 5, 7, 11, \ldots$, and we often denote them by $p_1, p_2, p_3, \ldots$. No practical formula is available for p_n, even though primes are the fundamental building blocks of the natural numbers in the sense that (except for order) every number n factors into a product of powers of distinct primes in exactly one way: $n = p_1^{k_1} p_2^{k_2} \ldots p_m^{k_m}$, say. This elementary (but nontrivial) fact is known as the *fundamental theorem of arithmetic* or the *unique factorization theorem*, and is proved in many textbooks. The earliest method for enumerating the primes is the well-known method of elimination commonly known as the *sieve of Eratosthenes*—the same Eratosthenes of Cyrene (276–194 BCE) who had served as a director of the Great Library at Alexandria, and who is known for having been the first person to accurately estimate the circumference of the Earth.

There are many elementary ways to show that there are infinitely many primes:

- The proof of Euclid of Alexandria, circa 300 BCE: For if not, then adding 1 to their product would produce a number which must have a new prime factor. □
- Another succinct proof: For any N, the smallest prime divisor of $N! + 1$ must exceed N. □
- A proof due to Kummer: If N were the product of all the primes, then N and $N+1$ must have a common prime divisor; this divisor must then also divide their difference which is $(N+1) - N = 1$. □
- In 1737 Euler showed that the infinitude of primes can also be deduced from (1.2) as follows: If their number were finite, then as $s \downarrow 1$ the product over primes there would remain finite, yet the sum $\zeta(s) \to \infty$. □
- A different proof, also based on (1.2): If the number of primes were finite, the value of the product when $s = 2$ would be rational, but Euler showed (in 1741) that $\zeta(2) \equiv \sum 1/n^2 = \pi^2/6$. □

Many other proofs for the infinitude of primes are known; see, for instance, Aigner & Ziegler (2014, [**9**]), Elsholtz (2021, [**329**]), Granville (2017, [**456**]), Meštrović (2018, [**754**]), Narkiewicz (2000, [**797**, Chapter 1]), and Ribenboim (1996, [**901**]), to name only a few.

It is, however, possible to specify the number of primes more precisely. In particular, there have been many results over the years that limit how far one has to go after seeing one prime before the next prime appears.

Upper bounds for gaps between primes. First, for all $n > 3$, there is always a prime p such that $n < p < 2n - 2$, a result known as *Bertrand's postulate*. It was conjectured by Joseph Bertrand in 1845 and was first rigorously proved by Chebyshev[3] in 1850, using a deduction we will give later in this chapter. Subsequently, Pál (or Paul) Erdős (1932, [**332**]) found a proof which is simpler than the one given by Chebyshev, and showed that between every number greater than 7 and its double there is always at least one prime that has a remainder of 1 when

[3]Alternately spelled Tchebyshev. Chebyshev—who himself used more than one version—has the distinction of a name that possesses many transliterations; see Narkiewicz (2000, [**797**, p. 13]).

divided by 4, and at least one that has a remainder of 3. (These, of course, are the only remainders possible for primes other than 2.) An immediate consequence of Bertrand's postulate is that consecutive primes satisfy

$$p_{n+1} \; < \; 2p_n \quad \text{or, equivalently,} \quad p_{n+1} - p_n \; < \; p_n,$$

so that every prime is less than twice its predecessor. Other results along such lines include

$$p_{n+2} < p_{n+1} + p_n, \quad p_{m+n} > p_m + p_n, \quad \text{and} \quad p_{mn} < p_m p_n,$$

the first two holding for $m, n \geq 2$, $(m, n) \neq (3, 4)$, $(4, 3)$, or $(4, 4)$.[4] One further consequence of Bertrand's postulate is that the sequence of primes—but with 1 included—is 'complete' in the sense that every n can be expressed as a *sum* of elements in this sequence (although not uniquely) using each one once at most. Proofs of Bertrand's postulate may be found in Hardy & Wright (2008, [**499**, p. 455]), as well as in de Koninck & Luca (2012, [**650**, Chapter 2, §7]), Overholt (2015, [**650**, Chapter 1, §2]), Rose (1994, [**914**]), and many other texts. Trost (1953, [**1066**]) is a noteworthy classic.

Next, for *any* $\epsilon > 0$, and provided n is large enough, there will always be a prime between n and $(1+\epsilon)n$. This is a consequence of the Prime Number Theorem which will be discussed further below. More incisively, Ingham (1937, [**581**]) proved that there is always a prime in $(n, n + n^{5/8+\epsilon})$, while Baker, Harman & Pintz (2001, [**70**]) proved that there is always a prime in $[n, n + n^{\theta}]$, or equivalently that $p_{n+1} \; - \; p_n \; \leq \; p_n^{\theta}$, for $\theta = 21/40 = 0.525$, all provided that n is large enough. These are very much deeper results. Under the Riemann hypothesis (RH, discussed in Chapters 5 and 6), any $\theta > \frac{1}{2}$ would do.

In fact, by assuming the Riemann hypothesis, far stronger results can be established. For example, Cramér (1920, [**242**]) showed (under RH) that there is a positive constant C such that

$$p_{n+1} - p_n \; \leq \; C\sqrt{p_n} \, \log p_n, \tag{2.1}$$

while Dudek et al. (2016, [**298**]), and Carnero et al. (2019, [**182**]) obtained the *explicit* versions

$$\limsup_{n\to\infty} \frac{p_{n+1} - p_n}{\sqrt{p_n} \, \log p_n} < \frac{21}{25}; \quad \text{and} \quad \frac{p_{n+1} - p_n}{\sqrt{p_n} \, \log p_n} \leq \frac{22}{25}, \quad \text{for } \; p_n > 3,$$

both under RH. Further, by assuming Montgomery's pair correlation conjecture (discussed in Chapter 6) *together* with the Riemann hypothesis, Heath-Brown (1982, [**529**]), and Heath-Brown & Goldston (1984, [**531**]) showed that one would then even have

$$\frac{p_{n+1} - p_n}{\sqrt{p_n} \log p_n} \to 0 \;\; \text{or, equivalently,} \;\; p_n - p_{n-1} = o\left(\sqrt{p_n \log p_n}\right).$$

And if an 1830 conjecture of Legendre held, about there always being a prime between consecutive squares, then we would further have $p_n - p_{n-1} \leq C\sqrt{p_n}$ for some $C > 0$.

---

[4]References for such results include Cohen (2022, [**212**]), Hansen (1972, [**484**]), Ishikawa (1934, [**585**]), Mitrinović & Popadić (1978, [**758**]), Nagura (1952, [**787**]), and Panaitopol (1998, [**830**]).

Smaller intervals are another matter. It is, for example, still unknown if there must eventually always be a prime between n^2 and $n^2 + n$, or even—as per Legendre's conjecture—between n^2 and $(n + 1)^2$. However, weaker versions of this conjecture have been established: for example, Jingrun Chen (1975, [**195**]) proved that for large enough integers there is always a prime, or a product of two primes, between consecutive squares, while Ingham (1937, [**581**]) had proved that there must always be a prime between n^3 and $(n + 1)^3$, once n is sufficiently large. Indeed, for the case of consecutive cubes, Dudek (2016, [**297**]) showed that $n \geq e^{e^{33.3}}$ would suffice. The result of Baker, Harman & Pintz mentioned above establishes that there is eventually always a prime between n^θ and $(n+1)^\theta$ for any $\theta > 40/19$; we demonstrate this 'corollary' of their result in Section 2.13 (Selected proofs).

As it happens, these results fall far short of what is widely believed—namely, that for any $\epsilon > 0$, every interval $[x, x + x^\epsilon]$ contains a prime once x is large enough. Furthermore, based on a probabilistic heuristic (which will be discussed in Chapter 4) Cramér (1936, [**245**]) had even conjectured that for some $C > 0$,

$$p_{n+1} - p_n \leq C \log^2 p_n.$$

The question of the *best* possible bound of the type $p_{n+1} - p_n \leq C p_n^\alpha \log^\beta p_n$ is thus completely open, although later in this chapter we will describe some results that limit how good such a bound can be.

Note that in statements like those above, the unspecified constants satisfy $C > 0$ and need not all be the same; likewise an ϵ invariably represents a constant $\epsilon > 0$ which is allowed to be arbitrarily small.

We have now seen several results that limit how large the gaps between consecutive primes can be; although modest in strength compared to what is believed to be true, each of these is a quantitative version of the statement that there are infinitely many primes. Still, we would like to have more precise quantitative information about the distribution of the primes. One significant achievement of nineteenth-century number theory was to show that such precision is available provide we look at weighted sums, or at products, involving primes.

Sums and products over primes. A manifestation of a different nature that the primes are infinite in number is the fact that the sum $\sum 1/p$ diverges. This may be seen from the straightforward inequalities

$$\sum_{n \leq x} \frac{1}{n} < \prod_{p \leq x} \left(1 - \frac{1}{p}\right)^{-1} < \prod_{p \leq x} e^{2/p} = \exp\left(2 \sum_{p \leq x} \frac{1}{p}\right),$$

together with the fact that $\sum 1/n$ diverges. (These inequalities may be taken as an exercise by the reader.) In fact, inspired by ideas introduced by Chebyshev, Mertens proved—*without appealing to the Prime Number Theorem*, which had not been established by the time of his 1874 papers(!)—that

$$\sum_{p \leq x} \frac{1}{p} = \log\log x + B + O\left(\frac{1}{\log x}\right), \quad \text{as } x \to \infty. \tag{2.2}$$

Here, B is the so-called *reciprocal prime* or *Meissel–Mertens constant*

$$B = \gamma + \sum_p \left(\frac{1}{p} + \log\left(1 - \frac{1}{p}\right)\right) = \gamma - \sum_{k=2}^{\infty} \frac{1}{k} \sum_p \frac{1}{p^k} \approx 0.261497\cdots, \tag{2.3}$$

and where $\gamma = 0.5772\ldots$ is the so-called Euler–Mascheroni constant—a constant (discussed in Appendix B) that will make many appearances in this book. (Observe that because $1/p + \log(1 - 1/p) = O(p^{-2})$, the first sum in (2.3) converges rapidly.) The result (2.2) is sometimes referred to as *Mertens's Second Theorem*. Mertens's result for $\sum 1/p$ may be contrasted with the well-known analogous result for $\sum 1/n$:

$$\sum_{n \le x} \frac{1}{n} = \log x + \gamma + \frac{\theta}{x}, \quad \text{with } |\theta| < 1. \tag{2.4}$$

Thus $\sum_{n \le x} 1/n$ behaves like $\log x$ while $\sum_{p \le x} 1/p$ behaves like $\log\log x$. A proof of (2.2) may be found in Section 2.13 (Selected proofs). An 'explicit bounds' version of (2.2) is given by

$$\log\log x < \sum_{p \le x} \frac{1}{p} < \log\log x + B + \frac{4}{\log^3 x}, \quad \text{for } x \ge e,$$

with B as in (2.3), while under the Riemann hypothesis Schoenfeld (1976, [**954**]) obtained

$$\left| \sum_{p \le x} \frac{1}{p} - \log\log x - B \right| < \frac{3\log x + 4}{8\pi\sqrt{x}} \quad \text{for } x > 13.5.$$

Numerical evidence shows the $O(1/\log x)$ error term in (2.2) to be predominantly positive, although Robin (1983, [**911**]) had shown that it does change signs infinitely often, and Büthe (2015, [**170**]) proved that there must be a sign change before the enormous value of 1.91×10^{215}.

As an aside, observe that since $\sum 1/p$ diverges while $\sum 1/n^2$ converges, one can say that primes are more numerous than squares. They are likewise more numerous than the integer parts of $n^{1+\epsilon}$ for any $\epsilon > 0$, and more numerous even than the integer parts of $n \log^k n$ for any $k > 1$. It will be worthwhile to keep in mind that $\sum 1/(n \log^k n)$ diverges when $k \le 1$ and converges when $k > 1$.[5] As a further aside, one may note that while $\sum 1/p$ diverges, the sum $\sum 1/p^\alpha$ converges for every $\alpha > 1$ since it is bounded by $\sum 1/n^\alpha$.

The result (2.2) of Mertens (1874, [**752**]) had been a major breakthrough, coming as it did before the Prime Number Theorem was established, and many useful consequences can be derived from it. In particular, each of the following four assertions is actually *equivalent* to (2.2) when taken together with the value of the constant B in (2.3): Thus we have, as $x \to \infty$,

$$\prod_{p \le x} \left(1 - \frac{1}{p}\right)^{-1} = e^\gamma \log x \left(1 + O\left(\frac{1}{\log x}\right)\right) \equiv e^\gamma \log x + O(1), \tag{2.5}$$

$$\prod_{p \le x} \left(1 - \frac{1}{p}\right) = \frac{e^{-\gamma}}{\log x} \left(1 + O\left(\frac{1}{\log x}\right)\right) \equiv \frac{e^{-\gamma}}{\log x} + O\left(\frac{1}{\log^2 x}\right), \tag{2.6}$$

$$\prod_{p \le x} \left(1 + \frac{1}{p}\right) = \frac{6e^\gamma}{\pi^2} \log x \left(1 + O\left(\frac{1}{\log x}\right)\right) \equiv \frac{6e^\gamma}{\pi^2} \log x + O(1), \tag{2.7}$$

[5] Divergence at $k = 1$ follows from $\int dx/(x \log x) = \log\log x$, while convergence for $k > 1$ follows because $\int dx/(x \log^k x) = (1-k)^{-1}(\log x)^{1-k}$.

$$\prod_{p\le x}\left(1+\frac{1}{p}\right)^{-1} = \frac{\pi^2 e^{-\gamma}}{6\log x} + O\left(\frac{1}{\log^2 x}\right). \tag{2.8}$$

Here, (2.6) and (2.8) approach 0, while (2.2), (2.5), and (2.7) approach $+\infty$. Proofs of these equivalences are also given at the end of this chapter. The error terms shown in (2.2) and (2.5)–(2.8) are not the best possible, although they were the best possible before the Prime Number Theorem was available; indeed, *any* improvement in any one of them—namely replacing the order $O(\cdot)$ shown by $o(\cdot)$—will render it equivalent to the Prime Number Theorem. The result (2.5)/(2.6) is variously referred to as *Mertens's formula* or as *Mertens's third theorem*. It is worth noting that (2.5) can be viewed as an approximation formula for the truncated Euler product (1.2) in the case where $s = 1$. Note too that relation (2.3) provides a way of relating sums involving $1/p$ to products involving $1 - 1/p$.

Numerous explicit bounds for the products appearing in (2.5)–(2.8) are also known. For example, Dusart (1998, [**301**]), (2018, [**304**]) showed that

$$\frac{e^{-\gamma}}{\log x}\left(1-\frac{0.2}{\log^2 x}\right) < \prod_{p\le x}\left(1-\frac{1}{p}\right) < \frac{e^{-\gamma}}{\log x}\left(1+\frac{0.2}{\log^2 x}\right),$$

where the inequality on the left holds for $x \ge 2973$, and the one on the right for $x > 1$; while under the Riemann hypothesis Schoenfeld (1976, [**954**]) showed that

$$\left|\prod_{p\le x}\left(1-\frac{1}{p}\right) - \frac{e^{-\gamma}}{\log x}\right| \le \frac{e^{-\gamma}}{\log x}\left(\frac{3\log x + 5}{8\pi\sqrt{x}}\right), \quad \text{for } x \ge 8,$$

and

$$\left|\prod_{p\le x}\left(1-\frac{1}{p}\right)^{-1} - e^{\gamma}\log x\right| \le e^{\gamma}\log x\left(\frac{3\log x + 5}{8\pi\sqrt{x}}\right), \quad \text{for } x \ge 13.1.$$

Robin (1983, [**911**]) proved that the order terms in (2.5)/(2.6) change signs infinitely often.

It is worth mentioning here that the factor $e^{-\gamma} \approx 0.56146$ appearing in (2.5)–(2.8) turns out to play an important role in the theory. This is because the product in (2.6) can be interpreted as being the proportion of all integers not divisible by any prime $\le x$, since (as we will see later) divisibility by any of a *finite* collection of primes are 'asymptotically independent events.' The result (2.6) shows that as x increases the proportion of integers having no prime factors $\le x$ approaches 0 at a rate proportional to $1/\log x$. But the factor $e^{-\gamma}$ (when viewed in light of the Prime Number Theorem) evidences the failure of independence when x is permitted to grow. (These points will come into clearer focus as our story unfolds.) In fact, as early as 1923 the famous mathematical duo Hardy and Littlewood[6] had already pointed out that the factor $e^{-\gamma}$ causes difficulties with certain arguments based purely on probability. Incidentally, either of (2.5) or (2.6) lead to the neat result

$$e^{\gamma} = \lim_{x\to\infty}\frac{1}{\log x}\prod_{p\le x}\left(1-\frac{1}{p}\right)^{-1}.$$

[6]Theirs was the longest-lasting research collaboration between two mathematicians of the highest rank.

The constants $e^{\pm\gamma}$ can be thought of as a deep property of the complete set of prime numbers. A handy and comprehensive summary of formulas for sums and products involving primes, natural numbers, and their logarithms, may be found in Appendix E.

Of course, the sum and the product of primes must also diverge and one can show that

$$\sum_{p\le x} p = \frac{1}{2}\frac{x^2}{\log x} + O\left(\frac{x^2}{\log^2 x}\right), \quad \text{and} \quad \prod_{p\le x} p = \exp\bigl(x(1+o(1))\bigr), \tag{2.9}$$

as well as

$$\sum_{p\le x} \log p \sim x, \quad \text{and} \quad \sum_{p\le x} \frac{\log p}{p} = \log x + O(1). \tag{2.10}$$

Of these four assertions the first three are each equivalent to the Prime Number Theorem (with the second and third really being identical statements). The fourth assertion, sometimes referred to as *Mertens's first theorem*, does not require the PNT; it is proved in Section 2.13. In passing, let us also mention the symmetrical results

$$\sum_{p\le x} \log^2 p + \sum_{pq\le x} (\log p)(\log q) = 2x\log x + O(x)$$

and

$$\sum_{p\le x} \log p + \sum_{pq\le x} \frac{(\log p)(\log q)}{\log pq} = 2x + O\left(\frac{x}{\log x}\right),$$

both due to Atle Selberg. The first of these played an essential role in the 'elementary' proof of the PNT by Selberg (1949, [**975**]) and by Erdős (1949, [**340**]); see Section 2.12 (Historical notes). Selberg also proved that there must exist constants A and B such that

$$\sum_{pq<x} \frac{1}{pq} = (\log\log x)^2 + A\log\log x + B + O\left(\frac{\log\log x}{\log x}\right).$$

In the last three displays, p and q both range over the primes.

So far, we have explored some results concerning pairs of consecutive primes, as well as some results that incorporate all of the primes up to some point. But (on account of their weighted formulations) these only give indirect information about the distribution of the primes themselves. Through the end of the nineteenth century, one mystery remained—a mystery that motivated a great deal of the development of pure mathematics at the time: Just how frequent are the primes, really?

2.2. The Prime Number Theorem (PNT)

Concerning the number of primes, there are more precise results that also align with intuition about how those numbers are distributed. Standard notation is to let $\pi(x)$ be the number of primes not exceeding x:

$$\pi(x) \equiv \sum_{p\le x} 1. \tag{2.11}$$

Thus $\pi(1.9) = 0$, $\pi(2) = \pi(2.5) = 1$, $\pi(10) = 4$, and so on. A celebrated result—which took many centuries to discover and to prove—is the *Prime Number Theorem*

(PNT). It states that $\pi(x)/(x/\log x) \to 1$ as $x \to \infty$ or, equivalently,

$$\pi(x) \sim \frac{x}{\log x}, \quad \text{as } x \to \infty. \tag{2.12}$$

Since $1/\log x$ decreases slowly, one sees from this that primes are plentiful.

Conjectured on numerical evidence by Adrien-Marie Legendre in 1798 (and published by him in 1808), and prior to that in 1791 by the then fourteen-year-old Carl Friedrich Gauss (1777–1855), the PNT thereafter became a 'holy grail'. The first *rigorous* result to establish the true asymptotic order of $\pi(x)$ is due to Chebyshev; in two pioneering papers (1852, [**192, 193**]) he showed there exist constants $0 < C_1 \leq 1 \leq C_2$ such that

$$C_1 \frac{x}{\log x} \leq \pi(x) \leq C_2 \frac{x}{\log x} \quad \text{for all } x, \tag{2.13}$$

and furthermore, that *if* $\pi(x)/(x/\log x)$ has an $x \to \infty$ limit, that limit must equal 1. Chebyshev proved these inequalities with $C_1 = \log\left(2^{1/2}3^{1/3}5^{1/5}/30^{1/30}\right) \approx 0.9213$ and $C_2 = 6C_1/5 \approx 1.1055$. His motivations were twofold: first, to obtain values of C_1 and C_2 as close to 1 as possible, and second, to use (2.13) to prove Bertrand's postulate.[7] In Section 2.13 we will demonstrate (2.13) in a simpler way, although with worse values for the constants. It is possible to improve upon Chebyshev's constants slightly by means of his approach, but it is generally conceded that the Prime Number Theorem cannot be established using the methods he introduced. For an attempt to do so, however, see Diamond & Erdős (1980, [**284**]).

In the only paper he ever wrote on number theory, Riemann (1859, [**906**])—a student of Gauss—laid out an ingenious plan for proving the Prime Number Theorem, and following up on Riemann's ideas, the PNT was finally established by Jacques Salomon Hadamard (1896, [**475**]), and independently by Charles Jean de la Vallée Poussin (1896, [**1079**]). The key step in their proofs involved showing that the Riemann zeta function $\zeta(s)$, considered as a function of a complex variable s, has no zeros on the vertical line $\Re s = 1$; that is, $\zeta(1 + it) \neq 0$ for all real t, to which (2.12) happens to be equivalent. The proof of the PNT was thus a pan-European endeavour: Riemann (1826–1866) was German, Hadamard (1865–1963) was French, and de la Vallée Poussin (1866–1962) was Belgian.

We will not prove the Prime Number Theorem in this book, although in Chapter 6 we will prove the crucial step on which it is based, namely that the zeta function $\zeta(s)$ has no zeros on the vertical line $\Re s = 1$. Proofs of the PNT can be found in many sources. Narkiewicz (2000, [**797**, Chapter 5]) offers a particularly thorough treatment.

Legendre's conjecture had been that $\pi(x) \sim x/(\log x + B)$, and using the best available tables of his day he estimated that $B = -1.08366$, although now it is known that the choice $B = -1$ is the one that correctly matches the second-order term of the expansion for $\pi(x)$. In fact, if one were to define $B(x)$ via $\pi(x) \equiv x/(\log x + B(x))$ it can be shown that $B(x) \to -1$ as $x \to \infty$. (Gauss knew of Legendre's conjecture but was reluctant to guess a value for B.) A set of explicit

[7]For proofs of Bertrand's postulate using (2.13) see, for instance, Chandrasekharan (1968, [**188**, Chapter VII, §3]), Fine & Rosenberger (2016, [**381**, Theorem 4.2.5]), or de Konink & Luca (2012, [**650**, Chapter 2, §7]).

inequalities for $\pi(x)$ based on Legendre's asymptotic form is given by

$$\frac{x}{\log x - \frac{1}{2}} < \pi(x) < \frac{x}{\log x - \frac{3}{2}},$$

with the left inequality holding for $x \geq 59$, and the one on the right holding for $x \geq 5$. Amiably, the average of the values $-\frac{1}{2}$ and $-\frac{3}{2}$ appearing here equals the mentioned optimal value of $B = -1$. Another set of bounds is given by

$$\frac{x}{\log x}\left(1 + \frac{1}{\log x}\right) \leq \pi(x) \leq \frac{x}{\log x}\left(1 + \frac{1.2762}{\log x}\right), \tag{2.14}$$

valid for $x \geq 599$ on the left, and $x > 1$ on the right. These and other such bounds may be found in various papers by Dusart, Ramaré, Robin, Rosser, and Schoenfeld, among others. Efficient *exact* computation of $\pi(x)$ for large x is described in Deléglise & Rivat (1996, [**272**]); for example, they determined that $\pi\left(10^{18}\right) =$ 24,739,954,287,740,860.

Now, if p_n is the nth prime, then $\pi(p_n) = n$. It follows that $\pi(x)$ as a function of x, and p_n as a function of n, are inversely related, so that results about $\pi(x)$—invariably asymptotic—can be translated into results about p_n, and conversely. One consequence is that statement (2.12) of the PNT is *equivalent* to the assertion

$$p_n \sim n \log n, \quad \text{as } n \to \infty. \tag{2.15}$$

A proof of this equivalence may be taken as an exercise, but it is also given in Section 2.13 (Selected proofs). Further, using more accurate formulas for $\pi(x)$, it can be shown that

$$p_n = n\big(\log n + \log\log n\big) + O(n). \tag{2.16}$$

A much more precise result is

$$p_n = n\left(\log n + \log\log n - 1 + \frac{\log\log n - 2}{\log n} + \frac{(\log\log n)^2 - 6\log\log n + 11}{2\log^2 n} + O(\cdots)\right),$$

with the indicated error term being $O\left(\left(\log\log n/\log n\right)^3\right)$. Such asymptotic results often possess known *explicit* versions. For example, it is known that

$$n\big(\log n + \log\log n - 1\big) < p_n < n\big(\log n + \log\log n - 1/2\big),$$

with the inequality on the left (due to Dusart, 1999, [**302**]) holding for $n \geq 2$, and the one on the right (due to Rosser & Schoenfeld, 1962, [**919**]) holding for $n \geq 20$. More incisively,

$$n\left(\log n + \log\log n - 1 + \frac{\log\log n - 2.1}{\log n}\right) < p_n < n\left(\log n + \log\log n - 1 + \frac{\log\log n - 2}{\log n}\right),$$

with the inequality on the left holding for $n \geq 3$ and the one on the right for $n > 688,383$. Dusart (2018, [**305**]) obtained sharper results by assuming RH. See also Robin (1983, [**911**]).

Arias de Renya & Toulisse (2013, [**39**]) provided a fascinating historical account of the development of asymptotic formulas for the nth prime, as well as of the role played by I. M. Pervushin, a Russian mathematician-priest who lived in Perm in the 1800s. One formula claimed by Pervushin—*clearly false!*—is

$$p_{n+1} - p_n = \log n + \log\log n + \frac{\log\log n - 1}{\log n} + O\left(\left(\frac{\log\log n}{\log n}\right)^2\right).$$

What should be noted about this formula is that had it been true it would not only have refuted the twin prime conjecture, but because its error term tends to zero, it would follow that prime gaps need not even be integers. As such, it calls attention to the fact that there are limits beyond which formulas for p_n cannot be expected to go. Additional bounds on $p_{n+1} - p_n$ will be discussed in Section 2.5.

The primes possess many amusing idiosyncrasies. One of the more curious ones, due to Golomb (1962, [**435**]), is that $n/\pi(n)$ eventually takes on every integer value greater than 1.

The Prime Number Theorem, as formulated in equation (2.12), is asymptotically correct but gives no indication of the quality of the approximation. Some of the inequalities in this section—such as (2.14)—provide a bit of information about how accurate the main term $x/\log x$ is as an approximation to $\pi(x)$. But it turns out, perhaps unexpectedly, that the best way to find more accurate asymptotic information about $\pi(x)$ is to replace $x/\log x$ by a different function.

2.3. Introducing the function $\mathrm{li}(x)$

A viewpoint alternate to (2.12), proposed by Gauss in the 1790s (although only published in 1849), and adopted by Peter Gustav Lejeune Dirichlet, is that the 'density' of primes near some t, namely the average number of primes per unit interval there, is $1/\log t$, so that

$$\pi(x) \sim \mathrm{li}(x) \equiv \int_0^x \frac{dt}{\log t} \equiv \lim_{\epsilon \searrow 0} \left(\int_0^{1-\epsilon} + \int_{1+\epsilon}^x\right) \frac{dt}{\log t}. \tag{2.17}$$

Since $\mathrm{li}(x) \sim x/\log x$ it follows that the approximations (2.12) and (2.17) are equivalent.[8] Nevertheless, it turns out that $\mathrm{li}(x)$ provides a far better approximation to $\pi(x)$ than $x/\log x$ does. The intrinsic reason for this is that the value $1/\log t = 1/\log x$ occurs only at the upper point of the integrand in (2.17), while in the formula $\pi(x) \sim x/\log x$, the value of $1/\log x$ has been applied to *all* of the numbers t below x.

The singularity in (2.17) at $t = 1$ can be bypassed by replacing the function $\mathrm{li}(x)$ by

$$\mathrm{Li}(x) \equiv \int_2^x \frac{dt}{\log t} = \mathrm{li}(x) - \mathrm{li}(2); \tag{2.18}$$

the difference between $\mathrm{li}(x)$ and $\mathrm{Li}(x)$ is only $\mathrm{li}(2) \approx 1.045$ which is usually inconsequential. One might equally consider using the *logarithmic sum*

$$\mathrm{ls}(x) \equiv \sum_{2 \le n \le x} \frac{1}{\log n};$$

by comparing the integral defining $\mathrm{Li}(x)$ with its Riemann sums, one can show that

$$\mathrm{Li}(x) \le \mathrm{ls}(x) \le \mathrm{Li}(x) + \frac{1}{\log 2}. \tag{2.19}$$

---

[8]An easy, worthwhile exercise: Using L'Hôpital's rule, show $(x/\log x)/\mathrm{li}(x) \to 1$ as $x \to \infty$.

It is worth taking a moment to notice that the logarithm function rises slowly enough for averages such as $(1/x)\int_1^x \log u\,du$ and $(1/x)\int_2^x (1/\log u)\,du$ to be asymptotic to their upper end-point values, namely $\log x$ and $1/\log x$, respectively, as $x \to \infty$. The forms $(1/x)\sum_{n\le x}\log n$ and $(1/x)\sum_{2\le n\le x}(1/\log n)$ are likewise asymptotic to $\log x$ and $1/\log x$. One consequence is that the *average spacing*, $\log x$, between primes *near* x is the same (asymptotically) as the average spacing between all primes *up to* x. Note too that even though one has the simple result $d\,\mathrm{li}(x)/dx = d\,\mathrm{Li}(x)/dx = 1/\log(x)$, the functions li and Li (and ls as well) do not correspond to any elementary formulas, and neither are there any attractive representations for their inverses $\mathrm{li}^{-1}(y)$, $\mathrm{Li}^{-1}(y)$, and $\mathrm{ls}^{-1}(y)$. However, using the formula $df^{-1}(x)/dx = 1/f'(f^{-1}(x))$, one may at least evaluate the derivative $d\,\mathrm{li}^{-1}(y)/dy = \log\left(\mathrm{li}^{-1}(y)\right)$, and likewise $d\,\mathrm{Li}^{-1}(y)/dy = \log\left(\mathrm{Li}^{-1}(y)\right)$.

Repeated integration by parts may be used to establish an asymptotic expansion for $\mathrm{li}(x)$; specifically, for any fixed $k \ge 1$ one can show that

$$\mathrm{li}(x) = \frac{x}{\log x}\left(1 + \sum_{j=1}^{k}\frac{j!}{\log^j x}\right) + O\left(\frac{x}{\log^{k+2} x}\right), \quad \text{as } x \to \infty. \tag{2.20}$$

There are also two nice convergent expansions for $\mathrm{li}(x)$, namely

$$\mathrm{li}(x) = \log\log x + \gamma + \sum_{n=1}^{\infty}\frac{\log^n x}{n!\,n} \quad \text{for } x > 1, \tag{2.21}$$

and

$$\mathrm{li}(x) = \log\log\frac{1}{x} + \gamma + \sum_{n=1}^{\infty}\frac{\log^n x}{n!\,n} \quad \text{for } 0 < x < 1. \tag{2.22}$$

So concludes our introduction to this unfamiliar yet surprisingly relevant function.

2.4. On the difference between $\pi(x)$ and $\mathrm{li}(x)$

We now come to one of the central questions driving the entire theory: How closely does $\mathrm{li}(x)$ approximate $\pi(x)$? Questions about the magnitude of this difference are motivated by statistical considerations regarding the extent to which the occurrence of primes behaves like a series of *independent events* with probabilities $1/\log n$, for each $n > 2$. (Appendix D lays out the essential ideas, although notice that we can already interpret the logarithmic sum $\mathrm{ls}(x)$ as the *expectation* of such a series.) As an initial calibration, in Chapter 4 we will see that a version of the Law of the Iterated Logarithm implies that under an independence heuristic, we would have

$$\left|\pi(x) - \mathrm{li}(x)\right| \overset{?}{=} O\left(\sqrt{\frac{x\log\log x}{\log x}}\right), \quad \text{as } x \to \infty, \tag{2.23}$$

with the order term shown being *exact* under the heuristic. (The $\overset{?}{=}$ means it's not a done deal, and indeed this order turns out not to be quite correct.)

In 1914, John Edensor Littlewood (1885–1977) showed *unconditionally* (that is, without assuming the Riemann hypothesis) that there are arbitrarily large values of x for which either

$$\pi(x) - \mathrm{li}(x) > \frac{1}{2}\frac{\sqrt{x}}{\log x}\log\log\log x \quad \text{or} \quad \pi(x) - \mathrm{li}(x) < -\frac{1}{2}\frac{\sqrt{x}}{\log x}\log\log\log x \tag{2.24}$$

holds. It is convenient to write such a pair of relations as

$$\pi(x) - \mathrm{li}(x) = \Omega_{\pm}\left(\frac{\sqrt{x}}{\log x}\log\log\log x\right);$$

it means that the inequalities in (2.24) each hold infinitely often, but not necessarily with the constants $\pm\frac{1}{2}$. Littlewood's result allows one to make the neater assertion

$$\limsup_{x\to\infty}\frac{|\pi(x) - \mathrm{li}(x)|}{\sqrt{x}\,/\log x} = \infty,$$

meaning that $\pi(x) - \mathrm{li}(x)$ must sometimes far exceed $C\sqrt{x}/\log x$ for any given value of C. (It consequently must also exceed $O(x^{1/2-\epsilon})$ for any $\epsilon > 0$.) A novelty of Littlewood's proof of (2.24) is that he first established it under the assumption that RH is true, and then separately under the assumption that RH is false. A simplified proof of Littlewood's result was given by Diamond (1975, [**282**]).

In the other direction, von Koch (1901, [**642**]) proved that if RH was true then

$$\limsup_{x\to\infty}\frac{|\pi(x) - \mathrm{li}(x)|}{\sqrt{x}\,\log x} < \infty$$

(note that this denominator is $\sqrt{x}\log x$ and not $\sqrt{x}/\log x$). It follows—if one assumes RH—that the true maximum order of $\pi(x) - \mathrm{li}(x)$ must lie somewhere between $\sqrt{x}/\log x$ and $\sqrt{x}\log x$. (The independence-based version (2.23) actually lies within this range.) Nevertheless, it is believed that the true maximum order of $\pi(x) - \mathrm{li}(x)$, while not as good as the one in (2.23), is nevertheless better than the one that has been proved under RH. In particular, Elliott (1972, [**315**]) had conjectured the true order to be

$$\left|\pi(x) - \mathrm{li}(x)\right| = O\left(\frac{\sqrt{x}}{\sqrt{\log x}}\log\log x\right),$$

while—based on tail estimates for infinite sums of random variables—Montgomery (1980, [**765**]) conjectured that

$$\limsup_{x\to\infty}\frac{|\pi(x) - \mathrm{li}(x)|}{(\sqrt{x}/\log x)(\log\log\log x)^2} = \frac{1}{2\pi}.$$

Littlewood's result, which forms a lower bound for the largest oscillations of the difference $\pi(x) - \mathrm{li}(x)$, has by now been complemented by unconditional upper bounds for the largest such oscillations. But to describe these results we need to make reference to the zeta function, $\zeta(s)$, where $s = \sigma + it$, as well as to certain of its so-called *zero-free regions*. These will be discussed in Chapters 5 and 6 and the reader need not concern themselves with those details here. For now, we will merely refer to three such regions, denoting them as

$$\text{(a) } \Re s = 1; \quad \text{(b) } \Re s > 1 - \frac{1}{C\log t}; \quad \text{and (c) } \Re s > 1 - \frac{1}{C(\log t)^{2/3}(\log\log t)^{1/3}}.$$

The first of these is the smallest zero-free region of any substantive value, namely the vertical line $\sigma = 1$. The second is the zero-free region of Hadamard and de la Vallée Poussin. And the last, due to Korobov and Vinogradov, is the best such region currently known. The zero-free region that would be *equivalent* to the Riemann hypothesis is $\Re s > \frac{1}{2}$. As mentioned, the details will be elaborated on in Chapters 5 and 6 and will not be needed here.

Remarkably, the true order of magnitude of the difference $\pi(x) - \mathrm{li}(x)$ depends on the location of the zeros of $\zeta(s)$. Given only the zero-free region (a), we can say no more than $\pi(x) \sim \mathrm{li}(x)$ as $x \to \infty$; this is the Prime Number Theorem to which the zero-free region (a) is in fact equivalent. Given the wider zero-free region (b), this result can be strengthened to

$$\pi(x) = \mathrm{li}(x) + O\left(xe^{-c\sqrt{\log x}}\right), \quad x \to \infty,$$

for some $c > 0$, while the zero-free region (c) leads to the best error term currently known:[9]

$$\pi(x) = \mathrm{li}(x) + O\left(xe^{-c(\log x)^{3/5}/(\log\log x)^{1/5}}\right). \tag{2.25}$$

None of the zero-free regions established so far allow one to conclude that there is an $\epsilon > 0$ such that $\pi(x) - \mathrm{li}(x) = O(x^{1-\epsilon})$, even though far more than this is believed to be true.[10]

Arias de Renya & Toulisse (2013, [**39**]) pointed out that if we have an approximation of the form

$$\pi(x) = \mathrm{li}(x) + O\big(r(x)\big), \tag{2.26}$$

then it can often be inverted to conclude that

$$p_n = \mathrm{li}^{-1}(n) + O\big(r(n\log n)\log n\big), \tag{2.27}$$

and conversely. We prove this under mild conditions in Section 2.13. In particular, a corollary to (2.25) (which actually is equivalent to it) is given by

$$p_n = \mathrm{li}^{-1}(n) + O\left(ne^{-c(\log n)^{3/5}/(\log\log n)^{1/5}}\right). \tag{2.28}$$

In contrast to the *unconditional* results stated above, the Riemann hypothesis (RH) is *equivalent* to *either one* of the considerably stronger assertions

$$\pi(x) = \mathrm{li}(x) + O\left(\sqrt{x}\,\log x\right), \quad \text{as } x \to \infty, \tag{2.29}$$

due to von Koch (mentioned previously), as well as a seemingly weaker version given by

$$\pi(x) = \mathrm{li}(x) + O\left(x^{1/2+\epsilon}\right), \quad \text{as } x \to \infty \ \text{ for all } \epsilon > 0. \tag{2.30}$$

It is an unusual feature of these two RH equivalences that if the weaker one in (2.30) were to hold, the stronger one in (2.29) would automatically follow. *Explicit* bound versions under the Riemann hypothesis include

$$\left|\pi(x) - \mathrm{li}(x)\right| < \frac{\sqrt{x}\,\log x}{8\pi}, \quad \text{for } x \ge 2657,$$

due to Schoenfeld (1976, [**954**]) and an inverted version

$$\left|p_n - \mathrm{li}^{-1}(n)\right| < \frac{\sqrt{n}\,\log^{5/2} n}{\pi}, \quad \text{for } n \ge 230,$$

---

[9] A version just slightly weaker than (2.25) but a bit easier on the eyes is

$$\left|\pi(x) - \mathrm{li}(x)\right| \le C_1\, x \exp\left(-C_2\,(\log x)^{3/5-\epsilon}\right).$$

[10] In particular, if $\frac{1}{2} \le \Theta \le 1$ is the supremum of the real parts of the zeros of $\zeta(s)$, it would follow that $\pi(x) - \mathrm{li}(x) = O\left(x^{\Theta+\epsilon}\right)$ and $\pi(x) - \mathrm{li}(x) = \Omega_\pm\left(x^{\Theta-\epsilon}\right)$, for all $\epsilon > 0$.

due to Robin (1983, [**911**]). These may be compared to unconditional bounds such as those in (2.14). Dusart (2018, [**304**]) provides some further results along such lines.

One curious fact—which also serves to caution that in this field numerical evidence can *never* be regarded as conclusive—is that $\pi(x) < \mathrm{li}(x)$ for *all* currently computable values of $\pi(x)$. Indeed, this has been verified numerically up to $x = 10^{19}$, as well as for selected values of x up to at least 10^{26}. (See, for example, Büthe, 2015, [**170**].) It became almost universally believed that this inequality must continue to hold for all sufficiently large values of x. Referring to this, Hardy (1937, [**488**]) said: "The conjecture is not only plausible but is supported by *all* the evidence of the facts" and he referred to it as "a striking example ... of a false conjecture which seems to have been endorsed even by Gauss and which took about 100 years to refute." Schmidt (1903, [**950**]) had shown that if $\pi(x) < \mathrm{li}(x)$ were true for all large enough x it would have implied RH. However, Littlewood's (1914, [**702**]) result (2.24) established that $\pi(x)$ and $\mathrm{li}(x)$ *must* switch positions infinitely often. Kaczorowski (1984, 1985, [**616**]) proved that for x large enough there are at least $c \log x$ sign changes below x, while arguments based on Cramér's heuristic model suggest that $\pi(x) - \mathrm{li}(x)$ should undergo at least one sign change on $(x, 2x)$ once x is large enough.

The first value of x at which $\pi(x)$ and $\mathrm{li}(x)$ switch positions is named after Stanley Skewes, a student of Littlewood, who in 1933, [**994**], showed, *assuming* RH, that this must occur at some point before

$$x = \exp(\exp(\exp(\exp 7.7))) < 10^{10^{10^{34}}}.$$

Commenting on this result at the time, Hardy (1937, [**488**]) said: "I think that this is the largest number which has ever served any definite purpose in mathematics."[11] Skewes (1955, [**995**]) later obtained the bound

$$x = \exp(\exp(\exp(\exp(\exp 7.705)))) < 10^{10^{10^{964}}},$$

this time *unconditionally* (that is, without assuming RH). Skewes's values were later successively reduced—sometimes with and sometimes without RH—by Lehman (1966, [**686**]), te Riele (1987, [**905**]), and others. In particular, Bays & Hudson (2000, [**84**]) established the bound $x = 1.4 \times 10^{316}$ unconditionally and showed that there are at least 10^{153} consecutive integers in that 'vicinity' for which $\pi(x) > \mathrm{li}(x)$, and likely a great many more.

To sum up, as of now the best available results indicate that the first reversal in the sign of $\pi(x) - \mathrm{li}(x)$ occurs after 10^{19} but before 1.4×10^{316}. That these numbers are so enormous stems from the need for large values of the $\log\log\log x$ which appears in Littlewood's proof. Determining the exact point at which the reversal first occurs lies well beyond currently foreseeable computational means, although Saouter et al. (2015, [**941**]) have reported some progress.

Having discussed the role that the size of $\pi(x) - \mathrm{li}(x)$ plays in the investigation of the *global* distribution of the primes, we will now return to looking at the gaps

[11] Hardy further remarked: "If the universe were the chess-board, the protons the [chess-pieces], and any interchange in the position of two protons a move, then the number of possible games would be something like the Skewes number. ... It does not seem at all likely that we shall ever know a single instance of the truth of Littlewood's theorem. ... This is an example in which the truth has defeated not only all the evidence of the facts and of common sense but even a mathematical imagination so powerful and profound as that of Gauss. ..."

between consecutive primes. Earlier (in Section 2.1) we were concerned with how large those gaps *cannot* be; next we will describe a series of significant results that probe the boundaries of how large, and how small, those gaps *can* be.

2.5. Large gaps, small gaps, and the twin primes conjecture

The Prime Number Theorem tells us that the average gap (that is, the average spacing) between consecutive primes near x is about $\log x$; it follows that the average of the *normalized spacings*, namely $(p_{n+1} - p_n)/\log p_n$, must equal 1, so that

$$\Delta \equiv \liminf_{n\to\infty} \frac{p_{n+1} - p_n}{\log p_n} \leq 1 \leq \limsup_{n\to\infty} \frac{p_{n+1} - p_n}{\log p_n} \equiv \Omega. \tag{2.31}$$

A class of natural and historically important questions concerns how much larger and how much smaller gaps between primes can be when compared to their mean spacing. The literature on *large gaps* and *small gaps* between consecutive primes is considerable, and in recent years significant progress has been made. Here we can trace only a few highlights, a key takeaway being just how wide the gulf is that separates the various results.

With respect to *large gaps*, that the lim sup value Ω is equal to $+\infty$ has been known for some time. It was first shown by Westzynthius (1931, [**1099**]) who established that

$$\limsup_{n\to\infty} \left(\frac{p_{n+1} - p_n}{\log p_n}\right) \left(\frac{\log\log\log\log p_n}{\log\log\log p_n}\right) > 0. \tag{2.32}$$

Following up on this Erdős (1935, [**334**]) proved the sharper version

$$\limsup_{n\to\infty} \left(\frac{p_{n+1} - p_n}{\log p_n}\right) \left(\frac{(\log\log\log p_n)^2}{\log\log p_n}\right) > 0, \tag{2.33}$$

while Rankin (1938, [**885**]) in a series of five papers showed that

$$\limsup_{n\to\infty} \left(\frac{p_{n+1} - p_n}{\log p_n}\right) \left(\frac{(\log\log\log p_n)^2}{(\log\log p_n)(\log\log\log\log p_n)}\right) \geq C \tag{2.34}$$

for $C = 1/3$. The results (2.33)/(2.34) allow one to say that infinitely many prime gaps exceed the average by a factor just under $\log\log p_n$. A series of subsequent efforts increased the value of C in (2.34) to $C = 2e^{\gamma}$. (Pintz (1997, [**848**]) traces the history of this bound.) For a long time, the asymptotic order of (2.34) had not been further improved although it was widely believed that much more must be true. In 1979, at a meeting in Durham, England, Erdős offered a prize of \$10,000—his largest ever—for a proof that the lim sup in (2.34) is actually infinite. This challenge was met some twenty-five years later when Maynard (2016, [**744**]), and independently (mere hours apart!) Ford, Green, Konyagin & Tao (2016, [**385**])—all of whom later combined efforts to produce the five-authored (2018, [**384**]) version—showed that not only is the lim sup in (2.34) infinite, but that furthermore

$$\limsup_{n\to\infty} \left(\frac{p_{n+1} - p_n}{\log p_n}\right) \left(\frac{\log\log\log p_n}{(\log\log p_n)(\log\log\log\log p_n)}\right) > 0; \tag{2.35}$$

these authors had expressed their result in the alternate but equivalent form

$$\max_{p_n \leq x} (p_{n+1} - p_n) \gg \frac{\log x \log\log x \log\log\log\log x}{\log\log\log x}.$$

Erdős's prize had to be split in two: \$5,000 went to Maynard, and the rest to the other four.

The result (2.35) remains far from what surely must be true. Indeed, based on his heuristic model (discussed in Chapter 4) Cramér (1936, [**245**]) had conjectured that

$$\limsup_{n\to\infty} \frac{p_{n+1}-p_n}{\log^2 p_n} \stackrel{?}{=} 1;$$

that this lim sup is nonzero is widely believed, although not the value of 1 on the right. This conjecture remains undecided; it is not even known whether or not $p_{n+1}-p_n$ can exceed $\log^\alpha p_n$ infinitely often for *any* value of $\alpha > 1$.

As for *small gaps* between primes, the history of

$$\Delta \equiv \liminf_{n\to\infty} \frac{p_{n+1}-p_n}{\log p_n}$$

divides sharply into the era prior to 2014 and the era thereafter. Summaries of the earlier history may be found in Goldston, Pintz & Yildirim (2007, [**428**]) and in Soundararajan (2007, [**1003**]), (2007, [**1004**]). Successively smaller bounds on Δ were obtained over nearly a century of work—both with and without unproven assumptions—by Erdős (1940, [**338**]), Bombieri & Davenport (1966, [**122**]), and by Huxley (1977, [**571**]), among others. One well-known result is that of Maier (1988, [**723**]) who established the bound $\Delta < 0.2485$. However, in an important breakthrough, Goldston, Pintz & Yidirim (2007, [**428**], (2009, [**429**]), (2010, [**430**]), (2013, [**431**]), (2014, [**432**]) proved, not only that $\Delta = 0$, but also the significantly sharper result

$$\liminf_{n\to\infty}\left(\frac{p_{n+1}-p_n}{\sqrt{\log p_n}\,(\log\log p_n)^2}\right) \equiv \liminf_{n\to\infty}\left(\frac{p_{n+1}-p_n}{\log p_n}\right)\left(\frac{\sqrt{\log p_n}}{(\log\log p_n)^2}\right) < \infty .$$

This allowed one to say not only that consecutive primes are infinitely often closer than any multiple of the average spacing, but also that there are infinitely many prime pairs which lie within about a square root of the average spacing. The best result of this type—up to the time of his paper—is due to Pintz (2014, [**852**]), who showed that

$$\liminf_{n\to\infty}\left(\frac{p_{n+1}-p_n}{\log p_n}\right)\left(\frac{\log p_n}{\log\log p_n}\right)^{4/7} < \infty.$$

Further details of this earlier work may be found in Heath-Brown (1988, [**427**]), Maier & Pomerance (1990, [**724**]), Granville (1995, [**427**]), Goldston & Motohashi & Yildirim (2006, [**427**]), and in references therein. Pintz (2007, [**850**]) gives a definitive review. Let us also mention here that the points of accumulation of $(p_{n+1}-p_n)/\log p_n$ are known to constitute a set of positive measure, but aside from 0 and ∞, no other member of this set has ever been established.

The era of small gaps *subsequent* to 2014 is bound up with the *twin primes conjecture* which asserts that there exist infinitely many prime pairs (p_n, p_{n+1}) for which $p_{n+1}-p_n = 2$, as well as with the question of *bounded gaps*, which asks only whether there is a constant K for which infinitely many prime pairs satisfy $p_{n+1}-p_n \le K$.[12] Any of these would, of course, imply that $\Delta = 0$ (although not conversely). In a stunning breakthrough—one which had eluded all of the experts and eclipsed all of the hard-won earlier results—an 'unknown' mathematician of

[12]The French aristocrat de Polignac appears to have been the first to raise such questions (1849, [**855**]), (1849, [**856**]), (1849, [**857**]), (1851, [**858**]). Alphonse de Polignac (1826–1862), also known as Prince de Polignac, was a mathematician who had graduated from l'École Polytechnique and had served as a captain in the French artillery. De Polignac had also conjectured that all large enough odd n were of the form $2^k + p$ but Erdős (1950, [**341**]) showed this to be false.

almost sixty years of age, Yitang Zhang,[13] proved *unconditionally* that there exists a finite bound K such that $p_{n+1} - p_n \leq K$ infinitely often, and furthermore that one can take $K = 70{,}000{,}000$. His paper was submitted in April of 2013, and its results subsequently announced at a seminar Zhang gave at Harvard. The paper was accepted one month after it was submitted, and published one year later. Some details of Zhang's 'lecturer-rags' to 'MacArthur Fellow–riches' story, as well as the 'shock-and-awe' refereeing of Zhang's Annals of Mathematics submission (John Friedlander and Henryk Iwaniec were the targeted referees), is described by Friedlander (2015, [**393**]) as well as by Wilkinson (2015, [**1109**]). See also the nice overview by Breitzman (2016, [**143**]), as well as the articles by Murty (2013, [**785**]), Granville (2015, [**455**]), Kowalski (2015, [**660**]), and Rezgui (2017, [**900**]).

In a subsequent breakthrough, based on a different method, James Maynard (2015, [**743**])—then a postdoc in Montreal, Canada—showed that this bound could be reduced to $K = 600$. Since then the bound has been further reduced to $K = 246$ via two 'Polymath projects'—organized around the results of Zhang, Maynard, and Terence Tao—with the second project bearing the title Polymath8b: "Bounded intervals with many primes." (Prior to the Polymath projects organized by Tao, Zhang's results were extensively discussed on a blog run by Scott Morrison, as well as on the MathOverflow website.) A fascinating account (in French) of Tao's project was given by de la Bretèche (2014, [**148**]). If a certain conjecture of Elliott & Halberstam (1970, [**323**]) were true it would allow one to further reduce this bound to $K = 12$, while a generalization of this conjecture would reduce it further to $K = 6$. One is thus left to wonder: Is the legendary twin primes conjecture now poised to fall?

Concerning twin primes, Brun (1915, [**159**]), (1919, [**160**]), (1919, [**161**]) proved that—unlike $\sum 1/p$ which diverges—the sum of twin prime reciprocals

$$B \equiv \sum_{\text{twins}} \left(\frac{1}{p} + \frac{1}{p+2}\right) = \left(\frac{1}{3} + \frac{1}{5}\right) + \left(\frac{1}{5} + \frac{1}{7}\right) + \left(\frac{1}{11} + \frac{1}{13}\right) + \cdots \tag{2.36}$$

is finite. Had it been infinite it would have immediately implied the truth of the twin prime conjecture; but, in any case, it does indicate that their number is substantially smaller than the number of primes. A heuristic argument for the finiteness of this constant is given in Section 2.13. However, (2.36) must be summed term-by-term and its convergence is exceedingly slow. Indeed, even after summing the first 10^9 twins it is believed that the relative error still exceeds 5%. The value of B, known as Brun's constant, has been estimated heuristically by Thomas Nicely (1996, [**806**]), (2001, [**807**]), among others, as $B \approx 1.90216058$, although aside from the unconditional bound $B < 2.347$, no *rigorous* evaluation of B is as yet available. (Some further details appear in Section 2.13.) Curiously, it was during the course of his computations in 1994 that Nicely discovered the notorious 'FDIV bug' in the Pentium processor's floating point division algorithm, a bug that ultimately

[13] Zhang was born in 1955 and had published only two articles prior to his 2013 submission. A comment by Andrew Granville sums up the universal initial reaction to Zhang's result: "Never heard of him. Absolutely never heard of him. There is no way that somebody I've never heard of has done this." On learning his paper had been accepted by the prestigious Annals of Mathematics, Zhang phoned his partner (the two were living at opposite ends of the US at the time) and told her: "Pay attention to the media and newspapers. You may see my name." Her reply was: "Are you drunk?" There is an engaging film about Yitang Zhang by George Csicseri that had been commissioned by MSRI.

cost the Intel company something in the order of a hundred million dollars. (Intel later claimed that it had become aware of this bug independently and its internal documents do not mention Nicely.) If there is a moral to this story, perhaps it is that large number-theoretic computations can sometimes provide checks against defects in computer hardware and software!

The qualitative twin primes conjecture also has a quantitative version analogous to the PNT: that conjecture states that the number of twin prime pairs up to x is asymptotic to

$$2\Pi_2 \frac{x}{(\log x)^2}, \quad \text{where } \Pi_2 = \prod_p \left(1 - \frac{1}{(p-1)^2}\right) = 0.6601618\cdots. \tag{2.37}$$

This formula provides an important input for the estimation of Brun's constant B.

The twin primes conjecture has been vastly generalized to the so-called *prime tuples conjecture* of Dickson (1904, [**287**]), and Hardy & Littlewood (1923, [**495**]). It asserts that for any integer $k \geq 2$, the set $\{n + a_1, n + a_2, \ldots, n + a_k\}$ contains only primes for infinitely many values of n, as long as the $a_1, a_2, \ldots, a_k$ do not range over all values modulo p of any prime p. This means, for instance, that there must be infinitely many triples $(p, p+2, p+6)$ of primes,[14] while the case $k = 2$ with $(a_1, a_2) = (0, 2)$ corresponds to the twin primes conjecture. These conjectures can be further extended to general linear polynomials $m_j n + a_j$. All of these have quantitative forms with asymptotic formulas analogous to equation (2.37), as well as to equations (2.38) and (2.40). Ribenboim (1996, [**901**, Chapter 4]) provides some background.

As will by now be evident, it is far easier to make plausible conjectures about the primes than it is to prove such statements. The twin primes conjecture (whose statement is both understandable and engaging—even to schoolchildren) is one case in point. However, even when proofs are scarce, it is a fitting part of our story to talk about a number of other celebrated open problems.

2.6. Goldbach's conjectures

Among the many still-open problems concerning the primes, and almost equal to the twin primes conjecture in renown, is the question of whether or not every *even* number $n > 2$ can be written as a sum of two primes. (Symbolically, this question asks if $\mathcal{P} + \mathcal{P} = 2\,\mathbb{N}$, but with the exception that 2 is not a sum of two primes.) This is the well-known *Goldbach conjecture*, which dates back to an exchange of letters between Christian Goldbach and Leonhard Euler in 1742, although a posthumously published manuscript of Descartes shows that in the 1630s or 1640s he had already conjectured that *every* integer $n > 1$ is the sum of 1, 2, or 3 primes. This 'binary' conjecture of Goldbach's further implies the *ternary Goldbach conjecture* (also mentioned in the 1742 letters) that every *odd* integer $n > 5$ is a sum of three primes.[15] The ternary conjecture does not imply the binary one, although it does obviously imply that every even integer $n > 8$ is a sum of 4 primes.[16]

---

[14] Of course, $(p, p+2, p+4)$ must have one term divisible by 3, and cannot satisfy the requirements of the conjecture.

[15] A linguistic footnote: *'ternary'* refers to a grouping of three; *'tertiary'* means ranking in third place.

[16] A possible exercise for an adventurous reader is to show that Bertrand's postulate would follow from the (binary) Goldbach conjecture (Ricardo (2005, [**902**])).

Schnirelmann (1930/1939, [**951**]) proved, unconditionally, that there is a fixed K such that every number $n \geq 1$ is the sum of at most K primes. Ramaré (1995, [**882**]) proved that every even integer is a sum of at most 6 primes, while under GRH (a generalized version of RH) the value 6 can be reduced to 4. Deshouillers, Effinger, et al. (1997, [**277**]) showed, under RH, that *every* odd number greater than 5 can be expressed as a sum of three primes, and Tao (2014, [**1042**]) proved—unconditionally—that every odd number is the sum of at most 5 primes.

The validity of Goldbach's conjecture has been verified numerically up to at least 4×10^{18} (Oliveira e Silva et al., 2014, [**825**]). It is also known that 'by far most' even numbers are a sum of two primes; see, for example, Montgomery & Vaughan (1975, [**772**]). Using sieve methods and a version of the Bombieri–Vinogradov theorem, Jingrun Chen (1966/1973, [**194**]) proved that every large enough even integer is the sum of a prime and a number that is product of at most two primes.[17] A simpler proof of Chen's result was given by Ross (1975, [**915**]).

By now, satisfactory progress has been made on the ternary Goldbach conjecture. Assuming GRH, Hardy & Littlewood (1923, [**495**]) showed that every sufficiently large odd number was the sum of three primes. In 1937 Ivan Matveyevich Vinogradov (1891–1983)[18] succeeded in showing *unconditionally* that not only can all *odd* n be written as a sum of three primes once n is sufficiently large, but that one can do so in $O\left(n^2/\log^3 n\right)$ ways, and more precisely in

$$\Theta(n) \times \frac{n^2}{2\log^3 n}\left(1 + O\left(\frac{\log\log n}{\log n}\right)\right) \tag{2.38}$$

ways, where

$$\Theta(n) = \prod_p \left(1 + \frac{1}{(p-1)^3}\right) \prod_{p|n}\left(1 - \frac{1}{p^2 - 3p + 3}\right).$$

Although Vinogradov's proof was technical, it started from the striking observation—based on Hardy & Littlewood's (1923, [**495**]) so-called "circle method" for additive arithmetical problems—that the number of ways n can be written as a sum of three primes is given by

$$\int_0^1 \sum_{p,q,r \leq n} e^{2\pi i t(p+q+r-n)}\, dt = \int_0^1 E_n^3(t) e^{-2\pi i n t}\, dt, \quad \text{where } E_n(t) = \sum_{p \leq n} e^{2\pi i p t}, \tag{2.39}$$

---

[17]A product of two primes is sometimes called an *almost prime*. In connection with almost primes, using difficult sieve-theory arguments, Jingrun Chen obtained results for *three* outstanding problems: (i) Between large enough consecutive squares there is always a prime or an almost prime (as was mentioned in Section 2.1); (ii) Every large enough even integer is a sum of two primes or of a prime and an almost prime; and (iii) There are infinitely many primes p such that $p + 2$ is either a prime or an almost prime. The references are Jingrun Chen (1973, [**194**]), (1975, [**195**]), (1978, [**196**]).

[18]The history of twentieth-century mathematics in the Soviet Union is less than exemplary. It is well documented that as director of the Steklov Institute over some fifty years, Vinogradov zealously pursued the anti-semitic policies of the Stalinist regime and Communist Party. He initiated rigorously discriminatory hiring practices, obstructed career promotions and election to the Academy of Sciences of even the most outstanding Jewish mathematicians, arranged for them to be secretly investigated by the KGB, and actively prevented them from attending international conferences—even for receiving a Fields Medal. A few sources include Sossinsky (2007, [**1000**]), Monastyrsky (2008, [**762**]), and Hargittai & Hargittai (2020, [**501**]).

using p, q, and r here for primes. Observe that each term on the left with $p+q+r-n=0$ contributes 1 to the integral, while the remaining terms each contribute 0. (Similar identities hold for twin primes and other additive problems.) A careful (though technical) exposition of Vinogradov's proof is laid out in the third chapter of the monograph by Estermann (1952, [**367**]). It had been known that $n > e^{3100}$ suffices for Vinogradov's result (Liu & Wang, 2002, [**708**]). But finally, Helfgott (2014, [**535**]), together with the remarkable computations of Helfgott & Platt (2013, [**536**]), succeeded in proving that *every* odd integer greater than 5 is the sum of three primes. Therefore the *ternary* Goldbach conjecture is now fully resolved.

A quantitative version of the binary Goldbach conjecture is that the number of representations of an even integer n as the sum of two primes is asymptotic to

$$2\Pi_2 \left(\prod_{p|n,\, p\geq 3} \frac{p-1}{p-2} \right) \times \frac{n}{\log^2 n} \tag{2.40}$$

where Π_2 is the constant from the quantitative twin primes conjecture (2.37). Of course, this conjecture also remains open. The identity (2.39) of Hardy and Littlewood for the number of representations of n as the sum of three primes adapts easily to the identity

$$\int_0^1 \sum_{p,q,r\leq n} e^{2\pi i t(p+q-n)}\, dt = \int_0^1 E_n^2(t) e^{-2\pi i n t}\, dt$$

for the number of representations of n as the sum of two primes; unfortunately, current understanding of the exponential sum $E_n(t)$ is not strong enough to carry through Vinogradov's proof here, even if the Riemann hypothesis is assumed.

2.7. Other open problems

In addition to the twin primes and Goldbach conjectures, a large number of other problems concerning the primes—many of which can be stated very simply—remain unsolved. Is there always a prime between consecutive squares? Are there infinitely many primes of the form n^2+1? (Iwaniec (1978, [**592**]) had proved that infinitely many n^2+1 are either prime or a product of two primes.) Are there infinitely many *Sophie Germain primes*,[19] namely primes of the form $2p+1$ where p is also prime? Is every odd number greater than 5 the sum of a prime and twice a prime? Is every large integer either a square or the sum of a prime and a square?[20] Does there exist a (nonlinear) polynomial $P(n)$ that takes on

[19]Sophie Germain (1776–1831) was a French mathematician who made significant contributions to number theory and mechanics. The primes named after her played an essential role in her incisive attempts to prove Fermat's Last Theorem. Among her many gems is the nifty formula

$$x^4 + 4y^4 = \Big((x+y)^2 + y^2\Big)\Big((x-y)^2 + y^2\Big).$$

Much has been written about the gender bias of her era. Musielak (2020, [**786**]) provides an authoritative biography.

[20]The conjecture is stated in this way because there are infinitely many squares that are not equal to the sum of a prime and a square. To see this, note that if $n^2 = p + m^2$ then $p = (n+m)(n-m)$, forcing $n+m=p$ and $n-m=1$, so that $p = 2n-1$; this is possible for some n, but not for all n.

prime values infinitely often?[21] The answer to all of these questions is believed to be 'yes'. For the question about polynomials, what is known is that there are no nonconstant polynomials which assume *only* prime values when evaluated at integer arguments. On the other hand Niven (1951, [**813**]) proved that for every real number $\beta > 8/3$ there is a real number $\alpha > 1$ such that $\lfloor \alpha^{\beta^n} \rfloor$ is prime for *every* $n \geq 1$; but hey—that's cheating!

Among the many other unproven conjectures, we will mention only three here: Brocard's conjecture that there are at least 4 primes in the interval (p_n^2, p_{n+1}^2), that is, between the squares of consecutive primes exceeding 2; *Andrica's conjecture,* that $\sqrt{p_{n+1}} - \sqrt{p_n} < 1$ for all n or equivalently $p_{n+1} - p_n < 2\sqrt{p_n} + 1$; and Firoozbakht's conjecture, that $p_n^{1/n}$ is a strictly decreasing function. The second of these, believed to be true, is just a variant of Legendre's conjecture on the existence of primes between consecutive squares (Andrica, 1986, [**22**]), while the third, believed to be false, is a stronger version of Cramér's conjecture (2.1). A conjecture related to Andrica's is $\lim_{n\to\infty}(\sqrt{p_{n+1}} - \sqrt{p_n}) = 0$ or, equivalently, $p_{n+1} - p_n = o\left(\sqrt{p_n}\right)$. In an influential address at the 1912 International Congress of Mathematicians held at Cambridge, England, Edmund Landau laid out four problems that he considered to be out of reach by then-current methods: Goldbach's conjecture, the twin primes conjecture, the existence of primes between n^2 and $(n+1)^2$, and the infinitude of primes of the form $n^2 + 1$. Now, well over one hundred years later, not a single one of these problems has been solved.

Hardy and Littlewood once conjectured that the prime counting function satisfies the convexity property

$$\pi(x+y) \leq \pi(x) + \pi(y) \quad \text{for } x, y > 1\,. \tag{2.41}$$

Viewed as $\pi(x+y) - \pi(y) \leq \pi(x)$, it says that no interval of length x contains more primes than the first x integers. While this conjecture has not been settled, Hensley & Richards (1973, [**541**]) have shown that—rather surprisingly—it is incompatible with the prime tuples conjecture mentioned previously. Since the prime tuples conjecture is widely believed to be true, conjecture (2.41) is thought to be false, but with counterexamples occurring only at enormous values of x, although possibly with moderate values for y. It is generally believed that a correct modification of this conjecture is

$$\pi(x+y) \leq \pi(x) + 2\pi(y/2) \quad \text{for } x, y > 1\,. \tag{2.42}$$

The special case $\pi(2x) \leq 2\pi(x)$, $x \geq 1$ of (2.41) had been established by Landau (1909, [**676**]), while Montgomery & Vaughan (1973, [**771**]) proved that $\pi(x+y) \leq \pi(x) + 2y/\log y$ for all $x, y > 1$. Rosser & Schoenfeld (1975, [**920**]) established the variant $\pi(x+y) < \pi(x) + 2\pi(y)$ for $x, y > 1$, while Mincu & Panaitopol (2008, [**757**]) showed that $\pi(x)\pi(y) < \pi(xy)$ for $x, y \geq \sqrt{53}$. Further details may be found in Richards (1974, [**903**]), Udrescu (1975, [**1076**]), Vehka (1979, [**1085**]), Panaitopol (2001, [**831**]), and Cohen (2022, [**212**]).

---

[21] While it is not known if there are nonlinear polynomials (in one variable) which take on prime values infinitely often, it is known that (except for trivial cases) quadratic and cubic polynomials take on *squarefree* values (values not divisible by the square of a prime) infinitely often (see Filaseta (1992, [**379**])).

Other additive problems concerning primes. Let us also comment on the long-standing question of whether or not the set of primes contains arbitrarily long sequences in arithmetic progression. Van der Corput (1937, [**236**]) had shown that there are infinitely many primes in nontrivial arithmetic progressions of length 3 or, equivalently, infinitely many n such that $p_n + p_{n+2} = 2p_{n+1}$. For arithmetic progressions of arbitrary length, this was settled in the affirmative only many years later by Green & Tao (2008, [**464**]), using methods first developed by Szemerédi and Gowers, together with subtle pseudorandomness properties of the primes. Blomer (2008, [**111**]) and Conlon et al. (2014, [**215**]) offer expositions of the Green–Tao result.

With respect to arithmetic progressions, in 1976, Erdős made what is perhaps his most famous conjecture, namely that arbitrarily long arithmetic progressions must occur in *any* set $\mathcal{A}$ of integers with the property $\sum_{n\in\mathcal{A}} 1/n = \infty$, and offered a prize of \$3,000 for its proof. (See also Erdős & Turán (1936, [**363**]).) This conjecture has not been settled, and until recently it was not even known if it held for progressions of length three—a question which dates back to Roth (1953, [**923**]). Behrend (1946, [**85**]) showed that for all N there are subsets $\mathcal{A} \subset [1, N]$ consisting of $\#\mathcal{A} = N\exp\left(-c\sqrt{\log N}\right)$ elements that do not contain any progressions of length 3, while in the opposite direction, Bloom (2016, [**112**]) showed that if $\mathcal{A} \subset [1, N]$ contains no nontrivial three-term arithmetic progressions, then

$$\#\mathcal{A} = O\left(\frac{N(\log\log N)^4}{\log N}\right).$$

The gap between these two results is considerable. The $(\log\log N)^4$ factor here represented a fundamental barrier, for if it could be removed then it would at least be possible to establish the simplest (namely the length 3) case of Erdős's conjecture. This barrier was overcome by Bloom & Sisask (2020, [**114**]) who proved that if $\mathcal{A} \subset [1, N]$ contains no nontrivial three-term arithmetic progressions then

$$\#\mathcal{A} = O\left(\frac{N}{(\log N)^{1+c}}\right).$$

Behrend's $N\exp\left(-c\sqrt{\log N}\right)$ lower bound—which has remained essentially unchallenged since 1946—is believed to be the one closer to the truth; the correct lower bound is commonly conjectured to be of the form $N\exp\left(-c(\log N)^c\right)$, although not necessarily with Behrend's value of $c = \frac{1}{2}$. In that direction, Kelley & Meka (2023, [**631**]) established upper bounds of this form with $c = \frac{1}{11}$, which was then refined by Bloom & Sisask (2023, [**115**]) to $c = \frac{1}{9}$.

The remarkable thing about Erdős's \$3,000 conjecture is that if it proved to be true, it would mean that—aside from the fact that $\sum 1/p = \infty$—there is nothing special about the primes so far as arithmetic progressions go! Incidentally, long progressions of primes occur only at very large numbers and with very wide gaps. Records of these are maintained in the "Primes in Arithmetic Progression Records" page of the `primerecords.dk` website maintained by J. K. Andersen.

Underpinning the results mentioned in the two previous paragraphs is a deep result due to Szemerédi referred to above. A subset $\mathcal{A}$ of the positive integers is said to have *positive upper density* if

$$\limsup_{n\to\infty} \frac{|\mathcal{A} \cap \{1, 2, 3, \ldots, n\}|}{n} > 0.$$

(Note that the primes do *not* have this property.) Szemerédi's theorem states that if $\mathcal{A}$ has positive upper density then $\mathcal{A}$ must contain (infinitely many) arithmetic progressions of length k, for every value of k. (The case $k = 3$ is the one that was considered by Roth.) Szemerédi's proof had been based on intricate combinatorial reasoning, although his result would also follow if Erdős's $3,000 conjecture was proved.

There is no shortage of open problems concerning the primes, particularly those that entwine the additive and multiplicative properties of the integers. Here is an unusually interesting one which dates back at least to Ostmann (1956, [**827**, vol. I, p. 13])—a kind of inverse to the Goldbach conjecture which Erdős (1986, [**353**]) once referred to as "one of the most interesting unconventional problems on primes." If $\mathcal{A}$ and $\mathcal{B}$ are subsets of the positive integers, the *sumset* $\mathcal{A} + \mathcal{B}$ is defined as the set of all numbers $a + b$ where $a \in \mathcal{A}$ and $b \in \mathcal{B}$. The question is this: Do there exist two such sets, $\mathcal{A}$ and $\mathcal{B}$, each containing at least two elements, such that beyond some value x the sumset $\mathcal{A} + \mathcal{B}$ is identical to the set $\mathcal{P}$ of primes—in other words, such that $(\mathcal{A} + \mathcal{B})) \cap [x, \infty)$ and $\mathcal{P} \cap [x, \infty)$ are identical once x is large enough? One thing known for certain is that for this to hold the number of elements in $\{n \le x \colon n \in \mathcal{A}\}$ would have to lie between $O\big(\sqrt{x}/(\log x \log\log x)\big)$ and $O\big(\sqrt{x} \log\log x\big)$ as $x \to \infty$, and likewise for $\mathcal{B}$. Ostmann had conjectured that for the primes no such decomposition is possible. Nonetheless, Terence Tao & Tamar Ziegler (2023, [**1051**]) proved the following scarcely believable result: There exist infinite sequences $a_1 < a_2 < \cdots$ and $b_1 < b_2 < \cdots$ of positive integers such that $a_i + b_j$ is a prime whenever $1 \le i < j$. (The reader may wish to reflect on the mechanisms by which such a result could possibly be true.)[22] Additional background may be found in Elsholtz (2001, [**326**]), (2006, [**327**]), (2017, [**328**]), Elsholtz & Harper (2015, [**330**]), and Tao & Vu (2006, [**1049**]).

Collections of many other open problems in number theory may be found in Guy (2004, [**474**]), Sárközy (2001, [**942**]), Shanks (1985, [**981**]), Sierpiński (1964, [**989**]), and in many other sources. Erdős was a particularly prolific poser of open problems, many of which can be found among the many hundreds of papers he has authored.

A great many open problems about primes come from simple human curiosity, observation, and pattern spotting. Others arise from deliberate experimentation—playing with the prime numbers themselves. As it happens, however, random models provide yet another prolific source of predictions, expectations, and open problems about the primes. We will next introduce one such perspective.

2.8. Poisson behaviour of the primes

One major theme of this book is that even though the sequence of prime numbers is deterministic, it is useful to define certain random processes in an attempt to model and try to understand their behaviour. This point of view will be explored in depth in Chapter 4, but the preliminary discussion here will help motivate some of the further properties conjectured to hold for the primes. If needed, Appendix D (or a textbook on probability and statistics) may be consulted for definitions and details of the basic terms to be used below.

[22]The key here is Dickson's prime tuples conjecture (Tamar Ziegler, personal communication).

A random variable X is said to have a *Poisson distribution* with mean $\theta \geq 0$ if it takes on the integer values $n = 0, 1, 2, \ldots$ with probabilities given by

$$P[X = n] = \frac{\theta^n}{n!} e^{-\theta}, \quad n = 0, 1, 2, \ldots.$$

(It is the limiting distribution, as $k \to \infty$, of the number of 'successes' in k independent trials, each of which has success probability θ/k.) A *Poisson point process* $N(t)$ *on the line*, with constant *intensity* λ, can be represented as a piecewise constant function with jumps of height 1 (one thinks of the 'jumps' as occurring at 'points') and can be unambiguously defined by the following two requirements:

(i) The number of 'points' (that is, jumps) $N(t+u) - N(u)$ in any interval of length t has a Poisson distribution with mean λt; and
(ii) The numbers of points in nonoverlapping intervals are statistically independent.

This definition extends in a natural way to allow the intensity $\lambda = \lambda(t)$ to vary. One may generate a Poisson process $M(t)$ with varying intensity $\lambda(t)$ from a Poisson process $N(t)$ with constant intensity 1, and conversely, via the 'time-transformations' $M(t) = N(m(t))$ and $N(t) = M\left(m^{-1}(t)\right)$, respectively, where $m(t) \equiv \int_0^t \lambda(u)\,du$. Poisson processes are discussed in Durrett (2016, [**300**]), Resnick (1992, [**898**]), Ross (1996, [**916**]), and many other texts.

Among the essential properties that a Poisson process—say, with constant intensity λ—can be shown to possess are the following:

(a) The gaps between consecutive points of the process are independent and exponentially distributed with mean $1/\lambda$ and density function given by $f(x) = \lambda e^{-\lambda x}$, $x \geq 0$;
(b) If a position on the line is selected independently of the process, the distance to the next point of the process also has an exponential distribution with mean $1/\lambda$;
(c) If two independent Poisson processes with respective intensities λ_1 and λ_2 are 'superimposed', the result is a Poisson process with intensity $\lambda_1 + \lambda_2$;
(d) If the points of a Poisson process of intensity λ are split into two processes by independently assigning its points with probabilities θ and $1-\theta$, the result is two independent Poisson processes with intensities given by $\theta\lambda$ and $(1-\theta)\lambda$;
(e) If a sequence of $k > 2$ consecutive points are selected at random, the $k-2$ intermediate points are distributed as if they were chosen uniformly and independently between the two endpoints. This also holds for the points in an arbitrary independently chosen interval of the process, when we *condition* on the number of points that lie in that interval.

Note that property (b) is a consequence of the *memoryless property* of the exponential distribution: specifically, if X has an exponential distribution, and $s, t > 0$, then

$$P[\,X > s+t \,|\, X > s\,] \equiv \frac{P[\,\{X > s+t\} \cap \{X > s\}\,]}{P[\,X > s\,]} = \frac{P[\,X > s+t\,]}{P[\,X > s\,]} = P[\,X > t\,].$$

Here the leftmost term denotes a conditional probability, and the rightmost equality follows by a simple computation. It is a standard exercise in probability to show that the property of independent and exponentially distributed gaps implies that the distance between the first and the last of a randomly selected sequence of $k+1$

points (equivalently, the sum of k consecutive gaps) has a gamma distribution with density $f(x) = \lambda^k x^{k-1} e^{-\lambda x} / \Gamma(k)$.

Turning now to their 'probabilistic' behaviour, the primes appear to play a game of chance. For if p and q are distinct primes, the proportions of integers divisible by p and by q are $1/p$ and $1/q$, respectively, while the proportion divisible by *both* p and q (and therefore by pq) is given by the *product*, $1/pq = 1/p \times 1/q$. (One demonstrates this by considering randomly chosen integers over an interval that is large compared to pq.) Divisibility by any of a *fixed* number of given distinct primes $p_1, p_2, \ldots, p_k$ can therefore be thought of as independently occurring events.[23] It follows that over a wide enough range of integers, the "probability" that an integer is *not* divisible by any of primes $p_1, p_2, \ldots, p_k$ can be taken as

$$\prod_{j=1}^{k} \left(1 - \frac{1}{p_j}\right). \tag{2.43}$$

To state things differently, in the prime factor representation

$$n = \prod_{\text{all primes}} p_j^{M_j}, \tag{2.44}$$

fixed and finite collections of the M_j become statistically independent when viewed over a wide enough range for n. (The exponents M_j here are nonnegative integers with all but finitely many of them being 0.) This means that for any fixed k, any distinct $j_1, \ldots, j_k$, and *any* subsets $A_1, A_2, \ldots, A_k$ of nonnegative integers,

$$\begin{aligned} \lim_{N\to\infty} \mathrm{P}_N\left[M_{j_1} \in A_1, \ldots, M_{j_k} \in A_k\right] \\ = \lim_{N\to\infty} \left\{\mathrm{P}_N\left[M_{j_1} \in A_1\right] \times \cdots \times \mathrm{P}_N\left[M_{j_k} \in A_k\right]\right\}. \end{aligned} \tag{2.45}$$

Here the probability $\mathrm{P}_N[\,\cdot\,]$ is based on choosing n uniformly at random from among the integers in $[1, N]$, and subsequently letting $N \to \infty$. When the subsets A_j are *finite*, result (2.45) is straightforward; a proof for *arbitrary* A_j is given in Section 2.13. This independence, however, does *not* carry over to the case where the number, k, of primes is allowed to increase arbitrarily with N. One may see this by comparing (2.43) with (2.6) and recalling the comments that were made there about the $e^{-\gamma}$ factor.

Arguments such as these *suggest* (but certainly do not *prove*) that far up the number line the occurrence of primes behaves like a Poisson process. For instance, it is widely believed that (for fixed λ) the number of primes in $[x, x + \lambda \log x]$ follows a Poisson(λ) distribution as $x \to \infty$, and that the numbers of primes in nonoverlapping intervals of this type become statistically independent. Remarkably, none of these seemingly plausible conjectures has yet to be proven. For the number of primes in the interval $(n, n + \lambda \log n]$, Gallagher (1976, [**405**]) showed that

$$\frac{1}{x}\left|\left\{n \le x : \pi(n + \lambda \log n) - \pi(n) = k\right\}\right| \longrightarrow \frac{\lambda^k e^{-\lambda}}{k!}, \quad \text{as } x \to \infty;$$

[23] A worthwhile exercise: Show that, as $N \to \infty$, the proportion of integers $n \le N$ which are coprime to m (that is, share no common factors with m) tends to

$$\prod_{p|m} \left(1 - \frac{1}{p}\right).$$

he also showed that the normalized spacings $(p_{n+1} - p_n)/\log p_n$ are exponentially distributed in the sense that for $0 < \alpha < \beta$

$$\frac{1}{\pi(x)} \left| \left\{ p_n \leq x : \alpha \leq \frac{p_{n+1} - p_n}{\log p_n} \leq \beta \right\} \right| \longrightarrow \int_\alpha^\beta e^{-t} dt\,, \quad \text{as } x \to \infty.$$

But Gallagher's proofs assumed not only the *prime tuples conjecture* of Dickson, Hardy, and Littlewood, but also that its conjectured error term satisfies a degree of uniformity in the shape of the tuples. Under the same assumptions, Freiberg (2018, [**392**]) proved that the numbers of primes in finite numbers of consecutive intervals are asymptotically independent. The prime tuples assumption is, in a sense, at least as strong as the results proved; however, they are widely believed and so lend credence to the Poisson nature of the primes. Empirical evidence also appears to support this property. The Poisson viewpoint also suggests that if $q_0 < q_1 < q_2 \cdots < q_{k+1}$ are large consecutive primes, their successive differences should behave like independent exponential variables, and the $q_1, q_2, \ldots, q_k$ should behave like points sampled uniformly and independently in the interval $[q_0, q_{k+1}]$.

Questions of Poisson behaviour also arise for the occurrence of twin primes, as well as in other analogous contexts. The Poisson behaviour of the primes was studied by Polvani (1933, [**860**]),[24] as well as by Kosambi (1963, [**655**]), Stein & Ulam (1967, [**1025**]), Good & Churchhouse (1968, [**442**]), and Mosteller (1972, [**780**]), (1972, [**781**]), among others. Khan (2010, [**633**]) showed that Gallagher's Poisson result would hold unconditionally if restricted to integers whose number of prime factors does not deviate too far from 'typical values'. We will revisit the topic of Poisson behaviour in Chapter 4.

Before we return to the Prime Number Theorem and to some matters related to it, let us take a brief detour to describe a simple application of elementary number theory that has made a truly profound difference in the 'real world'.

2.9. Primality testing, integer factorization, and cryptography

In Chapter 1 we mentioned the important role that primes play in modern cryptography. Here we will offer a few brief remarks on this topic. Primes are in ample supply and are easily multiplied together. In the vicinity of 10^{100}, for example, the proportion of primes is around $1/\log(10^{100}) \approx 1/230$, while the proportion remaining after numbers divisible by the first few primes are removed is considerably higher. At the same time, the security of many encryption systems (such as RSA—see [**5**]) depends on the difficulty of factoring numbers that are products of primes hundreds of digits long. There are thus two practical problems.

The first of these is to determine whether or not a given number n is prime. Naively, this can be done with no more than $\sqrt{n}$ trial divisions, but since $\sqrt{n} \equiv e^{(1/2)\log n}$, this approach is exponential in the number of digits of n (which is proportional to $\log n$). Ideally, one would like to have an algorithm which can make the determination in *polynomial time*, that is, an algorithm whose number of operations is bounded by a polynomial in $\log n$. The problem of finding such an algorithm,

[24]Giovanni Polvani (1892–1970), a contemporary of Enrico Fermi, was a physicist at the University of Milan, having been dismissed from the University of Pisa after refusing to acquiesce to the demands of the fascist regime. Leonardo Gariboldi offers an interesting history of the era in his 2023 article "L'Istituto di Fisica di Polvani a Milano..." which appeared in the Quaderni di Storia Della Fisica.

assuming one existed, remained unsolved for a considerable time. The first primality testing algorithm to require less than exponential time was found by Adleman, Pomerance & Rumely (1983, [**4**]) and required $(\log n)^{O(\log\log\log n)}$ operations. This almost-polynomial time algorithm is able to determine the primality of numbers consisting of some 300 digits within minutes. A true polynomial time algorithm (based in part on a theorem of Fermat) was ultimately found by Agrawal, Kayal & Saxena (2004, [**7**]); it requires no more than $(\log n)^{15/2}$ operations. These authors also provide an exposition (with many references) on the history of the search for a *deterministic* polynomial time algorithm for primality testing. Lenstra & Pomerance (2019, [**688**]) showed that this algorithm actually requires only $(\log n)^6$ operations. Curiously, the running time of this algorithm depends, in part, on the distribution of the Sophie Germain primes referred to previously. There are also *probabilistic* algorithms for primality testing which can only determine whether or not n is prime to within an arbitrarily small probability of error; see, for example, Rabin (1976, [**872**]), (1980, [**873**]). Remarkably, allowing a small amount of randomness dramatically increases the speed with which primality can be determined. However, none of the faster-than-exponential-time primality algorithms produce the actual factors for a number determined to be composite.

Of course, once numbers exceed a certain size, testing primality becomes onerous, even with the fastest algorithms. The Electronic Frontier Foundation (EFF)[25] offers a prize of \$150,000 US dollars to the first person or group who provably discovers a prime with at least 100 million digits, and \$250,000 for a prime with at least one billion digits. Smaller prizes for one million and ten million digits had been awarded in 2000 and in 2009. The winning primes were all *Mersenne numbers*, that is, numbers of the form $2^p - 1$ where p is a prime; as it happens, there exists a specialized algorithm (the "Lucas–Lehmer primality test") that determines the primality of Mersenne numbers much faster than available algorithms that work for all integers.

The second, and far more difficult problem, is that of *factoring* large integers. Naively, it too can be solved with no more than $\sqrt{n}$ divisions, but unlike primality testing, no factorization algorithm is known that runs in polynomial time in the number of digits, and it is widely believed that no such algorithm exists. The best-known factorization algorithms require something like $O\left(n^{1/4+o(1)}\right)$ steps; see for example Costa & Harvey (2014, [**237**]). Crandall & Pomerance (2005, [**248**]) may be consulted for further background. The 2010 article by Thorsten Kleinjung and twelve other authors titled "Factorization of a 768-bit RSA modulus," which appears in [**6**], exemplifies the type of Herculean efforts required to factor a 232-digit integer. A 250-digit number that formed part of the "RSA Factoring Challenge" was also finally factored some ten years later as were many other large numbers in the ensuing years. Such efforts are an essential part of ensuring the reliability of encryption standards, and RSA Labs used to offer large prizes to those who could factor certain 'challenge numbers'. Polynomial time factoring algorithms *might*

[25]The EFF is a nonprofit organization dedicated to "defending civil liberties in the digital world." It successfully represented Daniel Bernstein (then a graduate student at University of California, Berkeley) in his 1995 suit against the armaments traffic and export bureaucracy of the US government which was impeding his right to disseminate a mathematical paper he wrote about an encryption algorithm he called 'Snuffle'. The courts ultimately ruled that "software source code was speech protected by the First Amendment and that the government's regulations preventing its publication were unconstitutional."

be possible using quantum computers together with an ingenious Fourier-based algorithm devised by Peter Shor (1994, [**986**]), (1997, [**987**]), but these have so far not become a 'threat'.

As cryptographic methods are outside our scope, we offer only a skeletal description of Rivest, Shamir & Adleman's groundbreaking protocol now called the "RSA algorithm."[26] Set $n = pq$ where p and q are two *huge* primes. Choose a number $1 \leq e < n$ that is relatively prime to $\varphi(pq) = (p-1)(q-1)$. (Here $\varphi(n)$ is the Euler phi-function discussed in Chapter 3, the number of integers relatively prime to n.) This property of e guarantees that one can find the (essentially unique) d such that the product ed is congruent to 1 modulo $\varphi(pq)$; they are 'multiplicative inverses' to each other in this modular arithmetic. The 'public key' (n, e) is broadcast openly. The private key is (n, d), and the value of d is kept secret. A message is partitioned into pieces about the size of n; each piece is converted to a number, say, M, and is then encrypted as $m = M^e$ modulo n, and sent openly. When received it is then decrypted as m^d modulo n, which reproduces M. Finding the multiplicative inverse d to e is easy when $\varphi(pq) = (p-1)(q-1)$ is known. The security of the RSA system is based on the fact that—without knowing the values of p and q—it is extremely difficult to determine d knowing only e and n. A few additional details will appear in Chapter 3.

For large numbers, primality testing and integer factoring algorithms both involve nontrivial mathematics. A few references include Aumasson (2018, [**44**]), Barak (2021, [**76**]), Bressoud (1989, [**147**]), Crandall & Pomerance (2000, [**248**]), Guy (1975, [**473**]), Riesel (1994, [**907**]), Wagstaff (2013, [**1095**]), and H. C. Williams (1984, [**1112**]), (1984, [**1113**]), but there are many others. Blake, Seroussi & Smart (1999, [**109**]), for example, deals with the more powerful (and still more technical) methods of cryptography based on elliptic curves.

2.10. Why should the PNT be true?

We 'close' this chapter with some remarks on an obvious question: Why should the Prime Number Theorem, $\pi(x) \sim x/\log x$, be true? Gauss and Legendre had conjectured asymptotic equivalents to it in the 1790s, as had Dirichlet in the 1830s, and although Riemann had developed the extraordinary apparatus it required, even he was unable to prove this basic, fundamental result. And neither could Chebychev, Mertens, or von Mangoldt. Counterintuitive as it may seem, it is unlikely that any *genuinely simple* proof of the Prime Number Theorem will ever be found. However, the following heuristic considerations may be of interest.

As one progresses up the number line, the 'local' proportion of integers that are composite must be increasing, because there are ever more primes that have an opportunity to divide them. The remaining numbers, namely the primes, are therefore in locally decreasing proportion. Hence $\pi(x)/x$ 'should' be asymptotic to a decreasing function. Equivalently, the sequence of prime numbers must approach infinity more quickly than the sequence of integers, suggesting that $p_n \sim n\beta(n)$ for some increasing function $\beta(n)$. The 'simplest' increasing functions are of form $\beta(n) = Cn^\epsilon$, with $C > 0$ and $\epsilon > 0$. But these would imply that $p_n \sim Cn^{1+\epsilon}$, contradicting the 'elementary' fact that $\sum 1/p$ diverges. The next simplest increasing

[26] Immediately after Rivest, Shamir & Adleman's 1978 publication of their 'RSA algorithm', the US National Security Agency contacted the authors to advise them that if they mailed their paper abroad, they could be prosecuted for dealing in armaments illegally.

functions are of the form $\beta(n) = C\log^{\alpha} n$ with $\alpha > 0$. (Logarithms are the natural limits of functions such as $\beta(n) = Cn^{\epsilon}$ because $\lim_{\epsilon\downarrow 0}(x^{\epsilon}-1)/\epsilon = \log x$, for $x > 0$.) But $\sum 1/(n\log^{\alpha} n)$ converges when $\alpha > 1$, and diverges when $\alpha \le 1$; so one 'believable' candidate is the bordering value $\alpha = 1$, corresponding to $p_n \sim Cn\log n$.

Of course, even if one accepts this specious argument, it still leaves open the question of why $C = 1$ (or equivalently, fixing $C = 1$ say, why the logarithm should have base e). Chebyshev did prove that if $p_n \sim Cn\log n$ for some constant C, then that constant must be $C = 1$. The argument is that $p_n \sim Cn\log n$ is equivalent to $\pi(x) \sim \frac{1}{C}x/\log x$, which in turn implies (using a standard "partial summation" argument) that $\sum_{p\le x}\frac{1}{p} \sim \frac{1}{C}\log\log x$; but Mertens's result (2.2) obliges the coefficient $\frac{1}{C}$ of $\log\log x$ to equal 1.

In any case, a key flaw of the argument is that $\pi(x)/x$ decreasing over the long run does not guarantee that it is asymptotic to a decreasing function. (Consider, for instance, $A(x)/\log x$ with $A(x)$ wobbling between 1 and 2, say.) Concerning these points, Montgomery & Wagon (2006, [**775**]) have offered an elementary argument to show that *if* $\pi(x)/x$ was in fact asymptotic to a decreasing function, then necessarily $\pi(x) \sim x/\log_e x$. (One may compare this with the results of Chebyshev (1852, [**192**]), (1852, [**193**]) which were described earlier.)

Schröeder (2006, [**955**]) offered a different heuristic argument: Assuming divisibility by different primes to be independent events, the logarithm of the 'probability', $P(x)$, that x is not divisible by any prime $\le x$ satisfies

$$\begin{aligned}\log P(x) &\approx \log\prod_{p<x}\left(1-\frac{1}{p}\right)\\ &= \sum_{p<x}\log\left(1-\frac{1}{p}\right) \approx -\sum_{p<x}\frac{1}{p}\\ &\approx -\sum_{n<x}\frac{P(n)}{n} \approx -\int_2^x \frac{P(t)}{t}\,dt.\end{aligned}$$

The first (approximate) equality is just the assumed independence, while the third holds because the omitted terms sum to a finite constant and so are inconsequential. The fourth 'equality' just replaces a sum over primes by a probability-adjusted sum over the integers. Expressed in terms of $Q(x) \equiv 1/P(x)$, which represents the average distance between primes, the leftmost and rightmost expressions of the display can be recast in the form

$$\log Q(x) \approx \int_2^x \frac{dt}{t\,Q(t)}.$$

Differentiating this gives $Q'(x)/Q(x) = 1/[x\,Q(x)]$, or $Q'(x) = 1/x$, whose solution is $Q(x) = \log_e x + C$ or, equivalently, $P(x) = 1/(\log x + C)$. This conclusion is equivalent to the PNT, and for $C = 0$, it yields the PNT in the form commonly stated. Of course, there is a weak point in this heuristic argument as well: it is the assumption of independence in the divisibility by an infinitely increasing sequence of primes.

Two further heuristic arguments that lead one to conjecture the Prime Number Theorem were given by Hardy (1937, [**488**]).

2.11. Error terms

When discussing the error incurred by approximating $\pi(x)$ by $\text{li}(x)$, we were careful to indicate the best error terms currently known. Elsewhere, however, to streamline the discussion, we have not always stated the best available results. In the interest of providing a handy reference, we include a summary here—for some approximations involving the primes—of the best error terms currently known. As the relevant functions often involve compositions with exponential functions, it can be difficult to compare their orders of growth. Let us therefore recall the 'crucial lemma' that appeared in Chapter 1:

If $f(x)$ *and* $g(x)$ *both tend to* ∞, *and* $\log f(x) = o(\log g(x))$, *then* $f(x) = o(g(x))$.

Recall too that the converse of this lemma is false, and also that the two occurrences of little-o here may not be replaced by two occurrences of big-O.

It will also be convenient to make use of the notation

$$\lambda(x) = (\log x)^{3/5}(\log\log x)^{-1/5},$$

which is related to the Korobov–Vinogradov zero-free region (see Chapter 6). Using the lemma above (or otherwise) one may deduce that for any positive constant c, the function $e^{-c\lambda(x)}$ tends to zero more rapidly than any negative power of $\log x$ but more slowly than any negative power of x. It will also be worthwhile to observe at the outset that as $x \to \infty$ the following order terms (other than the first) all tend to 0, and at successively faster rates:

$$O(1), \quad o(1), \quad O\left(\frac{1}{\log^\alpha x}\right), \quad O\left(e^{-c\sqrt{\log x}}\right), \quad O\left(e^{-c\lambda(x)}\right), \quad O\left(\frac{1}{x^\alpha}\right).$$

(Here $\alpha > 0$ and $c > 0$ are arbitrary positive constants.) Thus every order term of this sequence may be replaced by any that come before. On the other hand, if a factor of x is introduced into each of these, they would all tend to ∞ (except, perhaps, for the last) at successively slower rates.

The multiple error terms appearing below are ordered in increasing strength. Subscripting—as in $o_{\text{PNT}}(\cdots)$—indicates order terms that are *equivalent* to the Prime Number Theorem, $O_{\text{KV}}(\cdots)$ indicates order terms based on the Korobov–Vinogradov zero-free region (the best currently known), while $O_{\text{RH}}(\cdots)$ indicates order terms that are conditional on the Riemann hypothesis. Unsubscripted order terms do not require the PNT. With these conventions, with B defined as in equation (2.3), and with the similar constant

$$A = -\gamma - \sum_p \frac{\log p}{p(p-1)},$$

we have, as $x \to \infty$, the following approximations:

$$\sum_{p\le x} p = \frac{x^2}{2\log x} + o_{\text{PNT}}\left(\frac{x^2}{\log x}\right), \; O_{\text{KV}}\left(x^2 e^{-c\lambda(x)}\right), \; O_{\text{RH}}\left(x^{3/2}\log^2 x\right),$$

$$\sum_{p\le x}\frac{1}{p} = \log\log x + B + O\left(\frac{1}{\log x}\right), \; o_{\text{PNT}}\left(\frac{1}{\log x}\right), \; O_{\text{KV}}\left(e^{-c\lambda(x)}\right), \; O_{\text{RH}}\left(x^{-1/2}\log x\right),$$

$$\sum_{p\le x}\frac{\log p}{p} = \log x + A + O(1), \; o_{\text{PNT}}(1), \; O_{\text{KV}}\left(e^{-c\lambda(x)}\right), \; O_{\text{RH}}\left(x^{-1/2}\log^2 x\right),$$

$$\prod_{p\le x}\left(1-\frac{1}{p}\right)^{-1} = e^{\gamma}\log x + O(1),\ o_{\rm PNT}(1),\ O_{\rm KV}\left(e^{-c\lambda(x)}\right),\ O_{\rm RH}\left(x^{-1/2}\log^2 x\right),$$

$$\prod_{p\le x}\left(1-\frac{1}{p}\right) = \frac{e^{-\gamma}}{\log x} + O\left(\frac{1}{\log^2 x}\right), o_{\rm PNT}\left(\frac{1}{\log^2 x}\right), O_{\rm KV}\left(e^{-c\lambda(x)}\right),\ O_{\rm RH}\left(x^{-1/2}\right),$$

$$\prod_{p\le x}\left(1+\frac{1}{p}\right) = \frac{6e^{\gamma}}{\pi^2}\log x + O(1),\ o_{\rm PNT}(1),\ O_{\rm KV}\left(e^{-c\lambda(x)}\right),\ O_{\rm RH}\left(x^{-1/2}\log^2 x\right),$$

$$\prod_{p\le x}\left(1+\frac{1}{p}\right)^{-1} = \frac{\pi^2}{6e^{\gamma}\log x} + O\left(\frac{1}{\log^2 x}\right), o_{\rm PNT}\left(\frac{1}{\log^2 x}\right), O_{\rm KV}\left(e^{-c\lambda(x)}\right), O_{\rm RH}\left(x^{-1/2}\right),$$

$$\vartheta(x) \equiv \sum_{p\le x}\log p = x + o_{\rm PNT}(x),\ O_{\rm KV}\left(xe^{-c\lambda(x)}\right),\ O_{\rm RH}\left(\sqrt{x}\log^2 x\right),$$

$$\psi(x) \equiv \sum_{p^k\le x}\log p = x + o_{\rm PNT}(x),\ O_{\rm KV}\left(xe^{-c\lambda(x)}\right),\ O_{\rm RH}\left(\sqrt{x}\log^2 x\right),$$

$$\pi(x) = {\rm li}(x) + o_{\rm PNT}\Big({\rm li}(x)\Big),\ O_{\rm KV}\left(xe^{-c\lambda(x)}\right),\ O_{\rm RH}\left(\sqrt{x}\log x\right),$$

$$p_n = {\rm li}^{-1}(n) + o_{\rm PNT}\Big({\rm li}^{-1}(n)\Big),\ O_{\rm KV}\left(ne^{-c\lambda(n)}\right),\ O_{\rm RH}\left(\sqrt{x}\log^{5/2} x\right).$$

The $\psi(x)$ sum is over all p and all k such that $p^k \le x$. The functions $\vartheta(x)$ and $\psi(x)$ are the Chebyshev functions to be discussed in Chapter 3. Note that any of the Korobov–Vinogradov order terms $O_{\rm KV}\left(xe^{-c\lambda(x)}\right)$ may be replaced by the weaker version $O\left(xe^{-c\sqrt{\log x}}\right)$ due to Hadamard and de la Vallée Poussin. One useful reference is Walfisz (1963, [**1096**]).

2.12. Historical notes

Here are a few historical notes to supplement any that have already made and to provide some additional perspective. Of particular importance are three highly significant precursors to the 1896 proof of Prime Number Theorem. These are due, respectively, to Dirichlet (1837, [**290**]), to Chebyshev (1852, [**192**]), (1852, [**193**]), and to Mertens (1874, [**751**]), (1874, [**752**]).

In the first of these, building on earlier work of Leonhard Euler (1707–1783), Peter Gustav Lejeune-Dirichlet (1805–1859) proved that every arithmetic progression $a + bn$, $n = 0, 1, 2 \ldots$, whose first term and common difference are relatively prime, contains an infinite number of primes. In this work—which marks the beginnings of rigorous analytic number theory—Dirichlet introduced the concepts now known as Dirichlet series, Dirichlet characters, and Dirichlet L-functions. The subject of primes in arithmetic progressions is not covered in this book but detailed treatments of this elegant theory can be found in many texts on analytic number theory. Here we limit ourselves to remarking that—for a given b—all progressions $a+bn$ generated by the relatively prime values of $1 \le a < b$ contain (asymptotically) the same number of primes (namely around $\pi(x)/\varphi(b)$ primes $\le x$, where φ is the Euler phi function) but with some interesting differences which give rise to the so-called 'prime number races'. (See, for example, Granville & Martin, (2006, [**458**]) or Feuerverger & Martin, (2000, [**378**]).)

The work of Pafnuty Lvovich Chebyshev (1821–1894)—which predates the PNT by almost fifty years—provided the first *rigorous* result concerning the growth rate of the primes counting function $\pi(x)$. As mentioned earlier in this chapter, in the context of proving Bertrand's postulate, Chebyshev established the result

$$0.92129 \leq \liminf_{x\to\infty} \frac{\pi(x)}{x/\log x} \leq \limsup_{x\to\infty} \frac{\pi(x)}{x/\log x} \leq 1.10555.$$

In other words, $\pi(x)/(x/\log x)$ stays between 0.92129 and 1.10555 in the $x \to \infty$ limit. Chebyshev's *tour-de-force* work of 1848 and 1850 is engagingly documented in Delone (2005, [**273**, Chapter 1]). Chebyshev was wealthy, and by no means immodest; Butzer & Jongmans (1999, [**175**]) provide some biographical details.

The third of these precursors to the PNT are the seminal results (2.2), (2.5), (2.6), and (2.7), all essentially due to Franz Mertens (1840–1927). While these results all follow from the PNT by elementary arguments (other than the value of B in (2.3)), the historical significance of Mertens's results stems from the fact that they predate the PNT by over twenty years, and that they had eluded all earlier attempts to prove them.

The 1896 proofs of the Prime Number Theorem made nontrivial use of complex analysis and it became widely believed that no 'elementary' proof of the PNT was possible. Here, 'elementary' is not synonymous with 'simple' but rather refers to a proof that does not use either complex analysis or Fourier analysis. In a 1921 lecture to the Mathematical Society of Copenhagen, Hardy summed up the prevailing opinion that a proof of the PNT "not fundamentally dependent [on complex analysis] seems to me extraordinarily unlikely", and he added: "If anyone produces an elementary proof ... [they] will show ... that the subject does not hang together in the way we have supposed, and that it is time for the books to be cast aside and for the theory to be rewritten" [**1012**]. Nevertheless, in 1949 both Atle Selberg and Pál Erdős stunned the mathematical world by publishing 'elementary' (but nontrivial) proofs of the PNT. These were based on some new results of Selberg and led immediately to a long standing dispute over priority, no doubt made all the more bitter by the radically different personalities and working styles of these two men. To quote Ernst G. Straus:

> *It has always been a source of great surprise and regret to me that two such superb minds and admirable human beings* [Erdős and Selberg] *have come to a permanent parting over a joint achievement that had been born with so much joy and hope...* [**1012**].

Details of this unfortunate feud may be found in Goldfeld (2003, [**419**]), and Spencer & Graham (2009, [**1012**]). The second of these includes comments by E. G. Straus and C. Pomerance. See also Baas & Skau (2008, [**51**]).

The 'elementary' proofs of Selberg and Erdős were later simplified somewhat by Newman (1980, [**802**]), Korevaar (1982, [**652**]), and Daboussi (1984, [**254**]). Daboussi's proof is described in Schwarz & Spilker (1994, [**960**, Chapter II, §9]), and is elegantly summarized in Tenenbaum & Mendès-France (2000, [**1055**, Chapter 4]). Other simplified versions were obtained by Hildebrand (1986, [**552**]), as well as by J. P. Kahane whose Fourier-based proof is described by J. B. Bost in the volume by Berline & Sabbah (2002, [**88**]). Diamond (1982, [**283**]) and de Koninck & Luca (2012, [**650**, Chapter 5]) give nice accounts of the ideas that are involved, as do Bateman & Diamond (2004, [**82**, Chapter 5]), Levinson (1969, [**694**]), Rose (1994, [**914**, Chapter 13]), Schwarz (1994, [**958**]), and Zagier (1997, [**1123**]). Not very well known is a proof by Norbert Wiener (1932, [**1104**]), whose only non-elementary use of the zeta function is that it does not vanish on the line $\Re s = 1$, but otherwise uses only real-variable (although nontrivial) methods. The 'elementary' proofs of the Prime Number Theorem are not easy reading but no 'simpler' proofs are known. Curiously, the 'elementary' proofs have so far not resulted in any significant innovations to the general theory.

The legend of Pál Erdős. Because the life and work of Pál Erdős (1913–1996) has been described in so many sources we will limit our comments about this remarkable man. A first remark is that he believed in 'The Book'—one in which God recorded the most elegant proofs possible. An avowed atheist, Erdős held that "You don't have to believe in God, but you should believe in The Book."[27] A second remark is that Erdős gave away essentially all of his earnings by funding prizes for solutions to unsolved problems of interest to him. These ranged from around \$25 for problems judged to be just outside current reach, to thousands of dollars for problems judged to be difficult and important, his largest prize having been \$10,000 for the result of Ford, Green, Konyagin, Maynard, and Tao described in equation (2.35).

Our more detailed comments about Erdős pertain to the fact that during the Hoover–McCarthy era, the FBI maintained a dossier on him consisting of hundreds of pages. All quotes below are from FBI files which were declassified in 2014. They had been redacted in order to protect their sources.

> In view of the information received from the Washington Field Office, it appears desirable to conduct further investigation as to this subject. ... Confidential informants who have furnished reliable data in the past, as well as other Government agencies, have provided voluminous data concerning ERDOS [sic]... [**783**].

The entries obsess about whether Erdős had any communist leanings. (He did not.) It notes, "Subject reportedly offered position at the University of California ... but declined to take loyalty oath [non-communist pledge] ... based on subject's policy of academic freedom rather than disloyalty." "He has repelled [sic] at being asked the question by any government as to whether he is or has been a member of the Communist Party (CP). He reacts violently to such questions, and refuses to comply with the various immigration requirements of certain governments." "It was T-1's impression that the subject was a communist at heart, and the fact that, as recently as the past year, he was granted a Hungarian passport despite his lengthy absence from Hungary, gives rise to the suspicion that there were strings attached to the granting of the document." "... the informant of the Washington Field Office expressed the opinion that it is indicated that ERDOS must have very influential connections in Budapest."

Eventually, all this was clarified: "While not identified as a communist, ERDOS has maintained his ties with communist Hungary, allegedly to protect the pension being paid to his aged mother by Hungary." Furthermore "... by virtue of his continuing ability to travel he serves as a walking propaganda advertisement for the

---

[27]It goes without saying that mere mortals cannot be the final arbiters or what belongs in "The Book." However, certain expositions, for example the one by Tenenbaum & Mendès-France (2000, [**1055**]) to name only one, approach it asymptotically.

Hungarian academic freedom." "...said that he considered ERDOS very reliable in every respect and an individual who has great affection for the United States."

Personal descriptions of Erdős which appear in the files include: "According to the informant, he originally left Hungary in 1934 due to provisions ... which restricted students of Jewish ancestry who could enter college ... The informant stated that ERDOS ... apparently was very much against the Hungarian Horthy Regime." "... advised that ERDOS ... has never held a permanent position for any period of time, as he prefers to travel around the United States and live off of his friends." "... advised that the subject is one of the top mathematicians in the world ... with typical atmospheric mind as related to factual things, that is, he is of the genius type who lives within his own mental scope, and that it is difficult to know him personally." "ERDOS is an outgoing, extroverted individual who is quite unconventional and has a wide range of knowledge on many subjects ... he delights in political discussions where he might take different positions just for the sake of argument." "... described ... as the type of person who, if engaged in research work of a secret nature, would probably divulge the results of his research to a foreign power in the misguided belief that his actions were serving the best interests of humanity as a whole."

From about 1954 until 1963, Erdős was denied entry to the US. The files state: "ERDOS, in 1954 ... was told by immigration authorities at that time he could remain in the United States but if he departed he could not come back." "... ERDOS had told him ... he considered this an unwarranted infringement on his liberties and that he would leave just because they said he couldn't come back." Erdős is also quoted as having said "I did not apply for U.S. citizenship since I am stateless by political conviction. I left (America) without a re-entry permit since I object to Iron Curtains—both Joe's (Stalin) and Sam's (U.S.)." Later it was revealed that "... his application for a re-entry permit was denied by the Immigration and Naturalization Service (INS) because his planned departure from the United States was deemed not in the best interest of the United States" [**783**].

Further biographical and other details about Erdős may be found in many sources, including Babai, Pomerance & Vertesi (1988, [**52**]), Baker, Bollobás & Hajnal (1990, [**68**]), Hoffman (1998, [**559**]), Rusza (1996, [**928**]), (1999, [**929**]), Schechter (1998, [**949**]), and Sós (2002, [**999**]). Additional historical information about number theory more generally may be found in Bateman & Diamond (1996, [**81**]), Goldfeld (2003, [**419**]), Goldstein (1973, [**420**]), and Narkiewicz (2000, [**797**]), among many other sources, not to mention the historical notes appearing in many textbooks.

2.13. Selected proofs

PROOF CONCERNING PRIMES BETWEEN n^θ AND $(n+1)^\theta$. The result of Baker, Harman & Pintz (2001, [**70**]) is that there is eventually always a prime in $[n, n+n^\theta]$, for $\theta = 21/40 = 0.525$. That this implies there is eventually always a prime between n^θ and $(n+1)^\theta$ for any $\theta > 40/19$ follows via Taylor approximation from the fact that if $x = n^{40/19}$, then

$$(n+1)^{40/19} \approx n^{40/19} + (40/19)n^{40/19-1} = x + (40/19)x^{21/40}. \qquad \square$$

PROOFS OF (2.2) AND THE SECOND EQUATION OF (2.10). Specifically, we will show that

$$\sum_{p\le x}\frac{\log p}{p}=\log x+O(1),\quad \text{and}\quad \sum_{p\le x}\frac{1}{p}=\log\log x+\text{Const}+O\left(\frac{1}{\log x}\right).$$

We first establish two preliminary results involving the function $\Lambda(n)$, defined as $\Lambda(n)=\log p$ if $n=p^k$ for some $k\ge 1$, and $\Lambda(n)=0$ otherwise. The two results are

$$\psi(x)\equiv\sum_{n\le x}\Lambda(n)=O(x), \tag{2.46}$$

and

$$\sum_{n\le x}\frac{\Lambda(n)}{n}=\log x+O(1). \tag{2.47}$$

To prove (2.46), we count occurrences of $\log p$ in $\psi(x)$ to obtain

$$\psi(x)\equiv\sum_{p\le x}\left\lfloor\frac{\log x}{\log p}\right\rfloor\log p\le\sum_{p\le x}\frac{\log x}{\log p}\log p=\pi(x)\log x=O(x).$$

Here $\lfloor\cdot\rfloor$ is the 'integer part' (or 'floor') function, and the last step uses Chebyshev's $\pi(x)=O(x/\log x)$ result proved further below; it does *not* require the PNT. To tackle (2.47), use integers N in place of x (for convenience). Estimating the sum $\sum_{n\le N}\log n$ by the corresponding integral $\int_1^N\log x\,dx$ of the increasing function $\log x$ quickly gives

$$\sum_{n\le N}\log n=N\log N-N+O(\log N). \tag{2.48}$$

However we alternately have, as a simple exercise,

$$\sum_{n\le N}\log n=\log N!=\sum_{p\le N}\left(\sum_{k\ge 1}\left\lfloor\frac{N}{p^k}\right\rfloor\right)\log p \tag{2.49}$$

(the inner sum is just the exponent of p in the prime factorization of $N!$) and this equals

$$\sum_{n\le N}\left\lfloor\frac{N}{n}\right\rfloor\Lambda(n)=\sum_{n\le N}\frac{N}{n}\Lambda(n)+O\left(\sum_{n\le N}\Lambda(n)\right)=N\sum_{n\le N}\frac{\Lambda(n)}{n}+O(N), \tag{2.50}$$

since the sum within the order term is just $\psi(N)=O(N)$ by (2.46). The result (2.47) follows on equating (2.48) with (2.50) and dividing by N.

We may now tackle $\sum_{p\le x}\log p/p$ by observing that

$$\sum_{p\le x}\frac{\log p}{p}=\sum_{n\le x}\frac{\Lambda(n)}{n}-\sum_{k\ge 2}\sum_{p^k\le x}\frac{\log p}{p^k}.$$

The required result follows by (2.47) since the double sum is bounded above by

$$\sum_p\left(\frac{1}{p^2}+\frac{1}{p^3}+\cdots\right)\log p=\sum_p\frac{\log p}{p(p-1)}\le\sum_{n>2}\frac{\log n}{n(n-1)}=O(1).$$

Finally, to tackle $\sum_{p\le x} 1/p$, first re-express the last result as

$$L(x) \equiv \sum_{p\le x} \frac{\log p}{p} = \log x + R(x). \text{ where } R(x) = O(1),$$

Then—using Riemann-Stieljes integration by parts (see Appendix F)—argue as follows:

$$\begin{aligned}
\sum_{p\le x} \frac{1}{p} &\equiv \int_{2-}^{x} \frac{dL(u)}{\log u} \\
&= \int_{2-}^{x} \frac{d\log(u)}{\log u} + \int_{2-}^{x} \frac{dR(u)}{\log u} \\
&= \int_{2-}^{x} \frac{du}{u\log u} + \left[\frac{R(u)}{\log u}\right]_{2-}^{x} - \int_{2-}^{x} R(u)\, d\left(\frac{1}{\log u}\right) \\
&= (\log\log x - \log\log 2) + \left(\frac{R(x)}{\log x} + 1\right) + \int_{2-}^{x} \frac{R(u)\, du}{u\log^2 u}
\end{aligned}$$

where the "+1" of the middle term arises from the fact that $R(2^-) = -\log 2$ by virtue of the definition of $R(\cdot)$. Using $R(x) = O(1)$, the result now follows because the $R(x)/\log x$ term is $O(1/\log x)$, while the last integral can be partitioned as $\int_2^\infty - \int_x^\infty$, in which the first piece is a constant, and the second is $O(1/\log x)$. □

REMARKS 2.1.

(i) The function $\Lambda(n)$ appearing here is the von Mangoldt function, while $\psi(x)$ is one of the Chebyshev functions; they are discussed in Chapter 3. It is in fact known that $\psi(x) \sim x$ and also that $\sum_{n\le x} \Lambda(n)/n = \log x - \gamma + o(1)$ with these each being equivalent to the PNT. The proof here, however, has used no more than (2.46) and (2.47).

(ii) The deduction of equation (2.48) can be phrased as a special case of Abel's summation formula (equation (F.3) of Appendix F) with $a(n) = 1$, $A(x) = \lfloor x \rfloor$, and $f(x) = \log x$:

$$\begin{aligned}
\sum_{n\le N} \log(n) &= A(N)\log N - \int_1^N A(y)\frac{1}{y}\, dy \\
&= N\log N - \int_1^N y\frac{1}{y}\, dy + \int_1^N (y - \lfloor y \rfloor)\frac{1}{y}\, dy \\
&= N\log N - (N-1) + O(\log N).
\end{aligned}$$

The order term arises from the last integral which is bounded by the integral $\int_1^N (1/y)\, dy$.

(iii) Mertens's result (2.2) for $\sum 1/p$ is historically important because it predates the PNT and is not based on it. However, if one assumes the PNT, so that $\pi(x) \sim x/\log x$ and $p_n \sim n\log n$, then Mertens's result follows very easily, for then, as $x \to \infty$

$$\sum_{p\le x} \frac{1}{p} \equiv \sum_{n\le\pi(x)} \frac{1}{p_n} \sim \sum_{n\le\pi(x)} \frac{1}{n\log n} \sim \log\log\pi(x) \sim \log\log x.$$

Here we have approximated $\sum 1/(n\log n)$ by $\int (1/x\log x)\, dx$ and used the fact that $\int_2^x dt/(t\log t) = \log\log x - \log\log 2$.

(iv) Alternate proofs of Mertens-type results may be found, for example, in Baker (2012, [**67**, §13.5, pp. 154–156]), de Koninck & Luca (2012, [**650**, §4.3 & 4.4, pp. 53–56]), and Stopple (2003, [**1034**, §7.1, pp. 162–174]).

(v) Evaluating the constant in the $\sum 1/p$ result (as shown in (2.2)) is considerably more difficult. Details may be found in Ingham (1932, [**580**, p. 23]), Hardy & Wright (2008, [**499**, §22.8]), Montgomery & Vaughan (2006, [**774**, Theorem 2.7]), and numerous other texts.

PARTIAL PROOFS OF (2.5), (2.6), (2.7) & (2.8) CONCERNING THE PRODUCT $\prod_{n\le x}(1\pm 1/p)^{\pm 1}$. We will assume the result (2.2) throughout, namely

$$\sum_{p\le x} 1/p = \log\log x + B + O(1/\log x)$$

with B as in (2.3). Using the Taylor expansion $\log(1-x) = -\sum_{k=1}^{\infty} x^k/k$, $|x|<1$, we get

$$\begin{aligned}
\log\left(\prod_{p\le x}\left(1-\frac{1}{p}\right)\right) &= \sum_{p\le x}\log\left(1-\frac{1}{p}\right)\\
&= -\sum_{p\le x}\sum_{k=1}^{\infty}\frac{1}{kp^k}\\
&= -\sum_{p\le x}\frac{1}{p} - \sum_{p\le x}\sum_{k=2}^{\infty}\frac{1}{kp^k}.\\
&= -\sum_{p\le x}\frac{1}{p} - \sum_{p}\sum_{k=2}^{\infty}\frac{1}{kp^k} + \sum_{p>x}\sum_{k=2}^{\infty}\frac{1}{kp^k}\\
&= -\log\log x - B - \sum_{p}\sum_{k=2}^{\infty}\frac{1}{kp^k} + O\left(\frac{1}{\log x}\right)\\
&= -\log\log x - \gamma + O\left(\frac{1}{\log x}\right).
\end{aligned}$$

Here have used the (2.2), the formula (2.3) for B, and the fact that

$$\sum_{p>x}\sum_{k=2}^{\infty}\frac{1}{kp^k} < \sum_{p>x}\sum_{k=2}^{\infty}\frac{1}{p^k} = \sum_{p>x}\frac{1}{p(p-1)} < \sum_{n>x}\frac{1}{n(n-1)}.$$

This converges to 0 at the rate $1/x$ (consult Appendix E), and therefore faster than $1/\log x$. On taking the exponential of both sides in the result just obtained we arrive at

$$\begin{aligned}
\prod_{p\le x}\left(1-\frac{1}{p}\right) &= \exp\left\{-\log\log x - \gamma + O\left(\frac{1}{\log x}\right)\right\}\\
&= \frac{e^{-\gamma}}{\log x}\times\exp\left\{O\left(\frac{1}{\log x}\right)\right\}\\
&= \frac{e^{-\gamma}}{\log x}\times\left\{1+O\left(\frac{1}{\log x}\right)\right\}\\
&= \frac{e^{-\gamma}}{\log x} + O\left(\frac{1}{\log^2 x}\right),
\end{aligned} \tag{2.51}$$

establishing (2.6). The equivalence of (2.5) and (2.6) follows as they are just reciprocal.[28]

To get (2.7), observe that

$$\prod_{p\le x}\left(1+\frac{1}{p}\right) = \left\{\prod_{p\le x}\left(1-\frac{1}{p}\right)^{-1}\right\}\left\{\prod_{p\le x}\left(1-\frac{1}{p^2}\right)\right\}$$

and that the last factor tends to $1/\zeta(2) = 6/\pi^2$. The equivalence of (2.7) and (2.8) follows as they are reciprocal as well. □

REMARK 2.2. The results (2.2), (2.5), (2.6), (2.7), and (2.8) are in fact all equivalent to one another. This may be established using a result such as

$$\sum_{p\le x}\left(\log\left(1-\frac{1}{p}\right)+\frac{1}{p}\right) = B-\gamma+O\left(\frac{1}{\log x}\right) \tag{2.52}$$

(compare with equation (2.3)) which allows one to relate sums in $1/p$ with products involving $(1-1/p)$. For instance, taking logarithms in (2.6) gives

$$\sum_{p\le x}\log\left(1-\frac{1}{p}\right) = -\gamma-\log\log x+O\left(\frac{1}{\log x}\right),$$

which, when combined with (2.52), yields

$$\sum_{p\le x}\frac{1}{p} = \log\log x + B + O\left(\frac{1}{\log x}\right).$$

PROOF OF CHEBYSHEV'S THEOREM (2.13). This theorem states that there exist constants $0 < C_1 \le 1 \le C_2 < \infty$ such that

$$C_1\frac{x}{\log x} \le \pi(x) \le C_1\frac{x}{\log x}, \quad \text{for } x > 1. \tag{2.53}$$

Chebyshev proved this result for $C_1 \approx 0.92129$ and $C_2 \approx 1.10555$. The proof given here is simpler; it yields worse values for C_1 and C_2, but it does establish the asymptotic rate.

An upper bound can be obtained using an inequality due to Erdős and Kalmar:

$$\prod_{p\le n} p \le 4^n.$$

It may be proved by induction. It certainly holds for $n = 1$, 2, and 3. Now if $n > 3$ is even, it cannot be prime, so $\prod_{p\le n} p = \prod_{p\le n-1} p$, and the result follows trivially. If n is odd, write $n = 2m+1$, and observe that

$$\prod_{m+1<p\le 2m+1} p \quad \text{divides} \quad \binom{2m+1}{m} \equiv \frac{(2m+1)!}{m!(m+1)!} \quad \text{evenly}$$

[28]In particular, taking reciprocals in (2.51) gives

$$\begin{aligned}\exp\left\{\log\log x+\gamma+O\left(\frac{1}{\log x}\right)\right\} &= e^{\log\log x}e^{\gamma}e^{O(1/\log x)}\\ &= e^{\gamma}\log x\left(1+O\left(\frac{1}{\log x}\right)\right)\\ &= e^{\gamma}\log x+O(1).\end{aligned}$$

because each of the indicated primes appears only in the numerator, and not in the denominator. But the binomial expansion of $(1+1)^{2m+1}$ contains $\binom{2m+1}{m}$ twice, once as $\binom{2m+1}{m}$, and once as $\binom{2m+1}{m+1}$, so

$$\binom{2m+1}{m} \leq \frac{1}{2}(1+1)^{2m+1} = 4^m.$$

Consequently, by induction,

$$\prod_{p\leq n} p = \prod_{p\leq m+1} p \times \prod_{m+1<p\leq 2m+1} p \leq 4^{m+1}4^m = 4^n,$$

thus proving the inequality.

Using this inequality and any r we immediately get

$$r^{\pi(n)-\pi(r)} \leq \prod_{r<p\leq n} p \leq 4^n,$$

and taking logarithms gives

$$\pi(n) \leq \frac{n\log 4}{\log r} + \pi(r) \leq \frac{n\log 4}{\log r} + r.$$

Choosing $r = \epsilon n/\log n$ for any $\epsilon > 0$, for example, then yields

$$\pi(n) \leq \frac{n}{\log n}\left(\log 4 \, \frac{\log n}{\log n - \log\log n + \log\epsilon} + \epsilon\right) \tag{2.54}$$

from which the upper bound in (2.53) follows.

To establish a lower bound we first define d_n to be the lowest common multiple of $\{1, 2, \ldots, n\}$, and then observe that $m\binom{n}{m}$ divides d_n. To see this, express these two integers in their prime factorizations:

$$d_n = p_1^{r_1}p_2^{r_2}\cdots p_k^{r_k} \quad\text{and}\quad m\binom{n}{m} = p_1^{s_1}p_2^{s_2}\cdots p_k^{s_k},$$

where the $p_1, p_2, \ldots, p_k$ represent all of the primes $p \leq n$. It suffices to show that the r and s exponents corresponding to any p satisfy $s \leq r$. Note that $r = \lfloor \log n/\log p\rfloor$. Now

$$m\binom{n}{m} = \frac{n\cdot(n-1)\cdots 2\cdot 1}{\{(m-1)\cdot(m-2)\cdots 2\cdot 1\}\times\{(n-m)\cdot(n-m-1)\cdots 2\cdot 1\}}.$$

The numerator contains $a = \lfloor n/p\rfloor$ terms divisible by p. Likewise, the factorial expansions in the denominator contain $b = \lfloor (m-1)/p\rfloor$ and $c = \lfloor (n-m)/p\rfloor$ terms divisible by p, respectively, for a total of $b+c$ factors in the denominator which are divisible by p. Observe, however, that $a-(b+c)$ equals either 0 or 1. Of course, we must also account for the additional powers of p that arise from terms that are divisible by p^2. An identical argument applies involving the counts $\lfloor n/p^2\rfloor$, $\lfloor (m-1)/p^2\rfloor$, and $\lfloor (n-m)/p^2\rfloor$. We continue in this way until we reach p^r and its corresponding counts $\lfloor n/p^r\rfloor$, $\lfloor (m-1)/p^r\rfloor$, and $\lfloor (n-m)/p^r\rfloor$. The result follows because the total number of 1's that will arise in this way (and therefore the total number of p-factors) will not exceed r.

We next use the established divisibility $m\binom{n}{m} \mid d_n$ to show that $d_n \geq 2^n$ provided $n \geq 7$. First note that

$$n\binom{2n}{n} \quad\text{and}\quad (2n+1)\binom{2n}{n}$$

both divide d_{2n+1}, the first because it divides d_{2n} which divides d_{2n+1}, and the second because it is the same as $(n+1)\binom{2n+1}{n+1}$. But because n and $2n+1$ are relatively prime it follows that $n(2n+1)\binom{2n}{n}$ also divides d_{2n+1} and therefore that

$$d_{2n+1} \geq n(2n+1)\binom{2n}{n} \geq n(1+1)^{2n}, \quad n \geq 1,$$

the last inequality following from the fact that $\binom{2n}{n}$ is the largest of the $2n+1$ terms in the binomial expansion of $(1+1)^{2n}$. Hence $d_{2n+1} \geq 2^{2n+1}$ if $n \geq 2$, and $d_{2n+2} \geq d_{2n+1} \geq 2^{2n+2}$ if $n \geq 4$, so the $d_n \geq 2^n$ result follows.

A lower bound result for (2.53) then follows immediately from the fact that

$$d_n = \prod_{p^r \| d_n} p^r \leq \prod_{p \leq n} n = n^{\pi(n)},$$

(because $p^r \leq n$ here) upon taking logarithms and using $d_n \geq 2^n$. □

REMARKS 2.3.

(i) Numerous proofs for this theorem are known yielding assorted upper and lower bounds $C_1 < 1 < C_2$. The proof here is modelled after the one in Tenenbaum (2015, [**1053**, Chapter 2, §6]); it gives $C_1 = \log 2$, while the upper bound (2.54) gives any value of C_2 greater than $\log 4$ once n is large enough. Alternate proofs can be found in many texts, for instance Baker (2012, [**67**, Chapter 13, §3]), Fine & Rosenberger (2007, [**381**, §42]), and Overholt (2015, [**828**, Chapteri]). de Konink & Luca (2012, [**650**, Chapter 2, §6]) obtain a lower bound by a method of Erdős which is simpler than the one used above, while Tenenbaum & Mendès-France (2000, [**1055**, Chapter 1]) obtain weaker constants but with a particularly elegant proof.

(ii) The proof here that $m\binom{n}{m}$ divides d_n is related to a formula of Legendre which states that the maximum power of a prime p that will divide $n!$ is given by the terminating sum $\sum_{k\geq 1}\lfloor n/p^k \rfloor$, where $\lfloor x \rfloor$ is the floor function (the integer part of x). (Indeed we had used this formula earlier in (2.49).)

(iii) We leave it as an exercise to determine the C_1, C_2 that follow from the given proof.

(iv) Bertrand's postulate is a direct consequence of Chebyshev's Theorem given sufficiently strong constants C_1 and C_2, since

$$\pi(2n) - \pi(n) \geq C_1\frac{2n}{\log(2n)} - C_2\frac{n}{\log n} = \left(2C_1 - C_2 - \frac{C_1 \log 4}{\log(2n)}\right)\frac{n}{\log n}.$$

In particular, the left-hand side must be positive once n is sufficiently large as long as $C_2/C_1 < 2$. For example, using the Chebyshev's constants $C_1 = 0.92129$ and $C_2 = 1.10555$ mentioned after equation (2.53), we obtain

$$\pi(2n) - \pi(n) \geq \left(0.73703 - \frac{1.27718}{\log(2n)}\right)\frac{n}{\log n}$$

which is positive for all $n \geq 3$. Additional references are Chandrasekharan (1968, [**188**, Chapter 7, §3]), and Hardy & Wright (2008, [**499**, §22.3]).

PROOF OF THE EQUIVALENCE OF (2.12) AND (2.15). This asserts that $\pi(x) \sim x/\log x$ and $p_n \sim n \log n$ are equivalent. Now from the first of these we have

$$n \equiv \pi(p_n) \sim \frac{p_n}{\log p_n}$$

so that

$$\log n \sim \log p_n - \log\log p_n \sim \log p_n.$$

Substituting this last into the preceding then yields

$$p_n \sim n \log p_n \sim n \log n.$$

Conversely, $p_n \sim n \log n$ implies on the one hand that

$$\pi(p_n) \equiv n \sim \frac{p_n}{\log n},$$

and on the other hand that

$$\log p_n \sim \log n + \log\log n \sim \log n;$$

combining these two gives

$$\pi(p_n) \sim \frac{p_n}{\log p_n}.$$

That p_n can be replaced by x here follows because the ratio p_n/p_{n+1} of consecutive primes tends to 1, which is, again, a consequence of $p_n \sim n \log n$. □

Note: This proof has shown that (2.15) and the PNT (2.12) are equivalent. Various other equivalences to the PNT are collected in Section 3.9.

PROOF OF THE $\pi(x) = \mathrm{li}(x) + O(r(x))$ AND $p_n = \mathrm{li}^{-1}(n) + O(r(n \log n) \log n)$ EQUIVALENCE. We can prove this under very mild assumptions on the function $r(x)$—say that it is positive and increasing and satisfies $r(y) \ll r(x)$ whenever $y \ll x$. (This last property holds if additionally $r'(x)$ is bounded, for example.)

First, assume that $\pi(x) = \mathrm{li}(x) + O(r(x))$. Define $E(x) = \pi(x) - \mathrm{li}(x)$, so that $E(x) = O(r(x))$ by assumption. Putting $x = p_n$ and rearranging gives $\mathrm{li}(p_n) = \pi(p_n) - E(p_n) = n - E(p_n)$, so that

$$p_n = \mathrm{li}^{-1}\big(\mathrm{li}(p_n)\big) = \mathrm{li}^{-1}\big(n - E(p_n)\big). \tag{2.55}$$

Now $\mathrm{li}'(x) = 1/\log x$, so applying the inverse function theorem $df^{-1}(x)/dx = 1/f'(f^{-1}(x))$ we get $d\mathrm{li}^{-1}(x)/dx = \log(\mathrm{li}^{-1}(x))$. Thus by the mean value theorem, there is some constant c between n and $n - E(p_n)$ such that

$$\begin{aligned} \mathrm{li}^{-1}(n) - \mathrm{li}^{-1}\big(n - E(p_n)\big) &= E(p_n) \log\big(\mathrm{li}^{-1}(c)\big) \\ &= O\big(r(p_n) \log(c \log c)\big) \\ &= O\big(r(n \log n) \log c\big) \end{aligned}$$

(where we have used the last assumption on $r(x)$), since $p_n \sim n \log n$ by the prime number theorem. Inserting this into equation (2.55) then gives

$$p_n = \mathrm{li}^{-1}(n) + O\big(r(n \log n) \log c\big) = \mathrm{li}^{-1}(n) + O\big(r(n \log n) \log n\big)$$

as claimed.

Similarly, define $F(n) = p_n - \mathrm{li}^{-1}(n)$. Putting $x = p_n$, so that $n = \pi(x)$, and rearranging gives $\mathrm{li}^{-1}\big(\pi(x)\big) = x - F\big(\pi(x)\big)$, so that

$$\pi(x) = \mathrm{li}\big(\mathrm{li}^{-1}\big(\pi(x)\big)\big) = \mathrm{li}\big(x - F\big(\pi(x)\big)\big). \tag{2.56}$$

Since $\mathrm{li}'(x) = 1/\log x$, the mean value theorem again implies that there is some constant c between x and $x - F(\pi(x))$ such that

$$\mathrm{li}(x) - \mathrm{li}\big(x - F(\pi(x))\big) = \frac{1}{\log c} F\big(\pi(x)\big) = O\Big(\frac{1}{\log x} F\big(\pi(x)\big)\Big),$$

so that equation (2.56) becomes

$$\pi(x) = \mathrm{li}(x) + O\Big(\frac{1}{\log x} F\big(\pi(x)\big)\Big).$$

Given the assumption that $F(n) = O\big(r(n\log n)\log n\big)$, we obtain

$$\begin{aligned}\pi(x) &= \mathrm{li}(x) + O\Big(\frac{1}{\log x} r\big(\pi(x)\log \pi(x)\big)\log \pi(x)\Big)\\ &= \mathrm{li}(x) + O\Big(r\Big(\frac{x}{\log x}\log x\Big)\Big)\\ &= \mathrm{li}(x) + O(r(x))\end{aligned}$$

as desired. □

Proof of the equivalence of (2.25) and (2.28). We need to show the equivalence between $\pi(x) = \mathrm{li}(x) + O(xe^{-c(\log x)^{3/5}/(\log\log x)^{1/5}})$ and $p_n = \mathrm{li}^{-1}(n) + O(ne^{-c(\log n)^{3/5}/(\log\log n)^{1/5}})$.

We use the result which has just been proved. Given the first of the two approximations we let $r(x) = x\exp\big(-c(\log x)^{3/5}/(\log\log x)^{1/5}\big)$ and then write down

$$r(n\log n)\log n = n(\log n)\exp\Big(-c(\log(n\log n))^{3/5}/(\log\log(n\log n))^{1/5}\Big)\log n.$$

Now $\log(n\log n) \sim \log n$ and $\log\log(n\log n) \sim \log\log n$, while the $\log^2 n$ factor can be absorbed into the exponential by reducing the value of c to a smaller positive number. So using the previous result leads to

$$p_n = \mathrm{li}^{-1}(n) + O(ne^{-c(\log n)^{3/5}/(\log\log n)^{1/5}}).$$

The reverse implication is similar. (Technically, what is true is that equation (2.28) implies equation (2.25) with the same constant c, while equation (2.25) implies equation (2.28) with any smaller value of c.) □

Heuristic argument for the finiteness of Brun's constant. Brun's constant is

$$B \equiv \sum_{\text{twins}} \Big(\frac{1}{p} + \frac{1}{p+2}\Big) = \Big(\frac{1}{3} + \frac{1}{5}\Big) + \Big(\frac{1}{5} + \frac{1}{7}\Big) + \Big(\frac{1}{11} + \frac{1}{13}\Big) + \cdots. \tag{2.57}$$

The PNT states that $\pi(x) \sim x/\log x$. In Chapter 4 it will be argued—by first assuming that primes occur independently with probabilities $1/\log n$, and then adjusting for dependences—that the number $\pi_2(x)$ of twin primes $\le x$ *should* be

$$\pi_2(x) \sim C\frac{x}{\log^2 x} \quad \text{as } x \to \infty, \tag{2.58}$$

for a certain conjectured constant $C \approx 1.3203$ (to be defined more precisely in equations (4.3)/(4.4)). Using the same argument as the one used to establish the equivalence of $p_n \sim n \log n$ with $\pi(x) \sim x/\log x$, (2.58) may be shown to be equivalent to $\widetilde{p}_n \sim \frac{1}{C} n \log^2 n$ where $\widetilde{p}_n$ denotes the nth twin prime. Since the sum $\sum_{n\geq 2} 1/n \log^2 n$ is finite the result $B < \infty$ would then follow from (2.58). □

REMARKS 2.4.

(i) Brun (1919, [**160**]), (1919, [**161**]) introduced sieve methods into number theory and used it to prove

$$\pi_2(x) = O\left(\frac{x(\log\log x)^2}{\log^2 x}\right).$$

This implies that $\widetilde{p}_n > C\big(n(\log n/\log\log n)^2\big)$ and consequently that $B < \infty$. Brun later established the stronger result $\pi_2(x) = O\left(x/\log^2 x\right)$ and also showed that $\pi_2(x) < 100x/\log^2 x$ for x large enough. Better explicit bounds (also based on sieve methods) were later obtained by Riesel & Vaughan (1983, [**889**]), Jie Wu (1990, [**1117**]), (2004, [**1118**]), and Cai & Lu (2003, [**177**]), among others. What is now known for certain is that for x large enough

$$\pi_2(x) \leq 4.5\frac{x}{\log^2 x} \tag{2.59}$$

(but see Lichtman (2023, [**698**])). If we put $x = \widetilde{p}_n$ here, so that $\pi_2(\widetilde{p}_n) = n$, we get

$$\widetilde{p}_n \geq \frac{1}{4.5} n \log^2 \widetilde{p}_n > \frac{1}{4.6} n \log^2 n,$$

once n is large enough. See also Cai (2002, [**176**]), Halberstam & Richert (1974, [**480**, p. 117]), LeVeque (1977/1996, [**692**, Chapter 6, §12]), Narkiewicz (2000, [**797**, pp. 339–340]).

(ii) Of course, no *lower* bound for $\pi_2(x)$ which increases to infinity can be expected whilst the twin prime conjecture remains unsettled.

(iii) The convergence of (2.57) to Brun's constant is exceedingly slow. Summing over all twin primes up to $p \leq 10^4$, 10^8, 10^{12}, and 10^{16}, respectively, yields the partial sums

$$1.61689356\ldots,\ 1.75881562\ldots,\ 1.80659242\ldots,\ \text{and } 1.83048442\ldots,$$

while the true value has been estimated by Nicely (1995, [**806**]), (1999, [**807**]) and others as $1.902160583\ldots$. Error bounds on such estimates are themselves estimated using Cramér-type heuristics. No rigorous upper bounds are known beyond the inequality $B < 2.347$. The thesis of Klyve (2007, [**637**, Chapter 3]) includes a comprehensive summary of known results for Brun's constant. Klyve conjectured that $1.90216054 < B < 1.90216063$, and assuming an extended form of the Riemann hypothesis, proved that $B < 2.17627$.

PROOF OF THE ASYMPTOTIC INDEPENDENCE RESULT (2.45). This asserts that in the prime factor representation

$$n = \prod_{\text{all primes}} p_j^{M_j} \quad \text{with integers } M_j \geq 0, \tag{2.60}$$

finite collections of the M_j become statistically independent. Specifically, for a fixed k, distinct $j_i, j_2, \ldots, j_k$, and *any* subsets $A_1, A_2, \ldots, A_k$ of nonnegative integers, it asserts that

$$\lim_{N\to\infty} \mathrm{P}_N\big[M_{j_1} \in A_1, \ldots, M_{j_k} \in A_k\big] = \lim_{N\to\infty} \big\{\mathrm{P}_N\left[M_{j_1} \in A_1\right]\times\cdots\times\mathrm{P}_N\left[M_{j_k} \in A_k\right]\big\};$$

here the probability $\mathrm{P}_N[\cdot]$ is based on choosing n uniformly at random from among the integers in $[1, N]$, and subsequently taking $N \to \infty$. Although the result is 'obvious', we will give a rigorous proof that does not require the sets A_j to be finite.

Let $\nu_p(n)$ be the power of p dividing n, and let $p_1, \ldots, p_k$ now represent any fixed primes. Given *finite* sets $A_j \subset \{0, 1, \ldots, t\}$, and with n implicitly restricted to $1 \le n \le N$, define

$$B_j \equiv \big\{n : \nu_{p_j}(n) \in A_j\big\} \quad\text{and}\quad B_j^* \equiv \big\{n : \nu_{p_j}(n) \in A_j,\ \text{or}\ \nu_{p_j}(n) > t\big\}.$$

In this case of finite A_j, it is fairly straightforward to see that

$$\lim_{N\to\infty} \mathrm{P}_N\big[B_1, \ldots, B_k\big] = \prod_{j=1}^{k}\left(\left(1 - \frac{1}{p_j}\right)\sum_{a\in A_j}\frac{1}{p_j^a}\right) = \lim_{N\to\infty}\prod_{j=1}^{k}\mathrm{P}_N\big[B_j\big]$$

and

$$\lim_{N\to\infty} \mathrm{P}_N\big[B_1^*, \ldots, B_k^*\big] = \prod_{j=1}^{k}\left(\left(1 - \frac{1}{p_j}\right)\sum_{a\in A_j}\frac{1}{p_j^a} + \frac{1}{p_j^{t+1}}\right) = \lim_{N\to\infty}\prod_{j=1}^{k}\mathrm{P}_N\big[B_j^*\big].$$

When N is a multiple of $p_1^{1+\max A_1} \times \cdots \times p_k^{1+\max A_k}$, these equalities are exact without the need to take limits; using limits allows N to be unrestricted.

But now let $A_1, \ldots, A_k$ be arbitrary—*possibly infinite*—subsets of $\{0, 1, 2, \ldots\}$, and for any integer $t \ge 0$ let $A_j(t) \equiv A_j \cap \{0, 1, \ldots, t\}$ and define the events $B_j(t)$ and $B_j^*(t)$ accordingly. Define events $C_j \equiv \{\nu_{p_j}(n) \in A_j\}$ and note that $B_j(t) \subseteq C_j \subseteq B_j^*(t)$, so that

$$\lim_{N\to\infty} \mathrm{P}_N\big[B_1(t), \ldots, B_k(t)\big] \le \lim_{N\to\infty} \mathrm{P}_N\big[C_1, \ldots, C_k\big] \le \lim_{N\to\infty} \mathrm{P}_N\big[B_1^*(t), \ldots, B_k^*(t)\big]$$

or, equivalently (by the calculation in the finite case),

$$\begin{aligned}\prod_{j=1}^{k}\left(\left(1 - \frac{1}{p_j}\right)\sum_{a\in A_j(t)}\frac{1}{p_j^a}\right) &\le \lim_{N\to\infty} \mathrm{P}_N\big[C_1, \ldots, C_k\big] \\ &\le \prod_{j=1}^{k}\left(\left(1 - \frac{1}{p_j}\right)\sum_{a\in A_j(t)}\frac{1}{p_j^a} + \frac{1}{p_j^{t+1}}\right).\end{aligned}$$

But both the leftmost and rightmost expressions have the same limit as $t \to \infty$, while the middle expression is independent of t; therefore the middle expression must equal the joint limit of the outer expressions, namely

$$\lim_{N\to\infty} \mathrm{P}_N\big[C_1, \ldots, C_k\big] = \prod_{j=1}^{k}\left(\left(1 - \frac{1}{p_j}\right)\sum_{a\in A_j}\frac{1}{p_j^a}\right).$$

In particular, this evaluation implies that

$$\lim_{N\to\infty} \mathrm{P}_N\big[C_1,\ldots,C_k\big] = \prod_{j=1}^{k} \lim_{N\to\infty} \mathrm{P}_N\big[C_j\big],$$

which confirms asymptotic independence result (2.45). □

CHAPTER 3

Interlude on Arithmetic Functions

These strange formulas attracted my attention, although incomprehensible to me at the time. I also heard about squaring the circle, asymptotes that curves approach without ever reaching them, and many similar things, the meaning of which I could not yet understand. But they acted on my imagination, instilling in me awe for mathematics as a higher and mysterious science, opening to those initiated into it a new and wonderful world, inaccessible to mere mortals.

. . . .

You are surprised that I can study both literature and mathematics at the same time. But in reality, mathematics requires the utmost imagination—one cannot be a mathematician without being at the same time a poet at heart. It seems to me that a poet must see what others do not see, and see more deeply. And a mathematician must do this as well.

Sofia Vasilyevna Kovalevskaya [**659**][1]

Arithmetic functions are central to all branches of number theory, and we devote this chapter to defining the most important ones, and to laying out their most important properties. The definitions for all of the functions to be considered are elementary, but their properties often lie very deep. Given the rather considerable detail we will include, we do not recommend that the reader go through the whole of this chapter on a first pass through. It will, however, prove very worthwhile to attentively read the first two sections. Unlike a textbook, the material of those sections is laid out in a compartmentalized way. Thus Section 3.1 deals almost exclusively with the general theory of arithmetic functions, while Section 3.2 introduces and describes the most important functions to which the theory is meant to apply. It is recommended to move freely between those two sections, at times referring to Section 3.2 whilst reading Section 3.1. Approaching the material in this way will prove helpful in forming an appreciation of the bigger picture that lies beneath.

The remainder of this chapter can then be read more selectively. Sections 3.3–3.8 treat properties of the functions introduced in Section 3.2. In each of these six sections the coverage focuses primarily on the statistical and asymptotic properties of the function being discussed and not on their so-called elementary or purely

[1] Based on pages 52–53 and 314 of [**659**]. (Translated by the authors; the assistance of Maria Tamboukou in locating source material is gratefully acknowledged.)

number-theoretic properties.[2] These sections offer a comprehensive resource (with detailed reference material) which can be turned to as and when information is desired about any one of the functions considered. Even so, these sections may well be of interest in their own right, and it will prove worthwhile to look over at least one or two of them at some point—for example, Section 3.3 which treats the Möbius function, or Section 3.4 which treats Euler's phi/totient function. This will lead to further appreciation for the methods and the concepts involved. The chapter concludes with a series of equivalences to the Prime Number Theorem, each formulated in terms of the arithmetic functions that have been introduced.

Some of the properties to be discussed will require passing references to the Riemann zeta function or to its zeros, and we will allow ourselves to do so freely. These will be discussed extensively in Chapters 5 and 6 (though of course there is nothing to prevent the reader from taking a sneak peek there). Proofs for many (certainly not all!) of the properties to be be mentioned will be found in Section 3.10, and ample references will be provided for proofs that are omitted.

3.1. Arithmetic functions

By an *arithmetic function*, say, $f(n)$—sometimes referred to as a *number-theoretic function*—we mean nothing more than a function[3] (usually real-valued or complex-valued) defined on the natural numbers $n = 1, 2, \ldots$. Typically, $f(n)$ will describe some arithmetical property of n. Such a function is said to be *additive* if $f(mn) = f(m) + f(n)$ *whenever m and n are relatively prime*, that is, whenever m and n have *greatest common divisor* $\gcd(m, n) = 1$. It is said to be *multiplicative* if $f(mn) = f(m)f(n)$ under the same condition on m and n. (By convention, $f(n) \equiv 0$ is not considered to be multiplicative.) If these properties were to hold for *all* m and n, then the function is said to be *completely additive* or *completely multiplicative*, as the case may be. (The terms *totally additive* and *totally multiplicative* are sometimes used instead.) Trivially, the sum of additive functions is additive, and the product of multiplicative functions is multiplicative. Also, if $f(n)$ is multiplicative and $f(n) \neq 0$ for all n, then its reciprocal $1/f(n)$ will also be multiplicative.

Note: It is recommended that the reader take a few moments out for a first look at the arithmetic functions that are introduced in Section 3.2.

In view of the definitions given above we find that if n has the 'canonical' prime factorization $n = p_1^{k_1} p_2^{k_2} \cdots p_m^{k_m}$, additive functions satisfy

$$f(1) = 0, \quad \text{and} \quad f(n) = \sum_{j=1}^{m} f\big(p_j^{k_j}\big) \equiv \sum_{p^k \,\|\, n} f\big(p^k\big), \tag{3.1}$$

and multiplicative functions satisfy

$$f(1) = 1, \quad \text{and} \quad f(n) = \prod_{j=1}^{m} f\big(p_j^{k_j}\big) \equiv \prod_{p^k \,\|\, n} f\big(p^k\big). \tag{3.2}$$

---

[2]The purely number-theoretic properties of the main arithmetic functions are extensively studied in the classic text of Hardy & Wright (2008, [**499**]), as well as in many other books on number theory.

[3]Of course 'f' is the function and not '$f(n)$'; we trust the reader is not troubled by this abuse of language.

The notation $p^k \| n$ means that p^k divides n but p^{k+1} does not. In the completely additive and completely multiplicative cases we would further have $f(p^k) = kf(p)$ and $f(p^k) = f(p)^k$, respectively, and the values that f takes on at the primes will determine f uniquely. An arithmetic function is called *strongly additive* or *strongly multiplicative*, respectively, if $f(p^k) = f(p)$ for all $k \geq 1$, and in these cases we will have

$$f(n) = \sum_{p|n} f(p) \quad \text{and} \quad f(n) = \prod_{p|n} f(p), \tag{3.3}$$

where $p \mid n$ refers to those primes that divide n. Note too that if $f(n)$ is additive (perhaps completely or strongly), then $g(n) \equiv \exp\bigl(cf(n)\bigr)$ will be multiplicative (also completely or strongly, as the case may be) for any value of c. A type of converse assertion is available as well but it requires taking the possibility that $g = 0$ into account, and in the complex-valued case, also requires taking into account the fact that the logarithm is not uniquely defined.

Now if one were to view n as having been selected randomly from $[1, N]$, with N large, then the expected proportion of integers divisible by p can for many purposes be reasonably approximated as $1/p$, and the expected proportion divisible by p^k but not by p^{k+1} as $(1/p^k)(1 - 1/p)$. (Show this.) Following up on this idea one can often justify approximations such as

$$\frac{1}{x}\sum_{n \leq x} f(n) \;\approx\; \sum_{p^k \leq x} f\left(p^k\right)\frac{1}{p^k}\left(1 - \frac{1}{p}\right)$$

for additive $f(n)$, and

$$\frac{1}{x}\sum_{n \leq x} f(n) \;\approx\; \sum_{p \leq x} \frac{f(p)}{p}$$

for strongly additive $f(n)$, when x is large. In the strongly additive case, for example, we have

$$f(n) = \sum_{p|n} f(p) = \sum_{p} f(p) I_p(n),$$

where $I_p(n)$ is an *indicator variable* which equals 1 when p divides n, and equals 0 otherwise. If one thinks of $I_p(n)$ as being a random variable, its expectation "$\mathrm{E}\bigl(I_p(n)\bigr)$" could be taken as $1/p$; one may use this idea to further motivate approximations of the type shown above.

Arithmetic functions and Dirichlet series. Every arithmetic function $f(n)$ corresponds—at least formally—to a 'generating function' of the form $F(s) = \sum_{n=1}^{\infty} f(n)/n^s$, $s \in \mathbb{C}$, called its *Dirichlet series*. Such series are discussed in Appendix C. A particularly useful property of multiplicative functions is that their corresponding Dirichlet series have *Euler product representations* analogous to (1.2). Taking the Unique Factorization Theorem into account, it is a standard exercise to show that the Dirichlet series $\sum_{n=1}^{\infty} f(n)/n^s$ of a multiplicative function converges absolutely for some $s \in \mathbb{C}$ if and only if $\sum_p \sum_{k \geq 1} f(p^k)/p^{ks}$ converges absolutely, and that one would then have[4]

$$\sum_{n=1}^{\infty} \frac{f(n)}{n^s} = \prod_p \left(1 + \frac{f(p)}{p^s} + \frac{f(p^2)}{p^{2s}} + \cdots\right) \tag{3.4}$$

with the product itself being convergent, and each of its factors being convergent Maclaurin series in p^{-s}. In the completely multiplicative case, (3.4) would reduce to

$$\sum_{n=1}^{\infty} \frac{f(n)}{n^s} = \prod_p \left(1 - \frac{f(p)}{p^s}\right)^{-1}. \tag{3.5}$$

These two relations remain *formally* valid regardless of any considerations of convergence. When $f(n) \equiv 1$ they reduce to relation (1.2), satisfied by the zeta function. To give but one insight into the value of such relations, consider the possibility of defining a *random* completely multiplicative function by viewing the $f(p)$ as independent random variables. The terms on the right in (3.5) would then be independently distributed, and (by taking logarithms, say) standard arguments for sums of independent random variables could be applied. (Appendices C and D provide further background for these remarks.)

Relations (3.4)/(3.5) also say something useful in the $s = 0$ case. For if f is multiplicative, say, (3.4) would become

$$\sum_{n=1}^{\infty} f(n) = \prod_p \left(1 + f(p) + f(p^2) + \cdots\right);$$

furthermore, if the sum on the left converges absolutely then each of the series $1 + f(p) + f(p^2) + \cdots$, as well as the product itself, will converge absolutely. (Lemma E.1 in Appendix E deals with convergence of infinite products.) If f is also completely multiplicative, each factor on the right-hand side will be a geometric series, resulting in the simplification

$$\sum_{n=1}^{\infty} f(n) = \prod_p (1 - f(p))^{-1}.$$

The convergence considerations for proving these last two relations *directly* are relatively straightforward. (See, for example, Chandrasekharan (1968, [**188**, Chapter VII, §4]).) In the converse direction, note that relations (3.4) and (3.5) can be recovered from these by simply substituting $f(n)/n^s$ for $f(n)$. Further properties of sums such as $\sum f(n)$ will be considered in Chapter 9.

Algebraic properties of arithmetic functions. Arithmetic functions can also be endowed with valuable algebraic properties. A *group* is a set $\mathcal{G}$ of elements $a, b, c, \ldots$ together with a binary operation "$\circ$" under which it is 'closed' (meaning that $a \circ b \in \mathcal{G}$ if $a, b \in \mathcal{G}$), provided that $\circ$ is 'associative' (meaning $a \circ (b \circ c) = (a \circ b) \circ c$), that there is an 'identity element' $e \in \mathcal{G}$ such that $a \circ e = e \circ a = a$ for all $a \in \mathcal{G}$, and that each $a \in \mathcal{G}$ has an inverse $a^{-1} \in \mathcal{G}$ such that $a^{-1} \circ a = a \circ a^{-1} = e$. The group $(\mathcal{G}, \circ)$ is called *abelian* if it also satisfies $a \circ b = b \circ a$ for all $a, b \in \mathcal{G}$.

EXERCISE 3.1. Show that each of the following is an abelian group: (i) the integers under addition; (ii) the rational numbers, but excluding 0, under multiplication; (iii) can you think of a few others?

[4]The convergence must be absolute for the rearrangements leading to (3.4) to be justifiable (although the formula could still hold beyond the region of absolute convergence).

Now let $\mathcal{A}$ be the set of all arithmetic functions, and for $f, g \in \mathcal{A}$, define $h = f + g$ by pointwise addition, that is, $h(n) = f(n) + g(n)$ for $n \geq 1$. Then $(\mathcal{A}, +)$ forms an abelian group in which the arithmetic function that is identically 0 is the identity.

An important binary operation, say "$*$", called *Dirichlet convolution*, can be defined on the group $\mathcal{A}$ of arithmetic function as follows: we write $h = f * g$ to mean

$$h(n) = \sum_{jk=n} f(j)g(k) \equiv \sum_{j|n} f(j)g\left(\frac{n}{j}\right) \equiv \sum_{j|n} f\left(\frac{n}{j}\right) g(j), \quad \text{for } n \geq 1.$$

(Sums such as $\sum_{d|n}$ are invariably over all positive divisors of n with 1 and n included.) This 'multiplicative' operation has the identity element

$$\mathbf{e}(n) = \begin{cases} 1, & \text{if } n = 1, \\ 0, & \text{if } n > 1, \end{cases}$$

so that $\mathbf{e} * f = f * \mathbf{e} = f$. (Dirichlet convolution is discussed further in Appendix C.) The structure $(\mathcal{A}, *)$ satisfies all of the requirements of an (abelian) group except for one: not every $f \in \mathcal{A}$ has a Dirichlet inverse, $f^{-1} \in \mathcal{A}$, such that $f^{-1} * f = f * f^{-1} = \mathbf{e}$. However, because $(\mathcal{A}, +, *)$ has all of the other properties mentioned above, as well as the 'distributive' property $f * (g + h) = (f * g) + (f * h)$, it constitutes what is known as a *ring*.

EXERCISE 3.2. Prove the associativity property by showing that $\big(f * (g * h)\big)(n)$ and $\big((f * g) * h\big)(n)$ are both equal to $\sum_{ijk=n} f(i)g(j)h(k)$.

The criterion for possessing an inverse under Dirichlet convolution is very simple: f has a Dirichlet inverse[5] if and only if $f(1) \neq 0$. If we let $\mathcal{A}' \subset \mathcal{A}$ be the subset of arithmetic functions that possess a Dirichlet inverse, then $(\mathcal{A}', *)$ will form an abelian group. This group contains an important subgroup $(\mathcal{A}^\dagger, *)$, where $\mathcal{A}^\dagger$ is the set of all *multiplicative* functions which have a Dirichlet inverse. That $\mathcal{A}^\dagger$ is actually a group follows because Dirichlet convolutions of multiplicative functions remain multiplicative, and less obviously, if a multiplicative function has a Dirichlet inverse it too will be multiplicative.[6] Further restricting $\mathcal{A}^\dagger$ to *completely* multiplicative or to *strongly* multiplicative functions does not result in subgroups as these are no longer closed under the $*$ operation; in fact, Dirichlet convolutions of completely multiplicative functions are rarely completely multiplicative.

A related fact is that if $f(n)$ is multiplicative, then so is its associated *divisor sum*

$$g(n) \equiv \sum_{d|n} f(d), \tag{3.6}$$

and conversely. In fact, the function defined as $\mathbf{1}(n) \equiv 1$ for all n is multiplicative, and (3.6) merely states that $g = f * \mathbf{1}$. On account of their group properties, and the fact that their Dirichlet series have Euler product representations, multiplicative functions are particularly important and can often be studied by way of their Dirichlet series.

---

[5]Also worth noting is that the Dirichlet inverse of a Dirichlet convolution is the Dirichlet convolution of the Dirichlet inverses: $(f * g)^{-1} = f^{-1} * g^{-1}$.

[6]The proofs of these assertions are somewhat involved. See Apostol (1976, [**25**, Chapter 2, §10]), or Niven, Zuckerman & Montgomery (1991, [**814**, Theorem 4.12]).

For later reference, we pause for a moment to record the following convolution identities:

$$\Big\{\mathbf{e}*\mathbf{1}=\mathbf{1},\ \ \mathbf{e}=\mu*\mathbf{1}\Big\},\ \Big\{\varphi*\mathbf{1}=\mathbf{n},\ \ \varphi=\mu*\mathbf{n}\Big\},\ \Big\{\mathbf{1}*\mathbf{1}=\tau,\ \ \mathbf{1}=\mu*\tau\Big\},$$

$$\Big\{\mathbf{n}*\mathbf{1}=\sigma,\ \ \mathbf{n}=\mu*\sigma\Big\},\ \ \Big\{\Lambda*\mathbf{1}=\mathcal{L},\ \ \Lambda=\mu*\mathcal{L}\Big\},$$

$$\Big\{-(\mu\,\mathcal{L})*1=\Lambda,\ \ -(\mu\mathcal{L})=\mu*\Lambda\Big\},\ \text{ and }\ \lambda*|\mu|=\mathbf{e}.$$

Here μ, φ, τ, σ, λ, and Λ represent the Möbius, Euler's phi, divisors, sum of divisors, Liouville, and von Mangoldt functions, all defined in Section 3.2. The identity element $\mathbf{e}$ and the function $\mathbf{1}$ are as defined above; $\mathbf{n}$ is defined as $\mathbf{n}(n)\equiv n$; $\mathcal{L}$ is defined as $\mathcal{L}(n)=\log n$; and the expression $\mu\,\mathcal{L}$ refers to pointwise multiplication of the two functions. The pairs of identities within each of the curly brackets are so-called *Möbius transform pairs*. They are based on the fact that $g=f*\mathbf{1}\iff f=g*\mu$; in other words, each of $\mathbf{1}$ and μ is the Möbius inverse of the other. We will return to this point and to these identities, and offer proofs, as we progress in this chapter.

As already mentioned, the properties of Dirichlet series and Dirichlet convolution are taken up in Appendix C. Here we merely note two important properties. The first concerns the connection between Dirichlet convolution and the multiplication of Dirichlet series. Thus suppose that

$$\left(\sum_{j=1}^{\infty}\frac{a(j)}{j^s}\right)\left(\sum_{k=1}^{\infty}\frac{b(k)}{k^s}\right)=\sum_{n=1}^{\infty}\frac{c(n)}{n^s}. \tag{3.7}$$

Then multiplying out and collecting terms we arrive at $c(n)=\sum_{jk=n}a(j)b(k)$ so that the arithmetic function $c(n)$ is the Dirichlet convolution of $a(n)$ and $b(n)$. Relation (3.7) will be valid in the region where both series on the left converge absolutely, but it will at least hold *formally* regardless of any considerations of convergence. The second important property concerns a relation between the summatory versions of arithmetic functions and Dirichlet convolution. Thus suppose that f and g are arithmetic functions, and that $F(x)=\sum_{n\le x}f(n)$ and $G(x)=\sum_{n\le x}g(n)$ are their *summatory* versions. Then the summatory version of the convolved function $f*g$ satisfies

$$\sum_{n\le x}(f*g)(n)=\sum_{n\le x}F\left(\frac{x}{n}\right)g(n)=\sum_{n\le x}G\left(\frac{x}{n}\right)f(n).$$

The proof may be taken as an exercise.

(Hint: Show that the left side equals $\sum_{jk\le x}f(j)g(k)$.)

More general properties of arithmetic functions can be somewhat involved. They are discussed extensively in Chandrasekharan (1970, [**189**]), McCarthy (1986, [**746**]), Schwarz & Spilker (1994, [**960**]), Sivaramakrishnan (1989, [**993**]), as well as in Vaidyanathaswamy (1931, [**1078**]) and numerous textbooks and other sources.

3.2. The usual suspects (definitions)

In Section 3.1 we dealt with general properties of arithmetic functions. Here we introduce the most important arithmetic functions along with a few of their relatives.

Möbius function $\mu(n)$: Defined as $\mu(1) = 1$, and as $\mu(n) = (-1)^k$ if n is a product of k *distinct* primes; otherwise (namely, if n is divisible by the square of any prime) $\mu(n) = 0$. Examples: $\mu(2) = -1$, $\mu(4) = 0$, and $\mu(6) = 1$. The summatory version $M(n) \equiv \sum_{k \le n} \mu(k)$ is known as the **Mertens function**. Note that $q(n) \equiv |\mu(n)| \equiv \mu^2(n)$, known as the **quadratfrei function** (or the **squarefree function**), serves as an indicator for numbers that are squarefree.

Euler totient or **phi function:** Usually denoted as $\varphi(n)$, but may also be written as $\phi(n)$. It is the number of integers in the range $[1, n]$ that are relatively prime to n (that is, have no common divisors with n). Examples: $\varphi(1) = \varphi(2) = 1$, and $\varphi(3) = \varphi(4) = 2$. Likewise $\varphi(9) = 6$, since 1, 2, 4, 5, 7, and 8 are relatively prime to 9, but 3, 6, and 9 are not.

Divisor function $\tau(n)$: The number of (positive) divisors of n, including 1 and n. This definition can also be stated as $\tau(n) \equiv \sum_{d|n} 1$. If n has the prime factorization $n = p_1^{k_1} p_2^{k_2} \cdots p_m^{k_m}$, then $\tau(n) = (k_1+1)(k_2+1) \cdots (k_m+1)$. (Remark: The notation $d(n)$ for this function is in common use as well.)

Sum of divisors function $\sigma(n)$: The sum of the (positive) divisors of n, including 1 and n. One can also write this as $\sigma(n) \equiv \sum_{d|n} d$. Example: $\sigma(12) = 1 + 2 + 3 + 4 + 6 + 12 = 28$.

Sum of kth powers of divisors function $\sigma_k(n)$: The sum of the kth powers of the divisors of n, including 1 and n. One may write this as $\sigma_k(n) \equiv \sum_{d|n} d^k$. The exponent k may be real or complex. Note that $\tau(n) = \sigma_0(n)$, $\sigma(n) = \sigma_1(n)$, and $\sigma_{-k}(n) = n^{-k}\sigma_k(n)$.

Number of distinct prime factors function $\omega(n)$: The number of *distinct* prime factors of n. Equivalently, $\omega(n) \equiv \sum_{p|n} 1$. Examples: $\omega(1) = 0$ and $\omega(3^2 \times 5^4) = 2$.

Number of prime factors function $\Omega(n)$: The *total* number of prime factors of n, counting with multiplicity. Equivalently, $\Omega(n) \equiv \sum_{p^k \| n} k$. If n has the prime factorization $n = p_1^{r_1} p_2^{r_2} \cdots p_k^{r_k}$, then $\Omega(n) = r_1 + \cdots + r_k$. Examples: $\Omega(1) = 0$ and $\Omega(3^2 \times 5^4) = 6$.

Liouville function $\lambda(n)$: Set $\lambda(1) = 1$, and $\lambda(n) = (-1)^{\Omega(n)}$. Alternately, $\lambda(n) \equiv (-1)^r$, where $r = r_1 + \cdots + r_k$ is sum of the powers in the prime factorization $n = p_1^{r_1} p_2^{r_2} \cdots p_k^{r_k}$. Thus $\lambda(n)$ equals $+1$ or -1 according as the *total* number of prime factors of n is even or odd. (We say that $\lambda(n)$ counts the *parity* of the number of prime factors of n.) Its summatory version, denoted by $L(n) \equiv \sum_{k \le n} \lambda(k)$, is the difference between the number of positive integers $\le n$ having an even, and an odd, number of prime factors. (Note that if n is squarefree, the functions $\lambda(n)$ and $\mu(n)$ will take on identical values.)

Von Mangoldt function $\Lambda(n)$: Defined as $\Lambda(n) = \log p$ whenever n is a prime power (that is, whenever $n = p^k$ for some prime p and some integer $k \ge 1$), and as $\Lambda(n) = 0$ otherwise.

Chebychev ϑ-function: This is the sum-over-primes function

$$\vartheta(x) \equiv \sum_{p \le x} \log p = \log \prod_{p \le x} p.$$

It is sometimes written as $\theta(x)$.

Chebychev ψ-function: This is the summatory version of the von Mangoldt function: $\psi(x) \equiv \sum_{n \le x} \Lambda(n) \equiv \sum_{p^k \le x} \log p$, where the last sum is over all primes p and all integers $k \ge 1$ such that $p^k \le x$.

It is customary to extend the domains of summatory functions such as $M(n)$ and $L(n)$ to real variables as follows:

$$M(x) \equiv \sum_{k \le x} \mu(k), \quad \text{and} \quad L(x) \equiv \sum_{k \le x} \lambda(k).$$

The functions $M(x)$ and $L(x)$—which are now step functions—are usually more convenient to use in analytical expressions. When viewed as functions of a real variable, however, the summatory functions $M(x)$ and $L(x)$ would not be considered to be arithmetic functions. Similar remarks apply to the Chebyshev functions $\vartheta(x)$ and $\psi(x)$.

Of the functions defined above, $\mu(n)$, $\varphi(n)$, $\tau(n)$, and $\sigma(n)$, as well as $\sigma_k(n)$, are all multiplicative. The function $\omega(n)$ is strongly additive, while $\Omega(n)$ is completely additive. Further examples of multiplicative functions include $f(n) = n^\alpha$ (for any $\alpha \in \mathbb{C}$) which is completely multiplicative, and $f(n) = (-1)^{n+1}$ which is strongly multiplicative. Note that the seemingly inconsequential variant $f(n) = (-1)^n$ is *not* multiplicative. The function $f(n) = \log n$ is completely additive, while the difference $\Omega(n) - \omega(n)$ (which can be viewed as a measure of how far n is from being squarefree) is merely additive. The von Mangoldt function $\Lambda(n)$ is neither additive nor multiplicative.

We have seen that the exponential of an additive function is multiplicative, and if the additive function is completely or strongly additive, its exponential will correspondingly be completely or strongly multiplicative. Thus since $\Omega(n)$ is completely additive the exponential $\lambda(n) = (-1)^{\Omega(n)}$ will be completely multiplicative. Likewise $2^{\omega(n)}$ will be strongly multiplicative, and $n^{it} = e^{it \log n}$ will be completely multiplicative. The proofs for all of the assertions that have been made here are easy exercises.

If p is a prime, these various functions take on the values $\mu(p) = -1$, $\varphi(p) = p - 1$, $\tau(p) = 2$, $\sigma(p) = p + 1$, $\sigma_k(p) = p^k + 1$, $\Omega(p) = 1$, $\omega(p) = 1$, $\lambda(p) = -1$, and $\Lambda(p) = \log p$, respectively. If $n = p^k$ is a prime power with $k > 1$, then $\mu(p^k) = 0$, $\varphi(p^k) = p^{k-1}(p-1)$, $\tau(p^k) = k + 1$, $\sigma(p^k) = (p^{k+1} - 1)/(p - 1)$, $\sigma_\alpha(p^k) = (p^{\alpha(k+1)} - 1)/(p^\alpha - 1)$, $\Omega(p^k) = k$, $\omega(p^k) = 1$, $\lambda(p^k) = (-1)^k$, and $\Lambda(p^k) = \log p$, respectively. The formulas for general n can be obtained either as immediate applications of the equations (3.1), (3.2), and (3.3), or more directly. The proofs for all of these assertions are easy exercises as well.

In the remainder of this chapter we will lay out some key properties for each of the arithmetic functions introduced above. Our focus will be more on their statistical and limiting behaviours than on their purely number-theoretic properties. A tremendous amount of material is available on the statistical properties of arithmetic functions. A selection of references includes Galambos (1970, [**402**]), Galambos, & Szüsz (1986, [**404**]), Indlekofer (1981, [**576**]), (1984, [**577**]), Kátai (1969, [**623**]), and de Koninck & Ivić (1980, [**649**]).

In this chapter, whenever the zeta function appears, the real part of its argument should be assumed to exceed 1 (so that the series defining $\zeta(s)$ is convergent) unless otherwise clear from the context. Also, except where explicitly stated otherwise, all error terms are *unconditional*, that is, they do not depend on any unproven hypotheses. Many error terms have significantly stronger versions if the Riemann hypothesis (or other unproven hypotheses) are assumed. References to original papers will be liberally provided for results not easily found in standard sources.

3.3. The Möbius function

We consider first the Möbius function $\mu(n)$ which was introduced by and named after August Ferdinand Möbius (1790–1868). This multiplicative function has the important property

$$\sum_{d|n} \mu(d) = \begin{cases} 1, & \text{if } n = 1, \\ 0, & \text{if } n > 1, \end{cases} \tag{3.8}$$

(or equivalently $\mu * \mathbf{1} = \mathbf{e}$) where, as always, the sum is over all positive divisors of n with 1 and n included. This fundamental property of μ is used extensively in applications, far more often than the definition of μ itself. One especially important application of it is to prove the *Möbius inversion formula* (due to Dedekind and Liouville rather than to Möbius). It asserts the equivalence

$$\begin{aligned} &g(n) = \sum_{d|n} f(d) \\ &\qquad \Longleftrightarrow f(n) = \sum_{d|n} \mu(d) g\left(\frac{n}{d}\right) \equiv \sum_{d|n} \mu\left(\frac{n}{d}\right) g(d) \equiv \sum_{ab=n} \mu(a) g(b), \end{aligned} \tag{3.9}$$

for $n \geq 1$. The arithmetic functions f and g are sometimes referred to as a Möbius transform pair. In terms of Dirichlet convolution this result can be stated concisely as

$$g = \mathbf{1} * f \iff f = \mu * g.$$

The Möbius inversion formula in effect asserts that $\mu(n)$ is the Dirichlet inverse of the function $\mathbf{1}(n)$ defined beneath (3.6). It is this fact that explains the collection of convolution identities that had been noted in Section 3.1. Proofs of (3.8) and of (3.9), as well as of (3.10), are given in Section 3.10.

It is worth pointing out here that the equivalence (3.9) has numerous noteworthy generalizations. For example, for *any* (not necessarily well behaved!) complex-valued functions F and G defined on $[1, \infty)$ we have

$$G(x) = \sum_{1 \leq n \leq x} F\left(\frac{x}{n}\right) \iff F(x) = \sum_{1 \leq n \leq x} \mu(n) G\left(\frac{x}{n}\right), \quad x \geq 1. \tag{3.10}$$

Two further examples, this time involving infinite sums, are: for F and G defined on $(0, \infty)$,

$$G(x) = \sum_{n=1}^{\infty} F(nx) \iff F(x) = \sum_{n=1}^{\infty} \mu(n) G(nx), \quad x > 0; \tag{3.11}$$

and for F and G defined either on $(0, 1)$, or on $[1, \infty)$,

$$F(x) = \sum_{n=1}^{\infty} G(x^n) \iff G(x) = \sum_{n=1}^{\infty} \mu(n) F(x^n), \quad x > 0, \tag{3.12}$$

the indicated $(0, 1)$ and $[1, \infty)$ domains being equivalent under the substitution of $F(1/t)$ and $G(1/t)$ for $F(t)$ and $G(t)$, respectively. The statements (3.11) and (3.12) are also equivalent, being related via transformations that replace $F(x)$ and $G(x)$

by $F(e^x)$ and $G(e^x)$. Remarkably, the *only* requirement for these results is the absolute convergence of the sums involved: One can show that either one of

$$\sum_{m,n=1}^{\infty} |F(mnx)| < \infty, \quad \text{or} \quad \sum_{m,n=1}^{\infty} |G(mnx)| < \infty$$

suffices for (3.11)—with either of these implying the other—and likewise that either one of

$$\sum_{m,n=1}^{\infty} |F(x^{mn})| < \infty, \quad \text{or} \quad \sum_{m,n=1}^{\infty} |G(x^{mn})| < \infty$$

suffices for (3.12).

Möbius inversion plays an important role not only in number theory, but also in combinatorics, group theory, and elsewhere. Further extensions (with applications in physics) may be found in the monograph by Nanxian Chen (2010, [**197**]). A selection of other references that deal with Möbius inversion include Hardy & Wright (2008, [**499**, §16.5]), Apostol (1976, [**25**, §2.7]), and Tenenbaum (2015, [**1053**, §2.5]), as well as Barnabei et al. (1986, [**77**]), Rota (1964, [**922**]), and the 1998–1999 Berkeley Math Circle notes by Stankova–Frenkel [**1019**].

Generating functions for μ, as well as for $|\mu|$, based on Dirichlet series are given by

$$\sum_{n=1}^{\infty} \frac{\mu(n)}{n^s} = \prod_p \left(1 - \frac{1}{p^s}\right) = \frac{1}{\zeta(s)}, \quad \text{and} \quad \sum_{n=1}^{\infty} \frac{|\mu(n)|}{n^s} = \frac{\zeta(s)}{\zeta(2s)}. \tag{3.13}$$

(Here and throughout we are using the standard notation $s = \sigma + it$.) The relations (3.13) both hold, not only for $\sigma > 1$, but also for $s = 1$ itself:

$$\sum_{n=1}^{\infty} \frac{\mu(n)}{n} = 0, \qquad \sum_{n=1}^{\infty} \frac{|\mu(n)|}{n} = \infty. \tag{3.14}$$

Of these two, the second is easy, while the first is equivalent to the Prime Number Theorem. The Riemann hypothesis is equivalent to the assertion that $\sum \mu(n)/n^s$ converges whenever $\sigma > \frac{1}{2}$, and if this convergence were to hold then the sum would continue to equal $1/\zeta(s)$ there. Proofs for Dirichlet generating function identities such as those in (3.13) may be found in Section 3.10. The importance of the Möbius function is already evident from its connections to the ζ function as seen in (3.13)

It is also known that the 'average value' of $\mu(n)$ is zero, that is,

$$\frac{1}{x}\sum_{n \le x} \mu(n) \to 0, \quad \text{or equivalently} \quad \sum_{n \le x} \mu(n) = o(x), \quad \text{as } x \to \infty, \tag{3.15}$$

a property which is also equivalent to the PNT. Further properties of $\mu(n)$ include:

$$\left|\sum_{n \le x} \frac{\mu(n)}{n}\right| \le 1; \quad \sum_{n \le x} \frac{\mu(n)}{n} = O\big(e^{-c\sqrt{\log x}}\big) = o(1); \tag{3.16}$$

$$\sum_{n \le x} \frac{\mu(n)}{n^2} = \frac{6}{\pi^2} + O\left(\frac{1}{x}\right); \quad \sum_{n=1}^{\infty} \frac{\mu(n)}{n^{1+it}} = \frac{1}{\zeta(1+it)}, \quad t \ne 0; \tag{3.17}$$

$$\sum_{n=1}^{\infty} \frac{\mu(n)\log n}{n} = -1; \qquad \sum_{n=1}^{\infty} \frac{\mu(n)\log^2 n}{n} = -2\gamma; \tag{3.18}$$

$$\sum_{n=1}^{\infty} \frac{\mu(n)x^n}{1-x^n} = x, \quad |x|<1; \qquad \sum_{n=1}^{\infty} \frac{\mu(n)x^n}{1+x^n} = x - 2x^2, \quad |x|<1; \tag{3.19}$$

and

$$\sum_{n\le x} \mu(n) \left\lfloor \frac{x}{n} \right\rfloor = 1, \quad \text{for all } x \ge 1; \tag{3.20}$$

(where $\lfloor \cdot \rfloor$ is the 'integer part' function). We also have the 'prime zeta function'

$$P(s) = \sum_p \frac{1}{p^s} = \sum_{n=1}^{\infty} \frac{\mu(n)}{n} \log \zeta(ns), \tag{3.21}$$

which will be discussed further in Chapter 5. A rather unusual identity involving the Möbius function, due to Davenport, is

$$\sum_{n=1}^{\infty} \frac{\mu(n)}{n} \left(\{n\theta\} - \frac{1}{2} \right) = -\frac{\sin(2\pi\theta)}{\pi},$$

where $\{\cdot\}$ is the 'fractional parts' function; it holds for every rational value of θ, and for almost all real values as well.

The value of $6/\pi^2$ in (3.17) stems from setting $s = 2$ at the left in (3.13), together with the evaluation $\zeta(2) = \pi^2/6$. The identities in (3.19) are examples of so-called Lambert series; proofs for an assortment of Lambert series identities are given in Appendix C. Putting, for example, $x = 1/2$ in these Lambert series leads to some nice identities involving $\mu(n)$. Furthermore, letting $x \uparrow 1$ in the second of these shows that μ has a negative bias, at least in that particular sense. The identity in (3.21) requires $\Re s > 1$.

REMARK 3.3. The second relation in (3.17) is a deep result and is not a special case of (3.13)/(3.14), where $s = 1 + it$ with $t \ne 0$ had not been included.

Turning to the quadratfrei function, we first mention the elementary property

$$\sum_{d|n} |\mu(d)| = 2^{\omega(n)}, \quad n > 1, \tag{3.22}$$

where $\omega(n)$ is the number of *distinct* prime factors of n. (Show this.) This property has the same flavour as the one in (3.8) and is proved similarly. For the summatory version of the quadratfrei function we have the following result:

$$Q(x) \equiv \sum_{n\le x} |\mu(n)| = \frac{6x}{\pi^2} + o\left(\sqrt{x}\right). \tag{3.23}$$

It incorporates the fact (explained in Chapter 4) that the asymptotic proportion of integers that are not divisible by a perfect square is $1/\zeta(2) = 6/\pi^2$. A proof of (3.23), but only with error $O(\sqrt{x})$, is given in Section 3.10. (The $o(\sqrt{x})$ error term requires the PNT.) Combining (3.23) with the fact that $\sum_{n\le x} \mu(n) = o(x)$, one finds that

$$\lim_{x\to\infty} \frac{1}{x} \Big| \{n \le x : \mu(n) = -1\} \Big| = \lim_{x\to\infty} \frac{1}{x} \Big| \{n \le x : \mu(n) = +1\} \Big| = \frac{3}{\pi^2},$$

so that among squarefree numbers the asymptotic proportions of those with even and odd numbers of prime factors are identical. (Show this.)

As is ubiquitous throughout, the *true* order of the approximation in (3.23) is unknown, and depends ultimately on the zero-free region of the zeta function. Assuming the Riemann hypothesis, Montgomery & Vaughan (1981, [**773**]) showed that the error in (3.23) can be reduced to $O(x^{9/28+\epsilon})$, while under the same assumption Chaohua Jia (1993, [**601**]) reduced this to $O\left(x^{17/54+\epsilon}\right)$; what is known for certain, however, is that it cannot be as low as $O\left(x^{1/4}\right)$. What is remarkable about these results is that the order of these errors is better than the $O\left(\sqrt{x}\log\log x\right)$ one might have expected from the Law of the Iterated Logarithm. The surprising take-away here is that the occurrence of squarefree integers exhibits far more regularity than that of a genuinely random sequence

As for the summatory version of μ, namely the Mertens function $M(x) = \sum_{n\le x}\mu(n)$, it has the trivial bound $|M(x)| \le x$, but even the slightest improvement $M(x) = o(x)$ is already equivalent to the PNT. The true order of $M(x)$ is bound up with the nonzero region of $\zeta(s)$. Under de la Vallée Poussin's zero-free region we would have $M(x) = O\left(x\exp\left(-c\sqrt{\log x}\right)\right)$ for some $c > 0$, while under the Korobov–Vinogradov zero-free region one obtains the best unconditional bound currently known: namely, there is a $c > 0$ such that

$$M(x) = O\left(x\exp\left(-c\frac{(\log x)^{3/5}}{(\log\log x)^{1/5}}\right)\right).$$

In contrast to these unconditional results, Littlewood (1912, [**701**]) proved that the Riemann hypothesis is *equivalent* to the stronger assertion $M(x) = O\left(x^{1/2+\epsilon}\right)$ for all $\epsilon > 0$, while Soundararajan (2009, [**1007**]) showed RH to be equivalent to[7]

$$M(x) \ll \sqrt{x}\exp\left((\log x)^{1/2}(\log\log x)^{14}\right).$$

Since it is known that $M(x) = \Omega_\pm\left(\sqrt{x}\right)$ (see the discussion which follows), the possible asymptotic behaviour of $M(x)$ is rather tightly bounded—at least if RH is assumed. If RH was false with Θ being the supremum of the real parts of the zeros of ζ, one could then show (essentially by using ((3.24) below) that $M(x) = \Omega_\pm(x^{\Theta-\epsilon})$ for any $\epsilon > 0$.

For purposes of comparison, a straightforward application of the Law of the Iterated Logarithm (see Appendix D) together with an independence assumption on the $\mu(n)$ would result in an almost sure growth rate for $M(x)$ of *exact* order $\sqrt{x\log\log x}$. However, based on Montgomery's pair correlation conjecture (discussed in Chapter 6) Gonek (unpublished) had conjectured that $M(x)$ has the still lesser *exact* order $O\left(\sqrt{x}\,(\log\log\log x)^{5/4}\right)$. Relying on numerical experiments, Kotnik & van de Lune (2004, [**657**]) conjectured that the order of $M(x)$ must exceed $O\left(\sqrt{x\log\log\log x}\right)$. Ng (2004, [**803**]) provides some additional details.

Based on numerical evidence, it was conjectured by Stieltjes (1905, [**1033**]) in 1885, as well as by Mertens (1897, [**753**])—and was widely believed at the time—that the Mertens function satisfies the inequality $|M(x)| \le \sqrt{x}$ for all x. This became known as the *Mertens conjecture* and, had it been true, it would have implied RH. The proof that the Mertens conjecture implies RH is a nice application of partial summation (see Appendix F) and is not difficult (Ingham (1942, [**583**]);

[7]Such bounds are ever subject to improvement. See, for example, Balazard & de Roton (2008, [**72**]).

Odlyzko & te Riele (1985, [**822**])). It follows from the computation

$$
\begin{aligned}
\frac{1}{\zeta(s)} &= \sum_{n=1}^{\infty} \frac{\mu(n)}{n^s} \\
&= \sum_{n=1}^{\infty} \frac{M(n) - M(n-1)}{n^s} \\
&= \sum_{n=1}^{\infty} M(n)\left(\frac{1}{n^s} - \frac{1}{(n+1)^s}\right) \\
&= s\sum_{n=1}^{\infty} M(n) \int_n^{n+1} \frac{dx}{x^{s+1}} \\
&= s\sum_{n=1}^{\infty} \int_n^{n+1} \frac{M(x)}{x^{s+1}}\, dx \\
&= s\int_1^{\infty} \frac{M(x)}{x^{s+1}}\, dx.
\end{aligned} \tag{3.24}
$$

These steps are valid when $\sigma > 1$. But if Mertens's conjecture were true then the last integral here, and therefore also $1/\zeta(s)$, would be analytic for all $\sigma > \frac{1}{2}$, and this would imply that $\zeta(s)$ cannot have zeros with $\sigma > \frac{1}{2}$. (The argument here has used analytic continuation, which is discussed in Appendix A.)

Ingham (1942, [**583**]) had proved that if the Mertens conjecture were true, it would follow that there must be infinitely many linear dependences[8] among the positive imaginary parts of the nontrivial zeros of $\zeta(s)$. Since there is no credible reason for this to hold in an object as transcendental as the zeta function, Ingham's result had already signalled that the Mertens conjecture was unlikely to be true. The conjecture was eventually shown to be false. By a method that involved computing the first 2000 complex zeros of ζ to some 100 decimal digits of accuracy, Odlyzko & te Riele (1985, [**822**]) were able to prove, unconditionally, that

$$
\liminf_{x\to\infty} \frac{M(x)}{\sqrt{x}} < -1.009, \quad \text{and} \quad \limsup_{x\to\infty} \frac{M(x)}{\sqrt{x}} > 1.06.
$$

(See also the historical notes by te Riele (1985, [**904**]), as well as the earlier result of Jurkat & Peyerimhoff (1976, [**609**]).) Note that it follows from this result that $M(x)$ changes sign infinitely often. Using similar methods, Hurst (2018, [**568**]) later improved these bounds to -1.83 and $+1.82$. See also Kotnik & te Riele (2006, [**656**]) and Best & Trudgian (2015, [**97**]). However, long before this, as a consequence of RH together with the widely believed 'linear independence assumption' alluded to above, Ingham (1942, [**583**]) had already established that

$$
\liminf_{x\to\infty} \frac{M(x)}{\sqrt{x}} = -\infty, \quad \text{and} \quad \limsup_{x\to\infty} \frac{M(x)}{\sqrt{x}} = +\infty. \tag{3.25}
$$

Result (3.25) is widely believed, but no unconditional proof of it is known.

In sum, hardly more than $M(x) = \Omega_\pm(\sqrt{x}\,)$ is known for certain. Whether or not a suitably normalized version of $M(x)$ possesses an asymptotic distribution is

[8]Such a linear dependence has the form $\sum_{j=1}^{k} a_j\gamma_j = 0$ for some integers a_j, not all of which are 0, where the γ_j are the imaginary parts of distinct complex zeros above the real axis.

also unknown, although using work of Heath-Brown and making certain assumptions, Ng (2004, [**803**]) argued that a normalized version of $e^{-y/2}M(e^y)$ should have an asymptotic distribution although not one that is Gaussian. No explicit counterexamples to the Mertens conjecture have been found although it is known that one must occur somewhere between 10^{14} and $10^{7\times 10^{39}}$. Odlyzko & te Riele (1985, [**822**]) give a nice historical survey.

Finally, we remark that the Mellin transform of $M(x)$ can be obtained from the generating function for μ. This follows from the integration by parts carried out in (3.24) which we restate succinctly as

$$\frac{1}{\zeta(s)} = \sum_{n=1}^{\infty} \frac{\mu(n)}{n^s} = \int_1^{\infty} \frac{1}{x^s}\, dM(x) = s\int_1^{\infty} \frac{M(x)}{x^{s+1}}\, dx\,, \qquad \Re s > 1.$$

This may be inverted using the Perron formula to give

$$M(x) = \frac{1}{2\pi i}\int_{c-i\infty}^{c+i\infty} \frac{x^s}{s\,\zeta(s)}\, ds\,, \quad \text{for any} \quad c > 1.$$

Mellin transforms and Perron's formula are discussed in Appendix C.

The Möbius function is discussed in many references including the classic text by Hardy & Wright (2008, [**499**]). Many analogues and generalizations of the Möbius function have also been proposed. For a comprehensive discussion of these, and many further references, see, for example, Sándor & Crstici (2004, [**939**, Chapter 2]), or McCarthy (1986, [**746**]).

3.4. Euler's phi function

We turn next to Euler's phi (or totient) function, $\varphi(n)$. It is defined as the number of positive integers $\le n$ relatively prime to n, and satisfies a basic formula of Euler-product type which may be proved using an inclusion-exclusion argument:

$$\varphi(n) = n\prod_{p|n}\left(1 - \frac{1}{p}\right). \tag{3.26}$$

From this one may see that $\varphi(n)$ is multiplicative and that $\varphi(n)/n$ is strongly multiplicative; correspondingly, $\log\varphi(n)$ is additive and $\log\bigl(\varphi(n)/n\bigr)$ strongly additive. Gauss proved that

$$\sum_{d|n} \varphi(d) = n, \quad n \ge 1; \tag{3.27}$$

applying the Möbius inversion formula (3.9) to this identity then gives

$$\varphi(n) = \sum_{ab=n} a\,\mu(b) \equiv \sum_{d|n} d\,\mu\left(\frac{n}{d}\right) \equiv \sum_{d|n} \frac{n}{d}\,\mu(d), \tag{3.28}$$

and one can recast this as

$$\frac{\varphi(n)}{n} = \sum_{d|n} \frac{\mu(d)}{d}.$$

In the notation introduced in Section 3.1, the relations in (3.27)/(3.28) can be represented as $\varphi * \mathbf{1} = \mathbf{n}$ and $\varphi = \mu * \mathbf{n}$. Euler's phi function has the Dirichlet generating function

$$\sum_{n=1}^{\infty} \frac{\varphi(n)}{n^s} = \frac{\zeta(s-1)}{\zeta(s)}, \quad \Re s > 2. \tag{3.29}$$

Proofs for all of the identities (3.26)–(3.29) are given in Section 3.10.

The phi function also satisfies the summatory relations

$$\sum_{n \le x} \varphi(n) = \frac{3x^2}{\pi^2} + O(x \log x), \qquad \sum_{n \le x} \frac{\varphi(n)}{n} = \frac{6x}{\pi^2} + O(\log x), \tag{3.30}$$

$$\sum_{2 \le n \le x} \frac{1}{\varphi(n)} = C \log x + o(\log x), \quad \text{and} \quad \sum_{2 \le n \le x} \frac{n}{\varphi(n)} = Cx + O(\log x), \tag{3.31}$$

where (at both occurrences here)

$$C = \prod_p \left(1 + \frac{1}{p(p-1)}\right) = \frac{\zeta(2)\zeta(3)}{\zeta(6)} \approx 1.943596.$$

Using the Korobov–Vinogradov zero-free region, the error terms in (3.30) can be taken as

$$O\left(x \, (\log x)^{2/3} (\log\log x)^{4/3} \right) \quad \text{and} \quad O\left((\log x)^{2/3} (\log\log x)^{4/3} \right),$$

respectively (Walfisz (1963, [**1096**])). Postnikov (1988, [**865**, p. 188]) observed they can even be taken as

$$O\left(x \, (\log x)^{2/3} (\log\log x)^{1+\epsilon}\right) \quad \text{and} \quad O\left((\log x)^{2/3} (\log\log x)^{1+\epsilon}\right),$$

a highly nontrivial fact. Erdős & Shapiro (1951, [**361**]) and Montgomery (1987, [**767**]) proved that

$$\sum_{n \le x} \varphi(n) - \frac{3x^2}{\pi^2} = \Omega_{\pm}\left(x\sqrt{\log\log x}\right), \quad \text{as} \quad x \to \infty,$$

so that the first error term in (3.30) cannot be as low as $o\left(x\sqrt{\log\log x}\,\right)$. A result which may be contrasted with the second approximation in (3.30) is

$$\sum_{n \le x} \log \frac{\varphi(n)}{n} = C'x + O(\log\log x), \quad \text{with} \quad C' = \sum_p \frac{\log(1 - 1/p)}{p} \approx -0.580058.$$

Concerning the *extremal* values of φ, it is known that

$$\liminf_{n \to \infty} \frac{\varphi(n)}{n/\log\log n} = e^{-\gamma} \quad \text{and} \quad \limsup_{n \to \infty} \frac{\varphi(n)}{n} = 1 \tag{3.32}$$

(with the second of these being trivial). From the first of these it follows that

$$\liminf_{n \to \infty} \frac{\varphi(n)}{n} = 0 \quad \text{and} \quad \lim_{n \to \infty} \frac{\varphi(n)}{n^{1-\epsilon}} = \infty.$$

In fact, $\varphi(n)/n = \prod_{p|n} (1 - 1/p)$, and as n ranges over $\{1, 2, \ldots, N\}$ with $N \to \infty$, the factors $1 - 1/p$ occur with probabilities $1/p$, and essentially independently of each other. So it turns out that the asymptotic distribution of the values of $\varphi(n)/n$ is the same as that of

$$\prod_p \left(1 - \frac{1}{p}\right)^{Z_p}$$

where the Z_p are independent 0–1 variables with $\mathrm{P}[Z_p = 1] = 1/p$. It can be shown that this product converges almost surely (see Appendix D) and that it corresponds to a distribution which—although dense in $(0,1)$—is also singular. Let

$$G(t) \equiv \lim_{N \to \infty} \frac{1}{N} \# \left\{ n \le N \colon \frac{n}{\varphi(n)} \ge t \right\} \tag{3.33}$$

be the 'natural density' of values of n for which $n/\varphi(n)$ exceeds t. Schoenberg (1928, [**952**]), (1936, [**953**])—the inventor of 'spline functions'—has shown that this limit exists (so, in particular, $\varphi(n)/n$ possesses a limiting distribution), is continuous, is strictly decreasing in t, and that it satisfies

$$\log G(t) = -e^{ct}(1 + o(t^{-2})), \quad \text{as} \quad t \to \infty,$$

with $c = e^{-\gamma}$. For additional details see Erdős (1939, [**337**]), (1946, [**339**]), Toulmonde (2006, [**1063**]), Weingartner (2007, [**1098**]), and Kowalski (2021, [**661**, §1.4]).

Among the various identities satisfied by Euler's phi function are

$$\sum_{n=1}^{\infty} \frac{\mu(n)}{\varphi(n)} = 0, \quad \sum_{n=1}^{\infty} \frac{\mu(n)\log n}{\varphi(n)} = 0,$$

$$\sum_{n=1}^{\infty} \frac{x^n}{1 - x^n} \varphi(n) = \frac{x}{(1-x)^2}, \quad \text{for } |x| < 1,$$

$$\sum_{n=1}^{\infty} \frac{x^n}{1 + x^n} \varphi(n) = x\frac{1 + x^2}{(1-x^2)^2}, \quad \text{for } |x| < 1,$$

as well as the Dirichlet convolution-based identities

$$\sum_{d|n} \frac{\mu^2(d)}{\varphi(d)} = \frac{n}{\varphi(n)}, \quad \text{and} \quad \sum_{d|n} \frac{n}{d} \varphi(d) = \sum_{j=1}^{n} \gcd(j,n)$$

the second of which is due to Cesàro; here gcd refers to the 'greatest common divisor'.

The phi function satisfies many inequalities. Among these are $\varphi(n) \ge \sqrt{n}$ except for $n = 2$ and $n = 6$, and more incisively $\varphi(n) > n^{2/3}$ for $n > 30$. We also have $\varphi(n) \le n - \sqrt{n}$ provided n is not a prime, with equality if and only if n is the square of a prime. Futhermore we have

$$\left| \sum_{n=1}^{N} \varphi(n) - \frac{3}{\pi^2} N^2 \right| \le \frac{3}{2} N \ln N + 3N,$$

from which simple computations yield bounds such as

$$0.3N^2 < \sum_{n=1}^{N} \varphi(n) \le 0.31N^2, \quad \text{for } N \ge 90.$$

Mitrinović & Popadić (1978, [**758**, §3]) provide additional inequalities and their sources.

Euler's phi plays an essential role in RSA public key cryptography. (Recall the discussion of RSA in Chapter 2.) If $n = pq$, where p and q are (large) primes, then $\varphi(n) = (p-1)(q-1)$. The essential result, due to Euler, which allows the RSA procedure to work is that $a^{\varphi(n)} \equiv 1 \,\mathrm{modulo}\, n$ whenever $\gcd(a,n) = 1$. (It has the well-known special case, due to Fermat, $a^{p-1} \equiv 1 \operatorname{modulo} p$ if p does not divide a.[9]) The security of RSA is based on the difficulty of factoring n, or equivalently on the difficulty of determining $\varphi(n)$, knowing only n.[10] For a proof and an insightful discussion of Euler's congruence Hardy & Wright (2008, [**499**, Chapter VI]) is recommended.[11]

Unlike many of the other functions discussed in this chapter, Euler's phi does not lend itself to any 'natural' equivalences to the PNT. However an interesting criterion for RH, due to Nicolas (1983, [**808**]), can be stated as follows. Let $N_k = p_1 p_2 \cdots p_k$ be the product of the first k primes. Then RH is *equivalent* to the inequality

$$\frac{N_k}{\varphi(N_k)} > e^\gamma \log\log N_k, \quad \text{for all } k \geq 1.$$

(Remark: This asserts that the lim inf in (3.32) is approached only from below.) In fact, to prove RH, it would suffice to show that this inequality holds for all but a finite number of k. A refinement of this condition, also due to Nicolas (2012, [**809**]) is that RH is equivalent to

$$\limsup_{n\to\infty} \sqrt{\log n}\left(\frac{n}{\varphi(n)} - e^\gamma \log\log n\right) = e^\gamma(4 + \gamma - \log 4\pi).$$

Concerning its connection to the PNT, Euler's phi function is useful primarily for expressing the asymptotics for primes in arithmetic progressions. Specifically, an arithmetic progression $a, a+n, a+2n, \ldots$ can contain more than one prime only if a and n are relatively prime, and it is a fundamental result of analytic number theory that the numbers of primes in the associated residue classes of n are asymptotically equal. Thus

$$\pi(x; n, a) \sim \frac{\pi(x)}{\varphi(n)} \sim \frac{1}{\varphi(n)} \frac{x}{\log x}$$

where $\pi(x; n, a)$ is the number of primes $\leq x$ which are congruent to $a \operatorname{modulo} n$, where a and n are relatively prime to each other.

Euler's phi function is discussed in Hardy & Wright (2008, [**499**, Chapter XVI]) and many other texts. See also Ford (1998, [**383**]) and the references therein. For generalizations and extensions of the phi function, see Mitrinović, Sándor & Crstici (1995, [**759**]), Sándor & Crstici (2004, [**939**, Chapter 3]), and Sándor, Mitrinović & Crstici (2006, [**940**, pp. 9–36]).

3.5. The divisor functions

The number of divisors function $\tau(n)$—known more simply as the 'divisor function'—as well as the sum of divisors and sum of kth powers of divisors functions, $\sigma(n)$ and $\sigma_k(n)$, are defined as $\tau(n) \equiv \sum_{d|n} 1$, $\sigma(n) \equiv \sum_{d|n} d$, and $\sigma_k(n) \equiv \sum_{d|n} d^k$. One can also state these definitions as Dirichlet convolutions: $\tau \equiv \mathbf{1} * \mathbf{1}$, $\sigma \equiv \mathbf{n} * \mathbf{1}$,

---

[9] The notation $a \equiv b \operatorname{modulo} n$ represents modular arithmetic and means that n divides $a - b$, or equivalently that dividing a by n leaves the same remainder as dividing b by n.

[10] Indeed, for a product $n = pq$ of two primes, knowing p and q yields $\varphi(n)$; while knowing pq and $\varphi(n)$ one may recover p and q using $n - \varphi(n) - 1 = p + q$ and the quadratic formula $(x-p)(x-q) = x^2 - x(p+q) + pq$.

[11] We cannot resist proposing two problems treated in Hardy & Wright (2008, [**499**, Chapter VI]):

(a) If p is a prime, show that *each* of $\binom{p}{1}, \binom{p}{2}, \ldots, \binom{p}{p-1}$ is divisible by p (Easy).

(b) Show that the product of *any* n successive positive integers is divisible by $n!$ (Not so easy).

and (in obvious notation) $\sigma_k \equiv \mathbf{n}^k * \mathbf{1}$. (Recall that $\mathbf{1}(n) \equiv 1$ and $\mathbf{n}(n) \equiv n$.) The Möbius inverse relations corresponding to these are $\tau * \mu = \mathbf{1}$, $\sigma * \mu = \mathbf{n}$, and $\sigma_k * \mu = \mathbf{n}^k$; with $n \geq 1$, one can write these as

$$\sum_{d|n} \tau(d)\mu\left(\frac{n}{d}\right) = 1, \quad \sum_{d|n} \sigma(d)\mu\left(\frac{n}{d}\right) = n, \quad \text{and} \quad \sum_{d|n} \sigma_k(d)\mu\left(\frac{n}{d}\right) = n^k,$$

respectively. The tau notation $\tau(\cdot)$ stems from the German word *teiler* meaning 'divisor'.

These divisor functions have Dirichlet generating functions given by

$$\sum_{n=1}^{\infty} \frac{\tau(n)}{n^s} = \zeta^2(s), \quad \sum_{n=1}^{\infty} \frac{\sigma(n)}{n^s} = \zeta(s)\zeta(s-1), \quad \text{and} \quad \sum_{n=1}^{\infty} \frac{\sigma_k(n)}{n^s} = \zeta(s)\zeta(s-k), \tag{3.34}$$

respectively. We mention also the less straightforward relations

$$\text{(a)} \sum_{n=1}^{\infty} \frac{\tau^2(n)}{n^s} = \frac{\zeta^4(s)}{\zeta(2s)}, \quad \text{and (b)} \sum_{n=1}^{\infty} \frac{\tau^3(n)}{n^s} = \frac{\zeta^4(s)\zeta(s-1)\zeta(s+1)}{\zeta^2(2s)}. \tag{3.35}$$

In these expressions, s, as well as k, may be complex; however all arguments of ζ must have real parts > 1. The values of $\tau(n)$ can also be generated by a *Lambert series*:

$$\sum_{n=1}^{\infty} \tau(n)x^n = \sum_{n=1}^{\infty} \frac{x^n}{1-x^n}, \quad |x| < 1. \tag{3.36}$$

The identities in (3.34) and (3.35)a are proved in Section 3.10, while (3.36) is proved in Appendix C.

The behaviour of $\tau(n)$ is highly erratic. Since $\tau(p^k) = k+1$ when p is prime, it equals 2 infinitely often and is otherwise unbounded. One can also show that $\tau(n) \leq 2\sqrt{n}$ for all n; $\varphi(n)\tau(n) > n$ for $n > 2$; and $\varphi(n)\tau^2(n) < n^2$ for $n > 12$. Of course $\liminf \tau(n) = 2$ while at the other extreme it is known that $\tau(n) = o(n^\epsilon)$ for all $\epsilon > 0$, and more precisely that

$$\limsup_{n\to\infty} \frac{\log \tau(n)}{\log n / \log\log n} = \log 2. \tag{3.37}$$

However, in spite of $\tau(n)$'s erratic behaviour, the values of $\log d/\log n$ as d ranges over the divisors of n, and when averaged over n in a certain natural way, stabilize to an arcsine distribution on $[0, 1]$. The details of this remarkable result may be found in Chapter 9.

The summatory version of $\tau(n)$ also stabilizes and satisfies

$$\sum_{n\leq x} \tau(n) = x\log x + (2\gamma - 1)x + \text{Error} \sim x\log x. \tag{3.38}$$

The question of the best error term here is known as the *Dirichlet divisor problem.* Dirichlet showed it can be taken as $O(\sqrt{x})$, while Hardy and Landau had each established that it cannot be less than $O(x^{1/4})$. Voronoi (1903, [**1093**]) obtained the error bound $O(x^{1/3}\log x)$. Subsequent efforts by many others have successively improved on this with the best known result being that of Huxley (2003, [**572**]) who established that (3.38) holds with error $O\left(x^{131/416}\right)$ where $131/416 \approx 0.3149$. See, for example, Titchmarsh (1986, [**1059**, §12.6]) or Ivić (1985, [**586**, §13.6]), as well as Diamond (1982, [**283**]) and Halberstam & Richert (1974, [**480**]). While the true error in (3.38) remains unknown, it is widely believed that $O(x^{1/4+\epsilon})$ is in fact the correct one. A proof of (3.38), but with Dirichlet's $O(\sqrt{x})$ error, is given in Section 3.10.

Results related to (3.38) include

$$\sum_{n \le x} \frac{\tau(n)}{n} = \frac{1}{2}\log^2 x + 2\gamma \log x + O(1),$$

and

$$\sum_{n \le x} \log \tau(n) = (\log 2) x \log\log x + Cx + o(x),$$

where

$$C = (\log 2)\left(\gamma + \sum_p \left[\log\left(1 - \frac{1}{p}\right) + \frac{1}{p}\right]\right) + \sum_p \sum_{k \ge 2} \frac{\log(1+1/p)}{p^k} \approx 0.830146.$$

It can also be shown that

$$\sum_{n \le x} \tau^2(n) \sim \frac{1}{\pi^2} x \log^3 x, \quad \sum_{n \le x} \tau^3(n) \sim c_3 x \log^7 x, \quad \text{and} \quad \sum_{n \le x} \tau^k(n) \sim c_k x (\log x)^{2^k - 1},$$

where

$$c_3 = \frac{1}{7!} \prod_p \left(1 - \frac{1}{p}\right)^4 \left(1 + \frac{4}{p} + \frac{1}{p^2}\right),$$

with the remaining constants c_k also known although more complicated. The asymptotics here, as well as in (3.38), are dominated by a small proportion of numbers for which $\tau(n)$ takes on significantly large values. In fact, most values of $\tau(n)$ are substantially smaller than $\log n$ as a consequence of the Erdős–Kac theorem as well as of the Hardy–Ramanujan inequalities for $\omega(n)$, both of which are discussed in Chapter 9.

Related to the divisor function $\tau(n)$ is the function $\tau_k(n)$ which counts the number of ways n can be written as a product $n = n_1 \times n_2 \times \cdots \times n_k$ of k positive integers.[12] It has the Dirichlet generating series

$$\zeta(s)^k = \sum_{n=1}^{\infty} \frac{\tau_k(n)}{n^s}, \quad \Re s > 1,$$

and summatory version that satisfies

$$\sum_{n \le x} \tau_k(n) = x P_{k-1}(\log x) + \text{Error}$$

where the P_{k-1} are certain degree $k-1$ polynomials. Note that $\tau_2(n) \equiv \tau(n)$. The relations mentioned here generalize the first identity in (3.34), and the result (3.38), respectively. The $\tau_k(n)$ are also commonly represented as $d_k(n)$. For background, the references in Keating et al. (2018, [**626**]) may be consulted.

For the sum of divisors function $\sigma(n)$ we have

$$\limsup_{n \to \infty} \frac{\sigma(n)}{n \log\log n} = e^\gamma, \quad \text{so that} \quad \limsup_{n \to \infty} \frac{\sigma(n)}{n} = \infty, \tag{3.39}$$

[12] Example: $\tau_3(25) = 6$, due to choices $(n_1, n_2, n_3) = (25,1,1), (1,25,1), (25,1,1), (5,5,1), (5,1,5), (1,5,5)$.

while for $\sigma_a(x)$ with real-valued a,

$$\liminf_{n\to\infty} \frac{\sigma_a(n)}{n^a} = 1, \text{ for } a > 0, \quad \text{and} \quad \limsup_{n\to\infty} \frac{\sigma_a(n)}{n^a} = \zeta(a), \text{ for } a > 1.$$

We also have the inequalities $\sigma(n) < 2.59\, n \log\log n$ for all $n \geq 7$, and $\sigma(n) > n + \sqrt{n}$ provided n is not a prime. Furthermore $\sigma(n)\varphi(n) \leq n^2 - 1$, with equality only if n is prime; $\sigma(n)/\tau(n) \leq (n+1)/2$ again with equality only if n is prime; and

$$\frac{6}{\pi^2} < \frac{\sigma(n)\varphi(n)}{n^2} < 1, \quad \text{for } n > 1. \tag{3.40}$$

In view of (3.40) the functions $\sigma(n)$ and $\varphi(n)$ can be expected to have complementary reciprocal properties. Indeed, $\sigma(n)/n$ and $n/\varphi(n)$ have the same asymptotic orders of magnitude; note the limiting bounds shown in (3.32) and (3.39).

A Dirichlet series result (due to Ramanujan) for the $\sigma_k(n)$ functions which generalizes (3.34) and (3.35) is given by

$$\sum_{n=1}^{\infty} \frac{\sigma_a(n)\sigma_b(n)}{n^s} = \frac{\zeta(s)\zeta(s-a)\zeta(s-b)\zeta(s-a-b)}{\zeta(2s-a-b)}$$

where s, a, and b may be complex, as long as all arguments of ζ have real part greater than 1. Meanwhile, a Lambert series identity generalizing (3.36) is given by

$$\sum_{n=1}^{\infty} \sigma_a(n) x^x = \sum_{n=1}^{\infty} \frac{n^a x^n}{1-x^n}.$$

For the summatory versions of σ and σ_k we have (see Apostol, 1985, [**26**, §3.6], for example)

$$\sum_{n\leq x} \sigma(n) = \frac{\pi^2}{12} x^2 + O(x \log^{2/3} x),$$

$$\text{and} \quad \sum_{n\leq x} \sigma_k(n) = \frac{\zeta(k+1)}{k+1} x^{n+1} + O(n^k) \quad \text{for } k > 1,$$

as well as

$$\sum_{n\leq x} \frac{\sigma(n)}{n} = \frac{\pi^2}{6} x - \frac{1}{2} \log x + O(\log^{2/3} x) = \frac{\pi^2}{6} x + O(\log x).$$

Furthermore, for all N we have

$$\frac{N-1}{2} < \frac{1}{N} \sum_{n=1}^{N} \sigma(n) \leq N, \quad \text{and} \quad \sum_{n=1}^{N} \frac{n}{\sigma(n)} > \frac{N+1}{2}.$$

The values of $\sigma(n)/n$ are known to be dense in $[1, \infty)$. Define the 'natural density' of the values of n for which $\sigma(n)/n$ exceeds t as the limiting proportion

$$H(t) \equiv \lim_{N\to\infty} \frac{1}{N} \left| \left\{ n \leq N : \frac{\sigma(n)}{n} \geq t \right\} \right|.$$

Davenport (1933, [**260**]) showed that this limit exists, is continuous and strictly decreasing, and that it satisfies

$$\log H(t) = -e^{ct}(1 + o(t^{-2})), \quad \text{as } t \to \infty,$$

with $c = e^{-\gamma}$. Like $G(t)$ defined in (3.33), the distribution $H(t)$ is also purely singular. Weingartner (2007, [**1098**]) pointed out that $H(t) \le G(t)$, a consequence of the fact that

$$\frac{\sigma(n)}{n} = \prod_{p^\nu \| n} \frac{1+p+p^2+\cdots p^\nu}{p^\nu} = \prod_{p^\nu \| n} \frac{1-p^{-\nu-1}}{1-p^{-1}} < \prod_{p|n} \frac{1}{1-p^{-1}} = \frac{n}{\varphi(n)}.$$

Recalling the complementary nature of $\sigma(n)$ and $\varphi(n)$ noted in (3.40), one may compare the density $H(t)$ for $\sigma(n)/n$ with the previously mentioned density $G(t)$ for $n/\varphi(n)$.

The lim sup in (3.39) is related to an inequality that is equivalent to the Riemann hypothesis. Specifically, Robin (1984, [**912**]) has shown that

$$\text{RH} \iff \sigma(n) < e^\gamma n \log\log n \text{ for } n > 5{,}040,$$

so that beyond 5,040 the lim sup value is approached only from below, although Ramanujan had already shown that under RH this inequality holds once n is large enough. Robin's criterion is particularly sharp for it is known *unconditionally* that

$$\sigma(n) < (e^\gamma n \log\log n)\,(1 + o(1));$$

indeed this is implied by the first inequality in (3.39). Robin's RH equivalence based on $\sigma(n)$ may be compared to Nicolas's previously mentioned RH equivalence based on $\varphi(n)$, while again recalling the complementary relation that exists between $\sigma(n)$ and $\varphi(n)$. For further discussion of Robin's RH equivalence, see Lagarius (2002, [**667**]).

3.6. The number of prime factors functions

The number of prime factors functions, ω and Ω, share very similar properties. For even though $\omega(n) \le \Omega(n)$ for all n, they nevertheless have the same asymptotic means

$$\text{(3.41)} \qquad \frac{1}{x}\sum_{n\le x} \omega(n) \sim \log\log x, \quad \frac{1}{x}\sum_{n\le x} \Omega(n) \sim \log\log x,$$

as $x \to \infty$, as well as the same asymptotic variances

$$\text{(3.42)} \quad \frac{1}{x}\sum_{n\le x} (\omega(n) - \log\log x)^2 \sim \log\log x, \quad \frac{1}{x}\sum_{n\le x} (\Omega(n) - \log\log x)^2 \sim \log\log x.$$

This occurs because $\omega(n)$ and $\Omega(n)$ are of similar size; specifically, one can show

$$\text{(3.43)} \quad \frac{1}{x}\sum_{n\le x} (\Omega(n) - \omega(n)) = C + O\left(\frac{1}{\sqrt{x}\log x}\right), \quad \text{for } C = \sum_p \frac{1}{p(p-1)} \approx 0.7732.$$

More precise versions of (3.41) are

$$\frac{1}{x}\sum_{n\le x} \omega(n) = \log\log x + B + O\left(\frac{1}{\log x}\right),$$

$$\text{and} \quad \frac{1}{x}\sum_{n\le x} \Omega(n) = \log\log x + A + O\left(\frac{1}{\log x}\right),$$

where B is the previously mentioned Meissel–Mertens constant

$$B=\gamma+\sum_p\left(\log\left(1-\frac{1}{p}\right)+\frac{1}{p}\right)=\gamma-\sum_p\sum_{k=2}^{\infty}\frac{1}{kp^k}=\gamma+\sum_{k=2}^{\infty}\mu(k)\frac{\zeta(k)}{k}\approx 0.261497,$$

and (with C as in (3.43))

$$A=B+C=\gamma+\sum_p\left(\log\left(1-\frac{1}{p}\right)+\frac{1}{p-1}\right)=\gamma+\sum_{k=2}^{\infty}\varphi(k)\frac{\zeta(k)}{k},$$

so that $A\approx 1.03465$. References to still more precise versions of these approximations may be found in the 'Distinct Prime Factors' page of `mathworld.wolfram.com`. The functions ω and Ω are considered in greater detail in Chapter 9 where proofs of the key results mentioned above may also be found.

On applying Chebychev's inequality (see Appendix D) using the mean and variance formulas (3.41) and (3.42), we find that the values of $\omega(n)$ stay very close to $\log\log n$ in a convergence in probability sense. Thus given any $\epsilon>0$, if n is chosen at random from among the integers $1,2,\ldots,N$, then the proportion of values of $\omega(n)/\log\log n$ which differ from 1 by more than ϵ tends to 0 as $N\to\infty$. The same result holds for $\Omega(n)$ as well. In this sense one can say that the average number of prime factors (whether distinct or not) of an integer n is $\log\log n$. (It does *not* mean that $\omega(n)$ or $\Omega(n)$ must eventually stay close to $\log\log n$.) These results are due to Hardy & Ramanujan (1917, [**497**]), but see also Turán (1934, [**1069**]) who showed further that

$$\limsup_{n\to\infty}\frac{\omega(n)}{\log n/\log\log n}=1,\quad\text{and}\quad\limsup_{n\to\infty}\frac{\Omega(n)}{\log n/\log 2}=1.$$

Of these, the second is trivial since $\Omega(n)\le(\log n)/(\log 2)$ with equality if and only if n is a power of 2. Also trivial is that $\omega(n)$ and $\Omega(n)$ take on their lower bound value of 1 infinitely often. Erdős & Kac (1939, [**357**]) proved that $\omega(n)$ and $\Omega(n)$ are asymptotically normally distributed with mean and variance both equal to $\log\log n$. The details are in Chapter 9.

Using $\omega(n)$, one can define a function which represents the number of integers $n\le x$ which have exactly k prime factors:

$$\pi_k(x)\equiv\#\{n\le x\colon \omega(n)=k\}.$$

In terms of this function, the Prime Number Theorem can be stated as $\pi_1(x)\sim x/\log x$. Landau established that for each *fixed* $k\ge 1$,

$$\pi_k(x)\sim x\frac{(\log\log x)^{k-1}}{(k-1)!\log x},\quad\text{as}\quad x\to\infty.$$

The values on the right correspond to the probabilities of a Poisson distribution whose mean is $\log\log x$. The same result continues to hold if $\pi_k(x)$ is based on Ω instead of ω. However, these results are not valid as stated if k is allowed to increase with x. Further details of Landau's result appear in Chapter 9.

Other results involving the number of prime factors functions include

$$\sum_{n\le x}2^{\omega(n)}=\frac{6}{\pi^2}x\log x+O(x),\quad\sum_{n\le x}2^{\Omega(n)}=Cx\log^2 x+O(x\log x),$$

for some constant C (not the same one as in (3.43)), as well as

$$\sum_{1<n\le x} \frac{1}{\omega(n)} \sim \frac{x}{\log\log x}, \quad\text{and}\quad \sum_{1<n\le x} \frac{\Omega(n)}{\omega(n)} = x + O\left(\frac{x}{\log\log x}\right).$$

Generating functions involving ω and Ω include

$$\sum_{n=1}^{\infty} \frac{\omega(n)}{n^s} = \zeta(s)P(s), \quad \sum_{n=1}^{\infty} \frac{\Omega(n)}{n^s} = \zeta(s)\sum_{n=1}^{\infty} P(ns), \quad\text{and}\quad \sum_{n=1}^{\infty} \frac{2^{\omega(n)}}{n^s} = \frac{\zeta^2(s)}{\zeta(2s)},$$

where $P(s) = \sum_p 1/p^s$ is the 'prime zeta function' (defined in (3.21) and discussed further in Chapter 5), and $\Re s > 1$. We also have

$$\sum_{d|n} |\mu(d)| = 2^{\omega(n)}, \quad\text{and}\quad \sum_{n=1}^{\infty} \frac{\mu(n)\omega(n)}{n} = 0.$$

Also worth mentioning is that the number of divisors function $\tau(n)$ satisfies

$$2^{\omega(n)} \le \tau(n) \le 2^{\Omega(n)}, \quad\text{so that}\quad \omega(n) \le \frac{\log \tau(n)}{\log 2} \le \Omega(n).$$

(Indeed, the first inequality holds because $2^{\omega(n)}$ is precisely the number of squarefree divisors of n.) As a consequence, the additive function $\log \tau(n)/\log 2$ acts like a 'number-of-prime-factors' function and so equals $(1+o(1))\log\log n$ except on a set of integers of zero density.

3.7. The Liouville function

The Liouville function, $\lambda(n)$, is the unique completely multiplicative function which assigns the values $\lambda(p) = -1$ at the primes. It is closely associated with the Möbius function since $\lambda(n) = \mu(n)$ for squarefree n. In fact, the Dirichlet inverse of $\lambda(n)$ happens to be $|\mu(n)|$:

$$\sum_{ab=n} \lambda(a)|\mu(b)| \equiv \sum_{d|n} \lambda(d)\left|\mu\left(\frac{n}{d}\right)\right| = \mathbf{e}(n) = \begin{cases} 1, & \text{if } n = 1, \\ 0, & \text{if } n > 1. \end{cases} \tag{3.44}$$

This identity, which can also be written as $\lambda * |\mu| = \mathbf{e}$, is equivalent to the fact that the Dirichlet series for λ, shown in (3.47), is the reciprocal of the one for $|\mu|$, shown previously in (3.13). A further connection between λ and μ is the identity

$$\lambda(n) = \sum_{d^2|n} \mu\left(\frac{n}{d^2}\right), \tag{3.45}$$

which can be proved by applying Dirichlet convolution to generating function (3.47).

The Liouville function has the elementary property

$$\sum_{d|n} \lambda(d) = \begin{cases} 1, & \text{if } n \text{ is a perfect square}, \\ 0, & \text{otherwise}, \end{cases} \tag{3.46}$$

and also satisfies

$$\sum_{n=1}^{\infty} \frac{\lambda(n)}{n} = 0, \quad \sum_{n=1}^{\infty} \frac{\lambda(n)\log n}{n} = -\zeta(2) = -\frac{\pi^2}{6}, \quad\text{and}\quad \sum_{n\le x} \frac{\lambda(n)}{n} \le 1 + \frac{1}{\sqrt{x}},$$

the first two of these being equivalent to the PNT. It has the Dirichlet generating function

$$\sum_{n=1}^{\infty} \frac{\lambda(n)}{n^s} = \frac{\zeta(2s)}{\zeta(s)} = \prod_p \left(1 + p^{-s}\right)^{-1}, \quad \text{for } \Re s > 1, \tag{3.47}$$

and the Lambert generating function

$$\sum_{n=1}^{\infty} \frac{\lambda(n) x^n}{1 - x^n} = \sum_{n=1}^{\infty} x^{n^2}, \quad |x| < 1.$$

The properties (3.44), (3.45), (3.46), and (3.47) are proved in Section 3.10.

At least as important as $\lambda(n)$ is its summatory version

$$L(x) \equiv \sum_{n \le x} \lambda(n),$$

which is the difference between the number of integers $n \le x$ having even versus odd numbers of prime factors (when counted with multiplicity). What is known for certain is that the order of $L(x)$ cannot be less than $O(\sqrt{x}\,)$, while the PNT is equivalent to the assertion $L(x) = o(x)$. It follows that half of the integers have an even number of prime factors and half have an odd number of prime factors, asymptotically. Under the de la Vallée Poussin zero-free region one can show that $\sum_{n \le x} \lambda(n) = O\big(xe^{-c\sqrt{\log x}}\big)$, while the best unconditional result currently known (based on the Korobov–Vinogradov zero-free region) is given by

$$L(x) = \sum_{n \le x} \lambda(n) = O\big(x \exp\big(-c(\log x)^{3/5} (\log \log x)^{1/5}\big)\big).$$

The Riemann hypothesis is equivalent to $L(n) = O(x^{1/2+\epsilon})$—we prove this in Section 3.10—although it is conjectured that the stronger result

$$L(x) = \sum_{n \le x} \lambda(n) = O(\sqrt{x} \log x)$$

should hold as well.

A famous question raised by Pólya (1919, [**861**]), whose answer turned out to be negative, is whether or not the summatory function satisfies $L(n) \le 0$ for all $n > 1$. Pólya showed that if one could establish that $L(n)$ does not change sign once n becomes large enough, then it would imply RH (and would also imply that all of the zeros must be simple). His argument was based on applying partial summation to the relation $\sum \lambda(n)/n^s = \zeta(2s)/\zeta(s)$ to get

$$\frac{\zeta(2s)}{\zeta(s)} = \int_1^{\infty} \frac{1}{x^s}\, dL(x) = s \int_1^{\infty} \frac{L(x)}{x^{s+1}}\, dx.$$

Now since $1/\zeta(s)$ has no *real* zeros, the rightmost *real-valued* singularity of the left side occurs at $s = \frac{1}{2}$, where $\zeta(2s)$ becomes $\zeta(1)$. On the other hand, if $L(x)$ is never positive, a theorem of Landau concerning singularities of Laplace transforms of nonnegative functions implies that the rightmost singularity of the right side must be real-valued. Because the functions involved are analytic, these two singularities must agree, so it follows that the right and left sides must then both be convergent for $\Re s > \frac{1}{2}$, which in turn implies that $\zeta(s) \ne 0$ for $\Re s > \frac{1}{2}$ as has been claimed. In the 1942 paper of Ingham referred to previously—the one in which he dealt with the Mertens conjecture—Ingham obtained analogous results for the function $L(x)$, showing that if either $L(n) < c\sqrt{n}$, or $L(n) > -c\sqrt{n}$, for some $c > 0$, and all n, then not only would RH follow with all zeros being simple, but also that the imaginary parts of the complex zeros would then satisfy the unlikely condition of being linearly dependent.

All such speculations were dashed, however, when Haselgrove (1958, [**516**]) proved that $L(n)$ changes sign infinitely often, and that there should be a counterexample to $L(n) \leq 0$ in the vicinity of $n = 1.845 \times 10^{361}$. Tanaka (1980, [**1039**]) later found the smallest counterexample; it occurs when $n =$ 906,150,257 at which $L(n) = +1$. It had also been conjectured that $\sum_{k=1}^{n} \lambda(k)/k$ must eventually remain nonnegative, but this too was disproved by Haselgrove, while Borwein, Ferguson & Mossinghoff (2008, [**135**]) later found that its first counterexample occurs at 72,185,376,951,205. Had that conjecture been true, Turán (1948, [**1071**]), (1959, [**1073**]) showed that it too would have implied RH. For some further details, see Lehman (1960, [**685**]), Tanaka (1980, [**1039**]), Mossinghoff & Trudgian (2012, [**778**]), (2017, [**779**]), Humphries (2012, [**567**]), and Alkan (2021, [**15**]). In particular, Borwein, Ferguson & Mossinghoff (2008, [**135**]) give a detailed historical account of Pólya's question. See also Wiener & Wintner (1957, [**1105**]).

There is, however, a sense in which one can say that the values of $\lambda(n)$ are skewed towards its negative value. Unlike $M(x)$ which changes sign rather frequently, the function $L(x)$ appears to take on negative values more often than positive values. Brent & van de Lune (2011, [**145**]) showed that there is a constant $c > 0$ such that, for any fixed $\alpha \geq 1$,

$$\sum_{n=1}^{\infty} \frac{\lambda(n)}{e^{n\pi x} + 1} = -\frac{c}{\sqrt{x}} + \frac{1}{2} + O\left(x^{\alpha}\right), \quad \text{as } x \downarrow 0.$$

The point here is that the left side is a weighted sum of the $\lambda(n)$ with positive weights that (for $n \ll 1/x$) are initially close to $1/2$, and then become small as n increases, while the right side becomes negative and tends to $-\infty$ as $x \downarrow 0$. This phenomenon is discussed in Martin, Mossinghoff, & Trudgian (2023, [**733**]) where further references may be found.

Pseudorandom sequences such as the Liouville function $\{\lambda(n)\}$ or the Möbius function $\{\mu(n)\}$ manifest themselves as interesting 'time series', and one can ask how their 'autocorrelation functions' behave, whether or not they tend towards 'stationarity', and so on. Such questions are related to conjectures of Chowla and of Elliot, as well as to recent work by Matomäki, Radziwiłł, Tao and others—matters that will be taken up in Chapter 12.

3.8. The von Mangoldt and Chebychev functions

We turn finally to the von Mangoldt function $\Lambda(n)$ and to the closely related Chebychev $\vartheta(x)$ and $\psi(x)$ functions, all three of which play an important role in the proof of the Prime Number Theorem. Recall first that the von Mangoldt function $\Lambda(n)$ is defined as

$$\Lambda(n) = \sum_{d|n} \lambda(d) = \begin{cases} \log p, & \text{if } n = p^k \text{ is a prime power,} \\ 0, & \text{otherwise.} \end{cases} \tag{3.48}$$

This function satisfies the simple relation

$$\sum_{d|n} \Lambda(d) = \log n, \tag{3.49}$$

and applying Möbius inversion to this relation gives

$$\Lambda(n) = \sum_{d|n} \mu\left(\frac{n}{d}\right) \log d \equiv \sum_{d|n} \mu(d) \log\left(\frac{n}{d}\right) \equiv -\sum_{d|n} \mu(d) \log d. \tag{3.50}$$

(The last equality used (3.8).) A second application of Möbius inversion then gives

$$\sum_{d|n} \mu\left(\frac{n}{d}\right) \Lambda(d) = -\mu(n) \log n. \tag{3.51}$$

(To see this use $\Lambda = -(\mu\mathcal{L}) * \mathbf{1}$ from (3.50) and convolve both sides by μ.) Recalling that $\mathbf{1}(n) \equiv n$ and $\mathcal{L}(n) \equiv \log n$, the relations (3.49), (3.50), and (3.51) can also be stated as

$$\Lambda * \mathbf{1} = \mathcal{L}, \quad \Lambda = \mu * \mathcal{L} = -(\mu\mathcal{L}) * \mathbf{1}, \quad \text{and} \quad \mu * \Lambda = -\mu\mathcal{L}.$$

The function Λ satisfies the Dirichlet series relations (for $\sigma > 1$)

$$\sum_{n=1}^{\infty} \frac{\Lambda(n)}{n^s} = -\frac{\zeta'(s)}{\zeta(s)}, \quad \text{and} \quad \sum_{n=2}^{\infty} \frac{\Lambda(n)/\log n}{n^s} = \sum_{p} \sum_{k=1}^{\infty} \frac{1/k}{p^{ks}} = \log \zeta(s), \tag{3.52}$$

the last sum being over all prime powers. These two are related, with the first identity being the derivative of the second (up to a negative sign). In fact, for Dirichlet series more generally,

$$f(s) = \sum_{n=1}^{\infty} \frac{a(n)}{n^s} \iff f'(s) = -\sum_{n=1}^{\infty} \frac{a(n) \log n}{n^s} \tag{3.53}$$

(aside from the constant $a(1)$); and provided $a(n)$ is completely multiplicative and the sums shown converge absolutely,

$$\log f(s) \equiv \log\left(\sum_{n=1}^{\infty} \frac{a(n)}{n^s}\right) = \sum_{n=2}^{\infty} \frac{a(n)\Lambda(n)/\log n}{n^s}. \tag{3.54}$$

Thus, for $a(n)$ completely multiplicative, (3.53) and (3.54) combine to yield

$$f(s) = \sum_{n=1}^{\infty} \frac{a(n)}{n^s} \iff \frac{d \log f(s)}{ds} \equiv \frac{f'(s)}{f(s)} = -\sum_{n=1}^{\infty} \frac{a(n)\Lambda(n)}{n^s}, \tag{3.55}$$

and the relations in (3.52) follows from these. (Of course, as in (3.52), the right sides of (3.54) and (3.55) can also be expressed as sums over prime powers, with similar remarks applying to many of the expressions occurring below.) The relation (3.53) only involves differentiation. The identities (3.49), (3.50), (3.51), (3.52), (3.54), and (3.55) are proved in Section 3.10. The formula (3.55) highlights the important role that the von Mangoldt function plays in relating Dirichlet series that are connected through logarithmic differentiation. The relation (3.54) will find an application in Chapter 10.

Versions related to (3.52) for the case $s = 1$ are given by

$$\sum_{n \le x} \frac{\Lambda(n)}{n} \equiv \int_1^x \frac{d\psi(t)}{t} = \log x - \gamma + o(1), \quad \text{and} \quad \sum_{n \le x} \frac{\Lambda(n) \log n}{n} = \frac{1}{2} \log^2 x + O(\log x).$$

The $o(1)$ term in the first of these is equivalent to the PNT; under RH it would be $O(x^{-1/2}\log^2 x)$. The second relation in (3.52) has the $s = 1$ version

$$\sum_{2\le n\le x} \frac{\Lambda(n)/\log n}{n} = \log\log x + \gamma + O\left(\frac{1}{\log x}\right),$$

while on the $\sigma = 1$ line one has

$$\sum_{2\le n\le x} \frac{\Lambda(n)/\log n}{n^{1+it}} = \log\zeta(1+it) + o(1), \quad t \ne 0.$$

Worth noting too is the close link between $\sum \Lambda(n)/n$ and $\sum \log p/p$: specifically

$$\sum_{n\le x} \frac{\Lambda(n)}{n} - \sum_{p\le x} \frac{\log p}{p} = \sum_{p^k\le x,\, k\ge 2} \frac{\log p}{p^k} = \sum_p \frac{\log p}{p(p-1)} + o(1)$$

with the sum on the right being a finite constant. The von Mangoldt function also satisfies the curious Fourier identity

$$\sum_{n=1}^{\infty} \frac{\Lambda(n)}{n}\{nx\} \;=\; -\frac{1}{\pi}\sum_{n=1}^{\infty} \frac{\log n}{n}\sin(2\pi nx), \quad x > 0,$$

where $\{\cdot\}$ is the fractional parts function.

In terms of $\Lambda(n)$, the PNT is equivalent to

$$\sum_{n\le x} \Lambda(n) = x + o(x),$$

while RH is equivalent to

$$\sum_{n\le x} \Lambda(n) = x + O\big(\sqrt{x}\log^2 x\big).$$

Formulas with multiple appearances of Λ include

$$\Lambda(n)\log n + \sum_{d|n} \Lambda(d)\Lambda\left(\frac{n}{d}\right) = \sum_{d|n} \mu(d)\log^2\left(\frac{n}{d}\right).$$

This result is due to Selberg and may be written alternately as

$$\Lambda(n)\log n + (\Lambda * \Lambda)(n) = (\mu * \log^2)(n).$$

We also have

$$\left(\sum_{n=2}^{\infty} \frac{\Lambda(n)}{n^s}\right)^2 = \sum_{n=2}^{\infty} \frac{(\Lambda*\Lambda)(n)}{n^s} = \left(\frac{\zeta'(s)}{\zeta(s)}\right)^2.$$

Also worth noting are the approximations

$$\sum_{1<mn\le x} \frac{\Lambda(m)\Lambda(n)}{mn} = \frac{1}{2}\log^2 x + O(\log x),$$

and

$$\sum_{1<mn\le x} \frac{\Lambda(m)\Lambda(n)}{mn\log(mn)} = \log x + O(\log\log x).$$

Selberg had also obtained the formula

$$\psi(x)\log x + \sum_{jk\le x} \Lambda(j)\Lambda(k) = 2x\log x + O(x). \tag{3.56}$$

It is the one that was used in the 1949 elementary proof of the PNT as well as in almost all subsequent elementary proofs. The sum in (3.56) has the alternate representations

$$\sum_{mn\le x}\Lambda(m)\Lambda(n)=\sum_{n\le x}\Lambda(n)\sum_{m\le x/n}\Lambda(m)=\sum_{n\le x}\Lambda(n)\psi\left(\frac{x}{n}\right)=\int_1^x\psi\left(\frac{x}{t}\right)d\psi(t).$$

A formula which is equivalent to (3.56) is given by

$$\sum_{n\le x}\Lambda(n)\log n+\sum_{jk\le x}\Lambda(j)\Lambda(k)=2x\log x+O(x). \tag{3.57}$$

It derives from a comparison of the initial terms in (3.56) and (3.57):

$$\left(\sum_{n\le x}\Lambda(n)\log n\right)-\psi(x)\log x=\sum_{n\le x}\Lambda(n)\log\left(\frac{n}{x}\right)=O(x).$$

The $O(x)$ error term here holds since, by partial summation, the difference on the left is exactly $-\int_2^x \psi(t)/t\,dt$, which is $O(x)$ by the Chebyshev bounds. (It is asymptotic to x itself under the PNT.) An 'elementary' proof of Selberg's formula may be found in Hardy & Wright (2008, [**499**, §22.14]). If one is 'allowed' to assume the PNT, Selberg's formula can be improved considerably. Specifically, Nevanlinna (1967, [**801**]) established that

$$\psi(x)\log x+\int_1^x\psi\left(\frac{x}{t}\right)d\psi(t)=2x\log x-(2\gamma+1)x+o(x),$$

while under RH the error term becomes $O\left(\sqrt{x}\log^3 x\right)$. Saidak (2008, [**931**]) discusses the correspondence between the error for this relation and the error for the PNT.

REMARKS 3.4.

(i) The formula (3.57) can be used to obtain the similar-looking formula

$$\sum_{p\le x}\log^2 p+\sum_{pq\le x}(\log p)(\log q)=2x\log x+O(x).$$

It follows from it because

$$\sum_{n\le x}\Lambda(n)\log n-\sum_{p\le x}\log^2 p=O\left(\sqrt{x}\log x\right),$$

and

$$\sum_{jk\le x}\Lambda(j)\Lambda(k)-\sum_{pq\le x}(\log p)(\log q)=O(x).$$

(ii) Under the PNT, $\psi(x)\sim x$, so that $\psi(x)\log x\sim x\log x$; using this, Selberg's formula (3.56) can be used to show that

$$\sum_{jk\le x}\Lambda(j)\Lambda(k)\sim\sum_{pq\le x}(\log p)(\log q)\sim x\log x.$$

(iii) Further relations of interest include

$$\sum_{n\le x}\frac{\Lambda(n)}{\log n}=\sum_{p^k\le x}\frac{1}{k}=\sum_{k=1}^{\infty}\frac{\pi(x^{1/k})}{k}=\frac{\psi(x)}{\log x}+\int_2^x\frac{\psi(t)}{t\log^2 t}\,dt.$$

(iv) An explicit bound involving $\Lambda(n)/n$ is

$$\left|\sum_{n\le x}\frac{\Lambda(n)}{n}-\log x+\gamma\right|\le\frac{2}{\log^2 x},\quad x>1.$$

Explicit bounds for $\psi(x)\equiv\sum_{n\le x}\Lambda(n)$ will be given below.

We consider next the Chebyshev functions

$$\vartheta(x)\equiv\sum_{p\le x}\log p=\log\prod_{p\le x}p,$$

and

$$\psi(x)\equiv\sum_{n\le x}\Lambda(n)\equiv\sum_{p^k\le x}\log p\equiv\sum_{p\le x}\left\lfloor\frac{\log x}{\log p}\right\rfloor\log p\equiv\sum_{k=1}^{\infty}\vartheta(x^{1/k}).$$

These are summatory over the primes and over the prime powers and are closely bound up with the Prime Number Theorem. In fact, unlike the proof of the PNT (which is difficult) it is comparatively easy to show that the three quantities

$$\frac{\pi(x)}{x/\log x},\quad\frac{\vartheta(x)}{x},\quad\text{and}\quad\frac{\psi(x)}{x}$$

all share the same lim sup as well as the same lim inf so that, in view of the PNT,

$$\pi(x)\sim\frac{\vartheta(x)}{\log x}\sim\frac{\psi(x)}{\log x},\qquad\text{as } x\to\infty,\tag{3.58}$$

or, equivalently,

$$\vartheta(x)\sim x,\quad\text{and}\quad\psi(x)\sim x,\qquad\text{as } x\to\infty.\tag{3.59}$$

(We prove the mentioned lim inf and lim sup equalities in Section 3.10.) In fact, the Prime Number Theorem is commonly proved by first proving either $\vartheta(x)\sim x$ or $\psi(x)\sim x$, and then applying the mentioned asymptotic equalities.

By way of intuition, the fact that $\vartheta(x)$ is asymptotic to x occurs because the $\log p$ in $\sum_{p\le x}\log p$ compensates for the density of primes among integers near p, and the fact that $\psi(x)$ is asymptotic to x occurs because the difference $\psi(x)-\vartheta(x)$ is merely $O(\sqrt{x})$:

$$\begin{aligned}\psi(x)-\vartheta(x)&=\sum_{p^k\le x}\log p-\sum_{p\le x}\log p\\&=\sum_{p^k\le x,\,k\ge 2}\log p\\&=\sum_{2\le k\le(\log x)/\log 2}\ \sum_{p\le x^{1/k}}\log p\\&=\sum_{2\le k\le(\log x)/\log 2}\vartheta(x^{1/k})\\&=\vartheta(x^{1/2})+\sum_{3\le k\le(\log x)/\log 2}\vartheta(x^{1/k})\\&\le\vartheta(x^{1/2})+\vartheta(x^{1/3})\sum_{3\le k\le(\log x)/\log 2}1\\&\ll x^{1/2}+x^{1/3}\log x\ll x^{1/2},\end{aligned}$$

where we have used the fact that the largest k that contributes must satisfy $2^k \leq x$. The $O(\sqrt{x})$ error for $\psi(x) - \vartheta(x)$ is in fact best possible. The fact is that prime powers p^k with $k \geq 2$ are much rarer than primes; indeed their number up to x is only

$$\sum_{p^k \leq x} 1 = \sum_{p^2 \leq x} 1 + O(x^{1/3}) = \pi(\sqrt{x}) + O(x^{1/3}) \sim \frac{\sqrt{x}}{\log \sqrt{x}} \sim \frac{2}{\sqrt{x}}\,\pi(x).$$

The relations in (3.59) can be envisioned graphically; the function $y = \psi(x)$ for example is piecewise constant, rising by an amount $\log p$ at prime powers $x = p^k$. The PNT is equivalent to the fact that this function is asymptotic to the 45-degree line $y = x$.

We mention here also the approximations

$$\int_0^x \psi(u)\,du \sim \frac{1}{2}x^2, \quad \text{and} \quad \int_1^x \frac{\psi(t)}{t^2}\,dt = \log x + O(1),$$

which were established by Mertens, as well as the much deeper result

$$\psi(x) = x - \sum_{|\rho| \leq x} \frac{x^\rho}{\rho} + O(\log^2 x) \tag{3.60}$$

involving a sum over the nontrivial zeros of $\zeta(s)$; it was conjectured by Riemann and proved by von Mangoldt. (See also formula (5.22) in Chapter 5.)

As regards the accuracy of the approximations to $\psi(x)$ and to $\vartheta(x)$, it can be shown—as a consequence of (3.60)—that any error of the form $O(x^\theta)$ that holds for any one of

$$\vartheta(x) = x + O(x^\theta), \quad \psi(x) = x + O(x^\theta), \quad \text{and} \quad \pi(x) = \operatorname{li}(x) + O(x^\theta) \tag{3.61}$$

must also hold for the other two. In fact, if any of these relations were to hold, it would follow that $\zeta(s)$ can have no zeros with $\sigma > \theta$. Conversely, if $\zeta(s)$ had no zeros with $\sigma > \theta$ the relations (3.61) would necessarily hold, except with error $O(x^{\theta+\epsilon})$ for any $\epsilon > 0$; in fact, one would then even have

$$\psi(x) - x = O(x^\theta \log^2 x), \quad \vartheta(x) - x = O(x^\theta \log^2 x), \quad \text{and} \quad \pi(x) - \operatorname{li}(x) = O(x^\theta \log x),$$

while if any one of these were to hold it would follow that $\zeta(s)$ has no zeros with $\sigma > \theta$. Consequently, either one of the error terms $O(x^{1/2+\epsilon})$ or $O(\sqrt{x}\log^2 x)$ for $\varphi(x) - x$ or for $\vartheta(x) - x$ would be equivalent to the Riemann hypothesis, with the same also being true for the error terms $O(x^{1/2+\epsilon})$ or $O(\sqrt{x}\log x)$ for $\pi(x) - \operatorname{li}(x)$. Unconditionally, however, the de la Vallée Poussin zero-free region allows us to replace the $O(x^\theta)$ in (3.61) by $O\big(x\exp(-c\sqrt{\log x}\,)\big)$, while the best unconditional results currently known (based on the Korobov–Vinogradov zero-free region) are given by

$$\psi(x) = x + O\left(x\exp\left(-c\,\frac{(\log x)^{3/5}}{(\log\log x)^{1/5}}\right)\right),$$

$$\vartheta(x) = x + O\left(x\exp\left(-c\,\frac{(\log x)^{3/5}}{(\log\log x)^{1/5}}\right)\right),$$

and

$$\pi(x) = \operatorname{li}(x) + O\left(x\exp\left(-c\,\frac{(\log x)^{3/5}}{(\log\log x)^{1/5}}\right)\right).$$

In the other direction, Schmidt (1903, [**950**]) had shown that $\psi(x) - x$ cannot be as small as $O(\sqrt{x})$, while Littlewood (1912, [**701**]), and Hardy & Littlewood (1916, [**492**]) showed that it must even exceed $O(\sqrt{x}\log\log\log x)$. These same results also hold with $\vartheta(x)$ in place of $\psi(x)$. Montgomery (1980, [**765**]) had even conjectured that

$$\liminf_{x\to\infty} \frac{\psi(x) - x}{\sqrt{x}\,(\log\log\log x)^2} = -\frac{1}{2\pi}, \quad \text{and} \quad \limsup_{x\to\infty} \frac{\psi(x) - x}{\sqrt{x}\,(\log\log\log x)^2} = +\frac{1}{2\pi}.$$

(See also Montgomery & Vaughan, (2006, [**774**, pp. 483–484].) We mention too that, under RH, a suitably normalized version of $e^{-y/2}\,(\psi(e^y) - e^y)$ will even possess an asymptotic distribution (Wintner (1935, [**1114**])).

A large number of *explicit bounds* for ϑ and ψ are available. These usually depend on combining known zero-free regions of $\zeta(s)$ with numerical verification of RH up to some given height. Typical examples of such bounds include

$$|\vartheta(x) - x| \le \min\left(\frac{1.83x}{\log^2 x}, \frac{21x}{\log^3 x}\right), \quad \text{for } x > 2;$$

$$|\psi(x) - x| \le 0.006409\,\frac{x}{\log x}, \quad \text{for } x \ge 1,515,000;$$

$$0 \le \psi(x) - \vartheta(x) \le 1.0012\sqrt{x} + 3\sqrt[3]{x}, \quad \text{for } x > 0.$$

Improvements to such bounds are under continual development. In fact, given any $k \ge 1$, it is typically possible to determine numerous explicit constants c_k and x_k such that

$$|\vartheta(x) - x| < \frac{c_k x}{\log^k x}, \quad \text{for } x > x_k.$$

Under RH, however, it can be shown that

$$|\vartheta(x) - x| < \frac{1}{8\pi}\sqrt{x}\log^2 x, \quad \text{and} \quad |\psi(x) - x| < \frac{1}{8\pi}\sqrt{x}\log^2 x,$$

the first holding for $x \ge 599$, and the second for $x \ge 74$. Many further results along such lines can be found in Rosser (1941, [**918**]), Rosser & Schoenfeld (1962, [**919**]), Diamond & Steinig (1970, [**285**]), Schoenfeld (1976, [**954**]), Robin (1983, [**911**]), Costa Pereira (1985, [**238**]), Massias & Robin (1996, [**736**]), Ramaré (2013, [**883**]), Dusart (1999, [**303**]), (2018, [**304**]), Büthe (2016, [**171**]), and Axler (2017, [**45**]).

Turning to some non-asymptotic results, in view of their definitions, $\psi(x) \ge \vartheta(x)$, and

$$\psi(x) = \vartheta(x) + \vartheta(x^{1/2}) + \vartheta(x^{1/3}) + \vartheta(x^{1/4}) + \cdots.$$

This sum terminates for any given x since the terms vanish once the $x^{1/k}$ arguments fall below 2. We mention also the relation

$$\log n! = \psi(n) + \psi\left(\frac{n}{2}\right) + \psi\left(\frac{n}{3}\right) + \psi\left(\frac{n}{4}\right) + \cdots$$

in which the sum also terminates for any given n, as well as the relation

$$\psi(n) = \log \operatorname{lcm}\{1, 2, \ldots, n\}$$

where lcm is the least common multiple function.

The connection between $\vartheta(x)$ and the prime counting function $\pi(x)$ is further evidenced by formulas such as

$$\pi(x) = \frac{\vartheta(x)}{\log x} + \int_2^x \frac{\vartheta(t)}{t\log^2 t}\,dt, \quad \text{and} \quad \vartheta(x) = \pi(x)\log x - \int_2^x \frac{\pi(t)}{t}\,dt, \quad x \ge 2.$$

Using (3.52) and partial summation we may also obtain the Mellin transform result

$$\frac{\zeta'(s)}{\zeta(s)} = -\sum_{n=1}^{\infty}\frac{\Lambda(n)}{n^s} = -\int_1^{\infty}\frac{d\psi(x)}{x^s} = -s\int_1^{\infty}\frac{\psi(x)}{x^{s+1}}\,dx, \quad \text{for } \sigma > 1,$$

and its corresponding inverse

$$\psi(x) = -\frac{1}{2\pi i}\int_{c-i\infty}^{c+i\infty}\frac{\zeta'(s)}{\zeta(s)}\frac{x^s}{s}\,ds, \quad c > 1,$$

with this integration being on any vertical line to the right of $s = 1$ in the complex plane. (Appendix C provides the needed background.)

In fact, owing to their close connections to the primes, the functions Λ, ϑ, and ψ are—like $\pi(x)$ itself—closely related to the zeta function, and can all be represented in terms of its zeros. Here we quote only one such result (due to von Mangoldt (1895, [**728**])):

$$\psi_0(x) = x - \sum_{\rho}\frac{x^\rho}{\rho} - \frac{\zeta'(0)}{\zeta(0)} - \frac{1}{2}\log(1 - x^{-2}), \quad x > 1, \tag{3.62}$$

where ρ ranges over all complex zeros of $\zeta(s)$, and $\psi_0(x) \equiv (\psi(x-) + \psi(x+))/2$ is the 'symmetrized' version of $\psi(x) \equiv \sum_{n \le x}\Lambda(n)$ in which values at jump points (namely at the prime powers) are replaced by the average of their left and right limits. Note too that $\zeta'(0)/\zeta(0) = \log 2\pi$ (see Chapter 5). The formula (3.62), and others like it, as well as their proper interpretations, will be discussed in Chapter 6. One further connection between ψ and the zeros of $\zeta(s)$ can be expressed in terms of the pair correlation conjecture. Goldston (1988, [**422**]) showed that the pair correlation conjecture holds if and only if

$$\int_T^{T^{1+\beta}}\left(\psi\left(x + \frac{x}{T}\right) - \psi(x) - \frac{x}{T}\right)^2\frac{dx}{x^2} \sim \beta\,\frac{\log^2 T}{T}, \quad \text{as } T \to \infty,$$

for every fixed $\beta > 0$. Discussion of this result is reserved for Chapter 6.

3.9. Equivalences to the Prime Number Theorem

In Chapter 2 the Prime Number Theorem was presented in the form $\pi(x) \sim x/\log x$, or as $\pi(x) \sim \mathrm{li}(x)$. However, there are many equivalent versions of the PNT and it is worthwhile for the sake of reference to summarize the more noteworthy ones here. Many of these can be stated in terms of the arithmetic functions discussed in this chapter, and some that we have already seen will be repeated here. Note that the first of the equivalences below was established in the proof of (2.15) in Chapter 2, while the last equivalence below is expressed in terms of the ζ function which is discussed only in Chapter 5.

$$\text{PNT} \iff p_n \sim n\log n,$$

$$\text{PNT} \iff \mathrm{li}(p_n) \sim n,$$

$$\text{PNT} \iff M(x) \equiv \sum_{n \le x}\mu(n) = o(x),$$

$$\text{PNT} \iff \sum_{n \le x}\frac{\mu(n)}{n} = o(1),$$

$$\begin{aligned}
\text{PNT} &\iff \sum_{n=1}^{\infty} \frac{\mu(n)\log n}{n} = -1, \\
\text{PNT} &\iff L(x) \equiv \sum_{n\le x} \lambda(n) = o(x), \\
\text{PNT} &\iff \sum_{n=1}^{\infty} \frac{\lambda(n)}{n} = 0, \\
\text{PNT} &\iff \sum_{n=1}^{\infty} \frac{\lambda(n)\log n}{n} = -\zeta(2) = -\frac{\pi^2}{6}, \\
\text{PNT} &\iff \sum_{p\le x} \frac{\log p}{p} = \log x + C + o(1), \\
\text{PNT} &\iff \vartheta(x) \equiv \sum_{p\le x} \log p \sim x, \\
\text{PNT} &\iff \psi(x) \equiv \sum_{n\le x} \Lambda(n) \equiv \sum_{p^k\le x} \log p \sim x, \\
\text{PNT} &\iff \sum_{n\le x} \frac{\Lambda(n)}{n} = \log x - \gamma + o(1), \\
\text{PNT} &\iff \int_1^{\infty} \frac{\psi(t)-t}{t^2}\,dt = -\gamma - 1, \\
\text{PNT} &\iff \zeta(s) \neq 0 \text{ for } \Re s = 1.
\end{aligned}$$

Of these equivalences, the one based on $\sum_{n\le x} \mu(n)$ roughly says that if a squarefree integer is chosen at random, it has an equal chance of having an even or an odd number of prime divisors, while the one based on $\sum_{n\le x} \lambda(n)$ says that an arbitrarily chosen integer has an equal chance of having (with multiplicity) an even and an odd number of prime divisors. However, these are rather strong assertions of 'equality' since under statistical independence these quantities would each have exact order $O\left(\sqrt{x \log\log x}\right)$—in accordance with the Law of the Iterated Logarithm (see Appendix D)—rather than $O\left(\sqrt{x}\right)$.

A collection of equivalences to the Riemann hypothesis, many of which can also be stated in terms of the arithmetic functions discussed in this chapter, may be found in Chapter 8.

3.10. Selected proofs

Notes: To aid in the logical flow among the proofs, their order here varies somewhat from that in the text. In particular, the proofs of the Dirichlet series generating function identities appear only in the following section, while proofs for the Lambert series identities are deferred to Appendix C.

PROOF OF THE FUNDAMENTAL MÖBIUS PROPERTY (3.8). This property asserts that

$$\sum_{d|n} \mu(d) = \begin{cases} 1, & \text{if } n = 1, \\ 0, & \text{if } n > 1. \end{cases}$$

Suppose $n > 1$ has the prime factorization $p_1^{m_1} p_2^{m_2} \cdots p_k^{m_k}$. Then we may as well suppose that $n = p_1 p_2 \cdots p_k$ because $\mu(d) = 0$ for any factors of n thereby forsaken. Thus n will have k divisors of the form p_j, $\binom{k}{2}$ divisors of the form $p_{j_1} p_{j_2}$, and so on. So altogether

$$\sum_{d|n} \mu(d) = \mu(1) - \binom{k}{1} + \binom{k}{2} - \binom{k}{3} + \cdots + (-1)^k \binom{k}{k}.$$

But $\mu(1) = 1$, and this sum is just the binomial expansion for $(1-1)^k = 0$. □

PROOF OF THE MÖBIUS INVERSION FORMULA (3.9). This formula states that

$$g(n) = \sum_{d|n} f(d) = \sum_{d|n} f\left(\frac{n}{d}\right) \iff f(n) = \sum_{d|n} \mu(d) g\left(\frac{n}{d}\right) = \sum_{d|n} \mu\left(\frac{n}{d}\right) g(d).$$

Assume that $g(n) = \sum_{d|n} f(d)$ holds for all n. Then

$$\sum_{d|n} \mu(d) g\left(\frac{n}{d}\right) = \sum_{d|n} \mu(d) \sum_{d'|\frac{n}{d}} f(d') = \sum_{dd'|n} \mu(d) f(d') = \sum_{d|n} f(d) \sum_{d'|\frac{n}{d}} \mu(d') = f(n).$$

The last equality used the fundamental Möbius property (3.8). Conversely, assume that $f(n) = \sum_{d|n} \mu(n/d) g(d)$ holds for all n. Then

$$\begin{aligned}
\sum_{d|n} f\left(\frac{n}{d}\right) &= \sum_{d|n} \sum_{d'|\frac{n}{d}} \mu\left(\frac{n}{dd'}\right) g(d') \\
&= \sum_{dd'|n} \mu\left(\frac{n}{dd'}\right) g(d') \\
&= \sum_{d|n} g(d) \sum_{d'|\frac{n}{d}} \mu\left(\frac{n}{dd'}\right) \\
&= g(n)
\end{aligned}$$

with the last equality here again using the fundamental Möbius property (3.8). □

A SYMBOLIC ALTERNATE PROOF. is $\mu * \mathbf{1} = \mathbf{1} * \mu = \mathbf{e}$. So if $g = (\mathbf{1} * f)$ then

$$\mu * g = \mu * (\mathbf{1} * f) = (\mu * \mathbf{1}) * f = e * f = f.$$

The converse is proved similarly. (Recall that Dirichlet convolution is both associative and commutative.) □

PROOF OF THE SECOND MÖBIUS INVERSION FORMULA (3.10). We need to show that

$$G(x) = \sum_{n \le x} F\left(\frac{x}{n}\right) \iff F(x) = \sum_{n \le x} \mu(n) G\left(\frac{x}{n}\right), \quad x > 0.$$

First note that $G(x) = F(x) = 0$ when $x < 1$ since the sums will then be empty; note also that *within the sums* the arguments of F and G are never less than 1. Now, in one direction—again using (3.8)—we have

$$\sum_{n \le x} \mu(n) G\left(\frac{x}{n}\right) = \sum_{n \le x} \sum_{m \le x/n} \mu(n) F\left(\frac{x}{mn}\right) = \sum_{k \le x} F\left(\frac{x}{k}\right) \sum_{n|k} \mu(n) = F(x).$$

(Note that k here is playing the role of mn.) The converse is proved similarly. □

PROOF OF (3.26), (3.27), AND (3.28) FOR EULER'S PHI FUNCTION. These formulas are, respectively:

$$\varphi(n) = n \prod_{p|n} \left(1 - \frac{1}{p}\right), \quad \sum_{d|n} \varphi(d) = n,$$

and

$$\varphi(n) = \sum_{d|n} \frac{n}{d} \mu(d) \equiv \sum_{d|n} d\,\mu\left(\frac{n}{d}\right) \equiv \sum_{ab=n} a\,\mu(b).$$

To prove the first of these, suppose that the primes dividing n are $p_1, p_2, \ldots, p_k$. Then multiplying out the product gives

$$n \prod_{j=1}^{k} \left(1 - \frac{1}{p_j}\right)$$
$$= n - \sum_{j} \frac{n}{p_j} + \sum_{j_1<j_2} \frac{n}{p_{j_1}p_{j_2}} - \sum_{j_1<j_2<j_3} \frac{n}{p_{j_1}p_{j_2}p_{j_3}} + \cdots + (-1)^k \frac{n}{p_1p_2\cdots p_k}.$$

The terms $\frac{n}{p_j}$ equal the number of values in $[1, n]$ divisible by p_j. The terms $\frac{n}{p_{j_1}p_{j_2}}$ equal the number divisible by $p_{j_1}p_{j_2}$, and so on. By the *inclusion–exclusion principle* (see Appendix D) we are left with the number $\varphi(n)$ of values in $[1, n]$ which are relatively prime to n.

To prove the second equality, suppose that n has the prime factorization $n = p_1^{m_1} p_2^{m_k}$, and note that the divisors of n range over $p_1^{r_1} p_2^{r_2} \cdots p_k^{r_k}$ for $0 \le r_j \le m_j$. Summing over these and using multiplicativity of φ gives

$$\sum_{d|n} \varphi(d) = \sum_{r_1, r_2, \ldots, r_k} \varphi\left(\prod_{j=1}^{k} p_j^{r_j}\right) = \sum_{r_1, r_2, \ldots, r_k} \prod_{j=1}^{k} \varphi\left(p_j^{r_j}\right).$$

Rearranging terms leads to

$$\prod_{j=1}^{k} \sum_{r_j=0}^{m_j} \varphi\left(p_j^{r_j}\right) = \prod_{j=1}^{k} (\varphi(1) + \varphi(p_j) + \cdots + \varphi(p_j^{m_j}))$$
$$= \prod_{j=1}^{k} (1 + (p_j - 1) + p_j(p_j - 1) + \cdots + p_j^{m_j - 1}(p_j - 1))$$
$$= \prod_{j=1}^{k} p_j^{m_j} = n.$$

Finally, the third equality follows on applying Möbius inversion to the second equality. □

PROOF OF (3.44), (3.45), AND (3.46) FOR THE LIOUVILLE FUNCTION. These formulas are:

$$\sum_{ab=n} \lambda(a)|\mu(b)| \equiv \sum_{d|n} \lambda(d) \left|\mu\left(\frac{n}{d}\right)\right| = \begin{cases} 1, & \text{if } n = 1, \\ 0, & \text{if } n > 1, \end{cases}$$

$$\lambda(n) = \sum_{d^2|n} \mu\left(\frac{n}{d^2}\right),$$

and

$$\sum_{d|n} \lambda(d) = \begin{cases} 1, & \text{if } n \text{ is a perfect square,} \\ 0, & \text{otherwise.} \end{cases}$$

To prove the first identity, observe that both sides are multiplicative functions since $\lambda(n)$ and $\mu(n)$ are. It therefore suffices to check the identity on prime powers $n = p^k$ (with $k \geq 1$). But in that case, the left-hand side becomes

$$\begin{aligned}\lambda(p^k)|\mu(1)| + \lambda(p^{k-1})|\mu(p)| + \lambda(p^{k-2})|\mu(p^2)| + \cdots + \lambda(p)|\mu(p^{k-1})| + \lambda(1)|\mu(p^k)| \\ = (-1)^k + (-1)^{k-1} + 0 + \cdots + 0 = 0.\end{aligned}$$

To prove the second identity, note likewise that $\nu(n) \equiv \sum_{d|n} \lambda(d)$ is multiplicative since $\lambda(n)$ is, so here too it suffices to check the identity on prime powers. But $\nu(p^k) = 1 - 1 + 1 - \cdots$, which equals 1 if k is even and 0 if k is odd. Thus $\nu(p_1^{k_1} \cdots p_m^{k_m}) = \prod_j \nu(p_j^{k_j})$ equals 1 if all k_j are even, and otherwise equals 0.

The third identity follows immediately from the second via Möbius inversion:

$$\lambda(n) = \sum_{c|n} \nu(c)\mu\left(\frac{n}{c}\right) = \sum_{c|n} \left\{\begin{matrix} 1, & \text{if } c \text{ is a perfect square,} \\ 0, & \text{otherwise} \end{matrix}\right\} \mu\left(\frac{n}{c}\right) = \sum_{d^2|n} \mu\left(\frac{n}{d^2}\right).$$

Alternately, note that $\nu(n) \equiv \sum_{d|n} \lambda(d)$ is multiplicative because $\lambda(n)$ is. But $\nu(p^k) = 1 - 1 + 1 - \cdots$, and this equals 1 if k is even and 0 if k is odd. Thus $\nu(p_1^{k_1} \cdots p_m^{k_m}) = \prod_j \nu(p_j^{k_j})$ equals 1 if all k_j are even, and otherwise equals 0. □

PROOFS OF (3.49), (3.50), AND (3.51) FOR THE VON MANGOLDT FUNCTION. These formulas are:

$$\sum_{d|n} \Lambda(d) = \log n,$$

$$\Lambda(n) = \sum_{d|n} \mu\left(\frac{n}{d}\right) \log d \equiv \sum_{d|n} \mu(d) \log\left(\frac{n}{d}\right) = -\sum_{d|n} \mu(d) \log d,$$

and

$$\sum_{d|n} \mu\left(\frac{n}{d}\right) \Lambda(d) = -\mu(n) \log n.$$

Writing $n = p_1^{m_1} \cdots p_k^{m_k}$, the first relation holds because only the divisors $d = p_j^{k_j}$ with $1 \leq k_j \leq m_j$ contribute to the sum and their contributions add up to

$$m_1 \log p_1 + m_2 \log p_2 + \cdots + m_k \log p_k = \log n.$$

The second relation follows on viewing the first relation as $\sum_{d|n} \Lambda(d)\mathbf{1}(n/d) = \log n$ and applying Möbius inversion. The last equality of the second relation follows on applying (3.8). The third relation follows upon viewing the last equality of the second relation as saying that $\Lambda(n) = -\sum_{d|n} (\mu \log)(d) \cdot \mathbf{1}(n/d)$, and applying Möbius inversion. □

REMARK 3.5. An alternate take on the third relation can be based on generating functions. So

$$-\frac{d}{ds}\frac{1}{\zeta(s)} = \frac{\zeta'(s)}{\zeta^2(s)} = -\frac{1}{\zeta(s)}\left(-\frac{\zeta'(s)}{\zeta(s)}\right).$$

In terms of the corresponding generating functions this gives

$$\sum_{n=1}^{\infty} \frac{\mu(n)\log n}{n^s} = \left(-\sum_{n=1}^{\infty} \frac{\mu(n)}{n^s}\right)\left(\sum_{n=1}^{\infty} \frac{\Lambda(n)}{n^s}\right).$$

It follows that $\mu(n)\log(n)$ must be the Dirichlet convolution of $-\mu(n)$ and $\Lambda(n)$. A similar treatment is possible for the second relation; see Hardy & Wright (2008, [**422**, Chapter XVII]).

PROOF OF (3.54) FOR THE VON MANGOLDT FUNCTION. This formula states that if $a(n)$ is completely multiplicative then

$$\sum_{n=1}^{\infty} \frac{a(n)}{n^s} = \exp\left(\sum_{n=2}^{\infty} \frac{a(n)\Lambda(n)/\log n}{n^s}\right)$$

within the half-plane of absolute convergence.

We start by writing the left-hand side as an Euler product, then using complete multiplicativity:

$$\sum_{n=1}^{\infty} \frac{a(n)}{n^s} = \prod_p \sum_{k=0}^{\infty} \frac{a(p^k)}{(p^k)^s} = \prod_p \sum_{k=0}^{\infty} \frac{a(p)^k}{(p^s)^k} = \prod_p \left(1 - \frac{a(p)}{p^s}\right)^{-1}$$

by the formula for a geometric series. Taking logarithms of both sides gives

$$\log\left(\sum_{n=1}^{\infty} \frac{a(n)}{n^s}\right) = \sum_p \log\left(1 - \frac{a(p)}{p^s}\right)^{-1} = \sum_p \sum_{k=1}^{\infty} \frac{1}{k}\left(\frac{a(p)}{p^s}\right)^k = \sum_p \sum_{k=1}^{\infty} \frac{a(p^k)/k}{(p^k)^s}$$

(again by complete multiplicativity), which implies that

$$\sum_{n=1}^{\infty} \frac{a(n)}{n^s} = \exp\left(\sum_p \sum_{k=1}^{\infty} \frac{a(p^k)/k}{(p^k)^s}\right) = \exp\left(\sum_{n=1}^{\infty} \frac{a(n)\Lambda(n)/\log n}{n^s}\right). \qquad \square$$

PROOF OF THE DIRICHLET DIVISOR APPROXIMATION (3.38). This states that

$$\sum_{n\le x} \tau(n) = x\log x + (2\gamma - 1)x + O(\sqrt{x}).$$

It will be instructive to begin a proof that is destined to not result in a sufficiently strong statement, because the idea will be salvaged in a moment. Note that

$$\sum_{n\le x} \tau(n) = \sum_{n\le x} \sum_{d\mid n} 1 = \sum_{d\le x} \sum_{\substack{n\le x \\ d\mid n}} 1 = \sum_{d\le x} \left\lfloor \frac{x}{d} \right\rfloor,$$

which uses the fact that the number of multiples of d not exceeding x equals $\lfloor \frac{x}{d} \rfloor$. Since $\lfloor \frac{x}{d} \rfloor = \frac{x}{d} + O(1)$, this approach would lead to

$$\sum_{n\le x} \tau(n) = \sum_{d\le x}\left(\frac{x}{d} + O(1)\right) = x\sum_{d\le x} \frac{1}{d} + O(x) = x\log x + O(x),$$

which is valid but not as accurate as we would like—the large number of summands caused a substantial error term. It is possible to establish the claimed $O(\sqrt{x})$ approximation from this starting point with a more complicated analysis, but we will use a different method.

To surmount the above difficulty, Dirichlet introduced a technique now called the 'hyperbola method', the name coming from the fact that we can write

$$\sum_{n\le x}\tau(n)=\sum_{uv\le x}1,$$

which is equal to the number of integer lattice points in the hyperbolic region of the 'uv-plane' bounded by $u>0$, $v>0$, and $uv\le x$. While it is known that somewhat more than $O(\sqrt{x})$ can be obtained using this approach, we will obtain the $O(\sqrt{x})$ using a simpler method due to Meissel; see Overholt (2015, [**828**, §1.6]).

If $d\mid n$, we will say that d is *small* if $d<\sqrt{n}$ and *large* if $d>\sqrt{n}$. Then because small divisors d are in one-to-one correspondence with their large counterparts n/d we will have

$$\tau(n)=2\sum_{\substack{d<\sqrt{n}\\ d\mid n}}1,$$

except that when n is a square one must add 1 to this result. Consequently

$$\sum_{n\le x}\tau(n)=\lfloor\sqrt{x}\rfloor+2\sum_{n\le x}\sum_{\substack{d<\sqrt{n}\\ d\mid n}}1=\lfloor\sqrt{x}\rfloor+2\sum_{d\le\sqrt{x}}\left(\left\lfloor\frac{x}{d}\right\rfloor-d\right).$$

The last step merely adds up the number of times each $d\le\sqrt{x}$ occurs as a small divisor among the integers $n\le x$: Each d divides $\left\lfloor\frac{x}{d}\right\rfloor$ of those integers, but d is not a small divisor in the d cases for which $n\le d^2$. The term $\lfloor\sqrt{x}\rfloor$ accounts for those cases $n\le x$ in which n is a square. So altogether, and bearing in mind that

$$2\sum_{d\le\sqrt{x}}d=\lfloor\sqrt{x}\rfloor^2+\lfloor\sqrt{x}\rfloor,$$

we have

$$\begin{aligned}\sum_{n\le x}\tau(n)&=2\sum_{d\le\sqrt{x}}\left\lfloor\frac{x}{d}\right\rfloor-\lfloor\sqrt{x}\rfloor^2\\&=2\sum_{d\le\sqrt{x}}\left(\frac{x}{d}+O(1)\right)-\left(\sqrt{x}+O(1)\right)^2\\&=2x\sum_{d\le\sqrt{x}}\frac{1}{d}+O\left(\sqrt{x}\right)-x+O\left(\sqrt{x}\right)\end{aligned}$$

and the result follows on approximating the sum using $\sum_{n\le x}1/n=\log x+\gamma+O(1/x)$; see "Sums in n" in Appendix E. (See also Apostol, 1985, [**26**, §3.5].) □

PROOF OF THE ASYMPTOTIC EQUIVALENCES BETWEEN $\pi(x)$, $\psi(x)$, AND $\vartheta(x)$. We will show that

$$\frac{\pi(x)}{x/\log x},\quad\frac{\vartheta(x)}{x},\quad\text{and}\quad\frac{\psi(x)}{x}$$

share identical values of lim inf and of lim sup as $x\to\infty$. Hence if any of these functions has a limit then so do the other two and their limiting values will be identical. It will also follow that either of $\vartheta(x)\sim x$ or $\psi(x)\sim x$ is equivalent to the Prime Number Theorem. The version $\psi(x)\sim x$ is arguably the most natural way to attack the PNT.

We follow Chandrasekharan (1968, [**188**, pp. 65–66]). From the definition of $\psi(x)$ we have

$$\psi(x) \equiv \sum_{p \le n} \left\lfloor \frac{\log x}{\log p} \right\rfloor \log p \le \sum_{p \le n} \left(\frac{\log x}{\log p} \right) \log p = \pi(x) \log x.$$

Therefore

$$\vartheta(x) \le \psi(x) \le \pi(x) \log x,$$

and consequently

$$\limsup_{x\to\infty} \frac{\vartheta(x)}{x} \le \limsup_{x\to\infty} \frac{\psi(x)}{x} \le \limsup_{x\to\infty} \frac{\pi(x)}{x/\log x}.$$

Next, fix $0 < \alpha < 1$. Then

$$\frac{\vartheta(x)}{x} \ge \frac{1}{x} \sum_{x^\alpha < p \le x} \log p \ge \frac{1}{x} \left(\alpha \log x\right) \left(\pi(x) - \pi(x^\alpha)\right) \ge \alpha \frac{\pi(x)}{x/\log x} - \alpha \frac{\log x}{x^{1-\alpha}},$$

first because $\log p > \log x^\alpha$ in the range $x^\alpha < p \le x$, and second because $\pi(x^\alpha) < x^\alpha$. But for any $0 < \alpha < 1$ the final term tends to 0 so that

$$\limsup_{x\to\infty} \frac{\vartheta(x)}{x} \ge \alpha \limsup_{x\to\infty} \frac{\pi(x)}{x/\log x}.$$

Since this holds for every $0 < \alpha < 1$ it also holds for $\alpha = 1$. Together with the previous inequalities for the lim sup, this establishes that the three lim sup values are identical. The equality of the three lim inf values is proved in an analogous way. □

3.11. Proofs of the generating function identities

PROOF OF THE IDENTITIES (3.13) FOR THE MÖBIUS AND QUADRATFREI FUNCTIONS. These identities are

$$\sum_{n=1}^{\infty} \frac{\mu(n)}{n^s} = \frac{1}{\zeta(s)}, \quad \text{and} \quad \sum_{n=1}^{\infty} \frac{|\mu(n)|}{n^s} = \frac{\zeta(s)}{\zeta(2s)}, \quad \Re s > 1.$$

The first identity follows from the definition of $\mu(n)$ and the expansion

$$\frac{1}{\zeta(s)} = \prod_p \left(1 - \frac{1}{p^s}\right) = 1 - \sum \frac{1}{p^s} + \sum \frac{1}{p^s q^s} - \sum \frac{1}{p^s q^s r^s} + \cdots \tag{3.63}$$

upon noting that the denominators on the right run over the values of n^s for which n is squarefree, and with signs determined by the parity of the number of primes in n. Next

$$\frac{\zeta(s)}{\zeta(2s)} = \prod_p \frac{1 - p^{-2s}}{1 - p^{-s}} = \prod_p \left(1 + p^{-s}\right) = \sum_{n=1}^{\infty} \frac{|\mu(n)|}{n^s},$$

since only squarefree n occur on multiplying out. □

REMARK 3.6. Strictly speaking (although it does give a correct answer) the expansion of the infinite product in (3.63) is not rigorous since it gives the terms containing *finitely* many p's but not those containing *infinitely* many p's. The situation parallels the Euler product argument given in Section 1.6 and can be handled similarly. An argument which is applicable here and which holds more generally may be found in Montgomery & Vaughan (2006, [**774**, Theorem 1.9, pp. 20–21]).

PROOF OF IDENTITIES (3.34), (3.35)(A), AND (3.36) FOR THE DIVISOR FUNCTIONS. These identities state that

$$\sum_{n=1}^{\infty}\frac{\tau(n)}{n^s}=\zeta^2(s),\quad \sum_{n=1}^{\infty}\frac{\sigma(n)}{n^s}=\zeta(s)\zeta(s-1),\quad \sum_{n=1}^{\infty}\frac{\sigma_k(n)}{n^s}=\zeta(s)\zeta(s-k),$$

$$\sum_{n=1}^{\infty}\frac{\tau^2(n)}{n^s}=\frac{\zeta^4(s)}{\zeta(2s)},\quad\text{and}\quad \sum_{n=1}^{\infty}\tau(n)x^n=\sum_{n=1}^{\infty}\frac{x^n}{1-x^n},\quad |x|<1.$$

The first of these follows from

$$\zeta^2(s)=\left(\sum_j\frac{1}{j^s}\right)\left(\sum_k\frac{1}{k^s}\right)=\sum_{j,k}\frac{1}{j^sk^s}=\sum_n\frac{\tau(n)}{n^s}.$$

In fact, this first identity as well as the second one both follow from the third identity which may be proved as follows:

$$\begin{aligned}\zeta(s)\zeta(s-k)&=\Big(\sum_{\alpha=1}^{\infty}\frac{1}{\alpha^s}\Big)\Big(\sum_{\beta=1}^{\infty}\frac{\beta^k}{\beta^s}\Big)\\&=\sum_{n=1}^{\infty}\frac{1}{n^s}\sum_{\alpha\beta=n}1\cdot\beta^k\\&=\sum_{n=1}^{\infty}\frac{1}{n^s}\sum_{d|n}d^k\\&=\sum_{n=1}^{\infty}\frac{\sigma_k(n)}{n^s}.\end{aligned}$$

For the fourth identity we have

$$\frac{\zeta^4(s)}{\zeta(2s)}=\prod_p\frac{1-p^{-2s}}{(1-p^{-s})^4}=\prod_p\frac{1+p^{-2s}}{(1-p^{-s})^3}=\prod_p\Big(\sum_{j=0}^{\infty}(j+1)^2p^{-js}\Big)=\sum_{n=1}^{\infty}\frac{\tau^2(n)}{n^s}.$$

Here we have used the peculiar identity

$$\frac{1+x}{(1-x)^3}=\sum_{j=0}^{\infty}(j+1)^2x^j,\quad |x|<1$$

which we leave for the reader.[13] The $\tau^2(n)$ at the last step arises because on viewing n as $n=p_1^{k_1}p_2^{k_2}\cdots$ the coefficient one obtains for n^{-s} is

$$(k_1+1)^2(k_2+1)^2\cdots=\big((k_1+1)(k_2+1)\cdots\big)^2=\tau^2(n).$$

The fifth identity is a Lambert series; these are proved in Appendix C. □

REMARK 3.7. The proofs of (3.34) and (3.35) can be viewed as special cases of the general formula (3.7) for Dirichlet series corresponding to a Dirichlet convolution.

---

[13] Hint, but only if needed: Differentiate $(1-x)^{-1}=\sum_{j=0}^{\infty}x^j$ twice and combine the results.

PROOF OF (3.29) FOR EULER'S PHI FUNCTION. This states that

$$\sum_{n=1}^{\infty} \frac{\varphi(n)}{n^s} = \frac{\zeta(s-1)}{\zeta(s)}, \quad \Re s > 2.$$

We offer two proofs. First, since $\varphi(n)$ is multiplicative we may apply (3.4) to obtain

$$\begin{aligned}
\sum_{n=1}^{\infty} \frac{\varphi(n)}{n^s} &= \prod_p \left(1 + \frac{\varphi(p)}{p^s} + \frac{\varphi(p^2)}{p^{2s}} + \cdots\right) \\
&= \prod_p \left(1 + \frac{p-1}{p^s} + \frac{p(p-1)}{p^{2s}} + \frac{p^2(p-1)}{p^{3s}} + \cdots\right) \\
&= \prod_p \frac{1 - 1/p^s}{1 - p/p^s} = \frac{\zeta(s-1)}{\zeta(s)}
\end{aligned}$$

after a little algebra. Alternately, the result follows from the computation

$$\zeta(s) \sum_{n=1}^{\infty} \frac{\varphi(n)}{n^s} = \left(\sum_{n=1}^{\infty} \frac{1}{n^s}\right)\left(\sum_{n=1}^{\infty} \frac{\varphi(n)}{n^s}\right) = \sum_{n=1}^{\infty} \frac{\Delta(n)}{n^s}$$

once $\Delta(n)$ is identified. For when the middle product is multiplied out the total contribution to $1/n^s$ can be seen to be $\sum_{d|n}(1 \times \varphi(d))$. Thus $\Delta(n) = \sum_{d|n} \varphi(d) = n$ by (3.27) so that

$$\sum_{n=1}^{\infty} \frac{\Delta(n)}{n^s} = \sum_{n=1}^{\infty} \frac{n}{n^s} = \zeta(s-1). \qquad \square$$

PROOF OF (3.47) FOR THE LIOUVILLE FUNCTION. This states that

$$\sum_{n=1}^{\infty} \frac{\lambda(n)}{n^s} = \frac{\zeta(2s)}{\zeta(s)} = \prod_p \left(1 + p^{-s}\right)^{-1}, \quad \text{for } \Re s > 1.$$

Using the Euler product $\zeta(s) = \prod_p \left(1 - p^{-s}\right)^{-1}$ we find that

$$\begin{aligned}
\frac{\zeta(2s)}{\zeta(s)} &= \prod_p \frac{1 - p^{-s}}{1 - p^{-2s}} \\
&= \prod_p \left(1 + \frac{1}{p^s}\right)^{-1} \\
&= \prod_p \left(1 - \frac{\lambda(p)}{p^s}\right)^{-1} \\
&= \prod_p \left(1 + \frac{\lambda(p)}{p^s} + \frac{\lambda(p)^2}{p^{2s}} + \frac{\lambda(p)^3}{p^{3s}} + \cdots\right) \\
&= \sum_{n=1}^{\infty} \frac{\lambda(n)}{n^s}.
\end{aligned}$$

In the last steps we used the fact that λ equals -1 on the primes, and then, after multiplying out the expanded product, the fact that λ is completely multiplicative.

Alternately, one could expand the second product, multiply out, and collect the terms to get

$$\prod_p \left(1 + \frac{1}{p^s}\right)^{-1} = \prod_p \left(1 - \frac{1}{p^s} + \frac{1}{p^{2s}} - \frac{1}{p^{3s}} + \cdots\right) = \sum_{n=1}^{\infty} \frac{\lambda(n)}{n^s}$$

using the definition of $\lambda(n)$. □

PROOF OF (3.52) FOR THE VON MANGOLDT FUNCTION. This states that

$$\sum_{n=1}^{\infty} \frac{\Lambda(n)}{n^s} = -\frac{\zeta'(s)}{\zeta(s)}, \quad \text{and} \quad \sum_{n=2}^{\infty} \frac{\Lambda(n)/\log n}{n^s} = \log \zeta(s).$$

From the Euler product for $\zeta(s)$ we have $\log \zeta(s) = -\sum_p \log(1 - p^{-s})$; differentiating, expanding the resulting $(1 - p^{-s})^{-1}$ into a geometric series, and using the definition of $\Lambda(n)$ then gives

$$-\frac{\zeta'(s)}{\zeta(s)} = \sum_p \frac{p^{-s} \log p}{1 - p^{-s}} = \sum_p \sum_{k=1}^{\infty} \frac{\log p}{p^{ks}} = \sum_{n=1}^{\infty} \frac{\Lambda(n)}{n^s},$$

thus establishing the first equality. To obtain the second equality we first use the Taylor expansion of $-\log(1-t)$, and then again use the definition of $\Lambda(n)$ to find that

$$\log \zeta(s) = -\sum_p \log(1 - p^{-s}) = \sum_p \sum_{k=1}^{\infty} \frac{1}{kp^{ks}} = \sum_{n=2}^{\infty} \frac{\Lambda(n)/\log n}{n^s}.$$ □

REMARK 3.8. The first equality results upon termwise differentiation of the second, while the second results upon termwise integration of the first. This exemplifies a general result for Dirichlet series proved in Appendix C: If f is multiplicative and $F(s) = \sum_n f(n)/n^s$ converges for $\sigma > \sigma_0$, then

$$\frac{d \log F(s)}{ds} \equiv \frac{F'(s)}{F(s)} = -\sum_n \frac{\Lambda(n) f(n)}{n^s},$$

and also converges for $\sigma > \sigma_0$. The converse result holds as well.

PARTIAL PROOF THAT $L(x) = O(x^{1/2+\epsilon})$ IS EQUIVALENT TO RH. In one direction, the proof resembles Pólya's argument. Using $\sum \lambda(n)/n^s = \zeta(2s)/\zeta(s)$ and partial summation we find

$$\frac{\zeta(2s)}{\zeta(s)} = \int_1^{\infty} \frac{1}{x^s}\, dL(x) = s \int_1^{\infty} \frac{L(x)}{x^{s+1}}\, dx.$$

If $L(x) = O(x^{1/2+\epsilon})$ for all $\epsilon > 0$ the right side will be finite when $s > 1/2$. It follows that the $\zeta(s)$ denominator on the left cannot have any zeros in that region. The converse argument is more difficult and requires the Landau theorem on singularities of Dirichlet series with nonnegative coefficients. □

CHAPTER 4

The Heuristic of Harald Cramér

It must tend to some end, or else our universe is ruled by chance, which is unthinkable. But what end? There is the great standing perennial problem to which human reason is as far from an answer as ever.

Arthur Conan Doyle [**295**]

The Swedish mathematician Harald Cramér (1893–1985) is widely acclaimed as one of the earliest pioneers of rigorous probability and statistical theory. He wrote the first significant book on probability to appear after Kolmogorov's seminal 1933 formulation of the theory (Cramér (1937, [**246**])), and under difficult wartime conditions he authored the first—and still influential—textbook on mathematical statistics (Cramér (1946, [**247**])). He is known for such fundamental contributions as the Cramér–Rao bound, the Cramér–Wold theorem, and the Cramér–von Mises goodness-of-fit test, as well as for extensive contributions to actuarial science, stochastic processes, and large deviation theory, among other areas. Less widely known outside of 'pure math' circles is that his original field of study was analytic number theory, and that his doctoral dissertation (completed in 1917 under Marcel Riesz) concerned Dirichlet L-series. In his earliest papers—dated between 1920 and 1936—Harald Cramér introduced what subsequently became a highly influential probability-based model for heuristic reasoning in number theory. The discussions in this chapter make substantive use of material from Appendix D.[1]

4.1. Cramér's model

Because the occurrence of primes among the integers is (at even a careful glance) hard to distinguish from a purely probabilistic phenomenon, it is natural to try to employ statistical means to describe their behaviour. In this chapter we will (at the outset) be concerned with the so-called *Cramér model* (sometimes also referred to as the *Gauss–Cramér model*) based on an "interesting and heuristic use of probability arguments" [**245**]. According to this model, primes (starting at 3) are viewed as occurring randomly and independently, with each integer n considered to be "prime" with probability $1/\log n$. A random sequence generated in this way can

[1] A namesake coincidence: Gabriel Cramer (1704–1752) was an accomplished Swiss mathematician who solved the "St. Petersburg paradox" associated with the Bernouilli clan: A game pays $\$2^n$ with probability 2^{-n}, $n = 1, 2\ldots$ Yet few would pay \$100,000 to play this game even though its expected return is infinite. G. Cramer's 'solution' included the notion of the *utility* of money as the square root (or the logarithm, say) of its amount. See Chapter 2 of the thought-provoking book by Diaconis & Skyrms (2018, [**280**]).

be represented by zero-one Bernoulli variables $U_1, U_2, U_3, \ldots$, with U_1 and U_2 chosen arbitrarily, while for $k \geq 3$,

$$\mathrm{P}\big[U_k = 1\big] = 1 - \mathrm{P}\big[U_k = 0\big] = \frac{1}{\log k}.$$

Cramér (1934–1936) denoted the space consisting of such sequences by $\mathcal{S}$, and proposed to view the sequence corresponding to the actual primes as being a typical member of $\mathcal{S}$. He then posited that if a property could be shown to hold *almost surely* (that is, *with probability* 1) for the elements of $\mathcal{S}$, then "we may take this as a *suggestion* that, for the particular sequence of ordinary prime numbers p_n, some similar relation may hold" [**245**]. In this heuristic, the actual primes are considered to have been generated from but a single realization of this probabilistic process, and the questions of interest centre on the "with probability 1" behaviour of the resulting *pseudoprime* sequences $\{n \geq 2 : U_n = 1\}$. Of course, the actual primes constitute a *deterministic* and not a *random* sequence, but the hope is that when viewed in this way it might allow for reasonable predictions to be made about properties of the primes which have yet to be established.

Cramér's proposal was not without controversy at the time. There is a historically interesting footnote, #4, in Cramér's (1936, [**245**]) paper which refers to "Arguments of this type being frequently misunderstood..." [**245**]. It seems likely that this footnote was meant to refute a dig by Hardy and Littlewood (1923, [**495**]) who had stated that "Probability is not a notion of pure mathematics, but of philosophy or physics." Hardy's disdain for probability, and how that led him to miss out on some extremely important results, is engagingly documented in Diaconis (2002, [**278**]). It is a germane point, however, that probability theory did not acquire its secure mathematical footing until the early 1930s when (piggybacking on the theory of measure developed by the French in the preceding decades) Andrey Kolmogorov (1903–1987) took up the challenge of Hilbert's sixth problem and published a rigorous formulation of probability in his ground-breaking 1933 monograph, *Grundbegriffe der Wahrscheinlichkeitsrechnung*.

We will now proceed to describe some properties that the sequence of "primes" arising from Cramér's model—sometimes referred to as "pseudoprimes" —possesses almost surely, properties which we might thus expect to also hold for the true primes—at least to some degree. In this chapter, p, p_n, $\pi(x)$, and $\zeta(s)$ take on their usual meanings; however, when dealing with their analogues for pseudoprime sequences—whether those of Cramér or of any other probability-based model—this will be indicated by adding a tilde: $\widetilde{p}$, $\widetilde{p}_n$, $\widetilde{\pi}(x)$, $\widetilde{\zeta}(s)$, and so on. In the arguments throughout this chapter, we will make frequent use of results appearing in Appendix D; it provides an in-depth discussion of probability theory, and in particular of the zero-one laws, almost sure convergence, the Law of the Iterated Logarithm, and so on.

4.2. Consequences of Cramér's model

As a first (and essentially tautological) example, consider the Prime Number Theorem (PNT). With the U_k defined as above, the number of 'primes' $\leq n$ is, according to Cramér's heuristic,

$$\widetilde{\pi}(n) \equiv \sum_{k=3}^{n} U_k. \tag{4.1}$$

We need to show that $\widetilde{\pi}(n) \sim n/\log n$, or $\widetilde{\pi}(n)/(n/\log n) \to 1$, *almost surely* (that is, with probability 1). Now the mean and variance of $\widetilde{\pi}(n)$, namely

$$\mathrm{E}\,\widetilde{\pi}(n) = \sum_{k=3}^{n} \frac{1}{\log k} \quad \text{and} \quad \mathrm{Var}\,\widetilde{\pi}(n) = \sum_{k=3}^{n} \frac{1}{\log k}\left(1 - \frac{1}{\log k}\right),$$

are both asymptotic to $n/\log n$ as $n \to \infty$.[2] Consequently, while the mean of $\widetilde{\pi}(n)/\mathrm{E}\,\widetilde{\pi}(n)$ equals 1, its variance tends to 0, so using Chebyshev's inequality (see Appendix D) we see that $\widetilde{\pi}(n)/\mathrm{E}\,\widetilde{\pi}(n)$ tends to 1 *in probability*. But the terms in (4.1) are *independent* so this convergence is also *almost sure* in view of a theorem of Lévy (see Appendix D) which states that for sums of *independent* variables, convergence in probability and almost sure convergence are equivalent. Consequently $\widetilde{\pi}(n) \sim \mathrm{E}\,\widetilde{\pi}(n) \sim n/\log n$ almost surely, and so the PNT is satisfied on $\mathcal{S}$. □

Cramér's heuristic is also consistent with the conjectured asymptotic Poisson behaviour of the primes considered in Chapter 2. For example, if n is large, the probability that the pseudoprime following n does not occur until after $n+\lambda\log n$ is

$$\sim \left(1 - \frac{1}{\log n}\right)^{\lambda \log n} \to e^{-\lambda}, \quad \text{as } n \to \infty,$$

since each of the approximately $\lambda\log n$ integers between n and $n+\lambda\log n$ have (independently of each other) an approximately $1-1/\log n$ probability of not being a pseudoprime. Similar computations 'establish' each of the (conditional) results of Gallagher which were discussed in Chapter 2, and in particular that

$$\lim_{x\to\infty} \frac{1}{x}\left|\left\{n \le x \colon \alpha \le \frac{\widetilde{p}_{n+1} - \widetilde{p}_n}{\log \widetilde{p}_n} \le \beta\right\}\right| = \int_\alpha^\beta e^{-t}\,dt\,, \quad 0 \le \alpha \le \beta \le \infty,$$

and

$$\lim_{x\to\infty} \frac{1}{x}\Big|\{n \le x \colon \widetilde{\pi}(n+\lambda\log n) - \widetilde{\pi}(n) = k\}\Big| = \frac{\lambda^k e^{-\lambda}}{k!}, \quad k = 0, 1, 2 \ldots.$$

Nonetheless, one should bear in mind that—in spite of seeming plausible, and in spite of the numerical evidence in its favour—the conjectured asymptotic Poisson behaviour for the actual primes has not been rigourously established.

As its mean increases, the Poisson distribution tends to a normal (or Gaussian) distribution. So over larger intervals, Cramér's model predicts that the number of primes should tend to a normal distribution as well. In particular, if $h/\log x \to \infty$ as $x \to \infty$ while $h/x \to 0$, the model predicts that the values of $\widetilde{\pi}(n+h) - \widetilde{\pi}(n)$, for $n \le x$, become normally distributed with mean and variance both asymptotic to $h/\log x$. However, while this mean does agree with the Prime Number Theorem, the corresponding value for the variance is not believed to be correct for the actual primes (although the asymptotic normality is believed to hold under the stated conditions). Granville (1995, [**451**]) and Soundararajan (2007, [**1003**]) offer some background on this.

Reasoning based on Cramér's heuristic does sometimes lead to fully correct results that can be rigorously confirmed. Consider for example the proportion of integers that are squarefree (that is, are not divisible by the square of any prime).

[2]Recall that the logarithm varies so slowly that, asymptotically, the terms in such sums can all be replaced by just their last term.

The proportion of integers not divisible by the square of the prime p is $1 - 1/p^2$, so assuming independence, the proportion of squarefree integers would be

$$\prod_p \left(1 - \frac{1}{p^2}\right) = \frac{1}{\zeta(2)} = \left(\sum_{n=1}^{\infty} \frac{1}{n^2}\right)^{-1} = \frac{6}{\pi^2}.$$

This answer is known to be correct (it was proved by Gegenbauer in 1885) even though the argument here has conveniently ignored the fact that—while true for *finite* collections—the independence assumption is not valid for *infinite* collection of events based on divisibility by different primes. Nevertheless, a corrected argument can be rigorously modelled after the proof of (2.45) in Chapter 2, with the underlying idea being that the contribution of the larger primes becomes asymptotically negligible. Arguing similarly one may show that the proportion of numbers that are cube-free is $1/\zeta(3)$, and so on.

Another example where the heuristic leads to correct results concerns the function $\omega(n)$ which counts the number of distinct prime factors of n. We may represent $\omega(n)$ as

$$\omega(n) = \sum_{p \le n} \mathrm{I}[p|n]$$

where the $\mathrm{I}[p|n]$ are zero-one variables indicating whether p divides n or not. For any *finite* number of primes the $\mathrm{I}[p|n]$ behave like random variables which are independent when $1 \le n \le N$ is selected randomly and N becomes large. If we assume full independence for all of the $\mathrm{I}[p|n]$ we find that $\omega(n)$ has mean and variance given by

$$\mathrm{E}\,\omega(n) = \sum_{p \le n} \mathrm{E}\,\mathrm{I}[p|n] = \sum_{p \le n} \frac{1}{p} \sim \log\log n,$$

and

$$\mathrm{Var}\,\omega(n) = \sum_{p \le n} \mathrm{Var}\,\mathrm{I}[p|n] = \sum_{p \le n} \frac{1}{p}\left(1 - \frac{1}{p}\right) \sim \sum_{p \le n} \frac{1}{p} \sim \log\log n,$$

by virtue of Mertens's result (2.2). (Remark: The variance computation uses the independence assumption; the mean computation does not.) These values are in fact correct and correspond to famous results of Hardy & Ramanujan and Turán, respectively. The independence part of the heuristic suggests further that the $\omega(n)$ should behave like asymptotically normal $N(\log\log n, \log\log n)$ variables and this too turns out to be correct. Chapter 9 provides a detailed discussion of these results for the actual primes.

To be fair, the argument we have just made was not based purely on the Cramér heuristic, depending as it did on Mertens's result. However, if the $\sum 1/p$ evaluation was based solely on Cramér primes that sum would be

$$\sum_{\tilde{p} \le n} \frac{1}{\tilde{p}} \sim \sum_{1 < m \log m \le n} \frac{1}{m \log m} \sim \int_2^{M(n)} \frac{dx}{x \log x} = \log\log x \Big|_2^{M(n)} \sim \log\log M(n) \sim \log\log n$$

since the Cramér primes satisfy $p_m \sim m \log m$. Here $M(n)$ is the largest integer m such that $m \log m \le n$, and the last equivalence is based on the fact that if $m \log m \approx n$ then $\log\log m \sim \log\log(n/\log m)$. One therefore does in fact get 'fairly' the results obtained above, using Cramér's heurstic alone.

Additive problems. Consider now the Goldbach conjecture, that every even number is a sum of two primes. Letting A_n be the event (in Cramér's model) that $2n$ is *not* a sum of two pseudoprimes, we will show that (with probability 1) no more than a finite number of the A_n can occur. By the first Borel–Cantelli lemma (see Appendix D) it will be enough to show that $\sum \mathrm{P}(A_n) < \infty$. Independence and elementary reasoning gives

$$\mathrm{P}(A_n) \le \prod_{m=3}^{n-1} \left(1 - \left(\frac{1}{\log m}\right)\left(\frac{1}{\log(2n-m)}\right)\right), \quad \text{for } n \ge 4,$$

and since $1 + x \le e^x$ for all real numbers x,

$$\mathrm{P}(A_n) < \prod_{m=3}^{n-1} \exp\left(-\frac{1}{\log m \times \log(2n-m)}\right) < \exp\left(-\frac{n-3}{\log^2 n}\right),$$

because the maximum of $\log m \times \log(2n-m)$ occurs when $m = n$. Consequently

$$\sum_{n=4}^{\infty} \mathrm{P}(A_n) < \sum_{n=4}^{\infty} \exp\left(-\frac{n-3}{\log^2 n}\right) < \infty,$$

so that (according to Cramér's heuristic) Goldbach's conjecture should hold, with at most a finite number of exceptions.

This argument can be considerably strengthened. Let n be an even integer. The number of unique ways to decompose n additively as $(n-m)+m$ is asymptotic to $n/2$, and the Cramér-based probability that $(n-m)$ and m are both primes is

$$\frac{1}{\log(n-m)} \times \frac{1}{\log n}.$$

An estimate for the number of ways n can be decomposed into a sum of two primes is then obtained by summing these terms over the values of $m \le n/2$: this sum is asymptotic to

$$\int_2^{n/2} \frac{du}{\log(n-u)\log u} \sim \frac{n}{2}\frac{1}{\log^2(n/2)} \sim \frac{Cn}{\log^2 n}$$

with the value of C resulting from this heuristic argument considered irrelevant. Hardy & Littlewood (1924, [**496**]) provided some evidence that this $n/\log^2 n$ asymptotic may actually be correct. Using an argument based on their 'circle method' they conjectured that the number of ways an even integer n can be written as a sum of two primes is asymptotic to

$$\left(\prod_{\substack{p|n \\ p\neq 2}} \frac{p-1}{p-2}\right) \prod_{p>2}\left(1 - \frac{1}{(p-1)^2}\right)\frac{2n}{\log^2 n} \quad \text{as } n \to \infty.$$

There is also a deterministic consideration (based on the fact that $p_n \sim n \log n$) that lends additional support to the Goldbach conjecture: Suppose we were to define the set S as $S = \{\lfloor n \log n \rfloor : n \ge 1\}$ (where $\lfloor \cdot \rfloor$ is the 'integer part' function). Then one can show that the sum set $S + S$ contains every integer greater than 25.

We may similarly consider the twin primes conjecture. Applying Cramér's heuristic naively, the divergence of the relevant sum $\sum(1/\log n)(1/\log(n+2))$ of probabilities—together with the second Borel–Cantelli lemma[3]—suggests first that the number of twin primes is infinite, and second that their number up to n should be about

$$\sum_{k=3}^{n} \frac{1}{\log k} \times \frac{1}{\log(k+2)} \approx \sum_{k=3}^{n} \frac{1}{\log^2 k} \approx \frac{n}{\log^2 n}, \tag{4.2}$$

and therefore asymptotic to $Cn/\log^2 n$ or $C\int_2^\infty \log^{-2} t\,dt$. That this value is as large as it is only strengthens the plausibility of the twin primes conjecture.

Of course, Cramér's heuristic cannot be used with abandon: for example, it makes the exact same prediction for the number of integers n for which n and $n+1$ are both prime. The model does not reflect number-theoretic facts about the actual integers, in which, given that $n > 2$ is prime, the 'probability' that $n+1$ is prime is not $1/\log(n+1)$ but 0, since $n+1$ must then be even; and likewise, the probability that $n+2$ is prime is not $1/\log(n+2)$ for numerous reasons, not least of which is that $n+2$ must then be odd and thus more likely to be prime than an integer of unknown parity.

Applied naively, Cramér's heuristic suggests that the value of C in (4.2) is 1, but as we saw in equation (2.37), Hardy & Littlewood (1923, [**495**]) conjectured the true value of this constant to be

$$C = \left(1-\frac{1}{2}\right)\left(1-\frac{1}{2}\right)^{-2} \prod_{p\geq 3}\left(1-\frac{2}{p}\right)\left(1-\frac{1}{p}\right)^{-2} \tag{4.3}$$

$$= 2\prod_{p\geq 3}\left(1-\frac{1}{(p-1)^2}\right) = 2\prod_{p\geq 3}\frac{p(p-2)}{(p-1)^2} \approx 1.3203\,. \tag{4.4}$$

The terms in (4.3) attempt to account for the dependences between n and $n+2$; The factors $(1-1/2)$ and $(1-1/2)^{-2}$ adjust for the fact that n and $n+2$ must both be odd; the heuristic evaluates this probability as $1/4$, although its true value is twice that. Likewise, for each prime $p \geq 3$, the p-term in the product corrects for the fact that the probability that neither n nor $n+2$ is divisible by p, is $1-2/p$, and not $(1-1/p)^2$ as determined by the heuristic. Similar reasoning leads to conjectures for prime triplets and other prime tuples. The values for the constants conjectured by Hardy & Littlewood agree with all numerical studies to date, and it is widely believed that, in some sense, dependences of the type just described account for all of the nonrandomness in the sequence of primes. Computations by Sabeh & Gourdon show the actual number of twin primes up to 10^{16} is 10,304,195,697,298 while the Hardy-Littlewood formula gives 10,304,192,554,496. Stronger *empirical* evidence in favour of the twin prime conjecture can scarcely be desired. For further details see, for example, Cherwell & Wright (1960, [**200**]), Bateman & Diamond (2004, [**82**, p. 335]), Iwaniec & Kowalski (2004, [**594**, p. 153]), or Hardy & Wright (2008, [**499**, §22.20]). We leave it to the interested reader to show that Dickson's prime tuples conjecture holds for the Cramér primes.

[3]Although events corresponding to twins whose endpoints touch are not independent (even under Cramér's model) one may split this sum in two to see that the mentioned Borel–Cantelli lemma can still be applied. Of course, for the real primes, aside from $\{3, 5, 7\}$ there are no other twins that touch since $\{n, n+2, n+4\}$ must always have one term divisible by 3.

Gaps between primes. Of particular interest is the application of Cramér's heuristic to questions of whether or not there are an infinite number of prime gaps substantially larger and substantially smaller than their average spacing. Concerning large gaps between primes, the heuristic leads to Cramér's widely cited 1936 conjecture

$$\limsup_{n\to\infty} \frac{p_{n+1}-p_n}{\log^2 p_n} = \text{Const} > 0, \tag{4.5}$$

which says that the order of the largest gaps is the square of the average gap size. To establish (4.5) for the pseudoprimes, consider the event A_n that n is prime and that the next prime exceeds $n+(1+\epsilon)\log^2 n$. Using $(1-1/x)^x \to e^{-1}$, the Cramér probability of this event is approximately

$$\mathrm{P}[A_n] \approx \frac{1}{\log n} \times \left(1-\frac{1}{\log n}\right)^{(1+\epsilon)\log^2 n} \approx \frac{e^{-(1+\epsilon)\log n}}{\log n} = \frac{1}{n^{1+\epsilon}\log n}.$$

Hence $\sum \mathrm{P}[A_n]$ is finite when $\epsilon > 0$, and infinite when $\epsilon \le 0$, so that gaps larger than $(1+\epsilon)\log^2 n$ can occur only a finite number of times if $\epsilon > 0$, but will occur infinitely often when $\epsilon \le 0$. This proves (4.5) for the Cramér primes, with Const $= 1$. Here we have used first "the (a) part" of the Borel–Cantelli lemma stated in Appendix D, and then "the (e) part" of the extended form of the lemma involving the requirement $\mathrm{P}[A_j \cap A_k] \le \mathrm{P}[A_j]\mathrm{P}[A_k]$. Taking $j < k$, the mentioned requirement holds here because A_j and A_k are independent, except when k falls in the range $\big(j,\, j+(1+\epsilon)\log^2 j\,\big]$ in which case $A_j \cap A_k$ will be null. Two alternate demonstrations of (4.5) for Cramér primes are given in Section 4.8 (Selected proofs).

Although far stronger than anything that has been proven so far, conjecture (4.5)—which originates in Cramér (1936, [**245**])—is widely believed to be true. However the value of 1 for the constant (implied by the given argument if left untweaked) is unlikely to be correct. The correct constant is sometimes referred to as the Cramér–Shanks–Granville ratio. Arguments accounting for divisibility by 'early' primes suggest that its value is at least $2e^{-\gamma} \approx 1.229$, although the largest currently known value of the ratio (which occurs at a certain 16-digits prime) is only about 0.92. See Granville (1995, [**451**]), (2009, [**454**]) for details. Recall from the discussion in (2.35) and (2.1) of Chapter 2 that what is currently known *unconditionally* is that there are infinitely many $(p_{n+1}-p_n)/\log p_n$ of size about

$$(\log\log p_n)(\log\log\log\log p_n/\log\log\log p_n),$$

(Ford et al. (2018, [**385**])), a value that is far less than the $\log p_n$ implied by (4.5). Cramér's conjecture does fall within the range of indeterminacy indicated in (2.1), yet curiously lies well beyond anything that even the Riemann hypothesis would imply.

As for small gaps, the implications of Cramér's heuristic were already evident from our discussion of twin primes in Chapter 2. In fact, Cramér primes take on the differences $\widetilde{p}_{n+1} - \widetilde{p}_n = k$ infinitely often with probability 1 for every $k \ge 1$, even though (as for the actual primes) the series of reciprocals $\sum_n (1/\widetilde{p}_n)\,\mathrm{I}[\widetilde{p}_{n+1} - \widetilde{p}_n = k]$ converges for every k. Thus every integer can be written as a difference of two

Cramér primes an infinite number of ways, consistent (aside from the natural difference between even and odd gaps) with the 1849 conjecture of de Polignac mentioned in Chapter 2. Straightforward arguments also show that Cramér primes contain arbitrarily long arithmetic progressions—consistent with the result of Green & Tao (2008, [**464**]). It can also be shown that in the region below x, and regardless of the value of k, the number of times $\widetilde{p}_{n+1} - \widetilde{p}_n = k$ occurs has the same order as $x/\log^2 x$, and that the number of times $\widetilde{p}_{n+m} - \widetilde{p}_n = k$ occurs has the same order as $x/\log^{m+1} x$, for $x \to \infty$.

Other heuristics. In addition to the examples already discussed, Cramér's heuristic accords with virtually all of the standard, well-known conjectures concerning the prime numbers. Thus, among the Cramér primes, there are (with probability 1) infinitely many Sophie Germain prime pairs $\widetilde{p}$ and $2\widetilde{p}+1$, as well as infinitely many primes of the form n^2+1. As for Legendre's unproven conjecture that there is always a prime between n^2 and $(n+1)^2$, for the Cramér primes one can show, far more generally, that all of the intervals $[n^{1+\epsilon}, (n+1)^{1+\epsilon}]$ contain primes once n is large enough, provided only that $\epsilon > 0$. A proof of this assertion is given in Section 4.8. Recall that for the actual primes, Baker, Harman & Pintz (2001, [**70**]) had established that (for n large enough) the interval $[n^\theta, (n+1)^\theta]$ must always contain a prime, but their proof only works for $\theta > 40/19 \approx 2.1$.

One of the more striking successes of the Cramér model is that it actually predicts the pair correlation conjecture of Montogomery & Dyson for the critical zeros of the zeta function (discussed in Chapter 6), and possibly even predicts the associated k-tuple correlation conjecture. Specifically, the results of Goldston (1988, [**422**]) imply that the pair-correlation conjecture holds if and only if

$$\int_T^{T^{1+\beta}} \left(\psi\left(x+\frac{x}{T}\right) - \psi(x) - \frac{x}{T}\right)^2 \frac{dx}{x^2} \sim \beta\,\frac{\log^2 T}{T}, \quad \text{as } T \to \infty,$$

for every fixed $\beta > 0$, where $\psi(x) \equiv \sum_{p^m \le x} \log p$ is the Chebyshev psi-function. It can be shown that Cramér's model actually satisfies this asymptotic (with probability 1) when ψ is replaced by the version $\widetilde{\psi}(x) \equiv \sum_{\widetilde{p}^m \le x} \log \widetilde{p}$ defined in terms of the Cramér primes.

Finally, Cramér's model for the primes can be shown to imply the Riemann hypothesis. Recall from Chapter 2 that the true maximum order of $\pi(x) - \mathrm{li}(x)$, though unknown, must exceed $O\left(\sqrt{x}/\log x\right)$, while under RH it would be $O\left(\sqrt{x}\log x\right)$. In fact RH is equivalent to the error term being $O(x^{1/2+\epsilon})$. Under Cramér's heuristic, however,

$$\liminf_{x\to\infty} \frac{\widetilde{\pi}(x) - \mathrm{li}(x)}{\sqrt{(2x\,\log\log x)/\log x}} = -1, \quad \text{and} \quad \limsup_{x\to\infty} \frac{\widetilde{\pi}(x) - \mathrm{li}(x)}{\sqrt{(2x\,\log\log x)/\log x}} = +1, \tag{4.6}$$

with probability 1, and this is substantially sharper than what is required for RH to hold. A proof for (4.6) is given in Section 4.8.

Denjoy (1931, [**274**]) proposed a different statistical argument in favour of the Riemann hypothesis based on the Möbius function. Recall that $\mu(n)$ is defined as 0 if n is divisible by a square, and otherwise as $+1$ or -1 when n has an even or an odd number of prime factors, respectively. RH is equivalent to the assertion $\sum_{n\le x} \mu(n) = O(x^{1/2+\epsilon})$ for all $\epsilon > 0$. Denjoy (1931, [**274**]) pointed out that if the ± 1 values of $\mu(n)$ behaved like a sequence of independent, identically distributed Bernoulli variables with probabilities $1/2$, then (as an immediate consequence of the Law of the Iterated Logarithm) that assertion, and consequently RH, would follow. Good & Churchhouse (1968, [**442**]) showed that this independence assumption agrees well with the data. Other probabilistic arguments favouring RH were given by Wintner (1944, [**1115**]), Denjoy (1964, [**275**]), and Ducray (1964, [**296**]). See also Wilf (1987, [**1107**]) and Steuding (2007, [**1027**, Chapter 3, §3]).

4.3. Maier's counterexample

In spite of the protestations of Hardy, Littlewood, and others, the influence of the Gauss–Cramér model on the development of number theory can scarcely be overstated. This model had, over a period of many decades, proven invaluable in guiding both theory and intuition about the primes. It had come to be trusted as an indispensable benchmark for suggesting the correct order of magnitude for many asymptotic results and for gauging the plausibility of various undecided conjectures. Indeed, no other method for conjuring up plausible predictions concerning the primes had ever been suggested. It therefore came as no small surprise when Helmut Maier (1985, [**722**]) provided a counterexample in which a straightforward prediction made by the model for the number of primes in certain long intervals fails completely, a counterexample which appears to suggest that no purely statistical model of the primes can capture all of their essential asymptotic properties.

The key issue is as follows: Cramér's model predicts that if we take $y = \log^\alpha x$, using any constant $\alpha > 2$, then with probability 1 we should have

$$\widetilde{\pi}(x+y) - \widetilde{\pi}(x) = \mathrm{li}(x+y) - \mathrm{li}(x) + O\left(\sqrt{y}\right), \quad \text{as } x \to \infty, \tag{4.7}$$

or, in view of the asymptotic equivalence $\mathrm{li}(x+y) - \mathrm{li}(x) \sim y/\log x$,

$$\widetilde{\pi}(x+y) - \widetilde{\pi}(x) = \frac{y}{\log x} + O\left(\sqrt{y}\right), \quad \text{as } x \to \infty. \tag{4.8}$$

A proof of these assertions is given in Section 4.8. Indeed, one may even allow $y = y(x)$ to vary freely in a range such as $10\log^2 x \le y \le \sqrt{x}$, say, and the error term in (4.7)/(4.8) would then be uniform in y. This is so because in this range, $\mathrm{li}(x+y) - \mathrm{li}(x) = y/\log t$ for some t between x and $x+y$ by the mean value theorem, and $\log t$ may be replaced by $\log x$ asymptotically, so that (4.8) will continue to hold for y in the wider range.

Selberg (1943, [**969**]) lent credence to the prediction (4.7)/(4.8) by proving for the actual primes (and $\alpha > 2$) that under the Riemann hypothesis,

$$\pi(x+y) - \pi(x) \sim \frac{y}{\log x}, \quad \text{as } x \to \infty, \tag{4.9}$$

for "almost all" integers x (that is, with a "0-proportion subset" of the integers removed). However, in his game-changing paper, Maier showed, unconditionally, that Selberg's result cannot hold for *all* integers x; specifically, he proved that the strict inequalities

$$\liminf_{x\to\infty} \frac{\pi(x+y) - \pi(x)}{y/\log x} < 1 < \limsup_{x\to\infty} \frac{\pi(x+y) - \pi(x)}{y/\log x}, \tag{4.10}$$

hold for *every* $\alpha > 0$ (and so in particular for every $\alpha > 2$), with either of these inequalities being in clear conflict with (4.7)/(4.8). (Bear in mind that $y \equiv y(x) \equiv \log^\alpha x$ is a function of x here.) Among other innovations, Maier's argument exploited the fact that the probabilistic model doesn't account for the factor $e^{-\gamma}$ in Mertens's formula (2.6), that is, it doesn't account correctly for divisibility by the small primes. Details of Maier's proof will be provided later in this chapter.

Stating things differently, Cramér's heuristic makes the straightforward prediction that

$$\pi\left(x + \log^{\lambda} x\right) - \pi(x) \sim \log^{\lambda-1} x, \quad \text{as } x \to \infty, \quad \text{for all } \lambda > 1, \tag{4.11}$$

and it is this prediction that fails for $\lambda > 2$. Maier's result (4.10) implies that for the actual primes, and for *all* $\lambda > 0$,

$$\frac{\pi\left(x + \log^{\lambda} x\right) - \pi(x)}{\log^{\lambda-1}(x)} \not\to 1, \quad \text{as } x \to \infty. \tag{4.12}$$

(It is, however, still possible that (4.9) might hold for all x if, for example, $y > x^{\epsilon}$.)

Subsequent to Maier's work, other 'irregularities' were unearthed in the distribution of the primes. Pintz (2007, [**849**]), for instance, provided an example of a fully *global* nature for which it seems that *any* model based on independent variables must fail: he showed that all such models incorrectly predict the asymptotic value of $\frac{1}{x}\int_0^x (\psi(y) - y)^2\, dy$ (this being related to the PNT error term). Other difficulties with Cramér's model were pointed out in Hildebrand & Maier (1989, [**554**]). Nevertherless, even with such shortcomings, Cramér's heuristic still remains an invaluable tool for guiding intuition and for developing plausible conjectures.

REMARK 4.1. One may ask: How large does $y \equiv y(x)$ need to be to guarantee the behaviour

$$\pi(x + y(x)) - \pi(x) \sim \frac{y(x)}{\log x}, \quad \text{as } x \to \infty,$$

expected under Cramér's model? Huxley (1972, [**571**]) and Heath-Brown (1980, [**527**]) showed that $y(x) = x^{7/12}$ would work, while under RH $y(x) = x^{1/2+\epsilon}$ would work. Huxley (1972, [**569**]) proved that $y(x) = x^{1/6+\epsilon}$ works for almost all x, while (as mentioned above) Selberg (1943, [**969**]) had shown, under RH, that $y(x)/\log^2 x \to \infty$ works for almost all x.

4.4. The Hawkins model

Numerous statistically based alternatives to Cramér's model have been proposed. Of these the most notable is a randomized version of the sieve of Eratosthenes due to David Hawkins (1957, [**518**]), (1974, [**519**]), (1987, [**520**]) a mathematical philosopher and graduate of Stanford and University of California Berkeley who had worked alongside Robert Oppenheimer and Stan Ulam at Los Almos. In Hawkins's model, 2 is declared to be prime, and then—independently of each other—each of the integers exceeding 2 is either discarded or kept with probabilities $\frac{1}{2}$. If $N > 2$ is the next number in the (undiscarded) subsequence that remains, it is declared to be prime, and each integer exceeding N—again, independently of the others—is discarded or kept with probabilities $\frac{1}{N}$ and $1 - \frac{1}{N}$, respectively. This process is continued indefinitely. In the Eratosthenes sieve, the deletions occur deterministically; nevertheless the deletion rates of the two sieving methods turn out to be asymptotically identical. In other words, the 'probability' that n is a Hawkins prime is (with probability 1) asymptotic to $1/\log n$ even though (unlike Cramér's model) in Hawkins's model the PNT was not "built-in." One can show that the sequence of *actual* primes belongs to the set of *most probable* sequences generated by the Hawkins model. Noteworthy too is that (unlike Cramér's model) Hawkins's model incorporates some degree of dependence amongst its 'primes' since the probability of an integer being declared prime now depends on the sequence of 'primes' that came before it.

Proving properties for Hawkins primes is more difficult than for those of Cramér, with such proofs typically based on detailed combinatorial arguments—as in Hawkins (1957, [**518**]), (1974, [**519**]), or Wunderlich (1974, [**1119**]), (1976, [**1120**])—or based on martingale methods—as as in Neudecker & Williams (1974, [**800**]), Neudecker (1975, [**799**]), or Heyde (1976, [**543**]), (1978, [**544**]). However, one helpful feature of the Hawkins model is that testing for divisibility only involves random elimination. For example, conditioning on earlier values of the sequence, one easily obtains the handy geometric distribution result

$$\mathrm{P}\left[\widetilde{p}_{n+1}-\widetilde{p}_n=j \mid \widetilde{p}_1,\ldots,\widetilde{p}_{n-1}\right] = \frac{1}{M_n}\left(1-\frac{1}{M_n}\right)^{j-1}, \quad j\geq 1,\ n\geq 1,$$

where M_n, an analogue of the Mertens product, is defined as

$$M_n = \prod_{1\leq k\leq n}\left(1-\frac{1}{\widetilde{p}_k}\right)^{-1}. \tag{4.13}$$

The quantity M_n will appear again in the discussion which follows.

The Prime Number Theorem for Hawkins primes, $\widetilde{p}_n$, was established by Wunderlich (1974, [**1119**]), (1976, [**1120**]) who showed that

$$\lim_{n\to\infty}\frac{\widetilde{p}_n}{n\log n} = 1 \text{ a.s.},$$

that is, almost surely (or with probability 1). An elementary proof of this result appears in Lorch & Okten (2006, [**711**]), (see also Lorch (2007, [**710**])), while a proof of this result (as well as of the next) based on martingales is given in Hall & Heyde (1980, [**481**, pp. 226–234]). Wunderlich also proved that

$$\lim_{n\to\infty}\frac{1}{\log n}\prod_{1\leq k\leq n}\left(1-\frac{1}{\widetilde{p}_k}\right)^{-1} \equiv \lim_{n\to\infty}\frac{M_n}{\log n} = 1 \text{ a.s.}, \tag{4.14}$$

this being an analogue of Mertens's theorem, except that here (as in Cramér's model) the limiting value obtained is 1 and not e^{γ}. Using martingale methods Heyde (1976, [**543**]), (1978, [**544**]) also established the further almost sure convergences

$$\lim_{n\to\infty}\frac{1}{\log\log n}\left(\frac{p_n}{n}-\log n\right)=1 \quad\text{and}\quad \lim_{n\to\infty}\frac{1}{\log\log n}\left(M_n-\log n\right)=1;$$

these may alternately be stated as

$$\widetilde{p}_n - n\log n \sim n\log\log n \quad\text{and}\quad M_n-\log n\sim\log\log n, \quad\text{as } n\to\infty, \text{ a.s.}$$

The Cramér version of (4.14) is much easier to prove, because in Cramér's case

$$\log\widetilde{M}_n = -\sum_{k=2}^{n} X_k\log\left(1-\frac{1}{k}\right), \quad\text{where } \mathrm{P}[X_k=1]=1-\mathrm{P}[X_k=0]=\frac{1}{\log k},$$

and this has mean asymptotic to $\sum_{k=2}^{n} 1/(k\log k)\sim\log\log n$, but variance which converges to a constant, from which it follows that $\widetilde{M}_n\sim\log n$, almost surely. (The required argument is identical to the one used earlier to prove the PNT for the Cramér primes.)

Many of the other properties of Hawkins primes turn out to be identical to those of the Cramér primes. For example, the sequences of Hawkins primes almost surely contain arbitrarily long arithmetic progressions. Neudecker (1975, [**799**]), Deheuvels (1985, [**265**]), and Bui & Keating (2006, [**166**]) showed that (with probability 1) the difference $\widetilde{p}_{n+1} - \widetilde{p}_n = k$ occurs infinitely often for every $k \geq 1$. Furthermore (with probability 1) the number of times that $\widetilde{p}_{k+1} - \widetilde{p}_n = k$ occurs in the region below x is asymptotic to $x/\log^2 x$, regardless of the value of k, which implies that the series of reciprocals $\sum(1/\widetilde{p}_n)\mathrm{I}[\,\widetilde{p}_{n+1} - \widetilde{p}_n = k\,]$ converges. Rivoal (2008, [**910**]) showed the related result that the number of integers $j \leq n$ such that $\widetilde{p}_{j+m} - \widetilde{p}_j = m$ (indicating a run of $m+1$ consecutive Hawkins primes) is almost surely asymptotic to $n/\log^m n$.

Deheuvels (1985, [**265**]) proved (again, with probability 1) that the differences $\widetilde{p}_{n+1} - \widetilde{p}_n$ exceed $(1-\epsilon)\log^2 n$ infinitely often, but exceed $(1+\epsilon)\log^2 n$ only a finite number of times for any $\epsilon > 0$, thus establishing that the Hawkins primes satisfy the Cramér conjecture

$$\limsup_{n\to\infty} \frac{\widetilde{p}_{n+1} - \widetilde{p}_n}{\log^2 \widetilde{p}_n} = 1, \text{ a.s.}$$

It is also known that for any $\epsilon > 0$, and once n is large enough, there is almost surely always a Hawkins prime between $n^{1+\epsilon}$ and $(n+1)^{1+\epsilon}$.

The Hawkins primes also satisfy an analogue of the Riemann hypothesis which mirrors the known RH equivalence $\mathrm{li}\,(p_n) = n + O(n^{1/2+\epsilon})$. Specifically, Neudecker & Williams (1974, [**800**]) showed first, that the limit

$$J \;\equiv\; \lim_{n\to\infty} \widetilde{p}_n \exp\big(-M_n\big) \;\equiv\; \lim_{n\to\infty} \widetilde{p}_n \exp\Bigg(-\prod_{1\leq k\leq n}\left(1 - \frac{1}{\widetilde{p}_k}\right)^{-1}\Bigg),$$

although random, exists and is positive with probability 1, and second, that

$$J \cdot \mathrm{li}\,(\widetilde{p}_n/J) \;=\; n + O\big(n^{1/2+\epsilon}\big), \quad \text{as } n \to \infty, \text{ a.s.}$$

(Here $J \cdot \mathrm{li}(\widetilde{p}_n/J) - n$ serves as an analogue for the error in the Prime Number Theorem.) Heyde (1978, [**545**]) later strengthened this result to

$$\limsup_{n\to\infty} \frac{|\, J \cdot \mathrm{li}\,(\widetilde{p}_n/J) - n\,|}{\sqrt{2n\log\log n}} \;\leq\; 3, \quad \text{a.s.},$$

thus effectively replacing the n^ϵ part of the $O(n^{1/2+\epsilon})$ error term by $\log\log n$, while Foster & Williams (1978, [**390**]) established that

$$\frac{1}{\sqrt{n}}\big(J \cdot \mathrm{li}\,(\widetilde{p}_n/J) - n\big) \;\xrightarrow{d}\; N(0,1), \quad \text{as} \quad n \to \infty,$$

as well as

$$\liminf_{n\to\infty} \frac{J \cdot \mathrm{li}\,(\widetilde{p}_n/J) - n}{\sqrt{2n\log\log n}} = -1 \quad \text{and} \quad \limsup_{n\to\infty} \frac{J \cdot \mathrm{li}\,(\widetilde{p}_n/J) - n}{\sqrt{2n\log\log n}} = 1, \quad \text{a.s.}$$

The results cited in this paragraph are all nontrivial.

At least one further nontrivial result concerning the Hawkins primes is worth noting since it represents an analogue to the Poisson distribution of the primes. Deheuvels (1985, [**265**]) established that there exists an independent and identically distributed sequence E_n of exponential random variables with mean 1 such that, with M_n defined as in (4.13), the distributional equality

$$\widetilde{p}_{n+1} - \widetilde{p}_n \stackrel{d}{=} \left\lfloor \frac{E_n}{-\log(1 - M_n^{-1})} \right\rfloor + 1$$

holds, where $\lfloor x \rfloor$ represents the integer part of x.

Further details concerning the Hawkins primes may be found in the references already cited, as well as in Williams (1974, [**1110**]), (1990, [**1111**]).

4.5. Other models for the primes

Many variations are possible for producing random sequences of pseudoprimes with reasonable properties. For example, declare $\widetilde{p}_1 = 2$ to be a prime; after the nth prime $\widetilde{p}_n$ has been established, the next one is defined to be $\widetilde{p}_{n+1} = \widetilde{p}_n + R_n$, where R_n has a geometric distribution over the integers ≥ 1, with mean $\operatorname{E} R_n = \log \widetilde{p}_n$. A different possibility is to declare $\widetilde{p}_1 = 2$ to be prime; then (independently of each other) the integers exceeding $\widetilde{p}_1$ are tested for primality using the probability $(1 - 1/\widetilde{p}_1) \equiv 1/2$, until one of them, namely $\widetilde{p}_2$, is declared to be prime. Then, after $\widetilde{p}_1, \ldots, \widetilde{p}_k$ have been declared prime, the integers greater than $\widetilde{p}_k$ are tested for primality using probabilities $(1 - 1/\widetilde{p}_1) \times \cdots \times (1 - 1/\widetilde{p}_k)$ until one of them, namely $\widetilde{p}_{k+1}$, is declared to be the next prime. Other statistical models have been proposed, for example, by Granville (1995, [**450**]), Garcia-Pérez, Serrano & Boguñá (2014, [**409**]), and by Banks, Ford & Tao (2023, [**74**]). Such methods for generating pseudoprime sequences can, in principle, all be studied for properties analogous to those of the actual primes. See also Gardiner et al. (1956, [**410**]), and Erdős & Jabotinsky (1958, [**355**]) and the references therein.

An interesting variation on the Hawkins sieve was proposed by Sanjoy Mahajan (1998, [**720**]) (author of the amusing and free-to-download book "Street Fighting Mathematics"). Recall from Chapter 2 that although Gauss and Riemann both believed that $\pi(n) < \operatorname{li}(n)$ for all n, Littlewood proved that it changes sign infinitely often. In the Hawkins model, however, $\widetilde{\pi}(n)$ is eventually always below $\operatorname{li}(n)$. Mahajan proposed a 'square root variation' of Hawkins's model in which, after n is declared to be a prime, only numbers below n^2 are subjected to the (ongoing) random elimination process. This somewhat increases the number of 'primes', and Mahajan showed that the probability of n being 'prime' will now oscillate around $1/\log n$, and he conjectured that for his version the first sign change of $\widetilde{\pi}(n) - \operatorname{li}(n)$ should occur around $n = 10^{27}$.

Of course, the principal shortcoming of the heuristic models proposed so far is that they do not account properly for the effects of divisibility by the *early* primes (and so do not account for the e^γ factor in Mertens's result). Determining when predictions of statistical models can be considered reliable is a difficult question, as is also the question of whether such models can be modified to more accurately reflect the properties of the actual primes and to not fail in constructions such as the one of Maier (and, in particular, to reproduce the e^γ 'small-primes factor' of Mertens's formula). Banks, Ford & Tao (2023, [**74**]) offer some progress along these lines.

An amusing take on such questions was once ventured by Pólya (1959, [**862**]). (See also Schröeder (2006, [**955**, pp. 54–55]).) Under independence, the probability that x is prime should be

$$\prod_{p \le x^*} \left(1 - \frac{1}{p}\right),$$

with $x^* = x^{1/2}$. In view of Mertens's result (2.6), this does not give the desired value of $1/\log x$; but it would give $1/\log x$ if we were to put $x^* = x^{\exp(-\gamma)} \approx x^{0.56}$ instead. Thus changing $x^* = x^{1/2}$ to $x^* = x^{\exp(-\gamma)}$ 'accounts' for dependences among divisibility by primes, at least in this instance. Curiously, this same adjustment leads to an estimate for the number of twin primes that is identical to the one that is widely conjectured. Specifically, the probability that x and $x+2$ are both prime is now estimated as

$$\text{Prob} = \frac{1}{2} \prod_{2 < p \le x^*} \left(1 - \frac{2}{p}\right), \quad x^* = x^{\exp(-\gamma)}.$$

The factor $1/2$ arises because x must be odd, while the remaining factors represent the probabilities that neither x nor $x+2$ is divisible p. If we rewrite this as

$$\text{Prob} = \left(\frac{1}{2} \frac{\prod_{2<p\le x^*} \left(1 - \frac{2}{p}\right)}{\prod_{2<p\le x^*} \left(1 - \frac{1}{p}\right)^2} \right) \left(\prod_{2<p\le x^*} \left(1 - \frac{1}{p}\right)^2 \right)$$

we find that the first component on the right converges to the usual twin-prime constant (4.3)/(4.4), while the second part converges to $1/\log^2 x$. Similar arguments also work for the commonly conjectured counts for prime triplets, quadruplets, and so on.

4.6. Beurling primes

Going further afield, we now describe a very different model intended to generalize the primes—a model with the unusual property that the "primes" are no longer restricted to be integers (and therefore neither are the "integers"!). This variant was proposed by Arne Carl-August Beurling (1905–1986), a Swedish mathematician analyst who single-handedly decrypted the cypher used by the German Luftwaffe and navy during WWII, an astonishing feat which gave Sweden advance knowledge of Operation Barbarossa. (The cyphers Turing had worked on were those used by the field units.)

A set of *Beurling generalized primes* consists of *any* sequence of *real* numbers $\mathcal{P} = \{\widetilde{p}_k\}_{k=1}^{\infty}$ tending to infinity which satisfies $1 < \widetilde{p}_1 \le \widetilde{p}_2 \le \ldots$. The $\widetilde{p}_k$ need not be integers, and multiplicity is allowed. The associated set of *Beurling generalized integers* $\mathcal{B}$ consists of *all numbers* that can be formed as a finite product $\widetilde{p}_{\nu_1}^{\alpha_1} \widetilde{p}_{\nu_2}^{\alpha_2} \cdots \widetilde{p}_{\nu_m}^{\alpha_m}$ of these Beurling primes (here the α_j are positive integers). These numbers are then ordered and labelled as $\widetilde{n}_1 \le \widetilde{n}_2 \le \cdots$. Some values of $\widetilde{n}$ can occur in multiple ways, and once again this multiplicity is allowed. One then defines the functions

$$\widetilde{\pi}(x) = \sum_{\widetilde{p}_k \le x} 1, \quad \text{and} \quad \widetilde{N}(x) = \sum_{\widetilde{n}_k \le x} 1,$$

which, respectively, count the number of Beurling primes up to x, and the number of Beurling integers up to x, and then asks for asymptotic conditions on $\widetilde{N}(x)$, as minimal as possible, under which the corresponding prime number theorem, namely $\widetilde{\pi}(x) \sim x/\log x$, will hold.

Beurling (1937, [**98**]) showed that the condition

$$\widetilde{N}(x) = Ax + O\left(\frac{x}{\log^\gamma x}\right), \quad \text{as} \quad x \to \infty,$$

where $A > 0$ and $\gamma > 3/2$, is sufficient for this PNT to hold, and that it need not hold when $\gamma \leq 3/2$. Kahane (1997, [**618**]) generalized Beurling's result and proved that

$$\int_1^\infty \left|\frac{(\widetilde{N}(t) - At)\log t}{t}\right|^2 \frac{dt}{t} < \infty$$

for some $A > 0$ would also suffice for the PNT. Curiously, the value of A in these statements has no effect on the main asymptotic for $\widetilde{\pi}(x)$: having more or fewer integers (which satisfy the conditions) does not affect the asymptotic number of 'primes'. This observation becomes more intuitive when we realize that inserting or deleting a single Beurling prime will not affect the asymptotic formula for $\widetilde{\pi}(x)$ but is already enough to change the leading constant A in the asymptotic formula for $\widetilde{N}(x)$. For example, if we take our Beurling primes to be the set of all *odd* primes (that is, all primes except 2), then we still have $\widetilde{\pi}(x) \sim x/\log x$, but now $\widetilde{N}(x) \sim x/2$ since the associated set of Beurling integers will consist of all the odd numbers.

Under the condition $\widetilde{N}(x) = Ax + o(x)$, Olofsson (2010, [**826**]) proved a striking result—unexpectedly similar to Mertens's theorem—namely

$$\lim_{x\to\infty} \frac{1}{\log x} \prod_{p_k \leq x} \left(1 - \frac{1}{\widetilde{p}_k}\right)^{-1} = Ae^\gamma.$$

(The appearance of the factor e^γ here is nothing short of astonishing!) One consequence of this is that under the stated condition the limit

$$M = \lim_{x\to\infty} \left(\sum_{\widetilde{p}_k \leq x} \frac{1}{\widetilde{p}_k} - \log\log x\right)$$

will exist, and Olofsson showed that this limit is given by

$$M = \lim_{\alpha \downarrow 0} \left(\sum_k \frac{1}{\widetilde{p}_k \log^\alpha \widetilde{p}_k} - \frac{1}{\alpha}\right).$$

The same result also holds for the ordinary primes, where it is known as Meissel's Theorem. As a consequence, Pollack (2013, [**859**]) showed that for the Beurling primes $\mathcal{P}$ we have

$$\sum_{\widetilde{p} \leq x} \frac{1}{\widetilde{p}} = \log\log x + \gamma + \log A - \sum_{\widetilde{p}} \sum_{k=2}^\infty \frac{1}{k\widetilde{p}^k} + o(1).$$

Remarkably, except for the term $\log A$, this result is identical to the one in (2.2)/(2.3) obtained by Mertens. Rupert (2013, [**927**]) showed that the Hardy–Ramanujan and Erdős–Kac theorems for $\omega(n)$ also continue to hold for the Beurling numbers.

Having discussed—in the previous sections—several random models of the primes, we will now return to the true primes to provide details of the proof of Maier's surprising result—that the number of primes near x in intervals of length $(\log x)^\alpha$ is not asymptotic to its expected value, in stark contrast to what the random models predict.

4.7. Details of Maier's proof

Maier's proof, which is based on a result of Buchstab (1951, [**162**]), is straightforward in principle but requires somewhat more intricate reasoning than we have encountered so far. It involves constructing a certain (*enormous*) $M \times N$ matrix, each of whose rows is a sequence of N consecutive integers, and each of whose columns is a sequence of M integers in an arithmetic progression. Let D denote the (common) difference in these N arithmetic progressions. Such a matrix can be produced with the property that the number of primes appearing in it, when counted column by column and summed, exceeds the number predicted by a purely probabilistic model such as Cramér's. Of course, only columns containing numbers relatively prime to D are eligible to contain primes, and (by the PNT for arithmetic progressions) each such column will contain the expected number of primes; it is the number of eligible columns that can be made to exceed the probabilistic prediction. As a consequence, at least one of the rows must then also exceed its probabilistically predicted primes count. By constructing the matrix appropriately, the required contradiction can be produced. The other inequality asserted in (4.10) can be established by a similar argument.

Buchstab's fundamental contribution was to generalize the formula

$$\prod_{p\le\sqrt{x}}\left(1-\frac{1}{p}\right) = \frac{e^{-\gamma}}{\frac{1}{2}\log x}+O\left(\frac{1}{\log^2 x}\right) \tag{4.15}$$

of Mertens. We have encountered this formula previously in (2.6), except here it is written using the argument $\sqrt{x}$ in lieu of x. If the factors $(1-1/p)$ are thought of as representing the proportions of integers not divisible by p, the product shown *should* give the proportion of integers up to x that are prime, since a number is composite if and only if it has at least one prime factor less than its square root. However, the PNT says that the number of primes up to x is about $x/\log x$, while (4.15) would appear to suggest their number to be about $e^{-\gamma}x/\frac{1}{2}\log x$, an apparent overcount by a factor of $2e^{-\gamma}\approx 1.12292$. This overcount must evidently stem from a lack of independence in divisibility by primes $\le\sqrt{x}$, over the interval $[1,x]$. The interval $[1,x]$ is simply too short (relative to this collection of primes) for an adequate degree of independence to hold; equivalently, this collection of primes is simply too large relative to that interval.

The assertions we have made about (4.15) can be phrased alternately in terms of the set $B(2)\equiv\{n\colon\ p\mid n\Rightarrow p>n^{1/2}\}$, which clearly is just the set of all primes. What had been stated is no more than that the probabilistic prediction

$$x\prod_{p\le x^{1/2}}\left(1-\frac{1}{p}\right)$$

for the number of integers in $B(2)$ not exceeding x is asymptotic to $2e^{-\gamma}x/\log x$, and not to the correct result $x/\log x$. More generally, however, Buchstab considered integers having only large prime factors; specifically, he defined the set $B(\alpha)$ of integers n such that every prime divisor of n exceeds $n^{1/\alpha}$. The probabilistic prediction for the number of integers in $B(\alpha)$ not exceeding x is evidently

$$x\prod_{p\le x^{1/\alpha}}\left(1-\frac{1}{p}\right);$$

nevertheless, Buchstab (1951, [**162**]) showed, for any $\alpha \geq 2$, that

$$\#\{n \in B(\alpha)\colon n \leq x\} \sim e^{\gamma}\,\omega(\alpha)\cdot x \prod_{p \leq x^{1/\alpha}} \left(1 - \frac{1}{p}\right), \tag{4.16}$$

where $\omega(\alpha)$ is a certain explicitly known function. The Buchstab function $\omega(u)$ is quite unusual: it is defined by the differential-difference equation $\frac{d}{du}(u\,\omega(u)) = \omega(u-1)$ for $u \geq 2$, with initial values given by $\omega(u) = 1/u$ for $1 \leq u \leq 2$. It is continuous everywhere and is real-analytic everywhere except at integer values of u. Maier (and Buchstab) showed that $e^{\gamma}\omega(\alpha)$ tends to 1 as α tends to infinity (as one would naturally expect), but that it oscillates continually in value, rising above 1 and falling below 1 repeatedly, while doing so.[4]

Given a *huge*[5] real number x, and a fixed real number $\alpha > 2$, the aforementioned $M \times N$ matrix can now be constructed as follows: the number of columns is chosen to be $N = \lfloor \log^{\alpha} x \rfloor$, and the common difference D in each column is set equal to the product of all the primes[6] up to $\log x$; the number of rows M is chosen to be an *enormous* power of D; and the first entry of the matrix is set to equal $MD+1$. The entries in the matrix will then all be determined, and will all lie between MD and $2MD$; furthermore, the probabilistic prediction would be that the number of primes in such a matrix should be asymptotic to $MN/\log MD$.

Now, the jth column, which contains only numbers congruent to j modulo D, is eligible to contain primes precisely when j is relatively prime to D, or equivalently when j is not divisible by any prime up to $\log x \sim N^{1/\alpha}$. Since j ranges from 1 to N, the number of eligible columns is

$$\#\{n \in B(\alpha)\colon n \leq N\} \sim e^{\gamma}\omega(\alpha)\cdot N \prod_{p \leq \log x} \left(1 - \frac{1}{p}\right).$$

However if j were chosen at random, then (according to the probabilistic model) the probability that the column would be eligible to contain primes would be $\prod_{p \leq N^{1/\alpha}}(1 - 1/p)$; the actual proportion of eligible columns thus differs from the probabilistic prediction by the factor $e^{\gamma}\omega(\alpha)$.

Meanwhile, the number of primes in each eligible column is approximately

$$\frac{D}{\varphi(D)}\,\frac{M}{\log MD}, \tag{4.17}$$

---

[4]This shows, in fact, that the probabilistic predictions for the sets $B(\alpha)$ fail as well—except for those coincidental values of α for which $\omega(\alpha) = e^{-\gamma}$. But (perhaps out of pity for the probabilistic model) Buchstab did show that if $y(x)$ tends to infinity with x slowly enough (namely, $\log y = o(\log x)$), then the number of integers up to x whose prime factors all exceed y will indeed be asymptotic to the probabilistic prediction $x\prod_{p \leq y}(1 - 1/p)$.

[5]So huge, in fact, that $\log x$ (and therefore N) will itself be *extremely* large.

[6]As a consequence of the PNT, the logarithm of the product of all the primes up to some point z is asymptotically z in size; here this implies that $\log D \sim \log x$.

where φ is the Euler phi-function, by virtue of the PNT for arithmetic progressions,[7,8,9] and this is the same as

$$= \prod_{p \le \log x} \left(1 - \frac{1}{p}\right)^{-1} \frac{M}{\log MD}. \tag{4.18}$$

In the last step we made use of the elementary formula

$$\varphi(n) = n \prod_{p|n} \left(1 - \frac{1}{p}\right), \quad \text{or} \quad \frac{n}{\varphi(n)} = \prod_{p|n} \left(1 - \frac{1}{p}\right)^{-1};$$

if $n = D$ is the product of all primes up to $\log x$, we obtain the product shown in (4.18). Note that the result (4.18) actually matches the probabilistic prediction, once one takes into account that all integers in each eligible column are automatically free of prime factors up to $\log x$.

Therefore, combining (4.16) and (4.18), the total number of primes in this matrix is

$$\left(e^{\gamma}\omega(\alpha)N \prod_{p \le \log x} \left(1 - \frac{1}{p}\right)\right) \times \left(\prod_{p \le \log x} \left(1 - \frac{1}{p}\right)^{-1} \frac{M}{\log MD}\right) = e^{\gamma}\omega(\alpha)\frac{MN}{\log MD},$$

which we see differs from the predicted number of primes by the factor $e^{\gamma}\omega(\alpha)$. In particular, this means that one of the M rows must contain at least

$$e^{\gamma}\omega(\alpha)N/\log MD$$

primes, and one of the M rows must contain at most $e^{\gamma}\omega(\alpha)N/\log MD$ primes (in contrast to the probabilistic prediction of $N/\log MD$ primes). Since every row is an interval of integers of length $\sim \log^{\alpha} x$, Maier's result follows.[10]

Further details regarding Maier's proof may be found in Granville (1995, [**451**]) and Soundararajan (2007, [**1004**]). See also Maier (1981, [**721**]), (1985, [**722**]), (1988, [**723**]) Maier & Pomerance (1990, [**724**]), Maier & Stuart (2007, [**725**]), and Montgomery & Vaughan (2006, [**774**, Chapter 7]). Using Maier's method, Granville and Soundararajan (2007, [**461**]) showed that many arithmetic sets of integers, not

---

[7] Stated loosely, the PNT for arithmetic progressions says that primes are apportioned equally among the eligible progressions; here $\varphi(n)$ is the Euler function—the number of integers in $[1, n]$ coprime to n.

[8] Of the M numbers in the column, ordinarily about $M/\log MD$ would be prime; but of the D residue classes modulo D, only $\varphi(D)$ contain primes. Hence (4.17) follows.

[9] There is a rather technical point here, namely that without assuming GRH, the PNT for arithmetic progressions in the form used by Maier is not known to hold for every modulus D; there is an obstacle related to "exceptional zeros" of Dirichlet L-functions modulo D. However, such moduli are rare, and Maier showed that this obstacle is absent on a subsequence of values of D tending to infinity. In the definition of M as a "huge power of D", the precise nature of "huge" must also account for this technicality.

[10] An astute reader may have noticed two inconsistencies in this account. First, while every row is indeed an interval of integers, those integers have size around MD rather than x; therefore for Maier's result to be established, the length of the interval needs to be $\log^{\alpha} MD$, which is a huge constant factor larger than $\log x$. Second, we have claimed that there are intervals of length $\log^{\alpha} x$ that contain more primes than expected, and other intervals of the same length that contain fewer primes than expected; yet the argument above produced only a single constant $e^{\gamma}\omega(\alpha)$, which cannot be both greater than and less than 1. Both inconsistencies can be resolved by increasing the number of columns suitably, running the argument through for the still larger matrix, and then using a pigeonhole principle to select an appropriate subset of adjacent columns.

just the primes, exhibit an "uncertainty principle" in that they cannot simultaneously be well distributed in arithmetic progressions and also well distributed in short intervals; discrepancies such as these violate the corresponding Cramér-type probabilistic models for such arithmetic sets.

4.8. Selected proofs

We first provide a lemma that consists of a number of standard inequalities, stated without proofs. Some of these inequalities will be used in the proofs below.

LEMMA 4.2.

(i) *We have* $1+x \le e^x$ *and* $1-x \le e^{-x}$, *for all real* x. *It follows that for any* $r>0$

$$(1+x)^r \le e^{rx}, \quad \text{for } x \ge -1; \quad (1-x)^r \le e^{-rx}, \quad \text{for } x \le 1;$$

and

$$e^{-rx} \le (1+x)^{-r}, \quad \text{for } x \ge -1; \quad e^{rx} \le (1-x)^{-r}, \quad \text{for } x \le 1.$$

(ii) *The Bernoulli inequalities: For* $x \ge -1$ *we have*

$$1+rx \le (1+x)^r, \quad \text{for } r \ge 1; \quad (1+x)^r \le 1+rx, \quad \text{for } 0 \le r \le 1.$$

(iii) *For* $x \le 1$ *and* $r>0$ *we have*

$$(1-x)^r \le 1/(1+ra), \quad \text{where } a = \log 1/(1-x).$$

(iv) *We have* $e^{-x} \le 1-\frac{x}{2}$, *for* $0 \le x \le 1$.

(v) *The sequence* $\left(1+\frac{x}{n}\right)^n \to e^x$ *as* $n \to \infty$, *and for* $x>-1$ *is monotonically increasing.*

Note that replacing x by x/r at (i), (ii), or (iii) yields other valid identities.

AN ALTERNATE PROOF OF (4.5) FOR THE CRAMÉR PRIMES. This conjecture stated that

$$\limsup_{n\to\infty} \frac{p_{n+1}-p_n}{\log^2 p_n} = \text{Const} > 0. \tag{4.19}$$

The following proof for the Cramér primes is based on Heath-Brown (1988, [**530**]). Write $d_n \equiv p_{n+1}-p_n$ for consecutive prime differences, and consider an interval $\mathcal{I}(x) \equiv (x,2x]$. For a prime in $\mathcal{I}(x)$, the probability that d_n exceeds N is about $(1-1/\log x)^N$, and since there are about $x/\log x$ primes in $\mathcal{I}(x)$ the probability that $d_n \le N$ for all of them is about

$$\left\{1-\left(1-\frac{1}{\log x}\right)^N\right\}^{x/\log x}.$$

If $N = C\log^2 x$ with $C>0$, then using $(1-1/\log x)^{\log x} \to e^{-1}$ and part (i) of the lemma stated above, this will be approximately

$$\left(1-x^{-C}\right)^{x/\log x} \le e^{-x^{1-C}/\log x}. \tag{4.20}$$

Now the intervals $\mathcal{I}_2(m) \equiv \left[2^m, 2^{m+1}\right)$, with $m=1,2,\ldots$, partition the integers $n>1$. Summing the terms (4.20) over the values $x=2^m$ amounts to summing the probabilities over the partition; when $0 < C < 1$ this yields

$$\sum_{x=2^m} e^{-x^{1-C}/\log x} \le \sum_m \left(1 + \frac{\left(2^{1-C}\right)^m}{(m \log 2)}\right)^{-1} < \infty.$$

Therefore by the Borel–Cantelli lemma the upper bound $N = C\log^2 x$ is violated in at most a finite number of the intervals. When $C > 1$, $x^{1-C} \to 0$, and part (iii) of the lemma gives

$$\left(1 - x^{-C}\right)^{x/\log x} \le \frac{1}{1 - \frac{x}{\log x}\log(1 - x^{-C})} \sim \frac{1}{1 - \frac{x^{1-C}}{\log x}} \sim 1 - \frac{x^{1-C}}{\log x},$$

so $x^{1-C}/\log x$ approximates the probability that N is exceeded for some d_n in $\mathcal{I}(x)$. But

$$\sum_{x=2^m} \frac{x^{1-C}}{\log x} = \sum_{m=1}^{\infty} \frac{1}{(m \log 2)\left(2^{C-1}\right)^m} < \infty,$$

so the lower bound $N = C\log^2 x$ is violated in at most a finite number of intervals. □

Notes:

(i) The proof shows that under Cramér's model (4.5) or (4.19) holds with Const $= 1$.

(ii) The proof also shows that for the Cramér primes $\max_{p_n \le x}(p_{n+1} - p_n) \sim \log^2 x$.

(iii) Heath-Brown (1988, [**530**]) noted that a more careful argument based on Cramér's heuristic would allow one to strengthen the conjecture for Cramér primes to

$$\max_{p_n \le x}(p_{n+1} - p_n) = (\log x)(\log x + \log\log\log x) + \text{smaller terms}.$$

One consequence of Heath-Brown's observation is that the pseudoprime differences $\widetilde{p}_{n+1} - \widetilde{p}_n$ will then exceed $\log^2 n$ infinitely often.

A THIRD PROOF OF (4.5) FOR THE CRAMÉR PRIMES. Let

$$X_n \equiv \frac{p_{n+1} - p_n}{\log^2 p_n}, \quad \text{and} \quad Y_k \equiv \max_{n \ge k}\{X_n\}.$$

According to Cramér's heuristic, the X_n become exponentially distributed with mean $1/\log p_n$. Consequently we would have

$$\mathrm{P}[X_n \le x] \sim 1 - e^{-x \log p_n} = 1 - p_n^{-x} \sim 1 - (n \log n)^{-x}.$$

Because the X_n will also be independent, we have

$$\mathrm{P}[Y_k \le x] = \prod_{n=k}^{\infty} \mathrm{P}[X_n \le x] \sim \prod_{n=k}^{\infty}\left(1 - (n \log n)^{-x}\right).$$

We now apply the lemma in Appendix E for infinite products: Since

$$\sum_{n=k}^{\infty}(n \log n)^{-x}$$

diverges when $x \le 1$ and converges when $x > 1$, it follows that when $x \le 1$ the $\prod_{n=k}^{\infty}$ product equals 0 for every value of k, while when $x > 1$ the product $\prod_{n=k}^{\infty}$

will converge to 1 as $k \to \infty$. It follows that $Y_k = 1$ with probability 1 for every k and consequently that $\limsup X_n = 1$. Furthermore, since the lim sup is a 'tail event' the Kolmogorov zero-one law guarantees that this limit is almost sure. □

PROOF OF THE EXISTENCE OF CRAMÉR PRIMES IN $[(n-1)^\alpha, n^\alpha]$ FOR $\alpha > 1$. If A_n is the event that there is no prime in $[(n-1)^\alpha, n^\alpha]$ then for the Cramér primes

$$\begin{aligned}
\mathrm{P}[A_n] &= \prod_{(n-1)^\alpha \le m \le n^\alpha} \left(1 - \frac{1}{\log m}\right) \\
&\le \left(1 - \frac{1}{\log n^\alpha}\right)^{n^\alpha - (n-1)^\alpha} \\
&= \left(1 - \frac{1}{\alpha \log n}\right)^{(\alpha \log n)\{[n^\alpha - (n-1)^\alpha]/\alpha \log n\}} \\
&\sim e^{-[n^\alpha - (n-1)^\alpha]/\alpha \log n} \\
&\sim e^{-n^{\alpha - 1}/\log n}.
\end{aligned}$$

Here we have used $(1 - 1/\alpha \log n)^{\alpha \log n} \to e^{-1}$ and $n^\alpha - (n-1)^\alpha \sim \alpha n^{\alpha-1}$. Now $\sum e^{-n^\beta}$ converges for any $\beta > 0$ (because $e^{n^\beta} > n^2$ once $n^\beta > 2 \log n$). Therefore $\sum_n \mathrm{P}[A_n]$ converges for $\alpha > 1$ so by the Borel–Cantelli lemma A_n can occur no more than a finite number of times. The result fails trivially for $\alpha \le 1$. □

REMARK 4.3. The underlying reason that one can go as low as $\alpha > 1$ for Cramér primes is that the number of integers in the intervals $[(n-1)^\alpha, n^\alpha]$ scales as $n^{\alpha-1}$ which is much faster than $\log n$. (For the actual primes the result has been proved only for $\alpha > 40/19$.)

PROOF OF (4.6) FOR THE CRAMÉR PRIMES.
This result states that

$$\limsup_{x \to \infty} \frac{\widetilde{\pi}(x) - \mathrm{li}(x)}{\sqrt{(2x \log \log x)/\log x}} = 1.$$

Now $\widetilde{\pi}(x) = \sum_{2<n\le x} U_n$ where the U_n are independent 0–1 Bernoulli variables based on probabilities $1/\log n$. Recall that the logarithmic sum, defined as $\mathrm{ls}(x) = \sum_{2<n\le x} 1/\log n$, satisfies $\mathrm{li}(x) - \mathrm{ls}(x) = O(1)$ (see (2.19) in Chapter 2). Then consider

$$\widetilde{\pi}(x) - \mathrm{ls}(x) = \sum_{2<n\le x} (U_n - \mathrm{E}(U_n)). \tag{4.21}$$

The variance of this sum is

$$s_x^2 \equiv \sum_{2<n\le x} \frac{1}{\log n}\left(1 - \frac{1}{\log n}\right) \sim \sum_{2<n\le x} \frac{1}{\log n} \sim \frac{x}{\log x}.$$

Now apply Kolmogorov's version of the Law of the Iterated Logarithm (see Appendix D). The boundedness condition on the terms in (4.21) required by this Law of the Iterated Logarithm is satisfied trivially. We therefore get

$$\limsup_{n\to\infty} \frac{\widetilde{\pi}(x) - \mathrm{ls}(x)}{\sqrt{2 s_n^2 \log\log s_n^2}} = 1, \quad \text{a.s.}$$

But

$$\sqrt{2s_n^2 \log\log s_n^2} \sim \sqrt{\frac{2n}{\log n} \log\log \frac{n}{\log n}} \sim \sqrt{(2n \log\log n)/\log n}$$

and hence the result. □

The last result we will prove is that Cramér's model is inconsistent with the result of Maier. Two proofs will be given, each of which uses a nontrivial *large deviation bound*. Many such bounds can be found in the literature. The following ones, which we state without proof, suffice for our needs. They were developed by Herman Chernoff, Harald Cramér, Wassily Hoeffding, and Herman Rubin.

The Chernoff–Cramér and Hoeffding exponential bounds.

Chernoff's Inequality. Let $0<\theta_i<1$, and let $\theta = \frac{1}{n}\sum_{i=1}^n \theta_i$. Let $X_1, X_2, \ldots, X_n$ be independent Bernoulli variables with $\mathrm{P}[X_i = 1] = \theta_i = 1 - \mathrm{P}[X_i = 0]$, and let $\bar{X} = \frac{1}{n}\sum_{i=1}^n X_i$. Then for all $\delta \geq 0$,

$$\mathrm{P}\left[\bar{X} - \theta > \delta\right] \leq \mathrm{P}\left[\bar{X} - \theta > \delta\theta\right] \leq \left(\frac{e^\delta}{(1+\delta)^{1+\delta}}\right)^{n\theta} \leq e^{-2n\delta^2}.$$

Hoeffding's Inequality. Let $X_1, X_2, \ldots, X_n$ be independent random variables which satisfy $0 \leq X_i \leq 1$, and let $\bar{X} = \frac{1}{n}\sum_{i=1}^n X_i$ and $\mu = \mathrm{E}\,\bar{X}$. Then for all $\delta > 0$,

$$\mathrm{P}\left[\bar{X} - \mu \geq \delta\right] \leq \left(\left(\frac{\mu}{\mu+\delta}\right)^{\mu+\delta}\left(\frac{1-\mu}{1-\mu-\delta}\right)^{1-\mu-\delta}\right)^n = e^{-D(\mu+\delta\|\mu)n} \tag{4.22}$$

and

$$\mathrm{P}\left[\bar{X} - \mu \leq -\delta\right] \leq \left(\left(\frac{\mu}{\mu-\delta}\right)^{p-\epsilon}\left(\frac{1-\mu}{1-\mu+\delta}\right)^{1-\mu+\delta}\right)^n = e^{-D(\mu-\delta\|\mu)n}. \tag{4.23}$$

(Note that for $\delta > 1-\mu$ and $\delta < \mu$, respectively, these probabilities will be 0.) Here

$$D(q\|p) \equiv q \log \frac{q}{p} + (1-q)\log\frac{1-q}{1-p} \tag{4.24}$$

is the so-called *Kullback–Leibler divergence* between Bernoulli variables with probability parameters q and p. It satisfies the inequality $D(p+\delta\|p) \geq 2\delta^2$, so that the right-hand sides of (4.22) and (4.23) are each bounded above by $e^{-2n\delta^2}$. It also satisfies

$$D\Big((1+\epsilon)p \parallel p\Big) \geq \epsilon^2 p/4, \quad \frac{1}{2} \leq \epsilon \leq \frac{1}{2} \tag{4.25}$$

which yields a stronger bound when $p < 1/8$.

Proof of assertions (4.7)/(4.8). These assert that with $y = \log^\alpha x$ and $\alpha > 2$, we have (with probability 1)

$$\widetilde{\pi}(x+y) - \widetilde{\pi}(x) = \mathrm{li}(x+y) - \mathrm{li}(x) + O\left(\sqrt{y}\right), \quad \text{as } x \to \infty.$$

One can replace the $\mathrm{li}(x+y) - \mathrm{li}(x)$ in this expression by its asymptotic equivalent $y/\log x$, but we will instead work with the version based on the *logarithmic sum*:

$$\sum_{x<n\leq x+y} U_n = \sum_{x<n\leq x+y} \frac{1}{\log n} + O(\sqrt{y}),$$

where the U_n are 0–1 independent Bernoulli variables with $\mathrm{P}[U_n = 1] = 1/\log n$. In this display the second sum is just the expected value of the first sum. Denote the (approximate) averages corresponding to these two sums by

$$\bar{Z}_x = \frac{1}{y} \sum_{x<n\leq x+y} U_n, \quad \text{and} \quad \mathrm{E}\,\bar{Z}_x = \frac{1}{y} \sum_{x<n\leq x+y} \frac{1}{\log n} \sim \frac{1}{\log x}$$

and consider the events $A_n \equiv \{\, \bar{Z}_n - \mathrm{E}\,\bar{Z}_n \geq C/\sqrt{y(n)}\, \}$. By the Borel–Cantelli lemma it will suffice to show that there exists a positive constant C such that $\sum_{n=1}^{\infty} \mathrm{P}\,[A_n] < \infty$. Keeping in mind that $\bar{Z}_n$ is an average of about $y(n)$ terms (and not n terms), Hoeffding's bound is essentially

$$\mathrm{P}\,[A_n] \leq e^{-D(\mu+\delta\|\mu)y(n)}, \quad \text{where } \mu = \frac{1}{\log x}, \quad \text{and} \quad \delta = \frac{C}{\log^{\alpha/2} x}. \tag{4.26}$$

We note that $\frac{\delta}{\mu} = C \log^{1-\alpha/2} x$, and that $\mu \to 0$ and $\delta \to 0$ as $x \to \infty$. We assume $\alpha > 2$ so that $\frac{\delta}{\mu} \to 0$ as well. Then using the definition in (4.24), the Taylor expansion $\log(1+x) = x - \frac{x^2}{2} + \cdots$, and dropping lower order terms along the way, we get

$$\begin{aligned}
D(\mu+\delta\|\mu) &= (\mu+\delta)\log\left(1+\frac{\delta}{\mu}\right) + (1-\mu-\delta)\log\left(1-\frac{\delta}{1-\mu}\right) \\
&\sim (\mu+\delta)\left(\frac{\delta}{\mu} - \frac{\delta^2}{2\mu^2}\right) - (1-\mu-\delta)\left(\frac{\delta}{1-\mu} + \frac{\delta^2}{2(1-\mu)^2}\right) \\
&= \frac{\delta^2}{2}\,\frac{1}{\mu(1-\mu)} - \frac{\delta^3}{2}\,\frac{1-2\mu}{[\mu(1-\mu)]^2} \\
&\sim \frac{C^2}{2}\log^{1-\alpha} x - \frac{C^3}{2}\log^{2-3\alpha/2} x.
\end{aligned}$$

Therefore, after multiplying by $y(x) = \log^{\alpha} x$ and substituting in (4.26) we find that

$$\mathrm{P}\,[A_n] \leq \exp\left(-\frac{C^2}{2}\log n\right) \times \exp\left(\frac{C^3}{2}\log^{1-\alpha/2} n\right).$$

The result $\sum \mathrm{P}\,[A_n] < \infty$ follows because the first factor on the right, which equals $n^{-C^2/2}$, is summable when $C^2/2 > 1$, while the second factor tends to 0 since $\alpha > 2$. The events $\left\{\, \bar{Z}_n - \mathrm{E}\,\bar{Z}_n \leq C/\sqrt{y(n)}\, \right\}$ are dealt with similarly. □

ALTERNATE PROOF OF ASSERTIONS (4.7) AND (4.8). We will again use the bound (4.22) of Chernoff and Hoeffding, but this time with the inequality

$$D((1+x)p\|p) \geq \frac{1}{4}x^2 p, \quad -\frac{1}{2} \leq x \leq \frac{1}{2}. \tag{4.27}$$

For our case $\mu \sim 1/\log n$, $\delta = C/\sqrt{y(n)}$, while $y(n) = \log^{\alpha} n$ plays the role of "n" in (4.22). The "x" in (4.27) is $x = \delta/\mu = C/\log^{-1+\alpha/2} n$. Consequently "$D$" here will be

$$\begin{aligned}
D(\mu+\delta \,\|\, \mu) &= D\left(\left(1+\frac{\delta}{\mu}\right)\mu \,\|\, \mu\right) \\
&= D\left(\left(1+\frac{C}{\log^{-1+\alpha/2} n}\right)\frac{1}{\log n} \,\|\, \frac{1}{\log n}\right)
\end{aligned}$$

$$
\begin{aligned}
&\geq \frac{1}{4} x^2 \mu \\
&= \frac{1}{4} \left(\frac{C}{\log^{-1+\alpha/2} n} \right)^2 \frac{1}{\log n} \\
&= \frac{C^2}{4 \log^{-1+\alpha} n}.
\end{aligned}
$$

So now, except for a finite number of the early terms, and with notation as before, we have

$$
\begin{aligned}
\mathrm{P}\left[\sum_{x<n\leq x+y} U_n - \sum_{x<n\leq x+y} \frac{1}{\log n} \geq C\sqrt{y(n)} \right] &= \mathrm{P}\left[\bar{Z}_n - \mathrm{E}\, \bar{Z}_n \geq \frac{C}{\sqrt{y(n)}} \right] \\
&\leq \exp\left(- \frac{C^2}{4 \log^{-1+\alpha} n} y(n) \right) \\
&= \exp\left(-\frac{1}{4} C^2 \log n \right) \\
&= n^{-C^2/4}.
\end{aligned}
$$

When summed over n, this converges if $C > 2$. A similar computation may be used for $\mathrm{P}\left[Z - EZ \leq -C\sqrt{y(n)} \right]$. That "$x$" eventually lies in $[-1/2, 1/2]$ is assured when $\alpha > 2$. □

CHAPTER 5

The Zeta Function

The natural development of such studies soon led [mathematicians] to embrace in their research the imaginary values of a variable as well as its real values ... and the vast doctrine of Cauchy greatly illuminated the fruitfulness of this generalization. It thus emerged that between two truths of the real domain, the easiest and the shortest path passes very frequently through the complex domain.

Paul Painlevé [**829**]

As mentioned in the introduction, in the theory of prime numbers an indispensable role is played by the zeta function

$$\zeta(s) \equiv \sum_{n=1}^{\infty} \frac{1}{n^s} = \prod_p \left(1 + \sum_{k=1}^{\infty} \frac{1}{p^{ks}}\right) = \prod_p \left(1 - p^{-s}\right)^{-1}, \quad \Re s > 1, \tag{5.1}$$

which Euler had introduced in 1737, and studied for real values of $s > 1$. Over a century later, in 1859, Riemann described how viewing $\zeta(s)$ as a function of a complex variable s had remarkable consequences. The significance of $\zeta(s)$, as well as its utility, stems from the fact that it links the *discrete sequence* of integers to a *continuous function* determined by the primes. It turns out that studying this function *analytically* opens a window onto a marvellous body of theory.

In this chapter we examine some basic theory for $\zeta(s)$ and indicate a number of its applications. This material is classical, but in its role as *deus ex machina*, it supplies context essential for other parts of this book. This chapter can be read casually if one wishes to avoid its full mathematical underpinnings, but ideally it is best read in conjunction with Appendix A which provides background to the elegant concepts which underlie the discussion, and in particular to the idea of analytic continuation which we will make use of several times. Occasional reference will also be made to Appendices B and C. Appendix A provides an informal exposition of key results from the theory of functions of a complex variable; Appendix B describes a number of such functions, but most notably the Gamma function which turns out to be bound up with the zeta function in deep and unexpected ways; and Appendix C discusses Dirichlet series of which the zeta function is a special although particularly important case. When taken together with these appendices, the material of this chapter is essentially self-contained, although as always, the reader can skim over parts that seem difficult and return to them at a later time.

5.1. Analytic continuation

Initially, the zeta function is defined only for complex numbers whose real part exceeds 1. In particular, since

$$\left|\sum_{n=1}^{\infty}\frac{1}{n^s}\right| \le \sum_{n=1}^{\infty}\frac{1}{|n^s|} = \sum_{n=1}^{\infty}\frac{1}{n^\sigma} < 1+\int_1^\infty \frac{1}{x^\sigma}\,dx = \frac{\sigma}{\sigma-1}, \quad \sigma>1,$$

the first sum in (5.1) converges for any $s=\sigma+it$ provided $\sigma>1$, and furthermore is analytic there—that is, all its complex derivatives exist (see Appendix A).[1] A vital fact about the zeta function, which was established by Riemann (1859, [**906**]), is that $\zeta(s)$ has an *analytic continuation* to the entire complex plane $\mathbb{C}$, with the exception of a simple pole with residue 1 at $s=1$. In other words, near $s=1$ it behaves like $\frac{1}{s-1}$, while $\zeta(s)-\frac{1}{s-1}$ is an *entire function*, that is, a function which is analytic and which has no singularities throughout all of the complex plane.[2] It is this resulting complex function of a complex variable—still denoted by $\zeta(s)$—that is known as the *Riemann zeta function*. In homage to the admittedly peculiar notation used by Riemann, the complex argument s is represented multilingually as $s=\sigma+it$.

The analytic continuation of ζ into the vertical region $0<\sigma\le 1$ of the complex plane is not particularly difficult to establish. For example, consider the identity

$$\zeta(s) = \frac{1}{1-2^{1-s}}\sum_{n=1}^{\infty}\frac{(-1)^{n+1}}{n^s} \tag{5.2}$$

due to Euler and to Dirichlet; it holds because

$$\eta(s) \equiv \sum_{n=1}^{\infty}\frac{(-1)^{n+1}}{n^s} = \sum_{n=1}^{\infty}\frac{1}{n^s} - 2\sum_{n=1}^{\infty}\frac{1}{(2n)^s} = \zeta(s)-\frac{2}{2^s}\zeta(s).$$

What is remarkable about (5.2) is that the sum $\sum(-1)^{n+1}/n^s$ on the right converges, not only for $\Re(s)>1$, but also for $\Re(s)>0$. When s is real this convergence is obvious by the alternating series test; a proof for complex s is given in Section 5.9. It follows that (5.2) provides an analytic continuation for $\zeta(s)$ valid right down to $\sigma>0$ except for the singularity at $s=1$ which is attested to by the term $(1-2^{1-s})^{-1}$ in (5.2).[3,4] The interval $0\le\sigma\le 1$, with the left boundary now included, is known as the *critical strip*, although this terminology is sometimes also used for the open region $0<\sigma<1$.

The fact that $1-2^{1-s}$ in (5.2) is real and negative on the segment $0<\sigma<1$ of the real line, while the alternating sum is clearly real and positive there, establishes that on that segment $\zeta(s)<0$. In fact—and to anticipate a little—if $\zeta(s)$ were to equal 0 for some $s=\sigma+it$ within the vertical strip $0<\sigma<1$ of $\mathbb{C}$, then on account

---

[1] We mention in passing that the Euler product on the right-hand side of (5.1) can be shown to remain valid even for $\sigma=1$, as long as $t\ne 0$.

[2] For the casual reader: *Entire* means that the complex-valued function of a complex variable is defined everywhere in $\mathbb{C}$, and is *very* smooth, so that it can be suitably differentiated an arbitrary number of times.

[3] An exercise: Actually, the function $1-2^{1-s}$ equals 0 for many values of s. Find all of them.

[4] Although we will not make explicit use of this fact, it is worth mentioning here that the function $\eta(s)$ can also be continued analytically—and to an *entire* function—as a consequence of relation (5.2) together with the analytic continuation of $\zeta(s)$ to be described below.

of (5.2), and using

$$n^s \equiv n^{\sigma+it} \equiv n^\sigma e^{it\log n} \equiv n^\sigma\{\cos(t\log n) + i\sin(t\log n)\},$$

it would provide a solution to the pair of simultaneous (real variable) equations

$$\sum_{n=1}^{\infty}\frac{(-1)^n}{n^\sigma}\cos(t\log n) = 0, \quad \text{and} \quad \sum_{n=1}^{\infty}\frac{(-1)^n}{n^\sigma}\sin(t\log n) = 0. \tag{5.3}$$

A prize of one million US dollars is available for a proof that all solutions (σ, t) of (5.3) with $\sigma > 0$ must satisfy $\sigma = \frac{1}{2}$. Details about how to collect this prize may be found in Chapter 8.

An alternate approach to the analytic continuation of ζ into the strip $0 < \sigma \leq 1$ can be based on the straightforward identity

$$\zeta(s) \;=\; \frac{s}{s-1} - s\int_1^\infty \frac{x - \lfloor x \rfloor}{x^{s+1}}\,dx, \tag{5.4}$$

also proved in Section 5.9. Here $\lfloor x \rfloor$ represents the integer part of x, so that the numerator term $0 \leq x - [x] < 1$ is just a 'sawtooth' function. Since the integral on the right is analytic for $\Re(s) > 0$, we again obtain an analytic continuation for ζ down to $\sigma > 0$, except for the (again explicitly visible) simple pole with residue 1 at $s = 1$. (Of course, since analytic continuations are unique, the representations (5.2) and (5.4) must take on identical values.) Equation (5.4) in fact coincides with the first iteration of an Euler–Maclaurin summation (this is shown in Appendix F), and repeated iterations can be used to extend the definition of ζ down the real axis by one unit at a time, thereby demonstrating the possibility of analytically continuing $\zeta(s)$ to the whole complex plane. However, the functional equation approach to the analytic continuation—to which we turn to next—is both more efficient and more elegant.

5.2. The functional equation

Although a little more difficult to establish, the analytic continuation of the function $\zeta(s)$ (defined initially for $\Re s > 1$) to the remainder of the complex plane can also be realized by means of the remarkable functional equation

$$\zeta(s) \;=\; 2(2\pi)^{s-1}\sin\frac{\pi s}{2}\,\Gamma(1-s)\zeta(1-s), \quad \text{for all } s \neq 0, 1, \tag{5.5}$$

due to Riemann (1859, [**906**]). (Here Γ is the Gamma function discussed extensively in Appendix B.) This equation allows one to obtain the values of $\zeta(s)$ for $\Re(s) \leq 1/2$ from its values for $\Re(s) \geq 1/2$, and also evidences a curious connection between the Gamma and zeta functions.[5] Alternate forms of this functional equation include

$$\zeta(1-s) = 2(2\pi)^{-s}\cos\frac{\pi s}{2}\,\Gamma(s)\zeta(s) \quad \text{and} \quad \zeta(s) = \frac{(2\pi)^s \sin\frac{\pi s}{2}\,\zeta(1-s)}{(\sin \pi s)\Gamma(s)}, \tag{5.6}$$

as well as an elegant symmetrical form which can be stated as

$$\Xi(s) \equiv \pi^{-s/2}\Gamma\left(\frac{s}{2}\right)\zeta(s) = \pi^{-(1-s)/2}\Gamma\left(\frac{1-s}{2}\right)\zeta(1-s) \equiv \Xi(1-s), \tag{5.7}$$

[5] Remark: Equation (5.5) allows one to compute $\zeta(s)$ for values of $\Re s < 0$ from its values for $\Re s > 1$. Its values in the strip $0 < \Re s < 1$ may be obtained using (5.2), although its values in $0 < \Re s \leq \frac{1}{2}$ will be related to its values in $\frac{1}{2} \leq \Re s < 1$ also in accordance with (5.5).

or, alternately, as

$$\xi(s) = \xi(1-s), \quad \text{where } \xi(s) = \frac{1}{2}s(s-1)\pi^{-s/2}\Gamma\left(\frac{s}{2}\right)\zeta(s). \tag{5.8}$$

Aside from factors that are well understood, versions (5.7)/(5.8)—which also appeared in Riemann's memoir—evidence a symmetry of the zeta function around the point $s = 1/2$. The function $\xi(s)$ is known as the *completed zeta function.*

Using well-known identities for the Gamma function (which may be found in Appendix B) the equivalence of the forms (5.5)–(5.8) is not difficult to establish and we leave this as a worthwhile exercise for the reader. In particular, (5.5) and the first equality in (5.6) are related via the reflection principle, while the two equalities in (5.6) are connected through the formula $\sin 2\theta = 2\sin\theta\cos\theta$. Finally, the equivalence of (5.5) with (5.7)/(5.8) follows by a combination of the reflection principle and the Legendre duplication formula.

Equations (5.7)/(5.8) say that $\xi(s)$, as well as $\Xi(s)$, are even functions of $s - \frac{1}{2}$. The factor $s(s-1)$ in (5.8) is not of essence; it serves merely to cancel the pole of $\zeta(s)$ at $s = 1$, and that of $\Gamma(s/2)$ at $s = 0$, in the definition of $\xi(s)$. (The function $\Xi(s)$, however, does have those two poles.) The values $\xi(0) = \xi(1) = \frac{1}{2}$ may be defined by requiring continuity. The function $\xi(s)$ is *entire* (namely, analytic on the whole complex plane) and, as we will see in Chapter 6, also has the advantage that its *only* zeros exactly match the so-called 'nontrivial zeros' of $\zeta(s)$. Furthermore, $\xi(s)$ is real-valued[6] on the 'critical line' $\Re s = \frac{1}{2}$, and so its sign changes on that line can help to locate the zeros that lie on it.[7] Finally, let us mention the rendition

$$\zeta(s) = \chi(s)\zeta(1-s) \tag{5.9}$$

of the functional equation, in which (essentially as above)

$$\chi(s) \equiv \pi^{s-1/2}\frac{\Gamma((1-s)/2)}{\Gamma(s/2)} = 2^s\pi^{s-1}\sin\left(\frac{\pi s}{2}\right)\Gamma(1-s) = \frac{(2\pi)^s}{2\,\Gamma(s)\cos(\pi s/2)}\,. \tag{5.10}$$

The function $\chi(s)$ satisfies the identity $\chi(s)\chi(1-s) = 1$, a fact that may be deduced trivially using only (5.9), although it can also be verified directly from its definition.

The analytic continuation of ζ, as well as the functional equation, can be established in numerous ways, and two classical proofs of (5.5) are provided in Section 5.9. Titchmarsh (1986, [**1059**, Chapter II]), for example, presents no fewer than seven different methods (and variations thereof) through which the functional equation can be derived.

The structure and behaviour of $\zeta(s)$ turn out to be of great interest. We begin by noting that $\zeta(s)$ essentially possesses two distinct symmetries, as illustrated in Figure 5.1. First, $\zeta(s)$ is *conjugate symmetric* about the real axis: from its definition in (5.1) we see that $\zeta(\overline{s}) = \overline{\zeta(s)}$, or $\zeta(\sigma - it) = \overline{\zeta(\sigma + it)}$ when $\sigma > 1$, and therefore for all s by a standard argument based on analytic continuation. This indicates, in particular, that $\zeta(s)$ must be real-valued on the *whole* real axis. Second, the functional equations (5.5)–(5.8) show that the 'completed zeta function' $\xi(s)$ is *radially symmetric* about the point $s = \frac{1}{2}$, so that $\xi(s) = \xi(1-s)$, or, equivalently, $\xi(\frac{1}{2} - s) = \xi(\frac{1}{2} + s)$. Figure 5.1 shows an arbitrary point s, together with its

[6] To see this, put $s = \frac{1}{2} + it$ in (5.8) to get $\xi(\frac{1}{2} + it) = \xi(\frac{1}{2} - it)$; being conjugates, these must be real.

[7] The critical line and the nontrivial zeros of $\zeta(s)$ will be discussed in detail in Section 5.4.

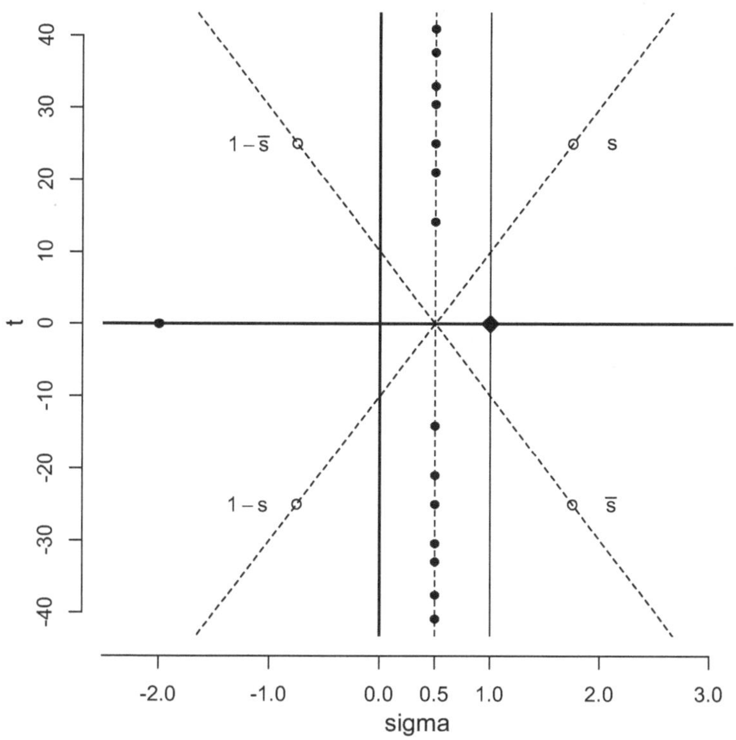

FIGURE 5.1. Symmetries of the zeta function. The dots indicate zeros of ζ, and the diamond indicates its singularity at $s = 1$. The point s reflects about the real axis to $\overline{s}$ and these two points reflect radially through the point $\frac{1}{2}$ (sloped lines) to $1 - s$ and $1 - \overline{s}$, respectively. The evaluations of ζ at the four points shown are related as in equation (5.11). In this figure s lies outside the critical strip, but the same relationships hold if s were to lie inside.

reflections, namely $\overline{s}$, $1 - s$, and $1 - \overline{s}$ (the four small circles). The evaluations of ζ at these respective points satisfy

$$\zeta(s) = \overline{\zeta(\overline{s})} = \chi(s)\zeta(1 - s) = \chi(s)\overline{\zeta(1 - \overline{s})}. \tag{5.11}$$

Note that the evaluation at $1 - \overline{s}$ can be obtained either by conjugate symmetry from the evaluation at $1 - s$, or by radial symmetry from the evaluation at $\overline{s}$. The dots in the figure show zeros in the critical strip, as well as the first of the "trivial zeros" which occurs at $s = -2$.

5.3. The behaviour of $\zeta(s)$ on vertical lines

Now that we know that $\zeta(s)$ is defined for all complex numbers s (except $s = 1$), we are led to inquire about its general behaviour—for instance, where it is growing, or bounded, or oscillatory. It turns out that a great many statements about its behaviour can be elegantly stated when we restrict to vertical lines $\Re(s) = \sigma$ in the complex plane.

By way of example, Figure 5.2 shows graphs of the zeta function along vertical line segments at position $\sigma = -1$, $\sigma = 1/2$, and $\sigma = 2$, with values of t ranging from 0 up to about 30 in each case. The solid and dashed lines correspond to the real and imaginary parts, respectively. At $\sigma = -1$ we have $\zeta(-1) = -1/12$, and the real and imaginary parts then oscillate at rapidly increasing frequencies. At $\sigma = 1/2$ we have $\zeta(1/2) \approx -1.460$; note the four locations (the first one at $t \approx 14.1$) where the real and imaginary parts simultaneously vanish, indicating zeros of $\zeta(s)$. On this "critical line" the real part initially appears to have a strong positive bias, but an asymptotic normality result due to Selberg (see Chapter 7) assures us that it will ultimately not have any such bias of sign. Note that the amplitudes of the real and imaginary parts rise only slowly in these first two graphs. Finally, at $\sigma = 2$ we have $\zeta(2) = \pi^2/6$, and the two functions are bounded with the real part always positive (as we will see momentarily). As will be explained in Chapter 10, the rescaled function $\zeta(2+it)/\zeta(2)$ is the characteristic function of a distribution.

On the real line itself, as σ increases without bound, we obviously have $\zeta(\sigma) \downarrow 1$ in view of its definition, and consequently also $\zeta(2\sigma)/\zeta(\sigma) \to 1$. As a general remark, one can also say that in any half-plane $\Re(s) > 1+\epsilon$, the function $\zeta(s)$ will be relatively well behaved on account of the convergence there of the series (5.1). (Here, as always, $\epsilon > 0$ and is arbitrary.) In any such half-plane $|\zeta(s)|$ will be bounded, and it is not hard to show that it will also be bounded away from zero. More precisely, along vertical lines for fixed values of $\sigma > 1$, one can show that

$$\liminf_{|t|\to\infty} |\zeta(s)| = \frac{\zeta(2\sigma)}{\zeta(\sigma)} < |\zeta(s)| \le \zeta(\sigma) = \limsup_{|t|\to\infty} |\zeta(s)|. \tag{5.12}$$

The supremum value is of course attained on the real axis, but once t is bounded away from 0, the indicated infimum and supremum values will be approached arbitrarily closely infinitely often as $|t| \to \infty$, but will never be attained. One may understand this behaviour from the representation

$$\zeta(\sigma+it) = \sum_{n=1}^{\infty} \frac{1}{n^{\sigma+it}} = \sum_{n=1}^{\infty} \frac{\cos\big((\log n)t\big)}{n^\sigma} - i\sum_{n=1}^{\infty} \frac{\sin\big((\log n)t\big)}{n^\sigma}, \quad \sigma > 1$$

involving infinite linear combinations of sinusoids most of which are not commensurate. For fixed $\sigma > 1$, the $\zeta(\sigma+it)$ are so-called 'almost periodic functions' as discussed at item (2) of Section 5.8 (Complements).

The zeta function is also relatively well behaved when $\Re(s) < -\epsilon$ on account of the functional equation linking its values for $\Re(s) < 0$ to its values for $\Re(s) > 1$. In particular, the functional equation together with Stirling's formula for the Gamma function (see Appendix B) can be used to show that for any fixed $\sigma < 0$,

$$\big|\zeta(\sigma+it)\big| = O\big(|t|^{1/2-\sigma}\big) \quad \text{and} \quad \frac{1}{\big|\zeta(\sigma+it)\big|} = O\big(|t|^{\sigma-1/2}\big), \quad |t| \ge \delta,$$

uniformly (in $|t| \ge \delta$), for any $\delta > 0$, and thus in particular that $\big|\zeta(\sigma+it)\big|$ is bounded away from 0 there.

Recall now that the vertical region $0 \le \sigma \le 1$ of the complex plane is known as the *critical strip*, and that the line $\sigma = \frac{1}{2}$ is called the *critical line*. As one approaches the critical strip from outside, things become complicated. First, within any vertical region $1 < \Re(s) < 1+\epsilon$ immediately to the *right* of the critical strip, regardless of how narrow, the function $\zeta(s)$ can be shown to take on *any* given *nonzero* complex value an infinite number of times, and with arbitrarily large $|t|$.

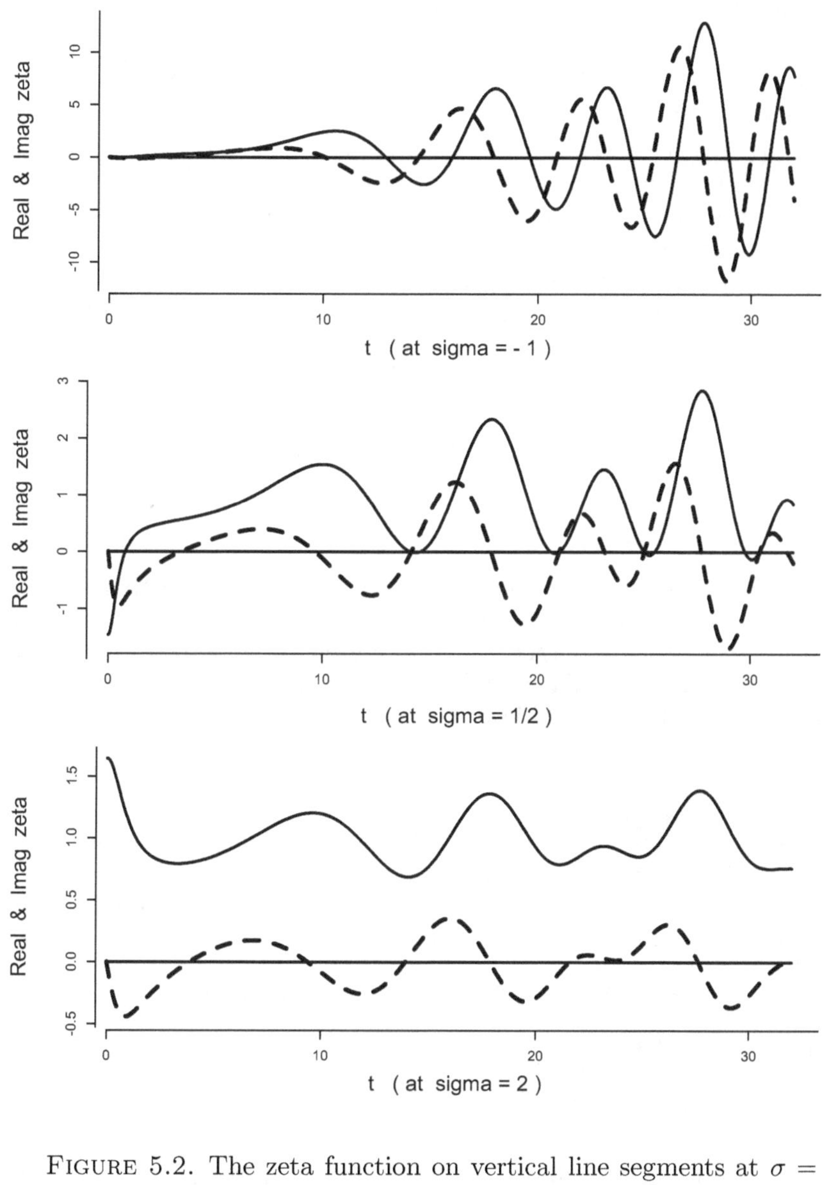

FIGURE 5.2. The zeta function on vertical line segments at $\sigma = -1$, $\sigma = 1/2$, and $\sigma = 2$. The solid and dashed lines correspond to the real and imaginary parts, respectively.

Thus, for example, it also approximates 0 there (but does not touch it) to within any specified accuracy, infinitely often.

Next, consider the behaviour of $\zeta(s)$ on the boundary lines $\sigma = 1$ and $\sigma = 0$. On the line $\sigma = 1$, as $|t| \to \infty$, both $\zeta(1+it)$ and $1/\zeta(1+it)$ are unbounded, and they approach (but never equal) 0 infinitely often. Unconditionally, the Korobov–Vinogradov zero-free region (which we will discuss later) leads to the bounds

$$\zeta(1+it) = O\left((\log t)^{2/3}\right) \quad \text{and} \quad \frac{1}{\zeta(1+it)} = O\left((\log t)^{2/3}(\log\log t)^{1/3}\right), \quad |t| \to \infty.$$

On the other hand, each of these functions is at least of order $O(\log\log t)$, since Littlewood showed that

$$e^{\gamma} \le \limsup_{t\to\infty} \frac{|\zeta(1+it)|}{\log\log t} \le 2e^{\gamma} \quad \text{and} \quad \frac{6}{\pi^2}e^{\gamma} \le \limsup_{t\to\infty} \frac{1/|\zeta(1+it)|}{\log\log t} \le \frac{12}{\pi^2}e^{\gamma};$$

here γ is the Euler–Mascheroni constant (discussed in Appendix B). The two upper bounds here (the right sides) were only established under the Riemann hypothesis (RH), while two lower bounds (the left sides)—of which the second is the harder to prove—were established without any unproven assumptions. Thus if RH were true, at most a factor of 2 would be in doubt in each case, and $O(\log\log t)$ would stand as the exact order of the two lim sup values. While RH by itself does not appear to be enough to determine the exact values of these limits, it is the lower bounds that are generally believed to be the true values. (See, for example, Granville & Soundararajan, (2006, [**459**]).) Explicit estimates, available under the Riemann hypothesis, include

$$\log\left|\zeta(1+it)\right| \le \log\log\log t + \gamma + \log 2 + o(1), \quad |t| \to \infty,$$

and

$$\max_{T\le t\le 2T} \log\left|\zeta(1+it)\right| \ge \log\log\log T + \gamma + o(1), \quad T \to \infty,$$

which together determine an *exact* $\log\log\log|t|$ order for the logarithm $\log|\zeta(1+it)|$ (under RH). Altogether then, one sees that on the line $\sigma = 1$ the largest values of the functions $\zeta(1+it)$ and $1/\zeta(1+it)$ both grow rather slowly. The functional equation and Stirling's formula can be used to obtain matching results for $\sigma = 0$; for example, under RH we would have

$$\zeta(0+it) = O\big(\sqrt{|t|}\log\log|t|\big), \quad \text{and} \quad \frac{1}{|\zeta(0+it)|} = O\left(\frac{\log\log|t|}{\sqrt{|t|}}\right), \quad |t| \to \infty.$$

As we step into the critical strip $0 < \Re(s) < 1$, understanding the behaviour of $\zeta(s)$ becomes far more complicated; within this region, the *true order* of $\zeta(s)$ is a deep problem and few complete answers are known. It is known *unconditionally* that there is an absolute constant $c > 0$ such that, for $|t| \to \infty$,

$$\zeta(\sigma+it) = O\big(|t|^{c(1-\sigma)^{3/2}}\log|t|\big), \quad \frac{1}{2} \le \sigma \le 1.$$

On the other hand, if RH can be assumed, this can be strengthened considerably to

$$\log\zeta(\sigma+it) = O\left(\frac{(\log|t|)^{2-2\sigma}}{\log\log|t|}\right), \quad \frac{1}{2} \le \sigma < 1. \tag{5.13}$$

In the other direction, it is known that for every $\frac{1}{2} < \sigma < 1$, there are positive constants $C_1 \equiv C_1(\sigma)$ and $C_2 \equiv C_2(\sigma)$ such that, as $|t| \to \infty$,

$$\begin{aligned} |\zeta(\sigma+it)| &= \Omega\left(\exp\left(C_1\frac{(\log|t|)^{1-\sigma}}{(\log\log|t|)^{\sigma}}\right)\right), \\ \text{and} \quad \frac{1}{|\zeta(\sigma+it)|} &= \Omega\left(\exp\left(C_2\frac{(\log|t|)^{1-\sigma}}{(\log\log|t|)^{\sigma}}\right)\right). \end{aligned} \tag{5.14}$$

These lower bounds hold unconditionally as well as under RH, the only difference being that the various explicitly known values for the constants C_1 and C_2 are larger under RH. Needless to say, these results are all nontrivial. Montgomery (1977, [**764**]) has conjectured that it is the lower bounds (5.14) rather than the upper ones (5.13) that represent the true orders of magnitude. Further details may be found in Aistleitner (2017, [**10**]), Granville & Soundararajan (2006, [**459**]), Ivić (1985, [**586**, Chapter 9]), Karatsuba & Voronin (1992, [**622**, Chapter VIII, §2, pp. 290–292]), Lamzouri (2008, [**674**]), (2011, [**675**]), Montgomery (1977, [**764**]), Soundararajan (2008, [**1005**]), and Titchmarsh (1986, [**1059**, Chapters VIII, XI, XIII]), as well as in the references which these works make to the contributions of Balasubramanian & Ramachandra, Bondarenko & Seip, and Levinson, among others.

The special case $\sigma = \frac{1}{2}$ is of very particular interest and merits a chapter of its own; it will be the subject of Chapter 7. However, before closing this section, we cannot refrain from mentioning a very unusual property related to traversing the values of the zeta function along vertical regions in the critical strip—a property known as *universality*.

Suppose that $\mathcal{D}$ is a closed disk contained in the right half, $\frac{1}{2} < \sigma < 1$, of the critical strip. If $g(s)$ is *any* nonvanishing function which is continuous on $\mathcal{D}$ and analytic in its interior, then $g(s)$ can be approximated—to within *any* desired precision—by the corresponding values of $\zeta(s)$ in some region $\mathcal{D} + it$ obtained by merely sliding $\mathcal{D}$ vertically far enough up the complex plane. (For a more precise statement, see item (5) of the Complements section below.) This surprising phenomenon, referred to as the universality property of the zeta function, was established by Voronin (1975, [**1092**]). In particular, it implies that for a fixed $\sigma \in (\frac{1}{2}, 1)$, the quantities $|\zeta(\sigma + it)|$ and $1/|\zeta(\sigma + it)|$ are both unbounded as $|t| \to \infty$. A precursor to Voronin's result was proved by Harald Bohr (brother of the physics Nobel laureate Niels Bohr) who showed that for any fixed $\frac{1}{2} < \sigma < 1$, the values of $\zeta(\sigma + it)$ as t ranges over all real numbers are dense over the whole complex plane, while Bohr and Jessen established that those values converged in distribution to a definite probability law corresponding to a definite probability density function. See, for example, Bagchi (1981, [**58**]), Titchmarsh (1986, Theorem 8.12 & §8.13), and Steuding (2007, [**1027**, Chapter 1, §3]).

On the critical line itself, a kind of opposite of this universality appears to be in effect. For while the equation $\zeta(\frac{1}{2} + it) = a$ is known to have an infinite number of solutions for t when $a = 0$, for other complex values of a, Selberg (1992, [**979**]) conjectured that this equation can have no more than a finite number of solutions. Later, Banks & Kang (2012, [**75**]) established that there are an infinite number of values of a for which the equation has only two solutions, and they conjectured that the equation can have at most two solutions for every nonzero value of a.

5.4. Zeros and the Riemann hypothesis

Of course, no discussion of the zeta function is complete without mention of the *Riemann hypothesis* (RH) which asserts that *all* of the *nontrivial zeros* of $\zeta(s)$, namely, the *complex-valued zeros* which lie in the *critical strip* $0 < \sigma < 1$, lie on the *critical line* $\sigma = \frac{1}{2}$. The background to this unsolved and much revered problem is as follows.

First, thanks to (5.12), we have already established that $\zeta(s)$ has no zeros when $\sigma > 1$. Using the functional equation (5.5) we find that $\zeta(s)$ equals 0 at the negative even integers. This happens because the sine function there has simple zeros at $s = 0, \pm 2, \pm 4, \ldots$ while the poles of the Gamma function cancel these zeros at the nonnegative even integers. (See the discussion of the $\Gamma(s)$ function in

Appendix B; it has no zeros, and has simple poles at $s = 0, 1, 2, \ldots$.) The zeros of $\zeta(s)$ at $s = -2, -4, -6, \ldots$ are referred to as its *trivial zeros*. It is a worthwhile exercise to match up the zeros and the poles of all of the components of (5.5), and in particular to verify that these trivial zeros are the only zeros of $\zeta(s)$ when $\sigma < 0$. That there are no zeros on the vertical lines $\sigma = 0$ and $\sigma = 1$ is a much deeper result which led to the first proofs of the Prime Number Theorem, and which is in fact equivalent to it; we will return to this point in Chapter 6.

In sum, all of the remaining zeros of $\zeta(s)$, namely the *nontrivial zeros*, are complex-valued[8] and must lie inside the *critical strip,* in other words, in the vertical region $0 < \sigma < 1$. Note, from its definition in (5.8), that the related function $\xi(s)$—which is real-valued on the $t = 0$ axis *as well as* on the $\sigma = \frac{1}{2}$ critical line—does not have any of the so-called trivial zeros but otherwise has zeros which exactly match those of $\zeta(s)$, both in location and in their multiplicity.

Now, on account of the two symmetries of $\zeta(s)$, as well as the third symmetry they induce (refer back to Figure 5.1), the *complex zeros*, that is, the zeros of $\zeta(s)$ inside the critical strip, $0 < \sigma < 1$ must lie symmetrically about the real axis $t = 0$, as well as symmetrically about the vertical line $\sigma = \frac{1}{2}$. It turns out that the exact locations of these zeros is of paramount importance in the study of the distributional properties of the prime numbers, with the locations of the zeros being the key to fully understanding the behaviour of the prime counting function $\pi(x)$. In particular, there are precise connections between zero-free regions of $\zeta(s)$ and the order of the error term in the Prime Number Theorem (as we had discussed in Chapter 2). These connections are described, for example, in Pintz (1984, [**847**]).

The *Riemann hypothesis* (RH) asserts that all of the nontrivial zeros of $\zeta(s)$ lie on the line $\sigma = \frac{1}{2}$. It was surmised by Riemann (1859, [**906**]) as being very likely ("*... es ist sehr wahrscheinlich, dass alle Wurzeln reell sind*") and is widely regarded as one of the most important unsolved problems in mathematics, one with far-reaching consequences throughout all of number theory and in some other areas as well. Although the question of whether or not RH is true remains unresolved, its truth is very often assumed when discussing the statistical properties of the zeros. Further details concerning these complex zeros will be provided in Chapter 6.

5.5. Special values of $\zeta(s)$

While the zeros of the Riemann zeta function are by far its most important values, it has quite a few other interesting special values which we now survey. Specifically, we will look at the values that $\zeta(\sigma)$ takes when σ is an integer. These values are certainly all real numbers, since from the functional equation (5.5), one sees that (except for its singularity at $s = 1$) the function $\zeta(s)$ is real-valued everywhere on the real axis.

On the interval $\sigma > 1$ of the real axis $\zeta(s)$ is real, positive, and decreasing. This is immediate from its infinite series definition (5.1), but it also follows from its product form since a convergent infinite product does not vanish (see Lemma E.1 in Appendix E). We have already seen that $\zeta(s)$ is strictly negative on the segment $0 < \sigma < 1$, but in fact it remains strictly negative on the wider segment $-2 < \sigma < 1$. Indeed, since $\zeta(s)$ has only simple zeros (and a simple pole) on the real axis, it must therefore change sign at every such point, and remain of constant sign between any

[8] We saw that $\zeta(s) \neq 0$ on the segment $0 < s < 1$, after equation (5.2) when we discussed the initial analytic continuation of $\zeta(s)$.

two such points. Our knowledge that the trivial zeros $-2, -4, -6, \ldots$ are all such zeros, allows us conclude that $\zeta(\sigma)$ is strictly positive for $-4 < \sigma < -2$, strictly negative for $-6 < \sigma < -4$, and so on.

Some explicitly known values of ζ include

$$\cdots \zeta(-5) = -\frac{1}{252}, \quad \zeta(-3) = \frac{1}{120}, \quad \zeta(-1) = -\frac{1}{12},$$

$$\zeta(0) = -\frac{1}{2}, \quad \zeta(2) = \frac{\pi^2}{6}, \quad \zeta(4) = \frac{\pi^4}{90}, \quad \zeta(6) = \frac{\pi^6}{945}, \quad \zeta(8) = \frac{\pi^8}{9450}, \quad \ldots.$$

The result that $\zeta(2) \equiv \sum 1/n^2$ equals $\pi^2/6$, once a long-standing question famously known throughout Europe as the Basel problem, was Euler's first big result and the one that secured his reputation.[9] The result for $\zeta(-1)$ can then be derived using $\sqrt{\pi}\,\Gamma(-1/2)\,\zeta(-1) = \pi^{-1}\zeta(2)$ obtained from the functional equation, together with the identity $-(1/2)\Gamma(-1/2) = \Gamma(1/2) = \sqrt{\pi}$. The result for $\zeta(0)$ is obtained from (5.5) using $\zeta(s) \sim 1/(s-1)$ near $s = 1$. For the sake of its interest let us note in passing that (5.2) may be used to evaluate

$$\zeta(1/2) = (1+\sqrt{2}) \sum_{n=1}^{\infty} (-1)^n/\sqrt{n} \approx -1.4603545$$

The remaining explicit values among those shown above can be explained via their connection to the *Bernoulli numbers*. These were introduced by Johann Bernoulli's older brother Jacob in the late 1600s and turn out to have a great many applications.[10] For integers $k \geq 0$, the B_k are defined via the generating relation

$$\frac{z}{e^z - 1} = \sum_{k=0}^{\infty} B_k \frac{z^k}{k!} = 1 - \frac{z}{2} + \sum_{k=1}^{\infty} B_{2k} \frac{z^{2k}}{(2k)!}, \quad |z| < 2\pi \tag{5.15}$$

where the radius of convergence 2π for the sums here originates from the fact that the ratio on the left becomes infinite as $z \to 2\pi i$. (The second of the equalities here will become transparent in a moment.) The Bernoulli numbers are rational, satisfy many identities, and occur in many classical Taylor series expansions. Their first few values are given by

$$B_0 = 1, B_1 = -\frac{1}{2}, B_2 = \frac{1}{6}, B_4 = -\frac{1}{30}, B_6 = \frac{1}{42},$$
$$B_8 = -\frac{1}{30}, B_{10} = \frac{5}{66}, B_{12} = -\frac{691}{2730};$$

other than B_1, the odd ones, namely $B_3, B_5, B_7, \ldots$, are all 0 (whence the second equality in (5.15)), while the even ones alternate in sign. For $n > 0$ the B_n can be generated by any of the equivalent recurrence relations

$$B_n = \sum_{k=0}^{n} \binom{n}{k} B_{n-k}, \quad \sum_{k=0}^{n} \binom{n+1}{k} B_k = 0, \quad \text{or } B_n = -\frac{1}{n+1} \sum_{k=0}^{n-1} \binom{n+1}{k} B_k.$$

---

[9]This problem had defeated Gottfried Leibniz as well as other great mathematicians of the time. Historical and other details concerning Euler's $\zeta(2)$ result may be found in Ayoub (1974, [**47**]), Lagarias (2013, [**668**]), and Sandifer (2007, [**938**]), as well as in Wikipedia's 'Basel Problem' entry.

[10]The three distinguished mathematicians of the Bernouilli clan are often a source of confusion: Jacob was the eldest, and Johann a younger brother, while Daniel was Johann's son.

The B_n possess many eccentricities. For example: there is a theorem of von Staudt which implies that there is no square in the denominator of any B_n (Hardy & Wright, (2008, [**499**, Chapter VII])).

It turns out that the even Bernoulli numbers are connected to the zeta function at its positive even and at its negative odd integer arguments via the relations

$$\begin{aligned} \zeta(2k) &= \sum_{n=1}^{\infty} \frac{1}{n^{2k}} = \frac{(-1)^{k+1} 2^{2k-1} B_{2k}}{(2k)!} \pi^{2k} \\ \text{and } \zeta(1-2k) &= \frac{-B_{2k}}{2k}, \qquad k = 1, 2, 3, \ldots; \end{aligned} \tag{5.16}$$

these relations can be inverted as

$$B_{2k} = \frac{(-1)^{k+1}(2k)!}{2^{2k-1}\pi^{2k}} \zeta(2k), \quad \text{and } B_{2k} = -2k\zeta(1-2k), \quad k = 1, 2, 3, \ldots. \tag{5.17}$$

Remarkably, $\zeta(s)$ takes on rational values at the *negative odd integers*, while at the *positive even integers* $2k$ it takes on rational multiples of π^{2k}. (Some of the assertions made here are proved in Appendix F.) At the *negative even integers* we have seen that $\zeta(s) = 0$. Of course, the results in (5.16) for $\zeta(1-2k)$ and $\zeta(2k)$ are linked via the functional equation. Also, because $\zeta(\sigma) \to 1$ for real $\sigma \to \infty$, we can show (using Stirling's formula) that the even Bernoulli numbers satisfy the asymptotic

$$B_{2k} \sim (-1)^{k+1} \frac{2(2k)!}{(2\pi)^{2k}} \sim (-1)^{k+1} 4\sqrt{\pi k} \left(\frac{k}{\pi e}\right)^{2k}, \quad \text{as } k \to \infty.$$

As for *positive odd integer* arguments, virtually nothing is known by way of exact results for $\zeta(s)$ other than the curious relation

$$\sum_{n=1}^{\infty} \frac{\zeta(2n+1)}{(2n+1)2^{2n}} = -\gamma + \log 2$$

(where γ is the Euler–Mascheroni constant), as well as the unusual identity

$$\zeta(3) \equiv \sum_{n=1}^{\infty} \frac{1}{n^3} = \frac{5}{2} \sum_{n=1}^{\infty} \frac{(-1)^{n+1}}{\binom{2n}{n} n^3} \approx 1.2020569$$

which is due to Apéry (1979, [**23**]), and from which he had established that $\zeta(3)$ is irrational. (It is a long-standing conjecture—supported by work of Beukers, Rivoal, Zudilin, and others—that the values of ζ at all positive odd integers are irrational.) Baker (2012, [**67**, §13.7]) provides an outline of Apéry's proof; see also van der Poorten (1978, [**864**]). Euler found the rapidly converging formula

$$\zeta(3) = \frac{\pi^2}{7} \left(1 - 4\sum_{n=1}^{\infty} \frac{\zeta(2n)}{(2n+1)(2n+2)2^{2n}}\right) = -\frac{4\pi^2}{7} \sum_{n=0}^{\infty} \frac{\zeta(2n)}{(2n+1)(2n+2)2^{2n}},$$

with the last equality following because $\zeta(0) = -1/2$. It is also known that[11]

$$\zeta(3) = \frac{5}{4} - \sum_{n=2}^{\infty} \frac{1}{n^3(n^2-1)}, \quad \text{and } \zeta(3) = \frac{1}{4} + \sum_{n=1}^{\infty} \frac{3n+2}{n^3(n+1)(n+2)}.$$

[11] Optional exercise: Show that the two representation for $\zeta(3)$ shown here are identical.

Information about the identities

$$\begin{aligned}\zeta(3) &= \frac{2}{7}\pi^2\log 2 - \frac{8}{7}\int_0^{\pi/2}\theta^2\cot(\theta)\,d\theta\\ &= \frac{2}{7}\pi^2\log 2 + \frac{16}{7}\int_0^{\pi/2}\theta\log(\sin(\theta))\,d\theta\\ &= \frac{2\pi^2}{9}\left(\log 2 + 2\sum_{n=0}^{\infty}\frac{\zeta(2n)}{(2n+3)2^{2n}}\right),\end{aligned}$$

and others like them—including their sources—may be found in Srivastava & Choi (2001, [**1017**]); see also Dabrowski (1996, [**256**]). For an engaging elementary treatment of "the $\zeta(3)$ problem" see Nahin (2021, [**789**]), where the surprising relation

$$\zeta(2) = \int_0^1\int_0^1 \frac{dx\,dy}{1-xy} = \frac{\pi^2}{6}$$

involving integration over the unit square is also discussed.[12] In passing, we mention that there are expressions for ζ at the even integers that mirror the result of Apéry, the first two of which are

$$\zeta(2) = 3\sum_{n=1}^{\infty}\frac{1}{\binom{2n}{n}n^2}, \quad \text{and} \quad \zeta(4) = \frac{36}{17}\sum_{n=1}^{\infty}\frac{1}{\binom{2n}{n}n^4}.$$

Other such relations can be found in Guy (2004, [**474**, F.16]).

For use in Appendices E and F we define here the *Bernoulli polynomials* $B_n(x)$, defined on $0 \le x \le 1$, for $n \ge 0$. They are obtained from their generating relation

$$\frac{te^{tx}}{e^t - 1} = \sum_{n=0}^{\infty}\frac{t^n}{n!}B_n(x)$$

(compare with (5.15)) and may be produced recursively using the identity

$$B_n(x) = \sum_{k=0}^{n}\binom{n}{k}B_{n-k}x^k.$$

The first few are given by $B_0(x) = 1$, $B_1(x) = x - \frac{1}{2}$, $B_2(x) = x^2 - x + \frac{1}{6}$, and $B_3(x) = x^3 - \frac{3}{2}x^2 + \frac{1}{2}x$. They may alternately be obtained by starting with $B_0(x) \equiv 1$, and then using the recursion formula $B_k'(x) = kB_{k-1}(x)$ together with $\int_0^1 B_k(x)\,dx = 0$ for $k \ge 1$. The Bernoulli numbers are special values of the Bernoulli polynomials, namely $B_n = B_n(0)$.

The Bernoulli numbers have a long history and were in fact discovered independently by the renowned but secretive Japanese mathematician Seki 'Kowa' Takakazu (1642–1708), known to his students as *Sansu* (the 'Arithmetical Sage'). These numbers satisfy many identities and have many applications in combinatorics and other areas. Although the links between Bernoulli numbers and values of $\zeta(s)$ hint at connections between the B_n and the primes, explicit results in that direction are distinctly limited. Proofs, and further details about the B_n may be found in Montgomery & Vaughan (2006, [**774**, p. 331 & Appendix B]), Vandiver (1961, [**1080**]), Rademacher (1964, [**874**, Chapters 1 & 2]), H. Cohen (2007, [**211**, vol. II, Chapter 9]), as well as in numerous other texts on number

[12] The $\int_0^1\int_0^1$ integral may be evaluated using $1/(1-u) = 1 + u + u^2 + u^3 + \cdots$ for $|u| < 1$.

theory and combinatorics. Dilcher, Skula & Slavutskii (1991, [**289**]), as well as a website maintained by Dilcher & Slavutskii (2007, [**289**]), provide *very* extensive bibliographies.

Also worth mentioning are the curious relations

$$\sum_{k=2}^{\infty}(\zeta(k)-1)=1,\ \sum_{k=2}^{\infty}(-1)^k(\zeta(k)-1)=\frac{1}{2},$$
$$\sum_{k=1}^{\infty}(\zeta(2k)-1)=\frac{3}{4},\ \sum_{k=1}^{\infty}(\zeta(2k+1)-1)=\frac{1}{4},$$

the last two of which are just sums over the even and the odd integers, and obtained by taking the sum and difference of the first two. Related results include the interrelated identities[13]

$$\sum_{k=2}^{\infty}\frac{1}{k}(\zeta(k)-1)=1-\gamma,\quad \sum_{k=2}^{\infty}\frac{(-1)^k}{k}(\zeta(k)-1)=\log 2+\gamma-1,$$
$$\sum_{k=1}^{\infty}\frac{1}{k}(\zeta(2k)-1)=\log 2,\quad \sum_{k=1}^{\infty}\frac{1}{2k+1}(\zeta(2k+1)-1)=1-\gamma-\log\sqrt{2},$$
$$\sum_{k=2}^{\infty}\frac{(-1)^k}{k}\zeta(k)=\gamma,\quad\text{and}\quad \sum_{k=2}^{\infty}\frac{(-1)^k}{k+1}\zeta(k)=1+\frac{\gamma}{2}-\frac{1}{2}\log 2\pi.$$

Numerous such relations date back to Goldbach; their proofs typically involve substituting $\zeta(k)=\sum 1/n^k$, arranging these sums by rows, and summing the resulting square array by columns. See, for example, Adamchik & Srivastava (1998, [**3**]), and Srivastava & Choi (2001, [**1017**]). Two relations of similar appearance are (Keiper (1992, [**630**]))

$$\sum_{k=2}^{\infty}\frac{\zeta(k)-1}{2^k}=\log 2-\frac{1}{2},\quad\text{and}\quad \sum_{k=2}^{\infty}\frac{\zeta(k)-1}{k2^k}=\frac{1-\gamma}{2}+\log\frac{\sqrt{\pi}}{2}.$$

Finally, we mention the recurrence relation

$$\zeta(2n)=\frac{2}{2n+1}\sum_{k=1}^{n-1}\zeta(2k)\zeta(2n-2k),\quad n\geq 2,$$

which mirrors the known identity

$$B_{2n}=-\frac{1}{2n+1}\sum_{k=1}^{n-1}\binom{2n}{2k}B_{2k}B_{2n-2k},\quad n\geq 2.$$

These are related through (5.16)/(5.17), from which one can also get identities for $\zeta(1-2k)$.

In addition to the values of $\zeta(s)$ noted above, various further explicit evaluations are known involving the behaviour of $\zeta(s)$ near $s=1$. These include

$$\lim_{s\to 1}(s-1)\zeta(s)=1,\quad \lim_{s\to 1}\left(\zeta(s)-\frac{1}{s-1}\right)=\gamma,\quad \lim_{s\to 1}(s-1)^2\zeta'(s)=-1,$$
$$\lim_{s\to 1}\left(\frac{\zeta'(s)}{\zeta(s)}+\frac{1}{s-1}\right)=\gamma,\quad \lim_{s\to 1}(\zeta(s)+(s-1)\zeta'(s))=\gamma,$$

[13] Note that in these relations one can swap between sums $\sum_{k=2}^{\infty}$ and $\sum_{k=1}^{\infty}$ provided one uses the principal value result $\zeta(1)=\gamma$ shown in (5.18).

as well as a *principal value* at $s = 1$:

$$\lim_{\epsilon\to 0} \frac{\zeta(1+\epsilon)+\zeta(1-\epsilon)}{2} = \gamma . \tag{5.18}$$

Derivatives at 0 and at the negative even integers are given by

$$\zeta'(0) = -\frac{1}{2}\log 2\pi, \quad \text{and } \zeta'(-2n) = (-1)^n \frac{(2n)!}{2(2\pi)^{2n}}\zeta(2n+1) \text{ for } n \geq 1.$$

For higher derivatives of $\zeta(s)$ at $s = 0$, see Apostol (1985, [**26**]).

An explicit *Laurent expansion* for $\zeta(s)$ about $s = 1$ also happens to be available. It was established by Stieltjes in a letter to Hermite in 1885, is valid for all s, and is given by

$$\zeta(s) = \frac{1}{s-1} + \sum_{k=0}^{\infty} \frac{(-1)^k}{k!}\gamma_k\,(s-1)^k \tag{5.19}$$

where the coefficients γ_k are the so-called *Stieltjes constants*,

$$\gamma_k \equiv \lim_{x\to\infty}\left(\sum_{1\leq m\leq x} \frac{\log^k m}{m} - \frac{\log^{k+1} x}{k+1}\right), \quad k = 0,1,2,\ldots. \tag{5.20}$$

(Here $\log^0 m$ means 1 for all m, including for $m = 1$.) These constants are essentially just the derivatives of $\zeta(s) - (s-1)^{-1}$ at $s = 1$:

$$\gamma_k = \lim_{s\to 1}\left((-1)^k\zeta^{(k)}(s) - \frac{k!}{(s-1)^{k+1}}\right).$$

The initial Stieltjes constant $\gamma_0 \equiv \gamma \approx 0.5772$ is just the Euler–Mascheroni constant, while the next few are $\gamma_1 \approx -0.0728$, $\gamma_2 \approx -0.00969$ and $\gamma_3 \approx 0.00205$. However, these early values are not representative of what ultimately happens to the coefficients: as n increases, the γ_n eventually take on enormous values with highly erratic signs, so that the numerical behaviour of the expansion (5.19) is hopelessly unstable. The Stieltjes constants are known to satisfy the bounds $|\gamma_k| \leq c_k(k-1)!/\pi^k$ where c_k equals 2 or 4 according as k is odd or even. An asymptotic formula for the γ_k may be found in Knessel & Coffey (2011, [**638**]). Surprisingly perhaps, the result (5.19)/(5.20) is not unduly difficult to establish and a proof is given in Section 5.9. Note that owing to some historical ambiguities the expansion (5.19)/(5.20) is sometimes written instead as

$$\zeta(s) = \frac{1}{s-1} + \widetilde{\gamma}_0 + \widetilde{\gamma}_1(s-1) + \widetilde{\gamma}_2(s-1)^2 + \cdots$$

with the $\widetilde{\gamma}_k$ given by

$$\widetilde{\gamma}_k \equiv \frac{(-1)^k}{k!}\lim_{x\to\infty}\left(\sum_{m\leq x} \frac{\log^k m}{m} - \frac{\log^{k+1} x}{k+1}\right), \quad k = 0,1,2,\ldots.$$

Further details concerning the Laurent expansion for $\zeta(s)$ may be found in Bohman & Froberg (1988, [**117**]), Briggs (1955, [**152**]), Briggs & Chowla (1955, [**153**]), Choudhury (1995, [**202**]), Keiper (1992, [**630**]), and Lagarias (2013, [**668**, §3.2 and Theorem 3.2.1]).

5.6. Connecting $\zeta(s)$ to the primes

The function $\zeta(s)$ encodes a great deal of information, and in particular *all* of the information about the primes. It is, for example, easy to show that the zeta function is connected to the prime counting function $\pi(x)$ according to the transformation

$$\log \zeta(s) = s \int_1^\infty \frac{\pi(x)}{x(x^s - 1)}\, dx, \quad \Re(s) > 1. \tag{5.21}$$

A proof of this relation is given in Section 5.9. A formula for $\pi(x)$ can, in principle, be obtained by 'inverting' this integral equation. Titchmarsh (1986, [**1059**, §3.7]), for example, carries out such an inversion. However, neither this nor any other way of deriving such a formula for $\pi(x)$ is at all straightforward.

In one of its most important applications, the zeta function gives rise to a remarkable and *explicit* formula for $\pi(x)$. Conjectured by Riemann in his 1859 memoir, and proved by von Mangoldt in 1895, it connects the primes, *in an exact count*, to a certain sum over the zeros of the ζ-function. The precise formula is somewhat involved (it is exhibited at item (13) in the complements near the end of this chapter) but a simpler, and equivalent form of it (also due to von Mangoldt) can be stated as

$$\psi^*(x) \;=\; x - \sum_\rho \frac{x^\rho}{\rho} - \frac{\zeta'(0)}{\zeta(0)} - \frac{1}{2}\log\left(1 - \frac{1}{x^2}\right). \tag{5.22}$$

Here $\psi^*(x)$ is the symmetrized version of the Chebychev function $\psi(x)$:

$$\psi^*(x) \equiv \frac{1}{2}\{\psi(x) + \psi(x-0)\}, \quad \text{where } \psi(x) \equiv \sum_{n \le x} \Lambda(n) \equiv \sum_{p^m \le x} \log p.$$

When $x = p^m$ is a prime power, $\psi^*(x)$ will include only half of the last term of the sum. The sum in (5.22) ranges over *all* of the nontrivial zeros ρ of ζ (with multiplicity), including those above as well as below the real axis, and must be taken in conjugate pairs ρ and $\overline{\rho}$:

$$\sum_\rho \frac{x^\rho}{\rho} \;\equiv\; \lim_{T\to\infty} \sum_{|\Im(\rho)| \le T} \frac{x^\rho}{\rho}; \tag{5.23}$$

the paired sum is convergent but is not absolutely convergent. The log-term in (5.22) originates from other "x^ρ/ρ" terms when taken over the trivial zeros $(-2, -4, -6, \ldots)$, as may be appreciated from its Taylor expansion

$$\frac{1}{2}\log\left(1 - \frac{1}{x^2}\right) = \sum_{k=1}^\infty \frac{x^{-2k}}{-2k},$$

while the term x in (5.22) originates from the pole of $\zeta(s)$ at $s = 1$. The term $\zeta'(0)/\zeta(0)$ can be replaced by its actual value, namely $(-1/2 \log 2\pi)/(-1/2) = \log 2\pi$. The result (5.22) is nontrivial. A proof can be found, for example, in Davenport (2000, [**262**, Chapter 17]).

When considered as a sum over conjugate pairs, (5.23) is a superposition of sinusoids of decreasing amplitude, linking the zeros of ζ to the primes in a kind of harmonic analysis. This has led to its being called the "music of the primes" (Berry & Keating (1992, [**93**]); du Sautoy (2003, [**947**]); Berry (2012, [**92**])). Specifically, with $\rho = \beta + i\gamma$, and keeping in mind that $|\rho| = \sqrt{\beta^2+\gamma^2}$ and $x^{\pm i\gamma} = e^{\pm i\gamma \log x} = \cos(\gamma \log x) \pm i \sin(\gamma \log x)$, we have

$$\begin{aligned}\frac{x^\rho}{\rho} + \frac{x^{\overline{\rho}}}{\overline{\rho}} &= x^\beta \left(\frac{x^{i\gamma}}{\beta + i\gamma} + \frac{x^{-i\gamma}}{\beta - i\gamma} \right) \\ &= \frac{2x^\beta}{|\rho|} \left(\frac{\beta}{|\rho|} \cos(\gamma \log x) + \frac{\gamma}{|\rho|} \sin(\gamma \log x) \right) \\ &= \frac{2x^\beta}{|\rho|} \cos(\gamma \log x - \theta),\end{aligned}$$

where $\theta = \arctan(\gamma/\beta)$. This is sinusoidal in $\log x$ with period $2\pi/\gamma$, phase shift θ, and amplitude envelope $2x^\beta/|\rho|$. Terms corresponding to higher zeros oscillate more rapidly and have smaller amplitude envelopes. Under the Riemannn hypothesis all values of β would equal $\frac{1}{2}$, so that the x-variation in the amplitudes would be the same for all of the 'vibrational' frequencies.

As a general rule, the Chebyshev function $\psi(x)$ is analytically easier to deal with than the prime counting function $\pi(x)$, and results about one of them can easily be converted into results about the other by elementary methods such as partial summation. The function $\psi(x)$ satisfies a transformation analogous to (5.21), namely

$$-\frac{\zeta'(s)}{\zeta(s)} = s \int_1^\infty \frac{\psi(x)}{x^{s+1}}\, dx, \quad \Re s > 1, \tag{5.24}$$

which like (5.21) also attests to the connection between $\zeta(s)$ and the primes. A proof of (5.24) also appears in Section 5.9. The inverse to this transformation can also be regarded as the one given by (5.22).

Estimating the sum over ρ in (5.22) is one of the central problems of the theory. If the Riemann hypothesis were true, all of the nontrivial zeros ρ would satisfy $\Re\rho = 1/2$ so that all $|x^\rho| = \sqrt{x}$. It would then follow from (5.22) that

$$\psi(x) = x + O\left(\sqrt{x}\log^2 x\right),$$

which in turn can be shown to imply that

$$\pi(x) - \mathrm{li}(x) = O\left(\sqrt{x}\log x\right), \quad \text{or } \pi(x) = \int_0^x \frac{dt}{\log t} + O\left(\sqrt{x}\log x\right).$$

Indeed, it would also follow that the difference between $\mathrm{li}(x)$ and $\pi(x)$ never exceeds $3\sqrt{x}\log x$, and that the difference between consecutive primes satisfies $p_{n+1} - p_n = O\left(\sqrt{p_n}\log^2 p_n\right)$. In fact, any of these various consequences is actually equivalent to RH. On the other hand, if RH were false, it would follow that the error of approximation of $\pi(x) - \mathrm{li}(x)$ must exceed $O\left(x^\Theta\right)$ where Θ is the supremum of the real parts of the violating zeros.

Two of the assertions just made require comment. The first is that the *infinitely many* $|x^\rho| = \sqrt{x}$, when summed as in (5.23), cannot end up larger than $O\left(\sqrt{x}\log^2 x\right)$. And the second, when $|x^\rho| = \sqrt{x}$ is violated, is that no fortuitous cancellation occurs among the *infinitely many* terms in the sum to result in an error smaller than $O(x^\Theta)$. These technical details are sorted out, for example, in Montgomery & Vaughan (2007, [**774**, Chapter 15]).

In the foregoing, the connection between $\zeta(s)$ and the primes was discussed primarily in terms of the Chebyshev function $\psi(x)$ which is the summatory version of the von Mangoldt function $\Lambda(n)$, but it is also possible to connect $\zeta(s)$ to $\Lambda(n)$

via a Dirichlet series. Thus taking logarithms and then differentiating the identity $\zeta(s) = \prod_p (1 - p^{-s})^{-1}$ gives, for $\Re s > 1$,

$$-\frac{\zeta'(s)}{\zeta(s)} = \sum_p \frac{d\log(1-p^{-s})}{ds} = \sum_p \frac{\log(p)p^{-s}}{1-p^{-s}} = \sum_p \log(p) \sum_{k=1}^{\infty} p^{-ks} = \sum_{n=1}^{\infty} \frac{\Lambda(n)}{n^s}.$$

Here the $1 - p^{-s}$ denominator was parlayed into a geometric series, while the last sum has nonzero contributions $(\log p)/p^{ks}$ only at prime powers $n = p^k$.

5.7. Arithmetic functions and $\zeta(s)$

In Chapter 3 we introduced numerous arithmetic functions, including $\mu(n)$, $\varphi(n)$, $\tau(n)$, $\sigma(n)$, $\sigma_k(n)$, $\omega(n)$, $\Omega(n)$, $\lambda(n)$, and $\Lambda(n)$. Many of these arise naturally as the coefficients of Dirichlet series expansions associated with simple analytic transformations of the function $\zeta(s)$. Many such relations were already noted in Chapter 3, but it is worthwhile to collect the better known ones and provide a summary of them here:

$$\frac{1}{\zeta(s)} = \prod_p \left(1 - \frac{1}{p^s}\right) = \sum_{n=1}^{\infty} \frac{\mu(n)}{n^s} = s\int_1^{\infty} \frac{M(x)}{x^{s+1}} dx,$$

$$\frac{\zeta(s)}{\zeta(2s)} = \sum_{n=1}^{\infty} \frac{|\mu(n)|}{n^s} \equiv \sum_{n=1}^{\infty} \frac{\mu^2(n)}{n^s} = \prod_p \left(1 + \frac{1}{p^s}\right),$$

$$\frac{\zeta(s-1)}{\zeta(s)} = \sum_{n=1}^{\infty} \frac{\varphi(n)}{n^s}, \quad \frac{\zeta(2s)}{\zeta(s)} = \sum_{n=1}^{\infty} \frac{\lambda(n)}{n^s},$$

$$-\frac{\zeta'(s)}{\zeta(s)} = \sum_p \frac{\log p}{p^s - 1} = \sum_{n=1}^{\infty} \frac{\Lambda(n)}{n^s} = \int_1^{\infty} \frac{d\psi(x)}{x^s} = s\int_1^{\infty} \psi(x) x^{-s-1}\, dx,$$

$$\log \zeta(s) = \sum_{k=1}^{\infty} \sum_p \frac{1}{k} \frac{1}{p^{ks}} = \sum_{n=2}^{\infty} \frac{\Lambda(n)/\log n}{n^s},$$

$$\zeta^2(s) = \sum_{n=1}^{\infty} \frac{\tau(n)}{n^s}, \quad \frac{\zeta^3(s)}{\zeta(2s)} = \sum_{n=1}^{\infty} \frac{\tau(n^2)}{n^s}, \quad \frac{\zeta^4(s)}{\zeta(2s)} = \sum_{n=1}^{\infty} \frac{\tau^2(n)}{n^s},$$

$$\zeta(s)\zeta(s-1) = \sum_{n=1}^{\infty} \frac{\sigma(n)}{n^s}, \quad \zeta(s)\zeta(s-a) = \sum_{n=1}^{\infty} \frac{\sigma_a(n)}{n^s}, \quad \frac{\zeta^2(s)}{\zeta(2s)} = \sum_{n=1}^{\infty} \frac{2^{\omega(n)}}{n^s},$$

$$\frac{\zeta(s+i)+\zeta(s-i)}{2} = \sum_{n=1}^{\infty} \frac{\cos(\log n)}{n^s}, \quad \text{and} \quad \frac{\zeta(s+i)-\zeta(s-i)}{2} = \sum_{n=1}^{\infty} \frac{\sin(\log n)}{n^s}.$$

The result for $\zeta^2(s)$ has the generalization

$$\zeta^k(s) = \sum_{n=1}^{\infty} \frac{\tau_k(n)}{n^s},$$

where $\tau_k(n) \equiv \#\{(n_1, n_2, \ldots, n_k) : n = n_1 n_2 \cdots n_k,\ n_j \geq 1\}$ is the number of ways n can be represented as a product of k positive integers.

In addition to these relations, straightforward differentiation of

$$\zeta(s) = \sum_{n=1}^{\infty} 1/n^s$$

yields

$$\zeta'(s) = -\sum_{n=2}^{\infty} \frac{\log n}{n^s}, \quad \text{and } \zeta^{(k)}(s) = (-1)^k \sum_{n=2}^{\infty} \frac{\log^k n}{n^s} \quad \text{for } k = 1, 2, \ldots$$

In fact, any of the relations above may be differentiated yielding results such as

$$-\left(\frac{\zeta(2s)}{\zeta(s)}\right)' = \sum_{n=1}^{\infty} \frac{\lambda(n)\log n}{n^s}, \quad \left(\frac{\zeta'(s)}{\zeta(s)}\right)' = \sum_{n=1}^{\infty} \frac{\Lambda(n)\log n}{n^s},$$

and so on. Note too that substituting selected values for s can sometimes lead to interesting identities. For example, taking $s = 2$ in the first of the formulas leads to $\sum_{n=1}^{\infty} \mu(n)/n^2 = 6/\pi^2$. Taking $s = 1$ there is not legal, although it does in that instance correctly give $\sum_{n=1}^{\infty} \mu(n)/n = 0$. In all of the identities above, the real parts of all arguments of ζ are assumed to exceed 1. Proofs for many of the identities appeared in Chapter 3, and proofs for the others can be found in Hardy & Wright (2008, [**499**, Chapter XVII]), and other texts. Appendix C provides general background on Dirichlet series.

5.8. Complements

There is so much more to tell—so many highways and byways along which to travel from here. But we will have to limit ourselves by pointing to only a few, and to close this chapter by offering the following extended series of remarks.

(1) There is a wealth of computational issues involved in the numerical evaluation of $\zeta(s)$ for general values of s. These are discussed, for example, in Borwein, Bradley & Crandall (2000, [**132**]), or Gourdon & Sebah (2003, [**446**]).

(2) If we write $\zeta(s)$ (with $s = \sigma + it$) in the form

$$\zeta(\sigma + it) = \sum_{n=1}^{\infty} \frac{1}{n^{\sigma+it}} = \sum_{n=1}^{\infty} \frac{e^{-it\log n}}{n^\sigma} = \sum_{n=1}^{\infty} \frac{\cos(t\log n) - i\sin(t\log n)}{n^\sigma}$$

we see that for each fixed $\sigma > 1$, $\zeta(s)$ is a superposition of sinusoidal waveforms in t having amplitudes $1/n^\sigma$ and frequencies $(\log n)/2\pi$. As such it is a special case of what is known as an *almost periodic function*. Stated loosely, almost periodic functions are those whose finite segments each repeat infinitely often, to within any given precision. Such functions were first introduced by Bohr in studies of the zeta function. Their theory is discussed in Besicovitch (1932, [**96**]), and Bohr (1947, [**118**]). A key result is that a bounded, continuous, complex-valued function $f(x)$, $x \in \mathbb{R}$, is almost periodic if and only if it is a uniform limit of functions of the form $\sum_{j=1}^{k} c_j e^{i\lambda_j x}$, $\lambda_j \in \mathbb{R}$.

(3) The Euler product formula (5.1) is not valid if $\sigma < 1$ but can be shown to remain valid on the line $\sigma = 1$, as long as $t \neq 0$:

$$\zeta(1 + it) = \sum_{n=1}^{\infty} \frac{1}{n^{1+it}} = \prod_p \left(1 - \frac{1}{p^{1+it}}\right)^{-1}, \quad t \neq 0.$$

For $t \neq 0$, convergence of the sum occurs due to the rotational nature of n^{it}.

(4) In the early 1920s Hamburger proved that, within a broad class of permissible solutions, the ζ function is actually characterized by its functional equation.

(5) The *universality* result of Voronin mentioned earlier can be stated more formally: If $g(s)$ is a non-vanishing function, continuous on the disk $|s| \le r < 1/4$, and analytic in its interior, then for any $\epsilon > 0$ there are arbitrarily large real values t such that

$$\sup_{|s|\le r} \left| \zeta\left(s + \frac{3}{4} + it\right) - g(s) \right| < \epsilon.$$

For a survey of this phenomenon, see Laurinčikas (2003, [**681**]).

(6) The functional equation was actually conjectured by Euler in 1761, almost one hundred years prior to Riemann's 1859 paper. Working with real-valued s, and the function

$$\eta(s) = \sum_{n=1}^{\infty} \frac{(-1)^{n+1}}{n^s} = \left(1 - 2^{1-s}\right)\zeta(s), \quad s > 0, \tag{5.25}$$

Euler foreshadowed the analytic continuation known today by conjecturing that

$$\frac{\eta(1-s)}{\eta(s)} = -\Gamma(s)\frac{2^s - 1}{2^{s-1} - 1}\pi^{-s}\cos\left(\frac{\pi s}{2}\right)$$

even though he knew that the left side diverges outside of $0 < s < 1$. For details, see Narkiewicz (2000, [**797**, §4.1]), as well as the entertaining article by Ayoub (1974, [**47**]).

The function defined in (5.25), sometimes called the *alternating zeta function*, has all the zeros $1 + 2\pi i k \log 2$ ($k \in \mathbb{Z}$) of $\left(1 - 2^{1-s}\right)$, other than the one at $s = 1$, in addition to those of $\zeta(s)$.

(7) In addition to the functional equations (5.5)–(5.8), there are many other links between the Gamma and zeta functions. For example, making the change of variables $y = nx$ gives

$$\frac{\Gamma(s)}{n^s} = \frac{1}{n^s}\int_0^\infty y^{s-1}e^{-y}\,dy = \int_0^\infty x^{s-1}e^{-nx}\,dx, \quad \text{for } \sigma > 0,$$

and summing over n then gives

$$\Gamma(s)\zeta(s) = \sum_{n=1}^{\infty}\int_0^\infty x^{s-1}e^{-nx}\,dx = \int_0^\infty \frac{x^{s-1}}{e^x - 1}\,dx\,, \quad \text{for } \sigma > 1.$$

This, in effect, says that $\Gamma(s)\zeta(s)$ is the Mellin transform of $1/(e^x - 1)$; see Appendix C. Two further identities linking the ζ and Γ functions are

$$\log\Gamma(1+z) = -\gamma z + \sum_{n=2}^{\infty}(-1)^n\frac{z^n}{n}\zeta(n)\,, \quad |z| \le 1\,,$$

and

$$\sum_{n=2}^{\infty}\zeta(n)x^{n-1} = -\Psi(1-x) - \gamma\,, \quad |x| < 1,$$

where $\Psi(x) \equiv \Gamma'(x)/\Gamma(x)$ is the digamma function. Note that setting $z = 1$ in the result for $\log\Gamma(1+z)$ yields an identity we had seen earlier:

$$\sum_{n=2}^{\infty}\frac{(-1)^n\,\zeta(n)}{n} = \gamma.$$

(8) Riemann obtained his functional equation for $\zeta(s)$ using the *theta function*

$$\theta(x) \equiv \sum_{n=-\infty}^{\infty} e^{-\pi n^2 x}$$

and its fundamental functional identity[14]

$$\theta(x) = \theta(1/x)/\sqrt{x}, \quad \text{or} \quad \theta(1/x) = \sqrt{x}\,\theta(x).$$

One begins with

$$\Gamma\left(\frac{s}{2}\right) \equiv \int_0^\infty e^{-x} x^{s/2-1}\, dx,$$

makes the change of variables $x = \pi n^2 u$ to get

$$n^{-s}\Gamma(s/2)\pi^{-s/2} = \int_0^\infty e^{-\pi n^2 u} u^{s/2-1}\, du,$$

and then sums over n to arrive at

$$\zeta(s)\Gamma(s/2)\pi^{-s/2} = \int_0^\infty \left(\sum_{n=1}^{\infty} e^{-\pi n^2 u}\right) u^{s/2-1}\, du.$$

By a further change of variables $v = 1/u$, followed by applying the theta functional identity, the right-hand side reduces ultimately to $\zeta(1-s)\Gamma((1-s)/2)\pi^{-(1-s)/2}$ thus establishing the functional equation. An elegant and concise version of this argument can be found in Tenenbaum & Mendès-France (2000, [**1055**, Chapter 2, §2]), and more complete versions can be found in Davenport (2000, [**262**, Chapter 8]), Iwaniec (2014, [**593**, Chapter 7]), or Montgomery & Vaughan (2006, [**774**, §10.2 & p. 457]), as well as in numerous other texts. An alternate proof of the functional equation is given in Section 5.9.

(9) There are a great many variations on the zeta function. One example is the *Hurwitz zeta function* defined as

$$\zeta(s,a) \equiv \sum_{n=0}^{\infty} \frac{1}{(n+a)^s}, \quad \text{for } \Re s > 1, \text{ with } 0 < a \le 1.$$

It too is holomorphic on $\Re s > 1$ and has an analytic continuation to the whole plane with a simple pole at $s = 1$ with residue 1. Its theory is roughly analogous to that of $\zeta(s) \equiv \zeta(s,1)$, although it has Euler product representations only when $a = 1$ and $a = \frac{1}{2}$. It has 'trivial' zeros on the negative real axis, is zero free for $\sigma \ge 1 + a$, and satisfies a functional equation. It also has a Laurent expansion analogous to that of $\zeta(s)$; Farr et al. (2018, [**370**]) may be consulted for details.

Another example is the *periodic zeta function*

$$\Omega(s,x) = \sum_{n=1}^{\infty} \frac{e^{2\pi i n x}}{n^s} = \sum_{n=1}^{\infty} \frac{\cos(2\pi n x) + i \sin(2\pi n x)}{n^s}$$

---

[14] This identity may be proved by applying the $x = 0$ case of the Poisson summation formula, namely

$$\sum_{n=-\infty}^{\infty} f(x+n) = \sum_{n=-\infty}^{\infty} \hat{f}(n) e^{2\pi i n x}, \quad \text{where } \hat{f}(y) \equiv \int_{-\infty}^{\infty} f(x) e^{-2\pi i x y}\, dx,$$

to the Fourier transform pair $f_t(x) = e^{-\pi x^2 t}$, $\hat{f}_t(y) = \frac{1}{\sqrt{t}} e^{-\pi y^2/t}$.

which has a period of 1 in the variable x. The case $x = \frac{1}{2}$ at which $e^{2\pi inx} = \pm 1$ has already been encountered. The Hurwitz and periodic zeta functions may be combined to form the *Lerch zeta function*

$$\Omega(s, a, x) = \sum_{n=1}^{\infty} \frac{e^{2\pi inx}}{(n+a)^s}.$$

These various functions were introduced by Lipschitz and Hurwitz and were later studied by many authors, including Lagarias & Li (2012, [**669**]). Vilaplana (1987, [**1086**]), and Saxena et al. (2011, [**948**]) considered the application of these functions to statistical inference. Other references include Apostol (1976, [**25**, Chapter 12]), Cohen (2007, [**211**, §9.6.1]), and Titchmarsh (1986, [**1059**, §2.17]).

(10) The function $P(s) \equiv \sum_p 1/p^s$ is known as the *prime zeta function*, a definition inspired by blending the forms $\sum 1/n^s$ and $\sum 1/p$. This series converges absolutely and is analytic for $\Re s > 1$. Using the product form

$$\zeta(s) = \prod_p (1 - p^{-s})^{-1}$$

and the expansion $-\log(1-z) = z + \frac{z^2}{2} + \frac{z^3}{3} + \cdots$ one finds, for $\Re s > 1$,

$$\log \zeta(s) = -\sum_p \log(1 - p^{-s}) = \sum_p \sum_{k=1}^{\infty} \frac{p^{-ks}}{k} = \sum_{k=1}^{\infty} \sum_p \frac{p^{-ks}}{k} = \sum_{k=1}^{\infty} \frac{P(ks)}{k}. \tag{5.26}$$

Applying to this the Möbius inversion formula in the form

$$f(x) = \sum_{k=1}^{\infty} \frac{g(kx)}{k} \iff g(x) = \sum_{k=1}^{\infty} \frac{\mu(k)}{k} f(kx)$$

(obtained on applying (3.11) to $f(x)/x$ and $g(x)/x$) then yields

$$P(s) = \sum_{k=1}^{\infty} \frac{\mu(k)}{k} \log \zeta(ks), \quad \Re s > 1, \tag{5.27}$$

a formula which allows for fast and accurate computation. Note that for s near 1 the first term of the sum in (5.27) will dominate so that

$$P(s) \equiv \sum_p \frac{1}{p^s} \sim \log \zeta(s) \sim \log 1/(s-1), \quad \text{as } s \downarrow 1. \tag{5.28}$$

The functions $P(s)$ and $\pi(x)$ can also be linked; thus

$$P(s) \equiv \sum_p \frac{1}{p^s} = \int_1^{\infty} \frac{1}{x^s}\, d\pi(x) = s \int_1^{\infty} \frac{\pi(x)}{x^{s+1}}\, dx, \quad \sigma > 1$$

where we have used integration by parts.

The relation (5.27) can be used to show that $P(s)$ cannot be analytically continued to the left of the imaginary axis. Indeed, there are also difficulties in the analytic continuation below $\sigma = 1$ owing to the logarithmic singularities arising from the $\log \zeta(ks)$ terms in (5.27). Due to the coefficients $\mu(k)$, only those $\log \zeta(ks)$ for which k is squarefree will be active. The singularity of $\zeta(s)$ at $s = 1$ will lead to a logarithmic singularity $P(s) \sim \log(1/(s-1))$ at $s = 1$, and this will be repeated at every point $s = 1/k$ on the real axis, for squarefree k, due to the remaining $\log \zeta(ks)$ terms. Furthermore, each of the nontrivial zeros

ρ of $\zeta(s)$ will lead to logarithmic singularities at the positions $s = \rho/k$ for squarefree k. These singularities will rain down kaleidoscopically and create a wall of densely clustered logarithmic singularities along all of the imaginary axis—a natural boundary that analytic continuation is ubable to breach.

For an auxiliary viewpoint, rewrite (5.26) as

$$\log \zeta(s) = \sum_p \sum_{k=1}^{\infty} \frac{1}{kp^{ks}} = P(s) + R(s), \quad \text{where } R(s) = \sum_p \sum_{k \geq 2} \frac{1}{kp^{ks}},$$

and observe that $R(s)$ converges for $\sigma > 1/2$. Now $\zeta(s)$ is analytic except at $s = 1$, so $\log \zeta(s)$ is analytic everywhere except at $s = 1$ and at the zeros of $\zeta(s)$. It follows that $P(s)$ can be analytically continued down to at least $\sigma > 1/2$ except for the singularities at $s = 1$ and at any zeros of $\zeta(s)$ in $1/2 < \sigma < 1$. RH is equivalent to asserting that $P(s) - \log(s-1)$ can be analytically continued down to $\sigma > \frac{1}{2}$.

The function $P(s)$ has no known closed forms. Further details about $P(s)$ may be found in Titchmarsh (1986, [**1059**, §1.6 & §9.5]), Fröberg (1968, [**395**]), and Chernoff (2000, [**199**]). Older references include Landau (1920, [**677**]) and Estermann (1928, [**365**]), (1928, [**366**]). See also Walfisz (1963, [**1096**]).

(11) A few other interesting identities connecting $P(s) \equiv \sum_p 1/p^s$ and $\zeta(s)$ include

$$\sum_{n=1}^{\infty} \frac{\omega(n)}{n^s} = \zeta(s)P(s), \quad \sum_{k-1}^{\infty} P(ks) = \sum_{k=1}^{\infty} \frac{\varphi(k)}{k} \log \zeta(ks), \quad \Re(s) > 1,$$

and an amusing identity due to Vassilev & Missana (2016, [**1082**]): for integers $k > 1$,

$$P(k) = 1 - \sqrt{\frac{2}{\zeta(k)} - \sqrt{\frac{2}{\zeta(2k)} - \sqrt{\frac{2}{\zeta(4k)} - \sqrt{\frac{2}{\zeta(8k)} - \cdots}}}}.$$

(12) Motivated by the product formula $\zeta(s) = \prod_p (1-p^{-s})^{-1}$, and the Prime Number Theorem $p_n \sim n \log n$, Paul Chernoff (2000, [**199**]) defined the analogous 'Euler product'

$$\kappa(s) = \prod_{n=2}^{\infty} \left(1 - \frac{1}{(n \log n)^s}\right)^{-1}, \quad \Re s > 1. \tag{5.29}$$

We refer to $\kappa(s)$ as *Chernoff's (pseudo) zeta function.*

Recall first, from item (10) above and equation (5.26) there, that for $\Re s > 1$,

$$\log \zeta(s) = \sum_{k-1}^{\infty} \frac{1}{k} P(ks), \tag{5.30}$$

where $P(s) \equiv \sum_p p^{-s}$ is the *prime zeta function*. Each term $P(ks)$ of this sum has a singularity at $s = 1/k$ (since $\sum_p 1/p$ diverges); however, if we omit the $k = 1$ term, the part $\sum_{k=2}^{\infty}$ that remains will be analytic all the way down to $\sigma > 1/2$. It is only the $k = 1$ term of (5.30), namely $P(s)$, which causes 'trouble'. So RH is equivalent to saying that (except for its singularity at $s = 1$) $P(s)$ can be analytically continued[15] into the strip $1/2 < \sigma < 1$.

These arguments carry over to the case where the primes p_n are replaced by $n \log n$, as in (5.29), leading to the question of whether or not the 'pseudo prime zeta function'

$$Q(s) \equiv \sum_{n=1}^{\infty} \frac{1}{(n \log n)^s}$$

can be analytically continued into the critical strip. Chernoff showed that, except for its singularity at $s = 1$, $Q(s)$ can be analytically continued[16] all the way down to $\sigma > 0$. (This is in contrast with $P(s)$ which must be singular at the complex zeros of $\zeta(s)$.) It follows that Chernoff's zeta function (5.29) has no zeros in the strip $1/2 < \sigma < 1$; below $\sigma = 1/2$ the $k > 1$ terms of the $\sum_{k=1}^{\infty} Q(ks)$ analogue of (5.30) can produce singularities, and there is no known analogue here of the functional equation for $\zeta(s)$ that would allow us to 'reflect' $\sigma > 1/2$ into $\sigma < 1/2$. As an aside, one could conclude that the zeros of $\zeta(s)$ somehow reflect the irregularities in the spacing of the primes.

(13) We had mentioned that the zeta function gives rise to an *explicit* formula for $\pi(x)$, but only displayed—in (5.22)—an equivalent form of it based on Chebyshev's $\psi(x)$. For anyone who insists on seeing the explicit formula for $\pi(x)$, we shall now state it here. First, define

$$\begin{aligned} \Pi(x) &\equiv \sum_{2 \le n \le x} \frac{\Lambda(n)}{\log n} \\ &= \pi(x) + \frac{1}{2}\pi(x^{1/2}) + \frac{1}{3}\pi(x^{1/3}) + \cdots \\ &= \sum_{k=1}^{\infty} \frac{1}{k}\pi(x^{1/k}) \\ &= \sum_{p^k \le x} \frac{1}{k}. \end{aligned} \tag{5.31}$$

These sums are finite and terminate once $x^{1/k} < 2$. The function $\Pi(x)$ is a weighted count of the primes and prime powers up to x. Since powers beyond the first become scarce, $\Pi(x)$ and $\pi(x)$ will be of similar size: $\Pi(x) = \pi(x) + O\left(\sqrt{x}/\log x\right)$ (Ingham, 1932, [**580**, p. 104]). The Mellin transform of $\Pi(x)$ can be evaluated as[17]

$$s \int_1^{\infty} \Pi(x) x^{-s-1}\, dx = \log \zeta(s) \tag{5.32}$$

---

[15] We have glossed over the point that the logarithmic singularity of $P(s)$ at $s = 1$ requires a branch cut. Hence the $\sigma \le 1$ segment of the real line must be omitted from the analytic continuation, which should be carried out separately above and below the real axis.

[16] As with $P(s)$, the $\sigma \le 1$ segment of the real line must be omitted from this analytic continuation due to the logarithmic singularities of $Q(ks)$ at $s = 1/k$.

[17] We may compare this Mellin transform with (5.21) to find that

$$\int_1^{\infty} \Pi(x) x^{-s-1}\, dx = \int_2^{\infty} \frac{\pi(x)}{x(x^s - 1)}\, dx, \quad \Re(s) > 1,$$

and this may be inverted using Perron's formula to get

$$\Pi(x) = \frac{1}{2\pi i}\int_{c-i\infty}^{c+i\infty} \log(\zeta(s))\frac{x^s}{s}\,ds. \tag{5.33}$$

This can be evaluated by inserting into it the Weierstrass–Hadamard factorization for $\zeta(s)$ (described in Chapter 6) and leads to the following explicit formula for $\Pi(x)$:

$$\Pi(x) = \mathrm{li}(x) - \sum_\rho \mathrm{li}\,(x^\rho) - \log 2 + \int_x^\infty \frac{du}{u(u^2-1)\log u}\,. \tag{5.34}$$

Here $x > 1$, ρ ranges over the complex zeros of $\zeta(s)$, and the sum is interpreted as the $T \to \infty$ limit of $\sum_{|\rho|\le T} \mathrm{li}(x^\rho)$. When $x = p^k$ is a prime power, one must add $1/2k$ to the right side. The integral here stems from the trivial zeros and is asymptotically negligible. A noteworthy feature of formula (5.34) is the duality it shows between the zeros of the zeta function and the primes. The steps to get from (5.33) to (5.34) are delicate. A proof of identity (5.32) is given in Section 5.9. A Möbius inversion formula similar to the one in item (10) above may be applied to (5.31) to find

$$\Pi(x) = \sum_{k=1}^\infty \frac{1}{k}\pi\left(x^{1/k}\right) \iff \pi(x) = \sum_{k+1}^\infty \frac{\mu(k)}{k}\Pi\left(x^{1/k}\right) \tag{5.35}$$

where μ is the Möbius function; the second sum here also terminates once $x^{1/k} < 2$. These results motivate the definition

$$R(x) \equiv \sum_{k=1}^\infty \frac{\mu(k)}{k}\,\mathrm{li}(x^{1/k}). \tag{5.36}$$

Straightforward combination of (5.36) with (5.31), (5.34), and (5.35) leads finally to

$$\pi(x) = R(x) + \sum_\rho R\,(x^\rho) + \sum_{k=1}^\infty \frac{\mu(k)}{k}\int_{x^{1/k}}^\infty \frac{dt}{t(t^2-1)\log t}\,. \tag{5.37}$$

And (5.37)—for anyone who had insisted to see it—is the explicit formula for $\pi(x)$.

REMARKS 5.1.

(i) The stated formulas hold for all x not of the form p^k, but hold without restriction if one uses the symmetrized form $\pi^*(x) \equiv \big(\pi(x-0) + \pi(x+0)\big)/2$.

(ii) The function $R(x)$ in (5.37) acts as a kind of 'principal harmonic' with the $R(x^\rho)$ comprising the 'higher harmonics'. Reisel & Gohl (1970, [**888**]) have observed that the influence on $\pi^*(x)$ of any pair ρ and $\overline{\rho}$ of zeros enters into formula (5.37) linearly, and can therefore be individually attributed to the $R(x^\rho) + R(x^{\overline{\rho}})$ components. They provide computations based on this observation and show (as perhaps expected) that the higher zeros of $\zeta(s)$ influence $\pi^*(x)$ only at the larger values of x.

(iii) The first term of (5.37), namely $R(x)$, provides an approximation to $\pi(x)$ that is considerably more accurate than $\mathrm{li}(x)$.

(iv) For complex z, evaluating $\mathrm{li}(z) \equiv \int_0^z (1/\log t)\,dt$ requires considerations involving integration in the complex plane. For details, see Ingham (1932, [**580**, p. 82]), or Narkiewicz (2000, [**797**, p. 146]).

(v) The explicit formula was stated by Riemann and proved by von Mangoldt. Stopple (2003, [**1034**, Chapter 10]) traces the genesis of this result in a not unduly technical exposition chock-full of riveting historical detail. The engaging monograph by Mazur & Stein (2016, [**745**]) provides a gentler introduction to the ideas. A rigorous and more technical treatment may be found in Borwein et al. (2008, [**133**]).

(14) A spectral interpretation: Assume RH and represent the complex zeros of $\zeta(s)$ as $\frac{1}{2} \pm i\gamma_j$ with $0 < \gamma_1 \le \gamma_2 \le \cdots$ (see Chapter 6). Mazur & Stein (2016, [**745**]) point out that the functions

$$f(t) = -\sum_{p^k \le x} \frac{\log p}{p^{k/2}} \cos(t \log(p^k)) = -\sum_{p^k \le x} \frac{\log p}{p^{k/2}} \cos(kt \log p)$$

and

$$\hat{f}(u) = 1 + \sum_{0 < \gamma_j \le T} \cos(\gamma_j \log u)$$

are related in a manner analogous to a Fourier transform pair. As x increases, $f(t)$ tends to a sum of Dirac delta functions whose peaks occur at the complex ordinates γ_j of the zeros. And as T increases, $\hat{f}(u)$ tends to a sum of Dirac delta functions whose peaks occur at the values of the prime powers p^k with heights proportional to $\log p$. The functions, $f(t)$ and $\hat{f}(u)$ are simplified and truncated versions of the symmetrized forms

$$f(s) = \sum_{n=1}^{\infty} \Lambda(n) \left(n^{-s} + n^{-\bar{s}}\right) = \sum_{p^k} \log p \left(p^{-ks} + p^{-k\bar{s}}\right), \quad s = \frac{1}{2} + it,$$

and

$$\hat{f}(u) = 1 + \sum_{\gamma} \left(u^{i\gamma} + u^{-i\gamma}\right),$$

where $\Lambda(n)$ is the von Mangoldt function. See also du Sautoy (2013, [**947**]). The fact that the primes and the complex zeros of $\zeta(s)$ are in some sense 'duals' of each other may also be understood from the fact that $\zeta(s)$ can be represented by an Euler product in the primes, *as well as* by a Weierstrass–Hadamard product in the zeros:

$$\prod_p \left(1 - \frac{1}{p^s}\right)^{-1} = \zeta(s) = C(s) \prod_\rho \left(1 - \frac{s}{\rho}\right) e^{s/\rho}$$

for a certain function $C(s)$. (See equation (6.16) of Chapter 6.)

(15) It is deceiving that $\sum_1^\infty 1/n^2 = \zeta(2) = \pi^2/6$ is irrational, while the similar sum $\sum_1^\infty 1/n(n+1) = 1$ is not. That the second sum equals $\sum_1^\infty (1/n - 1/(n+1))$ and can be summed telescopically was noted in a 1650 book by Pietro Mengoli (1628–1686). It was Mengoli who raised the question of $\sum_1^\infty 1/n^2$, later to become known as the *Basel problem* when it was again raised by Jacob Bernoulli, then a professor at the University of Basel. In addition to solving the Basel problem in 1734, Euler also obtained the related results

$$\sum_{n=0}^{\infty} \frac{(-1)^n}{2n+1} = 1 - \frac{1}{3} + \frac{1}{5} - \frac{1}{7} + \cdots = \frac{\pi}{4},$$

$$\sum_{n=0}^{\infty} \frac{(-1)^n}{(2n+1)^3} = \frac{\pi^3}{32}, \quad \text{and} \quad \sum_{n=0}^{\infty} \frac{(-1)^n}{(2n+1)^5} = \frac{5\pi^5}{1536},$$

as well as a great many others.

Here is another identity which is amusing on account of its surprisingly humble (although tricky) origins:

$$\prod_p \frac{p^2+1}{p^2-1} = \frac{5}{2}.$$

If you would like to try your hand at it don't look at the solution in the footnote![18] A more challenging identity for anyone who is game is

$$\sum_{n=1}^{\infty} \prod_{k=1}^{n} \frac{p_k - 1}{p_{k+1}} = 1.$$

5.9. Selected proofs

PROOF THAT $\sum_{n=1}^{\infty}(-1)^{n+1}n^{-s}$ OF (5.2) CONVERGES WHEN $\Re s > 0$. Take consecutive pairs

$$\frac{1}{(2n-1)^s} - \frac{1}{(2n)^s} = s\int_{2n-1}^{2n} \frac{1}{x^{s+1}} dx,$$

and (with $\sigma = \Re s$) observe that

$$\left|\frac{1}{(2n-1)^s} - \frac{1}{(2n)^s}\right| \leq |s| \int_{2n-1}^{2n} \left|\frac{1}{x^{s+1}}\right| dx = |s| \int_{2n-1}^{2n} \frac{1}{x^{\sigma+1}} dx \leq \frac{|s|}{(2n-1)^{\sigma+1}}$$

since the last integrand is highest at its left boundary. The result follows on noting that the sum (over n) of these terms converges. □

PROOF OF (5.4). This result states that

$$\zeta(s) = \frac{s}{s-1} - s\int_1^{\infty} \frac{x - \lfloor x \rfloor}{x^{s+1}} dx. \tag{5.38}$$

We exhibit two essentially equivalent proofs. The first is based on Riemann–Stieltjes integrals and integration by parts: For real u write $u = \lfloor u \rfloor + \{u\}$ where $\lfloor u \rfloor$ is the greatest integer $\leq u$, and $\{u\} \equiv u - \lfloor u \rfloor$ is the fractional part so that $0 \leq \{u\} < 1$. Then

$$\begin{aligned} \zeta(s) &= 1 + \sum_{n=2}^{\infty} n^{-s} \\ &= 1 + \int_{1+}^{\infty} u^{-s}\, d\lfloor u \rfloor \\ &= 1 + \int_1^{\infty} u^{-s}\, du - \int_{1+}^{\infty} u^{-s}\, d\{u\} \end{aligned}$$

---

[18] Solution:

$$\prod_p \frac{p^2+1}{p^2-1} = \prod_p \frac{p^4-1}{(p^2-1)^2} = \prod_p \frac{1-\frac{1}{p^4}}{\left(1-\frac{1}{p^2}\right)^2} = \frac{\zeta^2(2)}{\zeta(4)} = \frac{(\pi^2/6)^2}{\pi^4/90} = \frac{5}{2}.$$

The 'more challenging identity' does not require the ζ function (Molinari (2015, [**761**])).

$$= 1 + \frac{1}{s-1} - \frac{\{u\}}{u^s}\bigg|_1^\infty + \int_1^\infty \{u\}\, du^{-s}$$
$$= \frac{s}{s-1} - s\int_1^\infty \frac{\{u\}}{u^{s+1}}\, du.$$

For $\Re s > 0$ the final integral is bounded thus giving the analytic continuation.

The second proof uses partial summation (see Appendix F): We have

$$\zeta(s) = \sum_{n=1}^\infty \frac{1}{n^s} = \sum_{n=1}^\infty n\left(\frac{1}{n^s} - \frac{1}{(n+1)^s}\right)$$
$$= s\sum_{n=1}^\infty n\int_n^{n+1} x^{-s-1}\, dx$$
$$= s\int_{1-}^\infty \lfloor x\rfloor x^{-s-1}\, dx$$
$$= s\int_1^\infty x\, x^{-s-1}\, dx - s\int_{1-}^\infty \{x\}\, x^{-s-1}\, dx$$
$$= \frac{s}{s-1} - s\int_{1-}^\infty \{x\}\, x^{-s-1}\, dx.$$

Alternately, using integration by parts, we have

$$\zeta(s) = \int_{1-}^\infty \frac{d\lfloor t\rfloor}{t^s} = s\int_1^\infty \frac{\lfloor t\rfloor\, dt}{t^{s+1}} = s\int_1^\infty \left(\frac{1}{t^s} - \frac{\{t\}}{t^{s+1}}\right)dt = \frac{s}{s-1} - s\int_1^\infty \frac{\{t\}\, dt}{t^{s+1}}. \quad \square$$

REMARKS 5.2. (a) There is a minor variant of relation (5.38) given by

$$\zeta(s) = \frac{1}{2} + \frac{1}{s-1} - s\int_1^\infty \frac{S(x)}{x^{s+1}}dx,$$

where $S(x) = x - \lfloor x\rfloor - \frac{1}{2}$ is a sawtooth function. It is proved in (5.39). A further integration by parts (equivalently, one more step of the Euler–Maclaurin summation) shows that the right hand side is analytic all the way down to $\sigma > -1$.
(b) Other variants worth noting include

$$\zeta(s) = s\int_1^\infty \frac{\lfloor x\rfloor}{x^{s+1}}, \quad \sigma > 1; \quad \text{and} \quad \zeta(s) = s\int_1^\infty \frac{\lfloor x\rfloor - x}{x^{s+1}}, \quad 0 < \sigma < 1.$$

PROOF OF THE LAURENT EXPANSION (5.19)/(5.20). This result states that

$$\zeta(s) = \frac{1}{s-1} + \sum_{n=0}^\infty \frac{(-1)^n}{n!}\gamma_n (s-1)^n, \quad \text{where } \gamma_n \equiv \lim_{m\to\infty} \sum_{k\le m}\left(\frac{\log^n k}{k} - \frac{\log^{n+1} m}{n+1}\right).$$

Since the original proof by Stieltjes was rather involved we follow a method that appeared in Bohman & Fröberg (1988, [**117**]); see also Lagarias (2013, [**668**]), Narkiewicz (2000, [**797**, pp. 161–164]), and Baillaud & Bourget (1905, [**64**, p. 155]). We will use the trivial identity

$$(s-1)\zeta(s) = \sum_{k=1}^\infty (s-1)\frac{1}{k^s},$$

and the telescoping sum

$$\sum_{k=1}^\infty \left(\frac{1}{k^{s-1}} - \frac{1}{(k+1)^{s-1}}\right) = 1.$$

On combining them one obtains

$$(s-1)\zeta(s) = 1 + \sum_{k=1}^{\infty}\left(\frac{1}{(k+1)^{s-1}} - \frac{1}{k^{s-1}} + (s-1)\frac{1}{k^s}\right),$$

or equivalently,

$$(s-1)\zeta(s) = 1 + \sum_{k=1}^{\infty}\left(e^{-(s-1)\log(k+1)} - e^{-(s-1)\log k} + (s-1)\frac{1}{k}e^{-(s-1)\log k}\right).$$

Upon Taylor expanding the exponentials, the right hand side becomes

$$1+\sum_{k=1}^{\infty}\left(\sum_{n=0}^{\infty}\frac{(-1)^n(s-1)^n}{n!}\left(\log^n(k+1)-\log^n k\right)+\frac{s-1}{k}\sum_{n=0}^{\infty}\frac{(-1)^n(s-1)^n\log^n k}{n!}\right).$$

Dividing both sides by $(s-1)$ then gives

$$\zeta(s) = \frac{1}{s-1} + \sum_{n=0}^{\infty}\frac{(-1)^n}{n!}\gamma_n(s-1)^n$$

where

$$\gamma_n = \lim_{m\to\infty}\sum_{k\le m}\left(\frac{\log^n k}{k} - \frac{\log^{n+1}(k+1) - \log^{n+1} k}{n+1}\right).$$

Due to the telescopic nature of its second component, the sum $\sum_{k\le m}$ here is identical to the $\sum_{k\le m}$ sum in the definition of γ_n appearing above. □

Proof of functional equation (5.5) for $\zeta(s)$ (outline). This equation is

$$\zeta(s) = 2(2\pi)^{s-1}\sin\frac{\pi s}{2}\,\Gamma(1-s)\zeta(1-s), \quad \text{for all } s \neq 0, 1.$$

One proof was outlined at item (8) of the 'complements' appearing above. Here we outline a different proof. It involves the sawtooth function $S(x) \equiv \lfloor x\rfloor - x + \frac{1}{2}$ and the identities

$$\zeta(s) = s\int_0^{\infty}\frac{S(x)}{x^{s+1}}dx = s\int_1^{\infty}\frac{S(x)}{x^{s+1}}dx + \frac{1}{s-1} + \frac{1}{2}. \tag{5.39}$$

The proof of these identities just involves assembling the following three components:

$$\zeta(s) = \frac{s}{s-1} - s\int_1^{\infty}\frac{x-\lfloor x\rfloor}{x^{s+1}}\,dx,$$

$$\int_0^1\frac{\lfloor x\rfloor - x}{x^{s+1}}dx = -\int_0^1\frac{dx}{x^s} = \frac{1}{s-1}, \quad \text{and } s\int_1^{\infty}\frac{\frac{1}{2}}{x^{s+1}}\,dx = \frac{1}{2}.$$

The first of these is relation (5.4) which was proved above, while the second and third involve only straightforward integrations. Now the sawtooth function $S(x)$ has the well-known Fourier series representation

$$S(x) = \sum_{n=1}^{\infty}\frac{\sin 2\pi nx}{\pi n}$$

valid for non-integer values of x. Substituting into (5.39) and interchanging $\int$ and $\sum$ gives

$$\zeta(s) = \frac{s}{\pi}\sum_{n=1}^{\infty}\frac{1}{n}\int_0^{\infty}\frac{\sin 2\pi n x}{x^{s+1}}\,dx = \frac{s}{\pi}\sum_{n=1}^{\infty}\frac{(2\pi n)^s}{n}\int_0^{\infty}\frac{\sin y}{y^{s+1}}\,dy.$$

The integral here can be evaluated using the Mellin transform result

$$\int_0^{\infty}\frac{\sin y}{y^{s+1}}\,dy = \Gamma(s)\sin\frac{\pi s}{2}, \quad 0 \le \Re s < 1$$

(see Appendix C). This leads to

$$\zeta(s) = \frac{s}{\pi}(2\pi)^s\sum_{n=1}^{\infty}\frac{1}{n^{1-s}}\{-\Gamma(-s)\}\sin\frac{\pi s}{2} = \frac{s}{\pi}(2\pi)^s\Gamma(1-s)\sin\frac{\pi s}{2}\zeta(1-s)$$

which is the required functional equation. □

REMARK 5.3. This proof is the first of many appearing in Titchmarsh (1986, [**1059**, §2.1]) but we have been most cavalier with its technicalities. In particular, the $\int\sum$ to $\sum\int$ interchange is hardly immediate, and the fact that the result is valid for all s (other than $s = 1$) must be carefully justified using analytic continuation. See also Baker (2012, [**67**, Chapter 14, pp. 163–164]).

PROOF OF (5.21). This equality states that

$$\log\zeta(s) = s\int_1^{\infty}\frac{\pi(x)}{x(x^s-1)}\,dx, \quad \Re(s) > 1.$$

We will again exhibit two essentially equivalent proofs with the first based on integration by parts. Thus

$$\begin{aligned}\log\zeta(s) &= -\sum_p \log\left(1-\frac{1}{p^s}\right)\\ &= -\int_1^{\infty}\log\left(1-\frac{1}{x^s}\right)d\pi(x)\\ &= -\log\left(1-\frac{1}{x^s}\right)\pi(x)\Big|_1^{\infty} + \int_1^{\infty}\pi(x)\,d\log\left(1-\frac{1}{x^s}\right)\\ &= s\int_1^{\infty}\frac{\pi(x)}{x(x^s-1)}\,dx.\end{aligned}$$

The second proof is by partial summation. We have

$$\log\zeta(s) = -\sum_p \log(1-p^{-s}) = -\sum_{n=2}^{\infty}(\pi(n)-\pi(n-1))\log(1-n^{-s}).$$

The terms of the last sum can be rearranged as

$$\sum_{n=2}^{\infty}\pi(n)\Big(\log(1-(n+1)^{-s}) - \log(1-n^{-s})\Big).$$

However, the term within the large parentheses is just

$$\int_n^{n+1}\frac{d\log(1-x^{-s})}{dx}\,dx = s\int_n^{n+1}\frac{1}{x(x^s-1)}\,dx$$

hence yielding the equality. □

REMARK 5.4. Compare the representation for $\log\zeta(s)$ proved here with the representation (5.32) for $\log\zeta(s)$ proved earlier.

PROOF OF (5.24). This result states that

$$-\frac{\zeta'(s)}{\zeta(s)} = s\int_1^\infty \frac{\psi(x)}{x^{s+1}}\,dx, \quad \Re s > 1.$$

It follows from the computation

$$-\frac{\zeta'(s)}{\zeta(s)} = \sum_{n=1}^\infty \frac{\Lambda(n)}{n^s} = \int_1^\infty \frac{1}{x^s}\,d\psi(x) = s\int_1^\infty \frac{\psi(x)}{x^{s+1}}\,dx.$$

The first equality was demonstrated in Chapter 3, the second holds because $\psi(x) \equiv \sum_{n\le x}\Lambda(n)$, and the third follows upon integrating by parts. □

PROOF OF (5.32). This equality states that for $\Re(s) > 1$

$$\log\zeta(s) = s\int_1^\infty \Pi(x)x^{-s-1}\,dx, \quad \text{where } \Pi(x) = \sum_{k=1}^\infty \frac{1}{k}\pi\left(x^{1/k}\right).$$

Using the Euler product the left side is

$$\log\zeta(s) = -\sum_p \log\left(1-p^{-s}\right) = \sum_p\sum_{k=1}^\infty \frac{1}{k}p^{-ks}.$$

On the other hand, noting that $\Pi(x) = \sum_{p^k\le x} 1/k$, the right side is

$$\int_1^\infty \sum_{p^k\le x}\frac{s}{k}x^{-s-1}\,dx = \sum_{p^k}\frac{s}{k}\int_{p^k}^\infty x^{-s-1}\,dx = \sum_p\sum_{k=1}^\infty \frac{1}{k}p^{-ks}. \qquad □$$

PROOF THAT $\zeta(s)$ IS ZERO-FREE FOR $\Re s > 1$. The proof is just an application of the lemma in Appendix E concerning infinite products. When $\sigma > 1$ we have

$$\zeta(s) = \prod_p\left(1-\frac{1}{p^s}\right)^{-1} = \prod_p\left(1+a_p\right),$$

where

$$a_p = \frac{1}{p^s} + \frac{1}{p^{2s}} + \frac{1}{p^{3s}} + \cdots = \frac{1}{p^s}\,\frac{1}{1-1/p^s}.$$

But each a_p is bounded as

$$|a_p| = \left|\frac{1}{p^s}\right|\left|\frac{1}{1-1/p^s}\right| \le \frac{2}{p^\sigma},$$

so that $\sum_p a_p < \infty$ when $\sigma > 1$. Hence by the mentioned lemma $\zeta(s)$ is nonzero. □

CHAPTER 6

Statistics of the Zeros

Gauss spoke of the magical allure that made number theory the favourite science of the earliest mathematicians, not to mention the inexhaustible richness in which it so greatly surpasses all other branches of mathematics. Kronecker compared those who study number theory with the Lotus Eaters who—having once tasted such fare—become unable to live without it.

David Hilbert[1]

Aside from their mathematical importance and their connection to the primes and to the Riemann hypothesis, the zeros of the zeta function exhibit patterns of chaotic behaviour that differ in interesting ways from those of conventional models of randomness. They are therefore of interest from both a mathematical and a probabilistic point of view. We begin with a brief overview of what we have learned about them in Chapter 5.

First, as we have seen, $\zeta(s) = 0$ at the negative even integers $s = -2, -4, -6\ldots$. These are the *trivial zeros* which from here on will be completely ignored. Apart from those, $\zeta(s)$ has no zeros on the vertical lines $\Re s = 0$ and $\Re s = 1$ (a fact which is equivalent to the Prime Number Theorem) and indeed no other zeros outside of $0 < \Re(s) < 1$, a vertical region of the complex plane known as the *critical strip*. Within this region, the locations of these *nontrivial zeros* satisfy two key symmetries: they occur symmetrically about the real axis $t \equiv \Im s = 0$, and (as a consequence of a radial symmetry stemming from the functional equation) they occur symmetrically about the *critical line* $\sigma \equiv \Re(s) = \frac{1}{2}$. Thus, if a zero $\rho = \frac{1}{2} + i\gamma$ occurs on the critical line, there will be another at $\rho = \frac{1}{2} - i\gamma$, while if a zero occurs off the critical line, say at $\rho = (\frac{1}{2}+\delta)+i\gamma$, then it will be one of a set of four, namely $\rho = (\frac{1}{2}\pm\delta)\pm i\gamma$. As none of the zeros in the critical strip fall on the real axis, they are also referred to as the *complex-valued zeros* of ζ. As we shall see, there are an infinite number of complex zeros. It is traditional to denote them as $\rho = \beta + i\gamma$, and to arrange those in the upper half-plane into a sequence $\rho_n = \beta_n + i\gamma_n$, with $\gamma_{n+1} \geq \gamma_n$, for $n = 1, 2, 3, \cdots$, the zeros being counted with multiplicity. The first few of them occur at $\frac{1}{2} \pm it$ with $t \approx 14.13,\ 21.02,\ 25.01,\ 30.42,\ 32.93,\ 37.58,\ 40.9\ldots$

6.1. The number of zeros up to a given height and $S(T)$

The number of zeros of $\zeta(s)$ in the region $0 < \sigma < 1$, $0 < t \leq T$ (in other words, "up to height T") is denoted by $N(T)$, and the number of those which lie on the critical line $\sigma = \frac{1}{2}$ is often denoted by $N_0(T)$. With this notation, the

[1] From the 1930 radio address which Hilbert gave in German; translated by the authors. https://old.maa.org/press/periodicals/convergence/david-hilberts-radio-address

Riemann hypothesis is the assertion that $N(T) = N_0(T)$ for all T. Counts such as $N(T)$ and $N_0(T)$ are invariably meant to allow for multiplicity, although no zero of multiplicity greater than one has ever been found and it is widely believed that all of the zeros of $\zeta(s)$ are simple.

Cauchy's argument principle (see Appendix A) leads to the identity

$$N(T) = \frac{1}{2\pi i} \oint \frac{\xi'(s)}{\xi(s)}\, ds$$

where $\xi(s)$ is the function defined in (5.8), the contour is the boundary of the rectangle $0 \le \Re s \le 1$, $0 \le \Im s \le T$ oriented positively (meaning counterclockwise), and T is not the imaginary part of a zero. Arguing from this, a rather sharp estimate for $N(T)$ was conjectured by Riemann in 1859, and proved by von Mangoldt in 1895; namely, as $T \to \infty$,

$$N(T) = \frac{T}{2\pi} \log \frac{T}{2\pi} - \frac{T}{2\pi} + O(\log T) \equiv \frac{T}{2\pi} \log \frac{T}{2\pi e} + O(\log T). \tag{6.1}$$

The zeros are therefore infinite in number and slowly become more densely spaced as $T \to \infty$, with the average (vertical) gap around T being asymptotic to $2\pi/\log T$. One implication of equation (6.1) is that $N(T+1) - N(T) = O(\log T)$ as $T \to \infty$. Furthermore, Littlewood (1924, [**705**]) proved that $\gamma_{n+1} - \gamma_n \to 0$ as $n \to \infty$, which in particular shows that the distances between successive zeros is bounded.

A considerably sharper version of (6.1) is known, namely

$$N(T) = \frac{T}{2\pi} \log \frac{T}{2\pi} - \frac{T}{2\pi} + \frac{7}{8} + S(T) + O\left(\frac{1}{T}\right). \tag{6.2}$$

Here $S(T)$ is the *argument* of the zeta function, and specifically

$$S(T) = \frac{1}{\pi} \arg \zeta \left(\frac{1}{2} + iT\right),$$

this being the arg reached (starting with value 0) through continuous variation along the path consisting (for example) of the vertical line segment going from 2 to $2+iT$, and then continuing along a horizontal line segment to $\frac{1}{2}+iT$. The essential properties about this path are that it

(i) starts in the region $\sigma > 1$ where ζ has no zeros,
(ii) starts at a point where the arg is 0 (such as on the real axis there),
(iii) ends at the point $\frac{1}{2} + iT$, and
(iv) separates the zeros of ζ that lie below height T from those that lie above.

Any other path with these properties would also do—as for example the horizontal path that starts at $\infty + iT$, where arg $= 0$, and ends at $\frac{1}{2} + iT$. For the more precise formula (6.2) to hold for all T, we require that if T is the ordinate of a zero then $S(T) = S(T+0)$, so that $S(T)$ is right-continuous (as indeed $N(T)$ is as well). With this convention the $O(1/T)$ order term appearing in (6.2) represents a function which is actually continuous in T. Clearly, the mysterious function $S(T)$ is *extremely* important, encoding fine detail about the vertical placement of the zeros, so the remainder of this section will be devoted to it. The reason for the curious $\frac{7}{8}$ term (which in fact arises as $1 - \frac{1}{8}$ in the derivations) will also be explained here. Proofs of (6.1) and (6.2) may be found in Titchmarsh (1986, [**1059**, Chapters IX & X]), Montgomery & Vaughan (2007, [**774**, §14.1]), and elsewhere. See also Tenenbaum (2015, [**1053**, Chapter II.3]), and Borwein et al. (2008, [**133**, p. 66]).

The real-valued function $S(T)$ is extremely complicated. It jumps up by one unit (with multiplicity) for each zero that T bypasses, and between consecutive zeros it is continuous and monotonically decreasing with derivatives

$$S'(T) = -\frac{1}{2\pi}\log\frac{T}{2\pi} + O(T^{-2}), \quad \text{and} \quad S''(T) = -\frac{1}{2\pi T} + O(T^{-3}).$$

It also grows *very* slowly and changes sign infinitely often. However, if one were to remove the jumps from $S(T)$—by subtracting from it the function $N(T)$ say—one would be left with a function that is extremely smooth (and very slowly decreasing). Concerning the $S(T) + O(1/T)$ part of (6.2), Bäcklund (1918, [**55**]), and Rosser (1941, [**918**]), established unconditionally that

$$(6.3) \quad \left| N(T) - \frac{T}{2\pi}\log\frac{T}{2\pi} + \frac{T}{2\pi} - \frac{7}{8} \right| \\ \leq 0.137\log T + 0.443\log\log T + 1.588, \quad \text{for all} \quad T > e.$$

For $S(T)$ itself, Platt & Trudgian (2015, [**854**]) obtained the unconditional bound

$$(6.4) \quad |S(T)| \leq 0.11\log T + 0.29\log\log T + 2.29 = O(\log T), \quad \text{for all} \quad T > e.$$

Bounds such as these are under constant development.

The true order of $S(T)$ is not known, although it is widely believed to be $O\big((\log T)^{1/2+\epsilon}\big)$. von Mangoldt had shown that $S(T) = O(\log T)$, while Cramér (1918/1919, [**240, 241**]) showed that the Lindelöf hypothesis (LH, discussed in Chapter 7) implies $S(T) = o(\log T)$, and Littlewood (1924, [**705**]) showed that RH implies that $S(T) = O\big(\log T/\log\log T\big)$. One consequence is that, under LH and RH, the order term in (6.1) can be replaced by $o(\log T)$ and $O\big(\log T/\log\log T\big)$, respectively. (See Titchmarsh (1986, [**1059**, Chapter 14]).)

Ramachandra and Sankaranarayanan (1993, [**881**]) showed that under the Riemann hypothesis, and for T large enough, we would have

$$(6.5) \qquad \big|S(T)\big| \leq (3.516)\log T/\log\log T.$$

Improved inequalities were obtained by Fujii (2004, [**399**]), Goldston & Gonek (2007, [**424**]), and Carneiro, Chandee & Milinovich (2013, [**181**]); in particular, the latter authors established, under RH, that

$$|S(T)| \leq \left(\frac{1}{4} + O\left(\frac{\log\log\log T}{\log\log T}\right)\right)\frac{\log T}{\log\log T}, \quad \text{as} \quad T \to \infty.$$

However, aside from values of its implied constant, Littlewood's conditional bound $O(\log T/\log\log T)$ has never been improved. On introducing that result in his 1924 paper Littlewood wrote: "The proof of this theorem is long and difficult, and depends on a singularly varied set of ideas" [**705**]. Farmer, Gonek & Hughes (2007, [**369**]) have argued that the true maximal size of $S(T)$ could actually be even smaller—perhaps even as small as $O\left(\sqrt{\log T \log\log T}\,\right)$.

In the opposite direction, Selberg had shown, unconditionally, that

$$S(t) = \Omega_{\pm}\left(\frac{(\log t)^{1/3}}{(\log\log t)^{7/3}}\right), \quad t \to \infty,$$

or stated otherwise,

$$\liminf_{T\to\infty} \frac{S(T)}{(\log T)^{1/3}/(\log\log T)^{7/3}} < 0, \quad \text{and} \quad \limsup_{T\to\infty} \frac{S(T)}{(\log T)^{1/3}/(\log\log T)^{7/3}} > 0.$$

Later work of Tsang (1986, [**1067**]) improved these unconditional results by showing that the $(\log\log T)^{7/3}$ here may be reduced to $(\log\log T)^{1/3}$. On the other hand, if RH is assumed these lower bounds can be improved further to

$$\liminf_{T\to\infty} \frac{S(T)}{\sqrt{\log T/\log\log T}} < 0 \quad \text{and} \quad \limsup_{T\to\infty} \frac{S(T)}{\sqrt{\log T/\log\log T}} > 0, \tag{6.6}$$

as well as to the stronger statement

$$\limsup_{T\to\infty} \frac{|S(T)|}{\sqrt{\log T \log\log\log T/\log\log T}} > 0 \tag{6.7}$$

for its absolute value. (See Montgomery (1977, [**764**]), and Bondarenko & Seip (2018, [**129**]), respectively.) Relying on an analogy with random matrix theory, Farmer, Gonek & Hughes (2007, [**369**]) offered the explicit conjecture

$$\limsup_{T\to\infty} \frac{|S(T)|}{\sqrt{\log T \log\log T}} = \frac{1}{\pi\sqrt{2}}.$$

The results mentioned so far concern the *extreme* values of $S(T)$ but these are not typical. In fact Selberg proved that

$$\frac{1}{T}\int_0^T S^2(t)\,dt = \frac{1}{2\pi^2}\log\log T + o(1),$$

and more generally, for integers $k \geq 1$, that

$$\frac{1}{T}\int_0^T S^{2k}(t)dt = \frac{(2k)!}{k!(2\pi)^{2k}}(\log\log T)^k + O\left((\log\log T)^{k-1/2}\right), \quad \text{as } T\to\infty. \tag{6.8}$$

(Remark: If RH is assumed, one can reduce this error to $O\big(T(\log\log T)^{k-1}\big)$.) The leading term in (6.8) corresponds precisely to the moments of a normal distribution and allows one to show that $S(T)/\sqrt{\log\log T}$ behaves asymptotically like a Gaussian variable with mean 0 and variance $1/2\pi^2$. (This involves choosing T uniformly at random in $[0,t]$, and letting $t\to\infty$; see Ghosh (1983, [**414**]).) This normality implies that $S(T)$ is typically of size $O\left(\sqrt{\log\log T}\,\right)$ so that $S(T)$ cannot approach its extreme values too often. In fact, the largest value of $|S(T)|$ currently known is only about 2.3 and occurs in the vicinity of the 1.3-billionth zero. This fact, together with the bounds on $S(T)$ indicated above, means that the number $N(T)$ of zeros stays very close to the smooth function which remains in (6.2) after the $S(T) + O(1/T)$ component is removed. One implication is that the zeros of $\zeta(s)$ behave very unlike common statistical models for point process. Some references are Goldston (2005, [**423**]), Ivić (2013, [**590**, pp. 18–19]), Odlyzko (1987, [**818**]), and Titchmarsh (1986, [**1059**, §9.3 & §9.25]).

The term $\frac{7}{8}$ appearing in (6.2) warrants comment as it could obviously have been absorbed into $S(T)$. The reason for not doing so is that one then has the (unconditional) result $\int_0^T S(t)\,dt = O(\log T)$ due to Littlewood (1924, [**705**]). Without the $\frac{7}{8}$ this order would be $O(T)$ so the $\frac{7}{8}$ term has importance "on average." Even so, however, Littlewood's $O(\log T)$ result is very surprising since—in view of the mentioned asymptotic normality result of Selberg—one might have expected an order of about $O(T^{1/2+\epsilon})$. The much smaller order must evidently be the result of extensive jumps of $S(t)$ about 0. Note too that if one could make the smallest possible improvement for the value of the mentioned integral—from $O(\log T)$ to $o(\log T)$—it would become *equivalent* to the Lindelöf hypothesis, while under RH its order would become $\log T/\log\log T$. All of the results in this paragraph are due to Littlewood, although using other methods Cramér had already shown that $\int_0^T S(t)\,dt = O(T^\epsilon)$ for any $\epsilon > 0$. An explicit (unconditional) bound for the integral is

$$\left|\int_0^T S(t)\,dt\right| \le 0.059\log T + 3.054.$$

Simonič (2022, [**991**]) may be consulted for additional results of this type. Karatsuba & Korolev (2005, [**621**]) provide a particularly thorough study of the function $S(T)$.

6.2. Some properties of the zeros

For the discussions to follow, recall again that the nontrivial zeros are denoted as $\rho = \beta + i\gamma$, and that those in the upper half-plane are arranged in sequence as $\rho_n = \beta_n + i\gamma_n$, with $\gamma_{n+1} \ge \gamma_n$ for $n = 1, 2, 3\ldots$, and with multiplicity. The zeros in the lower half-plane may then be labelled as $\rho_{-n} = \beta_n - i\gamma_n$. Although we do not usually call attention to lower zeros individually, when sums $\sum_\rho$ or $\sum_\gamma$ over the complex zeros appear, they are to be understood as ranging over both the upper and the lower zeros unless indicated otherwise. Furthermore, they are then typically summed in corresponding pairs ρ_n, ρ_{-n} for purposes of convergence.

Now it follows from (6.1) that the average number of zeros in unit rectangles $0 \le \sigma \le 1$, $T \le t \le T+1$, is asymptotic to $(1/2\pi)\log T$, and likewise that for any fixed $h > 0$, the average value of $N(T+h) - N(T)$ is asymptotic to $(h/2\pi)\log T$. It also follows that

$$|\rho_n| \sim \gamma_n \sim \frac{2\pi n}{\log n}, \quad \text{as } n \to \infty, \tag{6.9}$$

so that $\sum_\rho |\rho|^{-\alpha}$ converges if $\alpha > 1$, and diverges if $\alpha \le 1$, while $\sum_\rho |\rho|^{-1}(\log|\rho|)^{-\alpha}$ converges if $\alpha > 2$, and diverges if $\alpha \le 2$. (See also "Sums in n and $\log n$" in Appendix E.) However, although $\sum |\rho|^{-1}$ diverges, the series $\sum \rho^{-1}$ itself converges provided that ρ and $\bar\rho$ terms are grouped together because, for $\rho = \beta + i\gamma$, we have

$$\frac{1}{\rho} + \frac{1}{\bar\rho} = \frac{2\beta}{\beta^2+\gamma^2} \le \frac{2}{|\rho|^2},$$

and $\sum |\rho|^{-2}$ converges. The value of the $\sum \rho^{-1}$ sum depends on the values of the β's and γ's; under RH all β's are $\frac{1}{2}$, but presumably the sum is still likely to be 'unreasonable' since it will depend on the exact values of the 'early' zeros.

Assertion (6.9) is weaker than (6.1). Applying partial summation to (6.2) one may derive the more precise formula

$$|\rho_n| \sim \gamma_n = \frac{2\pi n}{\log n}\left(1 - \frac{\log\log n}{\log n} + \frac{1 + 2\log 2\pi}{\log n} + O\left(\left(\frac{\log\log n}{\log n}\right)^2\right)\right), \tag{6.10}$$

as $n \to \infty$. It also follows—from either (6.1) or (6.9)—that as $T \to \infty$,

$$\sum_{0<\gamma\le T} \frac{1}{\gamma} = O\left(\log^2 T\right), \quad \sum_{0<\gamma\le T} \frac{1}{\gamma^2} = O(1), \quad \text{and} \quad \sum_{\gamma>T} \frac{1}{\gamma^2} = O\left(\frac{\log T}{T}\right), \tag{6.11}$$

these sums being over the complex ordinates γ of the upper zeros $\rho = \beta + i\gamma$. More precise approximations may also be obtained from (6.1)/(6.2); in particular,

$$\sum_{0<\gamma\leq T} \frac{1}{\gamma} = \frac{1}{2\pi}\left(\frac{1}{2}\log^2 T - (\log 2\pi)\log T\right) + O(1), \text{ as } T \to \infty, \tag{6.12}$$

and

$$\sum_{\gamma>T} \frac{1}{\gamma^2} = \frac{1}{2\pi T}\left(3\log\frac{T}{2\pi} - 1\right) + O\left(\frac{\log T}{T^2}\right), \text{ as } T \to \infty. \tag{6.13}$$

We will prove (6.12) and (6.13) in Section 6.11. Note that the approximations in (6.11), (6.12), and (6.13) would all remain valid if γ in these expressions is replaced by $|\rho|$.

6.3. The Weierstrass–Hadamard factorizations

Recall the 'completed zeta function' $\xi(s)$, which was defined in Chapter 5 as

$$\xi(s) \equiv \frac{1}{2}s(s-1)\pi^{-s/2}\Gamma\left(\frac{s}{2}\right)\zeta(s), \tag{6.14}$$

and its functional equation $\xi(s) = \xi(1-s)$. Recall too that the factor $(s-1)$ cancels the pole of $\zeta(s)$ at $s=1$, that the factor s cancels the pole of $\Gamma(s/2)$ at $s=0$, and that the trivial zeros of $\zeta(s)$ cancel the remaining poles of $\Gamma(s/2)$. The function $\xi(s)$ is therefore entire (that is, it has no singularities and is analytic on the whole complex plane), and furthermore, its zeros exactly match the nontrivial zeros of $\zeta(s)$.

Now it is possible to show that $\xi(s)$ satisfies the growth restriction $|\xi(s)| = O\left((\log|s|)^{c|s|}\right)$ as $|s| \to \infty$, for some $c > 0$. It turns out that this is more than enough to allow one to apply the product representation results of Weierstrass and Hadamard according to which an entire function can be factored in a manner that is analogous to how ordinary polynomials can be factored using their zeros[2] (See Appendix A). The leads to the factorization

$$\xi(s) = \frac{1}{2}e^{cs}\prod_{\rho}\left(1 - \frac{s}{\rho}\right)e^{s/\rho}, \text{ where } c = \frac{1}{2}\log 4\pi - 1 - \frac{\gamma}{2}, \tag{6.15}$$

valid for all s. The product here runs over all zeros of $\xi(s)$—equivalently, all nontrivial zeros of $\zeta(s)$—and is convergent if terms in ρ and $1-\rho$ are paired. Complete accounts of this result may be found, for example, in Davenport (2000, [**262**, §11 & 12]), as well as in Montgomery & Vaughan (2006, [**774**, §10.2]) and Titchmarsh (1986, [**1059**, §2.12]). Additional references are Iwaniec (2014, [**593**, Chapter 8]), and Tenenbaum (2015, [**1053**, p. 155]).

If we substitute (6.15) into (6.14), and use $(s/2)\Gamma(s/2) = \Gamma(s/2+1)$, we are led to

$$\zeta(s) = \frac{\pi^{s/2}e^{cs}}{2(s-1)\Gamma\left(\frac{1}{2}s+1\right)}\prod_{\rho}\left(1 - \frac{s}{\rho}\right)e^{s/\rho}, \quad s \neq 1, \tag{6.16}$$

---

[2]Recall that any degree-n polynomial $a_n z^n + a_{n-1}z^{n-1} + \cdots + a_0$ (with $a_n \neq 0$) having zeros $c_1, c_2, \ldots c_n$, can be written as a product $a_n \prod(z - c_n)$ over its zeros. Assuming $a_0 \neq 0$ for brevity, this can be written in the form $\prod\left(1 - \frac{z}{c_i}\right)$ times a constant. For entire functions that are not polynomials, certain additional factors are needed to ensure convergence.

and with c as in (6.15). Logarithmic differentiation then gives

$$\frac{d\log\zeta(s)}{ds} \equiv \frac{\zeta'(s)}{\zeta(s)} = \frac{1}{2}\log\pi + c - \frac{1}{s-1} - \frac{1}{2}\frac{\Gamma'(\frac{1}{2}s+1)}{\Gamma(\frac{1}{2}s+1)} + \sum_\rho \left(\frac{1}{s-\rho} + \frac{1}{\rho}\right),$$

provided s is not equal to 1 or to any zero of $\zeta(s)$. Notice that the trivial zeros of $\zeta(s)$ appear in this formula along with the nontrivial ones since

$$-\frac{1}{2}\frac{\Gamma'(\frac{1}{2}s+1)}{\Gamma(\frac{1}{2}s+1)} = \frac{\gamma}{2} + \sum_{n=1}^{\infty}\left(\frac{1}{s+2n} - \frac{1}{2n}\right);$$

this last identity follows from the Weirstrass–Hadamard product for the Gamma function in the form

$$\frac{1}{\Gamma(s+1)} = e^{\gamma s}\prod_{n=1}^{\infty}\left(1+\frac{s}{n}\right)e^{-s/n}$$

(see Appendix B), upon differentiating logarithmically and substituting $\frac{s}{2}$ for s. Combining these expressions we find that

$$\text{(6.17)}\quad \frac{d\log\zeta(s)}{ds} \equiv \frac{\zeta'(s)}{\zeta(s)} = -\frac{1}{s-1} + \sum_\rho\left(\frac{1}{s-\rho}+\frac{1}{\rho}\right) + \sum_{n=1}^{\infty}\left(\frac{1}{s+2n}-\frac{1}{2n}\right) + C$$

where

$$C = \frac{1}{2}\log\pi + c + \frac{\gamma}{2} = \log 2\pi - 1.$$

The representation (6.17) is of great value in applications of the theory. The Weierstrass–Hadamard factorization of $\zeta(s)$, or more precisely of the *entire* function $(s-1)\zeta(s)$, can be obtained by substituting the factorized form for Γ into (6.16):

$$(s-1)\zeta(s) = 2^{s-1}\pi^s e^{-s}\left(\prod_{n=1}^{\infty}\left(1+\frac{s}{2n}\right)e^{-s/2n}\right)\left(\prod_\rho\left(1-\frac{s}{\rho}\right)e^{s/\rho}\right), \quad s\in\mathbb{C}.$$

The two products represent contributions from the trivial and nontrivial zeros, respectively.

There is a further result worth mentioning that may be obtained on letting $s \to 1$ in (6.17). Recall first that

$$\lim_{s\to 1}\left(\frac{\zeta'(s)}{\zeta(s)} + \frac{1}{s-1}\right) = \gamma,$$

and secondly that $\sum_1^N \frac{1}{n} \sim \gamma + \log N$ as $N \to \infty$, and observe that the latter implies that

$$\begin{aligned}\sum_{n=1}^{N}\left(\frac{1}{1+2n}-\frac{1}{2n}\right) &= \sum_{n=1}^{N}\left(\frac{1}{1+2n}+\frac{1}{2n}\right) - \sum_{n=1}^{N}\frac{1}{n}\\ &= -1 + \sum_{n=1}^{2N+1}\frac{1}{n} - \sum_{n=1}^{N}\frac{1}{n}\\ &\sim -1+\gamma+\log(2N+1)-\gamma-\log N\\ &\sim -1+\log 2.\end{aligned}$$

In view of these two results, letting $s \to 1$ in (6.17) yields

$$\sum_\rho\left(\frac{1}{1-\rho}+\frac{1}{\rho}\right) = \gamma - C + 1 - \log 2 = 2 + \gamma - \log 4\pi;$$

one may state this alternately as

$$\sum_{\rho} \frac{1}{\rho(1-\rho)} \equiv \sum_{\rho} \left(\frac{1}{\rho} + \frac{1}{1-\rho} \right) \equiv 2 \sum_{\rho} \frac{1}{\rho} = 2 + \gamma - \log 4\pi, \tag{6.18}$$

since (on account of the symmetries) ρ and $1-\rho$ each run over all of the zeros. (These symmetries also result in the sum $\sum 1/\rho$ being real valued.) The result (6.18) could also be arrived at by substituting product representation (6.15) into the functional equation for $\xi(s)$ and then logarithmically differentiating. It is an interesting feature of formula (6.18) that all three of the constants γ, e, and π play a role within it.

The result (6.18) has a rather interesting consequence, namely that

$$\sum_{\rho} \frac{1}{|\rho|^2} \equiv \sum_{\rho} \left(\frac{1}{\rho} + \frac{1}{\bar{\rho}} \right) \geq 2 + \gamma - \log 4\pi \approx 0.046191417932242\ldots, \tag{6.19}$$

with equality if and only if the Riemann hypothesis is true. To understand why this is so consider a matching pair, $\rho_1 = \left(\frac{1}{2} - \epsilon\right) + i\gamma$, and $\rho_2 = \left(\frac{1}{2} + \epsilon\right) + i\gamma$, of zeros which lie off the critical line. Computations yield

$$\frac{1}{\rho_1(1-\rho_1)} + \frac{1}{\rho_2(1-\rho_2)} = \left(\frac{1}{2} - 2\epsilon^2 + 2\gamma^2 \right) \Big/ D, \tag{6.20}$$

and

$$\frac{1}{|\rho_1|^2} + \frac{1}{|\rho_2|^2} = \left(\frac{1}{2} + 2\epsilon^2 + 2\gamma^2 \right) \Big/ D, \tag{6.21}$$

where

$$D = \left(\frac{1}{4} - \epsilon^2 \right) + \gamma^4 + \gamma^2 \left(\frac{1}{2} + 2\epsilon^2 \right).$$

The claimed assertion follows on noting that the numerator in (6.21) exceeds the numerator in (6.20) if $\epsilon > 0$. Of course, the sum on the left in (6.19) can be computed to hundreds or even hundreds of thousands of decimal places, but needless to say not to *all* decimal places.

In (5.22) of Chapter 5 a formula was stated for the symmetrized version of $\psi(x)$:

$$\psi^*(x) = x - \sum_{\rho} \frac{x^\rho}{\rho} - \log 2\pi - \frac{1}{2} \log \left(1 - \frac{1}{x^2} \right). \tag{6.22}$$

We mention this here because it too, like (6.17), is an expression involving only the zeros of $\zeta(s)$. Here the "$\sum x^\rho/\rho$" analogue corresponding to the trivial zeros (namely, $-2, -4, \ldots$) is encoded in the term

$$\log \left(1 - \frac{1}{x^2} \right) = \sum_{k=1}^{\infty} \frac{x^{-2k}}{-2k},$$

while the "x" originates from the pole of $\zeta(s)$ at $s = 1$. In examining the formula (6.22) it is illuminating to bear in mind the equivalence between $\psi^*(x) \sim x$ and the Prime Number Theorem. Since the last two terms in (6.22) are inconsequential, the importance of the zeros of zeta, which are manifest in the $\sum x^\rho/\rho$ term, becomes evident.

Incidentally, the result (6.22) may also be proved by starting with

$$\sum_{n=1}^{\infty} \frac{\Lambda(n)}{n^s} = -\frac{\zeta'(s)}{\zeta(s)},$$

and invoking the Perron inversion formula (see (C.7)/(C.8) of Appendix C) to obtain

$$\psi^*(x) \equiv \sum_{n\leq x}{}^{*} \Lambda(n) = \frac{1}{2\pi i}\int_{c-i\infty}^{c+i\infty}\left(-\frac{\zeta'(s)}{\zeta(s)}\right)\frac{x^s}{s}\,ds, \quad \text{for any } c>1,$$

where the asterisk means to use only half of the last term of the sum when x is an integer. However, the remainder of this proof is nontrivial. If one could push the line of integration all the way leftwards ($c \to -\infty$) the integral would then equal the sum of the residues of the integrand at its poles, and the result would follow. However this step turns out to be extremely delicate. Davenport (2000, [**262**, §17]), Edwards (1974, [**310**, §3.2]), Ivić (1985, [**586**, Chapters 1 & 12]), and Montgomery & Vaughan (2006, [**774**, Chapter 12, pp. 397–401]), for example, may be consulted for details. We mention finally that there is also a formula for $\pi(x)$ based solely on the zeros of $\zeta(s)$. This rather complicated formula appeared in item (13) of Section 5.8.

6.4. The number of zeros on the critical line

Riemann was the first to compute (manually) a few nontrivial zeros and to prove that they lie on the critical line. Gram (1903, [**449**]) calculated the first 15 roots and proved there were no other complex roots 'below' them. Alan Turing (1943, [**1074**]), (1953, [**1075**])—the first to rigorously carry out such computations using a computer—proved the same for some 1104 zeros.[3] Computations by many others, including Lehmer (1956, [**687**]), Rosser, Yohe & Schoenfeld (1969, [**921**]), Brent (1979, [**144**]), Brent et al. (1982, [**146**]), van de Lune & others (1983, [**715**]), (1986, [**716**]), Odlyzko (1987, [**818**]), (1992, [**820**]), (1998, [**821**]), and Gourdon & Demichel (2004, [**445**]), have established that more than the first *ten trillion* (10^{13}) zeros lie on the critical line, as also do batches of zeros lying as high as 10^{24}. Platt (2016, [**853**]) provides some history of how such remarkable computations were done and provides further references. Needless to say, such computations do not establish the Riemann hypothesis, and sound arguments can be offered to suggest that if there were zeros off the critical line, they would occur only at enormous heights. Chapter 8 provides some additional detail.

Of the zeros found so far, all have been simple. The possibility of multiple zeros cannot be ruled out, although it is widely believed that there are none. If the pair correlation conjecture (discussed in Section 6.8) were true, it would follow that *almost all* of the zeros are simple.[4] What has been established rigorously so far is that of the zeros which lie on the critical line, no fewer than 19/27 of them

---

[3] Turing's computations were subsequent to manual computations of the first 1041 zeros by Titchmarsh. It involved building an analogue computer at Cambridge mirroring one used at Manchester for analyzing tidal phenomena, the essential difference being the trigonometric frequencies required for the computations. The 1939 grant application for the £40 Turing required for this purpose was supported by Hardy and Titchmarsh. Later on, in 1950, Turing used the 'Manchester Mark I' analogue computer to carry out similar computations until that machine finally gave out. Turing apparently believed RH to be false, and was trying to find a complex zero off the critical line, an application that was among the motivations for his pioneering work on electronic computers. Not well known is that his thesis at Cambridge had been on the Central Limit Theorem. See for example, Booker (2006, [**130**]), (2016, [**131**]), Hejhal & Odlyzko (2013, [**534**]), and Hodges (2014, [**557**]).

[4] Otherwise the cases of $\gamma_m - \gamma_n = 0$ with $m \neq n$ in the $N(\alpha, \beta : T)$ count in (6.38) would contribute a Dirac delta function on the right in (6.39).

(or just over 70%) are simple. See Bui & Heath-Brown (2013, [**165**]), and Conrey, Ghosh & Gonek (1998, [**228**]).

Concerning the question of whether all of the complex zeros of $\zeta(s)$ lie on the critical line, some partial results are known. In what was hailed as a remarkable achievement at the time, Hardy (1914, [**485**]) proved that $N_0(T) \to \infty$, meaning that there are infinitely many zeros on the critical line. Hardy & Littlewood (1921, [**493**]) later strengthened this result to $N_0(T) \gg O(T)$, meaning that $N_0(T)$ is *at least* of order T. In 1942, Selberg proved that $N_0(T) \gg O(T \log T)$, thereby establishing (in view of (6.1)) that a *positive proportion* of the zeros lie on the critical line. Levinson (1974, [**695**]) showed this proportion exceeds one third (more specifically, exceeds 0.342), while Conrey (1989, [**219**]) showed it exceeds two fifths:

$$\liminf_{T\to\infty} \frac{N_0(T)}{N(T)} > \frac{2}{5}$$

(and more precisely, exceeds 0.4088). In fact, the proportion two fifths holds even if one were to require those zeros to be simple. Small improvements were obtained, for example, by Feng (2012, [**374**]) (whose claimed value of 0.4128 was later reduced to 0.4107), and by Pratt et al. (2020, [**867**]) who obtained 0.41729. Further incremental improvements are to be expected.

Closely connected to questions concerning the proportion of zeros that lie on the critical line is a class of theorems which guarantee the *existence* of zeros on short intervals of that line. A typical theorem of this type asserts that, once T is large enough, there must be a zero of $\zeta(1/2 + it)$ in every interval of the form $(T, T + H)$ when $H \geq T^{5/32} \log^2 T$. A related result states that for every $\epsilon > 0$ there is a $C > 0$ such that, once T is large enough, and $H = T^{27/82+\epsilon}$, there must be at least $CH \log T$ zeros of $\zeta(\frac{1}{2} + it)$ in the interval $(T, T + H)$. The proofs for such results are difficult. They are extensively discussed in Karatsuba & Voronin (1992, [**622**, Chapter VI]); see their theorems on pages 168, 174, and 192 therein.

6.5. Zero density estimates

For $\sigma \geq \frac{1}{2}$, the number of zeros $\rho \equiv \beta + i\gamma$ of $\zeta(s)$ such that $\beta \geq \sigma$ and $0 < \gamma \leq T$ is commonly denoted by $N(\sigma, T)$. (Thus RH states that $N(\sigma, T) \equiv 0$ whenever $\sigma > 1/2$.) Bohr & Landau (1914, [**120**]) established the remarkable result that $N(\sigma, T) = O(T)$ for every (fixed) $\sigma > \frac{1}{2}$. Results of this type are known as *zero density estimates*. Since for any $\sigma > \frac{1}{2}$ the order of $N(\sigma, T)$ is less than that of $N(T) = O(T \log T)$, the striking conclusion is that for any $\epsilon > 0$ only an infinitesimal proportion of the zeros in the critical strip lie further than ϵ away from the critical line. One may thus say that *almost all of the nontrivial zeros lie arbitrarily close to the critical line*. (It does *not* follow, however, that only a zero proportion of the zeros lie off the line!)

Ingham (1940, [**582**]) improved on the result of Bohr and Landau by showing that

$$N(\sigma, T) \ll O\left(T^{\frac{3(1-\sigma)}{2-\sigma}} \log^5 T\right) \ll O\left(T^{\frac{3(1-\sigma)}{2-\sigma}+o(1)}\right), \quad \text{for } \frac{1}{2} \leq \sigma \leq 1, \tag{6.23}$$

while Huxley (1972, [**570**]) later proved that

$$N(\sigma, T) \ll O\left(T^{\frac{3(1-\sigma)}{3\sigma-1}+o(1)}\right), \quad \text{for } \frac{1}{2} \leq \sigma \leq 1. \tag{6.24}$$

These results are not uniform in σ, but both are better than Bohr & Landau's $O(T)$, especially for values of σ near 1.[5] The exponents of T in (6.23) and (6.24) cross at $\sigma = \frac{3}{4}$ (where both exponents equal 0.6), with Ingham's result superior in the region $\sigma < \frac{3}{4}$, and Huxley's superior when $\sigma > \frac{3}{4}$. While some incremental improvements are known for $\sigma > \frac{3}{4}$, for $\sigma < \frac{3}{4}$ no meaningful improvement on Ingham's result has yet been found, the main obstacle being the lack of sharp enough bounds for the moments of the ζ beyond the fourth. It is known, however, that

$$\int_{1/2}^{1} N(\sigma, T)\, d\sigma = O(T),$$

a result which is actually stronger than what the Ingham–Huxley bounds would imply. Other zero density estimates were derived by Carlson, Conrey, Gritsenko, Heath-Brown, Hoheisel, Hua, Jutila, Montgomery, Moser, and Selberg, among others. (See, for example, Simonič (2020, [**990**]).) However, none of these results suffice to prove that *almost all* of the complex zeros lie on the critical line.

In this context there is an important conjecture known as the *density hypothesis* which states that for any $\epsilon > 0$,

$$N(\sigma, T) = O\left(T^{2(1-\sigma)+\epsilon}\right), \tag{6.25}$$

uniformly for $\frac{1}{2} \leq \sigma \leq 1$. It is implied by the Lindelöf hypothesis (LH) (and so also by the Riemann hypothesis) and is far stronger than what either of Ingham's or Huxley's bounds would imply (although still weaker than either of RH and LH).[6] Its importance stems not only from the fact that it is a stepping stone towards proving RH, but also from the fact that if it were shown to be false it would invalidate LH as well as RH. Furthermore, the density hypothesis is strong enough to allow for some valuable results to be proved, particularly about primes in short intervals. In particular, (6.25) can be shown to be equivalent to

$$\psi(x+y) - \psi(x) \sim y, \quad \text{for } y \geq x^{\frac{1}{2}+\epsilon}, \text{ as } x \to \infty, \tag{6.26}$$

(where ψ is Chebyshev's ψ function) and this suffices to establish that

$$\pi(x+y) - \pi(x) \sim \frac{y}{\log x}, \quad \text{for } y \geq x^{\frac{1}{2}+\epsilon}, \text{ as } x \to \infty, \tag{6.27}$$

and that consecutive primes satisfy

$$p_{n+1} - p_n = O\left(p_n^{1/2+\epsilon}\right), \tag{6.28}$$

a result close to Legendre's conjecture about primes between consecutive squares.

Huxley was able to obtain the bound

$$N(\sigma, T) = O\left(T^{\frac{12}{5}(1-\sigma)+o(1)}\right), \tag{6.29}$$

---

[5]It is interesting to note, in passing, that one important step in establishing such results involves obtaining good bounds for Dirichlet polynomial moments of type

$$\int_0^T \left| \sum_{n \leq N} a_n n^{-it} \right|^k dt.$$

[6]Note that at $\sigma = \frac{3}{4}$ the exponent of T in (6.25) is $\frac{1}{2}$ as compared to Ingham and Huxley's 0.6.

and to show that it holds *uniformly* throughout $\frac{1}{2} \le \sigma \le 1$. To understand the significance of this, let us remark that any bound of the form

$$N(\sigma, T) = O\left(T^{(A+o(1))(1-\sigma)}\right), \tag{6.30}$$

will lead to the results shown in (6.26) and (6.27), but only when restricted to values of $y \ge x^{1-\frac{1}{A}+\epsilon}$. So under the "$\frac{12}{5}$ bound", bounds (6.26) and (6.27) would only hold for $y \ge x^{\frac{7}{12}+\epsilon}$, while the order in (6.28) would only be

$$O\left(p_n^{\frac{7}{12}+\epsilon}\right).$$

From this it is clear that the value $A = 2$ in (6.30) is the smallest one could hope for—even if RH were to be assumed. The density hypothesis thus emerged, together with the challenge to reduce the 1972 value $\frac{12}{5} = 2.4$ of (6.29) to the value of 2 in (6.25). More than fifty years later, Maynard & Guth (2024, [**472**]) succeeded in taking the next leap forward; they established (6.30) for the value of $A = \frac{30}{13} \approx 2.3077$. As a consequence we now know that (6.26) and (6.27) both holds for $y \ge x^{\frac{17}{30}+\epsilon}$, while (6.28) holds with order term

$$O\left(p_n^{\frac{17}{30}+\epsilon}\right).$$

Titchmarsh (1986, [**466**, §9.15–9.24 & 9.28–9.29]), and Karatsuba & Voronin (1992, [**622**, Chapter 4]), provide extensive discussions of zero-density estimates. See also Borwein et al. (2008, [**133**, p. 85]), Bourgain (2000, [**138**]), Gritsenko (1997, [**466**]), Ivić (1985, [**586**, Chapter 11]), and Iwaniec & Kowalski (2004, [**594**, Chapter 10]).

6.6. Zero-free regions

Of course, *zero density estimates* are arguably less glamorous than results which give *zero-free regions*. In this section we will elaborate on the discussion in Section 2.4.

The smallest zero-free region of any real worth is the boundary line $\sigma = 1$ (and correspondingly $\sigma = 0$), and it is *equivalent* to the Prime Number Theorem. The zero-free region obtained by de la Vallée Poussin (1899, [**1079**]) for his proof of the Prime Number Theorem was stronger than what was required; he proved that there is a constant $C > 0$ such that $\zeta(\sigma + it)$ has no zeros in the region defined by

$$\sigma \ge 1 - \frac{C}{\log|t|}. \tag{6.31}$$

This represents a (somewhat thick) sliver just left of the line $\sigma = 1$, whose width at height t is proportional to $1/\log|t|$. To be more mindful here, one should really require that $|t|$ be large enough (so that the logarithm not be negative) but if (6.31) were replaced by

$$\sigma \ge 1 - \frac{C}{\log(2 + |t|)}$$

no such proviso would be needed. (Similar remarks apply to (6.32) and (6.33).)

Littlewood (1922, [**703**]) and Chudakov (1936, [**205**]) improved result (6.31) to

$$\sigma \ge 1 - C\frac{\log\log|t|}{\log|t|}, \quad \text{and } \sigma \ge 1 - \frac{C}{(\log|t|)^{3/4+\epsilon}}, \tag{6.32}$$

(for all $\epsilon > 0$), respectively. The best zero-free region currently known was obtained independently by Korobov (1958, [**654**]), and Vinogradov (1958, [**1090**]), and has since not been further improved. It is given by

$$\sigma \geq 1 - \frac{C}{\log^{2/3}(|t|)\,(\log\log(|t|))^{1/3}}, \tag{6.33}$$

although a neater and just slightly weaker version can be stated as

$$\sigma \geq 1 - \frac{C}{(\log|t|)^{2/3+\epsilon}}.$$

It in effect reduces the $\log|t|$ of de la Vallée Poussin in (6.31) to $\log^{2/3}|t|$. Kadiri (2005, [**617**]) has shown that in (6.31) we may take $C = 1/5.69693$ with $|t| \geq 2$, and that in (6.33) we may take $C = 1/57.54$ with $|t| \geq 10$. Noteworthy about the boundaries of all known zero-free regions is that they invariably tend towards $\sigma = 1$ as $|t| \to \infty$. No zero-free region that includes $\sigma > 1 - \epsilon$ for *any* $\epsilon > 0$ has ever been established.

These results lie very deep and required many new techniques to prove. A proof of the Korobov–Vinogradov zero-free region can be found in Karatsuba & Voronin (1992, [**622**, IV.3, pp. 119–120]), and a proof outline appears in the nifty "Lent 2014" course notes posted by Adam Harper on the web. Walfisz (1963, [**1096**]) provides an extensive treatment. Montgomery & Vaughan (2007, [**774**, §6.3]) give a detailed historical timeline for the evolution of zero-free regions, and Diamond (1982, [**283**]) gives a comprehensive survey. Zero-free regions are also discussed in Ellison & Mendès-France (1975, [**325**]), Ivić (1985, [**586**]), and Titchmarsh (1986, [**1059**]).

Among their applications, zero-free regions directly impact the order of the error term in the PNT and are virtually in one-to-one correspondence with them, a fact that was established by Ingham (1932, [**580**]). (See also the related discussion in Section 2.4.) Thus the PNT, $\pi(x) = \mathrm{li}(x) + o(x/\log x)$, is equivalent to the zero-free region $\sigma = 1$. The de la Vallée Poussin region (6.31) corresponds to the stronger result

$$\pi(x) - \mathrm{li}(x) = O\left(x \exp\left(-c\sqrt{\log x}\right)\right),$$

while Littlewood's and Chudakov's zero-free regions correspond to

$$\pi(x) - \mathrm{li}(x) = O\left(x \exp\left(-c\sqrt{\log x \log\log x}\right)\right),$$

and

$$\pi(x) - \mathrm{li}(x) = O\left(x \exp\left(-c\,(\log x)^{4/7-\epsilon}\right)\right),$$

respectively. Finally, the Korobov–Vinogradov region (6.33) yields

$$\pi(x) - \mathrm{li}(x) = O\left(x \exp(-c\,(\log x)^{3/5}(\log\log x)^{-1/5})\right),$$

the strongest unconditional result for $\pi(x)$ currently known. Note that these error terms for $\pi(x) - \mathrm{li}(x)$ all derive from those for $\psi(x) - x$; they are are in fact identical. Under RH, whereby all of $\sigma > \frac{1}{2}$ would be zero-free, we would have $\pi(x) - \mathrm{li}(x) = O\left(\sqrt{x}\log x\right)$, and conversely.

To better appreciate the impact of the zeros on the order terms for $\pi(x)$, $\psi(x)$ and $\vartheta(x)$, let Θ be the supremum of the real parts of the zeros of $\zeta(s)$. This Θ can be shown to be identical to the one defined as

$$\Theta \equiv \inf\{\alpha : \psi(x) = x + O(x^\alpha)\},$$

and it can be shown that this assertion remains valid if $\psi(x)$ here is replaced by either $\vartheta(x)$ or $\pi(x)$. See, for example, Montgomery & Vaughan (2006, [**774**, p. 463]) for a proof. It follows from this that for any $\frac{1}{2} \le \alpha \le 1$ the three assertions

$$(6.34)\quad \pi(x) = \operatorname{li}(x) + O_\epsilon(x^{\alpha+\epsilon}), \quad \psi(x) = x + O_\epsilon(x^{\alpha+\epsilon}), \quad \text{and} \quad \vartheta(x) = x + O_\epsilon(x^{\alpha+\epsilon}),$$

(where $\epsilon > 0$ is arbitrary) are equivalent, and furthermore that if any of these were to hold it would imply that $\zeta(s)$ has no zeros with $\sigma > \alpha$. In the other direction, the restriction that $\zeta(s)$ has no zeros for $\sigma > \alpha$ implies the stronger results
(6.35)

$$\pi(x) = \operatorname{li}(x) + O(x^\alpha \log x), \quad \psi(x) = x + O(x^\alpha \log^2 x), \quad \text{and} \quad \vartheta(x) = x + O(x^\alpha \log^2 x),$$

although for the converse the weaker assertions (6.34)—which are implied by those in (6.35)—would suffice. These results are due primarily to von Koch. See also Ingham (1932, [**580**, Chapter 2ff]), Turán (1950, [**1072**]), Walfisz (1963, [**1096**]), Pintz (1980, [**846**]), and Ellison & Mendès-France (1975, [**325**]).

While on this topic, let us pay tribute to a neat result of Taylor (1945, [**1052**]) which had been intended as a possible step towards establishing RH. This result states that if $\xi(s)$ is the variant of the zeta function defined in (6.14), then all of the complex zeros of the function $\xi\left(s + \frac{1}{2}\right) - \xi\left(s - \frac{1}{2}\right)$ lie on the line $\sigma = \frac{1}{2}$. The mentioned paper was prepared by Rees and Titchmarsh from an original manuscript that had been left behind; Flight-Lieutenant P. T. Taylor was killed in active service in November 1943.

At the going down of the sun and in the morning
We will remember them [**107**].

6.7. Statistics of the zeros

When discussing statistical properties of the zeros, it is common to assume RH so that the complex zeros all take the form $\rho_n = \frac{1}{2} + i\gamma_n$, with those in the upper half-plane sequenced according to $0 < \gamma_1 \le \gamma_2 \le \cdots$, as before. The matching zeros in the lower half-plane are then $\rho_{-n} = \frac{1}{2} + i\gamma_{-n}$ with $\gamma_{-n} = -\gamma_n$, but will not be needed here.

We have seen—as a consequence of (6.1)—that $\gamma_n \sim 2\pi n/\log n$ as $n \to \infty$, so that (asymptotically) the differences $\gamma_{n+1} - \gamma_n$ are $2\pi/\log n$ on average; one can equivalently say that the differences are $2\pi/\log \gamma_n$ on average. Recall too that the differences satisfy $\gamma_{n+1} - \gamma_n \to 0$. A sample of these spacings—specifically those occurring between the $(10^{20} - 6)^{\text{th}}$ and the $(10^{20} + 5)^{\text{th}}$ zeros on the critical line (which occur in the vicinity of $t \approx 15.2 \times 10^{18}$ where the average spacing is $2\pi/\log(15.2 \times 10^{18}) \approx 0.1423$)—is given by 0.0943, 0.0886, 0.1807, 0.0789, 0.2262, 0.0657, 0.1996, 0.1435, 0.1503, 0.1190, and 0.0724.

A *normalized* version of the γ_n can be defined in various asymptotically equivalent ways:

$$(6.36)\qquad \eta_n = \frac{\gamma_n \log \gamma_n}{2\pi}, \qquad \eta_n = \frac{\gamma_n}{2\pi} \log\left(\frac{\gamma_n}{2\pi}\right), \qquad \text{or} \quad \eta_n = \frac{\gamma_n}{2\pi} \log\left(\frac{\gamma_n}{2\pi e}\right).$$

Any of these will have mean spacing 1, or equivalently will satisfy $\eta_n \sim n$ as $n \to \infty$, although the last of these, which one can also write as $\eta_n = (\gamma_n/2\pi)\left(\log(\gamma_n/2\pi) - 1\right)$, can be shown (via (6.1)/(6.2)) to result in a slightly more precise unit-space 'unfolding' of the zeros.

Questions motivated by statistical considerations can be formulated in terms of the spacings $\eta_{n+1} - \eta_n$ between the normalized zeros. These may be taken, for example, as

$$\frac{\gamma_{n+1} - \gamma_n}{2\pi} \log \gamma_n, \quad \text{or as} \quad \frac{\gamma_{n+1} - \gamma_n}{2\pi} \log\left(\frac{\gamma_n}{2\pi}\right),$$

with the second of these giving somewhat tighter asymptotic unit values. Empirically it is evident that they do not follow a Poisson process, and one may ask (as with the primes) how often the spacings can be arbitrarily large or arbitrarily small relative to their mean.

Selberg (in 1946) proved *unconditionally* the existence of values $\lambda < 1 < \lambda'$ such that

$$\text{(6.37)} \qquad \liminf_{n\to\infty} \frac{\gamma_{n+1} - \gamma_n}{2\pi/\log(\gamma_n/2\pi)} \leq \lambda, \quad \text{and} \quad \limsup_{n\to\infty} \frac{\gamma_{n+1} - \gamma_n}{2\pi/\log(\gamma_n/2\pi)} \geq \lambda'$$

(the denominators here representing the average spacing between the zeros), and Fujii (1975, [**396**]) proved that both of these inequalities hold for a positive proportion of the zeros. Subsequent work has invariably assumed either RH or GRH (a generalized version of RH). Bui, Milinovich & Ng (2010, [**169**]) showed, under RH, that one can take $\lambda = 0.5155$ and $\lambda' = 2.6950$, while Feng & Wu (2012, [**375**]) improved this to $\lambda = 0.514$ and $\lambda' = 2.7327$. Bredberg (2011, [**141**]) established $\lambda' \geq 2.766$, and Bui & Milinovich (2018, [**168**]) obtained $\lambda' > 3.18$ (all under RH). Other such results may be found in Bui (2011, [**163**]), (2014, [**164**]), Feng & Wu (2013, [**376**]), Ghosh & Gonek (1984, [**226**]), (1986, [**227**]), Goldston & Trudgian & Turnage-Butterbaugh (2023, [**433**]), Goldston & Turnage-Butterbaugh (2019, [**434**]), Hall (many papers including (1999, [**482**]) and (2008, [**483**]), Montgomery & Odlezko (1984, [**769**]), Mueller (1981, [**784**]), Ng (2008, [**804**]), Simonič et al. (2022, [**992**]), Slezeviciene-Steuding & Steuding (2007), and Soundararajan (1996, [**1002**]); see also Titchmarsh (1986, [**1059**, §14.34]).

However, if the pair correlation conjecture (discussed below) were to be established, it would automatically imply that $\lambda = 0$ and $\lambda' = \infty$, meaning that arbitrarily small and arbitrarily large *normalized* gaps would occur infinitely often. If, furthermore, the zeros behaved like the eigenvalues of large random GUE matrices (also discussed below), then it would follow from work of Ben Arous and Bourgade (2013, [**86**]) that

$$\limsup_{n\to\infty} (\gamma_{n+1} - \gamma_n)\sqrt{\frac{\log \gamma_n}{32}} = 1.$$

Of related interest, in his proof that $\gamma_{n+1} - \gamma_n \to 0$, Littlewood (1924, [**705**]) had established, *unconditionally*, that

$$\gamma_{n+1} - \gamma_n = O\left(\frac{1}{\log\log\log \gamma_n}\right), \quad n \to \infty.$$

Ivić (1995, [**588**]), however, pointed out that the formula (6.2) for $N(T)$, together with the bound (6.5) for $S(T)$ *under RH*, would suffice to establish that RH implies

$$\gamma_{n+1} - \gamma_n = O\left(\frac{1}{\log\log \gamma_n}\right);$$

Goldston & Gonek (2007, [**424**]) made this quantitative by showing (under RH) that

$$\gamma_{n+1} - \gamma_n \leq \frac{\pi}{\log\log \gamma_n}(1 + o(1)), \quad \text{as} \quad n \to \infty.$$

By assuming the GUE hypothesis, Odlyzko (1987, [**818**]) had posited that

$$\max_{\gamma_n \geq T} (\gamma_{n+1} - \gamma_n) \sim \frac{8}{\sqrt{2\log T}}, \quad \text{as} \quad T \to \infty,$$

which—since $\sqrt{32}$ and $8/\sqrt{2}$ are equal—is consistent with the conditional lim sup result of Ben Arous & Bourgade mentioned above. On the other hand, Joyner (1992, [**608**]) had shown that if the lim inf and lim sup in Montgomery's result (6.6) were finite, it would follow that

$$\max_{\gamma_n \geq T} (\gamma_{n+1} - \gamma_n) = O\left(\frac{1}{\sqrt{\log T \log\log T}}\right),$$

and this is not consistent with the Ben Arous and Bourgade result. Thanks to result (6.7) of Bondarenko and Seip it is now known that the hypothesis Joyner used does not hold. Simonić (2022, [**991**]) may be consulted for some additional details.

Concerning the quantities $\sum(\gamma_{n+1} - \gamma_n)^k$, Fujii (1975, [**396**]), (1981, [**397**]) proved unconditionally that for any $k > 1$ there exist constants $0 < C_1(k) < C_2(k)$ and $T(k)$ such that

$$C_1(k)\frac{N(T)}{(\log T)^k} \leq \sum_{0 \leq \gamma_n \leq T} (\gamma_{n+1} - \gamma_n)^k \leq C_2(k)\frac{N(T)}{(\log T)^k}, \quad \text{for} \quad T \geq T(k).$$

Since $N(T) \sim (T/2\pi)\log T/2\pi$, this implies that for any $k \geq 1$,

$$\sum_{0 \leq \gamma_n \leq T} (\gamma_{n+1} - \gamma_n)^k = O\left(\frac{T}{\log^{k-1} T}\right).$$

A more explicit version of this approximation was conjectured by Ivić (1995, [**588**, eqns. 10 & 11]), while assuming RH, Fujii (1990, [**398**]) established the explicit bound

$$\sum_{0 \leq \gamma_n \leq T} (\gamma_{n+1} - \gamma_n)^2 \leq \frac{18\pi T}{\log(T/2\pi)},$$

for T large enough. Further discussion of such properties may be found in Ivić (2018, [**591**]).

6.8. The pair correlation conjecture

To discuss the behaviour of the zeros in further detail requires a broader view of dependence than arises in typical statistical models, because the spacings of the zeros do not correspond to any of the standard time series or point process models. Of pivotal significance here is the *pair correlation conjecture* of Montgomery (1973, [**763**]), arrived at in early 1972 using reasoning involving assumptions which included RH. This conjecture states that if we fix $0 < \alpha < \beta$, and define

$$N(\alpha, \beta; T) = \#\left\{m, n \colon 0 < \gamma_m, \gamma_n \leq T, \ \alpha \leq \frac{\gamma_m - \gamma_n}{2\pi/\log T} \leq \beta\right\}, \tag{6.38}$$

then

$$N(\alpha, \beta; T) \sim N(T) \int_\alpha^\beta \left(1 - \left(\frac{\sin \pi u}{\pi u}\right)^2\right) du \,, \quad \text{as } T \to \infty. \tag{6.39}$$

REMARKS 6.1.
(1) The scaling factor $2\pi/\log T$ in the denominator in (6.38) represents the average spacing between *adjacent* zeros within the indicated $[0, T]$ range.
(2) The factor $N(T)$ in (6.39) may be replaced by an asymptotic value such as $(T/2\pi)\log T$.
(3) The normalization $N(T)$ in (6.39) is of smaller order than the number of γ_m, γ_n pairs involved in the definition in (6.38). Correspondingly, the integral—if taken over $(-\infty, \infty)$—is unbounded. If taken over $[0, x]$, the integral will be asymptotic to x as $x \to \infty$.
(4) If we allow $\alpha < 0 < \beta$, then a delta function term $(T/2\pi)\delta(\alpha, \beta)$—where $\delta(\alpha, \beta)$ is 1 if $0 \in (\alpha, \beta)$ and 0 otherwise—must be added to the right side of (6.39) to cover the additional $\gamma_m = \gamma_n$ cases which would then arise.
(5) The term 'pair correlation' as used here is a property of point processes which differs in meaning from the term 'correlation' as used in classical statistics.
(6) The (equivalent) form in which Montgomery initially stated his conjecture in 1972 is

$$\sum_{\substack{0<\gamma,\gamma'\leq T \\ 0<\gamma-\gamma'\leq 2\pi a/\log T}} 1 \sim \frac{T}{2\pi}\log T \int_0^a \left(1 - \left(\frac{\sin \pi u}{\pi u}\right)^2\right) du.$$

(7) One can rephrase (6.38) using the alternate version $\widetilde{\gamma}_j \equiv \gamma_j \log \gamma_j / 2\pi$, and the differences $\widetilde{\gamma}_j - \widetilde{\gamma}_k$.
(8) The pair correlation function $1 - \left(\sin(\pi r)/\pi r\right)^2$ equals 0 at $r = 0$, rises to 1 at $r = 1$, and thereafter oscillates, tending to 1 from below.
(9) Odlyzko (1987, [**818**]) found that the fit of the pair correlation conjecture to data was essentially perfect.

One way to view pair correlation (also known as 2-point correlation) is that if an element is selected at random from the ensemble, it gives the probability intensity for finding elements at points a given distance away. Alternately, it is the probability intensity for two points to be a given distance apart. (A more precise definition is given below.) Conjecture (6.38)/(6.39) states that the number of pairs of distinct zero ordinates in $[0, T]$, whose difference is less than α times the average gap between consecutive zero ordinates in $[0, T]$, is asymptotic to

$$N(T) \int_0^\alpha \left(1 - \left(\frac{\sin \pi u}{\pi u}\right)^2\right) du.$$

For small α the value of this integral is approximately $\pi^2\alpha^3/9$. (Exercise: Show this.) If the zeros were distributed randomly (that is, according to a Poisson process) the pair correlation would be constant and the value of the corresponding integral would be proportional to α. The fact that α^3 is much smaller than α when α is small is an indication that the zeros of the zeta function are repelling one another.

The concept of pair correlation also arises when studying the energy levels of heavy nuclei, the distribution of particles in gasses and colloids, and the cosmic distribution of galaxies. In complex systems the large number of interactions renders it infeasible to determine any of the levels or positions individually, but substantial information can still be obtained by understanding the distribution of the spacings between them. The mean number of galaxies within a given distance of a randomly chosen one, for instance, can be obtained by a suitable integral over a pair correlation function; see for example Peebles (1980, [**838**]), Springel et al. (2005, [**1015**]), or Coil (2013, [**213**]). As another application, Totsuji & Kihara (1969, [**1062**]) found that the pair correlation for the spatial distribution of galaxies in the universe satisfies an inverse power of distance law, indicating that the universe is in a state of "neutral stability."

6.9. The random matrix theory connection

An often repeated story concerns the serendipitous encounter at tea-time in the afternoon of April 6th, 1972, between Hugh Montgomery and Freeman Dyson (1923–2020) at the Institute for Advanced Studies in Princeton, New Jersey—an encounter that had a significant impact on the face of analytic number theory going forward.[7] Dyson was a British-born mathematician turned theoretical physicist who had learned about the zeta function from a course by G. H. Hardy, and had for a while studied number theory under Harold Davenport. (As an undergraduate he had even toyed with Waring's problem about whether every integer can be written as a sum of seven cubes.) He is known for major contributions to quantum electrodynamics, solid-state physics, and astronomy, as well as for a number of ideas bordering (quite literally) on science fiction. He also did not believe the universe was a product of chance and had once said: "The more I examine the universe and the details of its architecture, the more evidence I find that the universe must in some sense have known we were coming" [**308**].

The possible energy levels of a quantum mechanical system correspond to the eigenvalues of a Hermitian operator on an infinite-dimensional Hilbert space (known as the Hamiltonian of the system). The properties of such eigenvalues are often approximated by those of the eigenvalues of certain corresponding finite-dimensional matrices. In the early 1950s, in connection with studies on nuclear fission, Eugene Wigner[8] (1902–1995) suggested that the statistical properties of the discrete energy levels of heavy nuclei might be modelled by the eigenvalues of large random matrices, and he conjectured that these should be governed by 'universal' statistical limit laws.

[7]Montgomery, a graduate student at Cambridge under Harold Davenport at the time, was briefly visiting the IAS in order to ask Atle Selberg whether or not his pair correlation result might already be known. It was Sarvadaman Chowla who insisted on introducing Montgomery to Dyson. The 50th anniversary of the Montgomery–Dyson meeting was celebrated by a conference held at the Institute for Advanced Studies in June of 2022; the talks given at that conference may be accessed on the IAS website.

[8]Random matrix theory (RMT) owes much of its early history to the atom bomb. Along with fellow Hungarians Edward Teller and John von Neumann—not to mention other luminaries such as Robert Oppenheimer, Enrico Fermi, and Richard Feynman—Wigner was a leading figure in the Manhattan Project, the nuclear weapons undertaking of World War II. For an engaging biography of Wigner see Szanton (1992, [**1038**]). The account of the Manhattan Project in Bhattacharya (2022, [**99**]) is particularly worthwhile.

Wigner's conjecture turned out to be prescient. Dyson had worked on random matrix theory, and in the 1960s had studied three types of matrix 'ensembles' corresponding to different symmetry structures: the Gaussian orthogonal ensemble (GOE), Gaussian unitary ensemble (GUE), and Gaussian symplectic ensemble (GSE). Each of these corresponds to different assumptions on the physical system which they are meant to model, regarding time reversal invariance and even or odd spin invariance (spatial rotational symmetry). Upon hearing Montgomery's pair correlation formula $1 - (\sin \pi u/\pi u)^2$ for the distribution of the differences between the zeros of the zeta function, Dyson—without even a moment's hesitation—told a stunned Montgomery that it coincides precisely with the pair correlation of the eigenvalues of large random Hermitian or unitary matrices—a seeming "miracle" connecting prime number theory to theoretical physics and suggesting that the spacing statistics for (high) zeros of $\zeta(s)$ follow the same laws. Also fortuitous was that the result for random matrices had already been worked out—by Dyson and by Wigner—a mere ten years earlier. The foreword by Dyson in The Oxford Handbook of Random Matrix Theory (Akemann et al. (2011, [**11**])) is well worth reading for further background.

Odlyzko (1987, [**818**]) computed the empirical pair correlation function for some eight million zeros of the zeta function near the 10^{20}th zero, and found it to be in perfect statistical agreement with that of the GUE. Montgomery raised the possibility, and Hejhal, Rudnick, and Sarnak, and Katz and Sarnak later demonstrated (under assumptions similar to those used by Montgomery), that all of the *many-level correlation functions* of the zeta zeros may in fact be identical to those of the GUE. While these results do not prove, they certainly strongly hint, that the limiting distribution of the zeros of the zeta function is identical to the one that asymptotically governs the eigenvalues of the GUE.

The literature on random matrix theory is not always user-friendly, and it is difficult to find accounts, unencumbered by considerable technical detail, of those aspects of random matrix theory that we need. We therefore offer a brief outline here.[9] To begin with basic definitions, the matrices of interest, $X \equiv [x_{i,j}] \equiv [x_{i,j}]_{i,j=1}^N$, are all $N \times N$, typically with $N \to \infty$. If the entries of X are real, it is *symmetric* if it equals its transpose, $X' = X$, so that $x_{j,i} = x_{i,j}$, and *orthogonal* if $XX' = X'X = I$. If its entries are complex, it is *Hermitian* if equal to its conjugate transpose $X^* = X$, so that $x_{j,i} = \overline{x_{i,j}}$, and *unitary* if $XX^* = X^*X = I$. The *eigenvalues* of symmetric or of Hermitian matrices are real; those of orthogonal or of unitary matrices lie on the unit circle in the complex plane. (The eigenvalues of symmetric-orthogonal or Hermitian-unitary matrices can thus only take on values $+1$ or -1.) In the orthogonal real case, the eigenvalues occur in complex conjugate pairs. In all of these cases, the *eigenvectors* can be taken to be real-valued and orthogonal to one another. There are many books on matrix algebra that can be consulted for more detail.

The definitions of the GOE, GUE, and GSE are often stated incompletely. These matrix "ensembles" correspond to probability models involving specified sample spaces with specified probability structures. The ubiquity of normal (that is, Gaussian) variables in their various definitions stems from their orthogonality

[9] Some of the discussion that follows involves concepts from linear algebra and may be read more lightly by readers who do not have this background.

invariance properties. The *Gaussian Orthogonal Ensemble* (GOE) is based on real-valued $N \times N$ symmetric matrices whose $N(N+1)/2$ free entries are independent 0-mean Gaussian random variables. Its N diagonal entries have variance 1, and its (free) $N(N-1)/2$ off-diagonal entries have variance $\frac{1}{2}$. Samples from the GOE may be generated by noting that if Z is an $N \times N$ matrix with independent standard normal entries, then $X = \frac{1}{2}(Z + Z')$ will have the GOE distribution. If X follows the GOE distribution, the joint density function of its free elements is easily shown to be proportional to $\exp\left(-\frac{1}{2}\operatorname{Trace} X^2\right)$, a comparatively tractable form.[10] The symmetry property which characterizes this model is that if X is GOE then so is $O'XO$ for any orthogonal matrix O; this symmetry property, together with independence of its free elements, determines the GOE probability model to within a (common) constant factor for the variances. Note that the GOE does not consist of orthogonal matrices, and furthermore that the product of two GOE matrices is not GOE.

The *Gaussian Unitary Ensemble* (GUE) is based on complex-valued $N \times N$ Hermitian matrices whose free real and imaginary entries are independent 0-mean normal random variables. The N diagonal elements are real and have variances $\frac{1}{2}$; the $N(N-1)$ free real and imaginary off-diagonal elements all have variances $\frac{1}{4}$. A sample from the GUE may be generated by noting that if Z is $N \times N$, where its $2N^2$ real and imaginary components are independent 0-mean Gaussians with variance $\frac{1}{2}$, then $X = \frac{1}{2}(Z + Z^*)$ has the GUE distribution.[11] The joint density of the free elements of a GUE matrix is easily shown to be proportional to $\exp\left(-\operatorname{Trace} X^2\right)$. The symmetry property which characterizes the GUE model is that if X is GUE then so is U^*XU for any unitary matrix U. As in the case of the GOE, this symmetry property, together with independence of its free elements, determines the GUE probability model to within a (common) constant factor for the variances. Note that the GUE does not consist of unitary matrices, and that the product of two GUE matrices is not GUE. As the quaternion-based *Gaussian Symplectic Ensemble* (GSE) will not play a role below we do not discuss it here.

The eigenvalues, $\lambda_1, \lambda_2, \ldots, \lambda_N$, which arise from the $N \times N$ random matrices of the GOE, GUE, or GSE models are all real-valued, and their joint probability density functions are known. They may be determined by laborious transformation of variables methods as in Mehta (2004, [**749**, Theorem 3.3.1]), or Forrester (2010, [**388**, Proposition 1.3.4]). In all three cases their density functions have the form

$$(6.40) \qquad f_\beta(\lambda_1, \lambda_2, \ldots, \lambda_N) = \frac{1}{Z(N;\beta)} \exp\left(-\frac{\beta}{2}\sum_{j=1}^{N} \lambda_j^2\right) \prod_{1 \le j < k \le N} |\lambda_j - \lambda_k|^\beta .$$

The normalizing constants $Z(N;\beta)$—referred to as the *partition function* in the context of statistical mechanics—were originally evaluated by Selberg and are known

[10]This is an easy exercise: Since the free matrix elements are independent Gaussians, one only needs to multiply up their densities. The GUE case is similar.

[11]The entries of this Z are standard complex normal variables. The *standard complex normal distribution* is that of a complex-valued random variable $Z = U + iV$ with U and V independent $N\left(0, \frac{1}{2}\right)$ variables. Its probability density function is given by $f(z) = (1/\pi)e^{-|z|^2}$ where $z = u + iv$.

explicitly:

$$Z(N;\beta) = (2\pi)^{N/2}\beta^{-N/2-\beta N(N-1)/4}\prod_{j=1}^{N}\frac{\Gamma\left(1+\frac{j\beta}{2}\right)}{\Gamma\left(1+\frac{\beta}{2}\right)}. \tag{6.41}$$

The cases of interest are $\beta = 1$, which corresponds to the GOE, $\beta = 2$, which corresponds to the GUE, and $\beta = 4$, which corresponds to the GSE. Function (6.40) does correspond to a density for other values of $\beta > 0$, but owing to their physical symmetry relevance, it is the values 1, 2, and 4 for β that are of primary interest in random matrix theory. One may think of β as counting the number of real Gaussian variables constituting the matrix entries, namely 1, 2, or 4, according as the entries are real, complex, or quaternion, respectively. For a long time it was thought that the $\beta = 1, 2, 4$ densities were the only ones that could arise for eigenvalues of random matrices. However, Dumitriu & Edelman (2002, [**299**]) showed how to construct matrices resulting in eigenvalue distributions corresponding to any possible value of $\beta > 0$. These matrices are tri-diagonal, with standard Gaussian entries on the main diagonal, and various gamma distributed entries on the two off-diagonals (all independent).

Because (6.40) is symmetric in the λ's, it can represent their density whether or not the λ's are considered to be ordered, the only difference then being a factor of $N!$ in the normalization constant. We therefore state explicitly that (6.40)/(6.41), as written, are for the case of *unordered* eigenvalues, so that this density is nonzero over all of $\mathbb{R}^N$. In passing, note that (6.40) can be written alternately as

$$f_\beta(\lambda_1, \lambda_2, \ldots, \lambda_N) = \frac{1}{Z(N;\beta)}\exp\left(-\frac{\beta}{2}\sum_{j=1}^{N}\lambda_j^2 - \beta\sum_{j<k}\log\frac{1}{|\lambda_j - \lambda_k|}\right).$$

If the $\log\left(1/|\lambda_j - \lambda_k|\right)$ terms are interpreted as the interaction strengths between pairs of particles, the negative of the exponent will represent the Hamiltonian (or potential energy) of a so-called Coulomb gas, in which β is proportional to the inverse of the 'temperature' of the system.

The *n-point correlation function* corresponding to the point process determined by (6.40) (which actually is just a marginal density function) is defined, for $n < N$, as

$$R_n(\lambda_1, \ldots \lambda_n) = \frac{N!}{(N-n)!}\int_{-\infty}^{\infty}\cdots\int_{-\infty}^{\infty} f_\beta(\lambda_1, \lambda_2, \ldots, \lambda_N)\, d\lambda_{n+1}\cdots d\lambda_N. \tag{6.42}$$

(Note that a second subscript of R, namely N, has been suppressed.) Introduced by Dyson (1962, [**307**]), it may be interpreted as saying that $R_n(\lambda_1, \ldots, \lambda_n)\delta\lambda_1\cdots\delta\lambda_n$ is the probability that there is a point—regardless of how it happens to be indexed—in each of the nonoverlapping infinitesimal intervals $(\lambda_j, \lambda_j + \delta\lambda_j)$, for $j = 1, \ldots, n$. The probability of there being a point in each of the disjoint intervals $E_1, E_2, \ldots, E_n$ is thus given by

$$\int_{E_n}\cdots\int_{E_1} R_n(\lambda_1, \ldots, \lambda_n)\, d\lambda_1 d\lambda_2\cdots d\lambda_n.$$

Viewed this way, the concept of n-point correlation is useful for describing the stochastic properties of point processes with very general dependence structures. (In statistical parlance, the R_n are sometimes referred to as *joint intensity functions*.)

Consider next the $N \to \infty$ limiting behaviour of eigenvalues distributed according to (6.40) in the GUE case $\beta = 2$. To begin with, the one-point correlation function $R_1(\lambda)$ will just be the density function for the distribution of the eigenvalues on the real line. In papers between 1951 and 1968 (for example, see [**1106**]), Wigner showed that after suitable normalization this has limiting distribution given by the *semicircle law* with density

$$f(\lambda) = \frac{2}{\pi}\sqrt{1-\lambda^2}, \quad |\lambda| \le 1. \tag{6.43}$$

Specifically, if the eigenvalues $\lambda_1, \ldots, \lambda_N$ are normalized as $\tilde{\lambda}_j = \lambda_j / 2\sqrt{N}$, then the density of the limiting empirical distribution of the $\tilde{\lambda}_j$ is given by (6.43). This semicircle law is very general and holds for a wide variety of distributions on the matrix elements; it even holds for all other values of $\beta > 0$. Some properties of the density (6.43), details of how it is derived, and an explanation for the $\sqrt{N}$ normalization of the eigenvalues are given in the RMT complements (items (4) and (6)).

Dyson showed that, as $N \to \infty$, the pair correlation function associated with the $\beta = 2$ case of (6.40) tends to

$$R_2(\lambda_1, \lambda_2) \;=\; 1 - \left(\frac{\sin(\pi(\lambda_1 - \lambda_2))}{\pi\lambda_1 - \lambda_2}\right)^2, \tag{6.44}$$

this being precisely the same pair correlation as in Montgomery's conjecture. In fact, Dyson (1962, [**307**]) had shown more generally that the asymptotic density for the n-level correlations of the GUE are given by the determinant of an $n \times n$ matrix:

$$R_n(\lambda_1, \lambda_2, \ldots, \lambda_n) \;=\; \det\left(\left[K(\lambda_i - \lambda_j)\right]_{i,j=1}^{n}\right), \quad \text{where } K(\lambda) = \frac{\sin \pi\lambda}{\pi\lambda}.$$

A process of this type is called a *determinantal point process*, while $K(\lambda)$ is referred to as its *correlation kernel*. Note that for $n = 2$ this gives

$$R_2(\lambda_1, \lambda_2) = \det\begin{bmatrix} K(0) & K(\lambda_1 - \lambda_2) \\ K(\lambda_1 - \lambda_2) & K(0) \end{bmatrix} \;=\; 1 - \left(\frac{\sin \pi(\lambda_1 - \lambda_2)}{\pi(\lambda_1 - \lambda_2)}\right)^2,$$

in agreement with (6.44). It is a remarkable property of this process that all of its marginal densities are determinantal processes of precisely the same type. The literature on determinantal point processes is quite technical. They are discussed in Akemann (2011, [**11**]), Baik, Deift & Suidan (2016, [**63**]), Dyson (1962, [**307**]), Forrester (2010, [**388**]), Hough et al. (2006, [**562**]), (2009, [**563**]), Johansson (2006, [**603**]), Kulesza & Taskar (2012, [**665**]), Mehta (2004, [**749**], and Pastur & Shcherbina (2011, [**834**]).

Many other probability models involving random matrices arise in physics. Of particular interest are the *Circular Orthogonal Ensemble* (COE), the *Circular Unitary Ensemble* (CUE), and (of less interest here) the *Circular Symplectic Ensemble* (CSE). Unlike the GOE, GUE, and GSE, the circular models are based on groups—with matrix multiplication being the group operator—and because these groups are compact topological groups each is endowed with a unique uniform probability measure (the so-called Haar measure). This allows one to speak of sampling *uniformly* from the group in the sense that the probability $\mathrm{P}[A]$ of any (measurable) subset A of group elements is invariant under translation by a member X of the group: $\mathrm{P}[A] = \mathrm{P}[XA] = \mathrm{P}[AX]$. These groups can therefore be thought of as corresponding to probability spaces. The COE is based on the group $O(N)$ of (real) orthogonal matrices, and the CUE on the group $U(N)$ of (complex) unitary matrices. Uniform (meaning Haar-distributed) samples from the orthogonal group $O(N)$ can be constructed by Gram–Schmidt orthonormalization of $N \times N$ matrices consisting of independent 0-mean normal variables with common variances, while Haar-distributed samples from the unitary group $U(N)$ can be constructed by Gram–Schmidt orthonormalization of $N \times N$ matrices consisting of complex entries, with all $2N^2$ real and imaginary parts being independent 0-mean normal variables with common variance. Specifically, for real matrices, normalize the first row; at successive rows remove the linear combination of the previous rows leaving a vector orthogonal to them, and normalize it. The analogous procedure, but based on the complex form of inner product, is used for unitary matrices. Due to the normalization steps the (common) variance of the matrix elements may initially be set to any value. The CSE group is analogous but based on quaternions. For numerical work and for large matrices, faster and more stable procedures for generating such random matrices are available.

Since the eigenvalues of matrices from any of the circular groups all lie on the unit circle we may represent them as $\{e^{i\theta_1}, e^{i\theta_2}, \ldots e^{i\theta_N}\}$. (If required, they may be ordered—for example according to $0 \le \theta_1 \le \theta_2 \le \cdots \le \theta_N < 2\pi$.) For the orthogonal group, eigenvalues will occur in complex conjugate pairs. The joint probability densities of the phases $\theta_1, \ldots, \theta_N$ from all three of the circular models have been determined to be

$$f(\theta_1, \ldots, \theta_N) = \frac{1}{Z(N;\beta)} \prod_{1 \le j < k \le N} \left| e^{i\theta_j} - e^{i\theta_k} \right|^\beta, \tag{6.45}$$

with $\beta = 1$, 2, and 4 corresponding to the COE, CUE, and CSE cases, respectively. The normalizing constant $Z(N;\beta)$, conjectured by Dyson, and proved by Selberg, is given by

$$Z(N;\beta) = (2\pi)^N \frac{\Gamma\left(1 + \frac{N\beta}{2}\right)}{\Gamma\left(1 + \frac{\beta}{2}\right)^N}. \tag{6.46}$$

Here, as before, (6.45)/(6.46) represent the density for the *unordered* eigenvalues. The asymptotic distributional structures of the COE, CUE, and CSE eigenvalues turn out to be identical to those of the GOE, GUE, and GSE, respectively, Consequently the properties of the CUE and the GUE have equal relevance for the number theoretic context. In particular, if we normalize the eigenvalues from the CUE to have unit spacing by setting $\widetilde{\theta}_j = \frac{N}{2\pi}\theta_j$, then, as $N \to \infty$, their pair correlation function will satisfy

$$\frac{1}{N} \#\left\{1 \le j < k \le N\colon\ \alpha \le \widetilde{\theta}_k - \widetilde{\theta}_j \le \beta\right\} \sim \int_\alpha^\beta \left(1 - \left(\frac{\sin \pi u}{\pi u}\right)^2\right) du, \tag{6.47}$$

just as for the GUE, and just as for Montgomery's conjecture for the zeros of $\zeta(s)$.

6.10. RMT complements

As was the case at the end of Chapter 5, there remains a great deal more to tell. We therefore close this chapter—and our discussion of random matrix theory—with the following series of eighteen remarks.

REMARKS 6.2.

(1) The role of random matrix theory in suggesting properties of the zeros of $\zeta(s)$ mirrors that of Cramér's heuristic for the primes. However, the RMT predictions have turned out to be more reliable, with substantial numerical evidence supporting the conjecture that the statistical behaviour of the zeros *exactly* matches the asymptotic behaviour of eigenvalues of the GUE and CUE. Yet for the zeros, virtually none of their distributional properties has been proven *unconditionally*. Curiously, no compelling reason is known as to why the two should have the same distributional asymptotics beyond the fact that they may just be sharing some 'universal' statistical properties. (By way of analogy, the Gaussian distribution and Central Limit Theorem are another example of a 'universal' property applicable to a wide range of unrelated phenomena.)

(2) The fact that the pair correlation for the Riemann zeros matches the pair correlation for the GUE to high numerical accuracy gave a boost to an old proposal of David Hilbert and George Pólya that there may be a Hermitian or unitary linear operator $\mathcal{H}$ on some Hilbert space (whose eigenvalues are necessarily real-valued) such that the eigenvalues of $\frac{1}{2} + i\mathcal{H}$ exactly match the zeros of ζ. Indeed RH is equivalent to the existence of such an operator. Not much progress has resulted from this proposal. However, see Berry & Keating (1999, [**94**]).

(3) *Wigner–Dyson universality.* Although Gaussian entries are analytically the simplest to treat, the exact distribution of the matrix entries appears not to matter for the asymptotics of their eigenvalues. [László] Erdős, Schlein & Yau (2009, [**331**]), and independently Tao, Vu & Krishanpur (2010, [**1050**]), have shown that the asymptotic eigenvalue statistics of a matrix do not depend on the distribution of its elements as long as they are independent, have finite variance, and share the same distribution. In fact, it has long been known that many other types of matrices (for instance banded $N \times N$ random matrices with at least $\sqrt{N}$ bands) also share these properties. This provides yet another meaning for the term 'universality'.

(4) The semicircle density $f(x)$ shown in (6.43) has all odd moments (including its mean) equal to 0. Its variance is $\frac{1}{4}$, and its higher even moments are given by

$$EX^{2k} = 2\frac{(2k-1)!!}{(2k+2)!!}$$

where $n!! \equiv n(n-2)(n-4)\cdots(2 \text{ or } 1)$, according as n is even or odd. As this density has bounded support, it is determined by its moments. So Wigner's semicircle law can be proved by computing limiting moments (for example by using the trace formula in item (6)) and showing they match those of $f(x)$. See, for example, Tao (2012, [**1041**, §2.4]). See also Mehta & Gaudin (1960, [**750**]).

(5) An elegant way of proving the semicircle law can be based on the convergence of Stieltjes transforms. The Stieltjes transform of a distribution function F on $\mathbb{R}$ is

$$S_F(z) \equiv E_F\left(\frac{1}{X-z}\right) = \int_{-\infty}^{\infty} \frac{1}{x-z}\, dF(x) = -\frac{1}{z}\sum_{k=0}^{\infty} \frac{EX^k}{z^k}$$

where $z = u + iv$ is complex valued with $v \neq 0$, and X is a random variable distributed according to F. The last equality holds under suitable conditions.

The density function corresponding to F may be recovered from the imaginary part, $\Im S_F$, of the Stieltjes transform via

$$f_X(u) = \lim_{v \downarrow 0} \frac{1}{\pi} \Im S_F(u + iv).$$

The Stieltjes transform of the empirical distribution of the eigenvalues $\lambda_1, \ldots, \lambda_n$ of a matrix A is simply

$$S_n(z) = \frac{1}{n} \sum_{j=1}^{n} \frac{1}{\lambda_j - z} = \mathrm{Trace}(A - zI)^{-1},$$

and if a sequence of Stieltjes transforms converges pointwise then the corresponding convergence in distribution also holds. This method has been used in RMT theory by Zhedong Bai (1999, [**61**]), one advantage of the method being that moment computations can then be bypassed. Stieltjes transforms are discussed in Widder (1946, [**1101**, Chapter VIII]). See also Bose, Chatterjee & Gangopadhyay (2003, [**137**]).

(6) That $\sqrt{N}$ is the correct order of normalization for the eigenvalues in (6.43) may be intuited from the *trace formula*

$$\mathrm{Trace}\left(A^k\right) = \sum_{j=1}^{N} \lambda_j^k, \quad k = 1, 2, 3, \ldots,$$

which relates the elements $a_{i,j}$ of a matrix A to its eigenvalues λ_j. For $k = 2$, and when the $a_{i,j}$ are independent random variables with means 0 and common positive variance σ^2, this gives

$$\sum_{j=1}^{N} \lambda_j^2 = \mathrm{Trace}\left(A^2\right) = \sum_{i=1}^{N} \sum_{j=1}^{N} a_{i,j}^2 \sim \sigma^2 N^2,$$

so that the average λ_j^2 is of order N. So it is plausible that the average λ_j has order $\sqrt{N}$.

(7) Dyson's *'threefold way'*: In a series of five papers titled 'Statistical theory of the energy levels of complex systems'—published in the Journal of Mathematical Physics in 1962—Dyson proposed that there are only three ensembles of irreducible matrices that are physically realistic in the sense of being invariant under a symmetry group. Dyson classified them into three types: unitary symmetric, Hermitian, or symplectic, corresponding to real, complex, or quaternion variables. These are the $\beta = 1, 2$, or 4 cases, respectively. Following up on Dyson ideas it later emerged that there may actually be ten—not just three—symmetry types which may be relevant for physics and random matrix theory—the *tenfold way*—however the issues involved are complicated; see for example Baez (2020, [**56**]).

(8) The limiting density for normalized eigenvalues is given by semicircle law (6.43). For the limiting density of the *spacings* between consecutive normalized eigenvalues, Wigner, in the 1950s, *surmised* the (mean 1) density

$$\frac{\pi}{2} x \exp\left(-\frac{\pi x^2}{4}\right), \quad x > 0. \tag{6.48}$$

Although not correct, it is in fact remarkably close to the true density whose formula is far more complicated. Density (6.48) (like the true density) tends to 0 at the origin, again reflecting the repulsion property of the eigenvalues. 'Surmise' (6.48) is characterized by the assumption that (a) given a level (that is, an eigenvalue) at E, the probability of there being another level around $E+S$ is proportional to S for small S, together with (b) an assumption of independence for nonoverlapping intervals. Neither of these two assumptions is correct, but their two errors compensate each other extremely well. The exact distribution for the spacings was derived in a *tour de force* by Gaudin (1961, [**411**]). This exact law is also universal, holding for a wide range of distributions on matrices although it has been proven only for Gaussian ensembles. Mehta (2004, [**749**]) may be consulted for details.

(9) RMT eigenvalues repel each other (in Dyson's words: "exactly as if they were like charges, repelling each other with a force varying inversely with the first power of the distance" [**307**]). This may be seen from the fact that densities (6.40) and (6.45) tend to 0 when any pair of eigenvalues become close. It may also be seen from the fact that the pair correlation drops to 0 at the origin.

(10) The product involving the *differences* of eigenvalues in (6.40) constitutes a *repulsion factor*. It is this product that determines their asymptotic behaviour. The term involving

$$\exp\left(\sum_j \lambda_j^2\right)$$

may be replaced by other terms of the form

$$\exp\left(\sum_j g(\lambda_j)\right)$$

without affecting the asymptotic distribution of the spacings. A similar remark applies to (6.45).

(11) Products of paired differences can be expressed as Vandermonde determinants:

$$\Delta(x_1,\ldots,x_n) \equiv \det\left(\left[x_k^{j-1}\right]_{j,k=1}^n\right) = \prod_{1\le j<k\le n} (x_k - x_j)\,. \tag{6.49}$$

Thus, for example,

$$\prod_{1\le j<k\le n} \left|e^{i\theta_j} - e^{i\theta_k}\right|^\beta = \left|\Delta(e^{i\theta_1},\ldots,e^{i\theta_n})\right|^\beta,$$

which for $\beta = 2$ gives

$$\prod_{1\le j<k\le N} \left|e^{i\theta_j} - e^{i\theta_k}\right|^2 = \det\left(\left[S(\theta_k-\theta_j)\right]_{j,k=1}^n\right), \quad \text{where} \quad S(\theta) = \frac{\sin(n\theta/2)}{\sin(\theta/2)}.$$

An easy proof of (6.49) involves noting that both sides are homogeneous polynomials of degree $n(n-1)/2$. Since these both vanish whenever $x_j = x_k$, they must be identical up to a constant factor, but the coefficient of $x_n^{n-1}x_{n-1}^{n-2}\cdots x_2$ equals 1 in both cases.

(12) *Variance saturation.* Eigenvalues of the GUE and the GOE are spaced more rigidly than points that are independent and uniformly distributed, and over long intervals they pull towards more uniform spacing. Let $M_n(a,b)$ represent the number of eigenvalues of an $n\times n$ matrix from the GUE that fall in the arc $\left(e^{2\pi ia}, e^{2\pi ib}\right)$ of the unit circle. Here $0\le a<b\le 1$, and the proportion

$L \equiv b - a$ which this arc forms of the circle is assumed to be fixed but small. Then $\mathrm{E}\, M_n(a,b) = n(b-a) = nL$. However the variance of M_n does not scale as n (it would equal nL if the points were fully random) but rather it scales as $\log n$:

$$\operatorname{Var} M_n(a,b) \;=\; \frac{\log nL}{\pi^2} + O(1).$$

Under the GOE the result would be $(2/\pi^2)\log nL + O(1)$. The random sampling implicit in these results can be based on moving the arc around the circle using the eigenvalues of a single random matrix, or alternately fixing the arc, and sampling the matrices randomly—that is, spatial versus ensemble averaging. This phenomenon also occurs in eigenvalues stemming from other mathematical structures and had even been encountered in early work by Hardy and by Cramér.

Selberg (1944, [**970**]) and Fujii (1975, [**396**]) proved (unconditionally) that variance saturation also occurs for the Riemann zeros and matches that of the GUE. Using intervals for $\Im s$ in which the expected number of zeros is L, the number variance is asymptotic to $(1/\pi^2)\log L$ as $L \to \infty$. Further details may be found in Berry (1988, [**90**]), Coram & Diaconis (2003, [**234**]), Diaconis (2003, [**279**, §3]), Jain et al. (1972, [**598**]), Johannson (2004, [**602**]), Mehta (2004, [**749**, Chapter 12]), Montgomery (1973, [**763**]), Odlyzko (1989, [**819**]), and Wieand (2000, [**1103**, §5]).

(13) *Strange correlations.* Continuing with the GUE and the notation of the previous item, let $M_n(a,b)$ and $M_n(c,d)$ be the number of points falling in the non-identical arcs $(e^{2\pi ia}, e^{2\pi ib})$ and $(e^{2\pi ic}, e^{2\pi id})$, respectively. Then in the $n \to \infty$ limit, the correlation

$$\operatorname{Corr}\big(M_n(a,b), M_n(c,d)\big) \;\to\; +\frac{1}{2},\; -\frac{1}{2},\; \text{ or } 0, \tag{6.50}$$

according to the three cases $a = c$, $b = c$, or (a,b) and (c,d) share no boundary. (In the third case, the intervals may be fully disjoint or one may be properly contained within the other.) It is at first glance not intuitive that (6.50) complies with the nonnegative definiteness property that correlations must satisfy. See Coram & Diaconis (2003, [**234**]) and Diaconis (2003, [**279**, §3]).

(14) In RMT, the distribution of the elements on the diagonal can be more or less arbitrary without affecting the asymptotic behaviour of the eigenvalues. This is a consequence of the fact that the proportion of elements that are on the diagonal tends to 0.

(15) While computations by Odlyzko (1987, [**818**]) and Odlyzko and Schonhage (1988, [**823**]) of very high zeros do show them to be in statistical agreement with the limiting eigenvalue distribution of the GUE or CUE, the zeros of the zeta function approach their asymptotic behaviour *very* slowly. Nevertheless, the convergence is convincing by the 10^{13}th zero, and is virtually perfect by the 10^{20}th zero which occurs around $T \approx 0.15 \times 10^{20}$. By way of contrast, for $n \times n$ random matrices the limiting distribution is already 'reached' around $n = 20$. (This is sometimes viewed as a reason for *not* believing that the zeros of $\zeta(s)$ are the eigenvalues of some unitary or Hermitian operator.) Since zeros at height T have approximate density $\frac{1}{2\pi}\log\frac{T}{2\pi}$, while eigenvalues of $n \times n$ random unitary matrices have density $\frac{n}{2\pi}$ on the unit circle, Coram & Diaconis (2003, [**234**]) suggested equating n with $\log\frac{T}{2\pi}$ (so as to match the mean spacings) and chose

to compare the statistical properties of zeros at height $T = 0.15 \times 10^{20}$ with those of GUE eigenangles for $n = 42$ in their study.

Keating and Snaith suggested modelling the function $\zeta(s)$ itself at height T by the characteristic polynomial of a matrix of size $N \approx \log T$. Their numerical experiments based on Odlyzko's data are convincing.

(16) *The general circular law.* Let $X_n = [x_{j,k}]_{j,k=1}^{n}$ be an $n \times n$ matrix whose entries $x_{j,k}$ are independent, identically distributed mean 0 variance 1 complex-normal random variables. (Each $x_{j,k}$ consists of independent normal, mean 0, variance $\frac{1}{2}$, real and imaginary parts so that X_n is composed of $2n^2$ independent normals in all.) Note that this matrix is neither orthogonal nor Hermitian. Ginibre (1965, [**415**]) proved that the joint density of the (complex-valued) eigenvalues of X_n is given by

$$\text{Const} \times \left(\prod_{j \neq k} |\lambda_j - \lambda_k|^2 \right) \times \exp\left(-\frac{1}{2} \sum_{j=1}^{n} |\lambda_j|^2 \right).$$

Using this, Girko (1984, [**416**]) proved that, as $n \to \infty$, the empirical distribution of the eigenvalues of $\frac{1}{\sqrt{n}} X_n$ tends to the uniform distribution over the unit disc $\{|z| \leq 1\}$ of the complex plane. Bai (1997, [**60**]) substantially relaxed the normality requirement, while the general result requiring no more than a 0-mean was established by Tao, Vu & Krishanpur (2010, [**1050**]). In fact, the matrix may be real or complex valued. Tao (2012, [**1041**]), and Bose, Chatterejee & Gangopadhya (2003, [**137**]) provide details and references. The problem of relaxing the independence assumption has not been solved.

(17) Although random matrix theory owes its origins to the work of Eugene Wigner, Michel Gaudin, Madan Lal Mehta, and Freeman Dyson dating back to the 1950's, RMT first arose in the field of multivariate statistical analysis with Wishart's derivation in 1928 of the distribution of estimated covariance matrices for Gaussian finite samples (Such matrices are of the form $X'X$.) Marčenko & Pastur (1967, [**729**]) studied the empirical distribution of the eigenvalues in the limit of large covariance matrices. Theory for random matrices arising in statistics is reviewed for example in Dieng & Tracy (2011, [**288**]), Izenman (2020, [**596**]), Johnstone (2007, [**605**]), and Paul & Aue (2014, [**837**]).

(18) Formulas are available for the limiting densities of the largest and smallest eigenvalues of a matrix, referred to as 'edge effects'. For instance, the smallest eigenvalue of an $n \times n$ matrix consisting of independent standard complex Gaussian variables satisfies

$$\lim_{n\infty} \mathrm{P}\left[n \cdot \lambda_{\min} \geq x\right] = e^{-x-x^2/2}.$$

See Edelman (1989, [**309**]), Shen (2001, [**984**]), and Tracy & Widom (2000, [**1064**]).

The literature on random matrix theory is extensive but generally technical. Miller & Takloo-Bighash (2006, [**756**, Chapters 15, 16, and 17]) offer an amiable introduction. See also Diaconis (2003, [**279**]), and the articles in the volume by Forrester, Snaith & Verbaarschot (2003, [**389**]). The classical work by Mehta (2004, [**749**]) is harder to follow although other books on this subject are harder still. Important technical references include Anderson, Guionnet & Zeitouni (2010, [**21**]), and Forrester (2010, [**388**]). For an extensive review from a physics viewpoint together with many historical details see Brody et al. (1981, [**154**]). Hayes (2003, [**521**]) gives a nice non-technical overview. Further details and developments can be pursued by consulting Goldston, Gonek, Özlük & Snyder (2000, [**425**]), Conrey & Snaith (2008, [**230**]), (2014, [**231**]), Tao (2012, [**1041**]), and the references therein.

6.11. Selected proofs

PROOF THAT (6.2) IMPLIES (6.12) AND (6.13). We need to show that

$$N(T) = \frac{T}{2\pi}\log\frac{T}{2\pi} - \frac{T}{2\pi} + \frac{7}{8} + S(T) + O\left(\frac{1}{T}\right)$$

implies both

$$\sum_{0<\gamma\leq T}\frac{1}{\gamma} = \frac{1}{2\pi}\left(\frac{1}{2}\log^2 T - (\log 2\pi)\log T\right) + O(1), \text{ as } T\to\infty,$$

and

$$\sum_{\gamma>T}\frac{1}{\gamma^2} = \frac{1}{2\pi T}\left(3\log\frac{T}{2\pi} - 1\right) + O\left(\frac{\log T}{T^2}\right), \text{ as } T\to\infty.$$

Select any value of a greater than 0 but less than γ_1 (the height of the first zero). Then applying (6.2) and using Riemann–Stieltjes integration we obtain

$$\begin{aligned}
\sum_{0<\gamma\leq T}\frac{1}{\gamma} &= \int_a^T \frac{1}{u}\,dN(u) \\
&= \frac{N(u)}{u}\bigg|_a^T - \int_a^T N(u)\,d\left(\frac{1}{u}\right) \\
&= \frac{N(T)}{T} + \int_a^T \frac{N(u)}{u^2}\,du \\
&= \frac{1}{2\pi}\log\frac{T}{2\pi e} + O\left(\frac{\log T}{T}\right) + \frac{1}{2\pi}\int_a^T\left(\frac{1}{u}\log\frac{u}{2\pi e} + O\left(\frac{\log u}{u^2}\right)\right)du \\
&= \frac{1}{2\pi}\log\frac{T}{2\pi e} + \frac{1}{2\pi}\left[\frac{1}{2}\log^2 u - (\log 2\pi e)\log u\,\bigg|_a^T + O\left(\frac{\log T}{T}\right)\right. \\
&= \frac{1}{2\pi}\log\frac{T}{2\pi e} + \frac{1}{2\pi}\left[\frac{1}{2}\log^2 T - (\log 2\pi e)\log T\right] + O(1) \\
&= \frac{1}{4\pi}\log^2 T - \frac{\log 2\pi}{2\pi}\log T + O(1), \text{ as } T\to\infty,
\end{aligned}$$

which settles (6.12). For (6.13) we have:

$$\begin{aligned}
\sum_{\gamma>T}\frac{1}{\gamma^2} &= \int_T^\infty \frac{1}{u^2}\,dN(u) \\
&= \frac{N(u)}{u^2}\bigg|_T^\infty - \int_T^\infty N(u)\,d\left(\frac{1}{u^2}\right) \\
&= \frac{N(T)}{T^2} + 2\int_T^\infty \frac{N(u)}{u^3}\,du \\
&= \frac{1}{2\pi T}\log\frac{T}{2\pi e} + O\left(\frac{\log T}{T^2}\right) + \frac{2}{2\pi}\int_T^\infty\left(\frac{1}{u^2}\log\frac{u}{2\pi e} + O\left(\frac{\log u}{u^3}\right)\right)du
\end{aligned}$$

$$\begin{aligned}
&= \frac{1}{2\pi T}\log\frac{T}{2\pi e} + \frac{1}{\pi}\left[-\frac{1}{u}\log u - \frac{1}{u} + \frac{\log 2\pi e}{u}\Big|_T^\infty + O\left(\frac{\log T}{T^2}\right)\right.\\
&= \frac{1}{2\pi T}\log\frac{T}{2\pi e} + \frac{1}{\pi T}\log\frac{T}{2\pi} + O\left(\frac{\log T}{T^2}\right)\\
&= \frac{3}{2\pi T}\log\frac{T}{2\pi} - \frac{1}{2\pi T} + O\left(\frac{\log T}{T^2}\right), \text{ as } T\to\infty. \qquad \square
\end{aligned}$$

PROOF THAT $\zeta(s)$ HAS NO ZEROS ON THE LINE $\sigma = 1$. One of the simplest proofs known is based on a 'trick' due to Mertens (1898). It involves the inequality

$$(6.51)\qquad 3 + 4\cos\theta + \cos 2\theta = 2(1+\cos\theta)^2 \ \geq\ 0, \quad \text{for real } \theta,$$

but is otherwise straightforward. First note that for $\sigma > 1$

$$\log\zeta(s) = \log\prod_p \left(1-p^{-s}\right)^{-1} = -\sum_p \log\left(1-p^{-s}\right) = \sum_p\sum_{m=1}^\infty \frac{1}{p^{ms}} = \sum_{n=2}^\infty \frac{c_n}{n^s}$$

where $c_n = \frac{1}{m}$ if $n = p^m$ and 0 otherwise, and also note that

$$\frac{c_n}{n^s} = \frac{c_n}{n^{\sigma+it}} = \frac{c_n}{n^\sigma}\big(\cos(t\log n) - i\sin(t\log n)\big).$$

We therefore have

$$\log|\zeta(s)| = \Re\log\zeta(s) = \sum_{n=2}^\infty \frac{c_n}{n^\sigma}\cos(t\log n).$$

Consequently

$$\begin{aligned}
\log\left|\zeta^3(\sigma)\zeta^4(\sigma+it)\zeta(\sigma+2it)\right| &= 3\log|\zeta(\sigma)| + 4\log|\zeta(\sigma+it)| + \log|\zeta(\sigma+2it)|\\
&= \sum_{n=2}^\infty \frac{c_n}{n^s}(3 + 4\cos(t\log n) + \cos(2t\log n)) \ \geq\ 0
\end{aligned}$$

due to (6.51) and the fact that the $c_n \geq 0$. Therefore

$$\left|\zeta^3(\sigma)\zeta^4(\sigma+it)\zeta(\sigma+2it)\right| \geq 1, \quad \sigma > 1,$$

and we may rewrite this as

$$(6.52)\qquad |(\sigma-1)\zeta(\sigma)|^3 \left|\frac{\zeta(\sigma+it)}{\sigma-1}\right|^4 |\zeta(\sigma+2it)| \ \geq\ \frac{1}{\sigma-1}.$$

Now suppose $\sigma(1+it) = 0$ for some $t = t_0 \neq 0$. To see that this isn't possible, set $t = t_0$ in (6.52) and let $\sigma \downarrow 1$. Then the right side tends to $+\infty$, but the left side tends to

$$1\cdot|\zeta'(1+it_0)|^4\cdot|\zeta(1+2it_0)|$$

which is finite on account of the fact the $\zeta(s)$ is analytic when $s \neq 1$. $\square$

REMARK 6.3. This proof is modelled after Chandrasekharan (1968, [**188**, pp. 123–124 & pp. 1, 63–83, 122–130]). Montgomery & Vaughan (2006, [**774**, p. 193]) use the same trigonometric inequality and point out that it is equivalent to the identity $1 - \cos 2\theta \leq 4(1+\cos\theta)$ used by de la Vallée Poussin. Strömbergsson (2021, [**1036**, p. 123–124]) also offers a proof.

CHAPTER 7

Statistics of the Critical Line

If people do not believe that mathematics is simple, it is only because they do not realize how complicated life is.

John von Neumann [**20**]

In Section 5.3 we began to examine the behaviour of $\zeta(s)$ along vertical lines in the complex plane. In view of the functional equation—which allows us to compute the values of $\zeta(s)$ when $\sigma < 1/2$ from its corresponding values at $\sigma > 1/2$—we will confine our attention here to the half-plane $\sigma \geq 1/2$. To begin with, recall that if we fix $\sigma > 1$, the function $\zeta(\sigma + it)$ will be a linear combination of mostly noncommensurate sinusoids (in t), and bounded away from 0 and ∞ according to

$$\liminf_{|t|\to\infty} |\zeta(s)| = \frac{\zeta(2\sigma)}{\zeta(\sigma)} < |\zeta(s)| \leq \zeta(\sigma) = \limsup_{|t|\to\infty} |\zeta(s)|. \tag{7.1}$$

Furthermore, we will see in Chapter 10 that (for $\sigma > 1$) the rescaled version $\varphi_\sigma(t) \equiv \zeta(\sigma + it)/\zeta(\sigma)$ is the characteristic function of a discrete distribution which takes on its values at $\log n$, for $n = 0, 1, 2, \ldots$.

On the line $\sigma = 1$, as $t \to \infty$, both $\zeta(1 + it)$ and $1/\zeta(1 + it)$ are unbounded, never touch 0, and are each of order at most $O((\log t)^{2/3+\epsilon})$ unconditionally, and as low as $O(\log\log t)$ if the Riemann hypothesis is assumed. In the region $1/2 < \sigma < 1$ far less is known and some conditional as well as unconditional results for that case were considered in Section 5.3.

That leaves the case of the critical line (namely $\sigma = 1/2$) with that line being the one on which—if the Riemann hypothesis is true—all of the complex zeros are to be found. It is to this singularly important case that this chapter is devoted.

7.1. The case $\sigma = 1/2$ and the Lindelöf hypothesis

On the critical line itself, the true order of the zeta function is a deep mystery. Indeed, it is not even known whether or not its values on that line are dense in all of $\mathbb{C}$. In connection with $\zeta\left(\frac{1}{2} + it\right)$, there is an important consequence of the Riemann hypothesis that was recognized by Ernst Leonard Lindelöf (1870–1946) in the 1890s (and published by him in 1908), which led to the highly influential *Lindelöf hypothesis* (LH). This conjecture states, quite simply, that

$$\zeta\left(\frac{1}{2} + it\right) = O\left(|t|^\epsilon\right), \quad \text{as} \quad |t| \to \infty, \quad \text{for every } \epsilon > 0, \tag{7.2}$$

and is widely believed to be true. (Note that the order term here can also be written as $o\left(|t|^\epsilon\right)$ since ϵ is arbitrary.) While RH implies LH, there is no apparent implication in the opposite direction (although see Gonek, Graham & Lee, (2020, [**440**])). Nevertheless LH has not proven to be much easier to establish.

Bäcklund (1918/1919, [**55**]) had established that the Lindelöf hypothesis is *equivalent* to the assertion that for any $\sigma^* > \frac{1}{2}$ the number of zeros of $\zeta(s)$ in the region $\sigma > \sigma^*$, $T \le t \le T+1$ is $o(\log T)$. That result is one of the ways one can establish that RH implies LH; see Titchmarsh (1986, [**55**, §13.5]).

The truth of the Lindelöf hypothesis would have nontrivial consequences,[1] such as a strong form of the density hypothesis, and the existence of primes in all intervals $[x, x + x^{1/2+\epsilon}]$ for x large enough, as well as in almost all intervals of the form $[x, x + x^{\epsilon}]$. It would also imply that $\pi(x + x^{\delta}) - \pi(x) \sim x^{\delta}/\log x$ for every $\delta > 1/2$. Furthermore, any *explicit* bound of the type $\left|\zeta(\frac{1}{2} + it)\right| < Ct^{\alpha}$ for $t > t_0$ would lead to a bound for the zero-free region, as well as a bound on the arg function $S(t)$ associated with the zero-counting function of $\zeta(s)$, with better bounds resulting from better values of the constants C, α, and t_0. Even so, however, the values of $\left|\zeta(\frac{1}{2} + it)\right|$ are typically much smaller than the sharpest bounds known, although all bounds do need to account for the rare occasions when $\left|\zeta(\frac{1}{2} + it)\right|$ is suddenly comparatively large.

A more obvious reason for interest in LH is that if it were shown to be false it would contradict RH. However, in spite of many efforts since the 1890s to prove LH, all that is known for certain is that (7.2) holds for a value of ϵ just slightly less than $\epsilon = \frac{1}{6} = .1666\ldots$. In particular, in 1908 Lindelöf himself obtained the bound $O(|t|^{1/4+\epsilon})$, while Hardy & Littlewood (1923, [**494**]) established the highly nontrivial bound $O\left(|t|^{1/6}\log^{3/2}|t|\right)$. Later efforts by Kolesnik (1985, [**646**]), Bombieri & Iwaniec (1986, [**123**]), and Huxley (2005, [**573**]) successively reduced the value of ϵ in (7.2) to $\epsilon = \frac{139}{858} \approx .1620$, then $\frac{9}{56} \approx .1607$, and then more steeply to $\frac{32}{205} \approx .1561$. The best asymptotic bound currently known is due to Bourgain (2017, [**139**]) who established (7.2) for all values of ϵ which exceed $\frac{13}{84} \approx .1548$.

Based on work of Riemann and Siegel, Landau had obtained the *explicit* (unconditional) upper bound

$$\left|\zeta\left(\frac{1}{2} + it\right)\right| \le 4\left(\frac{t}{2\pi}\right)^{1/4}, \quad \text{for } t \ge \frac{32}{\pi},$$

while Cheng & Graham (2004, [**198**]) obtained the 'van der Corput–Weyl bound'

$$\left|\zeta\left(\frac{1}{2} + it\right)\right| \le 3t^{1/6}\log t, \quad \text{for } t \ge e.$$

The constants 3 and e here were later improved by Platt & Trudgian (2015, [**854**]) to 0.732 and 2, and then to 0.618 and 3 by Hiary, Patel, & Yang (2022, [**546**]). For additional details see Patel & Yang (2023, [**835**]).

A sharper version of the Lindelöf hypothesis was obtained by Littlewood (1924, [**704**]). Using a more careful argument he showed that RH implies there is a constant $C > 0$ such that

$$\left|\zeta\left(\frac{1}{2} + it\right)\right| \le \exp\left(C\frac{\log|t|}{\log\log|t|}\right). \tag{7.3}$$

[1] The density hypothesis $N(\sigma; T) = O(T^{2-2\sigma+\epsilon})$ which (7.2) implies was discussed in Chapter 6, as was the notion of zero-free regions, and the definition of the function $S(t)$.

(This, in fact, is the $\sigma = \frac{1}{2}$ case of (5.13).) Chandee & Soundararajan (2011, [**187**]) (also under RH) obtained the more explicit version

$$\left|\zeta\left(\frac{1}{2}+it\right)\right| \leq \exp\left(\left(\frac{1}{2}\log 2 + o(1)\right)\frac{\log|t|}{\log\log|t|}\right), \quad \text{as} \quad |t| \to \infty,$$

where the term $o(1) \to 0$ as $|t| \to \infty$, while Simonič (2022, [**991**]) obtained the fully *explicit*

$$\left|\zeta\left(\frac{1}{2}+it\right)\right| \leq \exp\left(0.995\frac{\log|t|}{\log\log|t|}\right), \quad \text{for} \quad |t| \geq 2\pi.$$

(See also Ramachandra & Sankaranarayanan (1993, [**881**]), and Soundararajan (2008, [**1005**]).) However, aside from such successive reductions in values for C, no sharpening of Littlewood's (7.3) version of LH is known. In the other direction—following earlier work by Montgomery (1977, [**764**]), Titchmarsh (1986, [**1059**, Theorem 8.12]), Soundararajan (2008, [**1005**]), and Aistleitner (2016, [**10**])—Bondarenko & Seip (2017, [**128**]), and de la Bretèche & Tenenbaum (2019, [**150**]) established—*unconditionally*—the lower bound

$$\max_{0\leq t\leq T}\left|\zeta\left(\frac{1}{2}+it\right)\right| \geq \exp\left(\left(\sqrt{2}+o(1)\right)\sqrt{\frac{\log T \log\log\log T}{\log\log T}}\right). \tag{7.4}$$

The gulf between the upper bound (7.2)/(7.3) and lower bound (7.4) is considerable. Even so, however, the true maximal size of $|\zeta(1/2+it)|$ is believed to be much smaller than (7.3); yet Littlewood's bound has withstood further improvement for more than 100 years.

REMARKS 7.1.

(i) If Littlewood's bound (7.3) were proven to hold, it would then also hold beyond $\sigma = \frac{1}{2}$, and, for $|t|$ large enough, we would have more generally

$$|\zeta(s)| \leq \exp\left(C\frac{\log|t|}{\log\log|t|}\right)$$

uniformly for all $\sigma \geq \frac{1}{2}$ (but note that the result (5.13) is stronger).

(ii) The Lindelöf hypothesis can be shown to be *equivalent* to the 'zeta sum' assertion

$$\text{LH} \iff \left|\sum_{n\leq x} n^{it}\right| = O\left(\sqrt{x}\,|t|^{\epsilon}\right), \quad 1 \leq x \leq \sqrt{|t|}, \ x \to \infty.$$

Details of this equivalence can be found in Karatsuba & Voronin (1992, [**622**, pp. 162–163]).

(iii) The Lindelöf hypothesis can be shown to imply, for any $\sigma^* > 1/2$, that $\zeta(s)$ can be approximated in the half-plane $\Re s \geq \sigma^*$ by short initial segments of its Dirichlet series; see Gonek (2012, [**439**]).

7.2. The maxima of $\zeta(1/2+it)$ over intervals

Work pertaining to the size of $\left|\zeta(\frac{1}{2}+it)\right|$ over intervals has followed two streams. In this section, as well as in the next, we will consider the behaviour of this function over the whole of $0 \leq t < \infty$, or over whole segments such as $0 \leq t \leq T$, or $T \leq t \leq 2T$, with $T \to \infty$. There is, however, important recent work on the behaviour of $\left|\zeta(\frac{1}{2}+it)\right|$ over short intervals such as $T \leq t \leq T+1$ for large values of T. This second stream will be taken up in Chapter 12.

Over large intervals $[0, T]$, Farmer, Gonek & Hughes (2007, [**369**]) have conjectured that

$$\max_{0 \le t \le T} \log \left| \zeta\left(\frac{1}{2} + it\right) \right| \sim \sqrt{\frac{1}{2} \log T \log\log T}, \quad \text{as} \quad T \to \infty. \tag{7.5}$$

They arrived at this conjecture by relying on an analogy with random matrix theory together with a randomness model for the primes (although they did not rule out the possibility that Littlewood's bound might be closer to the truth). Some support for conjecture (7.5) is offered by a central limit result of Selberg (to be described later in this section) which implies that the values of $\log|\zeta(1/2 + it)|$ on $0 \le t \le T$ behave like $N(0, (1/2) \log\log T)$ random variables. By considering the nature of the dependence amongst the points of $\log|\zeta(1/2 + it)|$ it has been argued that its maximum over $0 \le t \le T$ should behave like the maximum of a sample of T independent draws from this normal distribution. A probability calculation then leads to the conjecture

$$\max_{0 \le t \le T} \log \left| \zeta\left(\frac{1}{2} + it\right) \right| \approx \sqrt{2 \log T} \sqrt{(1/2) \log\log T} = \sqrt{(\log T)(\log\log T)},$$

with the point being that this result is similar to (7.5). Background concerning such arguments may be found in Arguin et al. (2017, [**30**]), Harper (2019, [**510**]), Ivić (1985, [**586**, pp. 191, 197]), Ivić (2013, [**590**, p. 13]), and Huxley & Ivić (2007, [**574**]). In particular, Harper discusses the role played by assumptions of independence when conjectures such as (7.5) are formulated; should those assumptions prove to be unreliable then the bound (7.3) of Littlewood might well turn out to be the one closer to the truth.

Further insight into LH can be formulated in terms of the *Lindelöf* $\mu(\sigma)$ *function*

$$\mu(\sigma) \equiv \inf\{\alpha \colon \zeta(\sigma + it) = O(|t|^{\alpha}) \text{ as } |t| \to \infty\}.$$

Trivially, $\mu(\sigma) = 0$ when $\sigma > 1$ because $|\zeta(\sigma + it)|$ is bounded by $\zeta(\sigma)$ there. The functional equation—together with the complex version of Stirling's formula for the Gamma function (Appendix B)—may be used to show that $\mu(\sigma) = \frac{1}{2} - \sigma$ for $\sigma \le 0$. In other words, $\mu(\sigma)$ starts out as $\frac{1}{2}$ when $\sigma = 0$, and then rises upward with slope 1 when heading left; see Figure 7.1.

Now, a theorem of Phragmen and Lindelöf (Appendix A) on the interpolation of growth rates on vertical lines of analytic functions can be used to show that the function $\mu(\sigma)$ is convex. Therefore—as indicated by the upper dashed line in Figure 7.1—we must have $\mu(\sigma) \le \frac{1}{2}(1 - \sigma)$ for $0 \le \sigma \le 1$. In particular, this means that $\mu\left(\frac{1}{2}\right)$ cannot exceed $\frac{1}{4}$. But (thanks to aforementioned result of Hardy & Littlewood) it is known that $\mu\left(\frac{1}{2}\right) < \frac{1}{6}$. Accordingly, the function μ must at the very least satisfy $\mu(\sigma) \le \frac{1}{6}(3 - 4\sigma)$ for $0 \le \sigma \le \frac{1}{2}$, and $\mu(\sigma) \le \frac{1}{3}(1 - \sigma)$ for $\frac{1}{2} \le \sigma \le 1$. This is indicated by the two conjoined dotted line segments lying between the two dashed lines. Any lower values for $\mu(\frac{1}{2})$ will obviously allow improvements to these bounds.

The lower dashed line in Figure 7.1 is the continuation of the solid leftward-heading line, and gives values for $\mu(\sigma)$ if the Lindelöf hypothesis were to hold. This is because the Lindelöf hypothesis asserts that $\mu\left(\frac{1}{2}\right) = 0$, which if true would imply (again by convexity) that $\mu(\sigma) = 0$ when $\sigma \ge \frac{1}{2}$, and $\mu(\sigma) = \frac{1}{2} - \sigma$ when $\sigma \le \frac{1}{2}$. Thus, under LH, the function $\mu(\sigma)$ would consist of two straight line segments meeting at the point $(\frac{1}{2}, 0)$, as shown in the figure. And if RH were also true, then

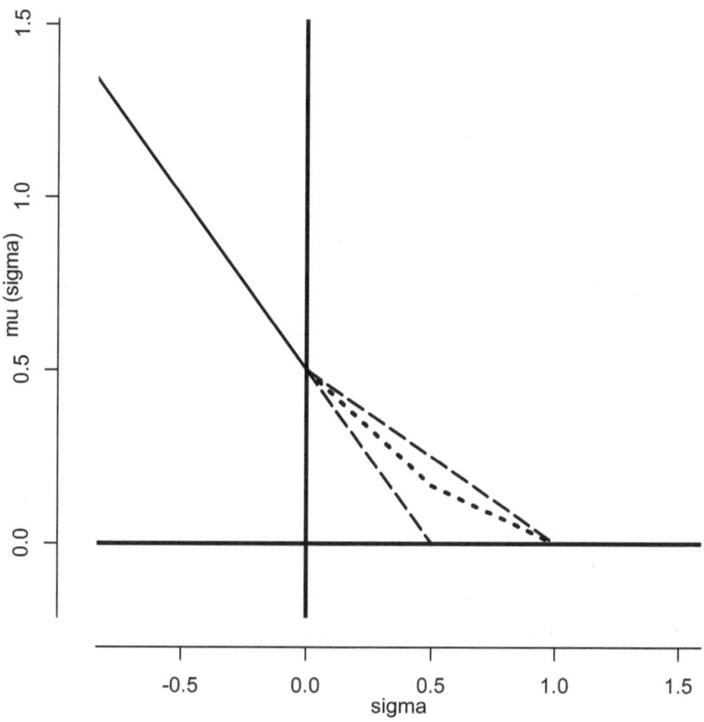

FIGURE 7.1. The Lindelöf function $\mu(\sigma)$: For $\sigma > 1$, $\mu(\sigma) = 0$, and for $\sigma \leq 0$, $\mu(\sigma) = \frac{1}{2} - \sigma$ (solid line). Convexity requires $\mu(\sigma) \leq \frac{1}{2}(1-\sigma)$ for $0 \leq \sigma \leq 1$ (upper dashed line). But since $\mu\left(\frac{1}{2}\right) < \frac{1}{6}$, $\mu(\sigma)$ is bounded above by the conjoined dotted line segments. LH asserts that $\mu\left(\frac{1}{2}\right) = 0$ and would lower the bound to the lower dashed line.

the more incisive version (5.13) of the assertion $\mu(\sigma) = 0$ for $\sigma \geq \frac{1}{2}$ would hold as well. Huxley & Ivić (2007, [**574**]) provide some further details.

7.3. Selberg's Central Limit Theorem

We now turn to a striking Central Limit Theorem (CLT) result for the value distribution of ζ on the critical line. It is stated in terms of

$$\log \zeta(1/2+it) \equiv \Re \log \zeta(1/2+it) + i\Im \log \zeta(1/2+it) \equiv \log|\zeta(1/2+it)| + i\pi S(t).$$

The arg here—namely the quantity $\Im \log \zeta(1/2+it)$—is taken to be $\pi S(t)$, where $S(T)$ is the function defined in Section 6.1 as $S(T) \equiv (1/\pi) \arg \zeta\left(\frac{1}{2}+iT\right)$ via continuous variation of the argument of the zeta function. (This arg is not restricted to lie in an interval of length 2π.) In 1946 Atle Selberg (in unpublished work) proved that if $E \subset \mathbb{C}$ is any (reasonable) subset of the complex plane, then

$$\lim_{T\to\infty}\frac{1}{T}\text{ measure}\left\{t\colon 0\le t\le T \text{ and } \frac{\log\zeta\left(\frac{1}{2}+it\right)}{\sqrt{\frac{1}{2}\log\log T}}\in E\right\}$$
$$=\frac{1}{2\pi}\iint_E e^{-(x^2+y^2)/2}\,dx\,dy.$$

This means, in essence, that the real and imaginary parts of $\log\zeta$ on the critical line behave like independent and normally distributed variables, whose means are 0 and whose variances are $\frac{1}{2}\log\log T$. This result follows from the convergence of moments to those of the mentioned Gaussian distribution; the relevant moments are given in (7.16) and (7.17) below. The singularities of $\log\zeta$ at the zeros do create technical problems, but as it happens, the (Lebesgue) measure of the problematical regions is not large enough to interfere with the result. Other unpublished probabilistic results due to Selberg are described by Hejhal in the Selberg volume [**532**]. A simplified proof of Selberg's result may be found in Radziwiłł & Soundararajan (2017, [**879**]).

Selberg's result implies, in particular, that over the interval $0\le t\le T$, the ratio

$$\log\left|\zeta(\frac{1}{2}+it)\right|/\sqrt{\frac{1}{2}\log\log T}$$

tends to a Gaussian random variable with mean 0 and variance 1, as $T\to\infty$. If we let τ be a random variable, and P_T be a probability measure under which τ has a uniform distribution on $[0,T]$, then writing $\sigma_T^2\equiv\frac{1}{2}\log\log T$ for the Selberg variance, the asymptotic normality result for the real part of the log can be expressed as

$$\mathrm{P}_T\left[\log\left|\zeta\left(\frac{1}{2}+i\tau\right)\right|\ge x\sigma_T\right]\ \to\ \frac{1}{\sqrt{2\pi}}\int_x^\infty e^{-\frac{1}{2}x^2}dx,\quad\text{as } T\to\infty. \tag{7.6}$$

Selberg's CLT mirrors a CLT result of Keating & Snaith (2000, [**627**]) for the characteristic polynomials of random matrices, further evidencing the link between RMT and properties of $\zeta(s)$ on the critical line. Specifically, these authors showed that if $Z(\theta;U)\equiv\det\left(I-Ue^{-i\theta}\right)=\prod_j\left(1-e^{i(\theta_j-\theta)}\right)$ is the characteristic polynomial of an $N\times N$ random unitary matrix with eigenvalues $e^{i\theta_j}$, then, under uniform (Haar) measure, and for fixed θ,

$$E\left(\left|U(\theta;U)\right|^{2k}\right)=\prod_{j=1}^{N}\frac{\Gamma(j)\Gamma(j+2k)}{\Gamma^2(j+k)},$$

and

$$\mathrm{P}\left(\frac{\log Z(\theta;U)}{\sqrt{\frac{1}{2}\log N}}\in E\right)\longrightarrow\frac{1}{2\pi}\iint_E e^{-(x^2+y^2)/2}\,dx\,dy,\quad\text{as } N\to\infty.$$

The distributional convergence promised by Selberg's CLT turns out to be extremely slow. Odlyzko (1997, [**821**]) found that its asymptotics are not reached, even by the 10^{20}th zero (where $T\approx1.52\times10^{19}$). Although the mean and variance for the (standardized) real part $\log|\zeta(\frac{1}{2}+it)|$ are already very close to 0 and 1 there, its 3rd and 4th moments are still only about -0.55 and 3.95 (instead of 0 and 3 for a standard Gaussian). The imaginary part however is substantially closer to its asymptotic distribution and has a more nearly symmetrical distribution than the real part in that vicinity. (Hints of this contrast can be seen in the $\sigma = 1/2$ curves back in Figure 5.2.)

It is worth taking a moment to intuit why Selberg's CLT result should be true. Although Euler's product for $\zeta(s)$ is not valid when $\Re s \le 1$, Hardy & Littlewood (1923, [**495**]) did nevertheless establish, *unconditionally*, that the zeta sum approximation

$$\zeta\left(\frac{1}{2}+it\right) = \sum_{n\le T} \frac{1}{n^{1/2+it}} + O\left(\frac{1}{\sqrt{T}}\right)$$

holds for all $T \le t \le 2T$, as $T \to \infty$. This in turn suggests a heuristic Euler product approximation such as

$$\zeta\left(\frac{1}{2}+it\right) \approx \prod_{p\le P}\left(1-\frac{1}{p^{1/2+it}}\right)^{-1} \approx \sum_{p\le P}\frac{1}{p^{1/2+it}} = \sum_{p\le P}\frac{e^{-it\log p}}{p^{1/2}}$$

for some suitable value of P. Now the exponentials $e^{-it\log p}$ in the numerator are 'rotational' terms whose frequencies, $\log p$, are not commensurate, so for most large values of t one reasonably expects the $e^{-it\log p}$ to behave like points that are independent and uniformly distributed on the unit circle. Such independence would induce the asymptotic normality of Selberg's result.[2] This heuristic will be taken up in greater detail in Chapter 12. A different heuristic involves identifying the distribution of $\zeta\left(\frac{1}{2}+it\right)$ with that of the characteristic polynomial of a matrix sampled randomly from the GUE. In random matrix theory it is known that if A is an $N \times N$ such matrix with N large, then

$$\frac{\log|\det(I-A)|}{\sqrt{\frac{1}{2}\log N}} \sim N(0,1).$$

Now in view of the Lindelöf hypothesis, large values of $\zeta\left(\frac{1}{2}+it\right)$ are of special interest. So it is natural to inquire if the normality result (7.6) holds 'in the tails', that is, when x is allowed to increase with T. (This is referred to as the case of *large deviations*.) If it did, then the so-called *Mills ratio* (also known as the Gaussian tail bound) would imply that[3]

$$\text{(7.7)} \qquad \mathrm{P}_T\left[\log\left|\zeta\left(\frac{1}{2}+i\tau\right)\right| \ge x\sigma_T\right] \sim \frac{1}{\sqrt{2\pi}}\frac{1}{x}e^{-\frac{1}{2}x^2}.$$

Here one first lets $T \to \infty$, and then lets $x \to \infty$. However, since (7.7) is not expected to hold if x is allowed to vary arbitrarily with T, it is natural to ask just how quickly x can be allowed to grow for (7.7) to continue to hold, and how quickly it can be allowed to grow so as to at least maintain the $O\big(\frac{1}{x}e^{-x^2/2}\big)$ order of magnitude. Following work by Jutila (1983, [**611**]), Radziwiłł (2011, [**876**]) established, unconditionally, that (7.7) continues to hold for $x \ll (\log\log T)^{1/10}$, a range that can be extended to $x \ll (\log\log T)^{1/6}$ if RH is assumed. Inoue (2019, [**584**, §2.3]) removed the assumption of RH from this latter result, but only for the upper bound

[2] This is a fairly straightforward exercise based on applying the CLT results of Appendix D.

[3] The *Mills ratio* (or tail bound) is the limit of the probability above x to the density at x. For the standard normal distribution it is given by

$$\left(\frac{1}{\sqrt{2\pi}}\int_x^\infty e^{-u^2/2}du\right)\Big/\left(\frac{1}{\sqrt{2\pi}}e^{-x^2/2}\right) \sim \frac{1}{x}, \quad \text{as } x\to\infty.$$

This result is easily proved using l'Hôpital's rule.

rather than for the actual asymptotic. Analogously, Soundararajan (2008, [**1005**]), (2009, [**1006**]) obtained upper bounds of the conjectured Gaussian shape for the case $x \gg \sqrt{\log\log T}$. See also Harper (2013, [**507**]), (2019, [**509**]), (2019, [**510**]). However, the question of how far into the tail one can go while still maintaining order $O\big(\frac{1}{x}e^{-x^2/2}\big)$ is unresolved.

Radziwiłł (2011, [**876**]) had conjectured that (7.7) should continue to hold even for x up to $o\big(\sqrt{\log\log T}\,\big)$, and that for x of size $O(\sqrt{\log\log T}\,)$, at least the $O\big(\frac{1}{x}e^{-x^2/2}\big)$ order which is implicit in (7.7) should hold. He further conjectured that for fixed $\alpha > 0$, and as $T \to \infty$,

$$(7.8)\quad \mathrm{P}_T\left[\log\left|\zeta\left(\frac{1}{2}+i\tau\right)\right| \geq \alpha\log\log T\right] \sim \frac{C_\alpha}{\sqrt{2\pi}}\int_{\alpha\sqrt{2\log\log T}}^{\infty} e^{-u^2/2}du \sim \frac{C_\alpha}{\sqrt{2\pi}}\cdot\frac{1}{\alpha(\log T)^{\alpha^2}\sqrt{2\log\log T}}.$$

The τ here is uniformly distributed on $[0,T]$ under P_T. The C_α is not equal to 1; instead (and somewhat unexpectedly) it is the leading coefficient of the 2αth moment of $\left|\zeta(\frac{1}{2}+it)\right|$. Radziwiłł's conjecture was based on calculations showing that, for a suitable choice for x (namely $x = T^{1/(\log\log T)^2}$), we have

$$(7.9)\quad \frac{1}{T}\,\text{measure}\left\{T \leq t \leq 2T\colon \Re\sum_{p\leq x}\frac{p^{-it}}{p^{1/2}} \geq \Delta\cdot\sqrt{\frac{1}{2}\sum_{p\leq x}\frac{1}{p}}\right\} \sim \frac{1}{\sqrt{2\pi}}\int_{\Delta}^{\infty} e^{-u^2/2}du,$$

uniformly in $\Delta = o\big(\sqrt{\log\log T}\,\big)$, and furthermore that this continues to hold for

$$\Delta \sim k\sqrt{\frac{1}{2}\log\log T}$$

as well, except that one must then attach a certain multiplicative constant to the right side. The $\Re\sum_{p\leq x}$ term in (7.9) is meant to approximate $\log|\zeta(1/2+it)|$, while the term under the square root gives the variance of the $\Re\sum_{p\leq x}$ term under the assumption that the p^{-it} are independent variables that are uniformly distributed on the unit circle. For subsequent extensions, see Fazzari (2021, [**371**]), (2021, [**372**]), as well as the more recent papers [**29**]–[**37**] by L.-P. Arguin and coauthors.

7.4. The moments of $\zeta\,(1/2+it)$

While questions about the *true order* of $\zeta(s)$ in the critical strip are extremely difficult, questions about their averages—quantities like $(1/T)\int_0^T|\zeta(\sigma+it)|^{2k}dt$—should (at least in principle) be a little easier to treat analytically, and somewhat more is known about them. Such results are referred to as *mean value theorems*. As in Section 7.3, the intervals for t considered here will be of the type $T \leq t \leq 2T$, with $T \to \infty$. (The results also hold for intervals $0 \leq t \leq T$, but the former allow for slightly greater generality.) Short intervals of the type $T \leq t \leq T+1$ will be taken up in Chapter 12.

To begin with, *off* the critical line it is known that[4]

$$\lim_{T\to\infty}\frac{1}{T}\int_T^{2T}|\zeta(\sigma+it)|^2\,dt=\zeta(2\sigma),\qquad \sigma>\frac{1}{2},$$

and

$$\lim_{T\to\infty}\frac{1}{T}\int_T^{2T}|\zeta(\sigma+it)|^4\,dt=\frac{\zeta^4(2\sigma)}{\zeta(4\sigma)},\qquad \sigma>\frac{1}{2}.$$

(Note that these also hold inside the critical region as long as $\sigma>1/2$.) At $\sigma=1$, the right hand sides here become $\pi^2/6$ and $5\pi^4/72$, respectively. We mention also the known identities

$$\lim_{T\to\infty}\frac{1}{T}\int_T^{2T}\Re\zeta(1+it)\,dt=1,\qquad \lim_{T\to\infty}\frac{1}{T}\int_T^{2T}\Im\zeta(1+it)\,dt=0,$$

$$\lim_{T\to\infty}\frac{1}{T}\int_T^{2T}\frac{1}{|\zeta(1+it)|^2}\,dt=\frac{\zeta(2)}{\zeta(4)}=\frac{15}{\pi^2},$$

and

$$\lim_{T\to\infty}\frac{1}{T}\int_T^{2T}\left|\frac{\zeta(2+2it)}{\zeta(1+it)}\right|^2dt=\zeta(2)=\frac{\pi^2}{6}.$$

Of course, it is the $\sigma=\frac{1}{2}$ case which is greatest interest. The Lindelöf hypothesis and Selberg's CLT result both lead to a focus on the *moments*

$$I_k(T)\equiv\frac{1}{T}\int_T^{2T}\left|\zeta\left(\frac{1}{2}+it\right)\right|^{2k}dt \tag{7.10}$$

of the zeta function on the critical line, since they potentially provide information on the extreme value behaviour of ζ on that line. In fact, Hardy & Littlewood (1916, [**492**]) had shown that the Lindelöf hypothesis is *equivalent* to the assertion

$$I_k(T)\equiv\frac{1}{T}\int_T^{2T}\left|\zeta\left(\frac{1}{2}+it\right)\right|^{2k}dt=O\left(T^{\epsilon}\right),\quad \text{for all } \epsilon>0, \text{ and all } k\geq 1, \tag{7.11}$$

and this, in turn, is *equivalent* to the assertion

$$\frac{1}{T}\int_T^{2T}|\zeta(\sigma+it)|^{2k}\,dt=O\left(T^{\epsilon}\right),\quad \text{for all } \sigma\geq\frac{1}{2},\ \epsilon>0, \text{ and all } k\geq 1.$$

These relations have been proven only for $k=1$ and $k=2$. We mention too that bounds such as these are also useful for establishing zero-density theorems. Altogether then, the moments of $|\zeta(\frac{1}{2}+it)|$ are of considerable importance.

Some insight into the utility of the moments $I_k(T)$ for studying the Lindelöf hypothesis can be obtained by replacing $2k$ by u in (7.10), and rewriting it as

$$\mathcal{M}(u)\equiv\frac{1}{T}\int_T^{2T}|\zeta(1/2+it)|^u\,dt=\frac{1}{T}\int_T^{2T}e^{u\log|\zeta(1/2+it)|}\,dt. \tag{7.12}$$

This is now the *moment generating function* of $\log|\zeta(1/2+it)|$ on $[T,2T]$, and as such is equivalent to the sequence of its moments as well as to its distribution there (see Appendix D). As u increases, the largest value of $\log|\zeta(1/2+it)|$ will dominate this expression because the exponential will magnify the relative difference between the maximum and any lesser values. In fact, $\mathcal{M}(u)^{1/u}$ is just the so-called '$\mathcal{L}^u$ norm'

[4]Were we to use intervals $[0,T]$ instead, these results would need need to be stated in the form $\lim(1/T)\int_1^T$ with the lower limit of 1 meant to bypass the singularity at $s=1$; but it may be taken as 0 when $\sigma\neq 1$.

of $|\zeta(1/2+it)|$; this will tend to its '$\mathcal{L}^\infty$ norm' as $u \to \infty$, and this is precisely the maximum value of $|\zeta(1/2+it)|$ on the $[T, 2T]$ interval. In particular,

$$\left(I_k(T)\right)^{1/2k},$$

as $k \to \infty$, must tend to this maximum value as well. By applying a certain 'multiscale decomposition' to $\log|\zeta(1/2+it)|$, Harper (2013, [**506**]) was able to establish (under RH) the exact order $I_k = O(T \log^{k^2} T)$; Soundararajan (2009, [**1006**]) had argued heuristically that this should hold regardless of whether or not RH was true. See also Heath-Brown (1981, [**528**]), Conrey & Ghosh (1984, [**224**]), Radziwiłł (2011, [**876**]), and Radziwiłł & Soundararajan (2013, [**878**]).

The emphasis on even moments in (7.10), at the expense of odd ones, is based in part on the fact that information on the even moments would suffice for settling Lindelöf's conjecture. But a further consideration is that computations involving even absolute moments can be simplified by means of the relation $|\zeta|^2 = \zeta\bar{\zeta}$, while no analogue is available for odd moments.

Notwithstanding previous remarks suggesting the contrary, the moments of the zeta function on the critical line—even their leading asymptotic behaviours—have proven very difficult to determine, and as mentioned, only two unconditional results are known. In the first of these, Hardy & Littlewood (1916, [**492**]) established the second moment result[5]

$$\int_T^{2T} \left|\zeta\left(\frac{1}{2}+it\right)\right|^2 dt \sim T\log T, \quad \text{as } T \to \infty, \tag{7.13}$$

which allows one to say that the average size of $\left|\zeta(\frac{1}{2}+it)\right|^2$ is $\log t$. It is now known more precisely that

$$\int_T^{2T} \left|\zeta\left(\frac{1}{2}+it\right)\right|^2 dt = T\left(\log\frac{T}{2\pi} + 2\gamma - 1\right) + \mathcal{E}(T), \tag{7.14}$$

with error term that satisfies $\mathcal{E}(T) = O\left(T^{1/3+\epsilon}\right)$, and even $\mathcal{E}(T) = O\left(T^{131/416+\epsilon}\right)$, although it has been conjectured that $\mathcal{E}(T) = O\left(T^{1/4+\epsilon}\right)$. Earlier contributions were made by Atkinson (1939, [**43**]) who obtained $\mathcal{E}(T) = O(\sqrt{T}\log T)$, and by Titchmarsh who in a series of papers obtained $O\left(T^{5/12}\log^2 T\right)$. (See Titchmarsh, 1986, [**1059**, pp. 145–146].) Further discussion of $\mathcal{E}(T)$ and additional references may be found in Ivić (1985, [**586**, pp. 83, 132, 443, & Chapter 15]), Iwaniec (2014, [**593**, p. 51]), Iwaniec & Kowalski (2004, [**594**, p. 557], and Titchmarsh (1986, [**1059**, p. 143, pp. 176ff]).

The other unconditional result is due to Ingham (1926, [**579**]) who established the fourth moment result

$$\frac{1}{T}\int_T^{2T} \left|\zeta\left(\frac{1}{2}+it\right)\right|^4 dt = \frac{1}{2\pi^2}\log^4 T + O\left(\log^3 T\right). \tag{7.15}$$

A much sharper version involving a fourth degree polynomial in $\log T$ was later obtained by Heath-Brown (1979, [**526**]):

$$\frac{1}{T}\int_T^{2T} \left|\zeta\left(\frac{1}{2}+it\right)\right|^4 dt = \sum_{k=0}^{4} c_k \log^{4-k} T + O\left(\frac{1}{T^{1/8-\epsilon}}\right).$$

[5]Note that in (7.13) and (7.14) we have not placed $1/T$ in front of the integrals.

Here $c_0 = 1/2\pi^2$ as in (7.15), while the remaining constant coefficients c_1, c_2, c_3, c_4 are explicitly known. An *exact* (although complicated) expression for the fourth moment was obtained, finally, by Motohashi (1997, [**782**]). See also Ivić (1991, [**587**, pp. 166, 203]), and Snaith (2000, [**996**, p. 88]).

There is also a class of results for moments of the logarithm of $\zeta(1/2+it)$. Specifically, assuming RH, Selberg (1946, [**970**]), (1946, [**972**]) had proved that

$$\begin{aligned}\frac{1}{T}\int_0^T (\Re\log\zeta(1/2+it))^{2k}\,dt &\equiv \frac{1}{T}\int_0^T \log^{2k}\left|\zeta(1/2+it)\right|dt \\ &= \frac{(2k)!}{k!2^{2k}}(\log\log T)^k + O\left((\log\log T)^{k-1}\right),\end{aligned} \tag{7.16}$$

and likewise that[6]

$$\begin{aligned}\frac{1}{T}\int_0^T (\Im\log\zeta(1/2+it))^{2k}\,dt &\equiv \frac{1}{T}\int_0^T (\arg\zeta(1/2+it))^{2k}\,dt \\ &= \frac{(2k)!}{k!2^{2k}}(\log\log T)^k + O\left((\log\log T)^{k-1}\right).\end{aligned} \tag{7.17}$$

These results also hold unconditionally, although with slightly weaker error terms. The leading terms on the right in (7.16) and (7.17) correspond to the moments of a Gaussian distribution and are what lie behind Selberg's asymptotic normality result for $\log\zeta(1/2+it)$. Note too that (7.17) has appeared previously in the discussion of the moments of $S(T)$ in Section 6.1.

For the case $k=1$ of (7.17) Goldston (1987, [**421**]) obtained the further result

$$\begin{aligned}\frac{1}{T}\int_0^T (\Im\log\zeta(1/2+it))^{2}\,dt &\equiv \frac{1}{T}\int_0^T (\arg\zeta(1/2+it))^{2}\,dt \\ &= \frac{1}{2}\log\log T + C + o(1),\end{aligned} \tag{7.18}$$

while for the case $k=1$ of (7.16), Lugar et al. (2023, [**713**]) found that

$$\begin{aligned}\frac{1}{T}\int_0^T (\Re\log\zeta(1/2+it))^{2}\,dt &\equiv \frac{1}{T}\int_0^T \log^{2}\left|\zeta(1/2+it)\right|dt \\ &= \frac{1}{2}\log\log T + C + o(1).\end{aligned} \tag{7.19}$$

In both of these cases, it is believed that

$$C = \frac{\gamma+1}{2} + \frac{1}{2}\sum_{m=2}^{\infty}\sum_p\left(\frac{1}{m^2}-\frac{1}{m}\right)\frac{1}{p^m}$$

with the inner sum being over the primes. The 1 which appears in the fraction $(\gamma+1)/2$ stems from a conjecture made by Goldston based on a strong version of the pair correlation conjecture; what is known for certain is that the true value must lie between 2/3 and 2. Using the relation $|x+iy|^2 = x^2+y^2$, (7.18) and (7.19) may be combined to give

$$\frac{1}{T}\int_0^T \left|\log\zeta\left(\frac{1}{2}+it\right)\right|^2 dt = \log\log T + 2C + o(1).$$

[6] Recall that to deal with the many-valuedness of the imaginary part of the logarithm (see Appendix B), its interpretation in (7.17) and (7.18) is based on the definition $S(T) \equiv (1/\pi)\arg\zeta(1/2+it)$ involving continuous variation of the arg, as described in Section 6.1.

As for the first moment of $\zeta(1/2+it)$, it is known (unconditionally) that

$$\int_T^{2T} \zeta\left(\frac{1}{2}+it\right) dt = T + O\left(T^{1/2}\right).$$

(Note that the leading term here is again real-valued.)[7] For the first *absolute* moment, only a conditional result is available: assuming RH we have

$$\int_T^{2T} \left|\zeta\left(\frac{1}{2}+it\right)\right| dt \sim CT\log^{1/4} T, \quad \text{as} \quad T \to \infty.$$

(The constant C is evaluated in Jabbarov & Hasanova, (2023, [**597**, Theorem 2]).) This is consistent with a result of Heap, Radziwiłł & Soundararajan (2019, [**523**]) who showed that for fixed real-valued $0 < \alpha \le 2$, the 2αth absolute moment satisfies the upper bound

$$I_\alpha(T) \equiv \int_T^{2T} \left|\zeta\left(\frac{1}{2}+it\right)\right|^{2\alpha} dt = O\big(T\log^{\alpha^2} T\big).$$

Results for moments beyond the fourth have proven still more difficult, with the problem of the sixth moment, for example, still unsolved. It is, however, generally believed that

$$\frac{1}{T}\int_T^{2T} |\zeta(1/2+it)|^{2k}\, dt = P(\log T) + o(1), \quad k = 1,2,3,\ldots, \tag{7.20}$$

where P is a polynomial of degree k^2. (The $k=1$ and $k=2$ cases are those of Hardy–Littlewood and Ingham, respectively.) It is also known that

$$\int_T^{2T} \left|\zeta\left(\frac{1}{2}+it\right)\right|^{2k} dt \;=\; O\big(T^{(k+2)/4}\log^{a(k)} T\big), \quad \text{for} \quad 2 \le k \le 6,$$

for some constants $a(k)$. In particular, Heath-Brown (1978, [**525**]) was able to establish the twelfth moment result

$$\int_T^{2T} \left|\zeta\left(\frac{1}{2}+it\right)\right|^{12} dt \;=\; O\left(T^2\log^{17} T\right).$$

Radziwiłł & Soundararajan (2013, [**878**]) and Heap & Soundararajan (2022, [**524**]), established the unconditional lower bound

$$I_\alpha(T) \equiv \frac{1}{T}\int_T^{2T} \left|\zeta\left(\frac{1}{2}+it\right)\right|^{2\alpha} dt \;\ge\; C_\alpha\,(\log T)^{\alpha^2},$$

for all real $\alpha \ge 0$, all T large enough, and certain constants C_α. Hence the order of the $2k$th moment cannot be less than $O\big(\log^{k^2} T\big)$. (Their paper gives references to earlier work on this result.) By assuming RH, Soundararajan (2009, [**1006**]) had obtained the upper bound

$$I_\alpha(T) \equiv \frac{1}{T}\int_T^{2T} \left|\zeta\left(\frac{1}{2}+it\right)\right|^{2\alpha} dt \;=\; O\big((\log T)^{\alpha^2+\epsilon}\big), \quad \text{for all real} \quad \alpha \ge 0,$$

and all $\epsilon > 0$, while Harper (2013, [**506**])—also under RH—established that the ϵ here may be replaced by 0. Consequently—under RH at least—the true order of $I_\alpha(T)$ is necessarily $O\big(\log^{\alpha^2} T\big)$, a result that is consistent with Selberg's CLT for

[7]Intuitively, the leading term, 1, of a (presumed and truncated) Dirichlet series would integrate to give the T, while all of the remaining terms would lead to oscillating integrals that cancel into the error term.

the critical line. (Of course, the cases $\alpha = 1$, $\alpha = 2$, and $0 < \alpha \leq 2$ all hold unconditionally by the results of Hardy and Littlewood, Ingham, and Heap, Radziwiłł, and Soundararajan, respectively.) Related references include Snaith (2000, [**996**]), Mezzadri & Snaith (2005, [**755**]), and Soundararajan (1995, [**1001**]). See also Beineke & Hughes (2009, [**87**]).

That there exist finite constants $C(\lambda) > 0$ for all real $\lambda \geq 0$ such that

$$I_\lambda(T) \equiv \frac{1}{T}\int_T^{2T} \left|\zeta\left(\frac{1}{2}+it\right)\right|^{2\lambda} dt \sim C(\lambda)(\log T)^{\lambda^2}, \quad \text{as } T \to \infty, \tag{7.21}$$

is a long-standing conjecture that is widely believed. It is based in part on the heuristic that on $t \in [0, T]$, the function $\zeta(1/2+it)$ is (at least for its non-small values) closely approximated by

$$\prod_{p<T} \left(1 - X_p/\sqrt{p}\right)^{-1}$$

where the X_p are independent and uniformly distributed on the unit circle in $\mathbb{C}$. (This heuristic is explained and dealt with extensively in Chapter 12.) The indicated moment should then be roughly of size

$$\mathrm{E}\left[\,\Big|\prod_{p<T}\left(1 - X_p/\sqrt{p}\right)^{-1}\Big|^{2\lambda}\right]$$

and this can be shown to be of order $(\log T)^{\lambda^2}$ as $T \to \infty$. (Compare also with the Hardy–Littlewood equivalence (7.11) for LH, and with (7.20).)

Work by Conrey & Ghosh (1998, [**225**]), Conrey & Gonek (2001, [**229**]), and Conrey & Farmer (2000, [**220**]), based on number theoretic arguments, as well as work by Keating & Snaith (2000, [**627**]), (2000, [**628**]) based on analogues to random matrix theory, have led to explicit conjectures regarding the value of $C(\lambda)$ in (7.21). For integer arguments, it is conjectured to be of the form $C(k) = a(k)g(k)/(k!)^2$. The so-called 'easy' term $a(k)$ here is an 'arithmetical factor' arising from the vagaries of the 'small primes', and is given by

$$a(k) = \prod_p \left(\left(1 - \frac{1}{p}\right)^{k^2} \sum_{m=0}^{\infty} \left(\frac{\Gamma(m+k)}{m!\,\Gamma(k)}\right)^2 \frac{1}{p^m}\right).$$

The 'difficult' term $g(k)$ is an 'RMT factor' conjectured from the moments of proxies to $\zeta\left(\frac{1}{2}+it\right)$ and is given by the integer-valued formula

$$g(k) = (k^2)!\prod_{j=0}^{k-1} \frac{j!}{(j+k)!}. \tag{7.22}$$

One such proxy is based on equating the behaviour of $\zeta\left(\frac{1}{2}+it\right)$ with that of characteristic polynomials of large random unitary matrices U; since its eigenvalues, θ_j, lie on the unit circle, this polynomial has the form $\det\left(I - Ue^{-i\theta}\right) = \prod_j \left(1 - e^{i(\theta_j - \theta)}\right)$. A different proxy is the presumed Euler product approximation alluded to above:

$$\zeta\left(\frac{1}{2}+it\right) \approx \prod_{p\leq x}\left(1 - \frac{1}{p^{1/2+it}}\right)^{-1} \approx \prod_{p \leq x}\left(1 - \frac{X(p)}{\sqrt{p}}\right)^{-1}.$$

Here $X(p) = p^{-it}$ is intended, but the $X(p)$ are instead treated as though they are independent random variables uniformly distributed on the unit circle in $\mathbb{C}$, while x is some suitably chosen value. (As mentioned, more about such reasoning will be found in Chapter 12.)

Result (7.13) of Hardy and Littlewood and (7.15) of Ingham had established that $g(1) = 1$ and $g(2) = 1/12$. While no other values of $g(k)$ have been proven, Conrey & Ghosh (1998, [**225**]) had conjectured that $g(3) = 42/9!$, and Conrey & Gonek (2001, [**229**]) had conjectured that $g(4) = 24024/16!$. The remarkable fact that this sparse data matched the values from RMT was instrumental in arriving at the conjecture (7.22) for $g(k)$. Conrey et al. (2005, [**222**]) have also conjectured lower order terms for these asymptotic expansions. Further detail may be found in Heath-Brown (1981, [**528**]), Snaith (2000, [**996**]), Keating & Snaith (2000, [**627**]), (2000, [**628**]), Diaconu et al. (2003, [**281**]), Gonek, Hughes & Keating (2007, [**441**]), Soundararajan (2009, [**1006**]), Harper (2013, [**506**]), Radziwiłł & Soundararajan (2013, [**878**]), and Heap, Radziwiłł & Soundararajan (2019, [**523**]). A key feature of these various conjectures is their exact match to corresponding properties of random matrix theory.

CHAPTER 8

Million Dollar Questions

Mathematicians have tried in vain to discover some order in the sequence of prime numbers, and we have reason to believe that it is a mystery which the human mind will never penetrate.

Leonhard Euler[1]

The Riemann hypothesis (RH) is widely considered to be one of the most important unsolved problems in mathematics. It attests to the far-reaching ways in which the zeta function and its zeros are linked to the chaotic and seemingly random character governing the distribution of the primes. The origin of this problem lies in two enduring sentences of Riemann's 1859 paper—the only one he ever wrote on number theory:

> *Man findet nun in der That etwa so viel reelle Wurzeln innerhalb dieser Grenzen, und es ist sehr wahrscheinlich, dass alle Wurzeln reell sind. Hiervon wäre allerdings ein strenger Beweis zu wünschen; ich habe indess die Aufsuchung desselben nach einigen flüchtigen vergeblichen Versuchen vorläufig bei Seite gelassen, da er für den nächsten Zweck meiner Untersuchung entbehrlich schien* [**906**].

Georg Friedrich Bernhard Riemann would certainly have chosen those words with care. In translating a passage such as this, one faces the difficulty of conveying subtlety and nuance across cultural, temporal and linguistic divides. The following translation is the authors' own and attempts to preserve the sense of the original; similar translations appear, for instance, in Edwards (1974, [**310**, Appendix]), Wilkins (2008, [**1108**]), and Mazur & Stein (2016, [**745**, §37]).

> *Indeed one now finds about that many real roots within these bounds, and it is very likely that all of the roots are real. One would admittedly wish for a stricter proof here; I have for the moment set these efforts aside after some fleeting vain attempts as it seemed to be unnecessary for the next objective of my investigation.*

The Riemann hypothesis is one of the seven "Problems of the Millennium" for which the Clay Mathematics Institute has offered a prize of one million US dollars. The current rules for this contest may be found on the Clay Institute's website.[2] The official description of this problem was formulated by Bombieri (2000, [**121**]),

[1]Euler went on to say: *'To convince oneself, one has only to glance at the tables of primes ... and one perceives that there is no order and no rule.'* For the source of this frequently cited quote, see Pólya (1959, [**862**]).

[2]`https://www.claymath.org/millenium-problems/`

and was later elaborated on and extended by Sarnak (2004, [**943**]). Note that it is a condition of the prize that RH be true; a counterexample would certainly lead to fame, but it would not collect.

The consequences of a resolution of RH would be far-reaching. The Riemann hypothesis had been one of the 23 problems (the eighth) posed by David Hilbert (1900, [**547**]) in his legendary address at the International Congress of Mathematicians held in Paris in 1900.[3] It was also the first of 18 problems for the twenty-first century proposed by Fields medallist Steven Smale, to coincide with the hundredth anniversary of Hilbert's Paris address. In 2008, the US Defense Advanced Research Projects Agency proposed its 23 "DARPA Mathematical Challenges" of which only 3, or possibly 4, can be considered to be purely mathematical in nature; one of these (problem number 19) is devoted to the Riemann hypothesis.

Many now legendary tales involving the Riemann hypothesis have been told. Hilbert once said: "If I were to awaken after having slept for a thousand years, my first question would be: Has the Riemann hypothesis been proven?"[4] And once, in July 1914, before a rough nighttime crossing of the English Channel to visit his collaborator Harald Bohr, Hardy sent Littlewood a postcard saying "Have proof of RH. Postcard too short for proof."[5] Hardy, an affirmed atheist, reasoned that if God really did exist, He would scarcely allow the ferry to sink in circumstances that would confer upon him such eternal glory.

Many false claims of a proof of the Riemann hypothesis have been made, including 'legitimately' erroneous ones by accomplished mathematicians such as Stieltjes,[6] Rademacher (who in 1945 claimed a *disproof* of RH), and several others. Among more contemporary claims that have not held up, that of Louis de Branges (from 2004 onwards) stands out because de Branges is well-known for having proved (in 1984) the 1916 Bieberbach conjecture of complex analysis. Ultimately, in 2010, de Branges issued an 'Apology' in an interesting (although rambling) essay posted on the web. XianJin Li, known for a criterion equivalent to RH (described below), and who had been a student of de Branges, also once claimed to have proved RH. In addition to these there are a great many false proofs in circulation by misguided individuals, at least one of which managed by some fluke to get published in a quasi-reputable journal (Polson, 2018, retracted 2019, Brazilian J. Probab. Statist.). On April 1, 1997, the highly respected Fields medalist Enrico Bombieri distributed an

---

[3]Hilbert's speech appears in the *Nachrichten von der Gesellschaft der Wissenschaften zu Göttingen, Mathematisch-Physikalische Klasse,* (1900, [**547**, pp. 253–297]). It begins as follows:

> Who among us would not gladly lift the veil beneath which the future lies hidden, to cast a glance at the future progress of our science and the secrets of its development in the coming centuries! What particular goals will the leading mathematical minds of future generations strive for? What new methods and new facts will the new centuries uncover in the vast and rich field of mathematical thought?

(Translated by the authors. See also Browder (1976, [**156**]).)

[4]This quote originates from a letter Hilbert wrote in 1900 to Ernst Zermelo: *"Ich würde gerne, wenn ich in tausend Jahren wieder aufwachte, als erstes nach dem Riemannschen Hypothesen fragen."* (See: *David Hilbert: Gesammelte Abhandlungen.*)

[5]This correspondence may be found in vol. 2 of The Collected Papers of G. H. Hardy.

[6]Stieltjes's claim had been that $\sum \mu(n)/n^s$ converges for all $\sigma > 1/2$; this implies RH by virtue of the fact that this sum would then equal $1/\zeta(s)$. Narkiewicz (2000, [**797**, pp. 158–160]) provides a riveting historical account of Stieltjes's persistence in his claim, and of its subsequent unravelling.

email stating that the Riemann hypothesis had just been proved by a young physicist. Great excitement ensued until it was realized this was no more than an 'April fools' prank.[7]

Of course, one should not to lose sight of the fact that even if the Riemann hypothesis were fully resolved, many questions concerning 'the finer shades in the behaviour of $\zeta(s)$,' [**1059**, Chapter 14, p. 336] as well as in the primes themselves, would still remain unanswered. It has been said that 'nothing is more precious than an unlit path', [**1088**] yet for anyone who may need it, we hasten to offer offer this fair warning. Quoting Pólya's advice to a student who was trying to prove RH:

> *If you want to climb the Matterhorn you might first wish to go to Zermatt where those who have tried are buried* [**863**].

8.1. Equivalences to the Riemann hypothesis

Many assertions equivalent to the Riemann Hypothesis are known, any one of which offers a path to the million dollar prize. One such equivalence was already noted in (5.3) of Chapter 5, and another in (6.19) of Chapter 6. A further instance will emerge in the discussion of infinitely divisible distributions in Chapter 10. We now present no fewer than twenty-four additional equivalences, selected from among the many that are known, focusing mainly on those that are most compelling, most simply stated, or most easily understood.

(1) The first equivalence is a trivial one. RH holds if and only if all nontrivial zeros satisfy

$$\left|\frac{\rho}{1-\rho}\right| = 1.$$

(2) The equivalence that may have led Riemann to his surmise in the first place:

$$\pi(x) = \mathrm{li}(x) + O\left(x^{\frac{1}{2}+\epsilon}\right), \quad \text{for all } \epsilon > 0 \iff \text{RH}$$

or the following equivalents to it:

$$p_n = \mathrm{li}^{-1}(n) + O\left(n^{\frac{1}{2}+\epsilon}\right), \quad \text{for all } \epsilon > 0 \iff \text{RH},$$

$$\mathrm{li}\left(p_n\right) = n + O\left(n^{\frac{1}{2}+\epsilon}\right), \quad \text{for all } \epsilon > 0 \iff \text{RH}.$$

[7] Excerpts from Enrico Bombieri's email of Tuesday April 1, 1997:

> There are fantastic developments to Alain Connes's lecture at IAS last Wednesday. Connes gave an account of how to obtain a trace formula involving zeroes of L-functions only on the critical line, and the hope was that one could obtain also Weil's explicit formula in the same context; this would solve the Riemann hypothesis for all L-functions at one stroke ... Well, a young physicist at the lecture saw in a flash that one could set the whole thing in a combinatorial setting using supersymmetric fermionic-bosonic systems (the physics corresponds to a near absolute zero ensemble of a mixture of anyons and morons with opposite spins) ... and, lo and behold, he got the required positivity of Weil's explicit formula! Wow! ... Please give this the highest diffusion.

(see `https://www.reddit.com/r/math/comments/b7xj23/bombieris_brilliant_april_fools_email_from_1997/`.)

Immediately afterwards, the US National Security Agency sent agents to Princeton to assess the resulting potential threats to national security. Among Bombieri's accomplishments is the landmark Bombieri–Vinogradov Theorem; less well known is that he is an artist of great versatility.

A version established by von Koch in 1901 states

$$\pi(x) = \mathrm{li}(x) + O\left(\sqrt{x}\,\log x\right) \iff \text{RH}.$$

And an explicit version due to Schoenfeld (1976, [**954**]) is give by:

$$|\pi(x) - \mathrm{li}(x)| \le \frac{\sqrt{x}\log x}{8\pi}, \quad x \ge 2657 \iff \text{RH}.$$

(3) Equivalences based on von Mangoldt's Λ, and Chebyshev's ψ function:

$$\psi(x) = \sum_{n\le x} \Lambda(n) = x + O\left(x^{\frac{1}{2}+\epsilon}\right), \quad \text{for all } \epsilon > 0 \iff \text{RH}$$

A version due to von Koch:

$$\psi(x) = x + O\left(\sqrt{x}\,\log^2 x\right) \iff \text{RH},$$

And an explicit version due to Schoenfeld (1976, [**954**]):

$$|\psi(x) - x| \le \frac{\sqrt{x}\log^2 x}{8\pi}, \quad x > 74 \iff \text{RH}.$$

(4) An equivalence due to Kunik & Lucht (2012, [**666**]) based on a Laplace transform of the von Mangoldt function:

$$\sum_{n=1}^{\infty} \Lambda(n)e^{-nx} = \frac{1}{x} + O\left(\frac{1}{\sqrt{x}}\right), \quad \text{as } x \downarrow 0 \iff \text{RH}.$$

(5) Equivalences based on the summatory version of the Mobius function $\mu(n)$, namely the Mertens function $M(x) \equiv \sum_{n\le x}\mu(n)$: Littlewood (1912, [**701**]) established that

$$|M(x)| \equiv \left|\sum_{n\le x} \mu(n)\right| = O\left(x^{\frac{1}{2}+\epsilon}\right), \quad \text{for all } \epsilon > 0 \iff \text{RH}.$$

REMARKS 8.1. Mertens had once conjectured that $|M(x)| \le \sqrt{x}$ but ths was disproved by Odlyzko & te Riele (1985, [**822**]). Unconditionally, there is no value of $\epsilon < \frac{1}{2}$ for which the above bound has been established. The best result to date is one based on the Korobov–Vinogradov bound:

$$M(x) = \sum_{n\le x} \mu(n) = O\left(x \exp\left(-C(\log x)^{3/5}/(\log\log x)^{1/5}\right)\right).$$

An alternate version of Littlewood's equivalence due to Soundararajan (2009, [**1007**]) is given by:

$$|M(x)| \equiv \left|\sum_{n\le x} \mu(n)\right| = O\left(\sqrt{x}\exp\left((\log x)^{1/2+\epsilon}\right)\right), \quad \text{for all } \epsilon > 0 \iff \text{RH}$$

while a related equivalence (the one Stieltjes once erroneously claimed to have proved) is:

$$\sum_{n=1}^{\infty} \frac{\mu(n)}{n^\sigma} \quad \text{converges for all } \sigma > 1/2 \iff \text{RH}.$$

(6) An equivalence based on the summatory version of the Liouville function:

$$|L(x)| \equiv \left|\sum_{n\le x}\lambda(n)\right| = O\left(x^{\frac{1}{2}+\epsilon}\right), \quad \text{for all } \epsilon > 0 \iff \text{RH}$$

This appears in Landau's 1899 thesis. (It would be implied if the occurrence of even and odd numbers of prime factors behaved like independent tosses of a fair coin.)

(7) Recall from equation (5.2) the alternating zeta function

$$\eta(s) \equiv \sum_{n=1}^{\infty}\frac{(-1)^{n-1}}{n^s} = \left(1-2^{1-s}\right)\zeta(s).$$

RH is equivalent to $\eta(s)$ having all its roots in $0 < \Re s < 1$ lie on $\Re s = \frac{1}{2}$.

(8) An equivalence based on the first derivative of the zeta function (Speiser (1934, [**1011**])):

$$\zeta'(s) \neq 0 \ \text{ for } 0 < \Re s < \frac{1}{2} \iff \text{RH}.$$

Remark 8.2. In 1996 Yildirim showed that under RH both ζ'' and ζ''' have no zeros in the region $0 \le \sigma < \frac{1}{2}$. They do however have zeros when $\sigma > \frac{1}{2}$; see Berndt (1970, [**89**]) and Binder, Pauli & Saidak (2015, [**105**]).

(9) An equivalence based on positivity of the derivatives of $\zeta(s)$ at the point $s = \frac{1}{2}$ (Pustylnikov (1999, [**871**])):

$$\left.\frac{d^n\zeta(s)}{ds^n}\right|_{s=1/2} > 0, \quad \text{for all } n \ge 1 \iff \text{RH}.$$

(10) Equivalences based on Farey 'series': The *Farey sequence* $\mathcal{F}_n$ *of order* n consists of the irreducible fractions j/k in $[0,1]$, for $0 \le k \le n$ (with j, k relatively prime), arranged in increasing order. The sequence $\mathcal{F}_5$, for example, is

$$\frac{0}{1}, \frac{1}{5}, \frac{1}{4}, \frac{1}{3}, \frac{2}{5}, \frac{1}{2}, \frac{3}{5}, \frac{2}{3}, \frac{3}{4}, \frac{4}{5}, \frac{1}{1}.$$

One property of $\mathcal{F}_n$ is that if u/v and u'/v' are successive terms, then $u'v - uv' = 1$. If m_n is the number of terms in $\mathcal{F}_n$, then $m_n \sim 3n^2/\pi^3$; see, for example, Hardy & Wright (2008, [**499**, Chapter III, & p. 354]), and Rademacher (1964, [**874**]). Let $\delta_{k,n} = a_{k,n} - k/m_n$, for $k = 1, 2, \ldots, m_n$, be the distance from the kth point of $\mathcal{F}_n$ (namely $a_{k,n}$) to the 'equidistributed' point k/m_n. Franel (1924, [**391**]) established that

$$\sum_{k=1}^{m_n}\delta_{k,n}^2 = O\left(n^{-1+\epsilon}\right), \quad \text{for all } \epsilon > 0 \iff \text{RH}$$

while Landau (following on Franel's article) established that

$$\sum_{k=1}^{m_n}|\delta_{k,n}| = O\left(n^{\frac{1}{2}+\epsilon}\right) \quad \text{for all } \epsilon > 0 \iff \text{RH}.$$

(Hardy & Wright (2008, [**499**, p. 44]), opined that Farey "seems to have been at the best an indifferent mathematician" who had little to do with the sequence named after him; Hardy (1929, [**487**]), provides some biographical detail of this amateur mathematician.)

(11) Redheffer (1977, [**887**]) introduced the $n \times n$ matrices $A(n)$ having entries $A_{i,j}$, with $A_{i,j} = 1$ if $j = 1$ or if i divides j, and $A_{i,j} = 0$ otherwise. It can be shown that the determinant of $A(n)$ equals $\sum_{k \le n} \mu(k)$. So it follows from item (5) above that RH is equivalent to

$$\det A(n) = O\left(\sqrt{n}\log^2 n\right), \quad \text{or} \quad O\left(n^{\frac{1}{2}+\epsilon}\right), \quad \text{for all } \epsilon > 0.$$

This equivalence leads to a interest in the eigenvalues of $A(n)$, a difficult problem studied by Vaughan (1993, [**1083**]), (1996, [**1084**]). A related matrix-based RH equivalence was proposed by Roesler (1986, [**913**]), (1987, [**913**]), (1990, [**913**]). See also Cardinal (2010, [**180**]).

(12) An equivalence based on lowest common multiples:

$$\left|\log\left(\operatorname{lcm}\{1,2,\ldots,n\}\right) - n\right| \le \frac{\sqrt{n}\log^2 n}{8\pi}, \quad n \ge 74 \iff \text{RH}.$$

This lcm is the smallest integer that is a multiple of each of $1, 2, \ldots, n$. Because $\log \operatorname{lcm}\{1, 2, \ldots, n\}$ is identical to the Chebychev function $\psi(n)$, this merely restates the Schoenfeld condition at item (3). (This inequality is sometimes stated as $< \sqrt{n}\log^2 n$ for $n \ge 3$ but this appears to have no source in the literature; Granville & Martin (2006, [**458**]) have $\le 2\sqrt{n}\log^2 n$ but without citation.)

(13) An equivalence due to Xianjin Li (1997, [**697**]) based on a positivity criterion involving the complex zeros, ρ, of $\zeta(s)$:

$$\lambda_n \equiv \sum_\rho \left(1 - \left(1 - \frac{1}{\rho}\right)^n\right) > 0, \quad \text{for all } n \ge 1 \iff \text{RH}.$$

Note that with $\xi(s) \equiv \frac{1}{2}s(s-1)\pi^{-s/2}\Gamma(\frac{s}{2})\zeta(s)$, Li's λ_n can also be expressed as

$$\lambda_n = \frac{1}{(n-1)!}\frac{d^n}{ds^n}\left(s^{n-1}\log\xi(s)\right)\Big|_{s=1} \ge 0, \quad n \ge 1.$$

Voros (2006, [**1094**]) reformulated Li's criterion as follows: RH holds if and only if

$$\lambda_n \sim \frac{1}{2}n(\log n - \log 2\pi - 1 + \gamma), \quad \text{as } n \to \infty.$$

Remark: For the function $\Xi(s) = s(s-1)\pi^{-s/2}\Gamma(\frac{s}{2})\zeta(s)$,

$$\frac{d}{ds}\log\Xi\left(\frac{1}{1-s}\right) = \sum_{n-1}^{\infty}\lambda_n s^{n-1}.$$

See also Bombieri & Lagarias (1999, [**124**]).

(14) An equivalence based on an *equality* involving the complex zeros:

$$\sum_\rho \frac{1}{|\rho|^2} = 2 + \gamma - \log 4\pi \iff \text{RH}$$

(As always, this sum is over *all* zeros in the critical strip with multiplicity.) The origin of this equivalence was discussed in (6.19).

(15) An equivalence based on an integral *equality*: Representing the zeros as $\rho = \beta + i\gamma$, Balazard, Saias & Yor (1999, [**73**]) proved that

$$\int_{\mathbb{R}} \log\left|\zeta\left(\frac{1}{2} + it\right)\right| d\mu(t) = \sum_{\beta > \frac{1}{2}} \log\left|\frac{\rho}{1-\rho}\right|, \quad \text{where } \mu(A) \equiv \frac{1}{2\pi}\int_A \frac{dt}{1/4 + t^2}.$$

Note that the measure $\mu(\cdot)$ represents the (scale-$\frac{1}{2}$) Cauchy density $f_Y(y) = \frac{2}{\pi}\frac{1}{1+(2y)^2}$, and that the sum is taken as zero if empty. It follows that RH is equivalent to

$$\int_{-\infty}^{\infty} \frac{\log|\zeta(1/2+it)|}{1/4+t^2}\,dt \equiv \int_{\Re(s)=1/2} \frac{\log|\zeta(s)|}{|s|^2}\,|ds| = 0. \tag{8.1}$$

This equivalence can also be stated as the expectation $E\,\log\left|\zeta(\frac{1}{2}+iY)\right| = 0$, where Y has the Cauchy density $f_Y(y)$ indicated above. A related equivalence due to Kondratyuk & Yatsulka (2008, [**648**]) is given by

$$\int_{\Re s=1/2} \frac{\log|\zeta(s)|}{|s|^4}\,|ds| = 2\pi(1-\gamma) \iff \text{RH}$$

For a version involving a denominator of $|s|^6$ in the integrand, see Basiuk & Tarasyuk (2016, [**80**]). It is natural to ask how quickly the integral in (8.1) converges. Bui, Lester & Milinovich (2013, [**167**]) have shown, under RH, that

$$\int_{-T}^{T} \frac{\log|\zeta(1/2+it)|}{1/4+t^2}\,dt = O\left(T^{-\alpha}\right), \text{ as } T\to\infty$$

holds for all $0<\alpha<2$, but fails for $\alpha=2$.

Remark: Singularities of $\log|\zeta(\frac{1}{2}+it)|$ are handled by means of principal values.

(16) Another equivalence involving an integral *equality*: Sekatskii et al. (2012, [**963**]) showed

$$\int_{-\infty}^{\infty} \frac{\log\zeta(1/2+it)}{(1/2+it)^2}\,dt = -2\pi \iff \text{RH}.$$

See also the related articles [**962**], [**964**], [**965**], [**966**], and [**967**].

(17) An equivalence based on an *equality* due to Volchkov (1995, [**1091**]):

$$\int_0^{\infty}\int_{1/2}^{\infty} \frac{1-12t^2}{(1+4t^2)^3}\log|\zeta(\sigma+it)|\,d\sigma\,dt = \frac{3-\gamma}{32}\pi \iff \text{RH}.$$

(The rationale for including this seemingly more complicated equivalence is that it, together with the three previous ones, are based on *equalities;* all of the other equivalences appearing on this list are based on asymptotic behaviours, or on *inequalities*.)

(18) Equivalences based on integrals involving $\pi(x)-\text{li}(x)$ and $\Pi(x)-\text{li}(x)$:

$$\int_1^{\infty}\left(\frac{\pi(x)-\text{li}(x)}{x}\right)^2 dx < \infty \iff \text{RH}.$$

(Compare with item (2).) A related equivalence is:

$$I = \int_1^{\infty}\left(\frac{\Pi(x)-\text{li}(x)}{x}\right)^2 dx < \infty \iff \text{RH}$$

where

$$\Pi(x) \equiv \sum_{p^m\le x}\frac{1}{m}.$$

(Recall from Chapter 5 that $\Pi(x)-\pi(x) = O\left(\sqrt{x}/\log x\right)$.) Interestingly, if the integral I is finite, it *must* be equal to the *computable* real number

$$J = \frac{1}{\pi}\int_0^{\infty}\left|\frac{\log\big((-1/2+it)\zeta(1/2+it)\big)}{1/2+it}\right|^2 dt$$

whose value lies somewhere between 1.266 and 1.273. See Arias de Renya (2008, [**38**]); the logarithm here must be understood in the sense explained in that reference. Associated with the last equivalence are the two functions

$$(\Pi(e^x) - \mathrm{li}(e^x))e^{-x/2} \quad \text{and} \quad \frac{\log((-1/2 + 2\pi it)\zeta(1/2 + 2\pi it))}{1/2 + 2\pi it}.$$

Under RH these functions are both in $\mathcal{L}^2(\mathbb{R})$, with the second being the Fourier transform of the first. Curiously, however, if RH is false, only the second will be in $\mathcal{L}^2(\mathbf{R})$.

(19) An equivalence based on the sum of divisors function: Lagarias (2002, [**667**]) showed

$$\sigma(n) < H_n + e^{H_n} \log H_n, \ \text{for all } n \geq 2 \iff \text{RH}.$$

Here $H_n \equiv \sum_{k=1}^{n} \frac{1}{k}$ is the nth harmonic number. This equivalence was derived from the very sharp one due to Robin (1984, [**912**]) that was discussed in Chapter 3:

$$\sigma(n) < e^{\gamma} n \log\log n, \ \text{for } n > 5040 \iff \text{RH}.$$

(Remark: These are true inequalities, not asymptotic ones.) Note that because $\gamma = \lim_{n\to\infty}(H_n - \log n)$, the criteria of Robin and Lagarias agree asymptotically. For further background, see Nicolas ([**808**], [**809**], [**810**]).

(20) Equivalences based on the values of ζ at positive integer arguments: It is known, unconditionally, that

$$R(x) \equiv \sum_{k=1}^{\infty} \frac{(-1)^{k+1}x^k}{(k-1)!\,\zeta(2k)} = O\left(x^{1/2+\epsilon}\right);$$

Riesz (1916, [**908**]) showed that RH is equivalent to the error term here being $O\left(x^{1/4+\epsilon}\right)$. (Recall from Chapter 5 that the $\zeta(2k)$ have *explicitly known* values.) Hardy & Littlewood (1916, [**492**]) showed that RH is also equivalent to

$$\sum_{k=1}^{\infty} \frac{(-1)^k x^k}{k!\,\zeta(2k+1)} = O\left(x^{-\frac{1}{4}+\epsilon}\right) \quad \text{as } x \to \infty.$$

Baez-Duarte (2003, [**57**]) obtained the RH equivalence

$$\sum_{j=0}^{k} (-1)^j \binom{k}{j} \frac{1}{\zeta(2j+2)} = O\left(k^{-3/4+\epsilon}\right).$$

Unconditionally this is known to be $O\left(k^{-1/2}\right)$. The fact that ζ enters in all of these only through its values on positive integers is a remarkable feature of these equivalences. However, these sums become extremely delicate as the value of x or k increases. To help appreciate the origins of such equivalences, we point out that (even for complex x)

$$R(x) = x \sum_{n=1}^{\infty} \frac{\mu(n)}{n^2} e^{-x/n^2}.$$

(21) An equivalence based on the Bernoulli numbers:

$$\sum_{k=1}^{\infty} \frac{k(2k-1)!x^k}{(2\pi)^{2k}(k-1)!B_{2k}} = O\left(x^{1/4+\epsilon}\right) \ \text{for all } \epsilon > 0 \iff \text{RH}.$$

This is essentially a restatement of Riesz's '$R(x)$-equivalence' and stems from the relation $B_{2k} = (-1)^{k+1}(2k)!\zeta(2k)/2^{2k-1}\pi^{2k}$, for $k \geq 1$ connecting the Bernoulli numbers to the zeta function.

(22) The following equivalence will be developed in Chapter 10: RH holds if and only if, for every $\frac{1}{2} < \theta < 1$, there exists a function $\nu_\theta(t)$ such that $\nu_\theta(t)\zeta(\theta + it)$ is the characteristic function (that is, the Fourier transform) of an infinitely divisible distribution.

(23) Recall the function $\xi(s)$ introduced in Chapter 5:

$$\xi(s) \equiv \frac{1}{2}s(s-1)\pi^{-s/2}\Gamma\left(\frac{s}{2}\right)\zeta(s), \quad \xi(s) = \xi(1-s).$$

Recall too that $\xi(s)$ is real valued on $\sigma = \frac{1}{2}$, and that its zeros are exactly those of the complex zeros of $\zeta(s)$. These zeros must in turn exactly match the singularities of $\log \xi(s)$ as well as of the logarithmic derivative $\xi'(s)/\xi(s)$. Now the mapping $s \to \frac{1}{s}$ maps the critical $\sigma = \frac{1}{2}$ line to the circle of radius 1 centred at $s = 1$. Consider then the power series expansions

$$\log \xi(1/s) + \log 2 = \sum_{k=0}^{\infty} \lambda_k (1-s)^k, \quad \text{and} \quad \frac{\xi'(1/s)}{\xi(1/s)} = \sum_{k=0}^{\infty} \tau_k (1-s)^k.$$

RH is equivalent to these power series having radius of convergence equal to 1. Formulas for the coefficients λ_k and τ_k may be found in Keiper (1992, [**630**]).

(24) Our last equivalence is more technical but is included because it is a favourite of the authors. Balazard (2003, [**71**]) reviews the connections between *completeness problems* and RH. An interesting one (which originates with Nyman and Beurling) can be formulated as follows: Let $\mathcal{L}^2(0,\infty)$ be the space of complex-valued functions on $(0,\infty)$ with respect to the measure dx/x^2, so that the inner product is given by

$$\langle u, v \rangle \equiv \int_0^\infty u(x)\,\overline{v(x)}\,\frac{dx}{x^2},$$

and $\mathcal{L}^2(0,\infty)$ consists of the functions $f(x)$ satisfying $\int_0^\infty |f(x)|^2 \frac{dx}{x^2} < \infty$. Since this space contains all bounded functions, it includes the indicator function, $g(x)$ say, of the interval $(1,\infty)$, as well as the sawtooth function $f(x) = \{x\}$ (the fractional part of x), along with all of its dilates $f(\frac{x}{c})$, for $c > 0$. Let $\mathcal{M}$ be the subspace of $\mathcal{L}^2(0,\infty)$ spanned by the dilates of f. Then RH is equivalent to the assertion that $g \in \mathcal{M}$, as well as to the assertion that $\mathcal{M} = \mathcal{L}^2(0,\infty)$. In fact, RH is even equivalent to the narrower assertion that g is contained in the subspace $\mathcal{H}$ spanned by the dilations $f(\frac{x}{c})$ restricted to $c = 1, 2, 3, \ldots$. (An extension of this result to spaces $\mathcal{L}^p(0,\infty)$ with $1 \leq p \leq 2$ yields equivalents to $\zeta(s)$ having no zeros above $\sigma = \frac{1}{p}$; we omit the details.) Relevant references include Nyman (1950, [**817**]), Alcantara-Bode (1993, [**13**]), Baez-Duarte (2003, [**57**]), and Bagchi (2006, [**59**]).

In addition to *equivalences* to RH, there are also propositions which provide *sufficient conditions* for RH to be true. Arguably, one of the most compelling among these (although shown eventually to be false) was that of Turán (1948, [**1071**]): If none of the partial zeta sums

$$\sum_{n=1}^{N} \frac{1}{n^s}, \quad N = 1, 2, 3, \ldots$$

vanish anywhere in the half-plane $\sigma > 1$, then RH is true. Turán showed that this implies

$$\sum_{n \leq x} \frac{\lambda(n)}{n} \stackrel{?}{\geq} 0$$

for all $x > 0$, so that by the result of Pólya (1919) discussed in Chapter 3, RH would be true. However Haselgrove (1958, [**516**]) proved that this condition is violated for some value of N. Spira (1968, [**1013**]) subsequently proved that Turán's partial zeta sum condition did hold for all $N \leq 9$ but then showed that the partial zeta sum does have a zero in the region $\sigma > 1$ for $N = 19$. Montgomery (1983, [**766**]) determined certain explicit regions in which zeros violating Turán's condition reside. See also Borwein et al. (2007, [**134**]) and Alkan (2021, [**15**]) for additional background.

Finally, it is worth mentioning that there are variations on the zeta function for which a version of RH can in fact be shown to hold. One of the more interesting among these is the one due to Taylor (1945, [**1052**]) mentioned in Chapter 6. Namely, that with $\xi(s)$ as defined in (6.14), all of the complex zeros of

$$\xi\left(s + \frac{1}{2}\right) - \xi\left(s - \frac{1}{2}\right)$$

lie on the line $\Re s = \frac{1}{2}$. Related examples may be found in Hejhal (1990, [**533**]), and in Lagarias & Suzuki (2006, [**670**]).

Further discussion of equivalences to RH may be found in Conrey & Farmer (2012, [**221**]), Borwein et al. (2008, [**133**, Chapter 5]), and in the two volume *tour de force* devoted to RH equivalences by Broughan (2017, [**157**]) which includes detailed proofs of the equivalences together with a great deal of historical information.

One other possibility which—for the sake of completeness—must be mentioned is that RH might be a formally undecidable proposition. Although the authors do not consider this likely, should that be the case, RH would be true but unprovable.[8] (A similar concern could likewise have been raised, at any point in its 356 year history, of Fermat's Last Theorem.)

8.2. Reasons why RH may be true

There are many reasons for believing that RH *should* be true. Here are nine compelling ones.

(1) Riemann (whose opinion must surely be taken seriously!) thought it likely to be true. Furthermore, many—*although certainly not all*—experts in the field believe it must be true, while a few remain genuinely undecided.

(2) If RH were false, many widely held beliefs about the distribution of the primes would descend into chaos. In his official problem description for the Clay Institute's prize, Enrico Bombieri put it this way:

> *In the opinion of many mathematicians the Riemann hypothesis ... is probably today the most important open problem in pure mathematics. ... The failure of the Riemann hypothesis would create havoc in the distribution of prime numbers. This fact alone singles out the Riemann hypothesis as the main open question of prime number theory. ... Notwithstanding some skepticism voiced in the past, based perhaps more on the number of failed attempts to a proof rather than on solid*

[8]The theorems of Kurt Gödel on the incompleteness and undecidability of common axiomatic systems do however allow for that possibility.

heuristics, it is fair to say that today there is quite a bit of evidence in its favour [**121**].

(3) According to Atle Selberg (whose opinion surely matters as well!):

...if one believes that there is something in this world that is as it should be, then I think that must be the truth of the Riemann hypothesis. It gives the best possible distribution of the prime numbers, and also what one would expect from a statistical point of view, namely that the deviation from [li(x)] *is not greater than the square root of* x. *It would entail an elegance that is striking. Besides, I must say that I trust Riemann's intuition very much* (Baas & Skau (2008, [**51**, p. 631])).

However, in a 1980 interview Selberg also said:

I think it is a good possibility that it will take a long time before it is decided [**51**, p. 634].

(4) Although numerical evidence is hardly conclusive the first hundred billion zeros have been shown to lie on the line, as well as millions of zeros in the vicinity of positions 10^{20}, 10^{21}, and 10^{22}. It is also known that at least 99% of zeros $\rho = \beta + i\gamma$ satisfy $|\beta - \frac{1}{2}| \le 8/\log|\gamma|$, and that at least 40% of them lie on $\sigma = \frac{1}{2}$ (and are simple).

(5) There are 'zero density theorems' from which one may conclude that the proportion zeros to the right of any fixed $\sigma > \frac{1}{2}$ (relative to all of the complex roots) is zero.

(6) The proof by Deligne of a conjecture of Weil concerning algebraic varieties over finite fields provides evidence in favour of RH. Weil's contribution to RH over finite fields may be mentioned here as well. (For details, see Bombieri's Clay prize article [**121**].)

(7) An argument proposed by Alain Connes (1999, [**216**]) involving Hamiltonian operators and statistical mechanics provides some evidence in favour of RH.

(8) Of the many results that have been proved assuming the validity of RH, none has so far been shown to be false. (Of course, many of these 'conditional' results seem almost as hard to prove unconditionally as is RH itself.)

(9) The lowest complex zeros whose real parts differ from $\frac{1}{2}$ must have real parts lying in $(0, 1)$. The 0 and 1 boundary values are precluded by the Prime Number Theorem. Yet no 'obvious' theoretical explanation suggests itself for any other values for these real parts.

8.3. Reasons why RH may be false

There are also many reasons for believing that RH *may* be false. Here are eleven compelling ones.

(1) Riemann's intuition was unparalleled but not infallible and his 1859 paper is not without false assertions. For instance, the order term he stated for the zeros counting function $N(T)$ is demonstrably false, as is also one of the identities he stated for $\pi(x)$. Furthermore, Riemann did not categorically assert the absence of zeros off the critical line but only that he thought it highly probable. In fact, of the six assertions in Riemann's 1859 paper for which no proofs have yet been offered, RH is the only one over which he had expressed even the slightest doubt. Edwards (1974, [**310**, §7.8])—after a careful and incisive analysis as to why Riemann may have been wrong—writes:

...although Riemann's insight was stupendous it was not supernatural, and what seemed 'probable' to him in 1859 might seem less so today.

(2) Every attempt to prove the Riemann hypothesis, from 1859 until now, has failed, and no one—after all this time, and all the efforts expended—has come up with a method through which the problem can be decisively approached. To quote Atle Selberg:

There have probably been very few attempts at proving the Riemann Hypothesis because, simply, no one has ever had any really good idea for how to go about it [**50**].

The one seemingly good idea that did arise—due to Landau, Pólya, and Hilbert—was to try to match the heights of the complex zeros with the eigenvalues of some unbounded self-adjoint operator in Hilbert space. A detailed discussion of this idea may be found in Berry & Keating (1999, [**94**]). While this approach has had some partial success (notably via random matrix theory) it has not really borne fruit (although some progress has been reported by Connes (1999, [**216**]), (2016, [**217**]).

(3) Selberg himself pointed out that there is plenty of cause for skepticism concerning the truth of RH; see Selberg (1946, [**971**]).

(4) No technical reasons are known as to why RH should be true. It had been conjectured largely on the basis of statistical arguments, but such arguments have not always proven reliable in number-theoretic contexts. Littlewood (who surely must also be taken seriously) believed RH to be false. To quote Littlewood:

I believe this to be false. There is no evidence whatever for it ... One should not believe things for which there is no evidence. ... I should also record here my feeling that there is no imaginable reason why it should be true [**707**].

Littlewood goes on to say:

...For a zero to be off the critical line a remarkable number of 'coincidences' have to happen. ... the chance of finding a zero off the line in a lifetime's calculations is millions to one against. It looks then as if we may never know [**707**].

(5) The existence of even the *slimmest* zero-free region of the form $1 - \epsilon < \sigma \leq 1$, for some $\epsilon > 0$, has still not been established. All known zero-free regions tend towards the $\sigma = 1$ boundary as the height $t \to \infty$, and even for regions of this type there has been essentially no progress since the work of Korobov and Vinogradov in 1958.

(6) If there were violations to RH they are expected occur so high up that detecting them computationally is unimaginable, and without an analytical way to establish their existence RH may never be resolved. Bombieri (2006, [**121**]) noted that if violations exist, they are likely to occur near unusually high peaks of $|\zeta(\frac{1}{2} + it)|$. Such peaks do occur at large t, generally where pairs of zeros are unusually close together. This is known as *Lehmer's phenomenon* and is described in Ivić (2013, [**590**, §2.3]). Referring to this Edwards (1974, [**310**, p. 178]) states that the Lehmer phenomenon "must give pause to even the most convinced believer of the Riemann hypothesis."

(7) The fact that so many zeros are known to lie on $\Re s = \frac{1}{2}$ cannot by itself be taken as convincing evidence for RH. One reason to be cautious is that $S(t)$—which measures the deviation of the number of zeros of height at most t from their expected number—has (asymptotic) mean square $(\log\log t)/2\pi^2$ and so increases very slowly. One can argue that unless $S(t)$ is reasonably large (say 100) one has not yet seen the true state of affairs as far as the behaviour of the zeros is concerned. However the height t for which $S(t)$ is of this size lies well beyond the capability of present-day computers.

(8) Many conjectures that are implied by and that are *far weaker* than RH have themselves proven enormously difficult and have not yielded to proof. Obvious examples include the Lindelöf hypothesis and the density conjecture. See, for example, Ivić (1985, [**586**, p. 47]), and Karatsuba & Voronin (1992, [**622**, §6]).

(9) False conjectures have been made by very able researchers, including Gauss, Riemann, Mertens, Rademacher, Stieltjes, Cramér, and others. RH may be no more than another false conjecture. That numerical evidence is unreliable and counterexamples hard to produce is amply demonstrated by Littlewood's (1914, [**702**]) result concerning sign changes of $\pi(x) - \mathrm{li}(x)$, or Odlyzko & te Riele's (1985, [**822**]) disproof of the Mertens conjecture. Indeed the enormity of the 'Skewes numbers' suggests that violations of RH may analogously occur so high up that they may not be detected for a very long time, if ever.

(10) RH is believed, in part, on the basis of statistical reasoning of Cramér-heuristic type. But, as Maier (1985, [**722**]) and others have shown, in number theory such reasoning does not guarantee correct conclusions. And yet RH says, in effect, that so far as fluctuations of the primes counting function $\pi(x)$ is concerned, the primes behave according to Cramér-type randomness.

(11) Ivić (2003, [**589**]) provides some mathematically deeper reasons for doubting RH, and proposes a possible approach for disproving it. Nevertheless, Ivić does believe that *almost all* of the complex zeros lie do on the critical line.

While RH can of course be disproved by disproving any of its *equivalences*, it can also be disproved by disproving any *necessary* but non-equivalent condition such as the Lindelöf hypothesis. Another nonequivalent implication is that under RH, $p_{n+1} - p_n = O(p_n^{1/2+\epsilon})$ for every $\epsilon > 0$. All that is currently known of this unconditionally is that it holds for every $\epsilon > 1/40$, as shown by Baker, Harmon & Pintz (2001, [**70**]). In fact, Goldston & Heath-Brown (1984, [**426**]) showed that RH would even imply $p_{n+1} - p_n = o\left(\sqrt{p_n \log p_n}\right)$—although this also assumes a conjecture of Montgomery associated with the pair correlation hypothesis. For purposes of comparison, recall the classical prediction of Cramér that $p_{n+1} - p_n = O\left(\sqrt{p_n}\, \log p_n\right)$. Many other implications of RH not equivalent to it are also known.

8.4. Other acclaimed problems

Although the main focus of this chapter has been on the Riemann hypothesis, many other noteworthy unsolved problems are known. The easily described twin primes conjecture and Goldbach conjecture were already discussed in Chapter 2. Here we mention a select number of more technical problems and conjectures any of whose solutions would lead to fame and fortune. Several of these have been encountered in earlier chapters.

(1) Does there exist a $\delta > 0$ such that all zeros of $\zeta(s)$ satisfy $\Re(s) < 1 - \delta$?

(2) Do all but an infinitesimal proportion of the complex zeros lie on the critical line?

(3) *The Lindelöf hypothesis* (LH) (described in Chapter 6): Show that

$$\zeta\left(\frac{1}{2}+it\right) = O\left(|t|^{\epsilon}\right), \quad \text{as } |t| \to \infty, \quad \text{for all } \epsilon > 0.$$

(4) Find an unconditional proof of Montgomery's *pair correlation conjecture*.

(5) Prove that all of the complex zeros of $\zeta(s)$ are simple.

(6) *The Linear Independence hypothesis* (LI): Are the imaginary parts of the zeros of the zeta function—that is, the real numbers $\gamma_n > 0$ such that $\zeta(\beta_n + i\gamma_n)$ equals 0—linearly independent over the rationals? If so, it would follow that all of the zeros are simple. In this connection recall that Ingham (1942, [**583**]) proved that had the Mertens conjecture been true then RH would follow, all zeros would be simple, and the complex zeros would satisfy infinitely many linear relations; see Odlyzko & te Riele (1985, [**822**]).

(7) *Primes in short intervals:* Determine the smallest value of θ such that every interval of the form $[x, x + x^{\theta}]$ contains a prime once x is large enough. The best value currently known, due to Baker, Harman & Pintz (2001, [**70**]), is $\theta = 21/40$ but it is believed that any $\theta > 0$ should do. A related condition can be stated as

$$\pi(x + x^{\theta}) - \pi(x) = \Big(1 + o(1)\Big)\frac{x^{\theta}}{\log x}, \quad \text{as } x \to \infty.$$

Under RH, the best known result of this type is that all intervals of the form $[x, x + C\sqrt{x}\log x]$ contain a prime for large x but even this is far from what is believed to be true.

(8) *Primes in logarithmically short intervals:* Cramér (1936, [**245**]) conjectured that for any $c > 2$ there will always be a prime in $[x, x + \log^c x]$ once x is large enough. This remains unknown even under RH. A stronger version posits the existence of a constant $C > 0$ such that all intervals $[x, x + C\log^2 x]$ contain a prime.

(9) *The Dirichlet divisor problem:* Let $\tau(n) \equiv \sum_{d|n} 1$ be the number of all divisors of n including 1 and n. For its summatory version Dirichlet proved that

$$D(x) \equiv \sum_{n \le x} \tau(n) = x\log x + x(2\gamma - 1) + O\left(\sqrt{x}\right).$$

The problem is to find the smallest value of θ for which the $O\left(\sqrt{x}\right)$ can be replaced by $O\left(x^{\theta+\epsilon}\right)$ for all $\epsilon > 0$. It is known that $\frac{1}{4} \le \theta < 0.315$, and conjectured that $\theta = \frac{1}{4}$.

(10) There are a great number of generalizations of the Riemann zeta function: for example, Dirichlet L-functions, Dedekind zeta functions, Artin and Hecke L-functions, L-functions associated with elliptic curves or more general varieties, L-functions associated to representations or automorphic forms, and an entire "Selberg class" of L-functions. All of these are expected to satisfy the analogous Generalized Riemann Hypothesis, and would have similarly impactful number-theoretic consequences.

(11) *The abc conjecture:* This famous and important problem—which originated with Masser and Oesterlé in 1985—concerns the deep conflict that exists between the additive and multiplicative properties of the integers. It is known

to imply Fermat's Last Theorem (with finitely many exceptions) and bears on many other arithmetical problems.

Define the *radical function* as $R(n) \equiv \sum_{p|n} p$. This function is multiplicative, and if n has unique factorization $n = p_1^{r_1} p_2^{r_2} \cdots p_k^{r_k}$, then $R(n) = p_1 p_2 \cdots p_k$. The abc conjecture states that for every $\epsilon > 0$, there are *at most* a finite number of *relatively prime* integer triplets satisfying $a + b = c$ such that $R(abc) < c^{1-\epsilon}$. Intuitively, this means that the product of all the distinct prime factors of the three integers should never be much less than c, or stated differently, that there cannot be many repeated prime factors among all three of a, b and c when $a + b = c$. We mention that the radical function $R(n)$ can be shown to satisfy

$$\sum_{n \leq x} \frac{R(n)}{n} = Cx + o(x), \quad \text{where } C = \prod_p \left(1 - \frac{1}{p(p+1)}\right).$$

See Masser (1985, [**735**]), Oesterlé (1988, [**824**]), Stewart & Tijdeman (1986, [**1031**]), Stewart & Yu (1996, [**1032**]), and Granville & Tucker (2002, [**462**]). Martin & Miao (2016, [**732**]) give a detailed discussion of this conjecture. See also Hardy & Wright (2008, [**499**, Chapter XIII, Notes]), and Tenenbaum (2015, [**1053**, pp. 209–210]).[9]

8.5. Equivalences to the Lindelöf hypothesis

Of the eleven problems just mentioned, the Lindelöf hypothesis (LH) is generally viewed as having the highest stature. For this reason, we close this chapter with some formulations which are *equivalent* to LH:

(1) An equivalence based on prime gaps:

$$\sum_{p_n \leq x} (p_{n+1} - p_n)^2 \; \leq \; x^{1+o(1)} \iff \text{LH}.$$

(2) An equivalence (see Section 7.4) based on the moments of $|\zeta(\frac{1}{2} + it)|$:

$$\frac{1}{T} \int_T^{2T} |\zeta(1/2 + it)|^{2k} \, dt = O\left(T^{\epsilon}\right), \text{ for all } \epsilon > 0, \text{ and all } k \geq 1 \iff \text{LH}.$$

(3) An equivalence (due to Bäcklund) based on the density of the zeros:

$$N(\sigma, T+1) - N(\sigma, T) = o(\log T), \quad \text{for every } \sigma > \frac{1}{2} \iff \text{LH}.$$

Recall that $N(\sigma, T)$ is the number of zeros of $\zeta(s)$ with $\Re s > \sigma$ and $0 < \Im s \leq T$. (This equivalence also shows that RH $\Rightarrow$ LH; Titchmarsh (1986, [**1059**, Theorem 13.5]).)

[9] **An update:** In 2012 Shinichi Mochizuki of Kyoto University released a 600-page proof of the abc conjecture written in an 'impenetrable idiosyncratic style' so laden with unfamiliar methods he had developed that it has not yet been possible to substantiate his results. The very few mathematicians who have been able to penetrate Mochizuki's work have expressed concerns that a certain key result—Corollary 3.12 on elliptic curves—may be wrong and irreparable. Nevertheless, Mochizuki's work was published in 2021 in the Publications of the Research Institute for Mathematical Sciences, Kyoto, to the consternation of many. Mochizuki (who received his PhD under Gerd Faltings at Princeton) is known for having solved a conjecture of Grothendieck on anabelian geometry. (See the articles in *Nature*, September 10, 2012 and April 3, 2020.)

(4) An equivalence based on the alternating zeta function,

$$\eta(s) \equiv \sum_{n=1}^{\infty} \frac{(-1)^n}{n^s} = \left(2^{1-s} - 1\right) \zeta(s),$$

discussed in Section 5.8. The Lindelöf hypothesis is equivalent to the condition

$$\lim_{T\to\infty} \frac{1}{2T} \int_{-T}^{T} |\eta(\sigma + it)|^{2k} \, dt < \infty$$

for all $\sigma > \frac{1}{2}$, and $k = 1, 2, \ldots$ (McCarthy (1986, [**746**, p. 80])).

(5) An equivalence which frames the Lindelöf hypothesis as *no more than an ordinary assertion in analysis*: The Lindelöf hypothesis is true if and only if, for every $\epsilon > 0$, and any $C > 0$ however large,

$$\sum_{n\leq x} n^{-it} = \frac{x^{1-it}}{1-it} + O\left(x^{1/2}|t|^{\epsilon}\right), \quad \text{for } 1 \leq x \leq |t|^C$$

(Gonek, Graham & Lee, (2020, [**440**, Theorem 3])). Note that the fractional term on the right is nothing more than the integral approximation $\int_0^x u^{-it} du$ of the sum on the left.

CHAPTER 9

Frontiers of Chance and Chaos

I know of scarcely anything so apt to impress the imagination as the wonderful form of cosmic order expressed by the "Law of Frequency of Error" [the Central Limit Theorem]. The law would have been personified by the Greeks and deified, if they had known of it. ...It is the supreme law of Unreason. Whenever a large sample of chaotic elements are taken in hand and marshalled in the order of their magnitude, an unsuspected and most beautiful form of regularity proves to have been latent all along.

Sir Francis Galton [**406**]

The discordance between the additive and multiplicative properties of the natural numbers gives rise to patterns which are situated somewhere between randomness and chaos. The topics considered in this chapter fall mainly under the rubric of *probabilistic number theory*, but strictly speaking, the description this implies is not entirely accurate. The term 'probabilistic number theory' hearkens back to the era in the early 1930s when Kolmogorov's formulation for probability (and its ensuing developments) were not yet widely entrenched, while number theorists of the time were discovering that the natural numbers possessed many limiting properties whose more complete descriptions mirrored notions of the LLN, CLT, and LIL of the then-evolving theory of probability.[1] But while many results of probabilistic number theory certainly borrow concepts from probability in their formulation, their proofs often also hinge on nontrivial number-theoretic considerations. There are however important exceptions to this, and efforts to delineate the purely probabilistic from the purely number-theoretic aspects of such results remains an ongoing area of investigation.

The key players here are the *arithmetic functions* $f(n)$—functions defined on the positive integers—which we introduced in Chapter 3. In particular, there are large classes of results which apply specifically to arithmetic functions that are additive, or apply specifically to arithmetic functions that are multiplicative. One may consider the properties of such a function when n is viewed as a random variable on the sample space $\{1, 2, \ldots, N\}$, typically under uniform probability, and typically in the $N \to \infty$ limit. (Indeed viewing functions in this way is what warrants the '*probabilistic* number theory' name.) Then even if $f(n)$ has erratic behaviour, natural questions can still be framed in terms of quantities such as its

[1] These abbreviations refer to the Law of Large Numbers, the Central Limit Theorem, and the Law of the Iterated Logarithm. In this chapter will often make use of concepts treated in Appendix D.

mean, its *variance*, and its *standard deviation*:

$$\mathrm{E}_N(f) \equiv \frac{1}{N}\sum_{n\le N} f(n), \quad \text{and} \quad \mathrm{Var}_N(f) \equiv \mathrm{StdDev}_N^2(f) \equiv \frac{1}{N}\sum_{n\le N}\left(f(n) - E_N f\right)^2.$$

Letting $N \to \infty$ one may also consider their limiting values (provided they exist)

$$\mathrm{E}(f) \equiv \lim_{N\to\infty} \mathrm{E}_N(f), \quad \text{and} \quad \mathrm{Var}(f) \equiv \mathrm{StdDev}^2(f) \equiv \lim_{N\to\infty} \mathrm{Var}_N(f).$$

The known results $\mathrm{E}(\mu) = 0$ (where μ is the Möbius function), which is equivalent to the Prime Number Theorem, and $\mathrm{Var}(\mu) = \mathrm{E}(\mu^2) = 6/\pi^2$, which gives the density of squarefree integers, already hint at the applicability of these concepts. In this connection a conjecture of Erdős & Wintner, later proved by Wirsing and Halász, is well worth noting: every multiplicative function f which takes on only the values -1, 0, and 1 possesses a mean value $\mathrm{E}\, f$.

One can likewise ask whether or not $f(n)$ possesses a *limiting distribution*, that is, given the 'finite sample' distributions $F_N(x) \equiv \frac{1}{N}\#\{n \le N\colon f(n) \le x\}$ of $f(n)$, does $F_N(x)$ converge to a distribution function $F(x)$ at all continuity points of F (so-called *weak convergence* $F \overset{w}{\to} F$)? By the Levy Continuity Theorem (see Appendix D), this question can equivalently be posed in the Fourier domain: do the "finite sample" *characteristic functions*[2]

$$\varphi_N(t) \equiv \frac{1}{N}\sum_{n\le N} e^{itf(n)}, \quad t \in \mathbb{R},$$

converge pointwise to a function $\varphi(t)$ which is continuous at $t = 0$? If they do, then φ is necessarily the characteristic function of a distribution F, that is, $\varphi(t) = \int e^{itx}\, dF(x)$, and F_n must converge weakly to F. Using this criterion, *convergence in distribution* translates into *convergence of mean values* for the arithmetic functions $g_t(n) = e^{itf(n)}$. More generally one can ask: when do there exist normalizing functions $\alpha(x)$ and $\beta(x) > 0$ such that

$$\frac{1}{x}\#\left\{n \le x\colon \frac{f(n) - \alpha(x)}{\beta(x)} \le t\right\}$$

converges (as $x \to \infty$) to a limiting distribution, $F(t)$ say, and what is that distribution?

In dealing with such questions, many methods from probability theory can arise, ranging from elementary results such as Chebyshev's inequality, to deeper ones such as Kolmogorov's Three-Series Criterion. Nevertheless, it is often—although not always—the number-theoretic considerations that provide the greater source of difficulty. Even so, however, statistical considerations have played an essential role in the development of the theory. While there is still much of the story of probabilistic number theory to be written, recent progress in this area has been considerable, and only a small part of it can be touched on here. More detail may be found in the texts by Kubilius (1964, [**663**]), Laurinčikas et al. (1977, [**682**]), Elliot (1979, [**318**]), (1980, [**319**]), Tenenbaum (2015, [**1053**]), and Kowalski (2021, [**661**]), and in the review articles by Rényi (1960, [**893**]), Galambos (1970, [**402**]), Schweiger & Manstavičius (1992, [**961**]), Schwarz (1994, [**958**]), (2007, [**959**]), Indlekofer (2002, [**578**]), (2002, [**579**]), Steuding (2002, [**1026**]), and Stakėnas (2015, [**1018**]), as well as in other references cited below.

[2] In some books the term 'characteristic function' denotes the 'indicator function' of a set; we use it exclusively to mean the Fourier transform $\varphi(t) = \int e^{itx}\, dF(x) = \mathrm{E}_F\, e^{itX}$ of a probability distribution $F(x)$.

Once adopted, a probabilistic viewpoint leads to several number-theoretic concepts. For example, the *natural density* $\nu(\mathcal{A})$ of some subset $\mathcal{A}$ of the natural numbers is defined as the limiting ratio

$$\nu(\mathcal{A}) = \lim_{x\to\infty} \frac{1}{x}\#\{n \le x\colon n \in \mathcal{A}\} \tag{9.1}$$

(provided the limit exists). Hence one can say that the even numbers have natural density $\frac{1}{2}$, while the primes have natural density 0. There are other versions of 'density' for integer sets, but here we will mention only one. The *logarithmic density*, say $\delta(\mathcal{A})$, of a subset $\mathcal{A}$ of positive integers can be defined in either of two equivalent ways:

$$\delta(\mathcal{A}) = \lim_{x\to\infty} \Bigg(\sum_{\substack{n\le x\\ n\in\mathcal{A}}} \frac{1}{n}\Bigg)\Bigg/\Bigg(\sum_{n\le x}\frac{1}{n}\Bigg), \quad\text{or}\quad \delta(\mathcal{A}) = \lim_{x\to\infty}\frac{1}{\log x}\sum_{\substack{n\le x\\ n\in\mathcal{A}}}\frac{1}{n}. \tag{9.2}$$

The rationale behind this definition involves the fact that $\sum_{n\le x}\frac{1}{n} \sim \log x$. It can be shown that whenever $\nu(\mathcal{A})$ exists, then so does $\delta(\mathcal{A})$ and the two are equal. But $\delta(\mathcal{A})$ can exist even when $\nu(\mathcal{A})$ does not. For example, if $\mathcal{A}$ is the set of all numbers consisting of an even number of digits, one can show that $\delta(\mathcal{A}) = \frac{1}{2}$ but $\nu(\mathcal{A})$ does not exist. Be that as it may, neither of the 'measures' $\nu(\cdot)$ or $\delta(\cdot)$ is countably additive, and numerous difficulties that arise in the theory can ultimately be traced back to this fact.

EXAMPLE 9.1. Let $g_k(n)$ be a zero-one indicator function for the *k-free* integers, that is, integers whose prime factors are all of multiplicity less than k:

$$g_k(n) = \begin{cases} 0, & \text{if } p^k \mid n \text{ for some prime } p,\\ 1, & \text{otherwise.}\end{cases}$$

(Note that $g_2(n) = |\mu(n)|$ where $\mu(n)$ is the Möbius function.) It can be shown that

$$g_k(n) = \sum_{d^k\mid n}\mu(d), \quad\text{and that}\quad \lim_{x\to\infty}\frac{1}{x}\sum_{n\le x} g_k(n) = \frac{1}{\zeta(k)}.$$

Thus the natural density of the k-free integers is equal to $1/\zeta(k)$. (See Apostol (1970, [**24**]).)

Next, one says that the arithmetic function $f(n)$ has *average order* $g(n)$ if

$$\sum_{n\le x} f(n) \sim \sum_{n\le x} g(n), \quad\text{as}\quad x\to\infty.$$

Such a function g obviously won't be unique; in typical usage g would be monotone, as well as 'simpler' and 'smoother' than f. One also says that $f(n)$ has *normal order* $g(n)$ if for every $\epsilon > 0$,

$$\left|\frac{f(n)}{g(n)} - 1\right| \le \epsilon$$

on a set of integers n whose natural density is 1; in other words, $g(n)$ approximates $f(n)$ arbitrarily well for almost all values of n. Typically, g would be 'simpler' than f and it too won't be unique. This concept, which dates to Hardy & Ramanujan (1917, [**497**]) and was new to number theory at the time, allows one to ignore a 'negligible' number of exceptional values and roughly corresponds to the notion of 'convergence in probability' (see Appendix D). If F_n is the distribution function corresponding to values of $f(n)/g(n)$ where n is chosen randomly from $\{1, 2, \ldots, n\}$, then normal order is equivalent to weak convergence of F_n to the degenerate distribution which assigns all of its mass to 1. General references include Tenenbaum (2015, [**1053**, pp. 454–455]), and Hardy & Wright (2008, [**499**, §22.10, 22.11, & 22.13]).

Remarks and examples:

- An arithmetic function can have an average order but not a normal order. For instance, if $f(n)$ equals 0 or 2 according as n is even or odd, then f has average order 1, but it does not have a normal order.
- A normal order and average order need not coincide. This will occur, for example, when the function is dominated by a small number of abnormally large values. An example is given by the divisor function $\tau(n)$ treated below.
- Normal order can be represented as $f(n) = g(n)\,(1 + o_{\text{a.e.}}(1))$. (a.e. $\equiv$ almost everywhere.)
- Let $f(n) = n$ if n is prime, and $f(n) = 1$ otherwise; and let $g(n) \equiv 1$. Then $f(n)$ has normal order $g(n)$ since the primes have density 0, but $g(n)$ is not the average order of $f(n)$ since

$$\sum_{n \le x} f(n) \sim \frac{x^2}{\log x}, \quad \text{while} \quad \sum_{n \le x} g(n) \sim x.$$

- With the exception of the powers of n, namely $f(n) = n^c$, an unbounded multiplicative function cannot have a *non-decreasing* normal order although it can have an average order (Birch, (1967, [**108**])).
- The Möbius function $\mu(n)$ has average value 0, and the Euler phi function $\varphi(n)$ has average order $6n/\pi^2$. These functions do not have normal orders.
- If $\tau(n)$ is the divisor function, then $\log \tau(n)$ has average order $(\log 2) \log\log n$, while (3.37) establishes that the normal order of $\log \tau(n)$ is also $(\log 2)(\log\log n)$. Underpinning these results is the fact that $2^{\omega(n)} \le \tau(n) \le 2^{\Omega(n)}$. For $\tau(n)$ itself, its average order is $\log n$. This follows from (3.38) since $\sum_{n \le x} \log(n) \sim \int_1^x \log t\, dt \sim x \log x$ which implies that $\sum_{n \le x} \tau(n) \sim \sum_{n \le x} \log n$. It turns out that the average order of $\tau(n)$ is determined by a small proportion of large values. Note that $\tau(n)$ does not have a normal order.
- Using $\sum_{m=1}^{n} m \sim \frac{1}{2} n^2$, we have

$$\sum_{n \le x} \sigma(n) \sim \frac{\pi^2}{12} x^2 \sim \sum_{n \le x} \frac{\pi^2}{6} n.$$

Consequently the average order of the sum of divisors function $\sigma(n)$ is $\frac{\pi^2}{6} n$.

- Hardy and Ramanujan proved (see below) that the prime divisors counting functions $\omega(n)$ and $\Omega(n)$ each have average order as well as normal order equal to $\log\log n$. The fact that $\omega(n) = \Omega(n) = 1$ infinitely often does not invalidate this result since this only happens on a set of density 0.

Continuing now, a *maximal order* of an arithmetic function $f(n)$ is a positive, nondecreasing function $M(n)$ such that

$$\limsup_{n \to \infty} \frac{f(n)}{M(n)} = 1.$$

When $f(n)$ is nonnegative this implies the weaker statement $f(n) = O(M(n))$. Likewise, for a positive function f, a *minimal order* $m(n)$ is defined via

$$\liminf_{n\to\infty} \frac{f(n)}{m(n)} = 1.$$

For example, the total number of prime factors function satisfies $\Omega(n) \geq 1$, for $n \geq 2$, and attains the value 1 at primes. Furthermore, $\Omega(n) \leq \log n/\log 2$, for $n \geq 1$, and attains this value when $n = 2^k$. Hence the minimal and maximal orders of $\Omega(n)$ are 1 and $\log n/\log 2$, respectively. Some arithmetic functions and their minimal and maximal orders are:

Function	Minimal order	Maximal order
$\Omega(n)$	1	$\log n/\log 2$
$\omega(n)$	1	$\log n/\log\log n$
$\log\tau(n)$	2	$(\log 2)\log n/\log\log n$
$\varphi(n)$	$e^{-\gamma}n/\log\log n$	n
$\sigma(n)$	n	$e^{\gamma}n\log\log n$

In the first half of this chapter, we give detailed expositions of some major theorems of probabilistic number theory: the Hardy–Ramanujan Theorem about the normal order of the prime factors counting functions $\omega(n)$ and $\Omega(n)$; the Turán–Kubilius inequalities that bound how far an additive function can deviate from its mean value on average; the Erdős–Kac Theorem, which shows how the functions $\omega(n)$ and $\Omega(n)$ can be said to have a normal limiting distribution; and the Erdős–Wintner and Delange Theorems that characterize when classes of additive or multiplicative functions have limiting distributions or limiting mean values. We then conclude the chapter with a series of complements along the same themes, followed by some historical remarks, as well as a selection of proofs of the results described.

9.1. The Hardy–Ramanujan Theorem

Arguably, the first major result of probabilistic number theory is due to Godfrey Harold Hardy (1877–1947) and Srinivasa Ramanujan (1887–1920),[3] appearing in a paper they published in 1917. Recall that $\omega(n)$ and $\Omega(n)$ represent, respectively, the number of *distinct* prime factors, and the *total* number of prime factors of n. Hardy and Ramanujan introduced the concept of normal order of an arithmetic function and proved that

$$\text{(9.3)} \qquad \frac{1}{x}\#\left\{n \leq x \colon \left|\frac{\omega(n) - \log\log n}{\sqrt{\log\log n}}\right| \leq \psi(n)\right\} \to 1, \quad \text{as } x \to \infty,$$

for *any* function $\psi(n)$ which tends to ∞ as $n \to \infty$. The key take-away from this is that $\omega(n)$ has normal order $\log\log n$, a statement that can be phrased in probabilistic terms as $\omega(U_n)/\log\log n \xrightarrow{p} 1$ as $n \to \infty$, where the U_n are random variables uniformly distributed over $\{1, 2, \ldots, n\}$. This allows one to say that the number of distinct prime divisors of an arbitrary number from 1 to n is (for large n) very likely to be close to $\log\log n$. Note that as a straightforward consequence of

[3]For an account of the astonishing and "romantic" story of Ramanujan, see Kanigel (1991, [**619**]). Accounts that also provide some technical detail include Hardy (1937, [**488**]), (1940, [**489**]), and Rao (2021, [**886**]). Accounts of Hardy himself include Albers et al. (2015, [**12**]), and Hardy (1967, [**491**]).

the fact that $\log\log x$ rises so slowly, one may replace either or both of the $\log\log n$ terms in (9.3) by $\log\log x$. One may demonstrate this using

$$\log\log x - 1 \le \log\log n \le \log\log x, \quad \text{for } x^{1/e} \le n \le x,$$

together with the fact that $x^{1/e} = o(x)$.

Although this result has been stated in terms of $\omega(n)$, the two functions $\omega(n)$ and $\Omega(n)$ typically have similar size, so the result (9.3) continues to hold if $\omega(n)$ is replaced by $\Omega(n)$. In Section 9.7 (Selected proofs) specifically, we will show that

$$\lim_{x\to\infty} \frac{1}{x} \sum_{n\le x} (\Omega(n) - \omega(n)) = \sum_p \frac{1}{p(p-1)}, \tag{9.4}$$

with the point being that this is a convergent sum. Consequently even though $\omega(n) \le \Omega(n)$, both turn out to have the same asymptotic means

$$\frac{1}{x}\sum_{n\le x} \omega(n) \sim \log\log x, \quad \text{and} \quad \frac{1}{x}\sum_{n\le x} \Omega(n) \sim \log\log x,$$

as well as asymptotic variances of the same order

$$\frac{1}{x}\sum_{n\le x} (\omega(n) - \log\log x)^2 = (1+o(1)) \log\log x,$$

and

$$\frac{1}{x}\sum_{n\le x} (\Omega(n) - \log\log x)^2 = (1+o(1)) \log\log x.$$

The $\log\log x$ centring in these variances are interchangeable with $\log\log n$ (as well as with the average $(1/x)\sum_{1<n\le x} \log\log n$), again because these values are virtually identical for all but the smallest x.

The variance results for ω and Ω were also established by Pál Turán (1934, [**1069**]) in an exemplary two-page paper which used only elementary methods. His purpose was to give a proof of (9.3) that was simpler than the rather complicated one of Hardy and Ramanujan. Turán used his variance results together with the Chebyshev inequality to prove

$$\mathrm{P}_n\Big[\ |\omega_n - m_n| \le \epsilon m_n\Big] \le \frac{\operatorname{Var} \omega_n}{\epsilon^2 \mu_n^2} = \frac{1+o(1)}{\epsilon^2 \log\log n} \to 0, \text{ as } n\to\infty, \text{ for every } \epsilon > 0,$$

this being, in fact, the Hardy–Ramanujan result. Here P_n represents the uniform probability on $\{1, 2, \ldots, n\}$, while ω_n represents ω chosen according to P_n, and m_n is its mean. Turán's proof was probabilistic in nature, but what is considered remarkable about his paper (aside from its elegance) is that Turán was not aware of the Chebyshev inequality and did not know any probability theory at the time. (Kolmogorov's theory was published only in 1933.) It was only after receiving a letter from Mark Kac, who had read Turán's 1934 paper, that Turán became aware of its probabilistic significance. (At the time Kac already suspected the Erdős–Kac normality result and wanted to know if Turán's method could be extended to evaluate the higher moments.) We will give Turán's proof in Section 9.7. See also Hardy & Wright (1986, [**499**, §22.11]), as well as the historical remarks in Section 9.6.

Incidentally, the Hardy–Ramanujan result appears to be fairly robust. For example, while it obviously cannot hold over the set of primes p, Erdős (1935, [**335**]) showed that it continues to hold over the set of shifted primes $(p-1)$'s; in other words, the normal orders of $\omega(p-1)$ and $\Omega(p-1)$ are also both $\log\log p$.

REMARK 9.2. Delange (1953, 1971, [**266, 270**]) refined Turán's result and proved that

$$\sum_{n\le x}(\omega(n)-\log\log x)^2 = x\log\log x + \Theta x + O\left(\frac{x\log\log x}{\log x}\right);$$

an expression for the value of the constant Θ may be found in Saidak (2002, [**930**]).

9.2. The Turán–Kubilius inequalities

Turán's (1934, [**1069**]) paper led to important generalizations (Turán (1936, [**1070**]), Kubilius (1956, [**663**])) and in particular led to the so-called Turán–Kubilius inequalities. These provide useful and computationally convenient bounds for the variances of additive arithmetic functions which—when used together with the Chebyshev inequality or with similar inequalities—often suffice to establish their normal orders. The Turán–Kubilius inequalities have also proven useful for establishing other asymptotic properties of additive functions. For example, they have been used in the proof of the Erdős–Wintner Theorem (discussed below); see for example Shapiro (1983, [**983**]). Two key references are Elliott (1979, [**318**, Chapter 4]), and Tenenbaum (2015, [**1053**, pp. 446–451]). Other references include Schwarz & Spilker (1994, [**960**, §1.4]), and Stakėnas (2015, [**1018**]).

Recall that if f is an additive arithmetic function, then $f(1)=0$, and if n has prime factorization $n = p_1^{k_1}\cdots p_m^{k_m}$, then

$$f(n) = \sum_{j=1}^{m} f(p_j^{k_j}) \equiv \sum_{p^k\| n} f(p^k) = \sum_{p^k\le n} f(p^k)\,\mathrm{I}\big[p^k \parallel n\big], \tag{9.5}$$

where $p^k \parallel n$ means that p^k divides n but p^{k+1} does not, and $\mathrm{I}[\,\cdot\,]$ is the 0–1 indicator function. Now as $N\to\infty$, the proportion of values of $n\le N$ divisible by p^k but not by p^{k+1} tends to $(1/p^k)(1-1/p)$, and this will be the long-run average of $\mathrm{I}\big[p^k \parallel n\big]$. This suggests approximating the mean $\frac{1}{N}\sum_{n\le N} f(n)$ by

$$E(x) = \sum_{p^k\le N}\frac{f(p^k)}{p^k}\left(1-\frac{1}{p}\right), \tag{9.6}$$

the sum here being over all prime powers $\le N$. (Note in passing that for a fixed p, the proportions $(1/p^k)(1-1/p)$ correspond to those of a geometric distribution.)

The representation in (9.5) also suggests a heuristic that allows one to gain some understanding of the variation of $f(n)$. Specifically, it suggests defining a *random variable* $\mathbf{f}$, based on the values of $f(n)$ for $1\le n\le N$, as follows:

$$\mathbf{f} \equiv \sum_{p^k\le N} f(p^k)X(p,k), \tag{9.7}$$

where $X(p,k)=1$ with probability $(1/p^k)(1-1/p)$, and $X(p,k)=0$ otherwise. Note that the expectation $\mathrm{E}\,\mathbf{f}$ coincides with the value $\mathrm{E}(x)$ shown in (9.6). If we regard the $X(p,k)$ as being mutually independent (we will examine this assumption later) then we will have

$$\operatorname{Var}\mathbf{f} = \sum_{p^k\le N} \operatorname{Var}\big(f(p^k)X(p,k)\big) \le \sum_{p^k\le N} \left|f(p^k)\right|^2 \frac{1}{p^k}\left(1-\frac{1}{p}\right).$$

Here the variances of the $X(p,k)$ have been bounded by their second moments. (It turns out that this restriction to second moments leads to bounds which are more convenient for computations, and yet remain adequate for most applications.) The Turán–Kubilius inequalities exploit this heuristic, and in one version state that if $f(n)$ is a (complex-valued) additive arithmetic function (not identically zero), then

$$\frac{1}{x}\sum_{n\le x}\left|f(n)-E(x)\right|^2 \le CD^2(x), \tag{9.8}$$

where

$$E(x) = \sum_{p^k\le x}\frac{f(p^k)}{p^k}\left(1-\frac{1}{p}\right), \quad D^2(x) = \sum_{p^k\le x}\left|f(p^k)\right|^2\frac{1}{p^k}\left(1-\frac{1}{p}\right) > 0, \tag{9.9}$$

and where $C>0$ is a *universal constant*—that is, a constant which does not depend on f, and holds for all $x\ge 2$. In the case that $f(n)$ is strongly additive, so that $f(p^k)=f(p)$, an entirely analogous heuristic argument can be made, and (9.8) will in fact hold with

$$E(x) = \sum_{p\le x}\frac{f(p)}{p}, \quad \text{and } D^2(x) = \sum_{p\le x}\frac{|f(p)|^2}{p} > 0. \tag{9.10}$$

These results can also be generalized to moments higher than the second. An essential point about these inequalities is that they hold regardless of any dependence structure among the terms of the additive function. A proof of (9.8)/(9.9) is given in Section 9.7 (Selected proofs).

In (9.9), either or both of the $(1-\frac{1}{p})$ factors appearing in $E(x)$ and $D^2(x)$ can typically be dropped. Also, $E(x)$ may be replaced by $\frac{1}{x}\sum_{n\le x}f(n)$ should one wish since this would merely reduce the left hand side of (9.8). The computations supporting these assertions are fairly straightforward applications of the Cauchy–Schwarz inequality (Hildebrand (1983, [**548**]) goes through the details) although such variations may affect the value of the constant C. Nevertheless, in all of the mentioned cases, and assuming only that f is additive, it is known that the value of C can be replaced by $C(x)=\frac{3}{2}+O\left(1/\sqrt{\log x}\right)$ universally, and that the value of $\frac{3}{2}$, as well as the indicated order term, cannot be further reduced. In the strongly additive case, C may actually be replaced by $\frac{3}{2}$ with the proviso that $x>x_0$, where x_0 is itself a universal constant. These results are due to Kubilius (1983, [**664**]), Hildebrand (1983, [**548**]), Stein (1984, [**1020**]), and Jungseob Lee (1989, [**683**]), (1992, [**684**]). See also Elliott (1987, [**320**]). We mention here that Charles Stein was a professor at Stanford who originated the 'Stein Method' for approximating distributions and made other original contributions to probability and statistics. Less well known was his long-time interest in analytic and probabilistic number theory.

Concerning the independence heuristic involving the $\mathrm{I}[p^k \parallel n]$ and $X(p,k)$ appearing in (9.5) and (9.7), the following approximations are worth noting: first,

$$\frac{1}{N}\sum_{n=1}^{N} \mathrm{I}\left[p^k \parallel n\right] = \frac{1}{N}\left(\left[\frac{N}{p^k}\right] - \left[\frac{N}{p^{k+1}}\right]\right) = \frac{1}{p^k}\left(1-\frac{1}{p}\right) + O\left(\frac{1}{N}\right),$$

and second, for primes $p \neq q$,

$$\begin{aligned}
&\frac{1}{N}\sum_{n=1}^{N}\left(\mathrm{I}\left[p^k \parallel n\right]\mathrm{I}\left[q^\ell \parallel n\right]\right)\\
&\quad = \frac{1}{N}\left(\left[\frac{N}{p^k q^\ell}\right] - \left[\frac{N}{p^{k+1}q^\ell}\right] - \left[\frac{N}{p^k q^{\ell+1}}\right] + \left[\frac{N}{p^{k+1}q^{\ell+1}}\right]\right)\\
&\quad = \frac{1}{p^k}\left(1-\frac{1}{p}\right)\frac{1}{q^\ell}\left(1-\frac{1}{q}\right) + O\left(\frac{1}{N}\right)\\
&\quad = \left(\frac{1}{N}\sum_{n=1}^{N}\mathrm{I}\left[p^k \parallel n\right]\right)\left(\frac{1}{N}\sum_{n=1}^{N}\mathrm{I}\left[q^\ell \parallel n\right]\right) + O\left(\frac{1}{N}\right).
\end{aligned}$$

Thus for fixed p^k and q^ℓ (and likewise for any finite collection of such) we observe behaviour consistent with independence in the $N \to \infty$ limit. It turns out that this degree of asymptotic independence—together with the Cauchy–Schwarz inequality—suffices to allow a proof of the Turán–Kubilius inequality to go through. A significant point is that the proof (see the one given in Section 9.7) requires only 'elementary' results from number theory.

9.3. The Erdős–Kac Theorem

Historically, the next major development in the probabilistic theory (after that of Hardy–Ramanujan) was due to Erdős & Kac (1939, [**357**]), (1940, [**358**]). Concerning their result, Erdős once ventured to say: "to be a little impudent and conceited, probabilistic number theory was born." While this distinction can also be claimed by Hardy–Ramanujan, (as well as by Turán—as interpreted by Kac), those results concerned only moments and not distributions.[4] This is a notable instance in which Hardy's disdain for probability had cost him a singularly important result. One may perhaps say this of Turán's indifference to probability as well.

Using number-theoretic methods Erdős and Kac obtained the distributional result

$$\lim_{x\to\infty}\frac{1}{x}\#\left\{n \leq x\colon \frac{\omega(n) - \log\log n}{\sqrt{\log\log n}} < \alpha\right\} = \frac{1}{\sqrt{2\pi}}\int_{-\infty}^{\alpha} e^{t^2/2}\,dt, \tag{9.11}$$

a striking result which links a fundamental property of the prime factorization of integers with the fundamental error distribution of statistical theory. A suggestive restatement of (9.11) is given by

$$\mathrm{P}\left(\frac{\omega\left(U_n\right) - \log\log n}{\sqrt{\log\log n}} < \alpha\right) \to \frac{1}{\sqrt{2\pi}}\int_{-\infty}^{\alpha} e^{t^2/2}\,dt, \quad \text{as } n \to \infty, \tag{9.12}$$

where U_n is uniformly distributed over $\{1, 2, \ldots, n\}$. One may describe this result succinctly by saying that $\omega(n)$ is distributed asymptotically as $N(\log\log n, \log\log n)$. The work of Erdős and Kac was based on Brun's method of sieves, although now

[4] To be fair, distributions did also appear in the work of Schoenberg (1928, [**952**]) concerning the function $\varphi(n)/n$.

it is known that sieve methods are not required for this result. Indeed what fundamentally underlies the result of Erdős and Kac are the two moment identities

$$\sum_{p\le x}\frac{\omega(p)}{p}=\log\log x+O(1),\quad\text{and}\quad \sum_{p<x}\frac{\omega^2(p)}{p}=\log\log x+O(1).$$

The equality of mean and variance for ω and for Ω hints at a Poisson distribution which lurks beneath, a point we will return to below. We remark that—as before—either or both of the $\log\log n$ terms appearing in (9.11) may be replaced by $\log\log x$.

One useful way to think of the Erdős–Kac result is to view $\omega(U_n)$ as a sum

$$\omega(U_n)=\sum_{p\le n}\mathrm{I}[p\mid U_n],$$

(where the $\mathrm{I}[p\mid U_n]$ are 1–0 indicators of whether p divides U_n or not), and to approximate the stochastic behaviour of $\omega(U_n)$ as that of

$$\widetilde{\omega}(U_n)=\sum_{p\le n}X_p\,,$$

where the X_p are *independent* Bernoulli random variables with probabilities $\frac{1}{p}$. The expected value of $\widetilde{\omega}(U_n)$ is $\mathrm{E}(\widetilde{\omega}(U_n))=\sum_{p\le n}\frac{1}{p}\sim\log\log n$, while its variance is $\sum_{p\le n}\frac{1}{p}(1-\frac{1}{p})\sim\log\log n$. (We are using the Mertens result (2.2) here.) Lindeberg's Central Limit Theorem then applies to $\widetilde{\omega}(U_n)$ to give the Erdős–Kac result for $\widetilde{\omega}$. Of course, this logic does not take correlations among the $\mathrm{I}[p\mid U_n]$ into account, but it turns out that the dependences among them are weak enough to allow the same normal limit result to hold. In fact, it was just such reasoning that led Kac to conjecture the asymptotic normality result (9.11) in the first place; at the time, Erdős did not know enough probability to have even been able to formulate it. Erdös & Kac (1940, [**358**]) begin by saying:

> *The present paper concerns itself with the applications of statistical methods to some number-theoretic problems. Recent investigations of Erdős and Wintner have shown the importance of the notion of statistical independence in number theory; the purpose of this paper is to emphasize this fact once again.*

They end with the remark:

> *We may perhaps point out that Lemma 2 (Brun) is the deepest part of the proof and that the statistical part is relatively superficial. However, the statistical considerations seemed to be suggestive and fruitful in leading to new and perhaps striking results* [**358**].

Just as in the Hardy–Ramanujan result, the similar sizes of $\omega(n)$ and $\Omega(n)$ imply that the asymptotic normality result (9.11)/(9.12) continues to hold if ω is replaced by Ω. Not surprisingly, however, the assured Gaussian behaviour of these functions kicks in very slowly; indeed $\log\log n$ does not attain a value of 4 until $n=\exp(\exp(4))\approx 5\times 10^{23}$, while numbers with 100 digits are on average expected to have only about 5.5 prime factors. Likewise, a straightforward computation based on (9.11) suggests that approximately 60% of the numbers around $n\approx 10^{70}$ should have somewhere between 3 and 7 prime divisors.

As already mentioned, the original proof of the Erdős and Kac result was based on complicated number-theoretic methods involving Brun's sieve. Kac (1949, [**612**]) suggested that a simpler proof should be possible using purely probabilistic methods, and in particular by establishing convergence of the moments. This program was successfully implemented by Halberstam (1955, [**479**]); Billingsley (1969, [**102**]) provides a detailed exposition; see also Granville & Soundararajan (2007, [**460**]). A proof of the Erdős–Kac result, relying on this moments method, will be outlined in Section 9.7.

It is natural to inquire about which other additive arithmetical functions have asymptotic distributions which are Gaussian. Such an extension of the Erdős–Kac asymptotic normality result ws obtained by Kubilius (1955, [**662**]) and by Shapiro (1956, [**982**]). They showed that if $f(n)$ is strongly additive, and if

$$D^2(x) \equiv \sum_{p \le x} \frac{f^2(p)}{p} \to \infty, \quad \text{with also} \quad \frac{1}{D^2(x)} \sum_{\substack{p \le x \\ |f(p)| > \epsilon D(x)}} \frac{f^2(p)}{p} \to 0 \tag{9.13}$$

for every $\epsilon > 0$, then the normalized $f(n)$ will tend to a Gaussian distribution. Specifically,

$$\lim_{x \to \infty} \frac{1}{x} \# \left\{ n \le x : \frac{f(n) - \mathrm{E}(x)}{D(x)} < t \right\} = \frac{1}{\sqrt{2\pi}} \int_{-\infty}^{t} e^{-u^2/2} \, du, \tag{9.14}$$

as $x \to \infty$, where

$$\mathrm{E}(x) \equiv \sum_{p \le x} \frac{f(p)}{p} .$$

In fact, this result continues to hold for a fairly wide class of additive functions which are not strongly additive. Hence, under the conditions in (9.13), the asymptotic behaviour of such arithmetic functions is in certain respects comparable to sums of independent, identically distributed random variables. We remark that the condition (9.13) originates from that of the Lindeberg CLT (see Appendix D).

Halberstam (1955, [**479**]) showed, in the strongly additive case, that if also $|f(p)| \le 1$ for all primes, then the weak convergence to normality can be strengthened considerably to include the convergence of all moments to those of a Gaussian:

$$\lim_{x \to \infty} \frac{1}{x} \sum_{n \le x} \left(\frac{f(n) - \mathrm{E}(x)}{D(x)} \right)^k = \frac{1}{\sqrt{2\pi}} \int_{-\infty}^{\infty} u^k e^{-u^2/2} \, du \,, \quad k > 0. \tag{9.15}$$

Remarkably, he obtained this result using only elementary (although computationally quite involved) methods. In assessing this result one should bear in mind that the weak convergence of a sequence of distributions does not require any of their moments to exist, but that the convergence of all moments to those of a distribution which is determined by its moments does imply the weak convergence of the distributions. Further details may be found in Kubilius (1964, [**663**]), Delange (1953, [**266**]), (1956, [**267**]), (1957, [**268**]), Delange & Halberstam (1957, [**271**]), Vilkas (1957, [**1087**]), and Galambos (1970, [**402**]).

The Erdős–Kac Theorem has also been generalized in various other directions. For example, Halberstam (1956, [**479**]) proved that the Erdős–Kac result holds for the shifted primes $p - 1$, and Alladi (1987, [**17**]) proved that it holds for integers without large prime factors. Rényi & Turán (1958, [**897**]) studied the error of the approximation in (9.11). Using the Berry–Esseen result of probability theory they established that the difference between the left and right sides of (9.11) is $O(1/\sqrt{\log\log x})$, uniformly for $\alpha \in \mathbb{R}$, and furthermore that this order is best possible. (Tenenbaum (2015, [**1053**, Chapter III.4, §4.4]) provides a proof of this result.) Delange (1956, [**267**]) decomposed this order term to obtain a still sharper result.

9.4. The Erdős–Wintner and Delange Theorems

An especially striking result of probabilistic number theory is a far-reaching companion to the Erdős–Kac Theorem which fully answers the question of when an additive arithmetic function possesses an asymptotic distribution—a result which doesn't restrict the resulting distributions to be Gaussian. It is due to Erdős (1935, [**336**]), (1937, [**336**]), (1938, [**336**]), who proved the sufficiency part, and Erdős & Wintner (1939, [**364**]), who proved the necessity. Like the Erdős–Kac Theorem, this foundational result has had enormous influence on the subsequent development of the theory.

THEOREM 9.3 (Erdős–Wintner). *Let f be a real-valued additive arithmetic function, and for each n, let F_n be the (finite-sample) distribution function based on the values that f takes on $\{1, 2, \ldots, n\}$. Then $f(n)$ will have an asymptotic distribution, that is, $F_n \overset{w}{\to} F$, for some distribution function F, if and only if each of the three series*

$$(9.16) \qquad (a) \sum_{|f(p)|>1} \frac{1}{p}, \qquad (b) \sum_{|f(p)|\le 1} \frac{f(p)}{p}, \quad \text{and} \quad (c) \sum_{|f(p)|\le 1} \frac{f^2(p)}{p}$$

converges; and when a limiting distribution exists, its characteristic function—namely its Fourier transform $\varphi(t) = \int e^{itx}\, dF(x)$—will be given by the convergent product

$$(9.17) \qquad \varphi(t) = \prod_p \left(1 - \frac{1}{p}\right)\left(1 + \sum_{k=1}^{\infty} \frac{\exp(itf(p^k))}{p^k}\right), \quad \text{for all } t \in \mathbb{R}.$$

Furthermore, this limiting distribution will be of pure type (meaning it will be purely discrete, purely absolutely continuous, or purely singular), and it will be one of the two continuous types if and only if the sum $\sum_{f(p)\neq 0} \frac{1}{p}$ diverges.

REMARKS 9.4.

(i) The criteria (9.16) are in effect a number-theoretic analogue of the Kolmogorov Three-Series Criterion (see Appendix D). The value of 1 used in (9.16) is arbitrary and may be replaced by any other finite positive value; if the three convergences hold for one such value, they will hold for all such values.

(ii) The characteristic function shown in (9.17) corresponds to the distribution of

$$\sum_p f(p^{\nu_p})$$

where the ν_p are independent geometric random variables corresponding to the proportion of integers n such that $p^k \parallel n$, namely $\mathrm{P}[\nu_p = k] = (1 - \frac{1}{p})p^{-k}$ for $k = 0, 1, 2, \ldots$.

(iii) Surprisingly, while the $k \geq 2$ values of $f(p^k)$ do affect the distribution result (9.17), they do not enter into the criterion (9.16) for the existence of a limiting distribution.

(iv) The theorem also holds for complex-valued additive function provided one replaces $f^2(p)$ by $|f(p)|^2$ in condition (c), and revises (9.17) in an appropriate way.

(v) Underlying the pure-types trichotomy is the Jessen & Wintner (1935, [**600**]) 'Law of Pure Types' for almost surely convergent sums of discrete-valued random variables. (See, for example, Breiman (1992, [**142**, Theorem 3.26])). No further conditions that help distinguish among the three cases are known.

(vi) Extensions of the Erdős–Wintner Theorem were developed by Kubilius, Elliot, Levin, Timofeev, and others; for these extended results and references see, for example, Schwartz (2007, [**959**]), Elliot (1980, [**319**, vol. II, Chapter 16]), or Narkiewicz (1983, [**796**, Chapter VII]).

(vii) Proofs of the Erdős–Wintner Theorem may be found in Tenenbaum (2015, [**1053**, Chapter III.4.1]), Novoselov (1964, [**816**]), Babu (1973, [**53**]), and Elliott (1979, [**318**, Chapter 5]). A partial proof is given in Section 9.7.

(viii) The approximation error for the Erdős–Wintner result was studied by Tenenbaum & Verwee (2021, [**1056**]).

One can further inquire as to when an additive arithmetic function will have a limiting distribution with a finite mean and variance. This question was studied by Elliott (1973, [**316**]) who found two sets of necessary and sufficient conditions. The first of these is that

$$\sum_p \frac{f(p)}{p} \quad \text{and} \quad \sum_p \frac{f^2(p)}{p}$$

both converge, while the second is that

$$\lim_{x\to\infty} \frac{1}{x} \sum_{n\leq x} f(n) \quad \text{exists, and} \quad \limsup_{x\to\infty} \frac{1}{x} \sum_{n\leq x} f^2(n) < \infty.$$

The first set of conditions is what one might expect if one could view $f(n)$ as a sum of independent variables, each of which has value $f(p)$ with probability $\frac{1}{p}$, and 0 otherwise. Less obvious is the second set of conditions which appear weaker, yet when combined with the assumption that f is additive still allow the result to go through. Indeed the limits of $\frac{1}{x}\sum_{n\leq x} f(n)$ and $\frac{1}{x}\sum_{n\leq x} f^2(n)$ will then both exist and will equal to the first two moments of the limiting distribution. Elliott's proof was based on the Turán–Kubilius inequality together with the Erdős–Wintner Theorem, but (surprisingly) also used a so-called Tauberian theorem for Lambert series.

The Erdős–Wintner Theorem has a distinctly probabilistic flavour, and its proofs are typically entwined with Kolmogorov's Three-Series Criterion in a essential way. Also worth noting is that a proof can be based on the fact that weak convergence $F_n \overset{w}{\to} F$ is equivalent to pointwise convergence of the corresponding characteristic functions. (See Appendix D for the probability background.) For as mentioned previously, an additive function $f(n)$ will have sample characteristic functions $\frac{1}{N}\sum_{n=1}^{N} e^{itf(n)}$ whose terms $g_t(n) = e^{itf(n)}$ are multiplicative (for any t), so that problems about its asymptotic distribution can be reduced to questions concerning the existence of means of the $g_t(n)$ for values of t about the origin. One can therefore prove the Erdős–Wintner Theorem with the help of the following multipart result for multiplicative functions. It is due to Delange (1961, [**269**]), Wirsing (1967, [**1116**]), and Halász (1968, [**476**]), (1971, [**477**]):

THEOREM 9.5 (Delange). *Let $g(n)$ be a complex-valued multiplicative arithmetic function such that $|g(n)| \leq 1$. Then it will have a nonzero limiting mean value*

$$M(g) \equiv \lim_{x\to\infty} \frac{1}{x} \sum_{n\leq x} g(n),$$

if and only if the following conditions (A) and (B) both hold:

$$\text{(9.18)}\qquad \begin{aligned} &\text{(A)} \quad \sum_p \frac{1-g(p)}{p} \quad \textit{is convergent, and} \\ &\text{(B)} \quad \sum_{\ell=1}^{\infty} \frac{g(p^\ell)}{p^\ell} \neq -1, \quad \textit{for all primes } p. \end{aligned}$$

In this case, the limiting mean value will equal

$$M(g) = \prod_p \left(1 - \frac{1}{p}\right)\left(1 + \sum_{\ell=1}^{\infty} \frac{g(p^\ell)}{p^\ell}\right). \qquad \text{(9.19)}$$

Furthermore, the limiting mean $M(g)$ will equal 0 if and only if the condition

$$\text{(C)} \quad \sum_p \frac{1 - \Re\left(g(p)p^{-it}\right)}{p} \quad \textit{diverges for every } t \in \mathbb{R}$$

also holds. If, in addition to $|g(n)| \leq 1$, the function $g(n)$ is real-valued, then its limiting mean $M(g)$ will always exist, and will still be given by (9.19); additionally its mean will equal 0 if the series (9.18) diverges.

REMARKS 9.6.

(i) Condition (B) is required only to ensure that the factors in the product for $M(g)$ are nonzero, and may altogether be replaced by the condition that $f(2^k) \neq -1$ for at least one integer $k \geq 1$). Condition (A) is the hypothesis that is used seriously in the proof.

(ii) When $g(n)$ is also strongly multiplicative, (9.19) may alternately be written as

$$M(g) = \prod_p \left(1 + \frac{g(p) - 1}{p}\right).$$

(iii) Note that when $g(n) = e^{f(n)}$, the convergence of (9.18) would imply the convergence of the three series in (9.16).

(iv) As mentioned earlier, Erdős and Wintner had conjectured that a multiplicative function which only takes on values -1, 0, or 1 must possess a mean value. Its truth is a corollary of Delange's theorem. Hence the Möbius function μ, for instance, must have a mean value. For μ, series (9.18) becomes $\sum(1-\mu(p))/p$ which equals $\sum \frac{2}{p}$ (because $\mu(p) = -1$ at the primes) and therefore diverges. Consequently the mean value of the Möbius function is 0. This fact is equivalent to the PNT, but isn't a proof of it, since the PNT is used in the proof of Delange's Theorem. The Erdős–Wintner conjecture had however implied the PNT.

(v) In the case when the multiplicative function g is real-valued with $|g(n)| \le 1$, and $\sum_p (1-g(p))/p$ is divergent, Wirsing and Hildebrand showed that there is a universal constant C such that

$$\left| \frac{1}{x} \sum_{n \le x} g(n) \right| \le C \left(1 + \sum_{p \le x} \frac{1-g(p)}{p} \right)^{-1/2}$$

holds for all such functions g, and for all $x \ge 2$. One consequence of this is that such functions g must have mean 0. (Schwarz & Spilker (1994, [**960**, Chapter IX, Theorem 1.4]).)

(vi) Unlike the real case, complex-valued g that satisfy $|g| \le 1$ need not have a mean value. An example is given by $g(n) = n^{it}$ which is multiplicative for any fixed real t. When $t \ne 0$, the sum $\frac{1}{x}\sum_{n \le x} g(n)$ does not converge: by approximating the sum by an integral one can show that $\frac{1}{x}\sum_{n \le x} g(n) \sim x^{it}/(1+it)$, which oscillates forever around a fixed circle.

(vii) For further generalizations and results, see Rényi (1965, [**895**]), (1965, [**896**]), Delange (1961, [**269**]), Daboussi (1981, [**252**]), (1989, [**253**]), Daboussi & Delange (1982, [**255**]), and Elliott (1975, [**317**]). Further discussion may also be found in Schwarz (2007, [**959**]), and in numerous articles by Galambos, In particular, Schwarz & Spilker (1994, [**960**, Chapter VII & IX]) provide an extensive treatment of the numerous components that comprise this theorem.

(viii) Tenenbaum (2015, [**1053**, Chapter III.4, §4.1]) spells out the details of how Delange's Theorem can be used to prove the Erdős–Wintner Theorem.

(ix) Some remarks on the proof of Delange's theorem may be found in Section 9.7. A complete proof may be found in Tenenbaum (2015, [**1053**, Chapter III.4, §4.2]).

The original proof of the Erdős–Wintner Theorem involved nontrivial number theory and even now no 'easy' proofs are known. Shapiro (1983, [**983**]) provides some relevant background. An alternate proof for the sufficiency was given by Rényi (1963, [**894**]). Remarkably, purely probabilistic proofs of the sufficiency (based on Kolmogorov's Three-Series Criterion) are possible and such a proof (which is nontrivial) was given by Novoselov (1964, [**816**]); see also Babu (1973, [**894**]). Billingsley (1974, [**104**, §7 & 8]) provides a (somewhat simplified, but still nontrivial) exposition of the original proof of the theorem, but expressed in purely probabilistic terms, and using only elementary results from number theory. A proof of the sufficiency is outlined in Section 9.7 (Selected proofs). For the necessity part of the theorem, all available proofs require nontrivial number theory. For additional details and discussion see, for example, Schoenberg (1936, [**953**]), Kubilius (1964, [**663**]), Delange & Halberstam (1957, [**271**]), Galambos (1970, [**402**]), Elliott (1979, [**318**, Chapter 5]), Sándor et al. (2006, [**940**, pp. 564–566]), and Tenenbaum (2015, [**1053**, Chapter III.4, §4.4 & Notes]).

9.5. Complements

Poisson behaviour of $\omega(n)$. The equality of the mean and variance in the Hardy–Ramanujan and Erdős–Kac Theorems is related to an underlying Poisson distribution. Let

$$\pi_k(x) \equiv \#\{n \le x \colon \omega(n) = k\}$$

be the number of integers $n \le x$ having exactly k distinct prime factors. Starting from the PNT and using an induction argument on k, Landau (1909, [**676**]) proved that *for a fixed* $k \ge 1$, and as $x \to \infty$,

$$(9.20) \qquad \pi_k(x) \sim \pi(x)\,\frac{(\log\log x)^{k-1}}{(k-1)!} \sim x\,\frac{\mu^{k-1}e^{-\mu}}{(k-1)!}, \quad \text{where } \mu = \log\log x$$

(although Gauss had already stated this in 1791). This same result also holds if $\pi_k(x)$ is based on the *total* number of prime factors $\Omega(n)$ counted with multiplicity. Hardy & Wright (2008, [**499**, §22.18]) show how this induction can be done; see also Montgomery & Vaughan (2006, [**774**, §7.4]). The proportions appearing in (9.20) correspond to a Poisson distribution with mean $\log\log x$.

While Poisson distributions tend to normality as their means increase, asymptotic normality cannot be used to obtain Poisson results. However, it is also the case that Landau's Poisson result cannot be used in any *straightforward* way to prove the Erdős–Kac asymptotic normality (although it does produce the correct result if used naively). This is because if k were allowed to grow with x, then the asymptotic behaviour of the proportions becomes more complicated. Harper (2009, [**503**]) showed how the Erdős–Kac Theorem can be established without appealing to Landau's result, using an approach based on Stein's method of Poisson approximation.

Sathe (1953–1954, [**945**]) and Selberg (1954, [**977**]) obtained sharp results for the case where k is allowed to increase with x. For $1 \le k \le A\log\log x$, with *any* $A > 0$, they proved

$$(9.21) \qquad \pi_k(x) \sim \pi(x)\,\frac{(\log\log x)^{k-1}}{(k-1)!}\,\lambda\!\left(\frac{k-1}{\log\log x}\right)$$

uniformly in k, where

$$\lambda(r) = \frac{1}{\Gamma(r+1)}\prod_p \left(1 + \frac{r}{p-1}\right)\left(1 - \frac{1}{p}\right)^r.$$

Furthermore, if $\omega(n)$ is replaced by $\Omega(n)$, the same result holds with λ replaced by

$$\lambda(r) = \frac{1}{\Gamma(r+1)}\prod_p \left(1 - \frac{r}{p}\right)^{-1}\left(1 - \frac{1}{p}\right)^r$$

but now uniformly only for $1 \le k \le (2-\delta)\log\log x$ for any $\delta > 0$. Noting that $\lambda(0) = 1$ we recover Landau's result (9.20) from (9.21) and see that it also holds for $k = o(\log\log x)$. However $\lambda(1) = 1$ as well, and consequently Landau's result also holds for $k \sim A\log\log x$ provided $A = 1$. For other values of A it will be off by a constant factor. Since we do know that $\sum_k \pi_k(x) = x$, it is somewhat curious (since the Poisson probabilities there do sum to 1) that the sum of the terms in the middle in (9.20) should exactly equal x as well. But (9.20) is not valid for all k, so this calculation is fallacious in spite of its outcome being correct. The explanation is that the bulk of this sum occurs in the vicinity of the central value $k \sim \log\log x$ of the Poisson distribution. For further details, see Tenenbaum (2015, [**1053**, §6.1, Thms. 4, 5 & Notes of Ch II.6]), Montgomery & Vaughan (2006, [**774**, §7.4]), and Kowalski (2021, [**661**, §2.3]).

Asymptotic behaviour of $\varphi(n)/n$ and $\sigma(n)/n$. (Here φ is Euler's phi-function and σ is the sum of divisors function.) In the 1920s and 1930s Schoenberg, Davenport, and others studied the asymptotic distributions of $\varphi(n)$ and $\sigma(n)$, and proved the existence of limiting distributions for the multiplicative functions $\varphi(n)/n$ and $\sigma(n)/n$. For example, the function $\log(\varphi(n)/n)$ is strongly additive and the Erdős–Wintner Theorem may be applied directly to show that $\log(\varphi(n)/n)$ converges in distribution. It turns out that the limiting distributions for $\varphi(n)/n$ and $\sigma(n)/n$ are both purely singular, with the first of these dense in $[0, 1]$, and the second dense in $[1, \infty]$. See Chapter 3 for further references and details. See also de Koninck & Luca (2012, [**650**, pp. 121–122]), and Tenenbaum (2015, [**1053**, p. 440]). Schwarz (1994, [**957**]) provides some historical details.

Asymptotic behaviour of $\tau(n)$, $d_j(n)$, and $p_j(n)$. (Here $\tau(n) = \sum_{d|n} 1$ is the number of divisors of n.) Methods similar to those of this chapter can be used to establish an asymptotic normality result for $\log_2 \tau(n)$, where $\log_2$ is the base 2 logarithm:

$$\frac{1}{x}\#\left\{n \le x\colon \frac{\log_2 \tau(n) - \log\log x}{\sqrt{\log\log x}} \le t\right\} \longrightarrow \frac{1}{\sqrt{2\pi}}\int_{-\infty}^{t} e^{-u^2/2}\,du\,, \quad \text{as } x \to \infty.$$

In particular, since $\log_2 \tau(n) \equiv \log \tau(n)/\log 2$ lies between $\omega(n)$ and $\Omega(n)$, it acts like them, and its moments can be obtained from those of $\omega(n)$ and $\Omega(n)$.

Write $\{d_j(n)\}_{1\le j\le \tau(n)}$ for the increasing sequence of divisors of n, and write $\{p_j(n)\}_{1\le j\le \omega(n)}$ for the increasing sequence of prime divisors of n. The study of $d_j(n)$ and $p_j(n)$ was undertaken by Erdős in the 1940s (see 1979, [**349**]), Erdős & Tenenbaum (1989, [**362**]), de Koninck & Tenenbaum (2002, [**651**]), and la Bretèche & Tenenbaum (2002, [**149**]). See also Tenenbaun (2015, [**1053**, III 3.6 & Notes]). Galambos (1976, [**403**]) obtained an asymptotic normality result for $\log\log p_j(n)$ as well as an asymptotic exponential distribution result for the gaps between the $\log\log p_j(n)$.

Stable law behaviour of additive functions. Some very unusual distributions can show up as the limit law of an arithmetic function. Tenenbaum (2015, [**1053**, p. 441, Example 259]) gives an example (due to Saffari) of an additive function whose limit law is a non-Gaussian stable law. Another such example was given by Levin & Fainleib (1966, [**693**]). In fact all of the stable laws can arise as limits of arithmetic functions (see Elliott ([**319**, Chapter 16])).

The arcsine law for $\tau(n)$. A particularly striking example of probabilistic behaviour occurs in connection with the divisors function $\tau(n)$ which counts the total number of divisors of n. We first recall that $\tau(n)$ has average

$$\frac{1}{x}\sum_{n\le x} \tau(n) \;=\; \log x + 2\gamma - 1 + O\left(\frac{1}{\sqrt{x}}\right);$$

and furthermore, that $\tau(n) = O(n^{\epsilon})$ for every $\epsilon > 0$, and more precisely, that

$$\limsup_{n\to\infty} \frac{\log_2 \tau(n)}{(\log n)/\log\log n} \;=\; 1.$$

Of course, $\tau(n)$ is never smaller than 2 (when $n > 1$), and equals 2 whenever n is prime.

Now suppose that $F_n(x)$ is the distribution function corresponding to uniform probability over the points $(\log d)/\log n$, where d ranges over the $\tau(n)$ divisors of n; in other words, $F_n(x)$ is the empirical distribution of the $(\log d)/\log n$. Clearly, the functions $F_n(x)$ are symmetric about $x = 1/2$ (since n/d divides n whenever d does) but otherwise the $F_n(x)$ vary erratically with n. Remarkably, however, when these functions are averaged, Deshouillers, Dress & Tenenbaum (1979, [**276**]) showed that they satisfy

$$\frac{1}{x}\sum_{n\leq x} F_n(x) \;=\; \frac{2}{\pi}\arcsin\sqrt{x} + O\left(\frac{1}{\sqrt{\log x}}\right)$$

uniformly for $0 \leq u \leq 1$, and with the order term being exact. The limit here corresponds to the celebrated arcsine law whose density is $1/\pi\sqrt{x(1-x)}$. Reflected in this density is the symmetry between the smaller divisors d and their corresponding larger partners $\frac{n}{d}$. Furthermore, this distribution places disproportionately high probability at the extremes, and means that integers typically have mostly small and mostly large divisors. The arcsine law is extensively discussed in Feller (1967, [**373**, vol. I, Chapter III]). (See also Appendix D.)

The smallest and largest prime divisors. Denote by $p(n)$ and $P(n)$ the smallest and largest prime factors of n, respectively. Then it is known that

$$\sum_{n\leq x} p(n) = \frac{1}{2}\frac{x^2}{\log x} + O\left(\frac{x^2}{\log^2 x}\right), \quad \text{and} \quad \sum_{n\leq x} P(n) = \frac{\pi^2}{12}\frac{x^2}{\log x} + O\left(\frac{x^2}{\log^2 x}\right).$$

It follows that the 'average orders' of $p(n)$ and $P(n)$ are $n/\log n$ and $\pi^2 n/6\log n$, respectively. That the two sums share the same order of magnitude may seem implausible at first. But this occurs because the first sum is dominated by contributions from n that are prime, while the second is dominated by contributions from n that are nearly prime. In any case, their leading terms do differ by a factor of $\pi^2/6$. Such results have been used to study the properties of certain cryptographic algorithms. It is also known that

$$\sum_{n\leq x}\frac{p(n)}{P(n)} = \frac{x}{\log x} + \frac{3x}{\log^2 x} + \frac{15x}{\log^3 x} + o\left(\frac{x}{\log^3 x}\right)$$

(Erdős & van Lindt (1982, [**356**]), and Huajia Chao (1987, [**190**])). Further results, due to de Bruijn (1951, [**158**]), include

$$\sum_{2\leq n\leq x} \log P(n) \;=\; Dx\log x + O(x\log\log x),$$

and

$$\lim_{n\to\infty}\frac{1}{x}\sum_{2\leq n\leq x}\frac{\log P(n)}{\log n} \;=\; \lim_{n\to\infty}\frac{1}{x\log x}\sum_{2\leq n\leq x}\log P(n) = D.$$

Note that the logarithmic ratios $(\log p)/\log n$ of the prime divisors of n all lie between 0 and 1 and are the natural quantities to look at when trying to understand the asymptotic behaviour of prime factors. The constant D indicated above is known as the *Golomb–Dickman constant* and it may be defined in equivalent ways as

$$D = 1 - \int_1^\infty \frac{\rho(v)}{v^2}dv = \int_0^\infty \frac{\rho(u)}{(1+u)^2}du = \int_0^1 e^{\mathrm{li}(t)}\,dt \approx 0.62433\ldots$$

where $\rho(u)$ is the Dickman–de Bruijn function to be defined below. A related result is

$$\sum_{2\le p\le x} \frac{1}{p(n)} = Cx + O\left(\frac{x}{(\log\log x)^2}\right), \quad \text{where } C = \sum_p \frac{1}{p(p-1)} \prod_{q<p}\left(1-\frac{1}{q}\right).$$

As it happens, the Golomb–Dickman constant D is also to equal to the asymptotic probability that the second largest prime factor of an integer is bounded above by the square root of its largest prime factor. It also arises in the study of random permutations where it is equal to the asymptotic value of the expectation $\mathrm{E}\,\ell(n)/n$ as $n \to \infty$, where $\ell(n)$ is the longest *cycle* of a permutation of length n, and expectation is calculated under a uniform distribution over the set of all possible permutations. See Shepp & Lloyd (1966, [**985**]) and Purdom & Williams (1968, [**870**]). For more detailed treatments of $\sum p(n)$ and $\sum P(n)$ see for example Alladi & Erdős (1977, [**18**]), (1979, [**19**]), or de Koninck & Luca (2012, [**650**, Chapter 9]).

The Dickman–de Bruijn distribution. Let

$$\Psi(x,y) \equiv \sum_{n\le x,\, P(n)\le y} 1 \;\equiv\; \#\{n \le x \colon P(n) \le y\}$$

be the number of integers $1 \le n \le x$ whose largest prime factor $P(n) \le y$. Such integers are called *y-friable* (or *y-smooth*). Estimates for $\Psi(x,y)$ are useful for analyzing factorization algorithms and related applications. The natural case to consider is when y is a power of x, namely $y = x^\alpha$, so that $\alpha = (\log y)/\log x$ with $0 < \alpha < 1$ fixed. Alternately one can take x to be a power of y as in $x = y^u$ so that $u = (\log x)/\log y$, with $u \ge 1$ fixed. Dickman (1930, [**286**]) proved the remarkable fact that for any fixed $u \ge 1$ the proportion of integers up to x which have all of their prime factors less than $x^{1/u}$ tends (as $x \to \infty$) to a nonzero limit, and furthermore he proved that this limit is given by

$$\frac{\Psi(x,y)}{x} \sim \rho(u), \quad u = (\log x)/\log y, \quad x \to \infty, \tag{9.22}$$

or equivalently

$$\lim_{x\to\infty} \frac{\Psi(x, x^{1/u})}{x} \sim \rho(u), \quad \text{or} \quad \lim_{y\to\infty} \frac{\Psi(y^u, y)}{y^u} \sim \rho(u),$$

where $\rho(u)$ is the so-called*Dickman–de Bruijn function.*[5] This is a continuous nonnegative function $\rho(u)$ on $0 \le u < \infty$, determined by the initial conditions

$$\rho(u) = 0, \;\; u < 0; \quad \text{and} \;\; \rho(u) = 1, \;\; 0 \le u \le 1;$$

together with the 'differential-delay equation'

$$u\rho'(u) + \rho(u-1) = 0, \quad u > 1. \tag{9.23}$$

[5]In a 2001 unpublished manuscript, Lars Holst of the Royal Institute of Technology in Stockholm makes the following observation:

> *Karl Dickman, born 1862, worked as an actuary in the end of the 19th and the beginning of the 20th century. Probably, he studied mathematics in the 1880's at Stockholm University, where the legendary Mittag-Leffler was professor. However, Dickman's 1930 paper seems to be his only publication in mathematics. The paper is a remarkable achievement of an old man* [**560**].

The function $\rho(u)$ thus satisfies (for $k \geq 1$) the recursive relation

$$\rho(u) = \rho(k) - \int_k^u \rho(t-1)\frac{dt}{t}, \quad k \leq u \leq k+1.$$

Note that (9.22) may be alternately stated as

$$\#\{n \leq x\colon P(n) \leq n^{1/u}\} \sim \rho(u)x.$$

For $u > 2$ there is no closed form for $\rho(u)$, but it is easy to establish that $\rho(u)$ is positive for $u > 0$, is strictly decreasing for $u > 1$, that $\rho(u) = 1 - \log u$ for $1 \leq u \leq 2$, and also that $u\rho(u) = \int_{u-1}^u \rho(v)\,dv$ for $u \geq 1$. It can further be shown that $\log \rho(u) \sim -u \log u$ as $u \to \infty$, consistent with a very fast decay of roughly $(u \log u)^{-u}$ for $\rho(u)$, and more precisely that

$$\log \rho(u) \;=\; -u \log u - u \log\log u + u + O\left(\frac{u \log\log u}{\log u}\right).$$

The function $\rho(u)$ also figures in the distribution, over the integers, of their largest prime factors $P(n)$: The quantities $(\log n)/\log P(n)$ live on $[1, \infty)$ and have limiting cumulative distribution function given by

$$\lim_{x\to\infty} \frac{1}{x}\#\left\{n \leq x\colon \frac{\log n}{\log P(n)} \leq u\right\} \to 1 - \rho(u),$$

with corresponding density function $-\rho'(u)$.

The error of the approximation $\Psi(x, y) \sim x\rho(u)$ is important for several reasons, and in particular for analyzing the computational complexity of 'subexponential-time' factoring algorithms in cryptography. This error was studied by Rankin (1938, [**885**]), and de Bruijn (1951, [**158**]) among others, and the approximation was generalized and extended by Canfield, Erdős & Pomerance (1983, [**179**]), Hildebrand & Tenenbaum (1986, [**555**]), (1993, [**556**]), and by Lichtman & Pomerance (2018, [**699**]). With $u = (\log x)/\log y$ it is now known that

$$\Psi(x, y) = x\rho(u) + O\left(\frac{x}{\log y}\right) = O(xe^{-u/2}), \quad \text{as } x \to \infty,$$

uniformly for $y \leq x$. Hildebrand (1984, [**549**]) showed that the Riemann hypothesis is *equivalent* to the assertion

$$\Psi(y^u, y) = y^u \rho(u) \exp\bigl(O(y^\epsilon)\bigr), \quad 1 \leq u \leq y^{1/2-\epsilon}, \quad \text{for all } \epsilon > 0.$$

de Koninck & Luca (2012, [**650**, Chapter 9]) provide a nice exposition for many of these and related results, while Dartyge (2019, [**259**]) provides some further details and references. See also Granville (2008, [**452**]) and Tenenbaum & Mendès-France (2000, [**1055**, Chapter 4, §6]).

The function $\rho(u)$ is of considerable interest in its own right. It can be shown that $\int_0^\infty \rho(u)du = e^\gamma$, and even that $\sum_{n=1}^\infty n\rho(n) = e^\gamma$ (with the value e^γ reminiscent of the formulas of Mertens). It follows that when $\rho(u)$ is rescaled as $e^{-\gamma}\rho(u)$ it becomes a probability density function which may be referred to as the *Dickman–de Bruijn density*. This density function has many interesting properties including the following: Let W_j be a sequence of 0-mean independent binary random variables distributed according to

$$\mathrm{P}\left[W_j = j\right] = \frac{1}{j}, \quad \mathrm{P}\left[W_j = 0\right] = 1 - \frac{1}{j}, \quad j = 1, 2, \ldots$$

Then $\frac{1}{n}\sum_1^n W_j$ has, asymptotically, the Dickman–de Bruijn distribution. It can also be shown that this density has the characteristic function

$$\varphi(t) \equiv \int_0^\infty \rho(u)e^{itu}\,du = \exp\left(\int_0^1 \frac{e^{itv}-1}{v}dv\right). \tag{9.24}$$

In Chapter 10 we will define the concept of infinitely divisible distributions and will consider some of their properties. Here we limit ourselves to remarking that the infinitely divisible distributions which also have finite variance are fully characterized by the fact that their characteristic functions $\varphi(t)$ are of the form

$$\log\varphi(t) = it\mu + \int_{-\infty}^{\infty}\left(e^{itu} - 1 - itu\right)\frac{dK(u)}{u}$$

where μ is a constant and $K(u)$ is a nondecreasing function of bounded variation. Such a distribution has mean μ, and if we take the left limit of K to be $K(-\infty) = 0$ (as we may) then the variance of the distribution will be its right limit $K(+\infty)$. The distribution defined by (9.24) is seen to be of this form, for if we set

$$K(u) = 0,\ \frac{u^2}{2},\ \frac{1}{2}, \quad \text{as} \quad u < 0,\ 0 \le u \le 1,\ u > 1,$$

then we obtain (9.24). So it follows that the Dickman–de Bruijn distribution is infinitely divisible and that it has mean 1 and variance $\frac{1}{2}$. Numerous properties of the Dickman–de Bruijn distribution have been explored by Cellarosi & Sinai (2011, [**185**]), (2013, [**186**]) where related references may also be found. See also Alladi (1982, [**16**]), Tenenbaum (2015, [**1053**, Chapter III.5]), as well as the mentioned paper by Holst (2001, [**560**]).

There is an interesting although obviously false heuristic argument worth noting which runs as follows: To obtain the count $\Psi(x,y)$ of integers $n \le x$ having no prime factors larger than y we need to remove from the set of integers $n \le x$ all those which are divisible by the larger prime factors. Assuming independence of the 'divisibility events' one might be tempted to estimate the number of integers that remain as

$$\Psi(x,y) \;\approx\; x \prod_{y<p\le x}\left(1-\frac{1}{p}\right).$$

Applying Mertens's Theorem to the product

$$\prod_{y<p\le x} = \prod_{p\le x}\Big/\prod_{p\le y}$$

then leads to $\Psi(x,y) \sim x/u$ where $u = (\log x)/\log y$. However in actual fact $\Psi(x,y)/x$ decreases exponentially fast in u. The difficulty of course is that divisibility by higher primes cannot be treated as asymptotically independent events. In particular, if one prime in the interval $(\sqrt{x}, x]$ were to divide some integer $n \le x$ then obviously that integer cannot be divided by any other prime in that interval. See Hidebrand & Tenenbaum (1993, [**556**]).

It is interesting to compare the Dickman function $\rho(u)$ with the Buchstab function $\omega(u)$ discussed in connection with Maier's counterexample in Section 4.7. The first of these estimates the proportion of integers n having no factors greater than $n^{1/u}$, while the second is the key term in the formula which estimates the proportion of integers n having no prime factors less than $n^{1/u}$. See equation (4.16). Buchstab's function is defined by the differential-delay equation

$$\frac{d}{du}\big(u\,\omega(u)\big) = \omega(u-1), \quad \text{or } \omega(u) + u\omega'(u) = \omega(u-1), \quad u \geq 2,$$

with initial values given by $\omega(u) = 1/u$ for $1 \leq u \leq 2$, and continuous at $u = 2$. This may be compared with the differential-delay equation (9.23) in the definition of Dickman's function.

Breaking sticks and the distribution of large prime factors. Let $n = p_1 p_2 \cdots p_k$ with $p_1 \geq p_2 \geq \cdots \geq p_k > 1$ be the prime factorization of n *with multiplicity* and in *descending order*, and adopt the convention that for this same value of n we have $p_j = 1$ whenever $j > k$. So 1, in effect, is being viewed as a prime, and the factorization of n viewed as an infinite sequence. Billingsley (1972, [**103**]) showed that if one fixed a value of $r > 1$ and chose n uniformly in $\{1, 2, \ldots, N\}$ then as $N \to \infty$ the distribution of the random r-vector

$$\left(\frac{\log p_1}{\log n}, \frac{\log p_2}{\log n}, \ldots, \frac{\log p_r}{\log n}\right)$$

converges to the marginal distribution (of an initial segment) of the *Poisson–Dirichlet distribution* with parameter 1. (Note that when $k < r$ the tail of this vector will consist of zeros.) The full (infinite) vector

$$\left(\frac{\log p_1}{\log n}, \frac{\log p_2}{\log n}, \ldots, \frac{\log p_k}{\log n}, 0, 0, \cdots\right),$$

where $k \equiv k(n)$ represents the total number of n's prime factors, will converge to the full Poisson–Dirichlet distribution. (Note that the terms of this vector sums to 1.)

We will not describe the general Poisson–Dirichlet distribution, but the case when the parameter equals 1—which is the one relevant here—can be imagined in the following way. Let $U_1, U_2, \ldots$ be an infinite sequence of independent random variables distributed uniformly on $[0, 1]$, and let

$$X_1 = U_1, \quad X_2 = (1 - U_1)U_2, \quad X_3 = (1 - U_1)(1 - U_2)U_3, \ \ldots.$$

Then $\sum_{j=1}^{\infty} X_j = 1$ (almost surely) and the distribution in question is that of the first r values of the X_j after the X_j have been rearranged into descending order. The process of generating the X_j values is sometimes referred to as the *stick-breaking* process because one may envision a stick of length 1 broken in two in accordance with a uniform distribution, setting the piece in the left hand aside, and breaking the piece in the right hand, with uniform distributions used to successively determine the break points.[6] The distribution which this describes yields sample points $(x_1, x_2, \ldots, x_r)$ say, with $\sum_{j=1}^{r} x_j \leq 1$; they may be viewed as arising from a distribution on the r-dimensional simplex

$$\left\{x_j \geq 0, \ j = 1, 2, \ldots, r+1; \ \sum_{j=1}^{r+1} x_j = 1\right\}.$$

Although not obvious, it is to be noted that the distribution in the $r = 1$ case here (which corresponds to the distribution of the largest X_j) is in fact the same

[6] For the sake of political correctness, let us note here that this procedure has an obvious modification for left-handed persons.

as the Dickman–de Bruijn distribution; a proof of this fact may be found in Holst (2001, [**560**]).

An interesting fact about the Dickman–de Bruijn distribution is that it is the unique 'law' $\mathcal{L}(V)$ which satisfies the distributional equality

$$\mathcal{L}(V) = \mathcal{L}\big(\max(U, (1-U)V)\big)$$

where U has a uniform distribution on $[0,1]$ and is independent of V. Furthermore, if the U_j are independent uniform random variables on $[0,1]$ then the series

$$Y \equiv U_1 + U_1U_2 + U_1U_2U_3 + \cdots$$

converges almost surely to a random variable which has the Dickman–de Bruijn distribution. For further details see Knuth & Trabb Pardo (1976, [**641**]), Lloyd (1984, [**709**]), Kingman (1993, [**636**]), Donnelly & Grimmett (1993, [**291**]), and Arratia, Barbour & Tavaré (1997, [**40**]), (1999, [**41**]), as well as the nice review article by Granville (2008, [**453**]).

Numbers composed of given primes. If $p_1, p_2, \ldots, p_k$ are a given set of distinct primes, then the number of $n \le x$ that can be produced using *only* those primes can be shown to be

$$\frac{\log^k x}{k! \prod_1^k \log p_j} + O(\log^{k-1} x).$$

This result is based on counting the number of solutions to $\prod_{j \le k} p_j^{\nu_j} \le x$, or equivalently to $\sum_{j \le k} \nu_j \log p_j \le \log x$, with integers $\nu_j \ge 0$; the volume of this simplex is the leading term shown above. (See Tenenbaum (2015, [**1053**, Chapter III.5, §2 & 3, and Theorem III.5.3, pp. 516–517].)[7]

9.6. Historical remarks

Paul (Pál) Turán (1910–1976) recalls:

> *When writing Hardy first in 1934 on my proof of Hardy–Ramanujan's theorem I did not know what Tchebycheff's inequality was and a fortiori of the central limit theorem. Erdős, to my best knowledge, was at that time not aware too. It was Mark Kac who wrote to me a few years later that he discovered when reading my proof in JLMS that this is basically probability and so his interest turned to this subject* (Elliott (1980, [**319**, p. 18])).

Since he was thinking of the Central Limit Theorem, the probabilist Mark Kac asked Turán if he could prove similar estimates for the higher moments

$$\sum_{n \le x} (\omega(n) - \log\log n)^k.$$

Knowing no probability theory, Turán replied saying that he "found absolutely no interest to do it actually" since he could foresee no applications for such a result! Pál Turán was married to Vera Sós, herself an outstanding mathematician. Sós (2002, [**999**]) provides a fascinating account of the interactions between Turán and Erdős.

[7]**Some further references:** Turán (1934, [**1069**]), (1936, [**1070**]), Kubilius (1964, [**663**]), Galambos (1970, [**402**]), Elliot (1979, [**318**]), Schwarz (2007, [**959**]), Cojocaru & Murty (2005, [**214**, Chapter 3 and exercises]), Narkiewicz (1983, [**796**, p. 243]), and Tenenbaum (2015, [**1053**]).

Commenting on the Erdős–Kac Theorem, Kac wrote:

> *If I remember it correctly I first stated (as a conjecture) the theorem on the normal distribution of the number of prime divisors during a lecture in Princeton in March 1939. ... Erdős was in thc audience, and he immediately perked up. Before the lecture was over he had completed the proof, which I could not have done not having been versed in the number-theoretic methods ...* (1985, [**615**, p. 24])

Kac further remarked:

> *The reader, I hope, will forgive my lack of modesty if I say that it is a beautiful theorem. It marked the entry of the normal law, hitherto the property of gamblers, statisticians and observateurs into number theory and ... gave birth to a new branch of this ancient discipline* (1985, [**615**, p. 24]).

9.7. Selected proofs

PROOF OF THE HARDY–RAMANUJAN THEOREM. Although it is an immediate consequence of the Turán–Kubilius inequality or of the Erdős–Kac Theorem (both proved below) it is instructive to prove this result separately. The proof here is modelled after the much simpler one of Turán which—as Erdős might have put it—was straight from "The Book." From number theory it uses (aside from unique factorization) only the elementary result of Mertens

(9.25) $$\sum_{p\le x}\frac{1}{p} = \log\log x + O(1), \quad \text{as} \quad x\to\infty,$$

which will get applied repeatedly.

We have, first, the approximation

(9.26) $$\begin{aligned}\sum_{n\le x}\omega(n) &= \sum_{n\le x}\sum_{p|n}1 = \sum_{p\le x}\left\lfloor\frac{x}{p}\right\rfloor = \sum_{p\le x}\frac{x}{p} + \sum_{p\le x}\left(\left\lfloor\frac{x}{p}\right\rfloor - \frac{x}{p}\right)\\ &= x\sum_{p\le x}\frac{1}{p} + O(x) = x\log\log x + O(x).\end{aligned}$$

Note that the $\left(\lfloor\frac{x}{p}\rfloor - \frac{x}{p}\right)$ terms which contribute to the $O(x)$ all lie in $[0,1)$, and although there are only $\pi(x)$ of them, the PNT has not been used here.

Secondly, let us establish the approximation

(9.27) $$\sum_{n\le x}\omega^2(n) = \sum_{\substack{pq\le x\\ p\ne q}}\left\lfloor\frac{x}{pq}\right\rfloor + \sum_{p\le x}\left\lfloor\frac{x}{p}\right\rfloor = x(\log\log x)^2 + O(x\log\log x).$$

To understand the source and the contributions of the $\sum\lfloor\frac{x}{pq}\rfloor$ and $\sum\lfloor\frac{x}{p}\rfloor$ components here, note first that

$$\omega^2(n) = \Big(\sum_{p|n}1\Big)\Big(\sum_{q|n}1\Big),$$

and that for a given $p\ne q$ pair, the total number of 1's that will occur, after multiplying out and summing over n, will be $\lfloor\frac{x}{pq}\rfloor$; likewise each given $p = q$

will contribute $\lfloor \frac{x}{p} \rfloor$ to the total. The impact of the $\sum \lfloor \frac{x}{p} \rfloor$ component in (9.27) is controlled by the computation in (9.26). To handle the $\sum \lfloor \frac{x}{pq} \rfloor$ term note that

$$\left(\sum_{p \le \sqrt{x}} \frac{1}{p} \right)^2 < \sum_{pq \le x} \frac{1}{pq} < \left(\sum_{p \le x} \frac{1}{p} \right)^2,$$

and then apply the Mertens result (9.25) to both sides to arrive at

$$\sum_{p \ne q} \left\lfloor \frac{x}{pq} \right\rfloor = x(\log\log x)^2 + O(x \log\log x).$$

Thus (9.27) is established.

Finally we may apply the Chebyshev inequality of probability theory. For this we need

$$\begin{aligned} V(x) &\equiv \frac{1}{x} \sum_{n \le x} \left(\omega(n) - \log\log x \right)^2 \\ &= \frac{1}{x} \sum_{n \le x} \omega^2(n) - 2 \log\log x \left(\frac{1}{x} \sum_{n \le x} \omega(n) \right) + \frac{\lfloor x \rfloor}{x} (\log\log x)^2 \\ &= O(\log\log x). \end{aligned}$$

The order term here follows after applying (9.26) and (9.27) and noting that the $(\log\log x)^2$ terms cancel. Now the Chebyshev inequality gives us

$$\frac{1}{x} \# \left\{ n \le x : |\omega(n) - \log\log n| \ge C(x) \sqrt{\log\log x} \right\} \le \frac{V(x)}{C^2(x) \log\log x} \to 0,$$

as long as $C(x) \to \infty$ however slowly, and this is the result of Hardy and Ramanujan. □

REMARKS 9.7.

(i) We could have proceeded alternately by setting

$$V(x) \equiv \frac{1}{x} \sum_{n \le x} \left(\omega(n) - \frac{1}{x} \sum_{n \le x} \omega(n) \right)^2$$

and then using (9.26) and (9.27) to also find

$$V(x) = \frac{1}{x} \sum_{n \le x} \omega^2(n) - \left(\frac{1}{x} \sum_{n \le x} \omega(n) \right)^2 = O(\log\log x).$$

The point here is that the choice of $\frac{1}{x} \sum_{n \le x} \omega(n)$, versus $\log\log x$, for 'centring' the variance is inconsequential because the difference is no more than $O(1)$.

(ii) Similar arguments based on the more precise approximation shown in (9.25) gives

$$V(x) = \log\log x + o(\log\log x).$$

In fact, it can be shown that

$$\frac{1}{x} \sum_{n \le x} \left(\omega(n) - \frac{1}{x} \sum_{n \le x} \omega(n) \right)^2 = \log\log x + B_2 + O\left(\frac{\log\log x}{\log x} \right) \tag{9.28}$$

where

$$B_2 = B - \frac{\pi^2}{6} - \sum_p \frac{1}{p^2} \approx -1.8357$$

with B as in (9.25).

(iii) A discussion of Turán's ideas may be found in Hardy (1940, [**489**, Lecture 3]).

PROOF OF BOUND (9.4) FOR $\sum(\Omega - \omega)$. The assertion is that

$$\lim_{x\to\infty} \frac{1}{x} \sum_{n\le x} (\Omega(n) - \omega(n)) = \sum_p \frac{1}{p(p-1)}. \tag{9.29}$$

Now the difference $\Omega(n) - \omega(n)$ equals

$$\Omega(n) - \omega(n) = \sum_{p^k|n} 1 - \sum_{p|n} 1 = \sum_{\substack{p^k|n\\ k\ge 2}} 1.$$

Therefore

$$\begin{aligned}\sum_{n\le x} (\Omega(n) - \omega(n)) &= \sum_{n\le x} \sum_{\substack{p^k|n\\ k\ge 2}} 1\\ &= \sum_{\substack{k\ge 2\\ p^k\le x}} \sum_{p^k|n} 1\\ &= \sum_{\substack{k\ge 2\\ p^k\le x}} \left\lfloor \frac{x}{p^k} \right\rfloor\\ &= \sum_{\substack{k\ge 2\\ p^k\le x}} \left(\frac{x}{p^k} + O(1)\right)\\ &= \sum_{\substack{k\ge 2\\ p^k\le x}} \frac{x}{p^k} + O(\sqrt{x}).\end{aligned}$$

For any given prime p, the sum over k is just a geometric series which equals $1/p(p-1) + O(1/x)$, so

$$\sum_{n\le x} (\Omega(n) - \omega(n)) = x \sum_{p\le\sqrt{x}} \left(\frac{1}{p(p-1)} + O\left(\frac{1}{x}\right)\right) + O(\sqrt{x}) = x \sum_p \frac{1}{p(p-1)} + O(\sqrt{x}),$$

whence the result follows. □

REMARKS 9.8.

(i) It follows from this result that $\omega(n)$ and $\Omega(n)$ have the same average order (namely $\log\log n$) and the same normal order (namely $\log\log n$).

(ii) Montgomery & Vaughan ([**774**, p. 65]) quote a remarkable result due to Rényi: The set of numbers $\{n : \Omega(n) - \omega(n) = k\}$ has density a_k where the a_k are the coefficients in

$$\sum_{k=0}^{\infty} a_k z^k = \prod_p \left(1 - \frac{1}{p}\right)\left(1 + \frac{1}{p-z}\right).$$

PROOF OF THE TURÁN–KUBILIUS INEQUALITY. The proof we give is modelled after that of Elliott (1979, [**318**, Chapter 4]). Assume first that f is real and nonnegative, and let

$$E(x) \equiv \sum_{p^k \le x} \frac{f(p^k)}{p^k}\left(1-\frac{1}{p}\right), \quad \text{and } D^2(x) \equiv \sum_{p^k \le x} \frac{f^2(p^k)}{p^k}.$$

We will use N instead of x (which causes no harm) and start by examining the expansion

$$\frac{1}{N}\sum_{n\le N}(f(n)-E(N))^2 = \frac{1}{N}\sum_{n\le N} f^2(n) - 2E(N)\left(\frac{1}{N}\sum_{n\le N} f(n)\right) + E^2(N). \tag{9.30}$$

We first upper-bound the quantity

$$\begin{aligned}\frac{1}{N}\sum_{n\le N} f^2(n) &= \frac{1}{N}\sum_{n\le N}\left(\sum_{p|n}\sum_{k} f(p^k)\,\mathrm{I}\left[p^k \parallel n\right]\right)\left(\sum_{q|n}\sum_{\ell} f(q^\ell)\,\mathrm{I}\left[q^\ell \parallel n\right]\right)\\ &= \frac{1}{N}\sum_{n\le N}\sum_{p|n}\sum_{q|n}\sum_{k}\sum_{\ell} f(p^k)f(q^\ell)\,\mathrm{I}\left[p^k \parallel n,\, q^\ell \parallel n\right].\end{aligned}$$

The contribution of the $p = q$ cases total to

$$\frac{1}{N}\sum_{p^k\le N} f^2(p^k) \times \sum_{\substack{n\le N\\ p^k\parallel n}} 1. \tag{9.31}$$

The $p \ne q$ cases contribute only when $p^k q^\ell \le N$, and their contributions total to

$$\frac{1}{N}\sum_{\substack{p\ne q\\ p^k q^\ell\le N}} f(p^k)f(q^\ell) \times \sum_{\substack{p^k\parallel n\\ q^\ell\parallel n\\ n\le N}} 1. \tag{9.32}$$

Now the $\sum 1$ term of (9.31) is bounded by N/p^k, and by an inclusion-exclusion argument the $\sum 1$ term of (9.32) equals

$$\left\lfloor\frac{N}{p^k q^\ell}\right\rfloor - \left\lfloor\frac{N}{p^{k+1} q^\ell}\right\rfloor - \left\lfloor\frac{N}{p^k q^{\ell+1}}\right\rfloor + \left\lfloor\frac{N}{p^{k+1} q^{\ell+1}}\right\rfloor \le \frac{N}{p^k q^\ell}\left(1-\frac{1}{p}\right)\left(1-\frac{1}{q}\right)+2.$$

Hence altogether (when f is real and nonnegative)

$$\begin{aligned}&\frac{1}{N}\sum_{n\le N} f^2(n)\\ &\le \sum_{p^k\le N}\frac{f^2(p^k)}{p^k} + \sum_{\substack{p\ne q\\ p^k q^\ell\le N}} \frac{f(p^k)}{p^k}\frac{f(q^\ell)}{q^\ell}\left(1-\frac{1}{p}\right)\left(1-\frac{1}{q}\right) + \frac{2}{N}\sum_{\substack{p\ne q\\ p^k q^\ell\le N}} f(p^k)f(q^\ell)\\ &\le D^2(N) + E^2(N) + \frac{2}{N}\sum_{p\ne q,\, p^k q^\ell\le N} f(p^k)f(q^\ell).\end{aligned} \tag{9.33}$$

Second, we have the lower-bound

$$\begin{aligned}\frac{1}{N}\sum_{n\le N} f(n) &= \frac{1}{N}\sum_{n\le N}\sum_{p|n}\sum_{k} f(p^k)\,\mathrm{I}\left[p^k \parallel n\right] = \frac{1}{N}\sum_{p^k\le N} f(p^k)\times \sum_{\substack{n\le N\\ p^k\parallel n}} 1 \\ &\ge E(N) - \frac{1}{N}\sum_{p^k\le N} f(p^k)\end{aligned} \tag{9.34}$$

(again assuming f real and nonnegative) since the $\sum 1$ term here is

$$\left\lfloor\frac{N}{p^k}\right\rfloor - \left\lfloor\frac{N}{p^{k+1}}\right\rfloor \ge \frac{n}{p^k}\left(1-\frac{1}{p}\right) - 1.$$

Inserting bounds (9.33) and (9.34) into (9.30) now gives

$$\begin{aligned}&\frac{1}{N}\sum_{n\le N}(f(n)-E(N))^2 \\ &\le D^2(N) + \frac{2}{N}\sum_{\substack{p\ne q\\ p^k q^\ell\le N}} f(p^k)f(q^\ell) + 2E(N)\frac{1}{N}\sum_{p^k\le N} f(p^k),\end{aligned} \tag{9.35}$$

in which the $E^2(N)$ term no longer appears.

To complete the proof we bound each of the components of (9.35) by multiples of $D^2(N)$. This will be based on three applications of the Cauchy–Schwarz inequality as follows:[8]

$$E(N) \le \sum_{p^k\le N} \frac{f(p^k)}{p^{k/2}}\cdot p^{k/2} \le D(N)\Bigg(\sum_{p^k\le N} p^k\Bigg)^{1/2},$$

$$\sum_{p^k\le N} f(p^k) = \sum_{p^k\le N} \frac{f(p^k)}{p^{k/2}}\cdot p^{-k/2} \le D(N)\Bigg(\sum_{p^k\le N} p^{-k}\Bigg)^{1/2},$$

and

$$\sum_{\substack{p\ne q\\ p^k q^\ell\le N}} f(p^k)f(q^\ell) = \sum_{\substack{p\ne q\\ p^k q^\ell\le N}} \frac{f(p^k)f(q^\ell)}{p^{k/2}q^{\ell/2}}\cdot p^{k/2}q^{\ell/2} \le D^2(N)\Bigg(\sum_{\substack{p\ne q\\ p^k q^\ell\le N}} p^k q^\ell\Bigg)^{1/2}.$$

Inserting these into (9.35) then gives

$$\frac{1}{N}\sum_{n\le N}(f(n)-E(N))^2 \le (1+\epsilon(N))D^2(N)),$$

where

$$\epsilon(N) = \frac{2}{N}\Bigg(\sum_{\substack{p\ne q\\ p^k q^\ell\le N}} p^k q^\ell\Bigg)^{1/2} + \frac{2}{N}\Bigg(\sum_{\substack{p^k\le N\\ q^\ell\le N}} p^{-k} q^\ell\Bigg)^{1/2}.$$

---

[8] The version of the Cauchy–Schwarz inequality being used three times here is

$$\left(\sum a_n b_n\right)^2 \le \left(\sum a_n^2\right)\left(\sum b_n^2\right).$$

It remains to show that $\epsilon(N) = o(1)$. For this, two further bounds are required. First

$$\sum_{p^k \leq N} p^{-k} = \sum_{p \leq N} \frac{1}{p} + \sum_{p^k \leq N,\, k \geq 2} \frac{1}{p^k} = \log\log N + O(1) \tag{9.36}$$

where Mertens's result has been used, and second

$$\sum_{q^\ell \leq N} q^\ell = O\Big(\sum_{q \leq N} q^{\log x/\log q}\Big) = O\Big(\sum_{q \leq N} x\Big) = O\Big(x \sum_{q \leq N} 1\Big) = O\Big(\frac{x^2}{\log x}\Big). \tag{9.37}$$

The first of the four order terms in (9.37) derives from the fact that each q generates a geometric series, $q^1, q^2, \ldots, q^\alpha$, where $\alpha = \left[\frac{\log x}{\log q}\right]$, and that this series sums to $(q/(q-1)) \cdot (q^\alpha - 1) < 2q^\alpha$. The last of the order terms in (9.37) follows from the PNT, although Chebychev's bound for $\pi(x)$ is all that is really needed, and would keep the proof 'elementary'. Using (9.36) and (9.37) we therefore see that

$$\sum_{\substack{p^k \leq N \\ q^\ell \leq N}} p^{-k} q^\ell = O\left(N^2 \frac{\log\log N}{\log N}\right)$$

and, reasoning as before,

$$\begin{aligned}
\sum_{\substack{p \neq q \\ p^k q^\ell \leq N}} p^k q^\ell &= 2 \sum_{\substack{p > q \\ p^k q^\ell \leq N}} p^k q^\ell \\
&= O\Big(\sum_{p^k \leq N} p^k \sum_{p < q \leq N/p^k} \sum_{\ell \leq \frac{\log(N/p^k)}{\log q}} q^\ell\Big) \\
&= O\Big(\sum_{p^k \leq N} p^k \sum_{p < q \leq N/p^k} \frac{N}{p^k}\Big) \\
&= O\Big(N \sum_{p^k \leq N} \pi\left(\frac{N}{p^k}\right)\Big) \\
&= O\Big(N^2 \sum_{p^k \leq N} \frac{1}{p^k \log(N/p^k)}\Big) \\
&= O\Big(N^2 \frac{\log\log N}{\log N}\Big).
\end{aligned}$$

It follows that $\epsilon(N) = o(1)$.

The restriction that f be nonnegative may be removed by applying the above to its positive and negative parts separately and combining the results, and the restriction that f be real may be removed by applying it to its real and imaginary parts separately and combining. We leave these details as an 'exercise' for interested readers; see, for example Steuding (2002, [**1026**, Chapter 6]), Elliott (1979, [**318**, Chapter 4]), or Tenenbaum (2015, [**1053**, §III.3]). □

REMARKS 9.9.

(i) Repeated application of the Cauchy–Schwarz inequality is the main technical device that has been used.

(ii) There are multiple versions of the Turán–Kubilius inequality—essentially all equivalent. See, for example, Hildebrand (1983, [**548**]), or Steuding (2002, [**1026**, Chapter 6]).

(iii) Postnikov (1987, [**865**, §4.4]) gives a particularly nice and carefully laid out proof.

On the proof of the Erdős–Kac Theorem. We follow Billingsley (1969, [**102**]), who in turn followed Halberstam (1955, [**479**]), to outline a proof using purely probability methods. From number theory, this proof again uses only unique factorization and Mertens's result (9.25).

It is convenient to first rewrite the Erdős–Kac result (9.11) in the form

$$\lim_{n\to\infty} \mathrm{P}_n\left[m : \frac{\omega(m) - \log\log m}{\sqrt{\log\log m}} < x\right] = \frac{1}{\sqrt{2\pi}}\int_{-\infty}^{x} e^{-t^2/2}\,dt,$$

where P_n is the uniform probability measure on $\{1, 2, \ldots, n\}$, and then to recast it as

$$\lim_{n\to\infty} \mathrm{P}_n\left[m : \frac{\omega_\alpha(m) - \log\log n}{\sqrt{\log\log n}} < x\right] = \frac{1}{\sqrt{2\pi}}\int_{-\infty}^{x} e^{-t^2/2}\,dt. \tag{9.38}$$

Here two alterations have occurred. The first is that the $\log\log m$ terms were changed to $\log\log n$; this is permissible because $\log\log m$ varies so slowly, and in this instance because

$$\lim_{n\to\infty} \mathrm{P}_n\left[m\colon \left|\frac{\omega(m) - \log\log m}{\sqrt{\log\log m}} - \frac{\omega(m) - \log\log n}{\sqrt{\log\log n}}\right| > \epsilon\right] = 0, \quad \text{for all } \epsilon > 0.$$

(This may be taken as an exercise by the interested reader.) The second alteration is that $\omega(m)$—which counts *all* prime divisors—has been replaced by a count, $\omega_\alpha(m)$—which omits prime divisors that are 'too large'. The idea is that divisibility by the smaller primes more closely resembles independent events. It turns out to be good enough to take

$$\omega_\alpha(m) = \sum_{p\le\alpha(n)} \mathrm{I}[p|m], \tag{9.39}$$

with $\alpha(n) = o(n^\epsilon)$ for all $\epsilon > 0$. (That this restriction will work is by no means obvious *a priori*.) However, in connection with a later use of Mertens's result, we also need

$$\sum_{\alpha(n)\le p\le n} \frac{1}{p} = o\left(\sqrt{\log\log n}\right). \tag{9.40}$$

The two competing requirements on $\alpha(n)$ can both be satisfied—for example by taking $\alpha(n) = n^{1/\log\log n}$. (That this choice of $\alpha(n)$ satisfies both requirements is again left as an exercise.) The fact that we may use $\omega_\alpha(m)$ in (9.38) in place of $\omega(m)$ follows by a standard probability argument, because (9.40) allows us to bound in probability the (average of) the terms which have been omitted.

The proof is based on comparing (9.39) to its independence-based counterpart

$$S_n \equiv \sum_{p\le\alpha(n)} X_p\,,$$

where the X_p are independent random variables such that $\mathrm{P}[X_p = 1] = \frac{1}{p}$ and $\mathrm{P}[X_p = 0] = 1 - \frac{1}{p}$. We first note that because the mean and variance of S_n, namely

$$c_n = \sum_{p \le \alpha(n)} \frac{1}{p}, \quad \text{and } s_n^2 = \sum_{p \le \alpha(n)} \frac{1}{p}\left(1 - \frac{1}{p}\right),$$

are (in view of (9.40)) each $\log\log n + o\left(\sqrt{\log\log n}\right)$, we may further recast (9.38) into

$$\lim_{n\to\infty} \mathrm{P}_n\left[m : \frac{\omega_\alpha(m) - c_n}{s_n} < x\right] = \frac{1}{\sqrt{2\pi}} \int_{-\infty}^{x} e^{-t^2/2}\, dt. \tag{9.41}$$

The proof now proceeds via the method of moments; we need to show that the moments

$$\mathrm{E}_n\left(\frac{\omega_\alpha(m) - c_n}{s_n}\right)^r, \quad r = 1, 2, \dots \tag{9.42}$$

converge to those of a standard normal distribution; here E_n is expectation under P_n, when m is uniformly distributed on $\{1, 2, \dots, n\}$. This is carried out in two steps. The first is to show that the moments

$$E_n\left(\frac{S_n - c_n}{s_n}\right)^r, \quad r = 1, 2, \dots \tag{9.43}$$

converge to those of a standard normal distribution. That $(S_n - c_n)/s_n$ converges to a standard normal follows from the Central Limit Theorem, but the required convergence of its moments is actually a consequence of the fact that the X_p are independent and bounded—they are just 0's and 1's. The proof is standard; it involves applying the binomial formula to (9.43) for every r, expanding out fully, and computing expectations.

The second step is to show that for every r, the difference between the moments (9.42) and (9.43) tends to 0 as $n \to \infty$. This can be proved by applying the binomial formula to expand out each of (9.42) and (9.43) fully, comparing expectations termwise, and checking that the total of the differences tends to 0, for every r. For the (elementary but messy) computational details of these two steps, we refer the reader to Billingsley (1969, [**102**]). □

REMARKS 9.10.

(i) Turán's formula for the asymptotic variance of ω can be proved using the same method.

(ii) The Hardy–Ramanujan result is an immediate consequence of the Erdős–Kac Theorem.

(iii) The method used in this proof can also be used to prove a Central Limit Theorem for completely additive functions f, provided that $f(p)$ is bounded and $\sum f^2(p)/p \to \infty$.

(iv) Granville & Soundararajan (2007, [**460**]) provide a simplified argument.

PARTIAL PROOF OF DELANGE'S THEOREM. The theorem states that if $g(n)$ is a complex-valued multiplicative function satisfying $|g(n)| \le 1$, then it will have a nonzero limiting mean value given by

$$\lim_{x\to\infty} \frac{1}{x} \sum_{n \le x} g(n) = \prod_p \left(1 - \frac{1}{p}\right)\left(1 + \sum_{\ell=1}^{\infty} \frac{g(p^\ell)}{p^\ell}\right), \tag{9.44}$$

if and only if the conditions (A) and (B) of that theorem are satisfied. Although we will not prove Delange's Theorem, we discuss it to see why it should be true.

We start by simplifying the problem by eliminating the action of large primes, say those primes $p > R$. Thus if n has the prime factorization $n = p_1^{k_1} p_2^{k_2} \cdots p_m^{k_m}$, we set to 1 any of the factors of $g(n) = g(p_1^{k_1})g(p_2^{k_2})\cdots g(p_m^{k_m})$ which involve a prime $p > R$. This amounts to approximating $g(n)$ by the multiplicative function $h(n)$ defined as

$$h(p^k) = \begin{cases} g(p^k), & \text{if } p \le R \\ 1, & \text{if } p > R. \end{cases}$$

If we use $h(p^\ell)$ in place of $g(p^\ell)$ on the right in (9.44) it will reduce to

$$\prod_{p \le R} \left(1 - \frac{1}{p}\right)\left(1 + \sum_{\ell=1}^{\infty} \frac{h(p^\ell)}{p^\ell}\right) \tag{9.45}$$

because any term in the product involving a prime $p > R$ will simply be

$$\left(1 - \frac{1}{p}\right)\left(1 + \frac{h(p)}{p} + \frac{h(p^2)}{p^2} + \cdots\right) = \left(1 - \frac{1}{p}\right)\left(1 + \frac{1}{p} + \frac{1}{p^2} + \cdots\right) = 1.$$

Now rewrite (9.45) as follows:

$$\prod_{p \le R} \left(\left(1 - \frac{1}{p}\right) + \left(1 - \frac{1}{p}\right)\frac{1}{p}h(p) + \left(1 - \frac{1}{p}\right)\frac{1}{p^2}h(p^2) + \cdots\right). \tag{9.46}$$

The factors $(1 - 1/p)(1/p^k)$ appearing here are just the proportions (when x is large) of those $n \le x$ for which $p^k \parallel n$ (that is, those n which are divisible by p^k but not by any higher power of p); the $k = 0$ factor gives the proportion of integers not divisible by p. Now if the product (9.46) were multiplied out, the typical term which results looks like

$$h(p_1^{k_1})h(p_2^{k_2})\cdots h(p_m^{k_m}) = h\left(p_1^{k_1} p_2^{k_2} \cdots p_m^{k_m}\right), \quad \text{where all } p_j \le R,$$

multiplied by a factor that equals the proportion of integers whose prime factorizations include $p_1^{k_1} p_2^{k_2} \cdots p_m^{k_m}$ and contain no additional prime factors $p < R$. Here $p_1, p_2, \ldots, p_k$ represent some subset of the primes $p < R$. It follows that (9.46) is identical to

$$\lim_{x \to \infty} \frac{1}{x} \sum_{n \le x} h(n)$$

since every integer n is accounted for in the implied partitioning. For the approximation of $g(n)$ by $h(n)$ to work the values of $g(p^k)$ for high primes cannot be too far from 1, lest the mean value of g be pushed towards 0. It turns out that the precise condition needed is that

$$\sum_p \sum_{k=1}^{\infty} \frac{1 - g(p^k)}{p^k}$$

must converge. But this is the same as Delange's condition (A) that $\sum_p (1 - g(p))/p$ converge, for it is easily seen that the $k \ge 2$ terms converge to a constant (since g is bounded).

A more complete proof of Delange's Theorem may be found, for example, in Tenenbaum (2015, [**1053**, Chapter III, Theorem 4.4 & 4.4]), or Postnikov (1988, [**865**, Chapter 4]). □

Partial proof of the Erdős–Wintner Theorem. We will deal only with the sufficiency part of this theorem. For the necessity, no proofs have yet been found which are able to avoid delicate number-theoretic reasoning.

We assume that the series (a), (b), and (c) in the statement of that theorem all converge; specifically, we assume that

(9.47)
$$\text{(a)} \sum_{|f(p)|>1} \frac{1}{p} < \infty, \quad \text{(b)} \sum_{|f(p)|\le 1} \frac{|f(p)|}{p} < \infty, \quad \text{and (c)} \sum_{|f(p)|\le 1} \frac{f^2(p)}{p} < \infty.$$

Setting $g(n) = e^{itf(n)}$ we will first verify that for arbitrary $t \in \mathbb{R}$ the condition (A) of Delange's Theorem, namely that $\sum_p (1 - g(p))/p$ converges, holds. To do this we split this condition into two components:

$$\sum_p \frac{1-g(p)}{p} = \sum_{|f(p)|>1} \frac{1-g(p)}{p} + \sum_{|f(p)|\le 1} \frac{1-g(p)}{p}. \tag{9.48}$$

Because $|1 - g(n)| \le 2$, the first component converges by the assumed convergence (a). To handle the second component, split the numerator term $1 - g(p)$ into three pieces as follows:

$$1 - e^{itf(p)} = -itf(p) + (1 - \cos(tf(p))) - i(\sin(tf(p)) - tf(p)).$$

When the first piece is used in the second component of (9.48) it converges due to the assumed convergence (b). The second piece has absolute value bounded by $\frac{1}{2}t^2 f^2(p)$; when used in the second component of (9.48) it converges due to the assumed convergence (c). Finally, when $|f(p)| \le 1$ the absolute value of the third piece is bounded by $\frac{1}{6}t^3|f^3(p)| \le \frac{1}{6}t^3 f^2(p)$ and when substituted into the second component of (9.48) it too converges by (c). Therefore Delange's condition is satisfied for every t. It follows that the characteristic functions $\varphi_n(t)$ corresponding to the finite-sample distribution of the partial sequences $\{f(1), f(2), \ldots, f(n)\}$ converge pointwise to the function

$$\varphi(t) \equiv \prod_p \left(1 - \frac{1}{p}\right)\left(1 + \sum_{\ell=1}^{\infty} \frac{\exp\{itf(p^\ell)\}}{p^\ell}\right). \tag{9.49}$$

To complete the proof that $f(n)$ has a limiting distribution it suffices to show that (9.49) is in fact a characteristic function. For this one may appeal to Levy's Continuity Theorem according to which this would follow if we could show that $\varphi(t)$ is continuous in an interval around the origin. This continuity would follow if the convergence of the partial products in (9.49) was *uniform* over such an interval. For details of such an *analytically* based approach see, for example, Tenenbaum (2015, [**1053**, §4.1, Lemma 4.3ff]).

We will prove that (9.49) is a characteristic function more probabilistically by appealing to the Kolmogorov Three-Series Criterion. First note that a typical term of (9.49), namely

$$\left(1-\frac{1}{p}\right) + \frac{1}{p}\left(1-\frac{1}{p}\right)e^{itf(p)} + \frac{1}{p^2}\left(1-\frac{1}{p}\right)e^{itf(p^2)} + \frac{1}{p^3}\left(1-\frac{1}{p}\right)e^{itf(p^3)} + \cdots$$

is the characteristic function of a discrete random variable X_p whose distribution is

$$\mathrm{P}\left[X_p = f(p^k)\right] = \left(1 - \frac{1}{p}\right)\frac{1}{p^k}, \quad \text{for } k = 0, 1, 2, \ldots$$

Positioning all X_p in the same probability space, the result would follow if we can establish that $\sum_p X_p$ is legitimately a random variable, for then (9.49) would be its characteristic function. In other words, we need to show that $\sum_p X_p$ converges (to a random variable) almost surely. According to the Three-Series Criterion this will follow if

$$\sum_p \mathrm{P}[\,|X_p| \geq A\,] < \infty, \qquad \sum_p \mathrm{E}\,|Y_p| < \infty, \qquad \text{and} \qquad \sum_p \operatorname{Var} Y_p < \infty, \tag{9.50}$$

where $Y_p \equiv X_p \mathrm{I}[\,|X_p| \leq A\,]$. We will show that they follow from the convergence assumptions (a), (b), and (c) of the Erdős–Wintner Theorem—see equation (9.47).

Taking into account of the distribution of X_p we have

$$\begin{aligned}
&\sum_p \mathrm{P}[\,|X_p| \geq A\,] \\
&\quad = \sum_{|f(p)|\geq A} \mathrm{P}[\,|X_p| \geq A\,] + \sum_{|f(p)|<A} \mathrm{P}[\,|X_p| \geq A\,] \\
&\quad \leq \sum_{|f(p)|\geq A} \left(1 - \left(1 - \frac{1}{p}\right)\right) + \sum_{|f(p)|<A} \left(1 - \left(1 - \frac{1}{p}\right) - \left(1 - \frac{1}{p}\right)\frac{1}{p}\right) \\
&\quad = \sum_{|f(p)|\geq A} \frac{1}{p} + \sum_{|f(p)|<A} \frac{1}{p^2},
\end{aligned}$$

with the first sum bounded by (a), and the second sum bounded by $\sum 1/n^2$. Next

$$\begin{aligned}
\sum_p \mathrm{E}\,|Y_p| &\equiv \sum_p \mathrm{E}(\,|X_p| \mathrm{I}[\,|X_p| \leq A\,]) \\
&= \sum_p \mathrm{E}(\,|X_p| \mathrm{I}[\,0 < |X_p| \leq A\,]) \\
&= \sum_{|f(p)|\geq A} \mathrm{E}(\,|X_p| \mathrm{I}[\,0 < |X_p| \leq A\,]) + \sum_{|f(p)|<A} \mathrm{E}(\,|X_p| \mathrm{I}[\,0 < |X_p| \leq A\,]).
\end{aligned}$$

Of these two sums, the first is bounded above by

$$\begin{aligned}
&A \sum_{|f(p)|\geq A} \mathrm{P}[\,0 < |X_p| \leq A\,] \\
&\quad \leq A \sum_{|f(p)|\geq A} \left(1 - \left(1 - \frac{1}{p}\right) - \left(1 - \frac{1}{p}\right)\frac{1}{p}\right) \leq A \sum_p \frac{1}{p^2} < \infty,
\end{aligned}$$

while the second is bounded above by

$$\sum_{|f(p)|<A} \left(1 - \frac{1}{p}\right)\frac{1}{p} f(p) + A \sum_{|f(p)|<A} \left(1 - \left(1 - \frac{1}{p}\right) - \left(1 - \frac{1}{p}\right)\frac{1}{p}\right),$$

with the first piece here bounded due to (b) and the second bounded by $\sum_p 1/p^2$. Finally,

$$\begin{aligned}
\sum_p \operatorname{Var} Y_p &\leq \sum_p \mathrm{E}\,|Y_p|^2 \\
&= \sum_{|f(p)|\geq A} X_p^2 \mathrm{I}[\,|X_p| \leq A\,] + \sum_{|f(p)|<A} X_p^2 \mathrm{I}[\,|X_p| \leq A\,],
\end{aligned}$$

and the two components here can be treated exactly as before except now using (c). It follows that $\sum_p X_p$ is legitimately a random variable and consequently that (9.49) is legitimately a characteristic function.

Finally, because $\varphi(t)$ is a convolution of discrete distributions corresponding to an almost surely converging sum of independent discrete random variables, the Jessen–Wintner Law of Pure Types applies. (See, for example, Breiman, (1992, [**142**, Theorem 3.26]).) It follows that $\varphi(t)$ corresponds to a distribution that is either discrete, absolutely continuous, or singular. This concludes the proof of the sufficiency. $\square$

CHAPTER 10

The Zeta Distributions

> Among the reminiscences of his days at the Manhattan Project in Los Almos, Otto Frisch recalls having met (around 1943) Stan Ulam, a brilliant Polish topologist. Frisch quotes Ulam as referring to himself as *"a pure mathematician who had sunk so low that his latest paper actually contained numbers with decimal points!"* [**394**], [**1077**].[1]

In this chapter we study the *distribution family* based on the Riemann zeta function, primarily—although not exclusively—in the context of *statistical inference*. In Section 10.1 we take an in-depth look at the probability properties of these distributions, while in Section 10.2 we focus rather comprehensively on the theory of statistical inference for this family. The zeta distributions constitute an *exponential family* which can be extended to a so-called *generalized linear model*, allowing one to accommodate predictor variables in a theoretically attractive way. That topic requires more involved statistical reasoning and is treated in Section 10.3. Finally, Section 10.4 lays out certain further details which—although arising from the probability and statistics context—turn out to have much wider interest. A number of proofs are also given at the end. The topics in this chapter—particularly those in Section 10.2, and still more so in Section 10.3—presuppose some background in theoretical statistics. Even so, however, this chapter includes much material that should prove of interest to readers without this background, and from which some interesting and surprising things can be learned. To quote Bradley Efron: "Those who ignore statistics are condemned to reinvent it!"[2] Our treatment in this chapter is purely theoretical; no numbers with decimal points will be found.[3]

10.1. Properties of the ζ-distributions

Because $\zeta(s) \equiv \sum_{n=1}^{\infty} 1/n^s$ consists of positive terms and converges for real values of $s > 1$, a parametric family of distributions on the integers $1, 2, 3 \ldots$ can be defined via the decreasing probability mass functions

$$f_\theta(x) \equiv \mathrm{P}_\theta[X = x] = \frac{1}{\zeta(\theta)} \frac{1}{x^\theta}, \quad x = 1, 2, 3, \ldots, \tag{10.1}$$

for parameter values $\theta > 1$, and corresponding (cumulative) distribution functions

$$F_\theta(x) \equiv \mathrm{P}_\theta[X \le x] = \frac{1}{\zeta(\theta)} \sum_{n \le x} \frac{1}{n^\theta}.$$

[1] An interesting article on Ulam's 'spiral of the primes' may be found on Wikipedia.

[2] `https://magazine.amstat.org/blog/2014/10/01/statview-oct14/`

[3] Other, of course, than in the section and equation numbers themselves!

Here X represents a random variable having the distribution (10.1). Rewriting (10.1) as

$$f_\theta(x) = \exp\big(-\theta \log x - \log \zeta(\theta)\big), \quad x = 1, 2, 3, \ldots, \tag{10.2}$$

we see that $\{f_\theta\}$ forms what is known as an *exponential family* of distributions with so-called *natural parameter* $\eta \equiv -\theta$ and *natural statistic* $Y \equiv +\log X$. Exponential families have invaluable theoretical properties, and are discussed, for example, in Efron (2023, [**311**]).

REMARKS 10.1.

(1) Alternate starting values for X can be accommodated, as for example by setting

$$f_\theta(x) \equiv \mathrm{P}_\theta[\, X = x\,] = \frac{1}{\zeta(\theta)}\,\frac{1}{(x+1)^\theta}, \quad x = 0, 1, 2, \ldots; \ \theta > 1. \tag{10.3}$$

(2) In the exponential family (10.2) we have taken $+\log x$ as the natural statistic and $-\theta$ as the natural parameter. One may instead take $-\log x$ as the natural statistic and $+\theta$ as the natural parameter. We have adopted the former of these conventions when the choice matters and shall then denote the corresponding natural parameter by $\eta \equiv -\theta$.

(3) The *zeta distribution* can also be thought of as being that of $\log X$ (or of $-\log X$) and thus supported on $\{\log n\}_{n=1}^\infty$ (or on $\{-\log n\}_{n=1}^\infty$). We will take the choice (10.1), supported on $\{1, 2, \ldots\}$, as defining the zeta distribution and will use the notation $Y \equiv \log X$ for the natural statistic, with X distributed according to (10.1). The distribution of Y is then simply

$$\mathrm{P}_\theta[Y = \log x] = \frac{1}{\zeta(\theta)}\,\frac{1}{x^\theta}, \quad x = 1, 2, 3, \ldots; \ \theta > 1. \tag{10.4}$$

(4) General background at various levels can be found in many statistics textbooks, including those authored by Cassella & Berger, Devore, Evans & Rosenthal (available free), Freund, de Groot & Schervish, Hogg, Knight, Wackerly & Mendenhall & Scheaffer, and Wasserman, to name a few. McCallagh & Nelder (1989, [**747**]) provides more specialized background.

Moments. The most basic descriptors of a probability distribution are its moments and (equivalently) its cumulants, and these are most easily obtained from their corresponding *generating functions*.[4] The natural statistic $Y \equiv \log X$ of the exponential family (10.2) has *moment generating function (mgf)* and *cumulant generating function (cgf)* given by

$$M_\theta(t) \equiv \mathrm{E}_\theta\, e^{tY} = \mathrm{E}_\theta\, e^{t \log X} = \mathrm{E}_\theta\, X^t = \frac{\zeta(\theta - t)}{\zeta(\theta)}, \quad t < \theta - 1, \tag{10.5}$$

$$\kappa_\theta(t) \equiv \log M_\theta(t) = \log \frac{\zeta(\theta - t)}{\zeta(\theta)}, \quad t < \theta - 1, \tag{10.6}$$

and corresponding *characteristic function (cf)* (namely, Fourier transform $\varphi(t) \equiv \int e^{ity}\, dF(y)$)

$$\varphi_\theta(t) \equiv \mathrm{E}_\theta\, e^{itY} = \frac{\zeta(\theta - it)}{\zeta(\theta)} = \frac{1}{\zeta(\theta)} \sum_{n=1}^\infty \frac{1}{n^{\theta - it}}$$

---

[4]The mgf (for instance) of a random variable X is $M(t) = \mathrm{E}\, e^{tX}$. Expanding the exponential leads to $\sum_{k=0}^\infty t^k\, \mathrm{E}\, X^k/n!$, and the kth moment $\mathrm{E}\, X^k$ may be obtained by differentiating k times and setting $t = 0$.

$$(10.7) \qquad = \frac{1}{\zeta(\theta)} \sum_{n=1}^{\infty} \frac{e^{it\log n}}{n^\theta} = \frac{1}{\zeta(\theta)} \sum_{n=1}^{\infty} \frac{\cos(t\log n) + i\sin(t\log n)}{n^\theta}, \quad t \in \mathbb{R}.$$

The *moments* of $Y \equiv \log X$ may be obtained by differentiating the mgf $M_\theta(t)$:

$$\mathrm{E}_\theta\, Y^k \equiv \mathrm{E}_\theta \log^k X = \frac{\partial^k}{\partial t^k} M_\theta(t)\bigg|_{t=0} = \frac{\partial^k}{\partial t^k} \frac{\zeta(\theta - t)}{\zeta(\theta)}\bigg|_{t=0} = (-1)^k \frac{\zeta^{(k)}(\theta)}{\zeta(\theta)}, \quad k \geq 0.$$

In particular, Y has *mean*

$$(10.8) \quad \mathrm{E}_\theta\, Y = -\frac{\zeta'(\theta)}{\zeta(\theta)} \equiv -\frac{d\log\zeta(\theta)}{d\theta} = \sum_{n=1}^{\infty} \frac{\Lambda(n)}{n^\theta} = \sum_p \sum_{k=1}^{\infty} \frac{\log p}{p^{k\theta}} = \sum_p \frac{\log p}{p^\theta - 1},$$

a positive, strictly decreasing, and convex function of $\theta > 1$, which ranges over the values $(1, \infty)$. The *variance* of Y may be computed (for example from the cgf) as

$$\begin{aligned}
\mathrm{Var}_\theta\, Y &= \frac{d^2\kappa_\theta(t)}{dt^2}\bigg|_{t=0} \\
&= \frac{\zeta''(\theta)\zeta(\theta) - (\zeta'(\theta))^2}{\zeta^2(\theta)} \\
&= \frac{d^2 \log\zeta(\theta)}{d\theta^2} \\
(10.9) \qquad &= \sum_{n=1}^{\infty} \frac{\Lambda(n)\log n}{n^\theta} \\
&= \sum_p \sum_{k=1}^{\infty} \frac{k\log^2 p}{p^{k\theta}} \\
&= \sum_p \frac{p^\theta \log^2 p}{(p^\theta - 1)^2}.
\end{aligned}$$

More generally the mth *cumulant* of Y is

$$\begin{aligned}
\mathrm{cum}_m\, Y &= \frac{d^m\kappa_\theta(t)}{dt^m}\bigg|_{t=0} \\
&= (-1)^m \frac{d^m \log\zeta(\theta)}{d\theta^m} \\
(10.10) \qquad &= \sum_{n=1}^{\infty} \frac{\Lambda(n)\log^{m-1} n}{n^\theta} \\
&= \sum_p \sum_{k=1}^{\infty} \frac{k^{m-1}\log^m p}{p^{k\theta}}, \quad m = 1, 2, 3, \ldots
\end{aligned}$$

In (10.8)–(10.10) $\Lambda(n)$ is the von Mangoldt function ($\Lambda(n) = \log p$ if $n = p^k$ with $k \geq 1$, and $\Lambda(n) = 0$ otherwise). As in (10.8), the functions in (10.9) and (10.10) are also positive, strictly decreasing, and convex. A proof of these properties is given in Section 10.5.

Following standard statistical theory, the quantity (10.9) must also equal the so-called *Fisher information* per observation

$$(10.11) \quad I(\theta) \equiv \mathrm{Var}_\theta \frac{\partial \log f_\theta(X)}{\partial\theta} \equiv -\mathrm{E}_\theta \frac{\partial^2 \log f_\theta(X)}{\partial\theta^2} = \frac{d^2\log\zeta(\theta)}{d\theta^2}, \quad \theta > 1;$$

it may also be expressed as

$$(10.12) \qquad I(\theta) = \operatorname{Var}_\theta(\log X) = \sum_{n=2}^{\infty} \frac{\Lambda(n)\log n}{n^\theta} = \sum_p \frac{p^\theta \log^2 p}{(p^\theta - 1)^2}, \quad \theta > 1.$$

Of course, the Fisher information for X, Y, or for any other one-to-one transformation of these variates will all be equal.

The computations so far were based on the moments and cumulants of the natural statistic $Y \equiv \log X$. For X itself, we have

$$(10.13) \qquad \mathrm{E}_\theta X^k = \frac{1}{\zeta(\theta)} \sum_{x=1}^{\infty} \frac{x^k}{x^\theta} = \frac{\zeta(\theta - k)}{\zeta(\theta)}, \qquad k < \theta - 1,$$

with all moments that exist being positive. Hence its variance, for instance, will be

$$(10.14) \qquad \operatorname{Var}_\theta X = \mathrm{E}_\theta X^2 - \left(\mathrm{E}_\theta X\right)^2 = \frac{\zeta(\theta-2)\zeta(\theta) - \zeta^2(\theta-1)}{\zeta^2(\theta)}, \qquad \theta > 3.$$

The numerator here must again be positive, and it is decreasing and convex. Section 10.5 includes a direct proof of this fact, and also shows that a generalized version of this numerator is positive as well. Finally, the characteristic function of X is given by

$$\mathrm{E}_\theta\, e^{itX} = \frac{1}{\zeta(\theta)} \sum_{n=1}^{\infty} \frac{e^{int}}{n^\theta} = \frac{1}{\zeta(\theta)} \sum_{n=1}^{\infty} \frac{\cos(nt) + i\sin(nt)}{n^\theta}$$

(compare with equation (10.7)). This is essentially a *periodic zeta function*, and an instance of what is known as a *polylogarithm*. It has a period of 2π in the variable t, which accords with the fact that it belongs to a distribution supported on the lattice of integers; see item (9) in the series of remarks in Section 5.9.

Since not all moments of X are finite it cannot have either a moment or a cumulant generating function. Nevertheless the number of its cumulants will equal the number of its moments, and those that exist can still be obtained by manipulating the formal generating series. Hence, for instance, the third and fourth cumulants of X (which will be finite when $\theta > 4$ and when $\theta > 5$, respectively) are given by

$$\begin{aligned}\operatorname{cum}_3 X &= \mathrm{E}\, X^3 - 3(\mathrm{E}\, X^2)\,\mathrm{E}\, X + 2(\mathrm{E}\, X)^3 \\ &= \frac{\zeta(\theta-3)\zeta^2(\theta) - 3\zeta(\theta-2)\zeta(\theta-1)\zeta(\theta) + \zeta^3(\theta-1)}{\zeta^3(\theta)}\end{aligned}$$

(which tends to -1 as $\theta \to \infty$) and

$$\operatorname{cum}_4 X = \mathrm{E}\, X^4 - 4(\mathrm{E}\, X^3)\,\mathrm{E}\, X - 3(\mathrm{E}\, X^2)^2 + 12(\mathrm{E}\, X^2)(\mathrm{E}\, X)^2 - 6(\mathrm{E}\, X)^4 = \mathrm{Num}/\zeta^4(\theta)$$

where

$$\begin{aligned}\mathrm{Num} = \zeta(\theta-4)\zeta^3(\theta) &- 4\zeta(\theta-3)\zeta(\theta-1)\zeta^2(\theta) - 3\zeta^2(\theta-2)\zeta^2(\theta) \\ &+ 12\zeta(\theta-2)\zeta^2(\theta-1)\zeta(\theta) - 6\zeta^4(\theta-1)\end{aligned}$$

(which is asymptotic to $2^{-\theta}$ as $\theta \to \infty$). One may also compute the *entropy* of the zeta distribution (Golomb (1970, [**436**])). This is given by

$$H(\theta) \equiv -\sum_{n=1}^{\infty} \mathrm{P}[X = n] \log \mathrm{P}[X = n]$$

$$= \frac{1}{\zeta(\theta)} \sum_{n=1}^{\infty} \Big(\theta \log n + \log \zeta(\theta)\Big) \Big/ n^{\theta}$$

$$= \log \zeta(\theta) + \frac{\theta}{\zeta(\theta)} \sum_{n=1}^{\infty} \frac{\log n}{n^{\theta}}$$

$$= \log \zeta(\theta) - \theta \frac{\zeta'(\theta)}{\zeta(\theta)},$$

a finite positive quantity which measures its 'statistical distance' from the (improper) uniform distribution on the positive integers. Since it depends only on probabilities, the entropy of X, Y, or of any other one-to-one transformation of these will all be equal. The concept of entropy arises in many fields, and particularly in information theory and thermodynamics.

Although we will bypass the details we mention that, for any fixed $\theta > 1$, the characteristic function $\varphi_{\theta}(t)$ in (10.7), and likewise each of $\zeta(\theta - it)$ and $\zeta(\theta + it)$, are nonnegative definite functions in t.[5] Furthermore, because they correspond to discrete distributions, these functions are *almost periodic*. One consequence is that for each fixed $\theta > 1$, the functions $\zeta(\theta - it)$ and $\zeta(\theta + it)$ over any finite t-interval will repeat to within any given degree of precision at arbitrarily high values of t. Note that the terms $n^{-\theta} e^{it \log n}$ of the ζ-sums have frequencies $\log n$, and that the frequencies $\log p$ from the primes are not commensurate. Almost periodic functions were studied by Besicovitch (1932, [**96**]) and Bohr (1947, [**118**]).

Infinite divisibility. It was observed by Alexandr Yakovlevich Khinchin (1894–1959)—in Khinchin (1938, [**635**, p. 35])—that the zeta distributions are *infinitely divisible*, a fact that turns out to have some bearing on the Riemann hypothesis. A little background is needed in order to describe this curious result. Infinite divisibility of a distribution, corresponding to a characteristic function $\varphi(t)$, means that for every $m = 1, 2, \ldots$, we have $\varphi(t) = \varphi_m^m(t)$, where each $\varphi_m(t)$ is itself the characteristic function of some distribution. This means that $\varphi(t)$ corresponds to an m-fold *convolution* for every m; that is, it corresponds to a sum of m independent and identically distributed random variables. The normal distributions and the Cauchy distributions (indeed any of the so-called 'stable laws') are examples of infinitely divisible distributions, and the Poisson and gamma distributions are as well.

It is an important characterization of the infinitely divisible distributions that they are precisely the same as the limiting distributions of weighted sums of independent Poisson distributed random variables. Infinitely divisible distributions are important in applications in part because they are precisely those distributions from which so-called *Lévy processes* can be constructed; in other words, they are the marginal distributions of stochastic processes which have 'independent and stationary increments' such as Brownian motion or homogeneous Poisson processes.

[5] A complex-valued function $\varphi(t)$ of a real variable t is said to be nonnegative definite if

$$\sum_{j=1}^{k} \sum_{\ell=1}^{k} c_j \, \overline{c_\ell} \, \varphi(t_j - t_\ell) \geq 0$$

for all $k = 1, 2 \ldots$, all real numbers $t_1, t_2 \ldots, t_k$, and all complex numbers $c_1, c_2, \ldots, c_k$. It is not hard to show that characteristic functions (of probability distributions) are nonnegative definite. There is a well-known characterization of Bochner which asserts (essentially) that the nonnegative definite functions correspond to the Fourier transforms of nonnegative functions.

A fine classical discussion of the theory of infinitely divisible distributions may be found in Gnedenko & Kolmogorov (1968, [**417**, Chapter 3]); additional references include Lukacs (1970, [**714**, Chapter 5]), Sato (1999, [**946**, Chapter 3, §7–8]), and Steutel & van Harn (2004, [**1029**]).

Before describing Khinchin's result, let us recall that the Poisson distribution with mean parameter $\lambda \geq 0$, to be denoted here by $\mathrm{P}(\lambda)$, has the probability mass function

$$\mathrm{P}_\lambda[\, X = x\,] = e^{-\lambda}\lambda^x/x!\,, \quad x = 0, 1, 2\ldots,$$

and has the readily computed characteristic function

$$\mathrm{E}_\lambda\, e^{itX} = e^{\lambda(e^{it}-1)}.$$

It is easily seen to be infinitely divisible since multiplying characteristic functions of this form results in a function of the same form: the λ's simply add.

We will also need to refer to the geometric distribution, $\mathrm{G}(\lambda)$, with parameter $0 < \lambda < 1$, which we will define here in the particular form

$$\mathrm{P}_\lambda\,[\, X = x\,] = (1-\lambda)\lambda^x, \quad x = 0, 1, 2\ldots \tag{10.15}$$

It has mean $\lambda/(1-\lambda)$, variance $\lambda/(1-\lambda)^2$, and the readily computed characteristic function

$$\mathrm{E}_\lambda\, e^{itX} = (1-\lambda)\left[1-\lambda e^{it}\right]^{-1}.$$

Now, using the Taylor expansion $\log(1-x) = -\sum_1^\infty x^k/k$, and combining terms, yields

$$\log \mathrm{E}_\lambda\, e^{itX} = \log(1-\lambda) - \log\left(1-\lambda e^{it}\right) = \sum_{k=1}^{\infty} \frac{\lambda^k}{k}\left(e^{ikt}-1\right).$$

From this one sees that the $\mathrm{G}(\lambda)$ distribution is the same as that of a linear combination of independent Poisson random variables whose means are λ^k/k, and which have been multiplied by k. We may represent this well-known distributional equality as

$$\mathrm{G}(\lambda) \overset{d}{=} \sum_{k=1}^{\infty} k\, \mathrm{P}\left(\lambda^k/k\right). \tag{10.16}$$

It follows that the geometric distribution is also infinitely divisible.

And one last item before we proceed. If N has a $P(\lambda)$ distribution, and if $X_1, X_2, \ldots$ are independent and identically distributed random variables which are independent of N and which have characteristic function $\varphi(t)$, then the sum

$$Y \equiv X_1 + X_2 + \cdots + X_N$$

(which consists of a random number of terms) is said to have a *compound Poisson distribution*. Using 'iterated expectation' its characteristic function is easily found to be

(10.17)

$$\mathrm{E}\left(e^{itY}\right) = \mathrm{E}\left[\mathrm{E}\left(e^{itY}\,\middle|\, N\right)\right] = \mathrm{E}\,\varphi^N(t) = \sum_{n=0}^{\infty} \frac{\lambda^n e^{-\lambda}}{n!}\,\varphi^n(t) = \exp\bigl(\lambda(\varphi(t)-1)\bigr).$$

This is clearly infinitely divisible (in the parameter λ). This form for the characteristic function is the distinguishing feature of compound Poisson distributions. It can be shown that a probability distribution on $\mathbb{R}$ is infinitely divisible if and only if it is a limit of compound Poisson distributions.

With these preliminaries out of the way, we now return to Khinchin's result. Using the Euler product representation, the characteristic function of the random variable $Y \equiv \log X$ based on the zeta distribution can be written as

$$\varphi_\theta(t) = \frac{\zeta(\theta - it)}{\zeta(\theta)} = \prod_p \left(1 - p^{-\theta}\right)\left(1 - p^{-\theta+it}\right)^{-1}. \tag{10.18}$$

Once again using the Taylor expansion $\log(1-x) = -\sum_1^\infty x^k/k$, $|x| < 1$ gives

$$\begin{aligned} \log \varphi_\theta(t) &= \sum_p \left(\log(1-p^{-\theta}) - \log(1-p^{-\theta+it})\right) \\ &= \sum_p \sum_{k=1}^\infty \frac{1}{k}\left(-p^{-k\theta} + p^{-k\theta+ikt}\right) \qquad (10.19) \\ &= \sum_p \sum_{k=1}^\infty \frac{p^{-k\theta}}{k}\left(e^{itk\log p} - 1\right). \end{aligned}$$

We therefore see that $\varphi_\theta(t)$ is a product of terms of the form $\exp\big(a(e^{ibt} - 1)\big)$. But each such term is infinitely divisible since it corresponds to a Poisson variable with mean parameter a, and multiplied by b; it takes its values on the lattice mb, $m = 0, 1, 2 \ldots$. For the Poisson components of (10.19), the a's range over the values $a = p^{-k\theta}/k$, while the b's range over the values $b = k \log p$. The result (10.19) can alternately be expressed as

$$Y \equiv \log X \overset{d}{=} \sum_p \sum_{k=1}^\infty (k \log p)\, \mathrm{P}\left(p^{-k\theta}/k\right) \tag{10.20}$$

where the $\mathrm{P}\left(p^{-k\theta}/k\right)$ are independent Poisson variables with parameters $p^{-k\theta}/k$. We mention, in passing, that result (10.19) can be also expressed in terms of the so-called *'Lévy–Khinchin representation'* for infinitely divisible distributions, namely

$$\varphi_\theta(t) = \exp\left(\int_0^\infty \left(e^{itx} - 1\right) dH_\theta(x)\right), \quad \text{with } H_\theta(x) = \sum_p \sum_{k=1}^\infty \frac{p^{-k\theta}}{k}\,\delta(x - k\log p),$$

where $\delta(x)$ is the Dirac delta function (a unit mass at $x = 0$). For details on this see Gnedenko & Kolmogorov (1968, [**417**, Chapter 3]), or one of the other references mentioned earlier.

An alternate approach to (10.18) is to observe that its factors

$$\left(1 - p^{-\theta}\right)\left(1 - p^{-\theta+it}\right)^{-1} = \left(1 - p^{-\theta}\right)\left(1 - p^{-\theta}e^{it\log p}\right)^{-1}$$

correspond to geometric distributions which have been multiplied by $\log p$ so that

$$Y \equiv \log X \overset{d}{=} \sum_p (\log p)\mathrm{G}(p^{-\theta}), \tag{10.21}$$

where the $\mathrm{G}(\cdot)$ here are independent geometric variables with parameters $\lambda = p^{-\theta}$. Since geometric distributions are also infinitely divisible, this again shows that the zeta distributions are infinitely divisible. Of course (10.20) and (10.21) must reconcile, so their coefficients of $\log p$, namely $\sum_{k=1}^\infty k\,\mathrm{P}\left(p^{-k\theta}/k\right)$ and $\mathrm{G}(p^{-\theta})$ must match. This merely corresponds to the distributional equality (10.16) with $p^{-\theta}$ in place of λ. The formal manipulations in (10.18)–(10.21) are justified more meticulously in Chung (2001, [**207**, §7.6]). Related references are Ehm (2001, [**312**]), (2007, [**313**]), Gut (2006, [**468**]), and Lin & Hu (2001, [**700**]).

REMARK 10.2. It is the distribution of $Y \equiv \log X$, not of X, that has been shown to be infinitely divisible although it does follow that X will have the corresponding property in a multiplicative sense. However, if one used the version of X that takes on the value 0, as in (10.3), then it can be shown that it too would be infinitely divisible in the standard sense. This happens to be a consequence of the fact that the probabilities $w_k \equiv \zeta^{-1}(\theta)(k+1)^{-\theta}$, $k = 0, 1, 2, \ldots$, have the *'completely monotone'* property $(-1)^n \Delta^n w_k \geq 0$ for $n \geq 0$, where $\Delta^0 w_k = w_k$, $\Delta w_k = w_{k+1} - w_k$, and $\Delta^n = \Delta\Delta^{n-1}$ when $n \geq 1$. Steutel & van Harn (2004, [**1029**, Chapter II, Example 11.6]) provides the details.[6]

The representation (10.21) allows for some further insights. Exponentiating both sides, it can be rewritten as

$$X \stackrel{d}{=} \prod_p p^{\mathrm{G}(p^{-\theta})}. \tag{10.22}$$

One may interpret this as saying that the zeta distributions generate integers X whose prime factors have independent and geometrically distributed exponents of the indicated form. The total number of these prime factors, namely $\Omega(X)$, is then $\sum_p \mathrm{G}(p^{-\theta})$, and because

$$\sum_p P[\,\mathrm{G}(p^{-\theta}) > 0\,] = \sum_p p^{-\theta} < \sum_n n^{-\theta} < \infty, \quad \text{for } \theta > 1,$$

the Borel–Cantelli Lemma (see Appendix D) guarantees that with probability 1, only a finite number of the $\mathrm{G}(p^{-\theta})$ exponents will be nonzero. Incidentally, the same can also be said for the $\mathrm{P}\left(p^{-k\theta}/k\right)$ Poisson variables appearing in (10.20): only a finite number of them can be nonzero. The distributions which are the factors that convolve to produce the distribution for Y can be determined from (10.22) by using the divisibility of the $\mathrm{G}(p^{-\theta})$ exponents there. This rather messy exercise is carried out by Sato (1999, [**946**]). See also Saito & Tanaka (2012, [**932**]).

A distribution related to the zeta can also be formed from the Dirichlet representation for $\log\zeta(s)$. By the definition of the von Mangoldt function, and for real θ, we have

$$\log\zeta(\theta) = \sum_{n=1}^{\infty} \frac{\Lambda(n)/\log n}{n^\theta} = \sum_p \sum_{k=1}^{\infty} \frac{1}{k}\frac{1}{p^{k\theta}}, \quad \theta > 1.$$

The terms of this Dirichlet series are nonnegative. So if U is a random variable which takes on the values $n = p^k$ with probabilities $1/\{\log\zeta(\theta)kp^{k\theta}\}$—in other words, if

$$\begin{aligned} \mathrm{P}\left[U = n \equiv p^k\right] &= \frac{1}{\log\zeta(\theta)}\,\frac{\Lambda(n)/\log n}{n^\theta} \\ &= \frac{1}{\log\zeta(\theta)}\,\frac{1}{kp^{k\theta}}, \quad \text{for primes } p, \text{ and } k = 1, 2, \ldots \end{aligned}$$

[6]An easy exercise: Show that an infinitely divisible distribution supported on the nonnegative integers must have positive probability at 0.

—then U has a well-defined probability distribution. The same of course is also true of the random variable $V = \log U$ with distribution

$$\mathrm{P}[V = \log n \equiv k \log p] = \frac{1}{\log \zeta(\theta)} \frac{1}{kp^{k\theta}}, \quad \text{for primes } p, \text{ and } k = 1, 2, \ldots$$

Interestingly, it turns out that if $V_1, V_2, \ldots$ are independent copies from this distribution, and if $Y = \log X$ has zeta distribution (10.1)/(10.4) as before, then these variables satisfy the distributional equality

$$Y = \log X \stackrel{d}{=} V_1 + V_2 + \cdots + V_N \tag{10.23}$$

where N is a Poisson random variable independent of the V_j and with mean $\log \zeta(\theta)$. This happens because, according to (10.21), for each prime p, the number of times $\log p$ occurs in Y is, given by $\mathrm{G}(p^{-\theta})$, while the number of times it occurs among the V_j in (10.23) is $\sum_{j=1}^{N} W_j^{(p)}$ where the $W_j^{(p)}$ are independent variables which, *for each fixed* p, have the probability structure

$$\mathrm{P}[W^{(p)} = k] = \frac{1}{\log \zeta(\theta)} \frac{1}{kp^{k\theta}}, \quad k = 1, 2, \ldots,$$

with $W^{(p)} = 0$ for the remaining probability. A direct proof of (10.23) using this fact is cumbersome; a simpler proof is given in Section 10.5. The representation in (10.23) is that of a *compound Poisson* and is, therefore, infinitely divisible.

The Riemann zeta distributions are discrete analogues of the so-called Pareto or Zipf distributions (Zipf (1949, [**1125**]); Johnson et al. (1992, [**604**, Chapter 11, §20])). In the continuous case, their basic characterizing property is the independence, for all n, of

$$\min\{X_1, X_2, \ldots, X_n\}, \quad \text{and} \quad \frac{X_1 + X_2 + \cdots X_n}{\min\{X_1, X_2, \ldots, X_n\}}.$$

In the discrete case, a basic defining property is the equality of the distribution of X and the conditional distribution of X/n *given* that n divides X.[7] Such distributions are commonly used in linguistics to model word frequencies, in geography to model the populations of communities, in insurance studies to model the numbers of claims, and in meteorology to model extreme events, as well as in other applications of this general kind. It is an empirical phenomenon for data of this type that the estimated values of θ are typically not very far from 1, thus corresponding to more heavily-tailed observations. A number of further properties of the zeta distributions may be found in Section 10.4.

10.2. Statistical inference

We turn next to the context of an *iid sample* $X_1, X_2, \ldots, X_n$ from a zeta distribution f_θ as defined in (10.1), with the parameter θ considered unknown.[8] A one-dimensional, *minimal sufficient statistic* for this exponential family of distributions is given by $\sum_{j=1}^{n} Y_j \equiv \sum_{j=1}^{n} \log X_j = \log \prod_{j=1}^{n} X_j$. (A sufficient statistic is a function of the data that carries all of the statistical information contained in the data

[7] We leave this assertion as an exercise; it is a consequence of the 'memoryless' property of geometric distributions. As the geometric is the only memoryless discrete distribution, this conditional distribution property of X/n characterizes the zeta distributions, although not quite completely.

[8] The abbreviation *iid* is widely used in statistics and stands for 'independent and identically distributed'.

under that model.) The *likelihood equation* for estimating θ, obtained by solving $d\big(\sum_j \log f_\theta(X_j)\big)/d\theta = 0$, is given by

$$-\frac{d\log\zeta(\theta)}{d\theta} = -\frac{\zeta'(\theta)}{\zeta(\theta)} = \frac{1}{n}\sum_{j=1}^{n}\log X_j,$$

which just equates the theoretical and empirical means. And because the required conditions are met, the *maximum likelihood estimator* (MLE) of θ, namely $\hat\theta_{\mathrm{MLE},n}$, is *asymptotically efficient* and *asymptotically normally distributed* as

$$\sqrt{n}(\hat\theta_{\mathrm{MLE},n} - \theta) \xrightarrow{d} N\left(0, \left[\frac{d^2\log\zeta(\theta)}{d\theta^2}\right]^{-1}\right).$$

Evaluations for the two derivatives of $\log\zeta(\theta)$ occurring here were given in (10.8) and (10.9). Alternate methods for estimating θ and goodness of fit testing were considered by Doray & Luong (1995, [**293**]), and Luong & Doray (1996, [**717**]).

For *hypothesis testing*, say for testing the *null hypothesis* $H_0 : \theta = \theta_0$ against the *alternative* $H_1 : \theta < \theta_0$ to be definite, the *likelihood ratio test* for a sample of size n is given by

(10.24) $$\text{Reject } H_0 \text{ if } \frac{1}{n}\sum_{j=1}^{n}\log X_j \text{ is large, or equivalently if } \prod_{j=1}^{n} X_j \text{ is large.}$$

In the context of 'finite sample size' such a test can be carried out by estimating the null distribution via Monte Carlo sampling. However, it is of interest to consider the null distribution analytically. Using $R \equiv \prod_{j=1}^{n} X_j$ as the test statistic, we know that R takes on integer values $m = 1, 2, \ldots$. So the event $R = m$ is assembled from the equiprobable ways in which m can be written as an ordered product of n integers. It follows that

(10.25) $$\mathrm{P}_\theta\,[\,R = m\,] = \exp\big(-\theta\log m - n\log\zeta(\theta) + \log\tau_n(m)\big),$$

where $\tau_n(m)$ is the number of distinct ways, when order matters, in which m can be factored into a product of n integers. For example, $\tau_2(3) = 2$, because 3 equals both 1×3 and 3×1; likewise $\tau_3(6) = 9$. And if $m = p_1^{k_1}p_2^{k_3}\cdots p_\ell^{k_\ell}$ is the prime factorization of m, then

$$\tau_n(m) = \prod_{j=1}^{\ell}\binom{k_j + n - 1}{n-1}.$$

Note that $\tau_2(m) = \tau(m)$ is the number of divisors function discussed in Section 3.5.

One way to proceed from here is to note that the terms (10.25) must sum to 1 so that

$$\zeta^n(\theta) = \sum_{m=1}^{\infty}\exp\left(-\theta\log m + \log\tau_n(m)\right) = \sum_{m=1}^{\infty}\frac{\tau_n(m)}{m^\theta}.$$

Hence the values of $\tau_n(m)$ needed for (10.25) can in principle be obtained as the coefficients in the Dirichlet series expansion of the nth power of the zeta function. They may also be obtained (using any $\sigma > 1$) via the formula (see Appendix C)

$$\tau_n(m) = \lim_{T\to\infty}\frac{1}{2T}\int_{-T}^{T}\zeta^n(\sigma + it)\,m^{\sigma+it}\,dt, \quad m = 1, 2, \ldots.$$

Although this quantity will not be required, we remark that the cumulative sum of the $\tau_n(m)$ may in principle be obtained by means of the Perron formula

$$\sum_{m\le x}{}^{*} \tau_n(m) = \frac{1}{2\pi i}\int_{c-i\infty}^{c+i\infty} \zeta^n(s)\frac{x^s}{s}\,ds\,, \quad \text{for } c>1,\ x>0$$

discussed in Appendix C. The asterisk on the sum indicates that the last term of the sum is multiplied by $\frac{1}{2}$ whenever x takes on an integer value.

An alternate approach is to note that because it is a product of n independent variates, R is essentially an n-fold convolution; inverting the characteristic function of $\log R$ then gives

$$\begin{aligned} \mathrm{P}_\theta[\,R=m\,] &= \mathrm{P}_\theta[\,\log R = \log m\,] \\ &= \lim_{T\to\infty}\frac{1}{2T}\int_{-T}^{T}\left[\frac{\zeta(\theta-it)}{\zeta(\theta)}\right]^n e^{-it\log m}\,dt \\ &= \lim_{T\to\infty}\frac{1}{2T}\int_{-T}^{T}\left[\frac{\zeta(\theta-it)}{\zeta(\theta)}\right]^n m^{-it}\,dt. \end{aligned}$$

The integrand here is an almost periodic function in t. However, neither of the last two approaches for obtaining a *p-value* for the test—namely for computing $\mathrm{P}_{\theta_0}\,[R\ge R_{\text{observed}}]$—is particularly tractable when the observed value of R is large.

To proceed asymptotically, one could use the version $\frac{1}{n}\sum_{j=1}^n \log X_j$ of the test statistic. Under H_0, this will be asymptotically normal with mean and variance given by (10.8) and (10.9). If *tail probabilities* are estimated using this normal distribution, the resulting test will be the same as the one based on Wilks's Theorem involving the asymptotic χ^2 distribution (under H_0) of the likelihood ratio statistic. A routine 'continuity correction' for this normal approximation could also be applied, since the values of $\sum_{j=1}^n \log X_j$ range over the discrete values $\log m$ where $m\ge 1$ is an integer. However, far more accurate values can usually be obtained using saddlepoint approximations. We will consider this now, but only briefly.

In general, if $Y_1, Y_2, \ldots, Y_n$ is an iid sample from a distribution whose cumulant generating function $\kappa(t)$ is finite in an interval about $t=0$, then one saddlepoint approximation for the distribution of $\bar{Y}\equiv(1/n)\sum_1^n Y_i$ is given by

$$\mathrm{P}[\,\bar{Y}>a\,] \sim 1-\Phi(r)+\varphi(r)\left(\frac{1}{u}-\frac{1}{r}\right), \tag{10.26}$$

and another is given by

$$\mathrm{P}[\,\bar{Y}>a\,] \sim 1-\Phi\left(r-\frac{1}{r}\log\frac{r}{u}\right). \tag{10.27}$$

Here Φ and φ are the cumulative distribution function and the density function of a standard normal variable,

$$r=\pm\sqrt{2n}\left(\hat{t}a-\kappa(\hat{t})\right)^{1/2}, \qquad u=\hat{t}\left(n\kappa''(\hat{t})\right)^{1/2}, \tag{10.28}$$

and $\hat{t}$ is the *saddlepoint* defined as the solution to

$$\kappa'(\hat{t})=a, \tag{10.29}$$

with the sign of r chosen to match that of $\hat{t}$. These nontrivial approximations were obtained by Lugannani and Rice (1980, [**712**]) and by Barndorff-Nielsen (1991, [**79**]) and are known to be highly accurate, even for very low values of n. (See also

Daniels (1954, [**258**])). In the present context κ_θ is as defined in (10.6), so that to approximate the tail probability

$$\mathrm{P}_\theta\left[\frac{1}{n}\sum_{j=1}^{n}\log X_j > a\right]$$

in the zeta distribution case one must first solve for the saddlepoint $\hat{t}$ using

$$-\frac{\zeta'(\theta-\hat{t})}{\zeta(\theta-\hat{t})} = a, \tag{10.30}$$

or equivalently

$$-\left.\frac{d\log\zeta(\sigma)}{d\sigma}\right|_{\sigma=\theta-\hat{t}} = \left.\sum_{n=1}^{\infty}\frac{\Lambda(n)}{n^\sigma}\right|_{\sigma=\theta-\hat{t}} = a,$$

and then compute the quantities

$$r = \pm\sqrt{2n}\big(\hat{t}a - \kappa_\theta(\hat{t})\big)^{1/2}, \quad \text{and} \quad u = \hat{t}\big(n\kappa''_\theta(\hat{t})\big)^{1/2},$$

using the definition of $\kappa_\theta(t)$ and the formula $\kappa''_\theta(t) = \sum\Lambda(n)\log n/n^{\theta-t}$. Substituting these values into (10.26) or (10.27) then leads to saddlepoint approximations for the p-value of the likelihood ratio test for the hypothesis testing problem (10.24). Greater accuracy can usually be attained by taking the discreteness of the distributions into account, but we will not pursue this here; details for how this may be done are given by Butler (2007, [**172**, §1.2.3–1.2.7, pp. 17–27]), or Kolassa (2006, [**644**, Chapter 5]).

As a final observation, note that the so-called *Kullback–Leibler divergence* (or 'relative entropy') of f_θ from f_{θ_0} in the family (10.1) is readily computed:

$$\mathrm{KL}(\theta_0,\theta) \equiv \mathrm{E}_{\theta_0}\log\frac{f_{\theta_0}(X)}{f_\theta(X)} = \log\zeta(\theta) - \log\zeta(\theta_0) - (\theta-\theta_0)\frac{\zeta'(\theta_0)}{\zeta(\theta_0)}. \tag{10.31}$$

This quantity must be positive when $\theta \neq \theta_0$. It measures the difficulty of detecting that θ is the true value for the parameter when it is thought initially that θ_0 is its true value. The so-called *Bahadur slope* for the likelihood ratio test of a hypothesis $H_0 : \theta = \theta_0$ against a true alternative value $\theta \neq \theta_0$ is, in general, given by twice the Kullback–Leibler divergence of f_θ from f_{θ_0}. This means that the 'p-value' of the likelihood ratio test converges to zero exponentially fast according to the formula

$$\lim_{n\to\infty}\left(-\frac{2}{n}\log\mathrm{P}_{H_0}\left[\frac{1}{n}\sum_{j=1}^{n}\log X_j > \mathrm{Obs}\right]\right) = 2\,\mathrm{KL}(\theta_0,\theta),$$

where 'Obs' is the (random) value of the test statistic when observed under the f_θ alternative, and P_{H_0} represents probability under f_{θ_0}. We remark that—as occurs in general—if $\log\zeta(\theta)$ in (10.31) were Taylor expanded about θ_0, the zeroth and first order terms would vanish, leaving a second order term equal to the per observation Fisher information at θ_0 (as in (10.11)), multiplied by $\frac{1}{2}(\theta-\theta_0)^2$, plus terms of higher order.

10.3. Generalized linear models

We next explore the possibility that the zeta distributions can provide the basis for a *generalized linear model* (GLM). GLMs are classes of natural and particularly tractable statistical models via which the parameters of a distribution family governing the observations are assumed to be a (given) function of a linear combination of predictor covariates associated with the observations. Estimating and testing the unknown coefficients of this linear combination is a primary goal of a GLM. For background see, for example, Jørgensen (1986, [**606**]), (1997, [**607**]), McCullagh & Nelder (1989, [**747**]), or Knight (1999, [**639**, Chapter 8]).

For it to be possible to construct a GLM from a family of densities $\{f_\theta\}$, a key requirement is that the variance $\sigma^2(\theta) \equiv \mathrm{Var}_\theta\, Y$ of an observation Y from that family must be viewed as being proportional to a function of its mean $\mu(\theta) \equiv \mathrm{E}_\theta\, Y$; we express this as

$$\sigma^2(\theta) \;\propto\; V(\mu(\theta)) \quad (\text{as } \theta \text{ varies})$$

for some suitable *variance function* $V(\mu)$. Here we are only concerned with the case where the model $\{f_\theta\}$ is a one-parameter exponential family of *univariate* distributions, with θ being a one-dimensional *natural parameter* of that family. This is the case that applies to the zeta distributions.

For a GLM, the (given) *covariates* x_j, assumed to be p-dimensional column vectors, must be connected, via a (given) *link function* $g(\cdot)$, to the means $\mu_j \equiv \mathrm{E}\, Y_j$ of the *observations*, Y_j, $j = 1, 2, \ldots, n$, by a *linear* relation as follows:

$$g(\mu_j) = x_j'\beta.$$

Here β is a column p-vector of *coefficients* which must be estimated (they are the 'GLM parameters') with $x_j'\beta$ denoting an inner product, while $j = 1, 2, \ldots, n$ ranges over the cases (that is, over the observations Y_j). To be a GLM, the predictor covariates may not enter into the model in any other way. (Note: Each observation Y_j corresponds to its own value, namely θ_j, of the distribution family's θ-parameter.) The so-called *canonical link function*, which is the one that relates the means μ_j of the observations Y_j to their corresponding θ_j parameter values by requiring that $g(\mu) = \theta$, is particularly convenient from a 'computational' standpoint (although it is neither obligatory nor necessarily of practical value). Under the canonical link we would have $\theta_j = x_j'\beta$, thus linking the covariates x_j to the parameters θ_j of the distribution family $\{f_\theta\}$ in a direct linear relationship. The one parameter zeta distributions (10.2) belong to the class of exponential families that admit a GLM model.

We have already seen—in (10.8) and (10.9)—that the natural statistic $Y \equiv \log X$ associated with the zeta distributions (10.1)/(10.4) satisfies

$$\mu = \mathrm{E}_\theta\, Y = -\,b'(\theta), \quad \text{and} \quad \sigma^2 = \mathrm{Var}_\theta\, Y = b''(\theta), \quad \text{where} \quad b(\theta) \equiv \log \zeta(\theta).$$

(Here primes represent differentiation.) The minus sign in front of $b'(\theta)$ is an artifact of our having chosen $+\log X$ rather than $-\log X$ as the 'natural statistic' in (10.2).

Using this notation, the 'variance function' for the zeta GLM is found to be

$$
\begin{aligned}
V(\mu) \propto \sigma^2 &= b''\left((-b')^{-1}(\mu)\right) \\
&= \left(\frac{d^2 \log \zeta(\theta)}{d\theta^2}\right)\Bigg|_{\theta=\theta(\mu)} \\
&= \left(\sum_{n=1}^{\infty} \frac{\Lambda(n)\log n}{n^{\theta}}\right)\Bigg|_{\theta=\theta(\mu)}
\end{aligned}
\tag{10.32}
$$

where

$$
\theta(\mu) = (-b')^{-1}(\mu) = \left(-\frac{d\log\zeta(\theta)}{d\theta}\right)^{-1}(\mu) = \left(\sum_{n=1}^{\infty}\frac{\Lambda(n)}{n^{\theta}}\right)^{-1}(\mu). \tag{10.33}
$$

For this expression $\zeta(\theta)$ must be regarded as a real-valued monotonically decreasing function of the real variable $\theta > 1$, with the -1 superscripts denoting function inverses. Substituting $f \equiv -b'$ in the formula $df^{-1}(y)/dy = 1/f'(f^{-1}(y))$ for the derivative of the function which is inverse to $y = f(x)$, leads to an alternate formula for the variance function (10.32):

$$
V(\mu) = -1\Bigg/\frac{d}{d\mu}\left(\left(-\frac{d\log\zeta(\theta)}{d\theta}\right)^{-1}(\mu)\right) \equiv -1\Bigg/\frac{d}{d\mu}\left(\left(\sum_{n=1}^{\infty}\frac{\Lambda(n)}{n^{\theta}}\right)^{-1}(\mu)\right), \tag{10.34}
$$

namely, the negative of the reciprocal of the derivative of the inverse of the negative logarithmic derivative of the zeta function. In view of all of the structure that a GLM must satisfy, only certain very special forms of variance functions $V(\mu)$ can arise in GLMs generated by exponential families. (See, for example, Jørgensen (1997, [**607**]), Letac & Mora (1986, [**690**]), and Letac (1992, [**689**]).) It is a consequence of what has just been shown that the particular form (10.32)/(10.34) belongs to the special class of permissible variance functions.

Following standard theory for GLMs—see for example, Efron (2023, [**311**]), Knight (1999, [**639**, Chapter 8]), or McCullagh & Nelder (1989, [**747**])—the likelihood equations for the (p-dimensional) parameter β associated with the (p-dimensional) covariates x_j will depend *only* on the link function $g(\mu)$, and on the variance function $V(\mu)$, and will given by the p equations in p unknowns

$$
\sum_{j=1}^{n} \frac{Y_j - \mu_j(\beta)}{g'(\mu_j(\beta))V(\mu_j(\beta))}\, x_j = 0, \tag{10.35}
$$

where $\mu_j(\beta) \equiv g^{-1}(x_j'\beta)$. (In (10.35), the x_j are vectors, the ratios preceding them are scalars, and the 0 is a vector.) Now the canonical link function is the one that satisfies $g(\mu) = \theta$. However $\mu = -b'(\theta)$, so if the canonical link is used we must have $g(\mu) = (-b')^{-1}(\mu)$. But then the $g'(\mu)$ in the denominator in (10.35) would become $d\left((-b')^{-1}(\mu)\right)/d\mu$ which, except for its sign, is precisely the reciprocal of the variance function in (10.34). This would cancel the V in the denominator in (10.35) and the likelihood equations (10.35) would simplify to

$$
\sum_{j=1}^{n} \left(Y_j - \mu_j(\beta)\right) x_j = 0, \quad \text{where} \quad \mu_j(\beta) = -\frac{d\log\zeta(\theta)}{d\theta}\Bigg|_{\theta = x_j'\beta}. \tag{10.36}
$$

Using the canonical link the likelihood function corresponding to (10.36) becomes

$$(10.37)\ L(\beta; Y_1, \ldots, Y_n) = \prod_{j=1}^{n} f_{\theta_j}(X_j) = \exp\Bigg(\Big(\sum_{j=1}^{n} Y_j x_j\Big)' \beta - \sum_{j=1}^{n} \log \zeta(-x_j'\beta)\Bigg),$$

where $Y_j \equiv \log X_j$, $j = 1, \ldots, n$. This is a p-dimensional exponential family with parameters β and (vector) sufficient statistic $\sum_{j=1}^{n} Y_j x_j \equiv \sum_{j=1}^{n} (\log X_j) x_j$. (Note: The 'raw observations' X_j from which the $Y_j \equiv \log X_j$ are obtained should not be confused with the predictors x_j.)

The covariance matrix for the MLE of β may be estimated from the known asymptotic formula for GLMs. It is given by $(X'WX)^{-1}$, where (in this formula) X is the $n \times p$ matrix whose jth row consists of the predictors x_j for the jth case, and W is an $n \times n$ diagonal matrix whose jth diagonal element is given by $(V(\mu_j))^{-1} (g'(\mu_j))^{-2}$. The μ_j may be estimated by replacing them by their MLEs, namely

$$\hat{\mu}_j = - \left. \frac{d \log \zeta(\theta)}{d\theta} \right|_{\theta = x_j'\hat{\beta}}$$

where $\hat{\beta}$ is the MLE of β. If the canonical link function is used, the estimated jth diagonal element of W would become simply $V(\hat{\mu}_j)$.

It should be stressed that the treatment in this section has been purely mathematical focusing on the novel variance function (10.34). In statistical applications the canonical link for the zeta GLM would be unsatisfactory as it requires the values of $x_j'\beta$ to stay within the range $(1, \infty)$ of θ. A more practical choice would be a link function such as $g(\mu) = \log(\mu)$ which maps the $(0, \infty)$ domain of μ onto $(-\infty, \infty)$. An alternate approach for dealing with covariates is considered by Doray & Arsenault (2002, [**292**]).

10.4. Complements

(1) In a former blog, Qiaochu Yuan had pointed out that if X is zeta distributed then: (a) The divisibility of X by distinct primes are independent events; (b) Exponents in its prime factorization are independent and geometrically distributed; (c) The *conditional* distribution of X/n, *given* that n divides X, is the same as the unconditional distribution of X; and (d) Together with monotonicity, such properties can be used to characterize the zeta distributions.

(2) Lin & Hu (2001, [**700**]) pointed out that if the coefficients $a(n)$ of a Dirichlet series $D(s) = \sum_{n=1}^{\infty} a(n)/n^s$ are nonnegative (and not all zero), and if σ is a real value inside the region of absolute convergence, then $D(\sigma + it)/D(\sigma)$ will also be a characteristic function (that is, the Fourier transform of a distribution), and will correspond to a discrete distribution supported on $\{-\log n \colon n = 1, 2, \ldots\}$. (Compare this with equation (10.7).) There are obviously *many* such Dirichlet series; to name just a few:

$$\frac{\zeta(s)}{\zeta(2s)}, \quad \frac{\zeta^2(s)}{\zeta(2s)}, \quad \frac{\zeta^3(s)}{\zeta(2s)}, \quad \frac{\zeta(s)\zeta(2s)\zeta(3s)}{\zeta(6s)}, \quad \text{and} \quad (-1)^k \zeta^{(k)}(s).$$

Lin & Hu also showed that the associated distributions are infinitely divisible if (but not only if) $a(n)$ is completely multiplicative (meaning that $a(mn) = a(m)a(n)$ for all m, n). Gut (2006, [**468**]) connected the infinite divisibility to the fact that the distributions involved are compound Poissons.

(3) The probabilities associated with (10.1) have a multiplicative property:

$$f_\theta(mn) = cf_\theta(m)f_\theta(n), \quad \text{or} \quad \mathrm{P}_\theta[X = mn] = c\mathrm{P}_\theta[X = m]\mathrm{P}_\theta[X = n], \tag{10.38}$$

for all $m, n = 1, 2, 3\ldots$, with the constant here being $c = \zeta(\theta)$. This property also holds for the probabilities associated with Dirichlet series which have nonnegative coefficients that are completely multiplicative.

(4) For $\theta > 1$, we have seen that $\zeta(\theta - it)/\zeta(\theta)$ is a characteristic function. It follows from Lukacs (1970, [**714**, Lemma 5.4.1, p. 111]) that the function $\chi(t) = \exp\big(a\,[\,\zeta(\theta - it)/\zeta(\theta)\; - 1\,]\big)$ is also a characteristic function for any $a > 0$, corresponding to a compound Poisson distribution that is infinitely divisible. (See also the earlier discussion around equation (10.17).)

(5) Gut (2006, [**468**]) showed that $\log\zeta(\theta - it)/\log\zeta(\theta)$ is also a characteristic function, and that it corresponds to a discrete distribution which assigns mass $\Lambda(n)\,/\,\big(\log\zeta(\theta)\cdot\log n\cdot n^\theta\big)$ to points $-\log n$, $n = 2, 3, \ldots$, where $\Lambda(n)$ is the von Mangoldt function. Taking the definition of $\Lambda(n)$ in account, this distribution assigns mass $(1/\log\zeta(\theta))\,\big(1/kp^{k\theta}\big)$ to points $-k\log p$ (exclusively).

(6) When $1/2 \le \theta < 1$, the function $\zeta(\theta - it)$ is unbounded so that $\varphi_\theta(t) = \zeta(\theta - it)/\zeta(\theta)$ cannot be the characteristic function of a distribution. Nakamura (2015, [**793**]), however, has shown that the modified functions

$$\frac{\zeta(\theta - it)/(\theta - it)}{\zeta(\theta)/\theta} \equiv \varphi_\theta(t)\,\frac{\theta}{\theta - it} \tag{10.39}$$

are, in fact, characteristic functions of distributions, provided only that $\theta > 0$, $\theta \neq 1$. These characteristic functions correspond to piecewise continuous and piecewise monotone (but otherwise highly unusual) density functions. The factor $\theta/(\theta - it)$ in (10.39) happens to be the characteristic function of an exponential distribution with mean $1/\theta$ (equivalently, a gamma distribution with parameters $\alpha = 1$ and $\beta = 1/\theta$) and so is infinitely divisible.[9] Therefore, when $\theta > 1$, (10.39) is infinitely divisible as well. (Remark: As in the $0 < \sigma < 1$ case of (10.39), characteristic functions may have factors that are themselves not characteristic functions.)

(7) It should not be surprising that Nakamura did not suggest that (10.39) is infinitely divisible when $0 < \theta < 1$, even though it is still a legitimate characteristic function. For if it were, then the Riemann hypothesis would follow because the characteristic function of an infinitely divisible distribution cannot vanish. (We prove this in Section 10.5.) Indeed, if functions could be found whose products with (10.39) were infinitely divisible for each $1/2 < \theta < 1$, then RH would follow. Nakamura mentions that if RH were true it may at least possible that convolutions represented by

$$\frac{\zeta(\theta - it)}{\zeta(\theta)}\left(\frac{\theta}{\theta - it}\right)^n$$

(which are in fact characteristic function for $0 < \sigma < 1$) might actually be infinitely divisible when $1/2 < \theta < 1$ for large enough n since the second factor

[9] For purposes of this discussion the density function of the gamma distribution is

$$f(x) = \Gamma^{-1}(\alpha)\beta^{-\alpha}x^{\alpha-1}e^{-x/\beta}, \quad x > 0, \quad (\alpha, \beta > 0).$$

Its characteristic function is $\phi(t) = (1 - i\beta t)^{-\alpha}$ which is obviously infinitely divisible in the parameter α.

of this product would then dominate. (Remark: A product of characteristic functions can be infinitely divisible even when its individual factors are not.)

A trivial observation: RH is true if and only if there exist functions $\psi_\theta(t)$ such that $\psi_\theta(t)\zeta(\theta+it)$ is an infinitely divisible characteristic function for every $1/2 < \theta < 1$.

(8) Recall the 'completed zeta function' $\xi(s) \equiv s(s-1)\pi^{-s/2}\Gamma(s/2)\zeta(s)$, and its functional equation $\xi(s) = \xi(1-s)$. Nakamura (2015, [**794**]) has shown that $\xi(\sigma - it)/\xi(\sigma)$ is a characteristic function for *every* value of $\sigma \in \Re$, even for $\sigma = 1/2$. He also showed that they are not infinitely divisible for *any* value of σ, and he evaluated their corresponding density functions as

$$f_\sigma(y) = \begin{cases} \frac{2}{\xi(\sigma)} \sum_{n=1}^{\infty} f(ne^{-y})e^{-\sigma y}, & y \le 0, \\ \frac{2}{\xi(\sigma)} \sum_{n=1}^{\infty} f(ne^{y})e^{(1-\sigma)y}, & y > 0, \end{cases}$$

where $f(x) = 2\pi\left(2\pi x^4 - 3x^2\right)e^{-\pi x^2}$. The case $\sigma = 1/2$ results in the density

$$f_{1/2}(y) = C \sum_{n=1}^{\infty} f(ne^{|y|})e^{|y|/2}, \; y \in \Re$$

where $C = 2/\xi(1/2) \approx -2.011584$. This function has the property that the zeros of its Fourier tansform are identical to those of ζ on the critical line.

(9) Owing to the singularity of $\zeta(s)$ at $s = 1$, the case $\theta = 1$ of $\zeta(\theta - it)/\zeta(\theta)$ is special. It can arguably be interpreted as representing either an 'improper' uniform, or an 'improper' harmonic distribution on all of the integers. Nanopoulos (1975, [**795**]) noted that there are subsets A of positive integers which have well-defined limiting probabilities under the zeta distribution when $\theta \downarrow 1$. For example, if A is the subset of integers divisible by m, then under a zeta distribution we would have

$$\mathrm{P}_\theta[A] = \sum_{k=1}^{\infty} \mathrm{P}[X = km] = \frac{1}{\zeta(\theta)} \sum_{k=1}^{\infty} \frac{1}{(km)^\theta} = \frac{1}{m^\theta}$$

and this has the well-defined limit $1/m$ as $\theta \downarrow 1$. In this way one can define (and then 'extend') a measure, say $\mu(A)$, over a class of subsets A of the positive integers. In this measure divisibility by different primes also represent independent events.

(10) Recall that the Hurwitz zeta function is a generalization of $\zeta(s)$ defined as

$$\zeta(\sigma + it, u) = \sum_{n=0}^{\infty} \frac{1}{(n+u)^{\sigma+it}}, \quad 0 < u \le 1, \quad \sigma > 1\,.$$

(See item (9) in the series of remarks in Section 5.8.) It too has an analytic continuation everywhere, except for a simple pole at $s = 1$. Hu et al. (2006, [**564**]) showed that the probability mass functions

$$\mathrm{P}_\theta[\,Y = \log(n+u)\,] = \frac{1}{\zeta(\theta, u)} \frac{1}{(n+u)^\theta}, \qquad n = 0, 1, 2, \ldots$$

have characteristic functions $\zeta(\theta - it, u)/\zeta(\theta, u)$, provided $\theta > 1$, but that these are infinitely divisible only when $u = \frac{1}{2}$ and $u = 1$. Curiously, however, the probability mass functions

$$\mathrm{P}[\,X = n+u\,] = \frac{1}{\zeta(\theta,u)}\,\frac{1}{(n+u)^{\theta}}\,, \qquad n = 0,1,2\ldots$$

are, in fact, infinitely divisible for any $0 < u \leq 1$, when $\theta > 1$. When $u \neq 1$ the infinite divisibility of either one of these two distributions seems surprising at first because this requires for it to be possible to produce all of the $Y = \log(n+u)$ values and all of the $X = n+u$ values (respectively) via convolutions of any order.

(11) Ehm (2001, [**312**]) introduced the function

$$\eta(s) \equiv \zeta(s)\,\frac{s-1}{s^2}\,, \tag{10.40}$$

and showed that it may be written as the Fourier–Laplace transform

$$\eta(\sigma+it) = \int_0^{\infty} e^{-(\sigma+it)x} g(x)\,dx\,, \quad \text{for} \quad \sigma > 0, \tag{10.41}$$

of the (real) function

$$g(x) = \sum_{1\leq n \leq e^x} (1 + \log n - x)\,, \qquad x \in \mathbb{R}.$$

The multiplier $(s-1)$ in (10.40) removes the pole of $\zeta(s)$ at $s = 1$, and the divisor s^2 ensures decay along vertical lines in the complex plane; however the zeros of ζ and η are identical. The function $g(x)$ equals 0 when $x < 0$, is strictly positive when $x \geq 0$, and is asymptotic to the linear function $x/2$ as $x \to \infty$. From (10.41) we see that, for values of $\sigma > 0$, the *tilted* forms $e^{-\sigma x} g(x)$ are proportional to densities whose characteristic functions are $\eta(\sigma - it)/\eta(\sigma)$, and that the functions

$$h_\sigma(x) = \exp(-\sigma x + \log g(x) - \log \eta(\sigma)) \tag{10.42}$$

comprise an exponential family of distributions which one may compare with the one in (10.2). The moments (although not the cumulants) for the distributions in (10.42) can be evaluated in the same *direct* way as for the zeta distributions. For the cumulants Ehm obtained formulas involving the nontrivial zeros of the zeta function, and showed that the Riemann hypothesis is equivalent to (for example) the integral condition

$$\frac{1}{\pi}\int_{-\infty}^{\infty} \log\left|\eta\left(\frac{1}{2}+it\right)\right| \frac{u-\frac{1}{2}}{\left(u-\frac{1}{2}\right)^2 + t^2}\,dt = \log \eta(u)\,, \quad \text{for all } u > \frac{1}{2},$$

and furthermore, that if this held for *any* $u > \frac{1}{2}$ then it must hold for *all* $u > \frac{1}{2}$. (The correct interpretation of such integrals requires using principal values around the zeros; see Ehm (2001, [**312**]), for details.) The cases $u = 1$ and $u = 2$ (as well as for other even integers) are particularly appealing:

$$\frac{1}{\pi}\int_{-\infty}^{\infty} \log\left|\eta\left(\frac{1}{2}+it\right)\right| \frac{1}{1+4t^2}\,dt = 0,$$

and

$$\frac{1}{\pi}\int_{-\infty}^{\infty} \log\left|\eta\left(\frac{1}{2}+it\right)\right| \frac{6}{9+4t^2}\,dt = \log\frac{\pi^2}{24}.$$

Unfortunately, since any zeros off the critical line have very high imaginary parts, numerical studies involving such criteria would be formidable. Ehm does not offer a probabilistic interpretation for the family of densities (10.42).

(12) Biane, Pitman & Yor (2001, [**101**]) review what is known about probabilistic interpretations of Riemann's function $\xi(s) \equiv \frac{1}{2}s(s-1)\pi^{-s/2}\Gamma(s/2)\zeta(s)$ in the context of random walks and Brownian motion. We will not define these terms but only mention that if

$$M = \sup_{0\le t\le 1} B(t), \quad \text{and} \quad m = \inf_{0\le t\le 1} B(t),$$

are the maximum and the minimum on $[0,1]$ of a Brownian bridge $B(t)$ (this is a Brownian motion conditioned on $B(1) = B(0) = 0$) then the random variable $Y = \sqrt{2/\pi}\,(M-n)$ has moments given by $\mathrm{E}\,Y^s = 2\xi(s)$ for all $s \in \mathbb{C}$. This random variable can be shown to satisfy the distributional equality

$$Y \stackrel{d}{=} \sqrt{\frac{1}{\pi}\sum_{n=1}^{\infty}\frac{V_n}{n^2}}$$

where the V_n are independent random variables with gamma density xe^{-x}, $x > 0$. See also Biane (2003, [**100**]).

(13) *An Erdős–Kac Theorem for the zeta distributions.* The zeta distributions satisfy an analogue of the Erdős–Kac Theorem: If $X \sim f_\theta(x)$ is distributed according to (10.1), then as $\theta \downarrow 1$, the number $\Omega_\theta(X)$ of its prime factors becomes normally distributed with mean and variance both equal to $\sum_p 1/p^\theta$, or equivalently equal to $\log 1/(\theta-1)$. A proof is given in Section 10.5.

REMARKS 10.3.

(a) In the classical Erdős–Kac Theorem, the mean and variance involve the partial sum $\sum_{p\le x} 1/p \sim \log\log x$, as $x \to \infty$, while in the present context it involves the convergent series $\sum_p 1/p^\theta$, as $\theta \downarrow 1$.

(b) As $\theta \downarrow 1$ the zeta distributions tend to an *improper harmonic distribution* on $\mathbb{N}$ whose probabilities for $X = n$ are proportional to $1/n$.

10.5. Selected proofs

PROOF FOR THE ASSERTIONS CONCERNING (10.8)–(10.10). These assertions concern the convexity and related properties of the mean, variance and other cumulants of the zeta distribution. We will show that if $D(s) = \sum_{n=1}^{\infty} a_n n^{-s}$ is any Dirichlet series with positive coefficients a_n that converges for $\Re s > \sigma_0$, then the function $D(x)$ evaluated at real numbers $x > \sigma_0$ is automatically positive, strictly decreasing, and convex. This implication applies, for example, to $\zeta(x)$ for $x > 1$.

Because it is valid to differentiate Dirichlet series term by term, and since $\frac{d}{dx}n^{-x} = (-\log n)n^{-x}$, we see that

$$D(x) = \sum_{n=1}^{\infty} a_n n^{-x}, \quad D'(x) = -\sum_{n=1}^{\infty} a_n(\log n)n^{-x},$$

and

$$D''(x) = \sum_{n=1}^{\infty} a_n(\log n)^2 n^{-x}.$$

In each series, all of the terms are positive by assumption (except for $\log 1$), and so $D(x)$ itself is positive, while $D'(x)$ is negative (meaning that $D(x)$ is strictly decreasing) and $D''(x)$ is positive (meaning that $D(x)$ is strictly convex).

The same argument works if the coefficients a_n are simply nonnegative, as long as there is some $n \geq 2$ for which $a_n > 0$ (that is, as long as $D(s)$ is not constant). In particular, the conclusion holds for the series

$$\sum_{n=1}^{\infty} \frac{\Lambda(n)(\log n)^{m-1}}{n^{\theta}}$$

for any integer $m \geq 1$ and any real number $\theta > 1$ that appears in equation (10.10) for $\operatorname{cum}_m Y$; and the Dirichlet series appearing in equations (10.8) and (10.9) for $\mathrm{E}_\theta(Y)$ and $\operatorname{Var}_\theta Y$ are just the special cases $m = 1$ and $m = 2$. □

REMARKS 10.4.

(1) The properties mentioned for $\mathrm{E}_\theta\, Y \equiv \mathrm{E}_\theta \log X$ are actually standard results for exponential families. Specifically, since (10.2) sums to 1, it follows that $\zeta(\theta)$ is the Laplace transform of a nonnegative function on the positive real axis and must therefore be strictly monotone and convex on $(1, \infty)$, as well as logarithmically convex.

(2) The inequality $\zeta''(s)\zeta(s) - (\zeta'(s))^2 \geq 0$ (from which it follows that $\operatorname{Var}_\theta Y$ is positive) can also be proved using the Cauchy–Schwarz inequality. A sharper inequality is given by

$$\zeta''(s)\zeta(s) - 2\left(\zeta'(s)\right)^2 > 0, \quad \text{on } (1, \infty);$$

it can be used to show that $1/\zeta(s)$ is strictly concave on $(1, \infty)$.

PROOF OF THE ASSERTION IN (10.14). More generally we will show that

$$\zeta(s - 2\delta)\zeta(s) - \zeta^2(s - \delta) > 0$$

when $\delta > 0$ and $s - 2\delta > 1$. To do this we first write out this quantity as

$$\left(\sum_{m=1}^{\infty} \frac{1}{m^{s-2\delta}}\right)\left(\sum_{n=1}^{\infty} \frac{1}{n^{s}}\right) - \left(\sum_{m=1}^{\infty} \frac{1}{m^{s-\delta}}\right)\left(\sum_{n=1}^{\infty} \frac{1}{n^{s-\delta}}\right).$$

The so-called AM-GM inequality can be then applied to matching pairs from each of the two products to yield

$$\left(\frac{1}{m^{s-2\delta}}\frac{1}{n^{s}} + \frac{1}{n^{s-2\delta}}\frac{1}{m^{s}}\right) - 2\left(\frac{1}{m^{s-\delta}}\frac{1}{n^{s-\delta}}\right) \geq 0. \qquad \square$$

PROOF OF DISTRIBUTIONAL EQUALITY (10.23). We need to show that if $Y = \log X$ is distributed according to the zeta distribution (10.1)/(10.4), then $Y \equiv \log X \stackrel{d}{=} V_1 + V_2 + \cdots + V_N$ where the $V_1, V_2, \ldots$ are independently distributed with $\mathrm{P}\left[V = \log p^k\right] = \left(kp^{k\theta} \log \zeta(\theta)\right)^{-1}$ for primes p and $k = 1, 2, \ldots$, and N is Poisson with mean $\log \zeta(\theta)$, and independent of the V_j.

Following Lin & Hu (2001, [**700**]) we prove a more general result. Suppose that $\Delta(s) \equiv \sum_{n=1}^{\infty} a(n)/n^s$ is a Dirichlet series whose coefficients $a(n)$ are completely multiplicative as well as nonnegative. Then in the same way as in (10.7)

$$\chi_\theta(t) \equiv \frac{\Delta(\theta + it)}{\Delta(\theta)} = \sum_{n=1}^{\infty} \frac{a(n)\, n^{-\theta}}{\Delta(\theta)}\, n^{-it} = \sum_{n=1}^{\infty} \frac{a(n)\, n^{-\theta}}{\Delta(\theta)}\, e^{-it \log n}$$

is the characteristic function of a discrete distribution that takes on the values $-\log n$ with probabilities $a(n)/\Delta(n)n^{\theta}$. Now by formula (3.54) and the multiplicativity of $a(n)$ we have

$$\log \Delta(\theta + it) = \sum_{n=2}^{\infty} \frac{a(n)\Lambda(n)/\log n}{n^{\theta+it}}$$

(where $\Lambda(n)$ is the von Mangoldt function). Therefore

$$\begin{aligned}\log \chi_\theta(t) &= \log \frac{\Delta(\theta+it)}{\Delta(\theta)} \\ &= \log \Delta(\theta) \left(\frac{\log \Delta(\theta+it)}{\log \Delta(\theta)} - 1 \right) \\ &= \log \Delta(\theta) \left(\frac{1}{\log \Delta(\theta)} \sum_{n=2}^{\infty} \frac{a(n)\Lambda(n)}{(\log n)\, n^{\theta}} e^{-it \log n} - 1 \right).\end{aligned}$$

We thus see that $\chi_\theta(t)$ has the compound Poisson form $\exp\big(\lambda(\varphi(t)-1)\big)$ as in (10.17), with $\lambda = 1/\log \Delta(\theta)$, and with $\varphi(t)$ the characteristic function of a distribution that takes values $-k \log p$ with probabilities $(1/\log \Delta(\theta))\,\big(a(n)/kp^{k\theta}\big)$; it represents the distribution of the V_j. The case $a(n) \equiv 1$ corresponds to (10.23). □

Proof that infinitely divisible characteristic functions (cf's) have no zeros. The proof uses several standard facts about cf's that can be found in many probability texts. If $\varphi(t)$ is an infinitely divisible cf then (by definition) we have that for each n, $\varphi(t) \equiv \varphi_n^n(t)$, where the $\varphi_n(t)$ are cf's. Trivially, we can therefore assert that $\varphi_n^n(t) \to \varphi(t)$ as $n \to \infty$, for every t. But $\varphi(t)$ equals 1 at $t = 0$ and is continuous around there, so for that convergence to be possible it is easy to see that we must then have $\varphi_n(t) \to 1$ for all t in an interval about 0. But the $\varphi_n(t)$ would thus converge to a function that is continuous about the origin (namely, identically equal to 1) so the function to which it converges must itself be a cf (this is a standard property of cf's). Furthermore, a cf which equals 1 in an interval about the origin must equal 1 everywhere. It follows that $\varphi_n(t) \to 1$ as $n \to \infty$ for *all* t, but since also $\varphi_n^n(t) \to \varphi(t)$, the conclusion that $\varphi(t)$ has no zeros follows. See also Lukacs (1970, [**714**, Chapter 5, Theorem 5.3.1]), or Gnedenko & Kolmogorov (1968, [**417**, Chapter 3, §17, Theorem 1]). □

Proof of the Erdős–Kac Theorem for the zeta distributions.
Our proof is based on the Lyapunov version of the Central Limit Theorem for 'triangular' arrays whose rows have infinite length (see Appendix D), and will use third central absolute moments. Now

(10.43) $$X \stackrel{d}{=} \prod_p p^{\mathrm{G}(p^{-\theta})}, \quad \text{so that } \Omega_\theta(X) \stackrel{d}{=} \sum_p \mathrm{G}(p^{-\theta}).$$

The notation $\Omega_\theta(X)$ in place of $\Omega(X)$ for the number of prime factors of X is meant to identify the distribution of X, and the $G(\lambda)$ are independent geometrically distributed random variables as in (10.15). The mean and variance of $G(\lambda)$ is $\mu \equiv \lambda/(1-\lambda)$, and $\sigma^2 \equiv \lambda/(1-\lambda)^2$, so that we have (exactly)

(10.44) $$\mathrm{E}\,\Omega_\theta(X) = \sum_p \frac{1}{p^{\theta}-1}, \quad \text{and } \operatorname{Var} \Omega_\theta(X) = \sum_p \frac{1}{p^{\theta}+p^{-\theta}-2}.$$

As $\theta \downarrow 1$, these are both asymptotic to $\sum_p 1/p^\theta$ which in turn is asymptotic to $\log 1/(\theta - 1)$ as a consequence of the identity (5.27) of Chapter 5; see also equation (E.4).

Now if $Y \sim G(\lambda)$, then since $\sum_{k=0}^\infty k^3 x^k = x(1+4x+x^2)/(1-x)^4$, we have

$$\mathrm{E}\, Y^3 \equiv (1-\lambda)\sum_{k=0}^\infty k^3 \lambda^x = \frac{\lambda(1+4\lambda+\lambda^2)}{(1-\lambda)^3},$$

and using the identity $\mathrm{E}(Y-\mu)^3 \equiv \mathrm{E}\, Y^3 - 3\mu\sigma^2 - \mu^3$ then gives

$$\begin{aligned} \mathrm{E}(Y-\mu)^3 &\equiv \sum_{k=0}^\infty \left(k - \frac{\lambda}{1-\lambda}\right)^3 (1-\lambda)\,\lambda^k \\ &= \frac{\lambda(1+4\lambda+\lambda^2)}{(1-\lambda)^3} - \left(\frac{3\lambda}{1-\lambda}\right)\frac{\lambda}{(1-\lambda)^2} - \left(\frac{\lambda}{1-\lambda}\right)^3 \\ &= \frac{\lambda(\lambda+1)}{(1-\lambda)^3}. \end{aligned} \tag{10.45}$$

Now for means $\mu \le 1$ (as will be our case) the absolute moments version of (10.45) will differ from (10.45) only in the first term. So when the mean $\mu \le 1$ we will have

$$\begin{aligned} \mathrm{E}\left|Y-\mu\right|^3 &\equiv \sum_{k=0}^\infty \left|k - \frac{\lambda}{1-\lambda}\right|^3 (1-\lambda)\,\lambda^k \\ &= \mathrm{E}(Y-\mu)^3 + 2(1-\lambda)\left(\frac{\lambda}{1-\lambda}\right)^3 \\ &= \frac{\lambda(\lambda+1)}{(1-\lambda)^3} + \frac{2\lambda^3}{(1-\lambda)^2} \\ &= \frac{\lambda(1+\lambda+2\lambda^2-2\lambda^3)}{(1-\lambda)^3}. \end{aligned}$$

Now if $Y \sim G(p^{-\theta})$, instead of $Y \sim G(\lambda)$, we would have

$$\mathrm{E}\left|Y-\mu\right|^3 = \frac{1}{p^\theta}\,\frac{1+p^{-\theta}+2p^{-2\theta}-2p^{-3\theta}}{(1-p^{-\theta})^3} \le \frac{16}{p^\theta}, \quad \text{since} \quad p \ge 2.$$

On the other hand the variance of $\Omega_\theta(X)$—shown in (10.44), let's call it s_θ^2 here—is bounded below as $s_\theta^2 > \sum_p 1/p^\theta$. Consequently, with $Y_{\theta,p} \sim G(p^{-\theta})$, the third moment Lyapunov criterion for X is

$$\frac{1}{s_\theta^3}\sum_p \mathrm{E}\left|Y_{\theta,p}\right|^3 \le \frac{\sum_p 16/p^\theta}{(\sum_p 1/p^\theta)^{3/2}} = \frac{16}{(\sum_p 1/p^\theta)^{1/2}} \to 0, \quad \text{as} \quad \theta \downarrow 1,$$

so that the required asymptotic normality follows. □

REMARKS 10.5.

(1) The Erdős–Kac result for the zeta distributions was proved independently by Cranston & Pelzer (2022, [**250**]) using a different method. Cranston & Mountford (2024, [**249**]) then showed how it may be used to prove the classical Erdős–Kac result wih the help of the following 'Tauberian theorem' due to Ikehara

and Wiener: If $\{c_n\}$ is a bounded sequence of complex numbers then

$$\lim_{\sigma \searrow 1} \frac{1}{\zeta(\sigma)} \sum_{n=1}^{\infty} \frac{c_n}{n^\sigma} = c \text{ if and only if } \lim_{N \to \infty} \sum_{n=1}^{N} c_n = c.$$

(2) Observe that the exponents in the prime factorization (10.43) of a zeta distributed random variable are fully mutually independent, even though there are an infinite number of such exponents. In contrast to this, divisibility of the natural numbers by *infinitely many* different primes is more delicate.

CHAPTER 11

Random Multiplicative Functions

Math is like ice cream, with more flavours than you can imagine.
Denise Gaskins[1]

Considerations of probability and statistics, as well as random and chaotic phenomena more generally, play important roles in many other parts of analytic number theory. In this chapter, as well as in Chapter 12, we discuss a number of topics linked to this theme, topics which give rise to questions that are not yet fully resolved. These include the random multiplicative functions known as *Radamacher chaos* and *Steinhaus chaos* which are the subject of this chapter. One key reason for interest in such questions is the hope that they may shed light on the Riemann hypothesis and on other unproven conjectures. In particular, these two 'chaos functions' were intended (at least originally) as models of the Möbius and the Liouville summatory functions, respectively. The proofs for some of the results to be discussed are lengthy and technical and have even required developing new results in probability theory. Such elaborate proofs have been omitted, but they may be found in the references that will be cited. Perret-Gentil (2019, [**840**]) provides a nice overview for some of the topics we will cover.

As they will frequently be referred to in the sequel, we recall the definitions of the Möbius function $\mu(n)$ and the Liouville function $\lambda(n)$, and their summatory versions $M(x) = \sum_{n\leq x}\mu(n)$ and $L(x) = \sum_{n\leq x}\lambda(n)$. The Möbius function $\mu(n)$ equals $(-1)^k$ if n is a product of k *distinct* primes, and otherwise is 0. The Liouville function $\lambda(n)$ is defined as the *parity* of the *total* number of prime factors of n, that is, $\lambda(n) = (-1)^k$, where k is the number of prime factors of n counted with multiplicity, or equivalently $\lambda(n) = (-1)^{\Omega(n)}$. A recurring theme which underlies the study of randomized versions of arithmetic functions concerns the degree to which their behaviour mimics that of *independent* random variables. But the bigger picture centres on the light which the behaviour of these functions can shed on the Riemann, Lindelöf, and other unproven hypotheses.

11.1. Random multiplicative functions

The asymptotic properties of arithmetic functions such as the Möbius function $\mu(n)$, and partial sums such as the Mertens function $M(x) = \sum_{n\leq x}\mu(n)$, have proven exceedingly difficult to discern. In an attempt to understand the behaviour of $\mu(n)$ and $M(x)$, as well as to help formulate conjectures about them, Aurel Friedrich Wintner (1903–1964) introduced a probabilistic model which, like $\mu(n)$,

[1]From her 'Let's Play Math' website: `https://denisegaskins.com/`. Quoted with permission.

has a multiplicative structure, but otherwise includes a significant component of genuine randomness.

Let $\{\epsilon_p\}$ be a sequence of independent ± 1 random variables which are indexed by the primes p, and which are distributed symmetrically as

$$\mathrm{P}\left[\epsilon_p = 1\right] = \mathrm{P}\left[\epsilon_p = -1\right] = \frac{1}{2}.$$

These are sometimes referred to as *Rademacher random variables*, and sometimes as *Bernoulli random variables*. Wintner (1944, [**1115**]) defined the random function

$$f(n) \equiv \begin{cases} \prod_{p|n} \epsilon_p, & \text{if } n \text{ is squarefree,} \\ 0, & \text{otherwise.} \end{cases} \tag{11.1}$$

Like $\mu(n)$, the function $f(n)$ is multiplicative, but unlike $\mu(n)$ it possesses probabilistic as well as number-theoretic attributes. It is commonly referred to as the *Rademacher random multiplicative function*—or as *Rademacher chaos* for short—and serves as a proxy for the Möbius function $\mu(n)$, while its summatory version

$$F(x) \equiv \sum_{n \le x} f(n)$$

serves as a proxy for the Mertens function $M(x) = \sum_{n \le x} \mu(n)$.

The historical context for Wintner's 1944 paper traces back to the recognition by Denjoy (1931, [**274**])—as well as by Cramér, Jessen, Paley, Wiener, and others at the time—that if the Dirichlet series $\sum \mu(n)/n^s$ for the Möbius function could be viewed as a 'typical' member of a random process of the form $\sum \epsilon(n)/n^s$ where the $\epsilon(n)$ were *independent*, then the series $\sum \mu(n)/n^s$ would converge almost surely for all $s > \frac{1}{2}$, and as a consequence the Riemann hypothesis would be true *with probability one*.[2] However, Paul Lévy (1931, [**696**])—and likely Hardy and Littlewood, among others—pointed out that this observation is meaningless from an arithmetical point of view if for no other reason than the fact that the probability that the independently distributed coefficients of $\sum \epsilon(n)/n^s$ would be *multiplicative* is zero. Thus $\sum \mu(n)/n^s$ ought, at the very least, to be thought of as having been selected from that zero-probability subset. Wintner's paper should be viewed as an attempt to address that one particular shortcoming of the independence-based model. Further reasons for studying random multiplicative functions will emerge as we progress.

It is worthwhile to take a moment to focus on the similarities and differences between $\mu(n)$ and Wintner's $f(n)$ as defined in (11.1). Both are supported (that is, are nonzero) only on square-free integers, and since they are each multiplicative, both are fully determined by their values on the primes. In the case of $\mu(n)$, its values on primes are deterministic, and are all equal to -1. For $f(n)$, its values on primes randomly take on $+1$ and -1 values, with these being independent across the primes. Injecting this randomness into the structure shifts some portion of the number-theoretic difficulties of $\mu(n)$ into the domain of probability, thus allowing its analysis to become somewhat more tractable. In part, this happens because the 'parity' of the Möbius function (namely, its dependence on whether the number of prime divisors is even or odd) gets broken. On the other hand, like $\mu(n)$, the

[2]Exercise: Using Kolmogorov's Three-Series Criterion or otherwise, show that if the $\epsilon(n)$ are independent with $\mathrm{P}[\epsilon(n) = -1] = \mathrm{P}[\epsilon(n) = +1] = \frac{1}{2}$, then $\sum_1^\infty \epsilon(n)/n^s$ converges a.s. to a random variable when $\Re s > \frac{1}{2}$.

function $f(n)$ still maintains a nontrivial dependence structure across the values of n (due essentially to the multiplicativity), and both functions are subject (although in different ways) to the number-theoretic nature of prime factorization across the sequence of integers.

A second random function, related in spirit to Rademacher chaos, can be defined as[3]

$$g(n) \equiv \prod_{p^k \| n} \epsilon_p^k = \prod_{p^k \| n} g^k(p),$$

where the random variables $\{\epsilon_p\}$ are again independent and indexed by the primes, but now are distributed uniformly over the complex unit circle $|z| = 1$. Such ϵ_p can be represented as $\epsilon_p \equiv g(p) \equiv \exp\{2\pi i\, \Theta_p\}$, where the Θ_p are independent and uniformly distributed on $[0, 1)$, so that

$$g(n) \equiv \exp\biggl(2\pi i \sum_{p^k \| n} k\, \Theta_p\biggr).$$

The function $g(n)$ is known as the *Steinhaus random multiplicative function*, or as *Steinhaus chaos* for short, while the ϵ_p are sometimes referred to as Steinhaus random variables. Steinhaus chaos is sometimes thought of as a proxy for the Liouville function $\lambda(n)$. We will denote its summatory version as $G(x) \equiv \sum_{n \le x} g(n)$.

Unlike Rademacher chaos which is supported on the square-free integers only, Steinhaus chaos is supported on all of the positive integers, and furthermore is a *completely* multiplicative function. An essential feature of these two *random* functions is that they are fully determined by their values on the primes. Furthermore, their moduli, $|f(n)|$ and $|g(n)|$, are both equal to 1 for all values of n over their supports, although the Steinhaus variables are complex-valued while the Rademacher ones are real. A significant technical difference is that the Steinhaus variables possesses a kind of 'translational invariance', in that for any fixed real t the probability distributions of the sequences $\{g(n)n^{it}\}$ and $\{g(n)\}$ are identical (which somewhat simplifies certain proofs) while no such result holds in the Rademacher case. Another important difference between the Rademacher and Steinhaus models (which mirrors the deterministic $\mu(n)$ and $\lambda(n)$ cases) is reflected in the Euler product factorizations of their associated Dirichlet series, namely

$$\sum_{n=1}^{\infty} \frac{f(n)}{n^s} = \prod_p \left(1 + \frac{f(p)}{p^s}\right), \quad \text{and} \quad \sum_{n=1}^{\infty} \frac{g(n)}{n^s} = \prod_p \left(1 - \frac{f(p)}{p^s}\right)^{-1},$$

respectively. (These products are valid for suitable values of s, and hold because $f(n)$ and $g(n)$ are each multiplicative.) The first product shown produces only 'square-free terms', while the second product expands out to include all 'integer terms'. However, in both cases, although the *coefficients* of their Dirichlet series have highly entangled dependences, the *factors* of their Euler products are statistically independent, a fact that is extensively exploited in proofs involving these multiplicative chaos models.

Of course, such models have numerous variants. For example, in the Steinhaus case, the ϵ_p can be taken to be Bernoulli variables. And in the Rademacher case the distribution of the ϵ_p can likewise be modified. One particularly natural variant for the ϵ_p involves random selection from the symmetrically located roots of

[3]Recall that $p^k \| n$ means that p^k divides n but p^{k+1} does not.

unity $e^{2\pi i q/r}$, $q = 0, 1, \ldots, r-1$, for some value of r. (See, for instance, Najnudel (2020, [**792**]).) The case $r = 2$ results in Rademacher variables (though supported on all integers rather than only squarefree ones), while the limit $r \to \infty$ tends to the Steinhaus case. However, in spite of their apparent similarities, the properties of the Rademacher and Steinhaus chaos models, as well as their proofs, while often similar, are not identical, in part because of the number-theoretic technicality that the first is only supported on integers that are squarefree.

When considering such 'chaos models', an obvious assortment of questions arise:

- Is there an (almost sure) limit value
$$\mathrm{E}\, f \equiv \lim_{x\to\infty} \frac{1}{x} \sum_{n\le x} f(n), \quad \text{or perhaps} \quad \lim_{x\to\infty} \frac{1}{\log x} \sum_{n\le x} \frac{f(n)}{n} \; ?$$
Furthermore, how large are $\sum_{n\le x} f^k(n)$, and $\mathrm{E}\big|\sum_{n\le x} f(n)\big|^r$?
- Do the empirical distributions of $\{f(n) \colon 1 \le n \le N\}$ converge weakly to a limiting distribution as $N \to \infty$?
- Is there a Central Limit result analogous to
$$\frac{\sum_{n\le x} f(n)}{\sqrt{\mathrm{E}\big|\sum_{n\le x} f(n)\big|^2}} \xrightarrow{d} N(0,1)\,?$$
Failing this, what is the correct normalization and limiting distribution?
- Is there an almost sure Law of Iterated Logarithm result analogous to
$$\limsup_{x\to\infty} \frac{|\sum_{n\le x} f(n)|}{\sqrt{2x \log\log x}} = 1 \;\; \text{a.s.?}$$
Failing this, what is the correct normalization and limiting constant?
- How rapidly do tail probabilities such as $\mathrm{P}\Big[\, \big|\sum_{n<x} f(n)\big| \ge \lambda\sqrt{x} \,\Big]$ decay as $\lambda \to \infty$?
- What can one say about the serial correlation structure of these series?

The interest in such questions stems not only from their number-theoretic origins but also from a desire to understand the impact of the subtly entangled dependences on the resulting probabilistic behaviour. Current understanding of the statistical properties of the Rademacher and Steinhaus models (as well as of randomness-based number-theoretic functions more generally) is not yet complete and has been an object of active research in recent years. See, for example, Harper, Nikeghbali & Radziwiłł (2015, [**514**]), and Harper (2020, [**511**]), (2024, [**513**]).

11.2. Properties of Rademacher and Steinhaus chaos, I

To distinguish between them when discussing their properties, we will use $f(n)$ and $g(n)$ to denote the Rademacher and Steinhaus chaoses, respectively, and will use $F(x) \equiv \sum_{n\le x} f(n)$ and $G(x) \equiv \sum_{n\le x} g(n)$ for their summatory versions. However, when stating a property that holds for both models, we will typically use just $f(n)$ and $F(x)$, but will make this clear on each such occasion.

Turning first to the Rademacher case, one finds immediately that the $f(n)$ have means $\mathrm{E}\, f(n) = 0$, and variances (for squarefree n) given by $\mathrm{E}\, f^2(n) = 1$.

Furthermore, since

$$f(m)f(n) = f\left(\frac{m}{(m,n)}\right) f\left(\frac{n}{(m,n)}\right)$$

when m and n are squarefree (or otherwise), we see that the $f(n)$ are uncorrelated:

$$\mathrm{E}\, f(m)f(n) = 0 \quad \text{for } m \neq n.$$

(Here (m,n) is the greatest common multiple.) In the Steinhaus case, noting the property

$$g(m)\overline{g(n)} = g\left(\frac{m}{(m,n)}\right) \overline{g\left(\frac{n}{(m,n)}\right)}$$

for all m and n (or again otherwise), we may see that

$$\mathrm{E}\, g(n) = 0, \quad \mathrm{E}\, |g(n)|^2 \equiv \mathrm{E}\, 1 \equiv 1, \quad \text{and} \quad \mathrm{E}\, g(m)\overline{g(n)} = 0 \quad \text{for} \quad m \neq n.$$

(Exercise: Demonstrate these assertions.) It follows from these results that

$$\mathrm{E}\, F^2(x) \equiv \mathrm{E}\left(\sum_{n\le x} f(n)\right)^2 = \frac{6x}{\pi^2} + O(\sqrt{x}\,),$$

and

$$\mathrm{E}\, |G(x)|^2 \equiv \mathrm{E}\left|\sum_{n\le x} g(n)\right|^2 = \lfloor x \rfloor = x + O(1).$$

The $6/\pi^2$ stems from the asymptotic proportion of squarefree numbers, and $\lfloor x \rfloor$ refers to the greatest integer $\le x$. (For the $O\left(\sqrt{x}\,\right)$ error term here see equation (3.23).)

The number-theoretic dependencies which remain among its terms results in asymptotic behaviour of $F(x) = \sum_{n\le x} f(n)$ which—although simpler than that of $M(x) = \sum_{n\le x} \mu(n)$, and now based as much on statistical as on number-theoretic considerations—is nevertheless still nontrivial. What does remain of interest, however, is to compare the behaviour of $F(x)$ with the version one would obtain if all of the non-zero $f(n)$ were simply independent ± 1 Bernoulli variables. Similar remarks apply to the function $G(x) = \sum_{n\le x} g(n)$ and its Liouville analogue, $L(x) = \sum_{n\le x} \lambda(n)$.

Using Kolmogorov's Three-Series Criterion (see Appendix D), one can establish that the Dirichlet series for Rademacher chaos, as well as its Euler product representation

$$\sum_{n=1}^{\infty} \frac{f(n)}{n^s} = \prod_p \left(1 + \frac{f(p)}{p^s}\right), \tag{11.2}$$

both converge almost surely whenever $\Re s > \frac{1}{2}$; a proof is given in Section 11.6. Were this result to hold for $\sum \mu(n)/n^s$ it would be equivalent to RH. In his elegant paper, Wintner (1944, [**1115**]) exploited the fact that the factors on the right in (11.2) are fully independent to show that for all $\epsilon > 0$ the Rademacher summatory function satisfies

$$F(x) = O(x^{1/2+\epsilon}), \quad \text{as well as } F(x) \neq O(x^{1/2-\epsilon}), \text{ almost surely.} \tag{11.3}$$

These assertions are proved in Section 11.6. For the actual Mertens function, $M(x)$, the first of the order assertions in (11.3) would again be equivalent to the Riemann hypothesis, while the second assertion there is known to hold because $M(x) >$

$x^{1/2-\epsilon}$ infinitely often. In this connection, recall from Section 3.3 the conjecture of Gonek that

$$\liminf_{x\to\infty} \frac{M(x)}{\sqrt{x}\,(\log\log\log x)^{5/4}} = -C, \quad \text{and} \quad \limsup_{x\to\infty} \frac{M(x)}{\sqrt{x}\,(\log\log\log x)^{5/4}} = +C,$$

for some positive constant C. (Unconditionally, little more than $M(x) = o(x)$ is known.) Clearly, it would be of interest to have sharp almost-sure bounds for $\sum_{n\le x} f(n)$ in the Rademacher case to help gauge Gonek's as well as other conjectures about $M(x)$.

Wintner's result (11.3) has subsequently been improved. Halász (1983, [**478**]) showed that there exist constants $0 < C_1, C_2 < \infty$ such that

$$|F(x)| = O\left(\sqrt{x}\,\exp\left(C_1\sqrt{\log\log x\,\log\log\log x}\right)\right), \tag{11.4}$$

and

$$|F(x)| \neq O\left(\sqrt{x}\,\exp\left(-C_2\sqrt{\log\log x\,\log\log\log x}\right)\right), \tag{11.5}$$

almost surely, as $x \to \infty$. Halász also showed that the upper bound (11.4) actually holds for *every* random multiplicative function—and therefore also for Steinhaus chaos—and with a common value for the constant C_1. Furthermore, when intended to apply to *all* multiplicative functions, the order term in (11.4) cannot be improved since it is actually attained for some multiplicative functions. (Remark: The upper and lower bounds (11.4) and (11.5) were obtained by different methods; the similarity of their shapes is no more than a pleasant coincidence.) We mention that independently of Halász, Erdős had already established the weaker upper and lower bounds $O(\sqrt{x}\log^A x)$ and $O(\sqrt{x}\log^{-B} x)$; see Erdős (1985, [**352**]).

In further sharpening of (11.4)/(11.5) for Rademacher chaos, Lau, Tenenbaum & Wu (2013, [**679**]) proved that for every $\epsilon > 0$ one has (with $x \to \infty$)

$$|F(x)| = O\left(\sqrt{x}\,(\log\log x)^{2+\epsilon}\right), \quad \text{almost surely,} \tag{11.6}$$

while Harper (2013, [**504**]) proved that as $x \to \infty$

$$|F(x)| \neq O\left(\sqrt{x}\Big/(\log\log x)^{5/2+\epsilon}\right), \quad \text{almost surely.} \tag{11.7}$$

It is interesting to note that Halász's lower bound result (11.5) depended in an essential way on obtaining good lower bounds for the tail probabilities of the supremum (over carefully restricted ranges for t and x) of the stochastic process

$$\Re\sum_{p\le x} \frac{\epsilon_p}{p^{1/2+1/\log x+it}} \equiv \sum_{p\le x} \frac{\epsilon_p \cos(t\log p)}{p^{1/2+1/\log x}}, \quad t\in\mathbb{R}, \tag{11.8}$$

where the ϵ_p are independent Rademacher random variables. (The purpose of the $1/\log x$ exponents was to assure that the sum of the variances of the terms in the sum—that is, the sum of the quantities $\cos^2(t\log p)/p^{1+2/\log x}$—would stay suitably bounded.) The result (11.7) likewise required developing lower bounds—still sharper than those of Halász—for the tail probabilities of the suprema of normalized versions of the process (11.8). (In his paper Harper also considered processes such as (11.8) in which the ϵ_p are independent standard normal variables.)

The bounds in (11.6) and (11.7) begin to resemble the scale of the Law of the Iterated Logarithm for independent random variables, namely $\sqrt{x\log\log x}$. In the

Rademacher case, for instance, were the $f(n)$ actually independent with ± 1 values each with probability $\frac{1}{2}$, the LIL would give

$$\limsup_{x\to\infty} \frac{|F(x)|}{\sqrt{2x\log\log x}} = \sqrt{\frac{6}{\pi^2}} \tag{11.9}$$

(the value $6/\pi^2$ being the asymptotic proportion of squarefree numbers). But (11.6) and (11.7) leave open the question of whether the number-theoretic dependences among the $f(n)$ might allow a bound *better* than that of the LIL to hold, perhaps even a bound like $o(\sqrt{x}\,)$, in spite of Erdős's (1985, [**352**]) conjecture that the larger bound $O(\sqrt{x}\,)$ would not hold.

In (2013, [**504**]) Harper argued that it seems unlikely that an almost sure result like $F(x) = O(\sqrt{x}\,)$ would hold, suggesting that $F(x)$ could have rare fluctuations not only of size $\sqrt{x\log\log x}$ (as per the LIL), and perhaps even larger, presumably corresponding to heavy tails in the distribution of $F(x)$. Harper (2023, [**512**]) then did finally resolve Erdős's conjecture in the affirmative by proving, for both Rademacher and Steinhaus chaos, that given *any* function $V(x)$ tending to infinity with x (however slowly) there almost surely exist arbitrarily large values of x for which

$$\left|\sum_{n\le x} f(n)\right| \ge \frac{\sqrt{x}\,(\log\log x)^{1/4}}{V(x)}. \tag{11.10}$$

Harper also suggested that the bound $O(\sqrt{x}\,(\log\log x)^{1/4+\epsilon})$ may plausibly be sharp for both chaos models, but cautioned that it may be a formidable task to prove it. Later—improving on (11.6)—Caich (2024, [**178**]) established the nontrivial *upper* bound

$$\left|\sum_{n\le x} f(n)\right| = O\left(\sqrt{x}(\log\log x)^{3/4+\epsilon}\right)$$

also applicable to both chaos models. Of course, these various results to not disprove Gonek's conjecture, but neither are they supportive of it.

The proof of a result such as (11.10) begins with ideas from the proof of the lower bound for the LIL, but then evolves into a *tour de force* to deal effectively with the complicated number-theoretic dependences that exist among the $f(n)$. Further details may be found in the various papers of Harper (including (2023, [**513**])), as well as in Wintner (1944, [**1115**]), Halász (1983, [**478**]), Hough (2011, [**561**]), Chatterjee & Soundararajan (2012, [**191**]), Lau, Tenenbaum & Wu (2013, [**679**]), and also the discussion in Erdős (1985, [**352**]), and in other references cited in these papers. In particular, Harper (2013, [**504**]) also provides a detailed overview of the various known methods for establishing probability bounds on the suprema of Gaussian processes. As work on these models continues, further results are to be expected.

11.3. Interlude: The symmetrical iid cases

Section 11.4 will deal with *moment properties* of the summatory versions of the Rademacher and Steinhaus chaos models. To provide a basis for comparisons, we take time out for a look at how symmetrical sums behave when based on identically and independently distributed variables.

Suppose then that the $\epsilon(n)$, $n = 1, 2, \ldots$, are independent and identically distributed symmetrical ± 1 Bernoulli variables:

$$\mathrm{P}\left[\epsilon(n) = -1\right] = \mathrm{P}\left[\epsilon(n) = +1\right] = \frac{1}{2}.$$

The $\epsilon(n)$ have mean 0 and variance 1. We will consider the asymptotic behaviour of

$$\mathrm{E}\left|\sum_{n \le N} \epsilon(n)\right|^q$$

for all $q > 0$. The following well-known result for $N(0, \sigma^2)$ Gaussian random variables will be useful:

$$\mathrm{E}\left|N(0, \sigma^2)\right|^q = \frac{\sigma^q\, 2^{q/2}}{\sqrt{\pi}} \Gamma\left(\frac{q+1}{2}\right), \quad q > -1.$$

Now, since the $\epsilon(n)$ are bounded, not only will their normalized partial sums, namely, $\sum_{n \le N} \epsilon(n)/\sqrt{N}$ converge in distribution to a standard normal, but all of their moments will converge to those of the standard normal as well (see Feller (1971, [**373**, vol. II, Chapter VIII, §1])). An identical argument applies to all of the absolute moments so that

$$\mathrm{E}\left|\sum_{n \le N} \epsilon(n)\right|^q \sim \mathrm{E}\left|N(0, N)\right|^q \sim \frac{(2N)^{q/2}}{\sqrt{\pi}} \Gamma\left(\frac{q+1}{2}\right), \quad \text{as } N \to \infty. \tag{11.11}$$

The complex-valued case, based on independent variables $\epsilon(n)$ which are uniformly distributed on the unit circle $|z| = 1$, is a bit more involved. Such $\epsilon(n)$ can be represented as $\epsilon(n) = \exp(2\pi i \theta(n))$ where the $\theta(n)$ are iid uniforms on $[0, 1)$, so that

$$\mathrm{E}\left|\sum_{n \le N} \epsilon(n)\right|^q = \mathrm{E}\left|\sum_{n \le N} \cos(2\pi\theta(n)) + i \sum_{n \le N} \sin(2\pi\theta(n))\right|^q. \tag{11.12}$$

One may think of the $\epsilon(n)$ as isotropically oriented unit vectors (that is, unit vectors with uniformly distributed orientations). Hence the quantity in (11.12) is the same as $\mathrm{E}\,|R|^q$ where R is the length of the resultant vector determined according to

$$R^2 = C^2 + S^2, \quad \text{where } C = \sum_{j=1}^{n} \cos(2\pi\theta(j)), \quad S = \sum_{j=1}^{n} \sin(2\pi\theta(j)). \tag{11.13}$$

To describe the distributions of C, S, and R, we need the Bessel functions of the first kind of orders 0 and 1. In general, the Bessel function $J_p(\cdot)$ of the first kind of order p is

$$J_p(x) = \sum_{k=0}^{\infty} \frac{(-1)^k}{\Gamma(p+k+1)\Gamma(k+1)} \left(\frac{x}{2}\right)^{2k+p},$$

so those of orders 0 and 1 are

$$J_0(x) = \sum_{k=0}^{\infty} \frac{(-1)^k}{(k!)^2} \left(\frac{x}{2}\right)^{2k}, \quad \text{and} \quad J_1(x) = \sum_{k=0}^{\infty} \frac{(-1)^k}{(k!(k+1)!} \left(\frac{x}{2}\right)^{2k+1}.$$

In terms of this notation, it is possible to evaluate the density function $h_n(r)$ and the distribution function $H_n(r)$ of R as

$$h_n(r) = r \int_0^\infty u J_0(ru) J_0^n(u)\, du, \quad \text{and } H_n(r) = r \int_0^\infty J_1(ru) J_0^n(u)\, du. \tag{11.14}$$

These functions rapidly become complicated with n although the case $n = 2$ is simple:[4]

$$h_2(r) = \frac{2}{\pi\sqrt{4-r^2}}, \quad 0 < r < 2.$$

A saddlepoint approximation for these distributions is however available. For background, see Mardia & Jupp (2000, [**730**, p. 67]).

Now the joint characteristic function for an individual unit vector, namely, $\big(\cos 2\pi\theta(j), \sin 2\pi\theta(j)\big)$, can be evaluated to be $J_0(\rho)$, and therefore the n-fold sum (C, S) will have characteristic function $J_0^n(\rho)$.[5] Using this, the joint density function of (C, S) can be evaluated as

$$f(c,s) = \frac{1}{2\pi}\frac{h_n(c^2+s^2)}{c^2+s^2},$$

with $h_n(\cdot)$ as in (11.14). The marginal densities of C and S then each turn out to be

$$f(c) = \frac{1}{\pi}\int_0^\infty \cos(c\rho)J_0^n(\rho)\,d\rho, \quad -n < c < n.$$

As $n \to \infty$, each of

$$\sqrt{\frac{2}{n}}C \text{ and } \sqrt{\frac{2}{n}}S$$

will tend in distribution to independent standard normal variables.[6] Consequently R^2 will tend to a chi-squared distribution according to

$$\frac{2}{n}R^2 \xrightarrow{d} \chi_2^2. \tag{11.15}$$

The χ_2^2 distribution has density $f(x) = \frac{1}{2}e^{-x/2}$, characteristic function $\varphi(t) = (1-2it)^{-1}$, and rth moments $2^r\Gamma(r+1)/\sqrt{\pi}$, for all $r > 0$. Here again, because the $\big(\cos 2\pi\theta(j), \sin 2\pi\theta(j)\big)$ are bounded, we have convergence of the moments, and it follows that for every $q > 0$

$$\mathrm{E}\Big|\sum_{n\le N}\epsilon(n)\Big|^q = \mathrm{E}\Big(\big(|R|^2\big)^{q/2}\Big) \sim N^{q/2}\,\Gamma\Big(\frac{q}{2}+1\Big)/\sqrt{\pi}, \quad \text{as } n \to \infty. \tag{11.16}$$

The results (11.11) and (11.16) may now be compared: The second is smaller than the first by a factor of $2^{q/2}$. This difference can be explained by the fact that the rotational (and indeed two-dimensional) nature of the complex-valued case allows more 'room' for cancellation to occur than the one-dimensional real-valued case. In fact, this phenomenon scales with dimension; in dimension d, the result (11.15) would become $(d/n)R^2 \xrightarrow{d} \chi_2^2$.

With these computations for comparisons behind us we now return to our main subject.

---

[4]Exercise: Derive the density $h_2(r)$ for the $n = 2$ case directly. (The derivation is very simple.)

[5]This characteristic function is expressed in polar coordinates. As the distribution is isotropic, only a radial coordinate ρ appears. These details will not concern us. (See Mardia & Jupp (2000, [**730**, Chapter 4)])).

[6]Exercise: Show that $\cos(2\pi\theta(j))$ and $\sin(2\pi\theta(j))$ are uncorrelated.

11.4. Properties of Rademacher and Steinhaus chaos, II

Turning now to *moments* of the chaos models, we seek to compare the asymptotics for expectations such as $\mathrm{E}\left|\sum_{n\le x} f(n)\right|^{2q}$ with what the asymptotics would be if the $f(n)$ were independent and identically distributed. And in particular, how big is $\mathrm{E}\left|\sum_{n\le x} f(n)\right|$? Due to the *orthogonality* of the Rademacher and Steinhaus processes, the second moment provides a natural point of contact with the independence-based model. In all of these cases $\mathrm{E}\left|\sum_{n\le x} f(n)\right|^{2}$ will have the precise order $O(x)$. (See (11.21).) In the independence based case (as for example if the terms were Gaussian) the size of the $2q$th moment would then be precisely $O(x^q)$. But this is not what happens—either below the second moment, or above it—under the dependence structure of the chaos models. Naively, say for the Steinhaus case, we have[7]

$$\mathrm{E}\Big|\sum_{n\le x} f(n)\Big| \le \Big(\mathrm{E}\Big|\sum_{n\le x} f(n)\Big|^2\Big)^{1/2} = \lfloor x\rfloor^{1/2} = O(\sqrt{x}),$$

but this estimate is crude and far from the truth. The proofs for most of the results in this section are long and involved. Some key references include Harper (2019, [**508**]), (2020, [**511**]), (2024, [**513**]), and Harper, Nikeghbali & Radziwiłł (2015, [**514**]).

REMARK 11.1. One should distinguish between results for $\mathrm{E}\left|\sum_{n\le x} f(n)\right|$ and results for $\sum_{n\le x} f(n)$; thus, for example, the former is an expected value, while the latter is subject to laws of iterated logarithm type.

A particularly curious fact in the Steinhaus case is that for all real $q>0$, and all $x>0$,

$$\mathrm{E}\Big|\sum_{n\le x} g(n)\Big|^q = \lim_{T\to\infty}\frac{1}{T}\int_0^T \Big|\sum_{n\le x} n^{-it}\Big|^q\,dt. \tag{11.17}$$

This "norm identity" evidences an unexpected link between Steinhaus chaos and partial sums associated with the zeta function. It also allows one to translate questions about moments associated with the Steinhaus function into questions about the limiting behaviour of the indicated zeta sums, and conversely. Result (11.17) is sometimes referred to as the *Bohr identification* or the *Bohr correspondence*. A proof of (11.17) is given in Section 11.6 (Selected proofs). In passing, let us mention also the related and equally curious equality

$$\mathrm{E}\Big|\sum_{n\le x} \frac{g(n)}{\sqrt{n}}\Big|^{2q} = \lim_{T\to\infty}\frac{1}{T}\int_0^T \Big|\sum_{n\le x} \frac{1}{n^{1/2+it}}\Big|^{2q}\,dt. \tag{11.18}$$

These so-called *pseudomoments* of the zeta function were studied by Conrey & Gamburd (2006, [**223**]), Bondarenko, Heap & Seip (2015, [**126**]), Bondarenko, Brevig, Sakesman, Seip & Zhao (2018, [**125**]), Heap & Lindqvist (2016, [**522**]), Brevig & Heap (2019, [**151**]), and Gerspach & Lamzouri (2022, [**413**])).

There is an interesting side story associated with (11.17) and with the lower order moments of the multiplicative chaos models. In connection with a problem

[7]We are using the well-known fact that $(\mathrm{E}\,|X|^r)^{1/r}$ is a nondecreasing function of r for $r>0$.

in harmonic analysis, Helson (2006, [**538**]), (2010, [**539**]) conjectured that for $q = 1$ there, the right-hand side of (11.17) is of order $o(\sqrt{x})$. In terms of Steinhaus variables this conjecture translates into the assertion

$$\mathrm{E}\left|\sum_{n\le x} g(n)\right| = o\left(\sqrt{x}\right), \quad \text{as } x \to \infty. \tag{11.19}$$

(The nature of the harmonic analysis connection is explained in Harper, Nikeghbali & Radziwiłł(2015, [**514**](, and Saksman & Seip (2016, [**933**]), but will not concern us here.) Now a result such as (11.19) is surprising—from both a number-theoretic as well as a statistical viewpoint—since better than square root cancellation occurs only rarely. Notwithstanding his conjecture, Helson did establish that $\mathrm{E}\left|\sum_{n\le x} g(n)\right| \ne O\left(\sqrt{x}/(\log x)^{1/4}\right)$, while Bondarenko & Seip (2016, [**127**]) strengthened this to

$$\mathrm{E}\left|\sum_{n\le x} g(n)\right| \ne O\Big(\sqrt{x}/(\log x)^{0.05616}\Big).$$

Bondarenko and Seip also proved that

$$\left(\mathrm{E}\left|\sum_{n\le x} g(n)\right|^q\right)^{1/q} \ne O\Big(\sqrt{x}/(\log x)^{0.07672}\Big)$$

for all $q > 0$, and they obtained somewhat better estimates for the cases $0 < q < 2$ which are the ones more closely associated with the Riemann hypothesis. Harper, Nikeghbali & Radziwiłł (2015, [**514**]) then established the still sharper bound

$$\mathrm{E}\left|\sum_{n\le x} f(n)\right| \ne O\left(\sqrt{x}/(\log\log x)^{3-\epsilon}\right), \quad x \to \infty,$$

for all $\epsilon > 0$, and also proved that for any $0 \le q \le 1$,

$$\mathrm{E}\left|\sum_{n\le x} f(n)\right|^{2q} \ne O\left(x^q/(\log\log x)^{6-\epsilon}\right), \quad x \to \infty,$$

with these two results holding in both the Steinhaus as well as the Rademacher cases. The 'trajectory' of these various results led to the belief that Helson's conjecture was unlikely to be true and that the exact order of (11.19) was actually $O(\sqrt{x})$. Additionally, a result of Chatterjee and Soundararajan (2012, [**191**]) proving that certain sums of the form $\sum_{x<n\le x+y} f(n)$ in the Rademacher case have a Gaussian limit showed that the analogue of (11.19) fails over "short intervals" thus also pointing against Helson's conjecture. (See also Soundararajan & Xu (2023, [**1009**]).)

On the other hand, so far as better than square-root cancellation goes, the related result

$$\lim_{t\to\infty} \frac{1}{T}\int_0^T \left|\sum_{n\le x} \frac{1}{n^{1/2+it}}\right| dt = O\left(\log^{1/4+\epsilon} x\right)$$

due to Bondarenko, Heap & Seip (2015, [**126**]) is of interest. It lent plausibility to Helson's conjecture since square-root cancellation would have implied an order more like $\log^{1/2} x$ for the left side of equation (11.18) when $2q = 1$. The matter was finally settled in Harper (2020, [**511**]) where it is proved (using decidedly intricate arguments) that for all real $0 \le q \le 1$, and uniformly for x large enough, the Rademacher and Steinhaus cases both satisfy

$$\mathrm{E}\Big|\sum_{n\le x} f(n)\Big|^{2q} \asymp \left(\frac{x}{1+(1-q)\sqrt{\log\log x}}\right)^q$$

(with $\asymp$ meaning that the left and right sides have the same order). These moments are smaller than for the independence case, and in particular we have

$$\mathrm{E}\Big|\sum_{n\le x} f(n)\Big| \asymp \frac{\sqrt{x}}{(\log\log x)^{1/4}}, \tag{11.20}$$

one conclusion from this being that Helson's surprising "better than square root cancellation" conjecture $o(\sqrt{x})$ is actually true. (See also Soundararajan & Xu (2023, [**1009**]).)

Turning now to higher moments, recall first our point of reference, namely that

$$\mathrm{E}\Big(\sum_{n\le x} f(n)\Big)^2 = \frac{6}{\pi^2}x + O(\sqrt{x}), \quad \text{and} \quad \mathrm{E}\Big|\sum_{n\le x} g(n)\Big|^2 = x + O(1). \tag{11.21}$$

For moments beyond the second, Harper, Nikeghbali & Radziwiłł (2015, [**514**]), and Heap & Lindqvist (2016, [**522**]) obtained the asymptotic formula

$$\mathrm{E}\Big(\sum_{n\le x} f(n)\Big)^k \sim C_k x^{k/2}(\log x)^{k(k-3)/2}, \quad \text{as } x\to\infty, \tag{11.22}$$

for the Rademacher case, valid for all *integer* moments $k \ge 3$. (The $k = 1$ and $k = 2$ cases do not follow this formula; the case $k = 2$ appears in (11.21), while the case $k = 1$ is trivial.) For Steinhaus chaos they obtained an asymptotic formula valid for the even-integer (absolute) moments:

$$\mathrm{E}\Big|\sum_{n\le x} g(n)\Big|^{2k} \sim D_k x^k (\log x)^{(k-1)^2}, \quad \text{as } x\to\infty, \quad k\ge 1. \tag{11.23}$$

These moments are all larger than those based on independence. The calculations behind these results involved expanding out the kth powers in the Rademacher case, and expanding out the $2k$th powers in the Steinhaus case (matching each factor with its conjugate), and then reducing the resulting sum of expectations to number-theoretic counting problems. When k is not a natural number (in the first case) or when $2k$ is not an even integer (in the second case) there is no obvious way to approach such calculations so less has been established, although Harper (2019, [**508**]) offers some results. The constants C_k and D_k appearing in (11.22) and (11.23) are known, though complicated, but can be shown to satisfy

$$C_k = \exp\left(-\frac{k^2}{2}\big(\log k + \log\log k\big) + O(k^2)\right), \quad k\to\infty,$$

and

$$D_k = \exp\big(-k^2\big(\log k + \log\log k\big) + O(k^2)\big), \quad k\to\infty.$$

Details about how expectations (11.22) and (11.23) can be viewed as a product of an 'arithmetic' factor and a 'geometric' factor can be found in the references cited. See also Harper (2024, [**513**]).

For the fourth Steinhaus moment, a rather precise result is available:

$$\mathrm{E}\left|\sum_{n\le x} g(n)\right|^4 = \frac{12}{\pi^2}x^2\log x + Cx^2 + O\left(x^{19/13}\log^{7/13} x\right),$$

where C is a certain number-theoretic constant (Ayyad, Cochrane & Zheng (1996, [**49**])). As it happens, this particular result is connected to the problem of counting the number of solutions to the congruence relation $x_1x_2 \equiv x_3x_4 \mod p$, and this is why this formula could be established.

There is a result of Conrey & Gamburd (2006, [**223**]) for the even integer moments of partial sums of the zeta function on the critical line which is of interest in connection with (11.18). These authors showed that for integers $k \ge 1$

$$\lim_{T\to\infty}\frac{1}{T}\int_0^T\left|\sum_{n\le x}\frac{1}{n^{1/2+it}}\right|^{2k} dt = B_k(\log x)^{k^2} + O\left((\log x)^{k^2-1}\right),$$

with constants B_k that are a product of two explicitly computed factors—an 'arithmetic' one and a 'geometric" one—just as for D_k in (11.23). What is of note here is how similar this result is to (11.23), that is, to the corresponding result for Steinhaus chaos; furthermore, the constants B_k and D_k have entirely analogous interpretations.

As one may see by comparing (11.22) and (11.23) the Rademacher moments are larger than the corresponding Steinhaus moments by some logarithmic factors. This is again due to the greater degree of cancellation which occurs among the complex-valued Steinhaus terms; because they are real-valued, the Rademacher variables experience less cancellation. Furthermore, on comparing these results with (11.11) and (11.16) of Section 11.3, we see that in both cases, the higher moments grow more rapidly (by factors involving logarithmic powers) than for sums of independent random variables.

11.5. Properties of Rademacher and Steinhaus chaos, III

Concerning distributional results, Harper (2020, [**511**]) proved that the convergence in probability

$$\frac{\sum_{n\le x} f(n)}{\sqrt{\mathrm{E}\left|\sum_{n\le x} f(n)\right|^2}} \xrightarrow{p} 0, \quad \text{as } x\to\infty, \tag{11.24}$$

holds for both Rademacher and Steinhaus chaos, with one conclusion from this being that the left side does not converge to the $N(0,1)$ expected under independence.[8] The result (11.24) is a straightforward consequence of the fact that for both models

$$\mathrm{E}\left|\sum_{n\le x} f(n)\right|^2 \asymp x$$

[8]Throughout this paragraph we will ignore the adjustments that are needed in the case of Steinhaus chaos to account for the fact that it is complex valued.

while (as noted in (11.20))

$$\mathrm{E}\left|\sum_{n\le x} f(n)\right| \asymp \sqrt{x}/(\log\log x)^{1/4},$$

so that the ratio in (11.24) cannot exceed $O\left(1/(\log\log x)^{1/4}\right)$. Harper (2013, [**505**]), however, showed that one would get convergence to a $N(0,1)$ if one restricted the two $\sum_{n\le x} f(n)$ sums in (11.24) to integers n having only a fixed number, say $\omega(n) = k$, of prime factors. In those restricted cases the resulting sums are a kind of number-theoretic generalization of so-called U-statistics sometimes referred to as chaoses of order k. (They are discussed, for example, in Blei & Janson (2004, [**110**]), Nourdin, Peccati & Reinert (2010, [**815**]), and Chatterjee & Soundararajan (2012, [**191**]).) Harper also showed that asymptotic normality would continue to hold if one restricted the number of factors as $\omega(n) \le k$, even if k were allowed to increase as fast as $o(\log\log x)$, but would no longer hold if k were allowed to increase as fast as $O(\log\log x)$. One may state this as

$$\frac{\sum_{n\le x;\ \omega(n)\le k} f(n)}{\sqrt{\mathrm{E}\left|\sum_{n\le x;\ \omega(n)\le k} f(n)\right|^2}} \longrightarrow N(0,1), \quad \text{as } x\to\infty, \quad \text{for } k = o(\log\log x).$$

(Recall from Chapter 9 that a typical integer up to x has about $\log\log x$ prime factors; so the integers up to x with only $o(\log\log x)$ factors comprise relatively sparse sets of integers.) Furthermore, Chatterjee and Soundararajan showed that

$$\frac{\sum_{x\le n\le x+y} f(n)}{\sqrt{\mathrm{E}\left|\sum_{x\le n\le x+y} f(n)\right|^2}} \longrightarrow N(0,1), \quad \text{as } x,y\to\infty, \quad \text{with } y = o(x/\log x).$$

See also Soundararajan & Xu (2023, [**1009**]), and Caich (2024, [**178**]). All of the results mentioned in this paragraph apply to both of the chaos models (except for adjustments to account for the fact that Steinhaus terms are complex-valued).

For the full sums which—in view of (11.20)—are properly normalized as

$$\frac{\sum_{n\le x} f(n)}{\sqrt{x}/(\log\log x)^{1/4}}, \tag{11.25}$$

it is not yet known whether or not limiting distributions exist nor what they might be. What is known for certain, however, is that those distributions cannot be $N(0,1)$ in the Rademacher case, or complex normal in the Steinhaus case. This is essentially a consequence of the persisting influence of 'early primes' on the random values of the $f(n)$. Relying on numerical studies (referred to in Harper (2013, [**505**])), Hough (2011, [**561**]) suggested that the limiting distributions—when *conditioned* on early primes—may well be Gaussian, although with differing variances, so that the *unconditional* distribution may be a limiting infinite *mixture* of Gaussians. Further background may be found in the references cited and in the related work of Gorodetsky & Wong (2024, [**443**]) and [Seth] Hardy (2025, [**500**]).[9]

The mentioned asymptotic normality results of Hough, Chatterjee & Soundararajan, and of Harper—for sums restricted to integers with not too many prime

---

[9]**Update:** Gorodetsky & Wong (2025, [**444**]) have shown, for the Steinhaus case, that (11.25) does have an asymptotic distribution, and that it is a certain scale-mixture of standard complex Gaussian distributions. An analagous result is expected for Rademacher chaos.

factors—relied on three distinct methods of proof, namely, the method of computing moments, Stein's method for approximating distributions, and a martingale approach. As one may appreciate from this, the range of probabilistic tools that have been brought to bear on establishing results of this nature is wide. However, in the moments method, for instance, where one needs to evaluate sums of products of the $f(n)$, some number-theoretic results must also be used, and number-theoretic considerations are needed for the other two methods as well. The proofs for all of these results are lengthy and technical. It is, however, worth mentioning that for the Rademacher case, many of the proofs—for the results here, as well as in the earlier chaos model sections (Sections 11.3 and 11.4)—do not require the ϵ_p variables that define the chaos to be ± 1 Bernoulli; it is often enough that they be independent, symmetric (so that ϵ_p and $-\epsilon_p$ have the same distribution), normalized (so that all $\mathrm{E}\,\epsilon_p^2 = 1$), and have bounded fourth moments ($\mathrm{E}\,\epsilon_p^4 \leq C$). Furthermore, many of the results continue to hold even if the ϵ_p are not identically distributed. The case where the ϵ_p are independent $N(0,1)$ variables is sometimes referred to as *Gaussian chaos*. Analogous remarks apply to the Steinhaus case.

Concerning the *tails* of the distributions associated with the summatory functions $F(x)$ and $G(x)$, we remark first that tail probabilities of sums of *uncorrelated* random variables (as opposed to *independent* ones) can sometimes help gauge the higher order dependences that exist among the variables. For example, larger tail probabilities tend to be associated with less randomness-based cancellation and with higher degrees of dependence, and conversely. By analogy with the case of independent variables one might at first expect for exponentially decaying tails to be possible, and indeed the 'close-to-LIL' results (11.6) and (11.7) lend some support to this. However this turns out not to be the case. Using Chebychev's inequality, one may obtain, for all $\lambda > 0$, the very crude bound

$$\mathrm{P}\left[\,\left|\sum_{n\leq x} f(n)\right| \geq \lambda\sqrt{x}\,\right] \;\leq\; \frac{\mathrm{E}\left|\sum_{n\leq x} f(n)\right|^2}{(\lambda\sqrt{x})^2} \;\leq\; \frac{1}{\lambda^2}.$$

Harper (2020, [**511**, Corollary 2]) showed how applying Chebychev's inequality to

$$\left|\sum_{n\leq x} f(n)\right|^2$$

(and using a fourth moment) allows one to improve this bound, but only slightly. Harper (2019, [**508**, Corollary 1])—for both Rademacher and Steinhaus chaos, and for all $2 \leq \lambda \leq \sqrt{\log x}$—then obtained the upper bound

$$\mathrm{P}\left[\,\left|\sum_{n\leq x} f(n)\right| \geq \lambda\sqrt{x}\,\right] = O\left(\frac{1}{\lambda^2} e^{-(\log^2 \lambda)/(\log\log x)}\right), \quad \text{as } x \to \infty.$$

But on the other hand, Harper (2023, [**512**]) established that for both chaos models there exist arbitrarily large valued of x for which $|\sum_{n\leq x} f(n)|$ exceeds $\sqrt{x}(\log\log x)^{1/4+o(1)}$ and he speculated that the exponent $1/4$ here may plausibly be sharp.

11.6. Selected proofs

PROOF OF THE ALMOST SURE CONVERGENCES (11.2). We need to show that the Dirichlet series and Euler product related to Rademacher chaos,

$$\sum_{n=1}^{\infty} \frac{f(n)}{n^s} = \prod_p \left(1 + \frac{f(p)}{p^s}\right) = \prod_p \left(1 + \frac{\epsilon(p)}{p^s}\right),$$

converge almost surely to a random variable when $\sigma \equiv \Re s > \frac{1}{2}$. Here the $\epsilon(p)$ are ± 1 Rademacher random variables. We will work with the logarithm of the Euler product and apply Kolmogorov's Three-Series Criterion. Thus, letting $X_p \equiv \log\left(1 + \epsilon(p)/p^s\right)$, we need to show that each of the three numerical series

$$\sum_p \mathrm{P}[\,|X_p| > c\,], \quad \sum_p \mathrm{E}(X_p \,\mathrm{I}[\,|X_p| \leq c\,]), \quad \text{and} \quad \sum_p \mathrm{Var}(X_p \,\mathrm{I}[\,|X_p| \leq c\,])$$

converges. The first convergence holds because $|X_p| \leq c$ once p is large enough so only a finite number of terms are nonzero. In the second and third sums the $\mathrm{I}[\cdots]$ indicator variables are identically 1 once p is large enough and so may be ignored. The middle sum can be controlled as

$$\begin{aligned}\left|\sum_p \mathrm{E}(X_p)\right| &\leq \sum_p |\,\mathrm{E}(X_p)| \\ &= \sum_p \left|\frac{1}{2}\log\left(1 + \frac{1}{p^\sigma}\right) + \frac{1}{2}\log\left(1 - \frac{1}{p^\sigma}\right)\right| \\ &= \sum_p \left|\frac{1}{2}\log\left(1 - \frac{1}{p^{2\sigma}}\right)\right| < C + \sum_p \frac{1}{p^{2\sigma}} < \infty.\end{aligned}$$

Here we used the simple bound $\left|\log(1+x)\right| < 2|x|$ which holds when x is close enough to 0, while the finite constant C stems from the early terms to which the bound does not apply. Finally, with $\mathrm{E}\,X_p = \frac{1}{2}\log(1 - 1/p^{2\sigma})$, we have

$$\sum_p \mathrm{Var}\,X_p = \sum_p \left(\frac{1}{2}\left(\log\left(1 - \frac{1}{p^\sigma}\right) - \mathrm{E}\,X_p\right)^2 + \frac{1}{2}\left(\log\left(1 + \frac{1}{p^\sigma}\right) - \mathrm{E}\,X_p\right)^2\right).$$

But each of the terms that result when the squares are multiplied out has a finite sum. For instance, arguing with the simple log-bound as before,

$$\sum_p \log^2\left(1 - \frac{1}{p^\sigma}\right) < C + \sum_p \frac{2}{p^{2\sigma}} < \infty,$$

and likewise for the other sums that arise. The Steinhaus case and the case where s is complex-valued follow along similar lines and are left as exercises. □

PROOF OF (11.3). The assertion is that the summatory version of the Rademacher chaos, $F(x) \equiv \sum_{n \leq x} f(n)$, satisfies the bounds

$$F(x) = O\left(x^{1/2+\epsilon}\right), \quad \text{and} \quad F(x) \neq O\left(x^{1/2-\epsilon}\right), \quad \text{almost surely.}$$

This result is just a consequence of the almost sure convergence of (11.2) for $\sigma > \frac{1}{2}$ (proved above) together with the following fact (Montgomery & Vaughan (2006, [**774**, Theorem 1.3])): If a Dirichlet series $\sum_1^\infty f(n)/n^s$ has a nonnegative abscissa of convergence β then this β will be the same as the greatest lower bound of those exponents β which satisfy the estimate

$$\sum_{k=1}^{n} f(k) = O\left(n^{\beta}\right). \qquad \square$$

PROOF OF BOHR IDENTIFICATION RESULT (11.17). This result was stated as

$$\mathrm{E}\left|\sum_{n\leq x} g(n)\right|^{2q} = \lim_{T\to\infty}\frac{1}{T}\int_0^T \left|\sum_{n\leq x} n^{-it}\right|^{2q} dt, \tag{11.26}$$

where $g(n)$ is the Steinhaus chaos. However in this form it masks the fact that the following more general result is true:

$$\mathrm{E}\, h\left(\sum_{n\leq x} g(n)\right) = \lim_{T\to\infty}\frac{1}{T}\int_0^T h\left(\sum_{n\leq x} n^{-it}\right) dt \tag{11.27}$$

for *any* smooth function h, including $h(x) = |x|^{2q}$. We first note that both sides of (11.27) represent expectations of certain random variables. On the left we have

$$\sum_{n\leq x} g(n) \equiv \sum_{n\leq x}\prod_{p^k \| n} \epsilon_p^k \equiv \sum_{n\leq x} \exp\left(2\pi i \sum_{p^k\|n} k\,\Theta_p\right)$$

where the random variables $\{\epsilon_p\}$, indexed by the primes, are independent and uniformly distributed on the complex unit circle $|z| = 1$, and the $\{\Theta_p\}$, also indexed by the primes, are independent uniforms on $[0, 1)$; the variables are connected as $\epsilon_p \equiv \exp(2\pi i\, \Theta_p)$. On the right in (11.27), the quantity t is to be thought of as a random variable which is uniformly distributed on $[0, T]$. The essential point is that for any given x, the limiting probability distribution of $\sum_{n\leq x} n^{-it}$, as $T \to \infty$, is the same as the distribution of $\sum_{n\leq x} g(n)$. (Note that these distributions are all on the complex plane.)

This convergence in distribution may be established as follows. For simplicity we will address only the case $x = 6$ but it covers all of the cases of a general proof. We thus have

$$\begin{aligned}\sum_{n\leq x} g(n) &= 1 + \epsilon_2 + \epsilon_3 + \epsilon_2^2 + \epsilon_5 + \epsilon_2\epsilon_3 \\ &= 1 + e^{2\pi i\Theta_2} + e^{2\pi i\Theta_3} + \left(e^{2\pi i\Theta_2}\right)^2 + e^{2\pi i\Theta_5} + e^{2\pi i\Theta_2}e^{2\pi i\Theta_3}\end{aligned}$$

and

$$\sum_{n\leq x} n^{-it} = 1 + e^{-it\log 2} + e^{-it\log 3} + (e^{-it\log 2})^2 + e^{-it\log 5} + e^{-it\log 2}e^{-it\log 3}.$$

But these two equations represent the same smooth function; in the first case it is evaluated at $\left(e^{2\pi i\Theta_2}, e^{2\pi i\Theta_3}, e^{2\pi i\Theta_5}\right)$, and in the second case it is evaluated at $\left(e^{-it\log 2}, e^{-it\log 3}, e^{-it\log 5}\right)$. What we now need is to show that the latter trio converges in distribution to the former trio when t is regarded as uniform on $[0, T]$, and $T \to \infty$. Since these random variables are all bounded this is equivalent to the convergence of all their joint moments. And since the variables are complex-valued this convergence of the moments must apply not only for the variables but also for their complex conjugates. This leads to consideration of

$$\mathrm{E}\left(\left(e^{-it\log 2}\right)^j\left(e^{-it\log 3}\right)^k\left(e^{-it\log 5}\right)^\ell\right)$$

for all integers j, k, ℓ (whether positive, negative, or zero). This is trivial when $j = k = \ell = 0$, but in all other cases it leads to

$$\mathrm{E}\big(e^{-it(j\log 2+k\log 3+\ell\log 5)}\big) = \mathrm{E}\big(e^{-itc}\big) \to 0$$

because c cannot be zero since logarithms of the primes are linearly independent over the rationals (an immediate consequence of the Unique Factorization Theorem). Hence all of the joint moments converge correctly thus leading to the convergence in distribution

$$\sum_{n\le x} n^{-it} \xrightarrow{d} \sum_{n\le x} g(n), \quad \text{as } T\to\infty,$$

and hence to the result. □

CHAPTER 12

Further Topics

When I go from hence let this be my parting word,
that what I have seen is unsurpassable.

I have tasted of the hidden honey of this lotus
that expands on the ocean of light,
and thus am I blessed –
let this be my parting word.

In this playhouse of infinite forms I have had my play...
... and if the end comes here, let it come –
let this be my parting word.

Rabindranath Tagore [**1040**]

In this chapter we pursue two broad themes. In the first three sections we continue with the subject of randomized number-theoretic models, along the lines of the Radamacher and Steinhaus chaos models studied in Chapter 11. However, we now shift gears—away from $\mu(n)$ and $\lambda(n)$—to study a randomized version of the zeta function, Kac's randomness model for additive arithmetic functions, and a randomized version of the Mertens function.

In the second theme, we explore the correlational and 'time series' behaviour of the (non-randomized) Möbius and Liouville functions. This concerns a class of difficult questions—raised by Sarvadaman Chowla in his memoir (1965, [**204**]) on the Riemann hypothesis—that have become known as the Chowla conjectures. Here too a recurring theme is the extent to which the behaviour of these functions mirrors that of *independent* random variables.

12.1. Randomized zeta functions

The motivation for studying randomized models of the zeta function stems from the Lindelöf hypothesis (LH) and related interest in the growth of the zeta function on the critical line. Recall from Chapter 7 that the Lindelöf hypothesis asserts that for every $\epsilon > 0$,

$$\max_{0\le t\le T}\left|\zeta\left(\frac{1}{2}+it\right)\right| = o\left(T^{\epsilon}\right), \quad \text{as } T\to\infty, \tag{12.1}$$

and that, unconditionally, all that is known is that the estimate (12.1) holds for any $\epsilon > 13/84 \approx 0.1548$ (Bourgain (2017, [**139**])). Relevant here as well is Selberg's CLT result which says that for t uniformly distributed in $[0, T]$, the real and imaginary parts of $\log\zeta(\frac{1}{2}+it)$ tend (in distribution, as $T\to\infty$) to independent $N\big(0, \frac{1}{2}\log\log T\big)$ random variables. Selberg's result suggests that most values of $\log|\zeta(\frac{1}{2}+it)|$ should be no larger than some multiple of $\sqrt{\log\log T}$. However, this reasoning does not take into account deviations in the tails of the distribution where the normal distribution is unlikely to hold. Large value of $\zeta(\frac{1}{2}+it)$ are thus of great interest, although rigorous results have been hard to come by and considerable uncertainty remains.

Given this dearth of information, several conjectures about the behaviour of $\zeta(\frac{1}{2}+it)$ have been advanced. In Section 7.2 we had indicated one such conjecture—due to Farmer, Gonek, Hughes & Keating (2007, [**441**])—that over the *full range*, $0 \le t \le T$, we should have

$$\max_{0\le t\le T} \log\left|\zeta\left(\frac{1}{2}+it\right)\right| \sim \sqrt{\frac{1}{2}\log T \log\log T}, \quad \text{as } T\to\infty \tag{12.2}$$

(whilst allowing for the possibility that a bound due to Littlewood might be closer to the truth). Subsequently, Fyodorov, Hiary & Keating (2012, [**400**]), and Fyodorov & Keating (2014, [**401**]), considered the maxima over randomly chosen *bounded intervals* of the critical line and they conjectured that for t sampled randomly on $[T, 2T]$ (and therefore also on $[0, T]$)

$$\max_{0\le h\le 1} \log\left|\zeta\left(\frac{1}{2}+i(t+h)\right)\right| = \log\log T - \frac{3}{4}\log\log\log T + O_P(1). \tag{12.3}$$

Here $O_P(1)$ is a term that is 'bounded in probability' (as $T\to\infty$), with many of its deeper properties having also been inferred by those authors. The inclusion of so precise a second-order term is a remarkable feature of this conjecture. It originates from the widely held belief that over finite intervals, far up on the critical line, the zeta function should behave like $Q(e^{i\theta})$ where Q is the characteristic polynomial of a large unitary matrix sampled randomly. The range $0 \le h \le 1$ in (12.3) can be replaced by any finite interval, for example $0 \le h \le 2\pi$, due (essentially) to the rotational nature of the ih component, a point that will become clearer as we proceed. It turns out that focusing on bounded intervals has technical advantages when dealing with the statistical dependences among the values of ζ at different points. A further advantage (in principle) is that bounded intervals are easier to investigate numerically, since one can then examine several randomly chosen finite intervals rather than just a single interval of length T.

The leading order term $\log\log T$ of (12.3) was established, at first conditionally on RH by Najnudel (2018, [**790**]), and then unconditionally by Arguin et al. (2019, [**31**]), in lengthy and technical papers. Harper (2019, [**509**]) subsequently conjectured a version of (12.3) which included a further term $\frac{3}{2}\log\log\log\log T$ on the right side. Finally, Arguin, Bourgade & Radziwiłł (2020, [**33**]) established the upper bound of the full Fyodorov–Hiary–Keating conjecture (12.3), unconditionally. However, the $O_P(1)$ term has not gone away, and since it allows for arbitrarily large values with ever smaller (although nonvanishing) probability, the Lindelöf hypothesis remains unsolved.

In this section we will examine certain randomization-based models that have been proposed in the attempt to assess and help corroborate conjectures such as (12.2) and (12.3). These are intended to model $\zeta(\frac{1}{2}+it)$ on finite intervals of t at high values, and are based on a heuristic motivated by straightforward moment computations.

PRINCIPLE 12.1 (Heuristic Principle). For t chosen randomly in $[0, T]$, with $T \to \infty$, any finite collection of 'rotational' components $p^{it} = e^{it \log p}$ (and likewise $p^{-it} = e^{-it \log p}$), based on distinct primes p, behave like independent random variables which are uniformly distributed on the complex unit circle $|z| = 1$. The real parts $\Re p^{it} \equiv \cos(t \log p)$ can also be viewed as independent amongst themselves, and likewise for the imaginary parts $\Im p^{it} \equiv \sin(t \log p)$. (Note, however, that *cos*$(t \log p)$ and $\sin(t \log p)$—whose squares sum to 1 if based on the same prime—cannot be regarded as being independent, although they will still be uncorrelated.)

EXERCISES 12.2.

(i) Show that the Unique Factorization Theorem is *equivalent* to *linear independence* of the $\log p$ in the sense that $\sum_j k_j \log p_j = 0$ for integers k_j (negative as well as positive), and distinct primes p_j, implies that all $k_j = 0$. (We say that the $\log p_j$ are not *commensurate*.)

(ii) Show that if $c \neq 0$ and t is sampled uniformly in $[0, T]$ then e^{itc} converges in distribution, as $T \to \infty$, to a uniform on the unit circle $|z| = 1$. (Hint: verify convergence of moments.)

(iii) Show that if p and q are distinct primes and t is sampled uniformly in $[0, T]$, then the vectors $\big(\cos(t \log p), \sin(t \log p)\big)$ and $\big(\cos(t \log q), \sin(t \log q)\big)$ become independent as $T \to \infty$. (Hint: it suffices to consider the joint moments.)

REMARK 12.3. The results at (ii) and (iii) above remain valid if one restricts t to a finite interval, say $T \leq t \leq T + 1$, with T tending to ∞.

Using Principle 12.1, we will now attempt to justify the following *randomness models for the zeta function on the critical line*: For t sampled randomly in $[0, T]$,

$$\zeta\left(\frac{1}{2} + it\right) \overset{d}{\approx} \prod_{p \leq P} \left(1 - \frac{U_p}{\sqrt{p}}\right)^{-1}, \quad \text{and } \log\left|\zeta\left(\frac{1}{2} + it\right)\right| \overset{d}{\approx} \Re \sum_{p \leq P} \frac{U_p}{\sqrt{p}}, \tag{12.4}$$

as $T \to \infty$. Here the U_p are independent random variables distributed uniformly on the unit circle $|z| = 1$, and the upper limit P depends on T in some suitable way. We will offer seven (interrelated) reasons as to why these models might be considered reasonable for trying to understand the behaviour of large values of $\zeta(\frac{1}{2} + it)$ on finite intervals.

First reason: Taking the logarithm of $\zeta(s) = \prod_p (1 - p^{-s})^{-1}$ and Taylor expanding gives

$$\begin{aligned} \log \zeta(s) &= -\sum_p \log\big(1 - \frac{1}{p^s}\big) \\ &= \sum_p \sum_{k=1}^{\infty} \frac{1}{kp^{ks}} \\ &= \sum_p \frac{1}{p^s} + \sum_{k=2}^{\infty} \frac{1}{k} \sum_p \frac{1}{p^{ks}} \\ &= \sum_p \frac{1}{p^s} + O(1), \end{aligned} \tag{12.5}$$

since the contribution of the prime powers p^{ks} with $k \geq 2$ is (uniformly) bounded. Of course, these steps are only valid when $\Re s > 1$, but one might expect that something resembling the Euler product would continue to have relevance below $\sigma = 1$, and in any case, the $O(1)$ component of (12.5) does remain valid down to $\Re s > \frac{1}{2}$. Assuming RH, and taking T large enough, it can be argued that the approximations *should* remain valid on the critical line for *most* values of t in high intervals of the type $[T, T+1]$, so that altogether

$$\zeta\left(\frac{1}{2} + it\right) \approx \prod_{p \leq P} \left(1 - \frac{1}{p^{1/2+it}}\right)^{-1}$$

(12.6)

$$\text{and } \log\left|\zeta\left(\frac{1}{2} + it\right)\right| \approx \Re \sum_{p \leq P} \frac{1}{p^{1/2+it}} + O(1)$$

for suitably chosen P. Applying the Heuristic Principle to the p^{it} components then results in the models in (12.4). (Exercise: justify the $O(1)$ in (12.5) for $\Re s > \frac{1}{2}$.)

Second reason: In a more careful treatment which included 'convergence factors' and assumed only the Riemann hypothesis, Harper (2013, [**507**]) proved that for T large enough,

$$\log\left|\zeta\left(\frac{1}{2} + it\right)\right| = \Re \sum_{p \leq T} \frac{1}{p^{1/2+it}} \frac{\log(T/p)}{\log T} + O(1)$$

for all $t \in [T, T+1]$ except on a set of arbitrarily small measure. (These small sets roughly consist of points in $[T, T+1]$ which are very close to ordinates of the zeros.) For typical values of T one expects the $1/p^{it}$ terms to act like independent uniforms on $|z| = 1$. The $\log(T/p)/\log T$ factors were introduced for purely technical reasons; we will simplify by ignoring them in the discussions which follow.

Third reason: In developing their conjecture Gonek, Hughes & Keating (2007, [**441**]) argued, again under RH, that on the critical line one has, roughly, and for suitable values of P,

$$\zeta\left(\frac{1}{2} + it\right) \approx \prod_{p \leq P} \left(1 - \frac{1}{p^{1/2+it}}\right)^{-1} \times \text{ Second Factor},$$

where the second factor is a kind of Hadamard product meant to account for zeros of zeta whose ordinates are close to t, and with those zeros modelled as the eigenvalues of large unitary matrices. For high values of t one would expect the two factors of this hybrid model to act independently, and for the second factor to have relatively limited influence. The Euler product for the zeta function should therefore be relevant even on the critical line.

Fourth reason: For large t, and any $x \leq t$, the main proposition of Soundararajan (2009, [**1006**]) can be used to argue that under RH we have the approximate upper bound

$$\log\left|\zeta\left(\frac{1}{2} + it\right)\right| \lesssim \Re \sum_{p \leq x} \frac{1}{p^{1/2+it}} + O\left(\frac{\log t}{\log x}\right). \tag{12.7}$$

(There cannot be *strict* inequality here due to the zeros of ζ; see Harper (2019, [**510**, p. 9]).) Now a crude upper bound for the sum here is $\sum_{p \leq x} 1/\sqrt{p} \sim 2\sqrt{x}/\log x$. (Exercise: explain this asymptotic using Cramér's heuristic as in Appendix E.)

Using this bound, and $C \log t / \log x$ to proxy the "big O" term, suggests minimizing (12.7) in x by minimizing

$$\frac{2\sqrt{x}}{\log x} + \frac{C \log t}{\log x}. \tag{12.8}$$

This has approximate optimum $x \asymp \log^2 t$, and substituting this into (12.8) recovers Littlewood's conditional bound $\log |\zeta(\frac{1}{2} + it)| \ll \log t / \log \log t$, lending credence to (12.7).

Fifth reason: Hardy and Littlewood proved—rigorously and unconditionally—that one can approximate $\zeta(\frac{1}{2} + it)$ by truncated sums of the zeta function: specifically, for $T \leq t \leq 2T$,

$$\zeta\left(\frac{1}{2} + it\right) = \sum_{n \leq T} \frac{1}{n^{1/2+it}} + O\left(\frac{1}{\sqrt{T}}\right), \quad \text{as } T \to \infty.$$

Representations (12.4) and (12.6) can be viewed as being approximations corresponding to this truncated zeta sum result.

Sixth reason: When $\sigma > 1$, we have the convergent sum

$$\zeta(\sigma + it) = \sum_{n=1}^{\infty} \frac{1}{n^{\sigma+it}} = \sum_{n=1}^{\infty} \frac{n^{-it}}{n^{\sigma}} = \sum_{n=1}^{\infty} \frac{e^{-it \log n}}{n^{\sigma}},$$

as well as its convergent Euler product equivalent

$$\zeta(\sigma + it) = \prod_p \left(1 - \frac{1}{p^{\sigma+it}}\right)^{-1} = \prod_p \left(1 - \frac{p^{-it}}{p^{\sigma}}\right)^{-1} = \prod_p \left(1 - \frac{e^{-it \log p}}{p^{\sigma}}\right)^{-1}.$$

Recognizing that for large values of t, finite collections of the $e^{-it \log p}$ behave like independent, uniformly distributed random variables on the unit circle, one is tempted to replace them by random variables U_p having precisely that distribution. Doing so leads to a proxy for the Euler product of $\zeta(\sigma + it)$ which, on multiplying out, can be reconstituted into a zeta sum:

$$\prod_p \left(1 - \frac{U_p}{p^{\sigma}}\right)^{-1} = \sum_{n=1}^{\infty} \frac{X(n)}{n^{\sigma}}.$$

Here $X(n)$ is related to the U_p as $X(n) = \prod_{j=1}^{m} U_{p_j}^{k_j}$, where the p_j and k_j range over the terms in the prime factorization $p_1^{k_1} \cdots p_m^{k_m}$ of n. So $\{X(n)\}$ is in fact the Steinhaus chaos. The equalities of the last three displays are rigorously valid when $\sigma > 1$, so aside from convergence issues, it seems plausible to assume that (for $\sigma > 1$) the distribution of $\zeta(\sigma + it)$—when t is sampled randomly in $[0, T]$ and $T \to \infty$—tends to a limiting distribution which is the same as that of $\sum_{n=1}^{\infty} X(n)/n^{\sigma}$. Bohr & Jessen (1930, 1932, [**119**]) showed that this same limiting distribution holds not only for $\sigma > 1$, but also for every value of $\sigma > \frac{1}{2}$. Now when σ varies over this range, the distribution of $\sum_{n=1}^{\infty} X(n)/n^{\sigma}$ varies continuously, and it converges to a normal distribution as $\sigma \downarrow \frac{1}{2}$; in fact, at $\sigma = \frac{1}{2}$ the distribution of $\sum_{n=1}^{\infty} X(n)/\sqrt{n}$ is exactly normal. It follows that the heuristic is consistent with Selberg's asymptotic normality result for the critical line. (Remark: In the range $\sigma \geq \frac{1}{2}$ the distribution of $\sum_{n=1}^{\infty} X(n)/n^{\sigma}$ is normal if and only if $\sigma = \frac{1}{2}$ by the Lyapounov CLT condition; see Appendix D.)

Seventh (and last) reason: Selberg's *unconditional* central limit result

$$\frac{\log|\zeta(1/2+it)|}{\sqrt{(1/2)\log\log T}} \xrightarrow{d} N\left(0, \frac{1}{2}\log\log T\right), \quad \text{as } T \to \infty, \tag{12.9}$$

when t is sampled randomly in $[T, 2T]$ (say) also lends credence to the heuristic

$$\log\left|\zeta\left(\frac{1}{2}+it\right)\right| \approx \Re \sum_{p \le T} \frac{1}{p^{1/2+it}} \overset{d}{\approx} \Re \sum_{p \le T} \frac{U_p}{p^{1/2}}. \tag{12.10}$$

For if the rotational components p^{-it} behave like independent random variables that are uniformly distributed on the unit circle then the rightmost sum in (12.10) would have mean 0 and variance $\sum_{p \le T} \frac{1}{2}\frac{1}{(\sqrt{p})^2} = \frac{1}{2}\sum_{p \le T}\frac{1}{p} \approx \frac{1}{2}\log\log T$, and furthermore would be asymptotically normal (on account of the independence) thus 'matching' Selberg's CLT result. Of course, this agreement between (12.9) and (12.10) is not quite precise as one expects about $T\log T$ zeros on the $[T, 2T]$ segment of the critical line where the logarithm won't be defined. However these zeros form a small set which (again) won't affect the result; Radziwiłł & Soundararajan (2017, [**879**]) show how that difficulty can be surmounted. (Here we have only considered $\log\left|\zeta\left(\frac{1}{2}+i\tau\right)\right|$; however the bivariate $\log\zeta\left(\frac{1}{2}+i\tau\right)$ case is analogous.)

Incidentally, in connection with Selberg's result (12.9), and the asymptotic normality of the sum in (12.10), if one could interpret those as also holding for larger deviations in the sense that the maximum of $\log|\zeta(\frac{1}{2}+it)|$ over $0 \le t \le T$ behaved like the maximum of (say) T independent trials from an $N\left(0, \frac{1}{2}\log\log T\right)$ distribution, then with $Z_1, Z_2, \ldots$, representing independent standard normal variables we would have

$$\begin{aligned}
\max_{0 \le t \le T} \log\left|\zeta\left(\frac{1}{2}+it\right)\right| &\overset{?}{\approx} \max_{0 \le j \le T} \log\left|\zeta\left(\frac{1}{2}+ij\right)\right| \\
&\overset{d}{\approx} \sqrt{(1/2)\log\log T}\max_{0 \le j \le T}\{Z_j\} \\
&\approx \sqrt{(1/2)\log\log T}\sqrt{2\log T} \\
&= \sqrt{\log T\log\log T}, \qquad T \to \infty.
\end{aligned}$$

Here we are using the fact that for standard normals $\max_{0 \le j \le T}\{Z_j\} \sim \sqrt{2\log T}$. This matches the conjecture of Farmer, Gonek & Hughes, while a more careful argument can be used to recover the constants in that conjecture as well. (Remark: We used a maximum of T trials, but any number CT^α of trials will give a result of the same order.)

A key feature in the reasoning throughout all of this section is the role played by assuming *some* independence *somewhere*. On a *global scale*, for instance, one might assume that for different primes the p^{it} behave independently once t is large, while on a *local scale* one might assume pointwise independence of quantities such as $\zeta(\frac{1}{2}+it_1)$ and $\zeta(\frac{1}{2}+it_2)$ once t_1 and t_2 are large and not too close together: $|t_1 - t_2| > 1$, say. Of course, such assumptions do discount the possibility of some underlying conspiracy among the variables which might invalidate the assumptions when dealing with more than a finite number of terms. For further development and justification for such heuristic reasoning see Gonek, Hughes & Keating (2007, [**441**]), Soundararajan (2009, [**1006**]), Harper (2013, [**507**]), Bourgain (2017, [**139**]), Arguin, Belius & Harper (2017, [**32**]), Bondarenko & Seip (2018, [**129**]), Najnudel (2018, [**790**]), and Arguin, Belius, Bourgade, Radziwiłł & Soundararajan (2019, [**31**]).

The upshot is that we will henceforth approximate $\log\left|\zeta(\frac{1}{2}+it)\right|$ as

$$\log\left|\zeta(\frac{1}{2}+it)\right| \equiv \Re\log\zeta(\frac{1}{2}+it) \approx -\Re\sum_{p\le T}\log\left(1-\frac{1}{p^{1/2+it}}\right) = \Re\sum_{p\le T}\frac{1}{p^{1/2+it}}+O(1),$$

and then approximate this distributionally as

$$\Re\sum_{p\le T}\frac{1}{p^{1/2+it}} = \Re\sum_{p\le T}\frac{p^{-it}}{p^{1/2}} \overset{d}{\approx} \Re\sum_{p\le T}\frac{U_p}{p^{1/2}} \overset{d}{=} \sum_{p\le T}\frac{\cos(\Theta_p)}{p^{1/2}} \tag{12.11}$$

where the U_p are independent random variables distributed uniformly on the unit circle, and the Θ_p are independent and distributed uniformly on $[0,2\pi]$. The quantity in (12.11) has mean 0, easily computed variance $\frac{1}{2}\sum_{p\le T}\frac{1}{p}\sim\frac{1}{2}\log\log T$, and behaves asymptotically like a Gaussian random variable.

We may now turn to the conjecture (12.3) of Fyodorov, Hiary, and Keating (FHK). Reasoning as above, for large t, and for $0\le h\le 1$ say, leads us to model $\log|\zeta(\frac{1}{2}+i(t+h))|$ as

$$\Re\sum_{p\le T}\frac{1}{p^{1/2+i(t+h)}} = \Re\sum_{p\le T}\frac{p^{-it}p^{-ih}}{p^{1/2}} \overset{d}{\approx} \sum_{p\le T}\frac{\Re\left(U_p\,p^{-ih}\right)}{p^{1/2}} \tag{12.12}$$

using some suitable value for T. With the understanding that t is large, we define

$$W_p(h)\equiv\frac{\Re\left(U_p\,p^{-ih}\right)}{p^{1/2}},\quad\text{and } X(h)\equiv\sum_{p\le T}W_p(h)=\sum_{p\le T}\frac{\Re\left(U_p\,p^{-ih}\right)}{p^{1/2}}.$$

REMARKS 12.4. For t large, and $0\le h_1,h_2\le 1$, we seek to understand the correlation between $\log|\zeta(\frac{1}{2}+i(t+h_1))|$ and $\log|\zeta(\frac{1}{2}+i(t+h_2))|$. When t is large, the heuristic principle suggests that the p^{-it} will approximate independent variables. However, since the values of h are not large, the p^{-ih} terms will induce correlations. The $p^{-ih}\equiv e^{-ih\log p}$ vary on a scale $\approx 1/\log p$, so the contributions from large primes should decorrelate quickly; however contributions from small primes will remain correlated over smaller intervals. An underlying multiscale structure will result in essentially logarithmic covariance.

Now, since $\Re U_p\,p^{-ih}=\cos(\Theta_p-ih)$ where the $U_p\equiv e^{i\Theta_p}$ are independent, we easily get

$$\mathrm{E}\left\{W_p(h)W_q(h')\right\}=\begin{cases}\frac{1}{2p}\cos(|h-h'|\log p), & \text{if } p=q,\\ 0, & \text{otherwise.}\end{cases}$$

Here we used the identity $2\cos(\alpha)\cos(\beta)=\cos(\alpha+\beta)+\cos(\alpha-\beta)$. Next, we have $\mathrm{E}\,X(h)=0$, while

$$\mathrm{E}\,X^2(h)=\sum_{p\le T}\frac{\mathrm{E}\cos^2\Theta}{p}=\frac{1}{2}\sum_{p\le T}\frac{1}{p}\approx\frac{1}{2}\log\log T. \tag{12.13}$$

by Mertens's Theorem. Furthermore, for $0 \le h_1, h_2 \le 1$ say, we will have

$$\begin{aligned} \mathrm{E}\, X(h_1)X(h_2) &= \sum_{p\le T}\sum_{q\le T} \frac{\mathrm{E}\, W_p(h_1)W_q(h_2)}{p^{1/2}q^{1/2}} \\ &= \sum_{p\le T} \frac{\mathrm{E}\cos\left(\Theta_p - h_1\log p\right)\cos\left(\Theta_p - h_2\log p\right)}{p} \qquad (12.14) \\ &= \frac{1}{2}\sum_{p\le T}\frac{\cos\left((h_1-h_2)\log p\right)}{p}. \end{aligned}$$

Now let $\alpha = h_1 - h_2$, and consider the values of (12.14) under various ranges for α. When $|\alpha|$ is much smaller than $1/\log T$ the cosine will not stray far from 1 for any of the $p \le T$, so (12.14) will nearly equal (12.13). The $X(h_1)$ and $X(h_2)$ will then be almost perfectly correlated, meaning that the values of $|\zeta(\frac{1}{2}+it+ih_1)|$ and $|\zeta(\frac{1}{2}+it+ih_2)|$ will be roughly equal for all values of t in $[T, 2T]$, say. When h_1 and h_2 are somewhat farther apart, but with $\alpha = |h_1 - h_2|$ still small enough compared with $1/\log T$ for the approximation $\cos x \approx 1 - x^2/2$ to be valid over the whole range of the sum, we will have[1]

$$\frac{1}{2}\sum_{p\le T}\frac{\cos(\alpha\log p)}{p} \approx \frac{1}{2}\sum_{p\le T}\frac{1}{p} - \frac{1}{4}\sum_{p\le T}\frac{\alpha^2\log^2 p}{p} \sim \frac{1}{2}\log\log T - \frac{1}{8}\alpha^2\log^2 T.$$

But for h_1 and h_2 still farther apart, with $|\alpha|$ substantially exceeding $1/\log T$, we may estimate (12.14) using the heuristic that n is prime with probability $1/\log n$. This gives

$$\begin{aligned} \frac{1}{2}\sum_{p\le T}\frac{\cos(\alpha\log p)}{p} &\approx \frac{1}{2}\sum_{2\le n\le T}\frac{\cos(\alpha\log n)}{n}\frac{1}{\log n} \\ &\approx \frac{1}{2}\int_2^T \frac{\cos(\alpha\log x)}{x\log x}\,dx \\ &= \frac{1}{2}\int_{\alpha\log 2}^{\alpha\log T}\frac{\cos(u)}{u}\,du \end{aligned}$$

by the change of variables $u = \alpha\log x$. Now the $u > c$ part of this integral is bounded for any $c > 0$ (one may show this using integration by parts or otherwise) so this becomes[2]

$$\begin{aligned} &= \frac{1}{2}\int_{\alpha\log 2}^{c}\frac{\cos(u)}{u}\,du + O(1) \\ &\approx \frac{1}{2}\int_{\alpha\log 2}^{c}\frac{1-u^2/2}{u}\,du + O(1) \\ &= \frac{1}{2}\int_{\alpha\log 2}^{c}\frac{du}{u} - \frac{1}{4}\int_{\alpha\log 2}^{c} u\,du + O(1) \end{aligned}$$

---

[1]Here we are using the results

$$\sum_{p\le x}\frac{1}{p} = \log\log x + C + O\left(\frac{1}{\log^A x}\right), \quad \text{and} \quad \sum_{p\le x}\frac{\log^2 p}{p} = \frac{1}{2}\log^2 x + O(\log x).$$

See for example Appendix E, or Montgomery & Vaughan (2006, [**774**, Theorem 2.7]).

[2]The constant c will be of order 1, selected only for the cosine approximation to hold. (In fact c may even be selected so as to exceed the $x^2/2$ part of the cosine expansion.)

$$= \frac{1}{2}\log u\Big|_{u=\alpha\log 2}^{u=c} - \frac{1}{4}\cdot\frac{u^2}{2}\Big|_{u=\alpha\log 2}^{u=c} + O(1)$$
$$= \frac{1}{2}\big(\log(c) - \log(\alpha\log 2)\big) - \frac{1}{8}\left(c^2 - (\alpha\log 2)^2\right) + O(1)$$
$$= -\frac{1}{2}\log\alpha + O(1).$$

Since this is dominated by $-\frac{1}{2}\log|\alpha|$, we are led to the *logarithmic form*

$$\mathrm{E}\, X(h_1)X(h_2) = \frac{1}{2}\log\left(|h_1 - h_2|^{-1}\right) + O(1).$$

Further details may be found in Harper (2013, [**504**, **506**, **507**]), Arguin, Belius & Harper (2017, [**32**]), and Harper (2019, [**510**]).

Now since $\mathrm{E}\, X(h) = 0$ for all h, while $\mathrm{E}\left(X(h_1)X(h_2)\right)$ depends only on $|h_1 - h_2|$, the process $X(h)$ is *'covariance stationary'*. Further, the joint distribution of the $W_p(h)$ terms is invariant under translation $h \to h + a$, and even under reflection $h \to -h$. However, in spite of these nice properties, the limiting behaviour of the $X(h)$ process is more complicated than the illustrative calculations shown above would suggest. The 36-page technical report and 75- and 24-page papers of Saksman & Webb (2016, [**934**]), (2020, [**935**]), (2020, [**936**]) can be consulted for a rigorous discussion. Briefly, this *log-correlated* process has Gaussian limits at *finite* collections of points, and it turns out that one can approximate its maximum by that of a stationary log-correlated Gaussian field. This involves a multi-scale decomposition based on an approximate hidden tree structure that governs the process. The details are technical and outside of our scope.

In a lengthy paper involving novel methods for estimating the maxima of branching random walks (meaning Brownian motions that keep splitting into two independent ones after exponential times) Arguin, Belius & Harper (2017, [**32**]) carried out a detailed study of the $X(h)$ process and proved that the heuristic model (12.12) satisfies

$$\max_{0\le h\le 1}\sum_{p\le T}\frac{\Re(U_p\, p^{-ih})}{p^{1/2}} = \log\log T - \frac{3}{4}\log\log\log T + o_P(\log\log\log T). \tag{12.15}$$

The significant (and indeed remarkable) point here is that random zeta function model (12.12) produces a result which coincides with that of random matrix theory, not only in its leading term, but far more strikingly, in its second order term as well, and furthermore that it matches the bounded-intervals conjecture (12.3) of Fyodorov, Hiary, and Keating.

REMARK 12.5. The maxima of identically distributed independent random variables with finite variance (when suitably normalized) can only converge to one of three distributional 'universality' classes: Gumbel, Frechet, or Weibull. See, for example, Gumbel (1954, [**467**]) or Resnick (1987, [**898**]). For correlated, and in particular for log-correlated processes, the extreme value theory is considerably more complicated; Remy (2020, [**892**]) and references therein may be consulted for some details.

It is worth mentioning that in earlier work based on Selberg's asymptotic normality result, Arguin, Belius & Harper (2017, [**32**]) had proposed that the order of the maximum of $\log|\zeta(\frac{1}{2} + it)|$ on $[T, T+1]$ might perhaps be estimated by the maximum of $\log T$ independent zero-mean Gaussians with variance $\frac{1}{2}\log\log T$. The underlying idea was that the maximum of a Gaussian process could be estimated by the maximum of a sample of points from that process, with those points chosen just the right distance apart for the dependence between them to have diminished in some optimal way. This maximum, however, turns out to behave as $\log\log T - \frac{1}{4}\log\log\log T + O_P(1)$, thus matching the first, but missing the second order term of the Fyadorov, Hiary, and Keating conjecture by an amount $\frac{1}{2}\log\log\log T$. These authors considered how the number of sampled points could be be modified so as to obtain a match to the second order term of the conjecture. However, it later emerged that the discrepancy could be resolved by assuming an underlying 'branching random walk'. One takeaway from those efforts is that Selberg's asymptotic normality result cannot be relied upon to shed light on the Lindelöf hypothesis.

In addition to providing a comprehensive overview of the methods underlying the results discussed here, Harper (2019, [**510**]) pointed out an important difference between the behaviour of a quantity such as $\Re\sum_{p\le T} 1/p^{1/2+it}$ on the critical line, versus the behaviour of the corresponding quantity $\Re\sum_{p\le T} 1/p^{\sigma+it}$ when $\sigma > \frac{1}{2}$. When T is large (so that the p^{it} behave like independent uniforms on the unit circle), the first of these has a mean square which is asymptotic to $\frac{1}{2}\log\log T$ as $T\to\infty$, while the mean square of the second converges to a finite value. Quoting Harper: "This is one of the key sources of difficulty and interest on the critical line, as compared with elsewhere in the complex plane ... $\log\log T$ is a very slowly growing function, which turns out to be key to the success of many of the arguments that we can implement" [**510**]. The idea is that the behaviour of ζ *on* $\sigma = \frac{1}{2}$ should not differ greatly from its behaviour *near* $\sigma = \frac{1}{2}$.

There is considerable ongoing work associated with the Lindelöf hypothesis, the maximum of the zeta function on the critical line, and the conjecture of Fyodorov, Hiary, and Keating. In particular—and as mentioned at the beginning of this section—the leading $\log\log T$ term of the conjecture (for the actual zeta function) was established by Arguin, Belius, Bourgade, Radziwiłł & Soundararajan (2019, [**31**]). They proved, unconditionally, that except for a vanishingly small proportion of the values of $t\in[T,2T]$, as $T\to\infty$,

$$\max_{|t-u|\le 1} \log\left|\zeta\left(\frac{1}{2}+iu\right)\right| \sim \log\log T.$$

(A conditional version of this result had been proved earlier by Najnudel (2018, [**790**]).) Furthermore, the upper bound part of the full conjecture (12.3) was established by Arguin et al. (2019, [**33**]), (2020, [**34**]). Their results also showed—again unconditionally, and for the actual zeta function—that for any function $h(T)$ tending to infinity however slowly, and with t sampled randomly in $[T,2T]$, the probability that

$$\max_{|t-u|\le 1} \log\left|\zeta\left(\frac{1}{2}+iu\right)\right| > \log\log T - \frac{3}{4}\log\log\log T + h(T)$$

tends to 1 as $T\to\infty$. Previous to this, with $T\le t\le 2T$ as before, Harper (2019, [**509**]) had established the upper bound

$$\max_{|t-u|\le 1} \log\left|\zeta\left(\frac{1}{2}+iu\right)\right| \le \log\log T - \frac{3}{4}\log\log\log T + \frac{3}{2}\log\log\log\log T + h(T)$$

with probability tending to 1, for any function $h(T) \to \infty$.

Additional details may be found in the references already mentioned as well as in Arguin et al. [**29**]–[**37**], Harper [**504**]–[**512**], Harper et al. [**514**], Fyodorov et al. [**400**], [**401**], Najnudel [**790**]–[**792**], Paquette & Zeitouni 2018, [**833**], Saksman & Webb [**934**]–[**936**], and Soundararajan [**1005**], [**1006**], and in more recent work by Arguin and coauthors.

12.2. Kac's randomness model for additive functions

Let g be a strongly additive function, so that $g(mn) = g(m) + g(n)$ when $(m, n) = 1$, and $g(p^k) = g(p)$ when p is prime. Such a function can be represented as

$$g(n) = \sum_p g(p)\ \mathrm{I}[p|n] \tag{12.16}$$

where $\mathrm{I}[p|n]$ is a 1–0 indicator of whether or not p divides n. As a heuristic for studying the asymptotic distribution of such $g(n)$ Kac (1959, [**613**])—and later Kubilius (1964, [**663**])—proposed the probabilistic model

$$g(n) \overset{d}{\approx} \sum_{p \le n} g(p) X(p) \tag{12.17}$$

where the $X(p)$ are independent random variables with

$$\mathrm{P}\big[X(p) = 1\big] = \frac{1}{p}, \quad \text{and } \mathrm{P}\big[X(p) = 0\big] = 1 - \frac{1}{p}.$$

According to this model, the mean and variance of $g(n)$, for $n \le x$ selected randomly, would (for large x) be estimated as

$$\mu(g, x) = \sum_{p \le x} \frac{g(p)}{p}, \quad \text{and } \sigma^2(g, x) = \sum_{p \le x} \frac{g^2(p)}{p} \left(1 - \frac{1}{p}\right).$$

These expressions are reminiscent of some we have seen in Chapter 9. Under mild conditions on the $g(p)$ (not growing too quickly, for example) we would also have

$$\sigma^2(g, x) = \sum_{p \le x} \frac{g^2(p)}{p} + O(1).$$

If $\sigma^2(g, x) \to \infty$ as $x \to \infty$, a Central Limit Theorem would apply, and Kac's model would lead to an asymptotic normality result: for any fixed value of u,

$$\frac{1}{x} \# \left\{ n \le x \colon \frac{g(n) - \mu(g, x)}{\sigma(g, x)} \le u \right\} \longrightarrow \frac{1}{\sqrt{2\pi}} \int_{-\infty}^{u} e^{-t^2/2}\, dt, \quad \text{as } x \to \infty. \tag{12.18}$$

A proof of this assertion is given in Section 12.5. Kac's heuristic is thus consistent with the theorem of Erdős & Wintner in the Gaussian case. Of course, this argument cannot be thought of as rigorous, since it has assumed independence across the entire set of primes, and not just over a finite subset of primes. Accounting fully for the dependence of the terms in sum (12.16) necessarily involves delicate number-theoretic considerations.

Returning to Kac's heuristic, it can be shown (although we omit the details) that it also correctly affirms that the left and right sides of (12.18) remain *asymptotic* to each other even if one allowed u to increase as fast as $u = o\left(\sigma^{1/3}(g, x)\right)$. Beyond this rate, however, and for general additive $g(n)$, the situation is more complicated and the normality-based asymptotic fails. (This is discussed in Mačiulis

(1983, [**718**, Theorem 2]); see also Ibragimov & Linnik (1971, [**575**, Chapter 6])). However Radziwiłł (2011, [**876**]) established an attractive alternate version to cover this case; it states that

$$\frac{1}{x}\#\left\{n\le x\colon \frac{g(n)-\mu(g,x)}{\sigma(g,x)}\ge u\right\}\sim \mathrm{P}\left[\frac{1}{\sigma(g,x)}\sum_{p\le x} g(p)\left(X(p)-\frac{1}{p}\right)\ge u\right]$$

uniformly for $1\le u\le o(\sigma(g,x))$, and with the $X(p)$ as before. It was also shown in that paper that if two arithmetic functions $f(n)$ and $g(n)$ were such that, as $x\to\infty$,

$$\frac{1}{x}\#\left\{n\le x\colon \frac{f(n)-\mu(g,x)}{\sigma(f,x)}\ge u\right\}\sim \frac{1}{x}\#\left\{n\le x\colon \frac{g(n)-\mu(g,x)}{\sigma(g,x)}\ge u\right\}$$

were to hold for the larger values of u as well, then under mild conditions $f(n)$ and $g(n)$ would necessarily be essentially the same function, that is, we would have $f(n)\equiv Cg(n)$ for some constant C. Stated otherwise, if two additive arithmetic functions take on large values with about the same probability then those two functions must be identical. For a purely probabilistic perspective on such results see Radziwiłł (2011, [**877**]).

An interesting question, raised by Elliott (1997, [**322**, p. 50]), concerns the extent to which the distribution of the values of an additive function on all of the integers determines (and is determined by) its distribution on the primes. Radziwiłł (2011, [**876**]) studies this problem and (among other results) shows that a strongly additive function is roughly Poisson distributed on the integers if and only if it tends to either 0 or 1 along the primes.

12.3. A randomized Mertens function

Recall that the Mertens function $M(x)\equiv\sum_{n\le x}\mu(x)$ is the summatory version of the Möbius function $\mu(n)$, and that $M(x)=o(x)$ is equivalent to the PNT, while $M(x)=O(x^{1/2+\epsilon})$ is equivalent to the Riemann hypothesis. If the nonzero values of $\mu(n)$ behaved like independent symmetric ±1 Bernoulli variables, then $M(x)$ would have mean 0 and variance $\sim 6x/\pi^2$, and $M(x)/\sqrt{6x/\pi^2}$ would tend to a $N(0,1)$ distribution. Denjoy (1931, [**274**]), (1964, [**275**]) noted that if this were the case then the Law of the Iterated Logarithm would imply that $M(x)=O\left(\sqrt{x}\log\log x\right)$ and RH would follow.

However, it is known that $M(x)$ does not follow an asymptotic normal distribution. In fact, assuming RH, and that all of the zeros are simple, and assuming further a conjecture of Gonek and Hejdal, it is possible to derive the explicit formula

$$\frac{M(x)}{\sqrt{x}}\equiv\frac{1}{\sqrt{x}}\sum_{n\le x}\mu(x)=2\Re\left(\sum_{\gamma>0}\frac{x^{i\gamma}}{\rho\zeta'(\rho)}\right)+o(1),$$

where $\rho\equiv\frac{1}{2}+i\gamma$ ranges over the nontrivial zeros of $\zeta(s)$ and ζ' is a derivative. (See Titchmarsh (1987, [**1059**, pp. 372–374]).) Arguing from this, together with an additional assumption, Ng (2004, [**803**]), (2008, [**805**]) showed that $M(e^y)/e^{y/2}$ has an asymptotic distribution as y varies, but not one that is Gaussian.[3]

---

[3]There are numerous results involving derivatives of $\zeta(s)$ at its zeros. These include

$$\sum_{0<\gamma\le T}\zeta'(\rho)\sim\frac{T}{4\pi}\log^2\frac{T}{2\pi},\qquad \sum_{0<\gamma\le T}\zeta^{(n)}(\rho)\sim\frac{(-1)^{n+1}}{n+1}\frac{T}{2\pi}\log^{n+1}\frac{T}{2\pi},$$

Under the widely believed assumption that the imaginary parts of the zeros are linearly independent over the rationals, Ng showed that the mentioned asymptotic distribution associated with $M(x)$ is the same as the distribution of the quantity

$$2\Re\left(\sum_{\gamma>0}\frac{X(\gamma)}{\rho\,\zeta'(\rho)}\right)$$

where the $X(\gamma)$ are independent random variables distributed uniformly on the unit circle. This distribution is also not Gaussian. Large deviation properties for distributions of this type were studied by Montgomery & Odlyzko (1988, [**770**]).

On the other hand, over short intervals where one considers quantities such as

$$M(n+h)-M(n)=\sum_{0\le m\le h}\mu(n+m),$$

the situation appears to be different. From numerical evidence, Good & Churchhouse (1968, [**442**]) had conjectured that sums of $\mu(n)$ in blocks of length h, where h is large, are in fact asymptotically normally distributed with mean 0 and variance $6h/\pi^2$. Ng provided some theoretical evidence to support this conjecture by calculating the moments

$$\sum_{n\le N}\big(M(n+h)-M(n)\big)^k$$

under certain assumptions. He also suggested that the asymptotic normality should hold if $h\to\infty$ with $h\le N^{1-\epsilon}$. For the summatory version $L(x)\equiv\sum_{n\le x}\lambda(n)$ of the Liouville function, Ng provided similar theoretical evidence suggesting that $L(n+h)-L(n)$, for $1\le n\le N$, should be asymptotically normal with mean 0 and variance h when $h\to\infty$ with $\log h/\log N\to 0$. Ng also suggested that some of his heuristic arguments may apply to the wider class consisting of multiplicative functions which have mean 0 and are bounded by 1.

12.4. The Chowla conjectures

We begin by again recalling that the Möbius and Liouville functions satisfy $\sum_{n\le x}\mu(n)=O(x)$ and $\sum_{n\le x}\lambda(n)=O(x)$ trivially, that the minuscule improvements $\sum_{n\le x}\mu(n)=o(x)$ and $\sum_{n\le x}\lambda(n)=o(x)$ are equivalents to the Prime Number Theorem, and that each of the improvements $\sum_{n\le x}\mu(n)=O(x^{1/2+\epsilon})$ and $\sum_{n\le x}\lambda(n)=O(x^{1/2+\epsilon})$ is equivalent to the Riemann hypothesis. However, in spite of the obstacles these equivalences portend, it is believed that the Möbius and Liouville functions satisfy many further properties of genuinely random sequences. We had already pointed out that if they behaved like independent and identically distributed random variables then the Law of the Iterated Logarithm would apply, resulting in $O\left(\sqrt{x\log\log x}\right)$ error terms, so that the Riemann hypothesis would follow. However genuine randomness has still broader implications, for it is entirely possible for serially correlated sequences to satisfy this bound.

(note the real-value biases and the alternating signs) and

$$\sum_{0<\gamma\le T}\left|\zeta'(\rho)\right|^2\sim\frac{T}{24\pi}\log^4\frac{T}{2\pi}.$$

See Hughes & Pearce-Crump (2022, [**566**]), and Gonek (1984, [**438**]) for sources and background.

In this section we address recent efforts directed towards the so-called Chowla conjectures concerning the serial dependence structure of these sequences. In time-series parlance, these questions concern the extent to which the Möbius and Liouville functions emulate the properties of "white noise." Such matters have been under active development, and interested readers will wish to seek out the latest results, and consult the most recent works of the authors to be mentioned.

The term *Chowla conjecture* is associated with two brief items (#56 and #57, out of 70 items in all) in a chapter titled "Some unsolved problems and statements of known results" in the monograph by Chowla (1965, [**204**]). These two items read as follows:

> *56. Let $\epsilon_1, \epsilon_2, \ldots \epsilon_g$ be arbitrary numbers each equal to $+1$ or -1, where g is a fixed (but arbitrary) number. Then the equations (in n)*
>
> $$\lambda(n+m) = \epsilon_m \qquad (1 \leq m \leq g)$$
>
> *have infinitely many solutions. For $g \geq 3$ this seems an extremely hard conjecture.*
>
> *57. Let $f(x)$ be an arbitrary polynomial with integer coefficients, which is not, however, of the form $cg^2(x)$ where c is an integer and $g(x)$ is a polynomial with integer coefficients. Then*
>
> $$\sum_{n=1}^{x} \lambda(f(n)) = o(x).$$
>
> *If $f(x) = x$ this is equivalent to the Prime Number Theorem. If the degree of $f(x)$ is at least 2, this seems an extremely hard conjecture* [**204**].

The term 'Chowla conjecture' has also been used for other assertions concerning serial dependence in the series $\{\mu(n)\}$ and $\{\lambda(n)\}$. A correlational version states that for any fixed $m \geq 1$, and any distinct[4] fixed integers $k_1, \ldots, k_m$,

$$\frac{1}{x}\sum_{n \leq x} \lambda(n+k_1)\lambda(n+k_2)\cdots\lambda(n+k_m) \to 0, \quad \text{as } x \to \infty; \tag{12.19}$$

one may write this alternately as

$$\sum_{n \leq x} \lambda(n+k_1)\lambda(n+k_2)\cdots\lambda(n+k_m) = o(x).$$

These conditions would hold (with probability 1) if the $\lambda(n)$ were genuinely independent random variables with equal probabilities of taking on the values ± 1. (Analogous assertions can be made for μ as well.) The conjecture (12.19) may be viewed as a generalization of the Prime Number Theorem, which corresponds to the case $m = 1$. It has, however, not yet been resolved for *any* other value of m—not even for the simplest $m = 2$ two-point correlation of adjacent terms, namely

$$\sum_{n \leq x} \lambda(n)\lambda(n+1) = o(x), \quad \text{as } x \to \infty. \tag{12.20}$$

---

[4]If a value of k_j occurred multiple times those factors could be removed in pairs as they only contribute factors of 1. Hence the requirement that the k_j be distinct is inconsequential as long as at least one of the k_j only occurs an odd number of times.

It is, for instance, not unthinkable that $\lambda(n) = \lambda(n+1)$ and $\lambda(n) = -\lambda(n+1)$ occur with frequencies other than $\frac{1}{2}$, an imbalance which would violate (12.20).

Closely related to Chowla's problem #56 is a formulation which states that all *sign patterns* (that is, all sequences of -1's and $+1$'s) for λ occur in the *same proportions* as for the tosses of a fair coin. Thus if $m \geq 1$, and $\epsilon_1, \epsilon_2, \ldots, \epsilon_m$ is any sequence of -1 and $+1$ values, and with $k_1, k_2, \ldots, k_m$ being distinct integers, this version states that

$$\lim_{x\to\infty} \frac{1}{x} \# \left\{ n \leq x \colon \lambda(n+k_1) = \epsilon_1, \ \ldots, \ \lambda(n+k_m) = \epsilon_m \right\} = \frac{1}{2^m}. \tag{12.21}$$

(Of course, the case $m = 1$ is again equivalent to the PNT.) Formulations (12.21) and (12.19) are equivalent (when taken as assertions valid simultaneously for all m). To show that (12.21) implies (12.19) is fairly straightforward, but the converse is a bit more tricky; if you would like to try your hand at it do not look at the footnote.[5] An alternate proof is given in Section 12.5. As an instance of this equivalence, the condition (12.20) means that the pairing $\big(\lambda(n), \lambda(n+1)\big)$ takes on its four sign patterns $(\pm 1, \pm 1)$ with asymptotically equal probabilities. Since λ is completely multiplicative (meaning that $\lambda(mn) = \lambda(m)\lambda(n)$ for all m, n), one expects its values to have no additive structure and therefore to behave like a sequence of independent values.

The conjecture (12.21) remains far from having been proven. The fact that $\big(\lambda(n), \lambda(n+1)\big)$ takes on all four sign patterns infinitely often has an elementary proof; if you care to try it, don't look at the footnote.[6] Hildebrand (1986, [**550**]), (1986, [**551**]), (1997, [**553**]) showed that each of the eight sign combinations of the triplet $\big(\lambda(n),\lambda(n+1),\lambda(n+2)\big)$ also occurs infinitely often. His proof was based on elementary combinatorial arguments using the multiplicative properties of λ at the small primes 2, 3, and 5. (A first step involved showing that $\lambda(15n-1)$ and $\lambda(15n+1)$ simultaneously equal 1, and likewise equal -1, infinitely often.) Matomäki, Radziwiłł & Tao (2016, [**740**]) extended Hildebrand's result by showing that these eight cases each occur with positive density. Curiously, the cases $(+,+,+)$ and $(-,-,-)$ were technically the most troublesome ones to treat. Tao & Teräväinen (2019, [**1047**]) established that all sixteen four-term sign patterns also occur infinitely often, although in earlier work Buttkewitz & Elsholtz (2011, [**173**]) had shown that each of the sixteen sign patterns do occur in infinitely many four-term arithmetic progressions $(\lambda(n), \lambda(n+d), \lambda(n+2d), \lambda(n+3d))$ and that this also holds for many completely multiplicative arithmetic functions.

---

[5] To see that (12.19) implies (12.21) observe first that

$$\prod_{j=1}^{m} \left(1 + \epsilon_j \lambda(n+k_j)\right) = 2^m,$$

when all $\epsilon_j = \lambda(n+k_j)$, and equals 0 otherwise. The result follows on expanding the product and summing over n. An alternate approach is to substitute $\lambda(h+k_j) = \epsilon_j\left(2 \cdot \mathrm{I}[\lambda(n+k_j) = \epsilon_j] - 1\right)$ into (12.19) and then expand the product.

[6] The $(+1,-1)$ and $(-1,+1)$ cases are trivial since $\lambda(n)$ changes signs infinitely often. To handle $(+1,+1)$, pick any odd n with $\lambda(n) = +1$ (so it has an even number of prime factors). If either of $\lambda(n-1)$ or $\lambda(n+1)$ equal 1 then we are done. Otherwise both of $\lambda(n-1)$ and $\lambda(n+1)$ equal -1, but then $\lambda((n-1)/2) = \lambda((n+1)/2) = +1$ gives such a pair. The same proof works for $(-1,-1)$: We just start with an odd number n with $\lambda(n) = -1$. (Remark: The result can also be shown to follow from the fact that the summatory function $L(x)$ changes sign infinitely often.) See also Harman, Pintz & Wolke (1985, [**502**]).

A much weaker version of (12.20), namely the question of whether or not

$$\limsup_{x\to\infty} \frac{1}{x}\left|\sum_{n\le x}\lambda(n)\lambda(n+1)\right| < 1$$

has likewise turned out to be difficult, although in recent years some progress has been made. In what was hailed as a major breakthrough, Matomäki & Radziwiłł (2016, [**737**]) succeeded in showing that for every $k \ge 1$ there exists a $\delta \equiv \delta(k) > 0$ such that (say)

$$\limsup_{x\to\infty} \frac{1}{x}\left|\sum_{n\le x}\lambda(n)\lambda(n+k)\right| < 1-\delta,$$

thus forging a significant inroad into the Chowla conjecture (under which this value should be 0). Stated otherwise this result means that, for every $k \ge 1$. each of the four sign patterns of $\big(\lambda(n), \lambda(n+k)\big)$ occurs a positive proportion of times. Matomäki and Radziwiłł also showed that this holds for any completely multiplicative function $f(n)$ satisfying $-1 \le f(n) \le 1$, provided $f(n) < 0$ for *some* $n > 0$. Subsequently, Teräväinen (2023, [**1057**]) established a lim sup result in the context of the general form shown in (12.19):

$$\limsup_{x\to\infty} \frac{1}{x}\sum_{n\le x}\lambda(n+k_1)\lambda(n+k_2)\cdots\lambda(n+k_m) < 1.$$

As it happens, the odd order cases are somewhat easier technically and Elliot (1992, [**321**]) had previously shown that

$$\limsup_{x\to\infty} \frac{1}{x}\left|\sum_{n\le x}\lambda(n)\lambda(n+1)\lambda(n+2)\right| \le \frac{20}{21}.$$

Nevertheless, the question of (say) whether or not

$$\frac{1}{x}\sum_{n\le x}\lambda(n)\lambda(n+1)\lambda(n+2) \to 0, \quad \text{as} \quad x\to\infty,$$

remains open. Graham & Hensley (1986, [**448**]), Baker & Harmon (1991, [**69**]), Cassaigne et al. (1999, [**183**]), (2000, [**184**]), and Soundararajan (2017, [**1008**]) may be consulted for some additional details.

In a further breakthrough, Matomäki & Radziwiłł (2016, [**737**]), (2020, [**738**]) proved that the Liouville function exhibits cancellation in almost all *short* intervals. Specifically, this result states that

$$\int_X^{2X}\left|\sum_{x\le n\le x+H}\lambda(n)\right|dx = o(HX)$$

as $X \to \infty$, provided that $H = H(X) \to \infty$. Subsequently, Matomäki, Radziwiłł & Tao (2015, [**739**]) proved an *averaged* version of the full Chowla conjecture:

$$\sum_{k_1,\dots k_m\le K}\left|\sum_{n\le N}\lambda(n+k_1)\lambda(n+k_2)\cdots\lambda(n+k_m)\right| = o(K^m N)$$

for any fixed $m \ge 1$, as *both* $K, N \to \infty$ with $K \equiv K(N) \le N$. Of course, were this to hold for any *bounded* value of K, it would be equivalent to the m-point Chowla conjecture. They also showed that this holds for the Möbius and other multiplicative functions which satisfy $|f(n)| \le 1$. Further results along such lines may be found in Matomäki & Teräväinen (2023, [**742**]), as well as in the five-author paper Matomäki, Radziwiłł, Teräväinen, Tao & Ziegler (2023, [**741**]). For earlier results, see Schur & Schur (1973, [**956**]), Sudo (1984, [**1037**]), Hildebrand (1986, [**550, 551**]), Croot (2003, [**251**]), Green & Tao (2010, [**465**]), and Buttkewitz & Elsholtz (2011, [**173**]).

Matomäki, Radziwiłł & Tao (2015, [**739**]) also proved that for every $t \in \mathbb{R}$,

$$\int_X^{2X} \left| \sum_{x \le n \le x+H} \lambda(n) e^{itn} \right| dx = o(HX), \tag{12.22}$$

as $X \to \infty$, with $H \equiv H(X) \le X$ tending to infinity with X, however slowly. Furthermore, the supremum of the left side over all t satisfies the same bound. (See also Davenport (1937, [**261**]).) One interest in this result stems from spectral analysis in time series. If the $\lambda(n)$ were fully random, the real and imaginary parts of the sum—for values of t ranging over a suitable finite set of values—would approximate independent Gaussian random variables with zero-mean and constant variance. Such results can sometimes help detect correlations among the $\lambda(n)$. In a series of papers—for example, (2023, [**1097**])—Walsh proved a far-reaching extension of (12.22) in which the term e^{itn} is replaced by $e^{itP(n)}$, where the $P(n)$ are polynomials, and a supremum over all polynomials of degree $\le k$ is taken *inside* the integral. (Here k is a fixed positive integer.) It turns out that only a slight further strengthening of this "Fourier uniformity" result would suffice to establish the full logarithmic Chowla conjecture (discussed below) as well as Chowla's conjecture #56; however, the details are technical and are omitted.

For the weaker *logarithmically averaged* version[7] of the Chowla conjecture substantial progress has been made. For the two-point correlation Tao (2016, [**1043**]) proved that

$$\lim_{x\to\infty} \frac{1}{\log x} \sum_{n \le x} \frac{\lambda(n)\lambda(n+k)}{n} = 0, \quad k \ne 0,$$

and that the four 2-point Liouville sign patterns each occur with *logarithmic density* $\frac{1}{4}$. He also showed that this holds for the Möbius function as well:

$$\lim_{x\to\infty} \frac{1}{\log x} \sum_{n \le x} \frac{\mu(n)\mu(n+k)}{n} = 0, \quad k \ne 0.$$

It is a consequence of this result is that all nine admissible configurations of $\big(\mu(n), \mu(n+k)\big)$ occur infinitely often. Tao also conjectured that for all $m \ge 1$

$$\lim_{x\to\infty} \frac{1}{\log x} \sum_{n \le x} \frac{\mu(n+k_1)\cdots\mu(n+k_m)}{n} = 0$$

as long as at least one of the k_j is repeated only an odd number of times. Subsequently, Tao & Teräväinen (2018, [**1045**]), (2019, [**1046**]), (2019, [**1047**]), (2019, [**1048**]) proved that

$$\frac{1}{\log x} \sum_{n \le x} \frac{\lambda(n+k_1)\lambda(n+k_2)\cdots\lambda(n+k_m)}{n} \to 0, \quad \text{as } x \to \infty,$$

[7] Recall equations (9.1) and (9.2). If $(1/x)\sum_{n\le x} f(n)$ converges then so does the logarithmic version $(1/\log x)\sum_{n\le x} f(n)/n$, and to the same value, but the converse need not hold.

for all *odd* values of m and distinct integers $k_1, \dots k_m$, and furthermore that this holds for the Möbius function as well. However the *even* values of $m \geq 4$ have proven to be more difficult, and their logarithmic versions have not yet been decided. The rates at which the two point logarithmically averaged Chowla forms tend to 0 is studied in Helfgott & Radziwiłł (2021, [**537**]), and Pilatte (2023, [**845**]).

Conjectures analogous to (12.19) and (12.21) can also be made for the Möbius function. However, since $\mu(n)$ ranges over the larger set $\{-1, 0, +1\}$, one must take its null values into account and this complicates the analysis. For example, $\mu(n)$ and $\mu(n+4)$ cannot be independent because they are correlated through the event that n is divisible by 4 (in which cases $\mu = 0$). Likewise, the events $\mu(n) = 1$, $\mu(n+1) = 1$, $\mu(n+2) = 1$ and $\mu(n+3) = 1$ cannot be independent since at least one of these must be wrong. For consecutive triples, the method of Hildebrant (1986, [**550**]) establishes that $\big(\mu(n-1), \mu(n), \mu(n+1)\big)$ takes on each of the eight sign patterns $(\pm 1, \pm 1, \pm 1)$ infinitely often (with n obviously being even), however that method does not allow values of 0 to be accounted for. This restriction was removed by Matomäki, Radziwiłł & Tao (2016, [**740**]), and Tao & Teräväinen (2019, [**1047**]) who allowed for the zeros and showed that $\big(\mu(n),\ \mu(n+1),\ \mu(n+2)\big)$ takes on all of its permissible sign patterns infinitely often, and each with positive lower natural density.[8] See also Pilatte (2023, [**845**]).

Ramaré (2018, [**884**]) provides a detailed discussion of the connections between the various seemingly different versions of the Chowla conjectures, and the question of whether or not results for $\{\lambda(n)\}$ can be parlayed into results for $\{\mu(n)\}$ and conversely.

The correlation conjecture (12.19) can be specialized to allow the arguments of λ to range over arithmetic progressions, and (barring certain degenerate cases) can be generalized to allow the arguments to range over polynomial values. It is also to be expected that under certain conditions analogous results should continue to hold for general multiplicative functions. However we will not discuss those cases nor enter into discussion of the Chowla conjecture #57 here. For some details see Landau (1909, [**676**]), Harman, Pintz & Wolke (1985, [**502**]), Katai (1996, [**624**]), Cassaigne et al. (1999, [**183**]), (2000, [**184**]), Sárközy (2001, [**942**]), Ganguli (2013, [**408**]), Srinivasan (2022, [**1016**]), and Teräväinen (2023, [**1057**]). Ayoub, Huard & Williams (1998, [**48**]) give a nice account of the life and work of Sarvadaman Chowla (1907–1995).

Najnudel (2020, [**792**]) studied the equivalent of the Chowla conjectures for *randomized* versions of arithmetic functions. Among other results he showed that if $f(n)$ is a completely multiplicative random function whose values are iid at the primes and distributed uniformly on the unit circle (this is the Steinhaus case), or distributed uniformly on the roots of unity of a given order, then for any fixed $k \geq 1$, the terms of $\big(f(n), f(n+1), \dots, f(n+k)\big)$ will be mutually (that is, *fully*) independent once n is large enough. Why this happens can be explained in terms of the logarithms of the prime factorizations associated with the elements in this collection, but it is a bit subtle. The reader who is up for a challenge may wish

[8]The lower and upper densities of a set $\mathcal{A}$ of positive integers are defined as

$$\liminf_{x\to\infty} \frac{\#\{n \leq x \colon n \in \mathcal{A}\}}{x} \quad \text{and} \quad \limsup_{x\to\infty} \frac{\#\{n \leq x \colon n \in \mathcal{A}\}}{x};$$

the two need not be equal.

to play with this for a while, using ± 1 values at the primes. Najnudel provides a proof.

As the full Chowla conjecture remains undecided, Sarnak proposed a somewhat weaker version of the conjecture. It is formulated in the context of 'topological dynamical systems' (X, T) in which X is a 'compact topological space', while $T : X \to X$ is a continuous map from X into itself. Sarnak's conjecture is that if T is *deterministic* (which here means that it has *zero 'topological entropy'*) then it is *Möbius orthogonal* (or *Möbius disjoint*), meaning that for every continuous function $f : X \to \mathbb{R}$ which maps X into the reals, and every $x \in X$,

$$\lim_{N\to\infty} \frac{1}{N} \sum_{n\le N} f(T^n x)\mu(n) = 0. \tag{12.23}$$

Here $T^n(x) \equiv T(T^{n-1}(x))$ is the nth iterate of T after starting at x. Loosely stated, this asserts that $\mu(n)$ is uncorrelated with every sequence possessing some degee of deterministic structure (that is, predictability).

We will not elaborate on the technical terms and concepts associated with this conjecture; the interested reader may consult the overview by Ferenczi, Kułaga-Przymus & Lemańczyk (2018, [**377**]) where the references to Sarnak's work, as well as to the work of many others, may be found. Here we will limit ourselves to three clarifying examples:

(i) The case where X consists of a single point: In this case $f(T^n x)$ is constant and (12.23) becomes $\sum_{n=1}^{N} \mu(n) = o(n)$. It is therefore equivalent to the Prime Number Theorem.

(ii) The case where X consists of a finite number of points: In this case $T^n x$—and therefore also $f(T^n x)$ in (12.23)—will range over all of the periodic functions whose period is equal to or less than the number of points in X. The requirement (12.23) will then essentially be equivalent to Dirichlet's prime number theorem for arithmetic progressions.

(iii) The case where X is the unit circle and T is a rotation: This case includes the rotational functions $f(T^n x) = e^{2\pi i n\alpha}$ for which Davenport (1937, [**261**]) had shown that (12.23) holds. In fact, Davenport proved the stronger result that

$$\sup_{\alpha} \left| \sum_{n=1}^{N} e^{2\pi i \alpha n} \mu(n) \right| = O\left(\frac{N}{\log^k N} \right)$$

holds for every k. (The constants associated with the order term will of course depend on k.) The same result is also known to hold if $e^{2\pi i \alpha n}$ is replaced by $e^{2\pi i \alpha P(n)}$ for polynomials $P(n)$.

Using purely combinatorial methods, Sarnak (2011, [**944**]) showed that the Chowla conjecture implies the Sarnak conjecture, although the converse does not appear to be available. However, Tao (2017, [**1044**]) proved that the *logarithmically averaged* versions of the Chowla and Sarnak conjectures are equivalent, and also showed that the Sarnak conjecture with respect to λ is equivalent to the same conjecture with respect to μ. Gomilko et al. (2018, [**437**]) have shown that if either the Sarnak conjecture or the logarithmic Chowla conjecture was true then there exists an increasing subsequence of the integers along which the full Chowla conjecture would hold. Sarnak surmised that a general rigidity theorem due to Marina

Ratner could provide an important link for proving his conjecture. An exposition of Ratner's deep results may be found in Morris (2005, [**777**]) and in a 2007 blog-post by Terence Tao.[9] See also Cellarosi & Sinai (2011, [**185**]), (2013, [**186**]).

12.5. Selected proofs

PROOF OF ASYMPTOTIC NORMALITY RESULT (12.18). This result states that

$$\frac{1}{x}\,\#\left\{n \le x\colon \frac{g(n)-\mu(g,x)}{\sigma(g,x)} \le u\right\} \;\to\; \frac{1}{\sqrt{2\pi}}\int_{-\infty}^{u} e^{-t^2/2}\,dt, \quad \text{as} \quad x \to \infty,$$

where

$$g(n) = \sum_{p\le n} g(p)X(p), \quad \mu(g,x) = \sum_{p\le x}\frac{g(p)}{p}, \quad \text{and} \quad \sigma^2(g,x) = \sum_{p\le x}\frac{g^2(p)}{p} + O(1),$$

and where $X(p) = 1$ with probability $\frac{1}{p}$, and is 0 otherwise. To prove this it will suffice to establish Lyapunov's condition for asymptotic normality (see Appendix D) in the form

$$\frac{1}{\sigma^{2+2\delta}(g,n)}\sum_{p\le n}\mathrm{E}\,|g(p)X(p)|^{2+2\delta} \to 0, \quad \text{as} \quad n\to\infty$$

for some $\delta > 0$. The expression on the left may be taken as

$$\sum_{p\le n}\frac{|g(p)|^{2+2\delta}}{p} \Bigg/ \left(\sum_{p\le n}\frac{g^2(p)}{p}\right)^{1+\delta}.$$

Now put $a(p) \equiv g^2(p) \ge 0$ (initially) and use the asymptotic $p \sim n\log n$ for the nth prime to recast this into the asymptotically equivalent expression

$$\sum_{2\le n\le N}\frac{a^{1+\delta}(n)}{n\log n} \Bigg/ \left(\sum_{2\le n\le N}\frac{a(n)}{n\log n}\right)^{1+\delta}.$$

Here N corresponds to the number of primes $p \le n$, while $a(n)$ has been redefined so as to correspond to the value of $g^2(p)$ at the nth prime. Next let

$$P(n) \equiv \frac{1}{\log\log N}\times\frac{1}{n\log n}, \quad 3 \le n \le N;$$

by an integral comparison $\sum_{n=3}^{N} P(n) \to 1$ as $N \to \infty$ so that for N not too small $P(n)$ acts as a probability measure.[10] With this notation the criterion can be taken as

$$\left(\frac{1}{\log\log N}\right)^{\delta} \times \left(\sum_{3\le n\le N} a^{1+\delta}(n)P(n)\right) \Bigg/ \left(\sum_{3\le n\le N} a(n)P(n)\right)^{1+\delta}.$$

Under mild growth conditions on g the expectation in the middle braces will be finite. The expectation in the third set of braces will then also be finite by the well-known monotonicity property for moments: $\mathrm{E}(|Y|^r)^{1/r} \le \mathrm{E}(|Y|^s)^{1/s}$ when $r \le s$. The result then follows because the term $1/(\log\log N)^{\delta}$ tends to 0. □

[9] "Ratner's Theorems" `https://terrytao.wordpress.com/2007/09/29/ratners-theorems/`.

[10] In fact, one could arrange here for $P(n)$ to exactly be a probability measure for each N simply by defining it as $P(n) = (c(N)n\log n)^{-1}$, where $c(N) = \sum_{2\le n\le N}(n\log n)^{-1} \sim \log\log N$.

PROOF OF THE EQUIVALENCE OF CHOWLA CONDITIONS (12.19) AND (12.21). With $m \geq 1$, the $k_1, k_2, \ldots, k_m$ distinct integers, and $\epsilon_1, \epsilon_2, \ldots, \epsilon_m$ any sequence of -1 and $+1$ values, we need to show the equivalence of

$$\sum_{n \leq x} \lambda(n+k_1)\lambda(n+k_2)\cdots\lambda(n+k_m) = o(x), \quad \text{as } x \to \infty,$$

and

$$\lim_{x\to\infty} \frac{1}{x}\#\Big\{n \leq x\colon \lambda(n+k_1) = \epsilon_1, \ \ldots, \ \lambda(n+k_m) = \epsilon_m\Big\} \ = \ \frac{1}{2^m}.$$

We will prove the following somewhat more general

Basic Result: Let $x(1), x(2), \ldots$ be an infinite series of values ranging over a finite set $J \equiv \{j_1, j_2, \ldots j_k\}$ of k distinct numbers. Let f be a probability mass function on J, so that $\sum_{j\in J} f(j) = 1$, and suppose f has moments $\eta(r) \equiv \mathrm{E}\, X^r$ where the random variable X has the distribution f. Consider the following assertions:

(A) For all positive integers L, all integers $0 = t_1 < t_2 < \cdots < t_L$, and all positive integers $r_1, r_2, \ldots, r_L$,

$$\frac{1}{N}\sum_{n=1}^{N} x^{r_1}(n+t_1)x^{r_2}(n+t_2)\cdots x^{r_L}(n+t_L) \ \longrightarrow \ \eta(r_1)\eta(r_2)\cdots\eta(r_L).$$

(B) For all positive integers M, all integers $0 = t_1 < t_2 < \cdots < t_M$, and all choices $j_{i_1}, j_{i_2}, \ldots, j_{i_M}$ of values in J, and with $\mathrm{I}[\,\cdot\,]$ being the 0–1 indicator function,

$$\frac{1}{N}\sum_{n=1}^{N} \mathrm{I}\Big[x(n+t_1) = j_{i_1}, \, x(n+t_2) = j_{i_2}, \, \cdots, \, x(n+t_M) = j_{i_M}\Big]$$
$$\longrightarrow \ f(j_1)f(j_2)\cdots f(j_1).$$

Then under the stated conditions, the assertions (A) and (B) are equivalent.

REMARKS 12.6. Condition (A) asserts asymptotic independence for moments, while condition (B) asserts asymptotic independence for events. Both (A) and (B) involve only finite collections of $\{t_j\}$. Note too that since X is bounded, the distribution f is determined by its moments.

PROOF. Consider first the one-dimensional context in which (A) and (B) are just

$$\text{(A)} \ \frac{1}{N}\sum_{n=1}^{N} x^r(n) \ \to \ \eta(r), \quad \text{and} \quad \text{(B)} \ \frac{1}{N}\sum_{n=1}^{N} \mathrm{I}[\,x(n) = j\,] \ \to \ f(j),$$

for all $r \geq 1$, and for all $j \in J$, respectively. The characteristic function $\varphi_N(t)$ for a uniform distribution on $x(1), x(2), \ldots x(N)$ can be written as either

$$\varphi_N(t) = \sum_{r=1}^{\infty} \frac{(it)^r\left(\frac{1}{N}\sum_{j=1}^{N} x^r(n)\right)}{r!}, \quad \text{or as } \varphi_N(t) = \sum_{j\in J}\left(\frac{1}{N}\sum_{n=1}^{N}\mathrm{I}[\,x(n) = j\,]\right)e^{ijt}.$$

The two representations are equivalent because each is just the characteristic function of the 'sample', the first based on empirical moments, and the second on empirical probabilities. (They can also be shown to be equivalent by expanding out and comparing.) As $N \to \infty$, these converge to

$$\varphi(t) = \sum_{r=1}^{\infty} \frac{(it)^r \operatorname{E} X^r}{r!}, \quad \text{and} \quad \varphi(t) = \sum_{j \in J} f(j) e^{ijt},$$

respectively, the two representations again being equivalent since each is just the characteristic function of f. The $\varphi_N(t)$ converge pointwise to $\varphi(t)$ in both cases; in the first case on account of (A) and the Dominated Convergence Theorem (since the values in J are bounded), and in the second case on account of (B) and finiteness of the sum. Thus (A) and (B) are each equivalent to the pointwise convergence $\varphi_N(t) \to \varphi(t)$, and therefore equivalent to each other. The argument for the higher dimensional cases $L > 1$ and $M > 1$ is analogous; the sums involved just become finite-dimensional multivariate sums. □

REMARKS 12.7.

(i) In the proof, the set J can be allowed to be infinite, but one then needs to protect the Dominated Convergence part of the proof.
(ii) For a real-valued completely multiplicative arithmetic function to take on only a finite number of values, it can only equal -1, 0, or $+1$ on the primes.

APPENDIX A

Analytic Functions

This appendix provides a self-contained account of those elements of complex analysis useful for following the discussions in this book. Referring to this subject (in the context of number theory) Andrew Wiles once said that *"The more analytic parts ... did not really catch my imagination until I had learned some complex analysis. Only then could I appreciate the power of the zeta function."*[1] Our treatment is meant to be a concise but readable account of this remarkable and surprisingly elegant theory. By omitting proofs we are able to offer a transparent overview, asking only in return that the reader take some time to reflect on the *meaning* of what has been said after each segment of this appendix has been read (and as needed, reread). The goal is to try to understand three things:

- What is a *holomorphic function* (synonymously, an *analytic function*)?
- What does *analytic continuation* mean?
- What are the key properties of analytic functions?

We assume that the reader is familiar with basic analysis from a rigorous course in calculus, with the use of complex numbers, Euler's formula $e^{ix} = \cos(x) + i\sin(x)$, and with the polar representation $z \equiv x + iy \equiv re^{i\theta}$, where $r = |z| = \sqrt{x^2 + y^2}$ and $\theta = \arg z$. Many introductory textbooks cover these topics.[2]

Holomorphic functions. The objects of interest to us are complex-valued functions $f(z)$ of complex variables $z = x + iy \in \mathbb{C}$, and their derivatives

$$f'(z) \equiv \frac{df(z)}{dz} \equiv \lim_{h\to 0} \frac{f(z+h) - f(z)}{h}. \tag{A.1}$$

Here the crucial point is that *this limit is required to exist in the complex sense*, which means that *it must exist independently of the manner in which $h \to 0$ in the complex plane*. This requirement, which is deceptive in its apparent simplicity, places considerable restrictions on the functions and has surprising and far-reaching consequences.

If $f(z)$ is defined on an open[3] subset $U \subset \mathbb{C}$ of the complex plane, it is said to be *holomorphic* (or *analytic*) at a point $z \in U$ if it has a derivative there in this complex sense. If this holds for all $z \in U$ then f is said to be holomorphic

[1] In the foreword to Hardy & Wright (2008, [**499**]). Andrew Wiles is known for having proved the 350-year-old conjecture known as Fermat's Last Theorem.

[2] Throughout this appendix we will use the more traditional $z = x + iy \in \mathbb{C}$ to represent a complex variable rather than Riemann's form $s = \sigma + it$ used elsewhere in this book. Incidentally, there is more to $\sqrt{-1}$ than meets the i; see Nahin (1998, [**788**]) for an illuminating discussion.

[3] Let's keep this simple: By saying U is *open* we mean that it excludes its boundary, so that any point in U is surrounded by a neighbourhood which lies entirely in U. A *neighbourhood* of a point can be thought of as a small circular disc which contains the point but whose boundary is missing.

on U. Functions holomorphic on all of $\mathbb{C}$ are said to be *entire*. It is a deep result—which is far from obvious or trivial—that if f is holomorphic on U, then it can automatically be differentiated on U (in the complex sense) infinitely many times. It is, however, quite straightforward to show that the rules for differentiating sums, products, quotients, and composite functions in the complex sense remain the same as the familiar rules for functions of a real variable.

Suppose now that f is a function on an open set U. By a *primitive* (or *antiderivative*) of f we mean a function F such that $F' = f$ everywhere on U. It will be of interest to know whether a holomorphic function f on U has holomorphic primitives. It turns out—and indeed this is a highly nontrivial fact—that if we require our open set U to be simply connected,[4] then holomorphic functions on U will indeed have holomorphic primitives (of all orders) along with holomorphic derivatives.

EXAMPLE A.1. On the right half-plane $U = \{x + iy\colon x > 0\}$, the holomorphic function $f(z) = \frac{1}{z}$ does have the holomorphic primitive $F(z) = \log z \equiv \log|z| + i \arg z$; in particular, the argument $\arg z \in (-\frac{\pi}{2}, \frac{\pi}{2})$ can be defined continuously on this right half-plane. (The complex logarithm function is discussed further in Appendix B.) However, on the punctured plane $U = \{z \in \mathbb{C}\colon z \neq 0\}$ the function $f(z) = \frac{1}{z}$ is still holomorphic but it does not have a holomorphic primitive, essentially because $\arg z$ cannot be defined continuously on the entire punctured plane (no matter what multiples of 2π we try to add as it "wraps around").

Putting $u + iv \equiv f(z)$, where $z = x + iy$, and viewing each of $u \equiv u(x, y)$ and $v \equiv v(x, y)$ as real functions of two real variables, the requirement that f be complex-differentiable at z leads to the so-called *Cauchy–Riemann equations*

$$\frac{\partial u}{\partial x} = \frac{\partial v}{\partial y} \quad \text{and} \quad \frac{\partial u}{\partial y} = -\frac{\partial v}{\partial x}\,. \tag{A.2}$$

These partial derivative equations are an immediate consequence of definition (A.1). They are obtained by equating the limits one obtains when one first lets $h \to 0$ along the real axis, and then lets $h \to 0$ along the imaginary axis. In view of these equations, it also follows that

$$\frac{df(z)}{dz} = \frac{\partial u}{\partial x} - i\,\frac{\partial u}{\partial y} = \frac{\partial v}{\partial y} + i\,\frac{\partial v}{\partial x}\,. \tag{A.3}$$

Here is a recommended, reasonably straightforward exercise: Verify (A.2) and (A.3).

A sort of converse to result (A.2) is also available: if $u(x, y)$ and $v(x, y)$ are continuously differentiable functions which satisfy the Cauchy–Riemann equations (A.2) on U, then the function defined as $f = u + iv$ will be holomorphic on U. Altogether then, one can say that once differentiation of $f(z)$ along the real and complex axes leads to the same value, the differentiation of $f(z)$ along *any* path will also result in that value. A further essential point is that even though (A.2) (corresponding to a holomorphic function) only refers to the existence (in x and in y) of *first* partial derivatives for u and v, when those two equations are satisfied, the real-valued function $u(x, y)$ and $v(x, y)$ will automatically be differentiable an infinite number of times.

[4] Loosely speaking, a *simply connected* set is one that has no 'holes': thus every simple closed curve in the set is the boundary of an interior that is also wholly contained in the set. For example, the punctured plane is not simply connected because of the 'hole' at $z = 0$.

Although we will make no use of this fact, it is worth mentioning that upon differentiating the Cauchy–Riemann equations one finds that $u(x,y)$ and $v(x,y)$ are *harmonic* functions. This means that $\nabla^2 u = 0$ and $\nabla^2 v = 0$, where ∇^2 is the Laplacian operator

$$\nabla^2 \equiv \frac{\partial^2}{\partial x^2} + \frac{\partial^2}{\partial y^2}.$$

One can also show that any harmonic function $u(x,y)$ is the real part of a holomorphic function (at least on simply connected sets), and that it determines the imaginary part $v(x,y)$ of that function to within an additive constant. A similar remark applies to $v(x,y)$. One refers to $u(x,y)$ and $v(x,y)$ as *harmonic conjugates*. Even though they are real-valued, harmonic functions share many of the properties of holomorphic functions. Harmonic functions are sometimes called *potential* functions and arise in studies of electrostatics, gravitation, fluid dynamics, and other physical processes associated with 'conservation laws', although in such applications they typically arise in their three-dimensional analogues.

EXAMPLE A.2. The exponential function

$$f(z) \equiv e^z \equiv e^{x+iy} = e^x \cos y + ie^x \sin y$$

(discussed in Appendix B) is entire. The functions $u(x,y) = e^x \cos y$ and $v(x,y) = e^x \sin y$ are conjugate harmonic functions and satisfy (A.2), as well as $\nabla^2 u = 0$ and $\nabla^2 v = 0$.

Contour integration. An essential concept in complex analysis is *integration along a curve*—also called a *path*, or a *contour*. If $z(t)$ traces out a piecewise smooth curve γ in $\mathbb{C}$, as the parameter t varies continuously over $[a,b] \subset \mathbb{R}$, the *line (or contour) integral* $\int_\gamma f(z)\,dz$ of f along γ can be defined—in the manner familiar from Riemann integration—as the limit of sums such as $\sum_j f\big(z(t_j^*)\big)\big(z(t_{j+1}) - z(t_j)\big)$ based on ever-finer sequences $z(t_1), z(t_2), \ldots, z(t_n)$ of points which lie on γ, with intermediate positions $z(t_j^*)$. When this limit exists, and some mild conditions are met, this integral can alternately be evaluated as

$$\int_\gamma f(z)\,dz \equiv \int_a^b f(z(t))z'(t)\,dt. \tag{A.4}$$

It can be shown that the value of such an integral does not depend on the parameterization $z(t)$ of the curve γ. Of course, if the path is traversed in the opposite direction, the value of the integral will be the same, except with opposite sign.

If the beginning and the endpoint of the curve are identical, in other words, if $z(a) = z(b)$, the curve is said to be *closed*. An example of a closed curve is the counterclockwise-oriented circle of radius r centred at z_0: we may represent it as $z(t) = z_0 + re^{it}$, where $0 \le t \le 2\pi$. To simplify subsequent discussions, we will consider only 'simple' curves and so we rule out curves that wind around more than once, curves that cross over themselves, and closed curves without an obvious interior and exterior. Also, when integrating over a closed curve, we will always assume the integration is in the positive, namely in the counterclockwise, direction. Closed curves can more generally cross over themselves, and can wind in both clockwise as well as counterclockwise directions. Such more general curves are treated in many textbooks and possess results analogous to the simpler ones that are considered below.

A fundamental and very far-reaching result of the theory is *Cauchy's Theorem*.[5] It dates back to an 1814 memoire by Augustin-Louis Cauchy (1789–1857) and states that if $f(z)$ is holomorphic on an open set $U \subset \mathbb{C}$, and if γ is *any* piecewise smooth curve in U which is closed (that is it ends where it began), *and whose interior is also in* U, then

$$\oint_\gamma f(z)\,dz = 0. \tag{A.5}$$

The notation $\oint$ rather than $\int$ here calls attention to the 'circular' (that is, the closed path) nature of the integral. One can avoid having to say that the interior of γ must also lie in U by assuming that the region U is simply connected to begin with. An alternate and equivalent way of stating Cauchy's Theorem is that if a function f is holomorphic on a region U without holes, then the line integral of f from a point a to a point b is independent of the path of integration, as long as a and b, as well as the paths, are all in U. This nontrivial result is proved in most books on complex analysis and is essentially a consequence of the fact, mentioned earlier, that every holomorphic function has a (holomorphic) primitive on such a region. Stein & Shakarchi (2003, [**1021**]), for example, offer a particularly elegant treatment of this result.

An immediate consequence of Cauchy's Theorem, which turns out to be particularly important, is that if f is holomorphic on an open set U containing a simple closed contour γ together with its interior, and if z is any point in the interior of the contour, then

$$f(z) = \frac{1}{2\pi i}\oint_\gamma \frac{f(w)}{w-z}\,dw\,. \tag{A.6}$$

Using (A.4), this result may be proved by first observing that

$$\oint_S \frac{dz}{z} = \int_0^{2\pi} \frac{ie^{it}}{e^{it}}\,dt = 2\pi i,$$

where S is a circle of radius 1 about the origin; we have parametrized it here as $z(t) = e^{it}$ with $0 \le t \le 2\pi$. One would obtain the same result if the circle had a different radius or if it was centred around some other point, say z_0. (It requires only the use of Cauchy's Theorem to show this.) Combining this fact with (A.5) then allows us to conclude that

$$\oint_\gamma \frac{dz}{z-z_0} = 2\pi i,$$

where γ is *any* simple closed contour (in U) about z_0. The proof of (A.6) is then completed by means of a simple argument involving integration around a 'keyhole' contour. The details may be found in many textbooks.

Result (A.6) is known as the *Cauchy Integral Formula* and it is possible to justify repeated differentiation under the integral to obtain the following more general version of that formula:

$$f^{(n)}(z) = \frac{n!}{2\pi i}\oint_\gamma \frac{f(w)}{(w-z)^{n+1}}\,dw\,. \tag{A.7}$$

[5]Sometimes also referred to as the Cauchy–Goursat Theorem. In their book, *Methods of Mathematical Physics*, Sir Harold Jeffreys and his wife Bertha have called this and other results of the theory "so surprising at first sight that nothing short of a proof can make them credible."

In fact, it is by arguing from these identities that one can prove the remarkable fact that a holomorphic function can be differentiated (in the complex sense) an infinite number of times. A useful observation regarding identities (A.6) and (A.7) is to appreciate that their integrands (as functions of w) remain holomorphic, except for the introduction of the 'pole' at $w = z$. (The notion of poles and other singularities is discussed below.) A particularly valuable (as well as surprising) conclusion from (A.6) is that the values of f on the boundary γ determine all of its values in the interior of γ. Note too that if z were to lie outside of γ, the integrands in formulas (A.6) and (A.7) would be holomorphic and so those integrals would simply vanish on account of (A.5).

We remark, in passing, that there is a converse to Cauchy's Theorem known as *Morera's Theorem*. It states that if f is continuous on an open region, and if all of its integrals along closed piecewise continuous curves in the region are zero, then f must be holomorphic in that region. This can be shown to hold even if the class of curves were to be restricted, as for example, to triangles or to rectangles. A noteworthy aspect of this result is that it has no smoothness requirement on f, only that it be continuous, and yet it would then follow that f must be infinitely differentiable.

A further essential fact about holomorphic functions is that they are fully determined by their values on any (infinite) set of distinct points possessing a point of accumulation.[6] More precisely, if f and g are holomorphic functions on a connected open set U, and are equal on such a subset of points in U (where the point of accumulation is also in U), then f and g must be identical on all of U. This uniqueness property is the rationale for a process known as *analytic continuation* in which a function which is defined and holomorphic over some region can have its definition extended to a function which is holomorphic over a larger region, with the point being that *this extension is unique*. When this process is continued as far as possible,[7] the resulting region can be considered to be the natural domain of definition of the holomorphic function. The construction of the Riemann zeta function, in which one starts with a function which is 'analytic' in a right half-plane and ends up with a function which is analytic everywhere (except at one point), is a key example of an analytic continuation.

EXAMPLE A.3. Polynomials $f(z) = a_n z^n + \cdots + a_1 z + a_0$ are holomorphic functions. Functions such as $f(z) = \sin(z)$, $\cos(z)$, and e^z (discussed in Appendix B) are entire functions, and other functions which have power series expansions in z are also holomorphic within their regions of convergence. So are sums, products, and compositions of holomorphic functions. Ratios of holomorphic functions are also holomorphic except at zeros of their denominator. If $f(z)$ is nonzero and holomorphic on a simply connected open set, one can show that $\log f(z)$ is also

[6] We say z is a *point of accumulation* if any open set containing z contains an infinite number of the points.

[7] In many practical cases there is a single maximal region of continuation; but in other cases, there can be multiple maximal regions all containing the original one. For example, we saw that $\log z$ is holomorphic on the right half-plane; it can be analytically continued to any 'slit plane' of the form $\{z \in \mathbb{C},\ z \neq 0\colon\ \arg z \neq \theta\}$ given a constant $\theta \in [\frac{\pi}{2}, \frac{3\pi}{2}]$, but those analytic continuations are not mutually compatible.

holomorphic there. Examples of functions which are not holomorphic[8] include the complex conjugate $f(z) = \bar{z}$ and the modulus $f(z) = |z|$. On the other hand, if $f(z)$ is holomorphic, then so is $\overline{f(\bar{z})}$.

EXAMPLE A.4. Consider the function $f_1(z)$ defined by the infinite series $f_1(z) = \sum_{n=0}^{\infty} z^n$. Initially, we only know that f_1 is well-defined within its circle of convergence $|z| < 1$ (it is in fact holomorphic there), and that it is undefined for $|z| \geq 1$. Suppose now that 'by pure luck' we recognize that the function $f_2(z) = 1/(1-z)$ coincides with $f_1(z)$ on $|z| < 1$. Then since f_2 is holomorphic everywhere (except at $z = 1$) it can be viewed as providing the analytic continuation of f_1 to the entire complex plane (except for $z = 1$). A similar argument can be applied to the pair $\sum_{n=0}^{\infty}(-1)^n z^n$ and $1/(1+z)$. Likewise, consider the result

$$\int_1^{\infty} t^{-z}\, dt = \frac{1}{z-1}.$$

This integral makes sense (that is, it converges) only for $\Re z > 1$, but the expression on the right provides its analytic continuation.

EXAMPLE A.5. The series $\zeta(s) = \sum_{n=1}^{\infty} 1/n^s$ converges only when $\Re(s) > 1$ and is analytic in that right half-plane. However straightforward computation yields

$$\left(1 - 2^{1-s}\right)\zeta(s) = \sum_{n=1}^{\infty} \frac{1}{n^s} - 2\sum_{n=1}^{\infty} \frac{1}{(2n)^s} = \sum_{n=1}^{\infty} \frac{(-1)^{n+1}}{n^s}.$$

Since the alternating series on the right can be shown to converge for $\Re(s) > 0$, this relation supplies an analytic[9] continuation for $\zeta(s)$ over the same region, as long as $s \neq 1$. (At $s = 1$, the factor $\left(1 - 2^{1-s}\right)$ on the left equals 0.) This result—as well as the analytic continuation of $\zeta(s)$ to the remainder of the complex plane—is considered in Chapter 5.

EXAMPLE A.6. Using concepts from Appendix D, the power series defined as $f(z) = \sum_0^{\infty} A_n z^n$, where the A_n are independent standard normal random variables, can be shown to define a random function which is analytic inside the unit disc almost surely (that is, with probability 1). The radius of convergence of the series (see formula (A.9) below) can be shown to be

$$\limsup_{n\to\infty} |A_n|^{1/n} = 1, \quad \text{almost surely}.$$

It turns out that these functions (almost surely) cannot be analytically continued anywhere beyond $|z| < 1$. Peres & Virág (2004, [**839**]) studied the distribution of the zeros of such functions. A related example is that if coefficients δ_n are selected independently with probabilities $P[\delta_n = +1] = P[\delta_n = -1] = \frac{1}{2}$, then with probability 1 the functions $f(z) = \sum_{n=0}^{\infty} \delta_n z^n$ which result also cannot be analytically continued beyond $|z| < 1$ almost surely. This will also be the case if the coefficients δ_n are selected according to $\delta_n = e^{i\theta_n}$, with the θ_n independent and uniformly distributed on $[0, 2\pi]$.

[8] We remark that there is a sense in which it can be said that for a complex function to be holomorphic, it must be free of $\bar{z}$ and depend only on z. For example, the modulus function, which can also be written as $|z| = \sqrt{z\bar{z}}$, is not holomorphic.

[9] The fact that convergent Dirichlet series are always analytic functions is discussed in Appendix C.

Analytic functions. A *power series* about a point $z_0 \in \mathbb{C}$ is an expansion of the form

$$f(z) = \sum_{n=0}^{\infty} a_n (z - z_0)^n, \tag{A.8}$$

where the coefficients a_n are complex numbers. These play an essential role in the theory. The radius of convergence r of series (A.8) can be shown to satisfy

$$1/r = \limsup_{n\to\infty} |a_n|^{1/n}; \tag{A.9}$$

the series converges for $|z-z_0| < r$ and diverges for $|z-z_0| > r$. On $|z-z_0| = r$, when r is finite, arbitrary behaviour is possible (although this boundary must contain at least one 'singularity' in a sense we will discuss soon). Note that the radius of convergence determined by (A.9) is unchanged if, for example, a_n in (A.8) is replaced by $|a_n|$ or if the signs of the a_n are arbitrarily altered. A key property of power series is that they can be differentiated (in the complex sense), as well as integrated, any number of times. Furthermore, these operations may be performed term by term, and they leave the radius of convergence unchanged. From these facts one obtains, upon differentiating (A.8), the standard formula

$$a_n = f^{(n)}(z_0)/n!. \tag{A.10}$$

The coefficients a_n can alternately be recovered from $f(z)$ by applying Cauchy's formula:

$$a_n = \frac{1}{2\pi i} \oint_\gamma \frac{f(w)}{(w - z_0)^{n+1}}\, dw$$

with γ being a contour about z_0 in the usual way. (See formula (A.7).)

The importance of power series in complex analysis stems, first, from the fact that a power series defines a function which is holomorphic within its disk of convergence, and second, from the converse fact (which can be proved using Cauchy's formula) that if f is holomorphic at z_0, then f can be represented as a power series about z_0 whose disk of convergence extends from z_0 to as far as the nearest singularity. In particular, an *entire* function can be represented by a power series around *any* point, and such a series will have an infinite radius of convergence. If a function f, defined on an open set $U \subset \mathbb{C}$, coincides in a neighbourhood of $z_0 \in U$ with a power series about z_0 with a positive radius of convergence, the function is said to be *analytic* at z_0. If this holds for all points in U then f is said to be analytic in U. It is a key result in the theory of complex functions that *being analytic over U and being holomorphic over U are equivalent conditions*; the terms analytic and holomorphic can thus be used interchangeably. One consequence of this is that holomorphic functions can be arbitrarily well approximated in any finite neighbourhood by a polynomial.

Singularities, zeros, and poles. A *singularity* of an analytic function is a point at which it is not defined;[10] for example, the function $1/(z-a)$ has a singularity at $z = a$ for any constant a. However, since we often speak of analytic

[10] In some contexts, it is helpful to speak of $f(z)$ having a *removable singularity* at $z = z_0$, where $f(z_0)$ is initially undefined, but setting $f(z_0) = \lim_{z\to z_0} f(z)$ results in a function that is continuous and analytic in a neighbourhood of z_0. For example, $(z^2-9)/(z-3)$ has a removable singularity at $z = 3$, and $(\sin z)/z$ has a removable singularity at $z = 0$. However, in this book we will always use 'singularity' to refer to a nonremovable singularity.

continuations, it is useful to use 'singularity' to refer to a second, closely related phenomenon: so we will say that an analytic function $f(z)$ has a singularity at $z = z_0$ if it does not possess any analytic continuation that includes a neighbourhood of z_0. For example, $\log z$ has a singularity at $z = 0$.

The zeros and poles of analytic functions play an important role in the theory. If the function f (not identically zero) is analytic on an open set U, and if $f(z_0) = 0$ at some $z_0 \in U$, then z_0 is called a *zero* of f. In that case, we will have $a_0 = 0$ in Taylor expansion (A.8) of $f(z)$ about the point $z = z_0$. The smallest integer n for which $a_n \neq 0$ is then called the *order* or the *multiplicity* of that zero. If $n = 1$ the zero is said to be *simple*. If an analytic function has a set of zeros with an accumulation point inside its (connected) domain, then it must be identically zero on its domain (since it agrees with the analytic function $g(z) = 0$ on a set with an accumulation point). An equivalent phrasing is that the zeros of a nontrivial analytic function must be 'isolated' (meaning that each one has a neighbourhood containing no other zeros).

A *pole* of an analytic function is a type of singularity which is a kind of opposite of a zero. We say that z_0 is a *pole* of f, of *order*[11] or *multiplicity* m, if f can be represented as

$$f(z) = \sum_{n=-m}^{\infty} a_n (z - z_0)^n \tag{A.11}$$

in a 'punctured' neighbourhood of z_0, for some $m > 0$ with $a_{-m} \neq 0$. This means that f is equal, near z_0, to the sum of an analytic function—the $\sum_{n=0}^{\infty}$ part—and the part $\sum_{n=-m}^{-1}$ which defines its behaviour about the pole; the latter sum is referred to as the *principal part* of the pole. If $m = 1$ the pole is said to be *simple*. An equivalent definition would be that $f(z)$ is equal to $(z - z_0)^{-m}$ times an analytic function that does not vanish at $z = z_0$. Poles of analytic functions must also be isolated.

A pole is not the only type of isolated singularity an analytic function can possess; all other types of singularities are called *essential*. An example is the function $f(z) = e^{1/z}$ at $z = 0$. Although otherwise analytic, this function tends to ∞ as $z \to 0$ on the positive real axis; tends to 0 as $z \to 0$ on the negative real axis; and oscillates with increasing frequency (although remaining bounded) as $z \to 0$ along the imaginary axis. In the vicinity of an essential singularity, an otherwise analytic function can still be represented by a double-sided series as in (A.11), except that the lower limit of the sum will now be $n = -\infty$. A pole and an essential singularity can be distinguished by the fact that at a pole z_0 we have $\lim_{z \to z_0} |f(z)| = \infty$, regardless of how z approaches z_0, while this is not the case at an essential singularity. The behaviour of an analytic function in the vicinity of an essential singularity is peculiar: a theorem of Picard says that within any neighbourhood (however small) about such a point, not only will the function take on *all* complex values—with at most one exception—but will take on each of its complex values infinitely many times.

[11]One could use the language of removable singularities to define the order of a zero of $f(z)$ at $z = z_0$ to be the largest integer n for which $f(z)/(z - z_0)^n$ has a removable singularity at $z = z_0$, and similarly define the order of a pole of $f(z)$ at $z = z_0$ to be the smallest integer m such that $(z - z_0)^m f(z)$ has a removable singularity at $z = z_0$.

The expansions (A.11) are called *Laurent series*, and (as intimated) may have lower limit $n = -\infty$. And just as the part $\sum_{n=0}^{\infty}$ may converge *within* a radius of R (about z_0), the part $\sum_{n=-\infty}^{-1}$ may converge *outside* a radius of r. When $r < R$, expansion (A.11) will then coincide with a function f that is analytic in the ring or *annulus* $r < |z - z_0| < R$. Furthermore, the formula (A.10) for the coefficients of the expansion can be generalized to

$$a_n = \frac{1}{2\pi i} \oint_\gamma \frac{f(w)}{(w - z_0)^{n+1}}\, dw, \quad \text{where now } n = 0, \pm 1, \pm 2, \ldots, \tag{A.12}$$

and where γ is now any closed contour that travels once around in the (r, R) annulus.

The $n = -1$ term in (A.11) is the only one that does not possess a primitive. (This is because even though $d \log z/dz = 1/z$, the function $\log z$ cannot uninterruptedly be analytically continued around a punctured neighbourhood of 0; see Appendix B.) The coefficient a_{-1} therefore turns out to play a special role and is called the *residue* of f at z_0; we write this as

$$\operatorname{res}_{z_0} f = a_{-1}.$$

Thus, for example, if $f(z)$ is analytic, $f(z)/(z-a)$ will have residue $f(a)$ at $z = a$, and more generally $f(z)/(z-a)^n$ will have residue $f^{(n-1)}(a)/(n-1)!$ there if $n > 1$. Using Cauchy's Theorem one can show that the integral along a closed curve (in the positive sense, namely counterclockwise) of an otherwise analytic function f, but having a pole at z_0 in the interior, is equal to $2\pi i$ times the residue of f at z_0. (The argument supporting this result is essentially the one indicated after equation (A.6).) A consequence of this is the *Residue formula* which states that if f is analytic on an open set containing the closed contour γ together with its interior, except for having poles in the interior of the contour at $z_1, z_2, \ldots, z_m$, then

$$\oint_\gamma f(z)\, dz = 2\pi i \sum_{k=1}^{m} \operatorname{res}_{z_k} f,$$

where the sum is over those poles. This formula has many applications to evaluating integrals.

If the only singularities of an otherwise holomorphic function $f(z)$ on $U \subset \mathbb{C}$ are (isolated) poles, the function is said to be *meromorphic* on U. If f is meromorphic on all of $\mathbb{C}$, as well as at ∞ in the sense that $g(x) \equiv f(1/z)$ is holomorphic around $z = 0$ except possibly for a pole there, then $f(z)$ is said to be *meromorphic on the extended complex plane*. In this case $f(z)$ will have no more than a finite number of poles and a finite number of zeros since poles and zeros must be isolated both 'locally' as well as 'at ∞'.

A theorem of Weierstrass implies that if f is meromorphic on an open set $U \subset \mathbb{C}$ then f must be the ratio, say g/h, of two holomorphic functions on U, while if f is meromorphic on all of $\mathbb{C}$, then this ratio must consist of two entire functions. It can be assumed that the zeros and poles of f (together with their multiplicities) will coincide with the zeros of g and the zeros of h, respectively. Meromorphic functions are substantially—although certainly not completely—determined by their zeros and their poles. For example, in the simplest case, where the function is meromorphic on the *extended* complex plane, an argument based on Liouville's Theorem shows that it must exactly equal a ratio of polynomials. (See also the remarks after the theorems of Weierstrass and Hadamard noted below.)

Finally, let us note an interesting connection between the power series expansions of analytic functions and Fourier series on circles. Thus suppose that $f(z) = \sum_{n=0}^{\infty} a_n z^n$ is analytic for $|z| < R$, and substitute $z = re^{2\pi i\theta}$, where $0 < r < R$ so that

$$f(z) = \sum_{n=0}^{\infty} a_n r^n e^{2\pi i n\theta}.$$

Then the a_n may be recovered by means of the well-known Fourier inversion formula which can be expressed here in the form (see, for example, Stein & Shakarchi, 2003, [**1021**, p. 101])

$$a_n = \frac{1}{2\pi r^n} \int_0^{2\pi} f(re^{i\theta})\, e^{-in\theta}\, d\theta, \quad n \geq 0. \tag{A.13}$$

In the more general case, where $f(z) = \sum_{n=0}^{\infty} a_n (z - z_0)^n$ the formula becomes

$$a_n = \frac{1}{2\pi r^n} \int_0^{2\pi} f(z_0 + re^{i\theta})\, e^{-in\theta}\, d\theta, \quad n \geq 0. \tag{A.14}$$

In fact, this is essentially formula (A.12) applied to the case where $n \geq 0$ there, and γ is a circle of radius r about z_0. When $n < 0$ the integrals (A.13) and (A.14) will both equal 0.

Analytic functions satisfy many truly remarkable properties. We close this appendix by highlighting some of the more important ones and invite the reader to reflect mindfully on as many as possible of the rather abundant collection of properties we now state. It is of course to be understood that to appreciate the full range of these results does require a certain amount of time.

PROPERTY A.1 (Liouville's Theorem). *An entire function $f(z)$ which is bounded must be a constant. More generally, if $|f(z)| \leq A + B|z|^k$ for some A and for some $B > 0$, then f must be a polynomial of degree at most k.*

PROPERTY A.2 (A theorem concerning limits). *If a sequence of holomorphic functions converges uniformly on an open set U, then the function to which the sequence converges will again be holomorphic. Indeed, this will be true even if the sequence only converges uniformly on subsets of U, as long as the collection of such subsets spans all of U. In particular, if the partial sums of a series of holomorphic functions converges uniformly on (subsets of) U, then the series itself represents a holomorphic function on U. Under these conditions, the sequence of derivatives will also converge uniformly. (This is very unlike the real variable case, in which uniform limits of differentiable functions need not be differentiable.)*

PROPERTY A.3 (The Maximum Modulus Principle). *If f is a nonconstant function holomorphic on an open region U, then $|f|$ cannot attain a maximum inside U. A corollary of this result is that if $|f|$ has a minimum in U and is not constant, then the minimum must equal 0, because otherwise $1/f$ would be holomorphic and non-constant near that point but then $|f|$ would have a maximum there. (It is easy to argue that these results hold for local maxima and local minima as well.) If the closure of U is compact, and if f is continuous on its boundary, then $|f|$ will attain both a maximum and a minimum on that boundary. (These results must also hold for polynomials, but are nontrivial even in that case.)*

PROPERTY A.4 (Schwarz's Lemma). *Suppose $f(z)$ is analytic in the unit disc $|z| < 1$, and that $f(0) = 0$. If $|f(z)| \leq 1$ in that disk then also $|f(z)| \leq |z|$ there,* *and* $|f'(0)| \leq 1$. *Furthermore, if* $|f(z)| = |z|$ *holds at any point* $z_0 \neq 0$ *in the disk then* $f(z) = cz$ *for some* $|c| = 1$. *(This lemma is essentially a corollary to the Maximum Modulus Principle, applied to the function* $f(z)/z$ *with its singularity at* $z = 0$ *removed.)*

PROPERTY A.5 (Runge's Theorem). *If* $f(z)$ *is analytic on a bounded open region* $U \subset \mathbb{C}$ *which contains a closed (bounded) subset* $K \subset U$*, then* $f(z)$ *can be uniformly approximated arbitrarily well on* K *by rational functions which have no poles in* K. *If* K *is simply connected then polynomials suffice for the approximation. (Remark: There are generalizations of this result due to Mergelyan and others; Fornaess et al., 2020,* [**386**]*, provide a nice survey of approximation results.)*

PROPERTY A.6 (Cauchy's Argument Principle). *If* $f(x)$ *is meromorphic about and within a closed contour* γ *whose boundary has neither zeros nor poles, then its 'logarithmic residue' satisfies*

$$\frac{1}{2\pi i}\oint_\gamma \frac{d\log f(z)}{dz}\,dz \equiv \frac{1}{2\pi i}\oint_\gamma \frac{f'(z)}{f(z)}\,dz \equiv \frac{1}{2\pi i}\oint_\gamma d\log f(z) = N - P,$$

where N *is the number of zeros and* P *is the number of poles of* f *within* γ*, both counted with multiplicity. (This result is a consequence of the Residue Theorem.) Note that* $d\log f(z)/dz$ *is defined as* $f'(z)/f(z)$.

PROPERTY A.7 (The Mean Value Property). *If* f *is holomorphic on a disc* $\{z : |z - z_0| \leq R\}$ *of radius* R *centred at* z_0*, then*

$$f(z_0) = \frac{1}{2\pi}\int_0^{2\pi} f\left(z_0 + re^{i\theta}\right)d\theta, \quad \textit{for any } 0 < r < R. \tag{A.15}$$

Thus $f(z_0)$ *equals the average of* f *on the circles about* z_0. *This result essentially follows from the* $n = 0$ *case of* (A.12)/(A.13) *since* $a_0 = f(z_0)$. *It follows from* (A.15) *that* $f(z_0)$ *also equals its average over the disc:*

$$f(z_0) = \frac{1}{\pi R^2}\iint\limits_{|z-z_0|\leq R} f(z)\,dx\,dy \quad (\textit{where } z = x + iy),$$

and also that

$$|f(z_0)| \leq \frac{1}{2\pi}\int_0^{2\pi}\left|f(z_0 + re^{i\theta})\right|d\theta, \quad \textit{for any } 0 < r < R.$$

The mean value property has a type of converse: If a continuous function satisfies (A.15) *throughout an open set then its real and imaginary parts are harmonic. (Recall that an analytic function has harmonic real and imaginary parts, and that a complex-valued function is analytic only if its real and imaginary parts are harmonic conjugates.)*

PROPERTY A.8 (Parseval's Identity). *If* $f(z) = \sum_{n=0}^{\infty} a_n(z - z_0)^n$ *converges for* $|z - z_0| < R$*, then*

$$\frac{1}{2\pi}\int_0^{2\pi}\left|f(z_0 + re^{i\theta})\right|^2 d\theta = \sum_{n=0}^{\infty}|a_n|^2 r^{2n}, \quad \textit{for } 0 \leq r < R.$$

(This is just the Parseval formula of Fourier analysis applied to the function determined by the values of f *on a circle of radius* r *about* z_0*.)*

PROPERTY A.9 (The Poisson Integral Formula). *If f is analytic within a disk $\{z \colon |z - z_0| \leq R\}$ and continuous on the disc, and if $z = z_0 + re^{i\theta}$ with $0 \leq r < R$ is a point inside the disk, then*

$$f(z) = \frac{1}{2\pi} \int_0^{2\pi} P_{r,R}(\theta - t) f(z_0 + Re^{it})\, dt, \tag{A.16}$$

where (for x real)

$$P_{r,R}(x) = \frac{R^2 - r^2}{R^2 - 2rR\cos x + r^2}$$

is the so-called Poisson kernel. It satisfies $(1/2\pi)\int_0^{2\pi} P_{r,R}(\theta - t)\, dt \equiv 1$ since (A.16) *also holds for $f \equiv 1$. (Formula* (A.16) *is a special case of the result that says that the values of an analytic function at arbitrary points inside a region are determined by its values on the boundary of that region.)*

PROPERTY A.10 (Rouché's Theorem). *If f and g are holomorphic on a connected open region together with its boundary, with $|f| > |g|$ everywhere on the boundary, then f and $f + g$ have the same number of zeros (counting multiplicity) inside the region. (A consequence of this result is that small changes to an analytic function within some region does not alter the number of zeros it has there.)*

PROPERTY A.11 (Hurwitz's Theorem). *If a sequence $\{f_n(z)\}$ of analytic functions converges uniformly to an analytic function $f(z)$ in some open region containing z_0, and if $f(z)$ has a zero of multiplicity m at z_0, then there will be an $\epsilon > 0$ such that once n is large enough, each $f_n(z)$ will have exactly m zeros (counting multiplicity) in the disk $|z - z_0| < \epsilon$, all of which converge to z_0 as $n \to \infty$.*

PROPERTY A.12 (Jensen's Formula). *Suppose that a disk of radius R about the origin lies interior to a region on which f is holomorphic. If $f(0) \neq 0$, and if f has no zeros on the boundary of the disc, then*

$$\log|f(0)| = \sum_{k=1}^{N} \log\left(\frac{|z_k|}{R}\right) + \frac{1}{2\pi}\int_0^{2\pi} \log\left|f(Re^{i\theta})\right| d\theta,$$

where $z_1, z_2, \ldots, z_N$ are the zeros of f in the disc, counted with multiplicity. Note that the $\log|f(0)|$ on the left is just a scaling factor which will be 0 if $|f(0)| = 1$. One can write the formula alternately as

$$|f(0)| \prod_{k=1}^{N} \log\left(\frac{R}{|z_k|}\right) = \exp\left(\frac{1}{2\pi}\int_0^{2\pi} \log\left|f(Re^{i\theta})\right| d\theta\right)$$

with the right side now being the geometric mean of $|f|$ on the boundary. (If $f(0) = 0$ one can consider instead $z^{-m}f(z)$, where m is the multiplicity.) Jensen's formula shows that as the number of zeros inside the disc increases—so that the $\sum \log$ term decreases—the $\int \log f$ term will increase. It provides a link between the number of zeros of an analytic function and the rate at which it can grow. Loosely stated, the more zeros a function has, the faster it must grow. For example, if $P(z)$ is a polynomial of degree n, its exact growth is $|P(z)| = O\left(|z|^n\right)$ as $|z| \to \infty$, and this corresponds exactly to its number of zeros.

PROPERTY A.13 (The Borel–Carathéodory Theorem). *Suppose $f(z)$ is analytic on the closed disk $|z| \leq R$, and let $A(r) \equiv \max_{|z| \leq r} |f(z)| = \max_{|z|=r} |f(z)|$ (these* *being equal by the maximum modulus principle). Then for* $0 < r < R$, *we have*

$$A(r) \leq \frac{2r}{R-r} \sup_{|z| \leq R} \Re f(z) + \frac{R+r}{R-r} |f(0)|.$$

(The point here is that one can bound an analytic function using only its real part.)

PROPERTY A.14 (The Little Picard Theorem). *Except when it is a constant, an entire function takes on every finite value with at most one exception. If it is not a polynomial, it will take on every finite value an infinite number of times with at most one exception.*

PROPERTY A.15 (The Great Picard Theorem). *If* f *is analytic in a region, except for an essential singularity at a point, then in every neighbourhood of that point* f *will take on every complex value, with at most one exception, an infinite number of times.*

PROPERTY A.16 (Gaston Julia's Theorem). *With the exception of polynomials, every entire function possesses a ray through the origin (called a Julia ray) with the property that in every angle about this ray (however narrow) the function takes on all possible values, with at most one exception, an infinite number of times.*

PROPERTY A.17 (Nonvanishing entire functions). *An entire function* $f(z)$ *which has no zeros must be of the form* $f(z) = e^{g(z)}$, *where* $g(z)$ *is an entire function. Consequently its logarithm* $g(z) \equiv \log f(z)$ *will also be an entire function. (Recall that a function is entire if it is analytic on all of* $\mathbb{C}$ *and has no singularities except possibly 'at* ∞*.')*

PROPERTY A.18 (A theorem of Weierstrass). *Given any sequence of points* $\{z_n\}$ *that do not accumulate (except perhaps at* ∞*), and any sequence* $\{m_n\}$ *of positive integers, there exists an entire function* f *whose zeros are precisely those* z_n, *with corresponding multiplicities* m_n. *Furthermore, any other such function must be of the form* $f(z)e^{g(z)}$, *where* $g(z)$ *is entire.*

PROPERTY A.19 (A theorem of Hadamard). *In the context and notation of the theorem of Weierstrass just mentioned, Hadamard showed that if* $|f(z)|$ *has an upper bound of the form* $Ae^{B|z|^\rho}$ *then the function* $g(z)$ *must be a polynomial of degree* $\leq \rho$. *A function* f *bounded in this way is said to have finite order with order of growth given by the infimum of such values of* ρ. *(See, for example, Stein & Shakarchi (2003,* [**1021**, Chapter 5, §4 and §5]*).) Polynomials have order* 0 *in this sense. The higher the order of a function is, the more zeros it can have in any given region.*

One joint consequence of these results of Weierstrass and Hadamard is that, given any nonoverlapping sequences of isolated points $\{a_n\}$ *and* $\{b_n\}$ *with assigned multiplicities, there is a meromorphic function on* $\mathbb{C}$ *whose zeros and poles are given precisely by* $\{a_n\}$ *and* $\{b_n\}$ *respectively, and with the corresponding multiplicities.*

PROPERTY A.20 (A result of Mittag-Leffler). *Given any finite or countable set of isolated points* $\{z_k\}$ *in* $U \subset \mathbb{C}$, *there exist holomorphic functions* f *on* U *which take on arbitrarily prescribed values for* $f(z_k)$ *as well as finite numbers of arbitrarily prescribed derivatives at each of those points.*

PROPERTY A.21 (Hadamard's Three Circles Theorem). *Suppose f is analytic within an annulus about zero bounded by radii $0 \le r \le R$, and for $r \le \rho \le R$ let*

$$M(\rho) \equiv \max_{|z|=\rho} |f(z)| = \max\left\{\left|f(\rho e^{i\theta})\right| : 0 \le \theta \le 2\pi\right\}.$$

Then the function $M(\rho)$ is log-convex over $[r, R]$. Specifically,

$$\log M(\rho) \le \frac{\log(R/\rho)}{\log(R/r)} \log M(r) + \frac{\log(\rho/r)}{\log(R/r)} \log M(R).$$

(This result takes its name from the fact that $M(\cdot)$ is defined on circles, while convexity of a curve involves a relation among triplets on that curve.)

PROPERTY A.22 (The Riemann Mapping Theorem). *If D is a simply connected open subset of $\mathbb{C}$, but neither empty nor equal to all of $\mathbb{C}$, then there exist one-to-one analytic functions which map D onto the disk $\{z \colon |z| < 1\}$. Such a function f will be unique if we require, for some given value $a \in D$, that $f(a) = 0$ and that $f'(a)$ be real with $f'(a) > 0$. Since the inverse mapping will also be analytic, it follows that if D' is a second such region there will be one-to-one analytic functions mapping D onto D'. (More precise statements of this result may be found in many texts.)*

PROPERTY A.23 (The Phragmén–Lindelöf Theorem). *We state this result only for the case of vertical strips as the general result is rather involved. Suppose $f(z)$ is bounded and holomorphic on the vertical strip $U = \{x + iy \colon a \le x \le b\}$. Denote the maxima of $|f|$ on vertical lines by*

$$M(x) = \sup\{|f(x + iy)| \colon y \in \mathbb{R}\}, \quad a \le x \le b.$$

Then $M(x)$ is log-convex on $a \le x \le b$, so that $M(x)^{b-a} \le M(a)^{b-x} M(b)^{x-a}$; or equivalently,

$$\log M(x) \le \left(\frac{b-x}{b-a}\right) \log M(a) + \left(\frac{x-a}{b-a}\right) \log M(b), \quad a < x < b.$$

Overholt (2015, [**828**, p. 241]*) gives an alternate version for vertical strips. Rudin (1987,* [**925**, Theorem 12.9] *) treats the case case of horizontal strips. Titchmarsh (1939,* [**1058**, §5.6–§5.8]*) considers a more general case. These results are basically extensions of the Maximum Moduls Principle. See also Stein & Shakarchi (2003,* [**1021**, Chapter 4, Theorem 3.4]*).*

PROPERTY A.24 (Green's (or Stokes's) Theorem). *Strictly speaking, this result is not part of complex analysis; but it has a similar flavour (and can even be used to prove Cauchy's Theorem). If $P(x, y)$ and $Q(x, y)$ are real functions of two real variables with continuous partial derivatives, then*

$$\oint_C (P dx + Q dy) = \iint_R \left(\frac{\partial Q}{\partial x} - \frac{\partial P}{\partial y}\right) dx dy,$$

where C is a simple closed curve and R is the region in its interior.

Of course, there is more to complex analysis than has been summarized in this appendix. We have not touched on the conformal (that is the local angle-preserving) and other geometric properties of holomorphic mappings, or on the theories of harmonic functions, elliptic functions, theta functions, or Riemann surfaces, among other topics.

On account of its importance—and its elegance as well—a great many books have been written on the subject of complex analysis. A concise selection of good ones (at assorted levels) might include the following: Ahlfors (1979, [**8**])—a classic, although at a somewhat higher level, Bak & Newman (1997, [**65**]), Boas (2010, [**116**]), Fornberg & Piret (2020, [**387**])—nicely illustrated, Gamelin (2001, [**407**])—a particularly thorough treatment, Kaplan (1966, [**620**]), Marsden & Hoffman (1999, [**731**]), Moore & Hadlock (1991, [**776**]), Priestley (2003, [**869**]), Remmert (1991, [**890**])—which contains a great deal of interesting historical commentary, and Stein & Shakarchi (2003, [**1021**])—one of the authors' favourites. A fair number of the books on this subject can be downloaded freely from their websites.

APPENDIX B

The Gamma and Related Functions

The *exponential function* on the complex plane is defined as the analytic continuation of its real-variable Taylor series and can be represented as

$$e^z = \lim_{n\to\infty}\left(1+\frac{z}{n}\right)^n = \sum_{n=0}^{\infty}\frac{z^n}{n!}, \quad z\in\mathbb{C}.$$

It is intimately linked with the trigonometric functions. The *sine* and *cosine* functions on $\mathbb{C}$ may be defined—again by analytic continuation of their real-variable counterparts—as

$$\sin z = \sum_{n=0}^{\infty}(-1)^n\frac{z^{2n+1}}{(2n+1)!} = \frac{e^{iz}-e^{-iz}}{2i}, \qquad \cos z = \sum_{n=0}^{\infty}(-1)^n\frac{z^{2n}}{(2n)!} = \frac{e^{iz}+e^{-iz}}{2},$$

and they satisfy $e^z = e^x(\cos y + i\sin y)$, where $z = x+iy$. The three functions mentioned are *entire* (that is, they are analytic on the whole complex plane). The sine and cosine also have Weierstrass product representations stemming from their zeros:

$$\frac{\sin\pi z}{\pi z} = \prod_{n=1}^{\infty}\left(1-\frac{z^2}{n^2}\right), \quad \text{and } \cos\pi z = \prod_{n=1}^{\infty}\left(1-\frac{z^2}{\left(n-\frac{1}{2}\right)^2}\right), \quad z\in\mathbb{C}.$$

The exponential function has no zeros, while the zeros of the sine and cosine functions all occur on the real line (at their usual lattice positions) and all have multiplicity 1. The sine and cosine functions are essentially identical, being only translates $\cos(z) = \sin(z+\pi/2)$ of each other. In fact, all functions whose zeros are spaced uniformly along a line and have multiplicity 1 are equal to translated, rotated and dilated versions of these to within a multiplicative factor $e^{g(z)}$ with $g(z)$ entire. Classical formulas satisfied by the trigonometric functions include[1]

$$\frac{\pi}{\sin(\pi z)} = \lim_{m\to\infty}\sum_{n=-m}^{m}\frac{(-1)^n}{z-n} = \frac{1}{z} + 2z\sum_{n=1}^{\infty}\frac{(-1)^n}{z^2-n^2},$$

$$\left(\frac{\pi}{\sin\pi z}\right)^2 = \sum_{n=-\infty}^{\infty}\frac{1}{(z-n)^2}, \qquad \frac{\sin\pi z}{\pi z} = \prod_{n=1}^{\infty}\cos\left(\frac{\pi z}{2^n}\right),$$

[1]When combined with the Weierstrass product for the sine, the first of these gives the amusing identity:

$$1 + 2z^2\sum_{n=1}^{\infty}\frac{(-1)^n}{z^2-n^2} = \prod_{n=1}^{\infty}\left(1-\frac{z^2}{n^2}\right)^{-1}.$$

and

$$\pi \cot \pi z \equiv \frac{\pi}{\tan(\pi z)} \equiv \frac{\pi \cos(\pi z)}{\sin(\pi z)} = \frac{1}{z} + 2z \sum_{n=1}^{\infty} \frac{1}{z^2 - n^2} \equiv \lim_{m\to\infty} \sum_{n=-m}^{m} \frac{1}{y+n}.$$

The complex version of the (natural) *logarithm* continues to be an inverse of the exponential function. If $z = x + iy = re^{i\theta} = e^{\log r + i\theta}$, with $r > 0$ and θ both real, $r = |z| = \sqrt{x^2 + y^2}$ and $\theta = \arg z$, then

$$\log z = \log |z| + i \arg z \equiv \log r + i\theta.$$

However, because $e^{2\pi i} = 1$, the imaginary part of the logarithm is defined only up to integer multiples of $2\pi i$. One may eliminate this multiplicity by requiring $-\pi < \theta \leq \pi$. By further requiring z to lie in the "slit plane" $\mathbb{C}\backslash(-\infty, 0]$—in other words, by not allowing z to lie on the negative real axis or to equal 0—we obtain the so-called *principal branch* of the logarithm function. When restricted in this way the logarithm becomes an analytic function and satisfies the Taylor expansion

$$\log(1 - z) = -\sum_{n=1}^{\infty} \frac{z^n}{n}, \quad \text{for } |z| \leq 1,\ z \neq 1.$$

(This expansion is most easily established by noting that the derivatives of the left and right sides are both equal to $-1/(1-z)$.) At $z = 0$, $\log z$ is said to have a *logarithmic singularity*.

Arbitrary powers can be defined using the principal branch of the logarithm: for $z \in \mathbb{C}\backslash(-\infty, 0]$, and arbitrary c, one defines $z^c = e^{c \log z}$. In this way z^c can also be viewed as being an analytic function of z when restricted to the slit-plane. The *logarithmic derivative* is not affected by the issue of multiplicity and may be defined quite generally as

$$\frac{d \log f(z)}{dz} \equiv \frac{f'(z)}{f(z)}, \quad f(z) \neq 0.$$

EXAMPLE B.1. If $f(z) = c\prod_{j=1}^{n}(z - c_j)$ is a polynomial then

$$\frac{d \log f(z)}{dz} \equiv \frac{f'(z)}{f(z)} = \sum_{j=1}^{n} \frac{1}{z - c_j}.$$

The remainder of this appendix is devoted to a detailed study of the *Gamma function*, an important, and well-studied function which—given its surprising connections to the Riemann ζ function—makes its properties of considerable interest. It may, in the first instance, be defined as

$$\text{(B.1)} \qquad \Gamma(s) = \int_0^{\infty} t^{s-1} e^{-t}\, dt = \int_0^1 (\log 1/t)^{s-1}\, dt, \quad \text{for } s \in \mathbb{C},\ \Re(s) > 0.$$

Its analytic continuation is not difficult to establish: For example, one can split the first $\int_0^{\infty}$ here into $\int_0^1 + \int_1^{\infty}$, expand the e^{-t} within the $\int_0^1$, and then integrate term by term to get

$$\text{(B.2)} \qquad \Gamma(s) = \sum_{n=0}^{\infty} \frac{(-1)^n}{n!(n+s)} + \int_1^{\infty} t^{s-1} e^{-t}\, dt, \quad s \in \mathbb{C},\ s \neq 0, -1, -2 \ldots.$$

The $\int_1^\infty$ integral here is entire, and we may see from this that the function $\Gamma(s)$ is analytic everywhere except for simple poles at $s = -n$ for $n = 0, 1, 2, \ldots$, with residues $(-1)^n/n!$.

The reciprocal function $1/\Gamma(s)$ is entire, with simple zeros at $n = 0, 1, 2, \ldots$, and has the Weierstrass–Hadamard product representation

$$\text{(B.3)} \qquad \frac{1}{\Gamma(s)} = s\, e^{\gamma s} \prod_{n=1}^{\infty} \left(1 + \frac{s}{n}\right) e^{-s/n}, \quad s \in \mathbb{C},$$

where $\gamma = 0.5772157\ldots$ is the Euler–Mascheroni constant defined in (B.6) further below.[2] Using the well-known formula (B.7) below, this relation may be stated alternately as

$$\text{(B.4)} \qquad \Gamma(s+1) = e^{-\gamma s} \prod_{n=1}^{\infty} \frac{e^{s/n}}{1 + s/n}, \quad s \neq -1, -2, -3, \ldots.$$

Indeed, following Weierstrass, these formulas may be taken as the definition of $\Gamma(s)$, with (B.1) then derived from it. The relations (B.3)/(B.4) again show that $\Gamma(s)$ has simple poles at $n = 0, -1, -2, \ldots$, with corresponding residues $(-1)^n/n!$, while the known convergence of the product in (B.3) can be used to show that $\Gamma(s)$ *has no zeros*. The Gamma function is among the simplest possessing the mentioned properties, and these features turn out to be essential for demonstrating that the complex zeros of Riemann's zeta function are distributed symmetrically about the line $\Re s = \frac{1}{2}$.

A formula closely related to (B.3)/(B.4) is

$$\text{(B.5)} \qquad \Gamma(s) = \frac{1}{s} \prod_{n=1}^{\infty} \left(1 + \frac{1}{n}\right)^s \left(1 + \frac{s}{n}\right)^{-1}$$

and is valid everywhere, except at the poles $s = 0, -1, -2, \ldots$. This formula dates to Euler in 1729 and is the one he used to define $\Gamma(s)$. In fact $\Gamma(s)$ may be defined using any of (B.2)–(B.5), and (B.1) can be derived from any one of them. A proof of the equivalence of (B.3) and (B.5) may be found in Karatsuba & Voronin (1992, [**622**, Appendix 3]); comparing them leads to the observation that the expressions

$$\prod_{n=1}^{\infty} \left(1 + \frac{1}{n}\right)^{-s} \left(1 + \frac{s}{n}\right) \quad \text{and} \quad e^{\gamma s} \prod_{n=1}^{\infty} \left(1 + \frac{s}{n}\right) e^{-s/n}$$

must be equal. (Note, however, that these expressions would no longer converge if the $\left(1 + \frac{s}{n}\right)$ factors were removed from them.) This equality is related to properties of the ubiquitous *Euler–Mascheroni constant* γ whose definition can be taken as

[2] Italian mathematician Lorenzo Mascheroni (1750–1800) is known for having shown that any geometric construction made using two compasses and a straight edge can be made by using two compasses alone.

any one of:[3]

$$\gamma = \lim_{N\to\infty}\left(\sum_{n=1}^{N}\frac{1}{n} - \log N\right)$$
$$= 1 - \int_1^\infty \frac{t-\lfloor t\rfloor}{t^2}\,dt \equiv 1 - \int_1^\infty \frac{\{t\}}{t^2}\,dt = -\int_0^\infty e^{-t}\log t\,dt. \tag{B.6}$$

Here $\lfloor\cdot\rfloor$ and $\{\cdot\}$ are the integer and fractional parts functions. This constant was announced by Euler in 1776 and studied by Mascheroni in 1790. The delightful book by Havil (2003, [**517**]) and the detailed review by Lagarias (2013, [**668**]) elucidate the captivating origins of this constant. Like e and π, the fundamental constant γ seemingly stands alone, bearing no known relation to any other mathematical constants, but while e and π have been proven to be irrational no such proof for γ has yet been offered. Curiously, whenever e and γ occur *together* in a 'mathematically natural' result, they seem typically to occur in the form $e^{\pm\gamma}$.

Now taking logarithms in (B.4) one obtains

$$\log\Gamma(s+1) = -\gamma s + \sum_{n=1}^{\infty}\left(\frac{s}{n} - \log\left(1+\frac{s}{n}\right)\right),$$

and using (B.7) once again, this relation can be differentiated to obtain

$$\frac{d\log\Gamma(s+1)}{ds} \equiv \frac{\Gamma'(s+1)}{\Gamma(s+1)} = \frac{\Gamma'(s)}{\Gamma(s)} + \frac{1}{s} = -\gamma + \sum_{n=1}^{\infty}\left(\frac{1}{n} - \frac{1}{n+s}\right), \quad s \neq 0, -1, -2, \ldots.$$

If we set $s = 1$ here, the sum at the right collapses telescopically to 1, and we find that $\Gamma'(1) = -\gamma$. Differentiating once more gives

$$\frac{d^2\log\Gamma(s)}{ds^2} = \sum_{n=0}^{\infty}\frac{1}{(n+s)^2}, \quad s \neq 0, -1, -2, \ldots.$$

Note that for integer arguments $n > 1$ the first derivative above collapses to

$$\psi(n) \equiv \frac{\Gamma'(n)}{\Gamma(n)} = -\gamma + \sum_{k=1}^{n-1}\frac{1}{k},$$

whence $\psi(1) = -\gamma$ and $\psi(2) = -\gamma + 1$; one can also show that

$$\psi(n+1/2) = -\gamma - 2\log 2 + \sum_{k=1}^{n}\frac{2}{2k-1}.$$

[3]The constant γ can also be taken as the 'principal value' $\zeta(1) = \gamma$ of the zeta function at 1, or even as

$$\gamma = \sum_{n=2}^{\infty}(-1)^n\frac{\zeta(n)}{n}.$$

Lagarias (2013, [**668**]) has argued that the name for the constant γ should rightly be Euler's alone.

For its interest, note that the sum appearing within (B.6) can also be represented as

$$H_N = \sum_{n=1}^{N}\frac{1}{n} = \int_0^1 \frac{1-x^N}{1-x}\,dx.$$

The function $\psi(s) \equiv \Gamma'(s)/\Gamma(s)$ is known as the *digamma function*, and $\psi(s+1) = \psi(s) + 1/s$ is its functional equation. At $s = \frac{1}{2}$ and $s = \frac{3}{2}$ it equals $-\gamma - 2\log 2$ and $2 - \gamma - \log 4$ respectively, while at $s = \frac{1}{4}$ and $s = \frac{3}{4}$ it equals $-\gamma - 3\log 2 - \frac{\pi}{2}$ and $-\gamma - 3\log 2 + \frac{\pi}{2}$. It is also known that $\psi(\frac{1}{3}) = -\gamma - \frac{3}{2}\log 3 - \frac{\sqrt{3}\pi}{6}$, and $\psi(\frac{2}{3}) = -\gamma - \frac{3}{2}\log 3 + \frac{\pi}{2\sqrt{3}}$. Like $\Gamma(s)$, the digamma function $\psi(s)$ is analytic everywhere except for simple poles at $s = 0, -1, -2, \ldots$, with residues all equal to -1. The digamma function has the Taylor expansion

$$\psi(s+1) = -\gamma + \sum_{k=1}^{\infty} (-1)^{k+1}\zeta(k+1)s^k, \quad |s| < 1$$

involving zeta function coefficients. Furthermore, it satisfies the relation

$$\psi(s) = \lim_{N\to\infty} \left(\log N - \sum_{n=0}^{N} \frac{1}{n+s} \right) = -\gamma + \sum_{n=0}^{\infty} \left(\frac{1}{n+1} - \frac{1}{n+s} \right),$$

and for each $n \geq 1$ and $\epsilon > 0$, satisfies the approximations

$$\psi(s+1) = \log s + \sum_{k=1}^{n} \frac{(-1)^k \zeta(1-k)}{s^k} + O\left(\frac{1}{|s|^{n+1}} \right), \quad -\pi + \epsilon < \arg s < \pi - \epsilon$$

from which we see that $\psi(s) \sim \log s + \frac{1}{2s}$ as $|s| \to \infty$ under the same condition on $\arg s$.

The Gamma function satisfies a number of important functional identities. Integrating (B.1) by parts leads to the *functional equation for the Gamma function*[4]

$$\Gamma(s+1) = s\Gamma(s), \quad \text{for } s \neq 0, -1, -2, \ldots. \tag{B.7}$$

Because $\Gamma(1) = \Gamma(2) = 1$, we obtain the well-known result $\Gamma(n+1) = n!$ for $n = 1, 2, \ldots$. Using (B.7) we also see that $\lim_{s\to 0} s\Gamma(s) = \Gamma(1) = 1$ from which we recover the behaviour $\Gamma(s) \sim \frac{1}{s}$ in the vicinity of the pole at $s = 0$. Property (B.7) can be used to recover the behaviour of $\Gamma(s)$ at its other poles as well.

For real $x > 0$, the function $\Gamma(x)$ is convex. It approaches ∞ as x approaches 0 from the right. On $x > 0$, it decreases up to about $x \approx 1.46163$, reaching its sole minimum of approximately 0.88560 there, and thereafter it increases. However this convexity, together with property (B.7), are not sufficient to uniquely characterize Γ. For a unique characterization (to within a constant factor) a condition such as the convexity of $\log \Gamma(x)$ is required as well.[5] The logarithmic convexity of the Gamma function follows from the fact that $d^2 \log \Gamma(x)/dx^2 > 0$ for $x > 0$. As for complex-valued arguments, it can be shown that if we fix the value of σ, then $|\Gamma(\sigma + it)|$ is strictly decreasing on $0 < t < \infty$ and thus $\partial \log |\Gamma(\sigma + it)|/\partial t < 0$ there.

The analytic continuation of Γ can be carried out in a way alternate to (B.1) by successive application of functional equation (B.7). Specifically, writing it as $\Gamma(s) = \Gamma(s+1)/s$ allows one to define $\Gamma(s)$ for $\Re(s) > -1$ using its values for $\Re(s) > 0$ as given by (B.1); this again shows that $\Gamma(s)$ has a simple pole at $s = 0$ with residue 1. Then using $\Gamma(s) = \Gamma(s+2)/(s(s+1))$ gives the analytic continuation down to $\Re s > -2$, and we again find that $\Gamma(s)$ has a simple pole at $s = -1$ with residue -1. This process can be continued indefinitely.

---

[4]Note, in particular, that (B.7) implies that $0!$ should be taken to equal 1, not 0.

[5]An optional exercise: Show that a logarithmically convex function must be convex. Show that $f(x) = x^2$ is convex but not logarithmically convex.

The Gamma function also satisfies an identity known as *Euler's reflection formula*

$$\Gamma(s)\Gamma(1-s) = \frac{\pi}{\sin(\pi s)}, \quad \text{for all } s,\ s \neq 0, \pm 1, \pm 2, \dots, \tag{B.8}$$

showing that $\Gamma(s)$ possesses a symmetry about the line $\Re(s) = \frac{1}{2}$. Setting $s = \frac{1}{2}$ here yields $\Gamma(\frac{1}{2}) = \sqrt{\pi}$, while combining with (B.7) then yields $\Gamma(-\frac{1}{2}) = -2\sqrt{\pi}$. Identity (B.8) can also be used (by the symmetry) to establish the analytic continuation of Γ. Observe too that the functions in (B.8) have no zeros, while the fact that both sides of (B.8) have simple poles at precisely all integer values underscores the naturalness of the relation. Property (B.8) also reveals that there is a natural sense in which one can say that the sine function (and its zeros) is constructed out of two copies of the $1/\Gamma$ function.

A third important identity is the so-called *Legendre duplication formula*

$$\sqrt{\pi}\, 2^{1-2s}\, \Gamma(2s) \;=\; \Gamma(s)\Gamma\left(s + \frac{1}{2}\right). \tag{B.9}$$

Here too both sides have matching poles—namely at 0 and at negative half-integer values. The $\sqrt{\pi}$ in equation (B.9) may be replaced by $\Gamma(\frac{1}{2})$ to give this equation a heightened visual appeal. An extension of (B.9) known as *Gauss's multiplication formula* is given by

$$\Gamma(s)\Gamma\left(s+\frac{1}{n}\right)\Gamma\left(s+\frac{2}{n}\right)\cdots\Gamma\left(s+\frac{n-1}{n}\right) \;=\; (2\pi)^{(n-1)/2} n^{1/2-ns}\,\Gamma(ns),$$

and holds for all integers $n \geq 1$, and for all s.

Functional relations (B.7), (B.8), and (B.9) can be variously combined to yield an assortment of useful identities. For example, combining (B.7) and (B.8) yields

$$\Gamma(s)\Gamma(-s) = -\frac{\pi}{s \sin \pi s}, \tag{B.10}$$

as well as

$$\frac{1}{\Gamma(1+s)\Gamma(1-s)} = \frac{\sin(\pi s)}{\pi s} = \prod_{n=1}^{\infty}\left(1 - \frac{s^2}{n^2}\right). \tag{B.11}$$

Likewise, using $\sin(\theta + \frac{\pi}{2}) = \cos(\theta)$, one can manipulate (B.8) and (B.9) into

$$\frac{\Gamma(s)}{\Gamma(1/2 - s)} = 2^{1-2s}\pi^{-1/2}\cos(\pi s)\Gamma(2s), \tag{B.12}$$

and, using $\frac{1}{2}\sin(\pi s)/\cos(\pi s/2) = \sin(\pi s/2)$), into

$$\Gamma(1-s)\sin\frac{\pi s}{2} = \frac{2^{-s}\sqrt{\pi}\,\Gamma\left(\frac{1-s}{2}\right)}{\Gamma\left(\frac{s}{2}\right)}. \tag{B.13}$$

Relations of this type are useful for reexpressing the functional equation for Riemann's zeta function into a variety of alternate forms.

With $s = \sigma + it$, a few other interesting relations satisfied by the Gamma function include

$$\Gamma\left(\frac{s}{2}\right)\Gamma\left(\frac{1-s}{2}\right) = \frac{1}{\sqrt{\pi}}2^{1-s}\Gamma(s)\cos\frac{\pi s}{2}, \quad \Gamma\left(\frac{1}{2}+s\right)\Gamma\left(\frac{1}{2}-s\right) = \frac{\pi}{\cos \pi s},$$

$$|\Gamma(it)|^2 = \frac{\pi}{t\sinh(\pi t)} = \frac{2\pi/t}{e^{\pi t} - e^{-\pi t}}, \quad \left|\Gamma\left(\frac{1}{2}+it\right)\right|^2 = \frac{\pi}{\cosh(\pi t)} = \frac{2\pi}{e^{\pi t}+e^{-\pi t}},$$

$$\Gamma(1+it)\Gamma(1-it) = \frac{\pi t}{\sinh \pi t} = \frac{2\pi t}{e^{\pi t} - e^{-\pi t}}, \quad \frac{\Gamma(s-\frac{1}{2})}{\Gamma(s)} = \frac{1}{\sqrt{\pi}} \int_{-\infty}^{\infty} \frac{du}{(1+u^2)^s},$$

$$\frac{\Gamma'(1-s)}{\Gamma(1-s)} - \frac{\Gamma'(s)}{\Gamma(s)} = \pi \cot \pi s, \quad \frac{\Gamma^2(\sigma)}{|\Gamma(s)|^2} = \prod_{n=0}^{\infty} \left(1 + \frac{t^2}{(n+\sigma)^2}\right),$$

and

$$\int_s^{s+1} \log \Gamma(z)\, dz = s \log s - s + \frac{1}{2} \log 2\pi, \quad s \neq 0, -1, -2, \ldots.$$

Of these nine formulas, the first two require only the functional identities for Γ already mentioned. For real $0 < t < 1$ there is also a nice Fourier representation given by

$$\log \Gamma(t) = \log \sqrt{2\pi} + \sum_{n=1}^{\infty} \left(\frac{1}{2n} \cos(2\pi n t) + \frac{\gamma + \log(2\pi n)}{\pi n} \sin(2\pi n t) \right), \quad 0 < t < 1.$$

This relation was first given by Kummer in the alternate form

$$\log \Gamma(t) = (\gamma + \log 2)\Big(\frac{1}{2} - t\Big) + (1-t) \log \pi$$
$$- \frac{1}{2} \log \sin \pi t + \sum_{n=1}^{\infty} \frac{\log n}{\pi n} \sin(2\pi n t), \quad 0 < t < 1.$$

Also worth noting is the reexpressed version of (B.5) given by

$$\log \Gamma(s) = -\log(s) + \sum_{k=1}^{\infty} \big(s \log(1 + 1/k) - \log(1 + s/k)\big), \quad s \in \mathbb{C} \backslash (-\infty, 0],$$

as well as the relations

$$\Gamma(s)\zeta(s) = \int_0^{\infty} \frac{t^{s-1}}{e^t - 1}\, dt, \quad \Re s > 1,$$

$$\log \Gamma(s+1) = -\gamma s + \sum_{n=2}^{\infty} (-1)^n \frac{\zeta(n)}{n} s^n, \quad |s| < 1,$$

and

$$\log \Gamma(1+s) = -\log(1+s) - s(\gamma - 1) + \sum_{n=2}^{\infty} (-1)^n \frac{\zeta(n) - 1}{n!} s^n, \quad \text{for } |s| < 1,$$

which provide three further links between the Γ and ζ functions.

The asymptotic behaviour of the Gamma function has been extensively investigated. Among the many approximations it satisfies, the most well known is the Stirling approximation for integers, $n = 1, 2, \ldots$. It can be stated as

$$\text{(B.14)} \qquad n! = \sqrt{2\pi n} \left(\frac{n}{e}\right)^n \exp(\theta(n)/12n), \quad 0 < \theta(n) < 1,$$

or equivalently as

$$\text{(B.15)} \qquad \sqrt{2\pi n} \left(\frac{n}{e}\right)^n < n! < \sqrt{2\pi n} \left(\frac{n}{e}\right)^n e^{1/12n}.$$

To higher order one has

$$\text{(B.16)} \quad n! = \sqrt{2\pi n} \left(\frac{n}{e}\right)^n \left(1 + \frac{1}{12n} + \frac{1}{288n^2} - \frac{139}{51840n^3} - \frac{571}{2488320n^4} + O\Big(\frac{1}{n^5}\Big)\right),$$

and more generally

$$n! = \sqrt{2\pi n}\left(\frac{n}{e}\right)^n \exp\left(\sum_{k=1}^{m} \frac{B_{2k}}{2k(2k-1)n^{2k-1}} + O\left(\frac{1}{n^{2m+1}}\right)\right),$$

where the B_{2k} are the Bernoulli numbers of Chapter 5. From these one obtains

$$\log n! = \log\left(\sqrt{2\pi}\right) - n + \left(n + \frac{1}{2}\right)\log n + o(1),$$

and more precisely

$$\log n! = \sum_{m=1}^{n} \log m = \log\left(\sqrt{2\pi}\right) - (n+1) + \left(n + \frac{1}{2}\right)\log(n+1) + O\left(\frac{1}{n}\right).$$

These all remain valid if $n \to \infty$ is real-valued, and $n!$ is replaced by $\Gamma(n+1)$; thus

$$\Gamma(x) \sim \sqrt{2\pi}x^{x-1/2}e^{-x}, \quad x \in \mathbb{R}, \quad x \to \infty.$$

Various other approximations can also be found in the literature. A particularly accurate one, due to Ramanujan is

$$n! \sim \sqrt{\pi}\left(\frac{n}{e}\right)^n \left(8n^3 + 4n^2 + n + \frac{1}{30}\right)^{1/6}.$$

These approximations can be extended to the case where Γ is viewed as a function of a complex variable, but in this case one needs to bound the domain of Γ away from its poles on the negative axis. (Around $s = 0$ we have already seen that $\Gamma(s) \sim \frac{1}{s}$.) Thus for a fixed $\delta > 0$ we have

$$\text{(B.17)}\quad \Gamma(s) = \sqrt{2\pi}e^{-s}s^{s-1/2}\left(1 + O\left(\frac{1}{|s|}\right)\right), \quad \text{for } |\arg s| \le \pi - \delta, \quad \text{as } |s| \to \infty,$$

and more precisely (under the same conditions),

$$\text{(B.18)}\qquad \Gamma(s) = \sqrt{2\pi}e^{-s}s^{s-1/2}\left(1 + \sum_{j=1}^{m} a_j s^{-j} + O\left(|s|^{-m-1}\right)\right), \qquad |\arg z| < \pi$$

for every $m > 0$, where the constants a_j here match those of (B.16): $a_1 = 1/12$, $a_2 = 1/288$, and so on. When s is real and positive, the remainder of this series is known to be less than the last of the terms retained. In expressions such as these, terms like $s^{s-1/2}$ are to be interpreted in the usual way as $e^{(s-1/2)\log s}$ using the principal value of the logarithm. Note that to meet a condition such as $|s| \to \infty$ or $1/|s| \to 0$, it is enough for only one of $|\Re s| \to \infty$ or $|\Im s| \to \infty$ to hold.

From (B.17) and (B.18), other useful approximations can be derived. For example, using (B.17), we find that, with $s = \sigma + it$, σ fixed or uniformly bounded, and $|t| \to \infty$,

$$\Gamma(s) = \sqrt{2\pi}(it)^{\sigma-1/2}e^{-\pi|t|/2}\left(\frac{t}{e}\right)^{it}\left(1 + O\left(\frac{1}{|t|}\right)\right),$$

from which we also get

$$|\Gamma(s)| = \sqrt{2\pi}\,|t|^{\sigma-1/2}e^{-\pi|t|/2}\left(1 + O\left(\frac{1}{|t|}\right)\right).$$

Thus (unlike its behaviour on the positive real line) $\Gamma(s)$ decays to 0 exponentially fast as $|t| \to \infty$ along vertical lines, uniformly within any vertical strip of bounded width in the complex plane. Karatsuba & Voronin (1992, [**622**, p. 344]) obtain the alternate version

$$\Gamma(s) = \sqrt{2\pi}\, t^{\sigma+it-1/2} \exp\left(-\frac{\pi t}{2} - it + \frac{\pi i}{2}\left(\sigma - \frac{1}{2}\right)\right)\left(1 + O\left(\frac{1}{|t|}\right)\right).$$

Approximations to the logarithm of the Gamma function may likewise be derived. Thus, (B.17) implies that

$$\log\Gamma(s) = \left(s - \frac{1}{2}\right)\log s - s + \log\sqrt{2\pi} + O\left(\frac{1}{|s|}\right) \sim s\log s, \quad |s| \to \infty,$$

while (B.18) allows the more precise result

$$\log\Gamma(s) = \left(s - \frac{1}{2}\right)\log s - s + \log\sqrt{2\pi} + \frac{1}{12s} + O\left(\frac{1}{|s|^2}\right), \quad |s| \to \infty.$$

More generally we have

$$\begin{aligned}\log\Gamma(s) = {} & \left(s - \frac{1}{2}\right)\log s - s + \log\sqrt{2\pi} \\ & + \sum_{k=1}^{m} \frac{B_{2k}}{2k(2k-1)s^{2k-1}} + O\left(\frac{1}{|s|^{2m+1}}\right), \quad |s| \to \infty.\end{aligned}$$

These results all assume that $|\arg s| \leq \pi - \delta$, with $\delta > 0$ fixed, as indicated in (B.17), and the implied constants of the order terms may of course depend on δ. Two variants of these approximations, valid when σ is uniformly bounded and $|t| \to \infty$ were stated by Ramachandra (2007, [**880**, pp. 143–144]):

$$\log\Gamma(s) = \left(s - \frac{1}{2}\right)\log(it) - it + \log(\sqrt{2\pi}) + O\left(\frac{1}{|t|}\right),$$

and

$$\log\Gamma(s) = \sqrt{2\pi}\, t^{s-1/2} e^{-\pi t/2 - it + (i\pi/2)(\sigma - 1/2)}\left(1 + O\left(\frac{1}{|t|}\right)\right).$$

Still more precise approximations for $\log\Gamma(s)$, correct to any order, can be expressed in terms of the Bernoulli numbers; these may be found in Whittaker & Watson (1927, [**1100**, §12.33]). The various logarithmic formulas also yield valid identities when differentiated. For example, the digamma function satisfies the approximation

$$\psi(s) \equiv \frac{\log\Gamma(s)}{ds} \equiv \frac{\Gamma'(s)}{\Gamma(s)} = \log s - \frac{1}{2s} + O\left(|s|^{-2}\right) = \log s + O\left(|s|^{-1}\right).$$

A number of Mellin transform identities involving the Gamma function may be found in Appendix C. Here we state only the following Mellin transform pair:

$$\int_0^\infty e^{-y} y^{s-1}\, dy = \Gamma(s), \quad \Re s > 0,$$

and

$$\frac{1}{2\pi i}\int_{c-i\infty}^{c+i\infty} \Gamma(s) z^{-s}\, ds = e^{-z}, \quad \Re z > 0,\ c > 0.$$

Finally, we mention the closely related Beta function (also due to Euler) defined as

$$B(\alpha, \beta) \equiv \frac{\Gamma(\alpha)\Gamma(\beta)}{\Gamma(\alpha+\beta)} = \int_0^1 (1-t)^{\alpha-1} t^{\beta-1}\, dt,$$

which often arises in statistical applications. The integral here converges only when both $\Re(\alpha) > 0$ and $\Re(\beta) > 0$, but $B(\alpha, \beta)$ can otherwise be defined by analytic continuation. It is interesting to note that (on account of the layout of its poles in $\mathbb{C}^2$) the Beta function has played a role as a scattering amplitude in the early physics of string theory.

Further details about the Gamma function as well as its history may be found in many classical and modern sources. These include Abramowitz & Stegun (1964, [**2**, Chapter 6]), Artin (1964, [**42**]), H. Cohen (2007, [**211**, vol. II, §9.6.2–§9.6.6]), Davis (1959, [**264**]), Everest & Ward (2005, [**368**, Chapter 9]), Finch (2003, [**380**, §1.5 and §4.5]), Henrici (1977, [**540**]), Jensen (1916, [**599**]), Karatsuba & Voronin (1992, [**622**, Appendix §3]), Lagarias (2013, [**668**]), Montgomery & Vaughan (2007, [**774**, Appendix C]), Nielsen (1906, [**811**]), Ramachandra (2007, [**880**, Chapter 10]), Remmert (1998, [**891**]), Tenenbaum (2015, [**1053**, Chapter II.0]), and Whittaker and Watson (1927, [**1100**, Chapter XII]). Gradshteyn & Rizhik (2014, [**447**, § 8.31–§ 8.34]) provide a comprehensive collection of series, products, and other results for the Gamma function.

APPENDIX C

Dirichlet Series

This appendix is devoted to the specialized class of analytic functions known as Dirichlet series, which we will treat in considerable detail. Their continuous analogue, the Mellin transform, will be discussed as well. Along the way we will comment briefly on Lambert series to which Dirichlet series are also related. Only minimal details of proofs are included. The theory of Dirichlet series was pioneered by many people in addition to Dirichlet, including Bohr, Hadamard, Hardy, Landau, Riesz, and Schnee, and more recently by Helson and Kahane. A recommended reference with carefully laid out proofs is Titchmarsh (1939, [**1058**, Chapter IX]); additional references will be provided as we proceed.

Given an *arithmetic function* $a(n)$—in other words, a function defined on the positive integers—its corresponding *Dirichlet series* expansion is given formally by

$$f(s) \equiv \sum_{n=1}^{\infty} \frac{a(n)}{n^s} \equiv \sum_{n=1}^{\infty} a(n) e^{-s \log n} \equiv \sum_{n=1}^{\infty} a(n)\, n^{-\sigma} e^{-it \log n}, \tag{C.1}$$

where here, as throughout this appendix, the complex argument s is taken as $s = \sigma + it$. A prototypical such function, which corresponds to $a(n) \equiv 1$, is the Riemann ζ-function. When manipulated as generating functions, Dirichlet series can be used to prove numerous identities involving arithmetic functions. However, among its more important applications is to help uncover asymptotic and statistical information about the sequence of its coefficients $\{a(n)\}$, and their partial sums $A(x) \equiv \sum_{n \le x} a(n)$.

Dirichlet series $\sum_{n=1}^{\infty} a(n) n^{-s}$ stand in contrast to power series $\sum_{n=0}^{\infty} a(n) s^n$. Unlike power series which converge geometrically, Dirichlet series generally converge more slowly, and questions about their convergence involve more delicate considerations. However, both are special cases of *generalized Dirichlet series* which are series of the form $\sum a(n) e^{-\lambda_n s}$, where $0 < \lambda_1 < \lambda_2 < \cdots$, with $\lambda_n = \log n$ corresponding to ordinary Dirichlet series, and $\lambda_n = n$ corresponding to power series. As such, Dirichlet series share many similarities with power series, although many differences as well. The more general λ_n-sequences will not concern us here; they are treated in the classical book by Hardy & Riesz (1915, [**498**]). The more recent book by Mandelbrojt (1972, [**727**]) treats various special topics.

While power series have domains of convergence which are discs, the domains of convergence of Dirichlet series are right half-planes. In particular, if a Dirichlet series converges at a point $s_0 = \sigma_0 + it_0$ in $\mathbb{C}$, then it converges at $s = \sigma + it$ for all $\sigma > \sigma_0$ and all $t \in \mathbb{R}$. It will also be analytic throughout all of that half-plane. The smallest value of σ_0 for which the series converges for all $\sigma > \sigma_0$ is called the *abscissa of (conditional) convergence* of the series and is denoted by σ_c. To the left of this value, namely for $\sigma < \sigma_c$, the series will converge nowhere, while on the $\sigma = \sigma_c$ boundary anything can happen. In further contrast to power series (which always converge absolutely within their discs of convergence), for a Dirichlet series, its half-plane $\sigma > \sigma_c$ of *conditional convergence* need not be the same as its half-plane $\sigma > \sigma_a$ of *absolute convergence*—the half-plane in which $\sum |a(n)|/n^\sigma < \infty$. The latter can lie up to, but it cannot lie more than, one full unit to the right: $\sigma_c \leq \sigma_a \leq \sigma_c + 1$. Titchmarsh (1939, [**1058**, Chapter IX]), for example, has proofs for all of the results mentioned above (as well as for many of those mentioned below).

EXAMPLES C.1. In extreme cases, one can have $\sigma_c = -\infty$, as when $a(n) = n^{-n}$, 2^{-n}, or $1/n!$; and one can have $\sigma_c = +\infty$, as when $a(n) = n^n$, 2^n, or $n!$. However, cases where $\sigma_c = \pm\infty$ do not frequently arise in applications. An example of a series for which σ_c and σ_a differ is given by the alternating series $\sum(-1)^n/n^s$ examined in Chapter 5: for this series we have $\sigma_c = 0$ and $\sigma_a = 1$.

Dirichlet series have an analogue to the radius of convergence formula (A.9) for power series, but for Dirichlet series the formula depends on whether or not $\sum_{n=1}^\infty a(k)$ converges. Specifically, the boundary (that is, the abscissa) of the half-plane of (conditional) convergence can be shown to be

$$\sigma_c = \limsup_{n\to\infty} \frac{1}{\log n} \log \left| \sum_{k=n+1}^{\infty} a(k) \right|, \quad \text{if } \sum_{n=1}^{\infty} a(k) \text{ converges,}$$

and

$$\sigma_c = \limsup_{n\to\infty} \frac{1}{\log n} \log \left| \sum_{k=1}^{n} a(k) \right|, \quad \text{if } \sum_{n=1}^{\infty} a(k) \text{ diverges.}$$

As mentioned, left of this boundary, the series will converge nowhere. The boundary σ_a of the half-plane of absolute convergence is obtained by replacing $a(k)$ by $|a(k)|$ in these formulas. Of course, if all $a(n) \geq 0$, then σ_c and σ_a will be identical. The usual precautions must be heeded when convergence is not absolute—namely that the terms in such series may not be arbitrarily rearranged. In any case, a Dirichlet series will be analytic throughout *all* of the half-plane on which it converges; for this, absolute convergence is not required. A Dirichlet series can also admit *analytic continuation* to regions far to the left of σ_c. An obvious example is given by the zeta function $\zeta(s) = \sum_1^\infty 1/n^s$ (for which $\sigma_c = \sigma_a = 1$). Another example is given by the alternating zeta function $\sum_1^\infty (-1)^{n+1}/n^s$. It can actually be analytically continued to all of $\mathbb{C}$ by virtue of the identity

$$\sum_{n=1}^{\infty} \frac{(-1)^{n+1}}{n^s} = \left(1 - 2^{1-s}\right) \zeta(s)$$

seen in Chapter 5; the function on the right is *entire* because the zero of $(1 - 2^{1-s})$ at $s = 1$ cancels the pole of $\zeta(s)$ there. The behaviour of Dirichlet series thus differs markedly from power series whose radius of conditional convergence, radius of absolute convergence, and maximum radius within which the series is analytic, are all identical.

One further difference between Dirichlet series and power series is that while the latter must always have at least one point on its circle of convergence at which it is singular, a Dirichlet series need not have any singularities on its boundary of convergence. Nevertheless both types of series can have boundaries consisting entirely of singularities. In the case when all $a(n) \geq 0$, however, Landau proved that the Dirichlet series must have a singularity at the point $s = \sigma_c = \sigma_a$ of the real line whenever that value is finite.

Questions regarding the convergence of a Dirichlet series can be extremely difficult. An example is given by the series $\sum_{n=1}^{\infty} \mu(n)/n^{\sigma} = 1/\zeta(s)$ based on the Möbius function $\mu(n)$. Trivially, its region of absolute convergence is just that of the ζ-function, namely $\sigma_a = 1$. However, other than $\frac{1}{2} \le \sigma_c \le 1$, nothing at all is known about its abscissa of conditional convergence, not even whether or not $\sigma_c < 1$. In fact, if one knew that $\sigma_c = \frac{1}{2}$, it would be equivalent to RH, while $\sigma_c < 1$ would imply that $\zeta(s)$ has no zeros for $\sigma > \sigma_c$.

Dirichlet series are highly specialized analytic functions, and typical analytic functions cannot be expressed as Dirichlet series. Yet given a function which is analytic in a half-plane, determining whether or not it can be represented as a Dirichlet series is not straightforward. It is known, however, that if a Dirichlet series $f(s)$ is absolutely convergent on a half-plane, then $1/f(s)$ can also be represented by an absolutely convergent Dirichlet series on that same half-plane provided only that $f(s)$ is bounded away from 0 there (Hewitt & Williamson, 1957, [**542**]). Being analytic, however, and each of its individual terms being analytic as well, any function that can be represented as a Dirichlet series will have all of the usual methods of complex analysis available to it. So like power series, Dirichlet series possess all of their derivatives and integrals, and may be differentiated and integrated termwise without affecting their convergence regions. Thus if $f(s) = \sum a(n)/n^{-s}$, then $f(s)$ along with

$$\frac{df(s)}{ds} = -\sum_{n=1}^{\infty} \frac{a(n)\log n}{n^s}, \quad \text{and} \quad \frac{d^k f(s)}{ds^k} = (-1)^k \sum_{n=1}^{\infty} \frac{a(n)\log^k n}{n^s}, \quad k \ge 1,$$

all share the same half-planes of convergence, as well as the same half-planes of absolute convergence, and likewise for the integrals of $f(s)$. (The fact that $\log^k n$ grows more slowly than n^{δ} for any $\delta > 0$ lies at the root of this result.) Furthermore, like power series, Dirichlet series are uniquely determined by their coefficients: if on a half-plane of convergence two Dirichlet series $\sum a(n)/n^s$ and $\sum b(n)/n^s$ agree at a countable collection of point $s_n = \sigma_n + it_n$ whose real parts satisfy $\sigma_n \to \infty$, then the coefficients $a(n)$ and $b(n)$ of the two series must be identical for all n. Hardy & Wright (2008, [**499**, Chapter XVII]) provide an elementary treatment of this result. Note too that a Dirichlet series is an almost periodic function on any vertical line within its region of (conditional) convergence because it is a sum of periodic terms on any such line.

The coefficients $a(n)$ of a Dirichlet series will typically not have faster than polynomial growth, for otherwise the series will not converge. If its terms $|a(n)| \le C$ are bounded, or more generally if $|a(n)| \le C \log^k n$ for some constants k and C, we would then have $\sigma_a \le 1$. Two other properties worth mentioning are, first, that a Dirichlet series is bounded on any *closed* half-plane which lies *inside* its domain of absolute convergence, and second, that as long as the series is nontrivial and has a nonempty domain of convergence, it will have a half-plane on which it is entirely free of zeros.[1]

When products are taken, power series and Dirichlet series possess convolution properties that mirror each other, these being additive and multiplicative in nature,

---

[1] These two properties are actually easy exercises. (Hint: Far enough to the right, the first nonzero term of the series will dominate the sum that remains.)

respectively. When two power series are multiplied together an *additive convolution* results. Thus if

$$\sum_{n=0}^{\infty} a(n)z^n \times \sum_{n=0}^{\infty} b(n)z^n = \sum_{n=0}^{\infty} c(n)z^n,$$

the resulting series is convergent on whichever is the smaller of the two original radii of convergence, and the coefficients $c(n)$ are given by the *additive* form $(n = j + k)$

$$c(n) = \sum_{j+k=n} a(j)b(k) \equiv \sum_{j=0}^{n} a(j)b(n-j).$$

When two Dirichlet series are multiplied together, the result will again be a Dirichlet series and it will be absolutely convergent on the smaller of the original half-planes of *absolute* convergence. (Conditional convergence will not suffice here.) Writing the product as

$$\left(\sum_{n=1}^{\infty} \frac{a(n)}{n^s}\right) \times \left(\sum_{n=1}^{\infty} \frac{b(n)}{n^s}\right) = \sum_{n=1}^{\infty} \frac{c(n)}{n^s}, \tag{C.2}$$

the $c(n)$ will be given by the *multiplicative* form $(n = jk)$

$$c(n) = \sum_{jk=n} a(j)b(k) \equiv \sum_{j|n} a(k)b(n/j)$$

called a *Dirichlet convolution*. Here j, k ranges over the factor pairs of n, while $j|n$ ranges over its divisors, including 1 and n. This result is proved by multiplying out the series and rearranging the terms of the double sum:

$$\left(\sum_{j=1}^{\infty} \frac{a(j)}{j^s}\right) \times \left(\sum_{k=1}^{\infty} \frac{b(k)}{k^s}\right) = \sum_{j,k} \frac{a(j)b(k)}{(jk)^s} = \sum_{n=1}^{\infty} \frac{1}{n^s}\left(\sum_{jk=n} a(j)b(k)\right).$$

One may represent this relation as $c = a * b = b * a$. Dirichlet convolution is associative and commutative. It has the identity element e, defined as $e(1) = 1$ and $e(n) = 0$ for $n > 1$, and satisfying $a * e = e * a = a$. If $f(s) = \sum a(n)/n^s$ and $g(s) \equiv 1/f(s) = \sum b(n)/n^s$ each have a finite abscissa of convergence, then g will be a *convolutional inverse* of f, meaning that $a * b = e$. A Dirichlet series need not have a convolution inverse, but when it does it will be unique. An example is given by the series $\mathbf{1}(n) \equiv 1$ for $n \geq 1$ corresponding to the zeta function $\zeta(s) = \sum 1/n^s$. Its convolution inverse is given by $\sum \mu(n)/n^s$ where μ is the Möbius function; we may write this as $\mathbf{1} * \mu = \mu * \mathbf{1} = e$, or as $\mathbf{1}^{-1} \equiv \mu$ or $\mu^{-1} = \mathbf{1}$. These points are elaborated on in Chapter 3.

A number of other properties of Dirichlet series are also worth noting. Thus suppose that the sequences $\{a(n)\}$ and $\{b(n)\}$ are related as a Möbius transform pair (see Chapter 3):

$$b(n) = \sum_{d|n} a(d), \quad \text{so that } a(n) = \sum_{d|n} \mu\left(\frac{n}{d}\right) b(d), \quad n \geq 1. \tag{C.3}$$

Then their Dirichlet series will satisfy the equivalent relation

$$\sum_{n=1}^{\infty} \frac{b(n)}{n^s} = \zeta(s) \sum_{n=1}^{\infty} \frac{a(n)}{n^s} \tag{C.4}$$

in any half-plane $\Re s > 1$ on which both these series converge absolutely (although this relation will hold formally regardless of convergence considerations). Indeed, this identity is just the special case $b(n) \equiv 1$ of equation (C.2). Another result linking Möbius inversion with Dirichlet series is

$$g(x) = \sum_{n=1}^{\infty} \frac{f(nx)}{n^s} \quad \Longleftrightarrow \quad f(x) = \sum_{n=1}^{\infty} \mu(n) \frac{g(nx)}{n^s}, \quad x \geq 1.$$

This is just an extension (for each fixed s) of one of the Möbius inversion formulas—namely (3.11)—appearing in Chapter 3. Its validity only requires absolute convergence of the sums involved. Finally, note that if $f(n)$ is a multiplicative function then its Dirichlet series will have an *Euler product representation*

$$F(s) = \sum_{n=1}^{\infty} \frac{f(n)}{n^s} = \prod_p \left(\sum_{m=0}^{\infty} \frac{f(p^m)}{p^{ms}} \right)$$

which will be valid in the region of absolute convergence. If $f(n)$ is completely multiplicative (so that the terms $f(p^k) = f(p)^k$ are in geometric progression) one can represent this Euler product more neatly as

$$F(s) = \sum_{n=1}^{\infty} \frac{f(n)}{n^s} = \prod_p \left(1 - \frac{f(p)}{p^s}\right)^{-1}.$$

Note that if $a(n)$ and $b(n)$ are both multiplicative functions then their Dirichlet convolution $a * b$ will be as well; a proof is given at the end of this appendix.

Dirichlet series bear an interesting connection to so-called *Lambert series* which are series of the form

$$\sum_{n=1}^{\infty} \frac{a(n)x^n}{1 - x^n}.$$

Lambert series possess the key property (which is due to Cesàro) that

$$\text{(C.5)} \qquad \sum_{n=1}^{\infty} \frac{a(n)x^n}{1 - x^n} = \sum_{n=1}^{\infty} b(n)x^n$$

if and only if $\{a(n)\}$ and $\{b(n)\}$ are a Möbius transform pair as in (C.3). It follows that every *formal* power series can be expressed as a *formal* Lambert series and vice versa. It also follows that the Dirichlet series identity (C.4) and the Lambert series identity (C.5) are equivalent. This equivalence is also proved at the end of this appendix. Noteworthy identities involving Lambert series include

$$\sum_{n=1}^{\infty} \frac{\mu(n)x^n}{1 - x^n} = x, \quad \sum_{n=1}^{\infty} \frac{\varphi(n)x^n}{1 - x^n} = \frac{x}{(1 - x)^2}, \quad \sum_{n=1}^{\infty} \frac{\lambda(n)x^n}{1 - x^n} = \sum_{m=1}^{\infty} x^{m^2},$$

$$\sum_{n=1}^{\infty} \frac{|\mu(n)|x^n}{1 - x^n} = \sum_{n=1}^{\infty} 2^{\omega(n)}x^n, \quad \sum_{n=1}^{\infty} \frac{\Lambda(n)x^n}{1 - x^n} = \sum_{n=1}^{\infty} \log(n)\, x^n,$$

$$\sum_{n=1}^{\infty} \frac{x^n}{1 - x^n} = \sum_{n=1}^{\infty} \tau(n)x^n, \quad \sum_{n=1}^{\infty} \frac{nx^n}{1 - x^n} = \sum_{n=1}^{\infty} \sigma(n)\, x^n,$$

$$\text{and } \sum_{n=1}^{\infty} \frac{n^{\alpha}x^n}{1 - x^n} = \sum_{n=1}^{\infty} \sigma_{\alpha}(n)\, x^n.$$

All of these hold for $|x| < 1$ (Hardy & Wright (2008, [**499**, Chapter 17, §10])). A relation also worth mentioning due to its similarity to those above is

$$\sum_{n=1}^{\infty} \frac{\mu(n)x^n}{1+x^n} = x - 2x^2.$$

Note that the result involving $\tau(n)$ is just the $\alpha = 0$ case for that of $\sigma_\alpha(n)$. (The arithmetic functions μ, φ, τ, λ, Λ, ω, and σ_α are defined in Chapter 3.) Aside from their number-theoretic relations, Lambert series satisfy some unusual identities in their own right, including

$$\sum_{n=1}^{\infty}(-1)^{n+1}\frac{x^n}{1-x^n} = \sum_{n=1}^{\infty}\frac{x^n}{1+x^n}, \quad \text{and} \quad \sum_{n=1}^{\infty} n\frac{x^n}{1-x^n} = \sum_{n=1}^{\infty}\frac{x^n}{(1-x^n)^2}.$$

These and other such identities may be found in Knopp (1928, [**640**, Chapter XII, §58]).

Now let $a(n)$ again be an arithmetic function, and $f(s) = \sum_1^\infty a(n)/n^s$, with $s = \sigma + it$, be its corresponding Dirichlet series. We will describe two approaches for recovering (at least in principle) the coefficients $a(n)$ given the function $f(s)$. In the first method the individual terms $a(n)$ are recovered directly using

$$\begin{aligned} a(n) &= \lim_{T\to\infty} \frac{1}{2T}\int_{-T}^{T} f(s)n^s\,dt \\ &\equiv \lim_{T\to\infty} \frac{1}{2T}\int_{-T}^{T} f(\sigma+it)n^{\sigma+it}\,dt\,, \quad n = 1, 2, \ldots, \end{aligned} \tag{C.6}$$

which one may write alternately as

$$\frac{a(n)}{n^\sigma} = \lim_{T\to\infty}\frac{1}{2T}\int_{-T}^{T} f(s)n^{it}\,dt \;\equiv\; \lim_{T\to\infty}\frac{1}{2T}\int_{-T}^{T} f(\sigma+it)n^{it}\,dt\,, \quad n = 1, 2, \ldots.$$

Here $\sigma > \sigma_a$ is *any* value in the region of absolute convergence of the series. Interestingly, if n is replaced by any *real* value $x > 0$, these integrals will still converge, but for noninteger values of x they will converge to 0. The principle behind this formula is simply that for every fixed σ each term $a(n)/n^s$, when viewed as a function in t, is a sinusoid of period $2\pi/\log n$:

$$\frac{a(n)}{n^s} = \frac{a(n)}{n^\sigma}\big(\cos(t\log n) - i\sin(t\log n)\big).$$

Hence $f(\sigma + it)$ is a linear combination of sinusoids (and so an *almost periodic function*) and (C.6) is just a Fourier inversion method for recovering the coefficients.

The second approach to inversion is based on the *summatory function* of the coefficients, namely, $A(x) \equiv \sum_{n\le x} a(n)$, for $x > 0$. Applying *Abel summation formula* (see Appendix F) one may verify that the Dirichlet series can be written as

$$f(s) = \sum_{n=1}^{\infty}\frac{a(n)}{n^s} = s\int_1^\infty A(x)x^{-s-1}\,dx, \tag{C.7}$$

for any $\sigma > \max\{0, \sigma_c\}$ where σ_c is the abscissa of (conditional) convergence of the series. The $\int_1^\infty$ integral appearing here—which equals $f(s)/s$—is essentially just a Mellin transform and can be inverted by *Perron's Formula*:

$$A^*(x) \equiv {\sum_{n\le x}}^{*} a(n) = \frac{1}{2\pi i}\int_{c-i\infty}^{c+i\infty} f(s)\frac{x^s}{s}\,ds\,, \quad \text{for } c > \max\{0, \sigma_c\},\ x > 0, \tag{C.8}$$

from which the $a(n)$ can be recovered. (Mellin transforms and their inversion are discussed further below.) The asterisk here means that the last $a(n)$ is to be multiplied by $\frac{1}{2}$ when x takes on an integer value, while the right-hand side of (C.8) is meant to be interpreted as

$$\lim_{T\to\infty} \frac{1}{2\pi i}\int_{c-iT}^{c+iT} f(s)\frac{x^s}{s}\,ds.$$

Curiously, if c satisfies $\sigma_c < c < 0$, then (C.8) continues to hold, but with $\sum^*_{n\le x} a(n)$ replaced by $\sum^*_{n>x} a(n)$. In passing, we mention the related and more rapidly converging formula

$$\sum_{n\le x} a(n)\log\frac{x}{n} = \frac{1}{2\pi i}\int_{c-i\infty}^{c+i\infty} f(s)\frac{x^s}{s^2}\,ds.$$

Formula (C.8) is in fact the $s = 0$ case of the more general result

$$\text{(C.9)}\qquad \sum_{n\le x}{}^{*}\,\frac{a(n)}{n^s} = \frac{1}{2\pi i}\int_{c-i\infty}^{c+i\infty} f(s+w)\frac{x^w}{w}\,dw, \quad \text{for } c > \max\{0,\sigma_c\},\ x>0.$$

The proofs for (C.8) and (C.9) require taking some delicate convergence issues into account. And for applications, the integrals involved would need to be approximated. Additional details may be found, for example, in Montgomery & Vaughan (2006, [**774**, Chapter 5, §1]).

EXAMPLE C.2. The Abel summation formula applied to the Riemann zeta function gives

$$\zeta(s) = \sum_{n=1}^{\infty}\frac{1}{n^s} = s\int_1^\infty \lfloor x\rfloor x^{-s-1}\,dx\,,$$

where $\lfloor x\rfloor$ is the floor function (greatest integer $\le x$). While not a very important fact, applying the Perron inversion formula to this yields

$$\lfloor x\rfloor^* = \sum_{n\le x}{}^{*}\,1 = \frac{1}{2\pi i}\int_{c-i\infty}^{c+i\infty}\zeta(s)\frac{x^s}{s}\,ds\,, \quad \text{for } c>1,\ x>0\,.$$

EXAMPLE C.3. Recall that $\sum_1^\infty \Lambda(n)/n^s = -\zeta'(s)/\zeta(s)$, for $\sigma > 1$. Applying (C.8) gives

$$\psi^*(x) \equiv \sum_{n\le x}{}^{*}\,\Lambda(n) = \frac{1}{2\pi i}\int_{c-i\infty}^{c+i\infty}\left(-\frac{\zeta'(s)}{\zeta(s)}\right)\frac{x^s}{s}\,ds, \quad x>0,\ c>1.$$

This result is important since it links the ζ-function to counts involving the primes, and can be used to derive the prime number theory (PNT), although it is not trivial to do so. Similar inversions can also be obtained for any of the Dirichlet series expansions appearing in Chapter 5.

Being sums of sinusoids on vertical lines, Dirichlet series are related to Fourier series and as such they satisfy an analogue of Parseval's identity (see Appendix F). Such identities can be used to establish *mean value properties* for Dirichlet series. Letting $f(s) = \sum a(n)/n^s$ with $s = \sigma + it$ again, the main result (due to Fritz Carlson) can be stated as

$$\text{(C.10)}\quad \lim_{T\to\infty}\frac{1}{2T}\int_{-T}^{T}|f(\sigma+it)|^2\,dt \equiv \lim_{T\to\infty}\frac{1}{2T}\int_{-T}^{T}\left|\sum_{n=1}^{\infty}\frac{a(n)}{n^s}\right|^2 dt = \sum_{n=1}^{\infty}\frac{|a(n)|^2}{n^{2\sigma}},$$

and is valid whenever σ lies in the region of absolute convergence of $f(s)$; in fact, $\sigma > \sigma_c + \frac{1}{2}$ would actually suffice (Narkiewicz (2000, [**797**, p. 315])). The case $\sigma = 0$, when applicable, results in the neat formula

$$\lim_{T\to\infty} \frac{1}{2T} \int_{-T}^{T} \left| \sum_{n=1}^{\infty} \frac{a(n)}{n^{it}} \right|^2 dt = \sum_{n=1}^{\infty} |a(n)|^2, \tag{C.11}$$

which is valid whenever the series $\sum |a(n)|^2$ converges. Bearing in mind the sinusoidal nature of n^{-it} these results are clearly comparable to the analogous ones for Fourier series.

More generally, let $f(s) = \sum_1^{\infty} a(n)/n^s$ and $g(s) = \sum_1^{\infty} b(n)/n^s$ be absolutely convergent for $\sigma > \sigma_1$ and $\sigma > \sigma_2$ respectively. Then for $\alpha > \sigma_1$ and $\beta > \sigma_2$ we have

$$\lim_{T\to\infty} \frac{1}{2T} \int_{-T}^{T} f(\alpha + it) g(\beta - it)\, dt = \sum_{n=1}^{\infty} \frac{a(n)\overline{b(n)}}{n^{\alpha+\beta}}.$$

The case $\alpha = \beta = \sigma$ then gives

$$\lim_{T\to\infty} \frac{1}{2T} \int_{-T}^{T} f(\sigma + it) g(\sigma - it)\, dt = \sum_{n=1}^{\infty} \frac{a(n)\overline{b(n)}}{n^{2\sigma}}$$

while when $f = g$ we recover (C.10).

EXAMPLE C.4. Putting $a(n) \equiv 1$ and $f(s) = \zeta(s)$ in (C.10) gives the mean value result

$$\lim_{T\to\infty} \frac{1}{2T} \int_{-T}^{T} |\zeta(\sigma + it)|^2\, dt = \sum_{n=1}^{\infty} \frac{1}{n^{2\sigma}} = \zeta(2\sigma), \quad \sigma > 1.$$

With a more delicate argument, this can be shown to hold even for $\sigma > \frac{1}{2}$, the only exception being that for $\sigma = 1$ we must change $\frac{1}{2T}\int_{-T}^{T}$ to $\frac{1}{T}\int_{1}^{T}$ to avoid the pole of $\zeta(s)$ at $s = 1$.

EXAMPLE C.5. The number of ways, $\tau_k(n)$, to write n as a product of k factors satisfies

$$\zeta^k(s) = \sum_{n=1}^{\infty} \frac{\tau_k(n)}{n^s}.$$

Applying (C.10) gives

$$\lim_{T\to\infty} \frac{1}{T} \int_{1}^{T} |\zeta(\sigma + it)|^{2k}\, dt = \sum_{n=1}^{\infty} \frac{\tau_k^2(n)}{n^{2\sigma}}, \quad \sigma > 1.$$

For $k \geq 2$ it is known that this continues to hold down to $\sigma > 1 - \frac{1}{k}$. For $k = 2$ there is a neat expression for the right hand side which leads to

$$\lim_{T\to\infty} \frac{1}{2T} \int_{-T}^{T} |\zeta(\sigma + it)|^4\, dt = \frac{\zeta^4(2\sigma)}{\zeta(4\sigma)} \quad \text{for } \sigma > \frac{1}{2}.$$

Our last topic in this appendix is the *Mellin transform*. For a function $f(y)$ of a real variable $y > 0$, it is defined as[2]

$$\widetilde{f}(s) = \int_0^\infty f(y)y^{s-1}\,dy = \int_0^\infty f(y)y^s\,\frac{dy}{y}\,, \quad \text{with } s = \sigma + it. \tag{C.12}$$

This will typically be finite in a vertical strip, say $c_1 < \sigma < c_2$, of the complex plane—possibly including some of the boundary—within which it will be holomorphic. The Mellin transform bears the same relation to Dirichlet series as the Fourier transform does to Fourier series. In fact, a Dirichlet series $\sum a(n)/n^s$ is essentially just a Mellin transform of the Stieltjes form $\int x^{-s}\,dF(x)$, in which F is the discrete 'distribution' which places mass $a(n)$ at positive integers n. The properties of Mellin transforms and Dirichlet series are therefore similar although not identical. Mellin transforms are also related to Laplace transforms and to Fourier transforms via suitable changes in variables. For instance putting $y = e^x$ and $s = it$ in (C.12) results in $\widetilde{f}(it) = \int_{-\infty}^{\infty} f(e^x)\big(\cos(tx) + i\sin(tx)\big)\,dx$. For this reason the properties of Mellin transforms are essentially in one-to-one correspondence with those of Fourier transforms. An example of this correspondence is given by the convolution property: If $h = f*g$ is an ordinary convolution, so that $h(y) = \int_0^\infty f(x)g(y-x)\,dx$, their Mellin transforms will satisfy $\widetilde{h}(s) = \widetilde{f}(s)\widetilde{g}(s)$.

REMARKS C.6. The definition of the Mellin transform is sometimes given as

$$\widetilde{f}(s) = \int_1^\infty f(y)y^{s-1}\,dy = \int_1^\infty f(y)y^s\,\frac{dy}{y}\,, \quad \text{with } s = \sigma + it.$$

The reason for this is that the $\int_0^1$ lower segment of the integral requires additional convergence considerations since one must then also ensure that $f(y)y^{\sigma-1}$ remains convergent at $y = 0$. This typically involves lower bounds on the admissible values of σ related to the behaviour of $f(y)y^{\sigma-1}$ as $y \downarrow 0$. One also sometimes encounters versions that look like $\int f(y)y^{-s+1}$. These can be turned into Mellin transforms by variable changes such as $y = \frac{1}{x}$. (An example will appear below.) Quite generally, however, the domains of definition of Mellin transforms are vertical strips in the complex plane, and they are holomorphic throughout those strips.

If two functions have the same Mellin transform (it is enough for the transforms to match on a segment of the real line) they will be identical, and under suitable conditions (C.12) can be inverted by means of a vertical integration in the complex plane:

$$f(y) = \frac{1}{2\pi i}\int_{c-i\infty}^{c+i\infty} \widetilde{f}(s)y^{-s}\,ds = \frac{1}{2\pi i}\int_{-\infty}^{\infty} \widetilde{f}(c+it)y^{-c-it}\,dt. \tag{C.13}$$

Here. as usual, $\int_{c-i\infty}^{c+i\infty}$ is meant in the sense $\lim_{T\to\infty}\int_{c-iT}^{c+iT}$. This formula is obtained by a change of variables in the formula for inverting Fourier integrals. On account of Cauchy's Theorem (Appendix A) the choice of the real value c is arbitrary as long as it is in the region where $\widetilde{f}(s)$ is finite. Equations (C.12) and (C.13) are

[2]The form $\frac{dy}{y}$ is often used because this measure (a weighted version of Lebesgue measure) is the natural measure on the positive real numbers that is invariant under multiplicative shifts $y \mapsto \lambda y$ as well as under reciprocals $y \mapsto \frac{1}{y}$. Indeed, the change of variables $y = e^x$ shows that the measure dy/y on the positive real numbers is a disguised version of Lebesgue measure dx on the entire real line.

referred to as a Mellin transform pair and conditions for their validity (which we omit) are analogous to those required for Fourier transforms. See, for example, Montgomery & Vaughan (2006, [**774**, Chapter 5]).

As in the case of Fourier analysis, a Mellin transform pair will satisfy a Plancherel relation. It can be obtained by a change of variables in the Fourier version and is given by

$$\int_0^\infty |f(y)|^2 y^{2\sigma-1}\,dy = \frac{1}{2\pi}\int_{-\infty}^\infty \left|\widetilde{f}(\sigma+it)\right|^2 dt. \tag{C.14}$$

This relation will be valid if both

$$\int_0^\infty |f(t)|t^{\sigma-1}dt < \infty \quad \text{and} \quad \int_0^\infty |f(t)|^2 t^{2\sigma-1}dt < \infty.$$

More generally, if $\widetilde{f}(s)$ and $\widetilde{g}(s)$ are the Mellin transforms of $f(y)$ and $g(y)$, then

$$\int_0^\infty f(y)\overline{g(y)}y^{2\sigma-1}\,dy = \frac{1}{2\pi i}\int_{c-i\infty}^{c+i\infty} \widetilde{f}(s)\overline{\widetilde{g}(s)}\,ds. \tag{C.15}$$

EXAMPLE C.7. Equations (C.7) and (C.8) are not in the form of a Mellin transform pair (C.12)/(C.13) but they can be put into that form by the change of variables $y = \frac{1}{x}$. This transforms (C.7) and (C.8) into

$$\frac{f(s)}{s} = \int_0^1 A\left(\frac{1}{y}\right) y^{s-1}dy, \quad \text{and} \quad A\left(\frac{1}{y}\right) = \frac{1}{2\pi i}\int_{c-i\infty}^{c+i\infty} \frac{f(s)}{s} y^{-s}ds,$$

valid for $\sigma > \max\{0, \sigma_c\}$ and $c > \max\{0, \sigma_c\}$ respectively. Note that the $\int_0^1$ integral here can be replaced by $\int_0^\infty$.

EXAMPLE C.8. Setting $\sigma = 0$ in (C.14) gives

$$\int_0^\infty |f(y)|^2 \frac{dy}{y} = \frac{1}{2\pi}\int_{-\infty}^\infty \left|\widetilde{f}(it)\right|^2 dt.$$

EXAMPLE C.9. With $f(s) = \sum a(n)/n^s$ and $A(x) = \sum_{n\le x} a(n)$ again, if we write (C.7) as

$$\frac{f(s)}{s} = \int_1^\infty A(x)x^{-s-1}\,dx \tag{C.16}$$

and apply (C.14), we get

$$\int_0^\infty |A(x)|^2 x^{-2\sigma-1}\,dx = \frac{1}{2\pi}\int_{-\infty}^\infty \left|\frac{f(\sigma+it)}{\sigma+it}\right|^2 dt, \quad \sigma > \max\{0, \sigma_c\}.$$

EXAMPLE C.10. If $L(x) \equiv \sum_{n\le x} \lambda(n)$ is the summatory version of the Liouville function then

$$\frac{\zeta(2s)}{\zeta(s)} = s\int_1^\infty L(x)x^{-s}\frac{dx}{x}.$$

Using the obvious notation $L^*(x)$, it follows that

$$L^*(x) = \frac{1}{2\pi i}\int_{c-i\infty}^{c+i\infty} \frac{\zeta(2s)}{\zeta(s)}\frac{x^s}{s}\,ds, \quad \text{for } c > 1,\ x > 0.$$

EXAMPLE C.11. Consider the Mellin transform of $\zeta(s)/s$ in the form

$$\zeta(s) = -s \int_0^\infty \{x\} x^{-s-1}\, dx, \quad 0 < \sigma < 1,$$

where $\{x\}$ is the fractional part function. Applying the Plancherel relation (C.14) gives

$$\int_0^\infty \{x\}^2 x^{-2\sigma-1} dx = \frac{1}{2\pi} \int_{-\infty}^{\infty} \left| \frac{\zeta(\sigma+it)}{\sigma+it} \right|^2 dt, \quad 0 < \sigma < 1.$$

The simplest Mellin transform is that of $f(y) = e^{-y}$ whose transform is the Gamma function $\Gamma(s)$ for $\Re s > 0$. We end by stating various other well-known examples:

$$\int_0^\infty e^{-ay} y^{s-1}\, dy = \frac{\Gamma(s)}{a^s}, \quad a > 0,\ \Re(s) > 0,$$

$$\int_0^\infty e^{-y} (\log y)^k y^{s-1}\, dy = \Gamma^{(k)}(s), \ \Re(s) > 0,$$

$$\int_0^\infty e^{-y^2} y^{s-1}\, dy = \frac{1}{2}\Gamma\left(\frac{s}{2}\right), \ \Re(s) > 0,$$

$$\int_0^\infty (\sin ay) y^{s-1}\, dy = \frac{\Gamma(s)}{|a|^s} \sin \frac{\pi s}{2} \ \mathrm{sign}(a), \quad -1 < \Re(s) < 1,$$

$$\int_0^\infty (\cos ay) y^{s-1}\, dy = \frac{\Gamma(s)}{|a|^s} \cos \frac{\pi s}{2}, \quad 0 < \Re(s) < 1,$$

$$\int_0^\infty (1+y)^{-1} y^{s-1}\, dy \ = \ \Gamma(s)\Gamma(-s) = \frac{\pi}{\sin \pi s}, \quad 0 < \Re(s) < 1,$$

$$\int_0^\infty \log(1+y) y^{s-1}\, dy = \frac{\pi}{s \sin \pi s}, \quad \text{for } -1 < \Re(s) < 0,$$

$$\int_0^\infty \frac{y^{s-1}}{e^y - 1}\, dy = \sum_{n=1}^\infty \int_0^\infty e^{-ny} y^{s-1}\, dy = \sum_{n=1}^\infty \frac{\Gamma(s)}{n^s} = \Gamma(s)\zeta(s), \ \Re(s) > 1,$$

$$\int_0^\infty \left\{ \frac{1}{t} \right\} t^{s-1}\, dt \ = \ -\frac{\zeta(s)}{s}, \ 0 < \Re(s) < 1,$$

where $\{x\}$ is the fractional part function. Also worth noting are the $\int_1^\infty$ versions

$$\int_1^\infty C\, x^{s-1}\, dx = -\frac{C}{s}, \quad \Re s < 0,$$

$$\int_1^\infty x^\rho\, x^{s-1}\, dx = \frac{1}{s-\rho}, \quad \Re s < -\rho,$$

$$\int_1^\infty \mathrm{li}(x)\, x^{s-1}\, dx = \frac{\log(-s-1)}{s}, \quad \Re(s) < -1.$$

All of the indicated transforms may be inverted; for example, the first of these is inverted as

$$\frac{1}{2\pi i} \int_{c-i\infty}^{c+i\infty} \frac{\Gamma(s)}{a^s} z^{-s}\, ds = e^{-az}, \quad \text{for} \Re z > 0 \text{ and } c > 0.$$

Finally we mention the nice dual relations (Montgomery & Vaughan ([**774**, pp. 143–144]))

$$\left(\sum_{n=1}^{\infty}\frac{a(n)}{n^s}\right)\Gamma(s)=\int_0^{\infty}\left(\sum_{n=1}^{\infty}a(n)e^{-ny}\right)y^{s-1}\,dy$$

and

$$\sum_{n=1}^{\infty}a(n)e^{-ny}=\frac{1}{2\pi i}\int_{c-i\infty}^{c+i\infty}\left(\sum_{n=1}^{\infty}\frac{a(n)}{n^s}\right)\Gamma(s)y^{-s}\,ds$$

for $\sigma>\max\{0,\sigma_c\}$ and $\Re y>0$ respectively. They are related to a result of Ramanujan: If

$$\varphi(x)=\sum_{n=0}^{\infty}\frac{(-1)^n}{n!}\lambda(n)x^n$$

then

$$\int_0^{\infty}x^{s-1}\varphi(x)\,dx=\Gamma(s)\lambda(-s).$$

Further properties of Dirichlet series and their applications may be found in Apostol (1976, [**25**, Chapter 8]), (1990, [**27**, Chapter 11]), Hardy & Riesz (1915, [**498**]), Hardy & Wright (2008, [**499**, Chapter XVII, pp. 381ff]), Montgomery & Vaughan (2007, [**774**, Chapters 1 & 5]), Tenenbaum (2015, [**1053**, Chapter II.1]), Titchmarsh (1939, [**1058**, Chapter IX] and 1986, [**1059**, Chapter 11]), and Widder (1971, [**1102**, Chapter 2]).

Treatments of Lambert series may be found in Apostol (1976, [**25**]), (1990, [**27**]), Berry (2010, [**91**]), Bromwich (1926, [**155**]), Hardy & Wright (2008, [**499**, Chapter 17, §10, pp. 338–341]), Knopp (1990, [**640**]), McCarthy (1986, [**746**]), NIST (2010, [**812**]), Postnikov (1988, [**865**]), and Rademacher (1973, [**875**]).

References that deal with Mellin transforms include Bertrand et al. (2000, [**95**]), Butzer & Jansche (1997, [**174**]), Kowalski (2021, [**661**, Appendix A.3]), and Widder (1946, [**1101**]).

Selected proofs

PROOF THAT IF f AND g ARE MULTIPLICATIVE THEN SO IS $f*g$. Let $\gcd(m,n)=1$. If $d|mn$ it must have the (unique) form $d=ab$ with $a|m$ and $b|n$. Then

$$\begin{aligned}(f*g)(mn)&=\sum_{a|m,\,b|n}f(ab)g(mn/ab)\\&=\sum_{a|m,\,b|n}f(a)f(b)g(m/a)g(n/b)\\&=\left(\sum_{a|m}f(a)g(m/a)\right)\left(\sum_{b|n}f(b)g(n/b)\right)\\&=(f*g)(m)\times(f*g)(n)\end{aligned}$$

□

PROOF OF THE CONNECTION BETWEEN LAMBERT SERIES AND MÖBIUS INVERSION. We need to show the equivalence between

$$b(n)=\sum_{d|n}a(d)\,,\quad a(n)=\sum_{d|n}\mu\left(\frac{n}{d}\right)b(d),\quad n\geq 1,\tag{C.17}$$

and

$$\sum_{n=1}^{\infty} \frac{a(n)x^n}{1-x^n} = \sum_{n=1}^{\infty} b(n)x^n. \tag{C.18}$$

Letting $b(n)$ be defined as in equation (C.17), we have

$$\begin{aligned}
\sum_{n=1}^{\infty} b(n)x^n &= \sum_{n=1}^{\infty} \sum_{d|n} a(d)x^n \\
&= \sum_{d=1}^{\infty} a(d) \sum_{n\geq 1,\, d|n} x^n \\
&= \sum_{d=1}^{\infty} a(d) \sum_{m\geq 1} x^{dm} \\
&= \sum_{d=1}^{\infty} a(d) \frac{x^d}{1-x^d}
\end{aligned}$$

since the last series is a geometric series with common ratio x^d. This calculation holds for $|x| < 1$ whenever the left-hand series converges absolutely. Reversing the calculation establishes the converse implication as well. □

PROOF OF THE CONVOLUTION PROPERTY FOR MELLIN TRANSFORMS. Suppose that $h(y) = \int_0^\infty f(x)g(y/x)\,\frac{dx}{x}$ is the multiplicative convolution of $f(y)$ and $g(y)$. We need to show that their Mellin transforms satisfy $\widetilde{h}(s) = \widetilde{f}(s)\widetilde{g}(s)$.

By definition, the Mellin transform of h is

$$\begin{aligned}
\widetilde{h}(s) &= \int_0^\infty h(y)y^s \frac{dy}{y} \\
&= \int_0^\infty \left(\int_0^\infty f(x)g\Big(\frac{y}{x}\Big)\frac{dx}{x} \right) y^s \frac{dy}{y} \\
&= \int_0^\infty \left(\int_0^\infty g\Big(\frac{y}{x}\Big)y^s\frac{dy}{y} \right) f(x)\frac{dx}{x}.
\end{aligned}$$

In the inner integral (where x is fixed) we make the change of variables $y = xz$, leading to the desired result:

$$\widetilde{h}(s) = \int_0^\infty \left(\int_0^\infty g(z)(xz)^s \frac{dz}{z} \right) f(x)\frac{dx}{x} = \int_0^\infty \widetilde{g}(s)x^s f(x)\frac{dx}{x} = \widetilde{g}(s)\widetilde{f}(s). \quad □$$

APPENDIX D

The Statistical Limit Laws

This appendix deals with probability theory and the fundamental statistical limit laws. These are needed for the arguments of Chapter 4 involving Cramér's heuristic as well as elsewhere in this book. Our focus will be primarily on the 'almost sure behaviour' of sequences of independent (but not necessarily identically distributed) random variables; while there are a few exceptions, most of the proofs for the results to be stated can be found in standard texts on probability and statistics. This journey will at times require us to dip into waters loosely referred to as 'measure theory', a subject which normally assumes a mathematically more mature attitude on the part of the reader. However, in bypassing the proofs, and in the frugal but self-contained path that we have carved out, the conscientious reader will be able to focus on and appreciate—one by one—the essential ideas that lie beneath, while passages that may at first seem difficult can be pondered over and diligently reread over a period of time. The sheer elegance of the subject matter is certain to offer its own rewards. The first half of this appendix is meant to be savoured patiently and read with care; the second half can then be read more lightly, and returned to as and when required.

It is worthwhile to remark that the origins of these laws, as well as some of their earliest applications, were to no small extent associated with problems in number theory, an early example being some questions raised by Émile Borel (1871–1956) about the statistical properties of digits occurring in the decimal expansions of real numbers.[1] Number-theoretic questions were also a driving force behind the discovery of the Law of the Iterated Logarithm (LIL) and some of its generalizations. In the versions appearing in this appendix, the statistical limit laws took many decades to establish. In the case of the LIL, contributions to its development were made by such luminaries as Hardy and Littlewood, and particularly by Khinchin and by Kolmogorov, prior to the essentially final version for the case of identically and independently distributed variables obtained by Hartman & Wintner (1941, [**515**]). For variables that are not identically distributed, or for variables possessing some degree of dependence, results in anything resembling their 'final' forms remain under development.

The usual formulation of probability theory starts with a set, say Ω, whose elements represent all the possible *outcomes* of the random phenomenon being investigated. This set is called the *sample space*. Probabilities, $\mathrm{P}[A]$, are defined for subsets $A \subset \Omega$. If $\omega \in \Omega$ is the random outcome that actually happened and

[1]The main result is that, except for a set of '(Lebesgue) measure 0', all real numbers are *normal*. Roughly stated, this means that their decimal expansions behave *in every way* as if they arose from pure randomness. One would expect that numbers such as e, π, γ, or $\sqrt{2}$ are normal, but remarkably not a single nontrivial example of a normal number has ever been established. Borel, a contemporary of Henri Lebesgue, participated in the French Resistance during WW II.

$\omega \in A$, then we say that the event A has *occurred*. A subtlety of the theory is that when Ω is uncountable it is rarely possible to define the probability $\mathrm{P}[A]$ for *all* possible subsets of Ω in a mathematically consistent way.[2] Those subsets of Ω for which probability *can* be defined are called *events* and are themselves collected into a set $\mathcal{A}$ called the *set* or the *class of events*. The set $\mathcal{A}$ is required to contain at least the *null event* $\emptyset$ (which is the empty set and never occurs) and the *sure event* Ω (which always occurs). Events may be manipulated according to the usual set operations, with a good example being the de Morgan laws

$$\overline{\left(\bigcup_{k=1}^{n} A_k\right)} = \bigcap_{k=1}^{n} \overline{A_k}, \quad \text{and} \quad \overline{\left(\bigcap_{k=1}^{n} A_k\right)} = \bigcup_{k=1}^{n} \overline{A_k}.$$

The triplet $(\Omega, \mathcal{A}, \mathrm{P})$ is referred to as a *probability space*.

It was recognized by Borel in the early 1900s, and exploited by Kolmogorov (1903–1987) in the early 1930s, that for a consistent but sufficiently general theory, the class of events $\mathcal{A}$ must be closed under *countable* unions, *countable* intersections, and complements:

- If $A \in \mathcal{A}$ then its complement $\overline{A} \in \mathcal{A}$;
- If a finite or *countably* infinite collection of events $A_j \in \mathcal{A}$, then $\bigcup_j A_j \in \mathcal{A}$ and $\bigcap_j A_j \in \mathcal{A}$.

Furthermore, the probability measure P is required to satisfy:

- $\mathrm{P}[\emptyset] = 0$ and $\mathrm{P}[\Omega] = 1$.
- If the finite or *countably* infinite collection of events $A_j \in \mathcal{A}$ are *mutually exclusive* (that is, disjoint), that is, if $A_j \bigcap A_k = \emptyset$ for all $j \neq k$, then $\mathrm{P}\left[\bigcup_j A_j\right] = \sum_j \mathrm{P}\left[A_j\right]$.

It is typically not possible to produce a consistent theory when Ω is uncountable and uncountable operations involving events are allowed.

Many useful result follow from the definitions, including the *inclusion–exclusion principle* $\mathrm{P}[A \cup B] = \mathrm{P}[A] + \mathrm{P}[B] - \mathrm{P}[A \cap B]$, or more generally

$$\begin{aligned} \mathrm{P}[A_1 \cup A_2 \cup \cdots \cup A_n] &= \sum_{1 \le i \le n} \mathrm{P}[A_i] - \sum_{1 \le i < j \le n} \mathrm{P}[A_i \cap A_j] \\ &+ \sum_{1 \le i < j < k \le n} \mathrm{P}[A_i \cap A_j \cap A_k] - \cdots + (-1)^n \mathrm{P}[A_1 \cap A_2 \cap \cdots \cap A_n], \end{aligned}$$

as well as the *continuity property* for countable sequences of *mutually exclusive* (disjoint) events:

$$\mathrm{P}\left[\bigcup_{k=1}^{n} A_k\right] \equiv \sum_{k=1}^{n} \mathrm{P}[A_k] \to \mathrm{P}\left[\bigcup_{k=1}^{\infty} A_k\right], \quad \text{as } n \to \infty.$$

In courses on probability theory, students are often asked to prove results such as these.

The formulation for probability theory which has just been outlined is self-contained, although brief, and is well worth rereading. However, if the concepts

[2] *Countably infinite* sets, like the primes, are sets that can be put in one-to-one correspondence with the integers. The real numbers are an example of a set that is not countable; a famous 'diagonal argument' due to Georg Cantor (1845–1918) is often used to show this. Examples of 'events' whose probability cannot be defined may be found in many textbooks.

seem disquieting, the reader will find it rewarding to seek out an appropriate book for examples and for additional background on this remarkable theory. Suitable references may be found at the end of this appendix. The 'easy exercise' appearing further below may also help to 'fix' ideas.

A few further concepts are needed before we can proceed. The 'observable quantities' of a random phenomenon under investigation are called *random variables* and are to be thought of as functions on the sample space $X : \Omega \to \mathbb{R}$, say. Thus if $\omega \in \Omega$ is the outcome that occurred, then X will take on the value $X(\omega)$. There can be many random variables, as in the vector $(X_1, X_2, \ldots, X_n)$, or as in the infinite sequence $X_1, X_2, X_3, \ldots$. A statement such as $a < X \leq b$ for real numbers $a < b$ is meant to refer to the event A defined as $A \equiv \{w \in \Omega \colon a < X(\omega) \leq b\}$, or equivalently as $A \equiv X^{-1}((a,b])$.

However, it is the *distribution* of a random variable X that is of primary importance. The *distribution function* of X is defined as $F_X(x) \equiv \mathrm{P}[X \leq x]$. For a finite collection $X_1, X_2, \ldots, X_n$ it is defined analogously as

$$F_{X_1,X_2,\ldots,X_n}(x_1, x_2, \ldots, x_n) \equiv \mathrm{P}\big[X_1 \leq x_1, X_2 \leq x_2, \ldots, X_n \leq x_n\big];$$

this is the probability of an intersection of n events (thus of all of them occurring). If $F_X(x) \equiv \int_{-\infty}^{x} f_X(y)\,dy$, where $f_X(x) \geq 0$ we call $f_X(x)$ the *density function* of X. Likewise if

$$F_{X_1,X_2,\ldots,X_n}(x_1, x_2, \ldots, x_n) \equiv \int_{-\infty}^{x_1}\int_{-\infty}^{x_2}\cdots\int_{-\infty}^{x_n} f_{X_1,X_2,\ldots,X_n}(y_1, y_2, \ldots, y_n)\,dy_1 dy_2 \cdots dy_n$$

we say $f_{X_1,X_2,\ldots,X_n}(x_1, x_2, \ldots x_n)$ is the density function of the random vector $(X_1, X_2, \ldots, X_n)$. When X takes on only a finite or countable number of values, one may speak instead of its *probability mass function*, $f_X(x_j) \equiv \mathrm{P}[X = x_j]$, with $\sum_j f_X(x_j) = 1$. For details about the many well-known parameterized distributions (binomial, Poisson, normal, and so on) and for methods of dealing with transformations on random variables, we refer the reader to any standard textbook on this topic.

REMARKS D.1. A function $F(x) \equiv \mathrm{P}[X \leq x]$ can be a distribution function if and only if it is nondecreasing, right-continuous, and has limits $F(-x) \to 0$ and $F(x) \to 1$ as $x \to \infty$. Furthermore, every distribution function $F(x)$ can be decomposed uniquely as

$$F(x) = \alpha F_{ac}(x) + \beta F_d(x) + \gamma F_s(x), \quad \text{where } \alpha, \beta, \gamma \geq 0, \quad \text{and } \alpha + \beta + \gamma = 1,$$

where $F_{ac}(x)$ is *absolutely continuous*—a distribution that corresponds to a density function; $F_d(x)$ is *discrete*—a distribution that takes on a finite or countable number of values whose probabilities add up to 1; and $F_s(x)$ is *singular*—a continuous distribution (function) whose derivative is 0 except on a set of measure 0 (like the famous Cantor function).

The concept of a 'long run average' for a random variable X (under 'repeated trials') is captured by its *expected value*, $\mathrm{E}\,X$. It is defined as $\mathrm{E}\,X = \int_{-\infty}^{\infty} x f(x)\,dx$ if X has density $f(x)$, and more generally as the Riemann–Stieltjes integral $\mathrm{E}\,X = \int_{-\infty}^{\infty} x\,dF_X(x)$ (defined in Appendix F), provided these exist. Still more generally, the *transformed* random variable $Y \equiv g(X_1, X_2, \ldots, X_n)$ can (when the density function exists) be shown to have expectation

$$\begin{aligned} \mathrm{E}\, Y &\equiv \mathrm{E}\, g(X_1, X_2, \ldots, X_n) \\ &= \int_{-\infty}^{\infty} \cdots \int_{-\infty}^{\infty} g(x_1, x_2, \ldots, x_n) f(x_1, x_2, \ldots, x_n)\, dx_1\, dx_2 \cdots dx_n. \end{aligned}$$

Note that because integration is a linear operator we will have in general, for constants c_j,

$$\mathrm{E}\Big(c_0 + \sum_{j=1}^{n} c_j X_j\Big) = c_0 + \sum_{j=1}^{n} c_j\, \mathrm{E}\, X_j.$$

Properties of a random variable X that can be expressed in terms of expectation include its *moments* $\mu_k \equiv \mathrm{E}(X^k)$ and *absolute moments* $\nu_k \equiv \mathrm{E}(|X|^r)$, for integers k and real numbers r, as well as their *centred* versions $\mu_k' \equiv \mathrm{E}(X - \mu_1)^k$, and $\nu_k' \equiv \mathrm{E}\,|X - \mu_1|^r$. Writing μ for μ_1, and σ^2 for ν_2'—these are called the *mean* and *variance* respectively—the identity $\sigma^2 = \mu_2 - \mu^2$ is an easy exercise. We often denote σ^2 by $\operatorname{Var} X$; the square root of the variance, σ, is known as the *standard deviation*.

Also fairly easy to establish is the very useful *Markov inequality*

$$\mathrm{P}\big[|X| \geq a\big] \leq \frac{\mathrm{E}\,(|X|^r)}{a^r}, \quad \text{for } a > 0,\ r > 0;$$

and (with $\mu = \mathrm{E}\, X$, $\sigma^2 = \operatorname{Var} X$) the special case of it known as *Chebyshev's inequality*:

$$\mathrm{P}\big[|X - \mu| \geq a\big] \leq \frac{\sigma^2}{a^2}, \quad \text{for } a > 0.$$

Another special case which generally results in sharper bounds is *Chernoff's inequality*

$$\mathrm{P}\big[X \geq a\big] \leq \inf_{t>0} e^{-at} M(t);$$

it follows upon optimizing over real values of t in the computation

$$\mathrm{P}\big[X \geq a\big] = \mathrm{P}\left[e^{tX} \geq e^{at}\right] \leq \frac{\mathrm{E}\, e^{tX}}{e^{at}} \equiv e^{-at} M(t).$$

Here $M(t) \equiv \mathrm{E}\, e^{tX}$, $t \in \mathbb{R}$, is called the *moment generating function (mgf)* of X because

$$\mathrm{E}\, e^{tX} \equiv \int_{\infty}^{\infty} e^{tx} dF_X(x) = \int_{\infty}^{\infty} \sum_{n=0}^{\infty} \frac{t^n x^n}{n!} dF_X(x) = \sum_{n=0}^{\infty} \frac{t^n}{n!} \mu_k,$$

$$\text{and} \quad \frac{d^k M(t)}{dt^k}\bigg|_{t=0} = \mu_k.$$

An mgf is essentially a Laplace transform. When the argument of $M(\cdot)$ is purely imaginary one obtains the *characteristic function (cf)*:

$$\varphi_X(t) \equiv \mathrm{E}\, e^{itX} = \int_{\infty}^{\infty} e^{itx} dF_X(x), \quad t \in \mathbb{R}.$$

Because this is a Fourier transform which can be inverted, cfs are in one-to-one correspondence with distributions. This isn't so for mgfs, although if an mgf is bounded in an interval around 0, it will actually be analytic in the corresponding vertical strip of the complex plane. Its values on the imaginary axis will then be a cf, and the mgf would then correspond to a unique distribution. (Of course, all moments of the distribution would need to be finite—and not grow too quickly—for there to be any chance for this to occur.) The logarithm of an mgf is called a *cumulant generating function* (cgf), and when expanded as $K(t) \equiv \log M(t) \equiv \log \sum_{n=1}^{\infty} \mu_n t^n/n! = \sum_{n=1}^{\infty} \kappa_n t^n/n!$ it defines the cumulants κ_n of the distribution. When these exist, they are related to the moments as $\kappa_1 = \mu_1$, $\kappa_2 = \mu_2 - \mu_1^2$, $\kappa_3 = \mu_3' = \mu_3 - 3\mu_2\mu_1 + 2\mu_1^3$, and so on.[3]

The last of the concepts we need—one which plays a crucial role in the theory—is that of *independence*. Two events A_1 and A_2 are said to be *independent* if $\mathrm{P}[A_1 \cap A_2] = \mathrm{P}[A_1]\mathrm{P}[A_2]$. A finite collection A_j, $j = 1, 2, \ldots, n$ of events is said to be *mutually independent* if

$$\mathrm{P}\left[A_{j_1} \cap A_{j_2} \cap \cdots A_{j_k}\right] = \mathrm{P}\left[A_{j_1}\right]\mathrm{P}\left[A_{j_2}\right]\cdots\mathrm{P}\left[A_{j_k}\right]$$

for all subcollections of the events. In the case of random variables, independence of X_1 and X_2 can be taken to mean that $F_{X_1,X_2}(x_1, x_2) = F_{X_1}(x_1)F_{X_2}(x_2)$ for all values of x_1 and x_2, while for a finite collection $X_1, X_2, \ldots, X_n$, mutual independence can be defined as

$$F_{X_1,X_2,\ldots X_n}(x_1, x_2, \ldots, x_n) = F_{X_1}(x_1)F_{X_2}(x_2)\cdots F_{X_n}(x_n), \quad \text{for all } x_1, x_2, \ldots, x_n.$$

Independence of an infinite collection of events or random variables is taken to mean the independence of all finite subcollections. The intuitive meaning of independence is highlighted in many textbooks. In particular, the independence of two events, say A and B, means that the *conditional probability* of A occurring given that B has occurred, denoted as $\mathrm{P}[A|B]$, is the same as $\mathrm{P}[A]$. Heuristic considerations lead to the definition $\mathrm{P}[A|B] \equiv \mathrm{P}[A \cap B]/\mathrm{P}[B]$, provided that $\mathrm{P}[B] \neq 0$. The consequences of independence, particularly for infinite collections of events or random variables, are far-reaching and profound, and lend to probability theory its distinctive flavour.

Here is an easy exercise which is meant to help the reader 'fix' ideas, and gauge their understanding of the underlying elementary concepts:

EXERCISE D.2. A fair coin (labelled H and T) and a fair six-sided die (labelled 1 through 6) are tossed simultaneously. The sample space Ω may be thought of as consisting of twelve outcomes, $\{H1, H2, \ldots, T5, T6\}$ say, which are all equally likely due to the underlying symmetries. Describe the most detailed class, $\mathcal{A}$, of events for this setup. (Hint: There are 2^{12} elements in it.) Describe the probability function $\mathrm{P}[\cdot]$ for this setup. Show that the events "coin falls heads" and "die falls odd" are independent. Show that the random variable, X, which equals 0 or 1 according to how the coin falls, and the random variable, Y, which equals the number of points showing on the die, are independent. Explain how a random variable W, equalling the square of the number showing on the die, can be viewed as a function on Ω; determine its distribution, and show that $\mathrm{E}\,W = 15\,\frac{1}{6}$, and that W has first absolute moment $10\,\frac{1}{2}$, variance ≈ 149.14, and standard deviation ≈ 12.21.

EXAMPLES D.3.

(1) A weighted coin has probability θ of falling heads, and $1 - \theta$ of falling tails, where $0 < \theta < 1$. A random experiment consists of repeated independent tosses of the coin until a heads first turns up. Here the sample space is *countable* and can be identified with the set of positive integers, $\Omega \equiv \{1, 2, 3, \ldots\}$, with the outcome "$n$" having probability $\theta(1 - \theta)^{n-1}$ because it involves $n - 1$ tails

[3]Cumulants have the advantage that they are additive for sums of independent random variables.

followed by one head. The class $\mathcal{A}$ of events is very big because it consists of all the possible subsets of Ω. The probability of any of these events is just the sum of the probabilities of the outcomes within it.

(2) An example of an *uncountable* sample space, $\Omega \equiv [0,1]$ say, might involve selecting a real number, uniformly at random, between 0 and 1. It is a fact (discussed in many books) that not every subset of Ω can be measured. On the other hand, the class of events $\mathcal{A}$ must at least contain all *intervals* that lie in Ω. (Such intervals are sets of numbers between a and b, where $0 \le a \le b \le 1$, and are allowed to include as well as exclude either of their two endpoints. The probability of such an interval is $b - a$.) But from the fact that $\mathcal{A}$ must be closed under countable unions, countable intersections, and complements, one sees that the class $\mathcal{A}$ of events must be quite complicated.

(3) Sample spaces can also be of higher dimension, as for example when choosing a point at random in the unit square $[0,1]^2$, or in the unit cube $[0,1]^3$. In fact 'hypercubes' of any dimension can be sample spaces, even when their number of dimensions is infinite. The class $\mathcal{A}$ of events will be still more complicated in such cases.

(4) The *normal* or *Gaussian distribution*, $N(\mu, \sigma^2)$, is defined by its density

$$f(x) = \frac{1}{\sqrt{2\pi}\,\sigma}\, e^{-(x-\mu)^2/2\sigma^2}.$$

It has mean $\mu \in \mathbb{R}$, variance $\sigma^2 > 0$, mgf $M(t) = e^{\mu t + \sigma^2 t^2/2}$, cgf $K(t) = \mu t + \sigma^2 t^2/2$, and cf $\varphi(t) = e^{i\mu t - \sigma^2 t^2/2}$. The $N(0,1)$ distribution is referred to as the *standard normal*.

(5) Although it is rarely used in this book, for completeness we define here the *d-dimensional multivariate normal distribution* , denoted as $N_d(\mu, \Sigma)$—where μ is a d-dimensional column vector of means, and Σ is a $d \times d$ matrix of variances and covariances. It is given by the density function

$$f(x) = \frac{1}{(2\pi)^{k/2}\,|\Sigma|^{1/2}} \exp\Big(-\frac{1}{2}(x-\mu)'\,\Sigma^{-1}(x-\mu)\Big).$$

Here $x \equiv (x_1, x_2, \ldots, x_d)'$, $\mu \equiv (\mu_1, \mu_2, \ldots, \mu_d)'$, the primes represent transposition, $|\Sigma|$ denotes a determinant, and Σ^{-1} denotes a matrix inverse. If $X \equiv (X_1, X_2, \ldots, X_d)'$ with $X \sim N_d(\mu, \Sigma)$—meaning that X has the $N_d(\mu, \Sigma)$ distribution—then $\mathrm{E}\, X_j = \mu_j$, and the *covariance* $\sigma_{j,k} \equiv \mathrm{E}\,(X_j - \mu_j)\,(X_j - \mu_k) = \mathrm{E}\, X_j X_k - \mu_j \mu_k$ is the (j,k)th element of Σ. Expressed more succinctly: $\mathrm{E}\, X = \mu$, and $\mathrm{E}\big[(X-\mu)(X-\mu)'\big] = \Sigma$. The jth diagonal element $\sigma_{j,j}$ of Σ is the variance of X_j.

With much of the essential background having now been laid out, we turn next to what are known as the *zero–one laws*. These considerably deeper results deal with infinite series of events and infinite series of random variables, as well as with so-called *tail events*. Given an infinite sequence of events, A_n, $n = 1, 2, \ldots$, recall the set-theoretic definitions

$$\limsup_{n\to\infty} A_n \equiv \bigcap_{n=1}^{\infty} \bigcup_{k=n}^{\infty} A_k, \quad \text{and} \quad \liminf_{n\to\infty} A_n \equiv \bigcup_{n=1}^{\infty} \bigcap_{k=n}^{\infty} A_k.$$

The lim sup consists of those outcomes that occur in an infinite number of the events, while the lim inf consists only of those that eventually occur in all of them (or equivalently, occur in all but a finite number of them). The resulting sets are examples of *tail events*—events that depend only on the 'tail' of the sequence, thus remaining unchanged if any finite subset of the sequence is removed or altered. Tail events can also arise in the context of an infinite sequence $X_1, X_2, \ldots,$ of random variables; these are events such as $\{\omega \in \Omega\colon \sum_1^\infty X_n < \infty\}$ or $\{\omega \in \Omega\colon N^{-1}\sum_1^N X_n \to \frac{1}{2}\}$ which do not depend on any finite subset of the variables.

The first of the zero–one laws we state here is the following.

THEOREM D.4 (Kolmogorov zero–one law). *If A is a tail event of an infinite sequence of independent random variables, $X_1, X_2, \ldots,$ then either* $\mathrm{P}[A] = 0$ *or* $\mathrm{P}[A] = 1$.

For this remarkable result to hold the X_n must be independent, but they do not need to be identically distributed. One corollary to Kolmogorov's law is that a real-valued random quantity, say X, which is determined solely by the tail of a sequence of independent random variables, must satisfy $\mathrm{P}[\,X < a\,] = 0$ or 1 for every value of a, and must therefore be constant with probability 1, that is, $\mathrm{P}[\,X = c\,] = 1$ for some constant c (possibly $\pm\infty$). Of the next two results, the first is just a consequence of Kolmogorov's law, while the second is often useful for distinguishing between tail events whose probability is 0 and those whose probability is 1.

THEOREM D.5 (Borel zero–one law). *If $\{A_n\}$ is any sequence of* independent *events, then*

$$\mathrm{P}\left[\limsup_{n\to\infty} A_n\right] \equiv \mathrm{P}\left[\bigcap_{n=1}^{\infty}\bigcup_{k=n}^{\infty} A_k\right] = 0 \ \ or\ \ 1,$$

so the probability that infinitely many *of the A_n occur is either 0 or 1. Likewise*

$$\mathrm{P}\left[\liminf_{n\to\infty} A_n\right] \equiv \mathrm{P}\left[\bigcup_{n=1}^{\infty}\bigcap_{k=n}^{\infty} A_k\right] = 0 \ \ or\ \ 1,$$

so the probability that eventually all *A_n occur is either 0 or 1.*

LEMMA D.6 (The Borel–Cantelli Lemmas).

(a) *If* $\sum_{n=1}^{\infty} \mathrm{P}[A_n] < \infty$, *then*

$$\mathrm{P}\left[\limsup_{n\to\infty} A_n\right] \equiv \mathrm{P}\left[\bigcap_{n=1}^{\infty}\bigcup_{k=n}^{\infty} A_k\right] = 0\,;$$

thus with probability 1, only finitely many of the A_n can occur.

(b) *If* $\sum_{n=1}^{\infty} \mathrm{P}[A_n] = \infty$, *and if the A_n are also mutually independent, then*

$$\mathrm{P}\left[\limsup_{n\to\infty} A_n\right] \equiv \mathrm{P}\left[\bigcap_{n}\bigcup_{k\geq n} A_k\right] = 1\,;$$

so that with probability 1, infinitely many of the A_n occur.

REMARK D.7. The quantity $\sum_{n=1}^{\infty} \mathrm{P}\,[A_n]$ equals the expected number of A_n that will occur.

Note that part (a) of this lemma does not assume independence, while the independence requirement of part (b) cannot be lightly dispensed with. In the following generalizations, parts (c), (d), and (e) are due to Erdős & Rényi (1959, [**359**]); parts

(d) and (e) are corollaries to part (c), with (d) being a special case of (e). Part (f) is due to Shuster (1970, [**988**]) and is of theoretical interest because it makes no mention of independence. Part (g), due to Barndorff & Nielsen (1961, [**78**]), also has no independence requirement.

LEMMA D.8 (Further Borel-Cantelli type lemmas).

(c) *If* $\sum_{n=1}^{\infty} \mathrm{P}[A_n] = \infty$, *then a necessary and sufficient condition for infinitely many of the* A_n *to occur with probability 1 is that*

$$\liminf_{n\to\infty} \frac{\sum_{j=1}^{n}\sum_{k=1}^{n} \mathrm{P}[A_j \cap A_k]}{\left(\sum_{j=1}^{n} \mathrm{P}[A_j]\right)^2} = 1\,. \tag{D.1}$$

(d) *If* $\sum_{n=1}^{\infty} \mathrm{P}[A_n] = \infty$, *and the* A_n *are only pairwise independent, then with probability 1, infinitely many of the* A_n *will occur.*

(e) *If* $\sum_{n=1}^{\infty} \mathrm{P}[A_n] = \infty$, *and* $\mathrm{P}[A_j \cap A_k] \le \alpha \mathrm{P}[A_j]\mathrm{P}[A_k]$ *for some* $\alpha \ge 1$, *and for all distinct* j, k *large enough, then with probability 1, infinitely many of the* A_n *will occur.*

(f) *A necessary and sufficient condition for infinitely many of the* A_n *to occur with probability 1 is that* $\sum_{j=1}^{\infty} \mathrm{P}[A \cap A_j] = \infty$ *for every event* A *such that* $\mathrm{P}[A] > 0$.

(g) *If* $\mathrm{P}[A_n] \to 0$ *as* $n \to \infty$ *and*

$$\sum_{n=1}^{\infty} \mathrm{P}\left[\, A_n \cap \overline{A_{n+1}} \,\right] < \infty, \tag{D.2}$$

then with probability 1 only finitely many of the A_n *will occur.*

REMARKS D.9. The liminf in (D.1) can exceed 1 but it can never be less than 1, and if it takes the value $C > 1$ then the probability that infinitely many of the A_n will occur is $\ge \frac{1}{C}$. Note that (D.2) by itself implies that there can be no more than a finite number of 'switches' between the sequences $\{A_n\}$ and $\{\overline{A_n}\}$, so one (and only one) of these must occur infinitely often. See also Kochen & Stone (1964, [**643**]), Spitzer (1964, [**1014**]), Petrov (2002, [**843**], and 2004, [**844**]), and Martikainen & Petrov (1990, [**734**]).

Our final zero–one result is included mainly for the sake of completeness, but it is worth noting that when both are restricted to iid variables the Hewitt–Savage zero–one law is actually slightly stronger than Kolmogorov's. The abbreviation *iid* stands for *independent and identically distributed*—a concept that plays a *very* important role in the theory.

THEOREM D.10 (Hewitt–Savage zero–one law). *If* A *is an event defined* symmetrically *on an infinite sequence of* iid *random variables* $X_1, X_2, \ldots,$ *(that is,* A *does not depend on the ordering of the variables), then either* $\mathrm{P}[A] = 0$ *or* $\mathrm{P}[A] = 1$.

Arguments motivated by Cramér's heuristic (for instance) are typically based on what is called almost sure convergence, or convergence with probability 1. Details concerning modes of convergence, $X_n \to X$ for random variables (as $n \to \infty$) are elucidated in many texts. We will consider four basic (although very different) modes of convergence:[4]

[4]To readers for whom this may be very new: Although our discussion is self-contained, to fully appreciate this 'rigorous probability' material requires both reflection, and 'maturity', and

- *Convergence in distribution* or *weak convergence*, $X_n \xrightarrow{d} X$ or $X_n \xrightarrow{d} F$, means that $\mathrm{P}[X_n \leq x] \to \mathrm{P}[X \leq x]$ at all x for which $F(x) \equiv \mathrm{P}[X \leq x]$ is continuous,
- *Convergence in rth mean* means that $\mathrm{E}\,|X_n - X|^r \to 0$, and
- *Convergence in probability*, $X_n \xrightarrow{p} X$, means that

$$\mathrm{P}\left[\,|X_n - X| > \epsilon\,\right] \to 0\,, \quad \text{as } n \to \infty, \text{ for all } \epsilon > 0.$$

- The fourth and much stronger form of convergence—which also is technically more difficult to establish—is *almost sure convergence* (or convergence with probability 1), denoted as $X_n \to X$ a.s., or as $X_n \to X$ w.p.1, with X assumed (w.p.1) to be finite. It is defined as

$$\mathrm{P}\left[\lim_{n\to\infty} X_n = X\right] = 1,$$

meaning that $\lim X_n(\omega) \to X(\omega)$ for all $\omega \in \Omega$ except on an event of probability 0. It can be shown that a.s. convergence is equivalent to any of the following four conditions:

$$\lim_{n\to\infty} \mathrm{P}\Big[\bigcap_{j=n}^{\infty} \{\,|X_j - X| \;\leq \epsilon\,\}\Big] = 1, \qquad \lim_{n\to\infty} \mathrm{P}\Big[\bigcup_{j=n}^{\infty} \{\,|X_j - X| \geq \epsilon\,\}\,\Big] = 0,$$

$$\sup_{m>n} |X_m - X_n| \xrightarrow{p} 0 \text{ as } n \to \infty, \quad \text{or} \quad \mathrm{P}\left[\;|X_n - X| \geq \epsilon \text{ infinitely often }\right] = 0,$$

with the $\epsilon > 0$ in these being arbitrary.

The connections among the various modes of convergence can be fairly subtle.[5] That $X_n \to X$ a.s. implies $X_n \xrightarrow{p} X$, which in turn implies $X_n \xrightarrow{d} X$, are standard results. A partial converse to the first of these states that if $X_n \xrightarrow{p} X$, then there is a subsequence such that $X_{n_k} \to X$ a.s., as $k \to \infty$. As for convergence in rth mean, the following two results may be noted. On the one hand, $\mathrm{E}\,|X_n - X|^r \to 0$ for any $r > 0$ implies that $X_n \xrightarrow{p} X$. (The proof is a simple application of the Markov inequality.) But on the other hand, the almost sure convergence $X_n \to X$ a.s. does not imply that $\mathrm{E}\,|X_n - X|^r \to 0$ (indeed the moments need not even exist); although it does imply this if a uniform boundedness condition $|X_n|^r \leq Z$ with $\mathrm{E}\,Z < \infty$ is met. (The proof is an instance of what is known as the 'Dominated Convergence Theorem'.)

As has been mentioned, almost sure convergence—which is the one most relevant for us—is also the most demanding one to establish. However there are a number of results which are often helpful. First of all, either of $\sum_{n=1}^{\infty} \mathrm{E}\,|X_n - X|^r < \infty$ for some $r > 0$, or $\sum_{n=1}^{\infty} \mathrm{P}[\,|X_n - X| \geq \epsilon] < \infty$ for all $\epsilon > 0$, suffice to establish that $X_n \to X$, a.s. (These criteria, in effect, require X_n to converges to X *quickly enough* in the indicated modes.) Secondly, it can be shown that if $\sum_{n=1}^{\infty} \mathrm{E}\,|X_n|^r < \infty$ for some $0 < r \leq 1$, then the sum $\sum_{n=1}^{\infty} X_n$ will converge a.s., while if $\sum_{n=1}^{\infty} \mathrm{E}\,|X_n|^r < \infty$ for some $1 \leq r \leq 2$, but with the X_i now also being *independent*, then the sum $\sum_{n=1}^{\infty}(X_n - \mathrm{E}\,X_n)$ will converge a.s. However, a word of caution: it can be shown that $(\mathrm{E}\,|X|^r)^{1/r}$ is a nondecreasing function of r; this is known as the *monotonicity property for moments*. So $\mathrm{E}\,|X|^r < \infty$ implies that $\mathrm{E}\,|X|^s < \infty$ for $s < r$. But there is no corresponding result for infinite *sums*:

(to be fair) may need a certain amount ot time and study. But once mastered, you will have reached a most worthwhile milestone!

[5]This paragraph, and the next two, are decidedly intense! But the facts are there for when they may be needed.

$\sum_{n=1}^{\infty} \mathrm{E}\,|X_n|^r < \infty$ for one value of r does not imply its convergence for any other value of r.

Under assumptions of independence much stronger properties can be established. A celebrated *theorem of Paul Lévy* states that if the X_n are independent then the almost sure, in probability, and in distribution convergences of $\sum X_n$ are all equivalent to one another. Assorted proofs of this nontrivial result may be found, for example, in Breiman (1968, [**142**, V]), Chung (2001, [**207**, Theorem 5.3.4]), Gut (2013, [**470**, p. 292]), Stout (1974, [**1035**, Theorem 2.13.2]) or Tucker (1967, [**1068**, §5.2]). It follows, under independence, that $\sum \mathrm{Var}\, X_n < \infty$ suffices to establish that $\sum (X_n - \mathrm{E}\, X_n)$ converges almost surely (as well as in rth mean for $r = 2$), although the independence is actually not required for the almost sure convergence here.[6] (If also $\sum \mathrm{E}\, X_n$ converged then $\sum X_n$ would converge a.s. as well.[7]) This has a converse which does however require the independence as well as the uniform boundedness $|X_n| \le M$. When these two conditions hold the almost sure convergence of $\sum(X_n - \mathrm{E}\, X_n)$ will be *equivalent* to the convergence of $\sum \mathrm{Var}\, X_n$. (Of course, under independence, sums such as $\sum X_n$ must either converge w.p.1 or diverge w.p.1, owing to Kolmogorov's zero–one law.) If one assumes only the *orthogonality condition* $\mathrm{E}\, X_j X_k = 0$ for all $j \ne k$, then the condition $\sum_1^{\infty} (\log n)^2\, \mathrm{E}\, X_n^2 < \infty$ suffices for the almost sure convergence of $\sum_1^{\infty} X_n$, and consequently also its convergence in distribution. (This result is due to Menshov and Rademacher.) While each of the numerous results stated in this paragraph, and in the two previous ones, require no more than patient diligence to *understand*, to *prove* the full collection of these results would be a formidable undertaking!

EXAMPLE D.11. Suppose that the X_n are independent, and equal ± 1 with equal probabilities. Then the 'harmonic' quantity

$$\sum_{n=1}^{\infty} \frac{X_n}{n}$$

converges almost surely to a well-defined random variable. The *Law of Pure Types* due to Jessen and Wintner asserts that a convergent sum of independent discrete random variables must be of pure type—namely discrete, absolutely continuous, or singular. It can be shown that the distribution of $\sum_{n=1}^{\infty} X_n/n$ possesses a density function.

Finally, we state the incisive *Three-Series Criterion* of Kolmogorov which fully characterizes the almost sure convergence of a sum of *independent random variables* in terms of the convergence of three *numerical* series: The sum $\sum_{i=1}^{\infty} X_i$ of independent random variables converges to some random variable X almost surely, if and only if each of the three series[8]

$$\sum_{i=1}^{\infty} \mathrm{P}[\,|X_i| > c\,], \quad \sum_{i=1}^{\infty} \mathrm{E}(X_i\ \mathrm{I}[\,|X_i| \le c\,]), \quad \text{and} \quad \sum_{i=1}^{\infty} \mathrm{Var}(X_i\ \mathrm{I}[\,|X_i| \le c\,])$$

[6] In general, if $\sum \mathrm{Var}\, X_n$ converges then $\sum(X_n - \mathrm{E}\, X_n)$ converges a.s. Conversely if the X_n are *independent*, uniformly bounded $|X_n| < c$, and all $\mathrm{E}\, X_n = 0$, then the almost sure convergence of $\sum X_n$ implies the convergence of $\sum \mathrm{E}\, X_n^2$ (Breiman (1968, [**142**, pp. 46–47])).

[7] In fact, for independent variables X_n, the convergence of both $\sum \mathrm{E}\, X_n$ and $\sum \mathrm{Var}\, X_n$ implies the almost sure convergence of $\sum X_n$. Kowalski (2021, [**661**, Appendix B.10]) provides a nice and complete proof.

[8] The $\mathrm{I}[\,|X_i| \le c\,]$ are just 0–1 indicator functions for occurrence of the event $|X_i| \le c$.

converges for some $c > 0$. Furthermore, if these series converge for one value of $c > 0$, then they will converge for all values of $c > 0$. (The criterion is sometimes stated just using $c = 1$.) If we define X_i^* as X_i when $|X_i| \leq 1$ and as 0 otherwise, an arguably more elegant version of this criterion can be stated as

$$\sum_{i=1}^{\infty} \mathrm{P}[X_i \neq X_i^*] < \infty, \quad \sum_{i=1}^{\infty} \mathrm{E}\, X_i^* < \infty, \quad \text{and} \quad \sum_{i=1}^{\infty} \mathrm{Var}\, X_i^* < \infty.$$

A proof may be found in Stout (1974, [**1035**, Theorem 2.13.4]), as well as numerous other sources.

REMARK D.12. The three-series conditions would all be met if the X_i were independent, and each of the series $\sum \mathrm{E}\, X_i$ and $\sum \mathrm{Var}\, X_i$ converged.

EXAMPLE D.13. The asymptotic distribution of $\varphi(n)/n$ (discussed in Chapter 3) is that of

$$\prod_p \left(1 - \frac{1}{p}\right)^{Z_p},$$

where the Z_p are independent, with $\mathrm{P}[Z_p = 1] = \frac{1}{p}$ and $\mathrm{P}[Z_p = 0] = 1 - \frac{1}{p}$. On taking logarithms, Kolmogorov's Three-Series Criterion can be used to show that this converges to a random variable almost surely.

EXAMPLE D.14. Let $f(p)$ be independent and symmetric Bernoulli random variables with $\mathrm{P}[f(p) = 1] = \mathrm{P}[f(p) = -1] = \frac{1}{2}$. The Euler products over the primes p, defined as

$$F(s) = \prod_p \left(1 + \frac{f(p)}{p^s}\right), \quad \text{and} \quad G(s) = \prod_p \left(1 - \frac{f(p)}{p^s}\right)^{-1},$$

correspond to the Dirichlet series of random multiplicative arithmetic functions which have been proposed as a proxies for the analogous series based on the Möbius and Liouville functions. (Background and details may be found in Chapter 11.) On taking logarithms, Kolmogorov's Three-Series Criterion can be used to show that each of $F(s)$ and $G(s)$ converges almost surely, not only for $\Re s > 1$, but also for $\Re s > \frac{1}{2}$.

We now indicate a few methods which are useful for establishing the convergence in distribution/weak convergence $X_n \xrightarrow{d} X$ of a sequence of random variables X_n to a random variable X. The alternate notations $X_n \xrightarrow{d} F$ and $F_n \xrightarrow{d} F$, where $F_n(x) = \mathrm{P}[X_n \leq x]$ and $F(x) = \mathrm{P}[X \leq x]$ may also be used in this context.

- The convergence $X_n \xrightarrow{d} X$ is *equivalent* to the convergence $\mathrm{E}\, h(X_n) \to \mathrm{E}\, h(X)$ for all bounded continuous functions h.
- If random variables X_n and X have mgfs $M_n(t) = \mathrm{E}\, e^{tX_n}$ and $M(t) = \mathrm{E}\, e^{tX}$ all of which are finite in some interval $|t| < \epsilon$ $(\epsilon > 0)$ about the origin, and if $M_n(t) \to M(t)$ as $n \to \infty$ for all such t, then $X_n \xrightarrow{d} X$. (Remark: mgfs which are equal in an interval about the origin correspond to identical distributions.)
- Weak convergence $F_n \xrightarrow{d} F$ is *equivalent* to pointwise convergence $\varphi_n(t) \to \varphi(t)$ for all t of the corresponding characteristic functions (cfs). Further, if a sequence of cfs converges to a function that is continuous around the origin then that function must itself be a cf and the corresponding convergence in distribution holds.

(Recall that cfs are in one-to-one correspondence with distributions functions.) These results are due to Paul Lévy.

- If all moments of $\{X_n\}$ are finite and converge to those of X whose distribution is *determined* by its moments, then $X_n \xrightarrow{d} X$. (This result is due to Fréchet and Shohat.) A sufficient condition (due to Torsten Carleman) for a sequence $\{\mu_k\}$ of moments to correspond to a unique distribution is $\sum_1^\infty \mu_{2k}^{-1/2k} = \infty$. No useful necessary and sufficient condition is known.

REMARK D.15. If $X \sim N(\mu, \sigma^2)$, its odd central moments $\mathrm{E}(X-\mu)^{2n+1}$ are zero by symmetry, while its even central moments can be computed to be

$$\mathrm{E}(X-\mu)^{2k} = \sigma^{2k}\,\frac{(2k)!}{2^k\,k!}, \quad k = 1, 2, \ldots.$$

Its cumulants are $\kappa_1 = \mu$, $\kappa_2 = \sigma^2$, with $\kappa_j = 0$ for all $j \geq 3$. Its mgf is finite in an interval around 0 (indeed finite everywhere) so normal distributions are determined by their moments. (They also satisfy Carleman's condition.)

EXAMPLE D.16. Although we omit the details, it is a result of 'extreme value theory' that the maximum of N independent standard normal distributions tends to a so-called Gumbel distribution, and one can show that this maximum becomes sharply concentrated around

$$M = \sqrt{2\log N}\left(1 - \frac{\log\log N}{4\log N}\right) \sim \sqrt{2\log N}.$$

Selberg's CLT states that for t large, $\log|\zeta(\frac{1}{2}+it)|$ is approximately $N\big(0, \frac{1}{2}\log\log t\big)$. Now, in intervals of length 2π near height T, $\log|\zeta(\frac{1}{2}+it)|$ has approximately $\log T$ zeros and $\log T$ midpoints between them. So if one could approximate the maximum of $\log|\zeta(\frac{1}{2}+it)|$ in an interval of length 2π (at 'high T') by $\log T$ independent normal variables, it would concentrate around

$$\sqrt{\frac{1}{2}\log\log T} \times \sqrt{2\log\log T}.$$

It is amusing that this is the same as the leading term of the Fyodorov–Hiary–Keating conjecture.

The definitions of the various modes of convergence extend in natural ways to sequences of *random vectors*. Here we state only the so-called *Cramér–Wold device*: The k-dimensional convergence in distribution (weak convergence)

$$(X_{1,n}, X_{2,n}, \ldots, X_{k,n}) \xrightarrow{d} (X_1, X_2, \ldots, X_k), \quad \text{as } n \to \infty,$$

holds if and only if $\sum_{j=1}^k a_j X_{j,n} \xrightarrow{d} \sum_{j=1}^k a_j X_j$ holds for all linear combinations.

We now turn to the three core statistical limit laws—arguably among the more remarkable theorems in all of mathematics—starting with the Strong Law of Large Numbers (SLLN), followed by the Central Limit Theorem (CLT), and then the Law of the Iterated Logarithm (LIL). Although our focus will be primarily on sequences of random variables which are independent, these laws will each be presented in considerable generality. Some of these results are of comparatively recent vintage and are therefore not yet in their 'ultimate' forms. It is to be understood throughout that degenerate cases (such as when the random variables are all constants) are meant to be excluded, as appropriate. Also, bear in mind that *iid* stands for *independent and identically distributed*, and 'almost surely (a.s.)' and 'with probability 1 (w.p.1)' mean the same thing.

THEOREM D.17 (Strong Law of Large Numbers (SLLN)).

(i) *If X, and $X_1, X_2, \ldots,$ are iid with* $\mathrm{E}\,|X| < \infty$, *then as* $n \to \infty$

$$\frac{X_1 + X_2 + \cdots + X_n}{n} \longrightarrow \mathrm{E}\,X, \quad \textit{almost surely.}$$

(No other conditions are needed.) It suffices if the X_j are identically distributed, but only pairwise independent. If $\mathrm{E}\,|X| = \infty$, *the average will diverge with probability 1.*

(ii) *Conversely, if X and $X_1, X_2, \ldots,$ are iid, and $(X_1 + X_2 + \cdots + X_n)/n$ converges a.s. to some finite constant c, then $\mathrm{E}\,|X|$ is finite, and $c = \mathrm{E}\,X$.*

(iii) *If X and the X_j are iid, then $\mathrm{E}\,|X|^p < \infty$ for some $0 < p < 2$ if and only if*

$$\frac{X_1 + X_2 + \cdots + X_n - nc}{n^{1/p}} \longrightarrow 0 \quad \textit{almost surely,}$$

where $c = \mathrm{E}\,X$ if $1 \le p < 2$, and may be taken as any arbitrary value when $0 < p < 1$. If the X_j are iid with $\mathrm{E}\,|X|^2 < \infty$ then

$$\frac{X_1 + X_2 + \cdots + X_n - n\,\mathrm{E}\,X_1}{\sqrt{n}\log n} \longrightarrow 0 \quad \textit{almost surely.}$$

(iv) *If the X_j are independent and now* not *necessarily identically distributed, but with all $\mathrm{E}\,|X_j|^{1+\epsilon} \le K$, for some $\epsilon > 0$ and some $K > 0$, then*

$$\frac{X_1 + X_2 + \cdots + X_n - \mathrm{E}(X_1 + X_2 + \cdots + X_n)}{n} \longrightarrow 0 \quad \textit{almost surely.} \tag{D.3}$$

(v) *The result* (D.3) *holds if (a) the variable X_j are independent and also $\sum_{j=1}^{\infty} \mathrm{E}(X_j - \mathrm{E}\,X_i)^2/j^2 < \infty$, or more generally, if (b) the X_j are independent, and $\sum_{j=1}^{\infty} \mathrm{E}\,|X_j|^p/j^{1+p/2} < \infty$ for some $p \ge 2$, or if (c) the X_j are uncorrelated, and $\sum_{j=1}^{\infty} (\log j)^2\,\mathrm{E}(X_j - \mathrm{E}\,X_j)^2 < \infty$.*

(v′) *The result* (D.3) *also holds if $\{X_j\}$ is a covariance stationary sequence with summable covariance function. In fact, it is enough if*

$$\sum_{j=1}^{n} |\operatorname{Cov}(X_1, X_j)| \le Cn/(\log n \log\log n)^2$$

for all n for some $C < \infty$.

REMARKS D.18.

(1) The so-called *Weak Law of Large Numbers* (WLLN), which is an easy consequence of the Chebyshev inequality, assumes uncorrelated variables with finite and equal variances but only claims convergence in probability. Specifically, for every $\epsilon > 0$, we have

$$\mathrm{P}\left[\left|\frac{1}{n}\sum_{j=1}^{n}(X_j - \mathrm{E}(X_j))\right| \le \epsilon\right] \le \frac{\sigma^2/n}{\epsilon^2} \to 0, \quad \text{as } n \to \infty.$$

Here σ^2 is the common variance of the X_j, and elementary computations give σ^2/n as the variance for the $\frac{1}{n}\sum_1^n$ term. This argument also works for any

$\sum_{j=1}^{n} c_j (X_j - \mathrm{E}(X_j))$, with uncorrelated variables, as long as its variance $\to 0$ as $n \to \infty$.

(2) The first assertion at (iii) is due to Marcinkiewicz and Zygmund, and has a known converse: If X and the X_i are iid and $(X_1 + \cdots + X_n - c_n)/n^{1/r} \to 0$, a.s., for some $0 < r < 2$ and some constants c_n, then $\mathrm{E}|X|^r < \infty$. (See Stout (1974, [**1035**, p. 126]).)

(3) For a 'direct' proof of the last assertion at (iii), see Breiman (1992, [**142**, Chapter 3, §6]). In fact, the $\sqrt{n}\log n$ denominator can even be replaced by $c_n\sqrt{n}\log\log n$, where c_n is any sequence such that $c_n \to \infty$, however slowly. (This is a consequence of the LIL.)

(4) Item (v) is known as the Rademacher–Menchoff Theorem (Stout, 1974, [**1035**, p. 20]). Part (v′) was included here only for completeness; we omit defining the terms used.

THEOREM D.19 (Central Limit Theorem (CLT)).

(i) *If $X_1, X_2, \ldots,$ are iid with mean μ and finite variance σ^2, then we have the convergence in distribution*

$$\frac{X_1 + X_2 + \cdots + X_n - n\mu}{\sigma\sqrt{n}} \longrightarrow^d N(0,1),$$

where $N(0,1)$ represents the standard normal distribution.

(ii) Lyapunov's Condition: *Suppose $X_1, X_2, \ldots,$ are independent but not necessarily identically distributed, with means $\mu_j = \mathrm{E}\,X_j$ and variances $\sigma_j^2 = \mathrm{E}(X_j - \mu_i)^2$, and let $s_n^2 = \sum_{j=1}^{n} \sigma_j^2$. Then the condition*

$$\frac{1}{s_n^{2+\delta}} \sum_{j=1}^{n} \mathrm{E}\,|X_j - \mu_j|^{2+\delta} \to 0 \quad \text{as } n \to \infty, \quad \text{for some } \delta > 0, \tag{D.4}$$

suffices for the convergence in distribution

$$\frac{1}{s_n} \sum_{j=1}^{n} (X_j - \mu_j) \to^d N(0,1). \tag{D.5}$$

(iii) Lindeberg's Condition: *Given the conditions in (ii), the convergence to normality* (D.5) *continues to hold if condition* (D.4) *is replaced by*

$$\frac{1}{s_n^2} \sum_{j=1}^{n} \mathrm{E}\Big(|X_j - \mu_j|^2\, \mathrm{I}[\,|X_j - \mu_j| > \epsilon s_n] \Big) \to 0 \quad \text{as } n \to \infty, \quad \text{for all } \epsilon > 0, \tag{D.6}$$

where $\mathrm{I}[\cdots]$ represents the 0–1 indicator function.

(iv) Converse to Lindeberg's Theorem: *Lindeberg's condition* (D.6) *implies that*

$$\frac{\max\{\sigma_k : 1 \le k \le n\}}{s_n} \to 0. \tag{D.7}$$

When combined with (D.7), *the asymptotic normality* (D.5) *implies the Lindeberg condition* (D.6).

(v) Central Limit Theorem for (triangular) arrays: *For $n = 1, 2, 3, \ldots,$ assume the nth row of an array consists of jointly independent, zero-mean, finite variance, random variables $X_{n,1}, X_{n,2}, \ldots, X_{n,k_n}$. Independence is not assumed across rows, but the length of the rows $k_n \to \infty$. Let $s_n^2 \equiv \sum_{\ell=1}^{k_n} \mathrm{Var}\, X_{n,\ell}$ be the row-sum variances. Then the standardized row-sums* *will tend to the* $N(0,1)$ *distribution if either of the following conditions hold:*

$$\frac{1}{s_n^{2+\delta}} \sum_{\ell=1}^{k_n} \mathrm{E}\,|X_{n,\ell}|^{2+\delta} \to 0, \quad \text{as } n \to \infty, \quad \text{for some } \delta > 0,$$

$$\frac{1}{s_n^2} \sum_{\ell=1}^{k_n} \mathrm{E}\left(X_{n,\ell}^2 \cdot \mathrm{I}\,[\,|X_{n,\ell}| > \epsilon s_n]\right) \to 0, \quad \text{as } n \to \infty, \quad \text{for all } \epsilon > 0.$$

(These are the Lyapunov and the more general Lindeberg conditions, respectively.) The rows may be of infinite length ($k_n = \infty$) provided the sum of variances for each row is finite.

(vi) Central Limit Theorem for infinite variance: *If X has distribution $F(x) \equiv \mathrm{P}[X \le x]$, one says that it belongs to the domain of attraction of a normal law if*

$$\frac{x^2 \mathrm{P}[\,|X| > x\,]}{\int_{|y|<x} y^2\, dF(y)} \equiv \frac{x^2 \int_{|y|>x} dF(y)}{\int_{|y|<x} y^2\, dF(y)} \to 0 \quad \text{as} \quad x \to \infty. \tag{D.8}$$

If $X_1, X_2, \ldots,$ are iid and satisfy (D.8) *then there exist constants a_n and b_n such that*

$$\frac{\sum_{i=1}^{n} (X_i - a_n)}{b_n} \to^d N(0,1) \quad \text{as} \quad n \to \infty.$$

REMARKS D.20.

(1) In part (ii) of the Theorem, uniform boundedness $|X_j| < C$, together with the condition $s_n^2 \to \infty$, would suffices not only for the asymptotic normality (D.5) to hold, but also for the convergence of all moments to those of the standard normal.

(2) Lyapunov's condition (D.4) holds whenever Lindeberg's condition (D.6) is satisfied.

(3) The Lyapunov and Lindeberg conditions both imply divergence of the sum s_n of the variances; indeed the CLT could not hold without it.

(4) The independence assumptions throughout cannot be relaxed to pairwise independence.

(5) The *rate of convergence to normality* in the CLT is important for many applications. Such results were first obtained by Berry and Esseen in the early 1940s.

(6) Condition (D.8) was discovered independently by Feller, Lévy, and Khinchin in 1935. It is satisfied if $\mathrm{E}\,|X|^2 < \infty$; conversely, if (D.8) is satisfied then $\mathrm{E}\,|X|^{2-\epsilon} < \infty$ for all $\epsilon > 0$. The finite variance case is the only one whose normalization satisfies $b_n \sim c\sqrt{n}$. In infinite variance cases b_n will be of the form $b_n \sim g(n)\sqrt{n}$, where $g(n)$ is a certain *slowly varying function* (meaning that $g(\alpha x)/g(x) \to 1$ as $x \to \infty$ for every $\alpha > 0$).[9] More generally, if there is a normalization such that a sum of iid variables converges to a nondegenerate distribution, that distribution must be a *stable law*. Details may be found in Gnedenko & Kolmogorov (1954, [**417**]), as well as in Petrov (1975, [**842**]), and in Ibragimov & Linnik (1971, [**575**]).

[9]Examples: Powers and iterates of the logarithm are slowly varying functions; powers of x are not.

EXAMPLE D.21. Motivated by questions about the behaviour of certain Dirichlet series on the critical line, let X_n be independent, zero-mean, unit-variance random variables identically distributed as $\mathrm{P}[X_n = 1] = \mathrm{P}[X_n = -1] = \frac{1}{2}$, and consider the sum

$$S(T) = \sum_{n \le T} \frac{X_n}{\sqrt{n}}.$$

Then $s_T^2 \equiv \operatorname{Var} S(T) = \sum_{n \le T} \frac{1}{n} \sim \log T$, and the Lyapounov condition is essentially

$$\frac{1}{(\sqrt{\log T})^{2+\delta}} \sum_{n \le T} \frac{1}{(\sqrt{n})^{2+\delta}} = \frac{1}{(\log T)^{1+\delta/2}} \sum_{n \le T} \frac{1}{n^{1+\delta/2}}.$$

This converges to 0 for any $\delta > 0$ (since the last sum is finite) and therefore

$$\frac{1}{\sqrt{\log T}} \sum_{n \le T} \frac{X_n}{\sqrt{n}} \xrightarrow{d} N(0,1), \quad \text{as } T \to \infty.$$

If all we know is that the X_n are independent, zero-mean, unit-variance (and not necessarily identically distributed) then one might still have convergence to normality, however the Lyapunov condition would not be assured; for example, the $\mathrm{E}\,|X_j - \mu_j|^{2+\delta}$ terms of (D.4) could be infinite for all $\delta > 0$. The Lyapunov condition also fails if one instead considers $S(T) = \sum_{n \le T} X_n / n^{1/2+\epsilon}$ with some $\epsilon > 0$. This happens because the variances will now decrease too quickly for asymptotic normality to be possible.

As Bernoulli random variables frequently arise in arguments stemming from Cramér's heuristic (as well as elsewhere) the following special case of the CLT is of particular interest. We include a proof since this basic result seems not to appear elsewhere.

THEOREM D.22 (CLT for Bernoulli variables). *Let $X_1, X_2, \ldots,$ be independent Bernoulli variables with*

$$\mathrm{P}\,[X_i = 1] \;=\; 1 - \mathrm{P}\,[X_i = 0] \;=\; \theta_i.$$

A necessary and sufficient condition for the convergence in distribution

$$\text{(D.9)} \qquad \sum_{i=1}^{n} (X_i - \theta_i) \Big/ \left(\sum_{i=1}^{n} \theta_i (1 - \theta_i) \right)^{1/2} \xrightarrow{d} N(0,1)$$

is that

$$\text{(D.10)} \qquad \sum_{i=1}^{n} \theta_i (1 - \theta_i) \to \infty, \quad \textit{or equivalently,} \quad \sum_{i=1}^{n} \min\{\theta_i, 1 - \theta_i\} \to \infty.$$

Furthermore, when (D.9) *holds, all moments will also converge to those of the* $N(0,1)$ *and conversely. If $\theta_i \to 0$, the condition* (D.10) *may be taken as $\sum_1^n \theta_i \to \infty$, and the denominator in* (D.9) *may be replaced by $\sqrt{\sum_1^n \theta_i}$.*

PROOF. Assume (D.10), and observe that for *any* $\delta > 0$

$$\begin{aligned} \mathrm{E}\,|X_i - \theta_i|^{2+\delta} &= \theta_i (1 - \theta_i)^{2+\delta} + (1 - \theta_i) \theta_i^{2+\delta} \\ &= \theta_i (1 - \theta_i) \{ (1 - \theta_i)^{1+\delta_i} + \theta_i^{1+\delta} \} \le \theta_i (1 - \theta_i). \end{aligned}$$

The Lyapunov condition in (D.4) will therefore be met since its two component factors are

$$\sum_{i=1}^{n} \mathrm{E}\,|X_i - \theta_i| \le \sum_{i=}^{n} \theta_i(1-\theta_i), \quad \text{and} \quad s_n^{2+\delta} = \left(\sum_{i=1}^{n} \theta_i(1-\theta_i)\right)^{1+\delta/2}.$$

Conversely, assume that $\sum_{i=1}^{\infty} \theta_i(1-\theta_i) < \infty$, and consider the decomposition

$$\sum_{i=1}^{\infty}(X_i - \theta_i) = \sum_{i=1}^{L}(X_i - \theta_i) + \sum_{i=L+1}^{\infty}(X_i - \theta_i) \tag{D.11}$$

with $L < \infty$ chosen so that the $\sum_1^L$ sum (which cannot be Gaussian) has positive variance. Since the $\sum_{L+1}^{\infty}$ sum consists of independent variables, the sum of whose variances is finite, it converges to a random variable almost surely. But if the convolution in (D.11) were Gaussian then the sum $\sum_{i=1}^{L}(X_i - \theta_i)$ must itself be Gaussian, by a well-known result that states that if a sum of independent variables is Gaussian each term of the sum must be Gaussian as well. Hence the normality in (D.9) would be impossible. Finally, because the variables X_i are bounded, convergence of the moments must hold as well. (For the moment convergence, see for example, Feller, 1971, [**373**, vol. 2, Chapter VIII, p. 251].) □

REMARK D.23. The condition in (D.10) implies that $\sum \theta_i$ and $\sum(1-\theta_i)$ both diverge. The converse is not true.

Concerning fluctuations of the partial sums $S_n = X_1 + \cdots + X_n$ of independent random variables, Kolmogorov's zero–one law tells us that if $a_n \uparrow \infty$, then $\limsup_{n\to\infty} S_n/a_n$ converges almost surely to either a finite constant or to $\pm\infty$. If the X_i are iid with mean μ and variance $\sigma^2 > 0$, the CLT tells us that

$$\frac{1}{\sigma\sqrt{n}}\,(X_1 + \cdots + X_n - n\mu)$$

converges *in distribution* to $N(0,1)$ and so is neither bounded above nor bounded below. On the other hand, the SLLN tells us, not only that $\frac{1}{n}\,(X_1 + \cdots + X_n - n\mu)$ converges a.s. to 0, but that $\frac{1}{\sqrt{n}\log n}\,(X_1 + \cdots + X_n - n\mu)$ converges a.s. to 0 as well (Breiman (1992, [**142**, Chapter 3, §6])). The Law of the Iterated Logarithm (LIL) to be stated next is based on the dividing by $\sqrt{n \log\log n}$ and provides a fundamental boundary between the SLLN and CLT. Remarkably, only a finite variance is required for the LIL to hold. It is worthwhile to notice here that because $\frac{1}{\sigma\sqrt{n}}\,(X_1 + \cdots + X_n - n\mu)$ converges in distribution, its product with $1/\sqrt{\log\log n}$ must tend to 0. But this convergence is only in probability and not almost sure, and therefore does not contradict the LIL. The LIL results stated below (some of which are of relatively recent vintage) are due to Khinchin, Kolmogorov, Hartman & Wintner, Petrov, Strassen, and Tomkins, among others.

THEOREM D.24 (The Law of the Iterated Logarithm (LIL)).

(i) *If $X_1, X_2, \ldots,$ are iid with mean μ and finite variance σ^2, then*

$$\mathrm{P}\left[\limsup_{n\to\infty} \frac{X_1 + X_2 + \cdots + X_n - n\mu}{\sigma\sqrt{2n\log\log n}} = 1\right] = 1$$

and (by symmetry, on replacing X_j by $-X_j$)

$$\mathrm{P}\left[\liminf_{n\to\infty} \frac{X_1 + X_2 + \cdots + X_n - n\mu}{\sigma\sqrt{2n\log\log n}} = -1\right] = 1.$$

Furthermore, w.p.1, all values in $[-1,1]$ *are accumulation points of the indicated sequence.*

(ii) *Conversely, if the* X_j *are iid, and*

$$\mathrm{P}\left[\limsup_{n\to\infty}\frac{|X_1+\cdots+X_n|}{\sqrt{2n\log\log n}}<\infty\right]>0,$$

then $\mathrm{E}\,X_1^2<\infty$, *and* $\mathrm{E}\,X_1=0$.

(iii) *If* $Y(t)$, $0\le t<\infty$ *is a standard Brownian motion, then*

$$\limsup_{t\to\infty}\frac{Y(t)}{\sqrt{2t\log\log t}}=1,\ \ a.s.,\quad \textit{likewise}\ \limsup_{t\downarrow 0}\frac{Y(t)}{\sqrt{2t\log\log(1/t)}}=1,\ \ a.s.$$

Furthermore, w.p.1, the corresponding set of limit points is all of $[-1,1]$.

REMARKS D.25.

(1) The lim sup condition in (i) is equivalent to the assertions that, for every $\epsilon>0$,

$$\mathrm{P}\Big[X_1+X_2+\cdots+X_n-n\mu\ >\ (1-\epsilon)\sigma\sqrt{2n\log\log n}\ \text{ inf. often}\Big]\ =\ 1,$$

and

$$\mathrm{P}\Big[X_1+X_2+\cdots+X_n-n\mu\ >\ (1+\epsilon)\sigma\sqrt{2n\log\log n}\ \text{ inf. often}\Big]\ =\ 0,$$

with a similar equivalence for the lim inf. (Of course, the probabilities of such tail events must necessarily be 0's or 1's due to Kolmogorov's zero–one law.)

(2) The LIL as stated at (i) implies that, almost surely,

$$\limsup_{n\to\infty}(X_1+X_2+\cdots+X_n)=+\infty,\quad \liminf_{n\to\infty}(X_1+X_2+\cdots+X_n)=-\infty,$$

$$\text{and}\qquad \frac{X_1+X_2+\cdots+X_n-n\,\mathrm{E}(X_1)}{c_n\sqrt{n\log\log n}}\to 0,$$

where c_n is any sequence which approaches ∞, however slowly.

(3) The version of the LIL stated in part (i) of the theorem is due to Harman & Wintner, while the converse in part (ii) is due to Strassen.

(4) Part (iii) has been included for the sake of completeness. We will not define Brownian motion, but remark that (iii) is a standard result and essentially a corollary of part (i).

The case of iid variables in the LIL is usually insufficient for applications such as those associated with Cramér's heuristic. An extended version of the LIL—due to Kolmogorov (1929, [**647**]), but see also Khinchin (1927, [**634**])—is available for independent variables that are not identically distributed, but which satisfy a certain boundedness condition. We state this result here, but it is the immediate corollary pertaining to Bernoulli variables which follows from it that is of primary interest.

THEOREM D.26 (Kolmogorov's LIL for bounded variables). *Let the* X_n *be zero-mean independent variables with finite variances, and let* $s_n^2\equiv\mathrm{E}(X_1+X_2+\cdots+X_n)^2\to\infty$. *If*

$$|X_n|\ \le\ \frac{s_n}{\sqrt{2\log\log s_n^2}}\,\epsilon_n\,,\ \ a.s.$$

for a sequence of constants such that $\epsilon_n\to 0$, *then*

$$\limsup_{n\to\infty}\frac{X_1+X_2+\cdots+X_n}{\sqrt{2s_n^2\log\log s_n^2}}=1,\ \ a.s.$$

and likewise for the lim inf. *The ϵ_n sequence cannot be replaced by a constant.*

REMARK D.27. The boundedness condition of this result is rather delicate. It cannot, for instance, be replaced by the ever-so-slightly weaker condition

$$\sqrt{2\log\log s_n^2}\,\left(|X_n|/\,s_n\right) \to 0$$

almost surely. For background see, for example, Tomkins (1992, [**1061**]).

THEOREM D.28 (LIL for Bernoulli variables). *Suppose the X_i are independent Bernoulli variables with*

$$\mathrm{P}[\,X_i = 1\,] \;=\; 1 - \mathrm{P}[\,X_i = 0\,] \;=\; \theta_i,$$

and let $s_n^2 = \sum_{i=1}^n \sigma_i^2$, where $\sigma_i^2 = \theta_i(1-\theta_i)$. Then provided $s_n^2 \to \infty$, we have

$$\mathrm{P}\left[\limsup_{n\to\infty} \frac{X_1 + X_2 + \cdots + X_n - (\theta_1 + \theta_2 + \cdots + \theta_n)}{\sqrt{2s_n^2 \log\log s_n^2}} = 1\right] \;=\; 1$$

and (by symmetry)

$$\mathrm{P}\left[\liminf_{n\to\infty} \frac{X_1 + X_2 + \cdots + X_n - (\theta_1 + \theta_2 + \cdots + \theta_n)}{\sqrt{2s_n^2 \log\log s_n^2}} = -1\right] \;=\; 1.$$

If the $\theta_i \to 0$, then we may take $s_n^2 = \sum_{i=1}^n \theta_i$ here.

EXAMPLE D.29. Under the Gauss–Cramér model for the primes, $\widetilde{\pi}(n)$ is a sum $\sum_1^n X_j$ of Bernoulli variables with success probabilities $\theta_j = 1/\log j$. Thus

$$s_n^2 \;=\; \sum_{j=3}^{n} \frac{1}{\log j}\left(1 - \frac{1}{\log j}\right) \;\sim\; \sum_{j=3}^{n} 1/\log j \;\sim\; n/\log n,$$

and

$$s_n^2 \log\log s_n^2 \;\sim\; \frac{n}{\log n}\log\log\frac{n}{\log n} \;=\; \frac{n}{\log n}\log(\log n - \log\log n) \;\sim\; \frac{n\log\log n}{\log n}.$$

It follows from the Bernoulli LIL result that in the Gauss–Cramér model, the difference $\widetilde{\pi}(x) - \mathrm{li}(x)$ has, as $x \to \infty$, the *exact* order $O\left(\sqrt{x(\log\log x)/\log x}\right)$.

Extensions of the LIL to the case of independent variables which are not identically distributed and which are not required to be bounded, as well as extensions of the LIL to various contexts involving dependence, remain a subject of ongoing investigations. The three results which follow, due, respectively, to Petrov (1966, [**841**]), Tomkins (1971, [**1060**]), and Heyde (1969, [**543**]), and Kesten (1972, [**632**]) are worth noting. A good reference is Stout (1974, [**1035**]).

THEOREM D.30 (Petrov's LIL). *Suppose the X_i are independent random variables with zero means and variances σ_i^2, and let $s_n^2 = \sigma_1^2 + \cdots + \sigma_n^2$. If*

$$\liminf_{n\to\infty} s_n^2/n > 0, \quad \textit{and} \quad \mathrm{E}\,|X_i|^{2+\epsilon} \le C$$

for some $\epsilon > 0$ and finite C, then

$$\limsup_{n\to\infty} \frac{X_1 + \cdots + X_n}{\sqrt{2s_n^2 \log\log s_n^2}} = 1, \quad \textit{a.s.}$$

THEOREM D.31 (A theorem of Tomkins). *Let the X_i be independent, with zero means, and variances σ_i^2, and let Φ be the standard normal distribution function. Let $S_n = X_1 + \cdots + X_n$ and $s_n^2 = \sigma_1^2 + \cdots + \sigma_n^2$, and assume that $s_n^2 \to \infty$ but $\sigma_n^2/s_n^2 \to 0$. If the tail conditions*

$$\frac{\mathrm{P}\left[S_n/s_n < -xu_n\right]}{\Phi(-xu_n)} \to 1, \quad \textit{and} \quad \frac{\mathrm{P}\left[S_n/s_n > xu_n\right]}{1-\Phi(xu_n)} \to 1$$

are satisfied for all x in a neighbourhood of 1 for some sequence u_n, then

$$\limsup_{n\to\infty} \frac{S_n}{s_n u_n} = 1, \quad a.s.$$

REMARK D.32. This result links the rate of convergence to normality of the tails of the distribution of S_n to LIL type results. If the condition stated here is satisfied for $u_n = \sqrt{\log\log s_n^2}$ it leads to an iterated-log type of LIL. With this choice for u_n, Tomkins's Theorem also lets one establish the Hartman-Wintner LIL.

THEOREM D.33 (A theorem of Heyde & Kesten). *Suppose the X_i are iid. No moment conditions are assumed. Then there will exist a sequence of centring constants, a_n, and a sequence of scaling constants, $0 < b_n \uparrow \infty$, such that*

$$\limsup_{n\to\infty} \frac{X_1 + X_2 + \cdots + X_n - a_n}{b_n} = 1, \quad a.s.,$$

if and only if

$$\liminf_{x\to\infty} \frac{x^2 \mathrm{P}[\,|X_1| > x\,]}{\mathrm{E}\{X_1^2 \;\mathrm{I}[\,|X_1| \le x]\}} = 0. \tag{D.12}$$

REMARK D.34. Condition (D.12) is identical to the necessary and sufficient condition for X_1 to belong to the *'domain of partial attraction of the normal law'*. For background, see for example Maller (1980, [**726**]), and references therein, as well as Gnedenko & Kolmogorov (1954, [**417**]), and Petrov (1975, [**842**]). Note that it is entirely possible for a distribution to satisfy the condition (D.12) of this theorem and yet to have no finite moments at all.

We mention one last LIL result, this one due to Lai (1974, [**671**]), which deals with 'windowed' (or 'delayed') sums involving the terms of an iid sequence that lie between positions n and $\lfloor n + n^\alpha \rfloor$, where $\lfloor\cdot\rfloor$ is the 'integer part' function.

THEOREM D.35 (Law of the Single Logarithm). *Suppose $X_1, X_2, \ldots,$ are iid random variables with mean 0, and variance $\sigma^2 < \infty$. If*

$$\mathrm{E}\left(\frac{|X|^{2/\alpha}}{(\,\log^+|X|\,)^{1/\alpha}}\right) < \infty, \; \textit{for some } 0 < \alpha < 1,$$

then

$$\mathrm{P}\left[\limsup_{n\to\infty} \frac{X_{n+1} + X_{n+2} + \cdots + X_{\lfloor n+n^\alpha \rfloor}}{\sqrt{2n^\alpha \log n}} = \sigma\sqrt{1-\alpha}\right] = 1,$$

and similarly for the lim inf. *Here* $\log^+ x = \max\{\log x, 1\}$.

A delayed sums result involving shorter intervals was obtained by Erdős & Rényi (1970, [**360**]), and an LIL for the extrema of partial sums was obtained by Chung (1948, [**206**]). Further details concerning the LIL may be found in Bauer (2011, [**83**, §34]), Bingham (1986, [**106**]), Chung (2001, [**207**, §7.5]), Feller (1971, [**373**, Chapter VII, §5]), Gut (2013, [**470**, Chapter 8]), Lamperti (1996, [**673**, pp. 61–70]), Stout (1974, [**1035**, Chapter 5]), and in publications by V. V. Petrov, and R. J. Tomkins, among others. Gut & Spătaru (2000, [**471**]), and the references therein provide an interesting alternative view of LIL asymptotics.

There remains at least one further category of limit laws that is of value for understanding the behaviour of sums of independent variables. We state one key result here but without providing the definitions; they may be found elsewhere.

THEOREM D.36 (The Arcsine Laws). *Let $B(t)$ denote standard Brownian motion on $0 \le t \le 1$. Then the following three quantities are well defined and have identical distributions:*

(a) *the proportion of 'time' that $B(t) > 0$,*
(b) *the time at which $B(t)$ attains its maximum, and*
(c) *the last time at which $B(t)$ changes sign.*

Their common distribution is the arcsine law, with density function

$$f(x) = \frac{1}{\pi\sqrt{x(1-x)}}, \quad 0 < x < 1,$$

and distribution function

$$F(x) = \int_0^x f(x)\,dx = \frac{2}{\pi}\arcsin\sqrt{x} = \frac{1}{2} + \frac{\arcsin(2x-1)}{\pi}, \quad 0 < x < 1.$$

With natural large-sample reformulations, the same distribution also holds for random walks based on variables which are iid, 0-mean, and finite-variance.

REMARKS D.37.

(1) The arcsine law is the same as that of the Beta$\left(\frac{1}{2}, \frac{1}{2}\right)$ distribution.
(2) Feller (1970, [**373**, vol. 1, Chapter III]) comments on the behaviour of random walks and on the remarkably unintuitive aspects of the Arcsine Laws. Here is one example: A random walk based on a fair coin typically spends a highly disproportionate amount of time either above or below the horizontal axis, regardless of the number of tosses.
(3) Révész (2013, [**899**, Chapters 12 & 25]), in discussing 'returns to the origin' under tosses of a fair coin, states: *"the particle during its first n steps visits the origin practically $n^{1/2}$ times...* [however these] *...$n^{1/2}$ visits are distributed in $[0, n]$ in a very nonuniform way. ...the length of the longest excursion up to n ...for some n can be nearly as big as n... At the same time, for some other n it will be about $n/\log\log n$ only but it cannot be smaller than this ... if the length of the longest excursion is $n/\log\log n$ only then the length of the second, third ... longest excursion is about the same... As a consequence ... the sum of the lengths of the $\log\log n$ longest excursions is nearly n..."*
(4) Needless to say, the time required (under tosses of a fair coin) to travel between a position which is close to the top of the LIL boundary and a location which is close to the bottom of the LIL boundary can be remarkably long.

In the subject of random walks, many open problems remain. A classic work is Spitzer (1964, [**1014**]); for a more current treatment see Hughes (1996, [**565**]). Révész (2013, [**899**]) provides a fairly detailed and relatively more accessible treatment.

There are many outstanding books on probability theory. A few at a relatively advanced level that the authors like include Bauer (2011, [**83**]), Breiman (1992, [**142**]), Chow & Teicher (1997, [**203**]), Chung (2001, [**207**]), and Gut (2013, [**469**]); any from this *exceedingly* abbreviated list can be recommended.

Among more elementary books any of the following are good bets: Chung & AitSahlia (2003, [**208**]), Eisen (1969, [**314**]), and Hoel, Port & Stone (1971, [**558**]). There are, however, a great many other possibilities.

APPENDIX E

Approximations for Sums and Products in Powers of n, p, $\log n$, and $\log p$

Many applications require estimates for sums or products involving powers of n, p, $\log n$, and $\log p$. For convenient reference, we offer a collection of such results here. Those involving only n are typically proved by standard methods of analysis, while those involving p generally require number-theoretic considerations. Of particular interest is to compare results when p and $n \log n$ are interchanged. We comment on this below, and also include a comprehensive lemma on infinite products. Unless otherwise stated, all approximations involve $x \to \infty$ or its analogue, and order terms have implied constants that may be parameter-dependent.

We use the following notation throughout: $\mathbb{N}$ refers to the positive natural numbers $n = 1, 2, 3, \ldots$, and $\mathbb{N} \cup \{0\}$ is used if 0 is to be included. Unidentified constants, with or without subscripts, are typically denoted by C and usually differ between occurrences. The notations B_j and $B_j(\cdot)$ represent the Bernoulli numbers and the Bernoulli polynomials; they are defined in Chapter 5. We also sometimes refer to the Stieltjes constants γ_k (also defined in Chapter 5) which arise in the Laurent expansion of $\zeta(s)$ around $s = 1$: for $k \in \mathbb{N} \cup \{0\}$

$$\zeta(s) = \frac{1}{s-1} + \sum_{k=0}^{\infty} \frac{(-1)^k}{k!} \gamma_k (s-1)^k, \quad \gamma_k \equiv \lim_{x \to \infty} \left(\sum_{m \le x} \frac{\log^k m}{m} - \frac{\log^{k+1} x}{k+1} \right).$$

The first few are $\gamma_0 = \gamma$ (the Euler–Mascheroni constant), $\gamma_1 \approx -0.0728$, $\gamma_2 \approx -0.00969$, and $\gamma_3 \approx 0.00205$. Finally, we will also use the notation (appearing in Chapters 2 and 6)

$$\lambda(x) = (\log x)^{3/5} (\log \log x)^{-1/5}$$

associated with the Korobov-Vinogradov zero-free region.

Sums in n

We first consider sums of powers of n. For $N \in \mathbb{N}$ and $K \ge 2$ we have

$$\sum_{n=1}^{N} \frac{1}{n} = \log N + \gamma + \frac{1}{2N} - \sum_{k=1}^{K-1} \frac{B_{2k}}{2kN^{2k}} - \frac{B_{2K}}{2KN^{2K}} \theta_K, \quad \text{with } 0 \le \theta_K \le 1.$$

A special case is

$$\text{(E.1)} \qquad \sum_{n \le x} \frac{1}{n} = \log x + \gamma + \frac{1}{2x} - \frac{1}{12x^2} + \frac{\theta}{60x^4}, \quad 0 \le \theta \le 1.$$

For real-valued[1] $\alpha > 0$, $\alpha \neq 1$, with $\{x\} \equiv x - \lfloor x \rfloor$ the fractional parts function, and with $S(x) = x - \lfloor x \rfloor - \frac{1}{2}$ a sawtooth function, we have

$$\begin{aligned}\sum_{n\le x}\frac{1}{n^\alpha} &= \zeta(\alpha) + \frac{x^{1-\alpha}}{1-\alpha} - \frac{\{x\}}{x^\alpha} + \alpha\int_x^\infty \{u\}\frac{du}{u^{\alpha+1}}\\ &= \zeta(\alpha) + \frac{x^{1-\alpha}}{1-\alpha} - \frac{S(x)}{x^\alpha} + O\left(\frac{1}{x^{\alpha+1}}\right) = \zeta(\alpha) + \frac{x^{1-\alpha}}{1-\alpha} + O\left(\frac{1}{x^\alpha}\right)\end{aligned}$$

with the implied constant of the order terms independent of α. Two special cases are

$$\sum_{n\le x}\frac{1}{n^2} = \zeta(2) - \frac{1}{x} + O\left(\frac{1}{x^2}\right) = \frac{\pi^2}{6} - \frac{1}{x} + O\left(\frac{1}{x^2}\right)$$

and

$$\sum_{n\le x}\frac{1}{\sqrt{n}} = 2\sqrt{x} + \zeta\left(\frac{1}{2}\right) + O\left(\frac{1}{\sqrt{x}}\right).$$

In the divergent cases ($0 < \alpha < 1$) the dominant term is $x^{1-\alpha}/(1-\alpha)$; In the convergent cases ($\alpha > 1$) it is $\zeta(\alpha)$. Connected to this latter case, we also have

$$\lim_{x\to\infty}\sum_{n\le x}\frac{1}{n^\alpha} = \zeta(\alpha) \sim \frac{1}{\alpha-1}$$

as $\alpha \downarrow 1$.

For general real-valued $\alpha \geq 0$ we have

$$\sum_{n\le x} n^\alpha = \frac{x^{\alpha+1}}{\alpha+1} + O(x^\alpha), \quad \alpha \geq 0.$$

H. Cohen (2007, [**211**, vol. II, Chapter 9, §9.2.5]) gives some extended versions of this result. For the special case $\alpha = \frac{1}{2}$ we have the better approximation

$$\sum_{n\le x}\sqrt{n} = \frac{2}{3}x^{3/2} + \frac{1}{2}\sqrt{x} + \zeta\left(-\frac{1}{2}\right) + O\left(\frac{1}{\sqrt{x}}\right).$$

A related sum for $k \in \mathbb{N}\cup\{0\}$, $m > k$, and $N \in \mathbb{N}$ is

$$\sum_{n=1}^{N}\frac{1}{(n+k)(n+k+1)\cdots(n+m)} = \frac{1}{m-k}\left(\frac{k!}{m!} - \frac{(N+k)!}{(N+m)!}\right).$$

We also give Faulhaber's formula (in three versions): For $N \in \mathbb{N}$ and $k \in \mathbb{N}\cup\{0\}$,

$$\begin{aligned}\sum_{n\le N} n^k = \frac{B_{k+1}(N+1) - B_{k+1}}{k+1} &= \frac{1}{k}\sum_{j=0}^{k}(-1)^j\binom{k+1}{j}B_j\cdot N^{k+1-j}\\ &= \frac{1}{k+1}\sum_{j=1}^{k}\binom{k+1}{j}B_j\cdot(N+1)^{k+1-j}.\end{aligned}$$

Special cases are

$$\sum_{n\le N} n = \frac{1}{2}N(N+1),\ \sum_{n\le N} n^2 = \frac{1}{6}N(N+1)(2N+1),\ \sum_{n\le N} n^3 = \frac{1}{4}N^2(N+1)^2.$$

[1] These results continue to hold when α is a complex number with $\Re\alpha > 0$ (and $\alpha \neq 1$ still), provided that we replace x^α with $x^{\Re\alpha}$ in the error term.

A few sums in n which are not of the form $\sum n^\alpha$ include

$$\sum_{n=1}^{\infty} \frac{(-1)^n}{n} = -\log 2, \quad \sum_{n=0}^{\infty} \frac{(-1)^n}{2n+1} = \frac{\pi}{4}, \quad \sum_{n=1}^{\infty} \frac{(-1)^{n+1}}{n^s} = \left(1 - 2^{1-s}\right)\zeta(s),$$

$$\sum_{n=1}^{\infty} \frac{1}{(n+k)(n+k+1)(n+k+2)\cdots(n+m)} = \frac{1}{m-k}\,\frac{k!}{m!}, \quad k \in \mathbb{N}\cup\{0\},\ m > k.$$

In particular,

$$\sum_{n=1}^{\infty} \frac{1}{n(n+1)} = 1, \quad \sum_{n=1}^{\infty} \frac{1}{n(n+1)(n+2)} = \frac{1}{4}, \quad \sum_{n=1}^{\infty} \frac{1}{n(n+1)(n+2)(n+3)} = \frac{1}{18},$$

and

$$\sum_{n=1}^{\infty} \frac{1}{n(n+1)(n+2)\cdots(n+k)} = \frac{1}{k\cdot k!}, \quad k \in \mathbb{N}.$$

We also have

$$\sum_{n=1}^{\infty} \frac{1}{(2n-1)^3} = \frac{7}{8}\zeta(3), \quad \sum_{n=1}^{\infty} \frac{1}{n(n+k)} = \frac{H_k}{k}, \quad \text{where } H_k = \sum_{n=1}^{k} \frac{1}{n}.$$

(Such sums can sometimes be evaluated by partial fraction decomposition and recognizing the telescoping series that result. No comprehensive table of such results is available.)

Sums in $\log n$

Using $\int \log^k x\,dx = C + x \sum_{j=0}^{k} (-1)^{k-j} \frac{k!}{j!} \log^j x$ one may derive:

$$\sum_{n\le x} \log^k n = x \sum_{j=0}^{k} (-1)^{k-j} \frac{k!}{j!} \log^j x + O(\log^k x), \quad k \in \mathbb{N}.$$

The case $k = 1$ (corresponding to the Stirling formulas —see Appendix B) admits the improved version

$$\sum_{n\le x} \log n = x\log x - x + \log\sqrt{2\pi x} + O\left(\frac{1}{x}\right)$$

and even

$$\sum_{k=1}^{n} \log k = n\log n - n + \log\sqrt{2\pi n} + \frac{1}{12n} - \frac{1}{360n^3} + \frac{1}{1260n^5} + O\left(\frac{1}{n^7}\right),$$

while for $k = 2$ we have

$$\sum_{n\le x} \log^2 n = x\log^2 x - 2x\log x + 2x + O(\log^2 x).$$

More generally, for any real β (but omitting the $n = 1$ term if $\beta < 0$ since $\log 1 = 0$):

$$\sum_{n\le x} \log^\beta n = x\log^\beta x + O(x\log^{\beta-1} x),$$

with the special cases $\beta = -1$ and $\beta = -2$ satisfying

$$\sum_{2 \le n \le x} \frac{1}{\log n} = \mathrm{li}(x) + O(1) = \frac{x}{\log x} + O\left(\frac{x}{\log^2 x}\right),$$

$$\sum_{2 \le n \le x} \frac{1}{\log n} = \frac{x}{\log x}\left(\sum_{k=0}^{K-1} \frac{k!}{\log^k x} + O_K\left(\frac{1}{\log^K x}\right)\right), \quad K \in \mathbb{N},$$

and

$$\sum_{2 \le n \le x} \frac{1}{\log^2 n} = \int_2^x \frac{du}{\log^2 u} + O(1) = \frac{x}{\log^2 x} + O\left(\frac{x}{\log^3 x}\right).$$

Other results, with $x \to \infty$, include

$$\sum_{2 \le n \le x} \log\log n = x \log\log x - \mathrm{li}(x) + O(\log\log x) \sim x \log\log x$$

and

$$\sum_{16 \le n \le x} \log\log\log n \sim x \log\log\log x.$$

The last two equivalences reflects the fact, for large n, that $\log\log n \approx \log\log x$ for almost all $n \le x$, and likewise for $\log\log\log x$.[2]

Sums in n and $\log n$

We first consider the general form $\sum_{n \le x} n^\alpha \log^\beta n$ where α and β are any real numbers. (When $\beta < 0$ the $n = 1$ term must be omitted since $\log 1 = 0$.) The cases $\alpha = 0$ and $\beta = 0$ were already treated. The divergent cases $\alpha > -1$, the convergent cases $\alpha < -1$, and the case $\alpha = -1$ will be treated separately.

The divergent $\alpha > -1$ case: We have in general:

$$\sum_{n \le x} n^\alpha \log^\beta n = \frac{x^{\alpha+1}}{\alpha+1} \log^\beta x + O\left(x^{\alpha+1} \log^{\beta-1} x\right), \quad \alpha > -1.$$

When β is a positive integer, $x^\alpha \log^\beta x$ has an exact antiderivative (not available for general β) allowing for a better error term. Its exact form depends on the range of α; for $\alpha \ge 0$, $\beta \in \mathbb{N}$ we have

$$\sum_{n \le x} n^\alpha \log^\beta n = x^{1+\alpha}\left(\sum_{k=0}^{\beta} \frac{(-1)^{\beta-k}}{(\alpha+1)^{\beta-k+1}} \frac{\beta!}{k!} \log^k x\right) + O\left(x^\alpha \log^\beta x\right),$$

while for $-1 < \alpha < 0$, $\beta \in \mathbb{N}$ we have

$$\sum_{n \le x} n^\alpha \log^\beta n = x^{1+\alpha}\left(\sum_{k=0}^{\beta} \frac{(-1)^{\beta-k}}{(\alpha+1)^{\beta-k+1}} \frac{\beta!}{k!} \log^k x\right) + C + O\left(x^\alpha \log^\beta x\right).$$

For the special case $\alpha > 0$ and $\beta = 1$ we have

$$\sum_{n \le x} n^\alpha \log n = \frac{x^{\alpha+1}}{\alpha+1}\left(\log x - \frac{1}{\alpha+1}\right) + O\left(x^\alpha \log x\right), \quad \alpha > 0.$$

[2]Folklore quips: It has been said that iterated logarithms do increase to infinity, but so slowly that no one has yet observed them doing so. On the other hand, "log log log log" might be the sound of an analytic number theorist drowning.

The convergent $\alpha < -1$ case: We have in general,

$$\sum_{n \le x} n^\alpha \log^\beta n = C + O\left(x^{\alpha+1} \log^\beta x\right), \quad \alpha < -1.$$

When β is a nonnegative integer, we can identify the constant:

$$\sum_{n \le x} n^\alpha \log^\beta n = (-1)^\beta \zeta^{(\beta)}(\alpha) + O\left(x^{\alpha+1} \log^\beta x\right), \quad \alpha < -1, \beta \in \mathbb{N} \cup \{0\},$$

where $\zeta^{(\beta)}(s)$ is the βth derivative of $\zeta(s)$ (with $\zeta^{(0)}(s) = \zeta(s)$). Special cases include

$$\sum_{n=1}^{\infty} \frac{\log n}{n^2} = -\zeta'(2), \quad \sum_{n=1}^{\infty} \frac{\log n}{n^s} = -\zeta'(s), \quad \sum_{n=1}^{\infty} \frac{\log^k n}{n^s} = (-1)^k \zeta^{(k)}(s), \quad \Re s > 1.$$

Finally, we consider $\sum n^\alpha \log^\beta n$ for $\alpha = -1$ according to three cases: $\beta > -1$ (divergent), $\beta = -1$ (also divergent), and $\beta < -1$ (convergent).

The divergent $\alpha = -1$, $\beta > -1$ case: We have in general

$$\sum_{n \le x} \frac{\log^\beta n}{n} = \frac{\log^{\beta+1} x}{\beta+1} + C + O\left(\frac{\log^\beta x}{x}\right), \quad \beta > -1.$$

(The fact that the main term is an antiderivative of $(\log^\beta x)/x$ has allowed for a very good error term.) When β is a nonnegative integer, we can identify the constant that appears:

$$\sum_{n \le x} \frac{\log^\beta n}{n} = \frac{\log^{\beta+1} x}{\beta+1} + (-1)^\beta \gamma_\beta \, \beta! + O\left(\frac{\log^\beta x}{x}\right), \quad \beta \in \mathbb{N} \cup \{0\},$$

where γ_k are the Stieltjes constants and γ_0 the Euler–Mascheroni constant. For $\beta = 1$ this gives

$$\sum_{n \le x} \frac{\log n}{n} = \frac{1}{2} \log^2 x - \gamma_1 + O\left(\frac{\log x}{x}\right).$$

The divergent $\alpha = -1$, $\beta = -1$ case: This is the special case

$$\sum_{2 \le n \le x} \frac{1}{n \log n} = \log \log x + C + O\left(\frac{1}{x \log x}\right). \tag{E.2}$$

The convergent $\alpha = -1$, $\beta < -1$ case: In this case we have

$$\sum_{2 \le n \le x} \frac{\log^\beta n}{n} = C - \frac{\log^{\beta+1} x}{\beta+1} + O\left(\frac{\log^\beta x}{x}\right), \quad \beta < -1.$$

A few sums involving n and $\log n$ not of the form $n^\alpha \log^\beta n$ include

$$\sum_{n \le x} \frac{\log(x/n)}{n} = \frac{1}{2} \log^2 x + C \log x + C'x + O(1),$$

$$\sum_{n \le x} \frac{\log^k(x/n)}{n} = \frac{\log^{k+1} x}{k+1} + O(1), \quad k \ge 1,$$

$$\sum_{n \le x} \log^k \left(\frac{x}{n}\right) = k!\,x + O\left(\log^k x\right), \quad k \in \mathbb{N},$$

with special cases

$$\sum_{n\le x}\log\left(\frac{x}{n}\right)=x+O(\log x),\quad\text{and}\quad\sum_{n\le x}\log^2\left(\frac{x}{n}\right)=2x+O(\log^2 x).$$

We also have

$$\sum_{n=1}^{\infty}\frac{(-1)^n\log n}{n}=\gamma\log 2-\frac{1}{2}(\log 2)^2.$$

Sums over p and Cramér's heuristic

For sums involving p and $\log p$ the following should be noted: Provided the sequence $a(n)$ corresponds to a 'well-behaved' function $a(x)$, the asymptotic properties of a sum $\sum a(p)$ over primes can often be guessed as

$$\sum_{p\le x}a(p)\sim\sum_{2\le n\le x}\frac{a(n)}{\log n},\quad\text{or}\quad\sum_{p\le x}a(p)\sim\int_2^x a(x)\frac{dx}{\log x}.$$

Here the sum over n and the integral over x have been probability-adjusted using Cramér's heuristic that n is prime with probability $1/\log n$. For example, with this heuristic

$$\sum_{p\le x}\frac{1}{p}\sim\sum_{n\le x}\frac{1}{n\log n}\sim\log\log x+C,$$

which happens to be correct, although owing to vagaries of the early primes, any value for C obtained in this way would not be correct. Similarly one might argue that

$$\sum_{p\le x}\frac{\log p}{p}\sim\sum_{2\le n\le x}\frac{\log n}{n}\frac{1}{\log n}=\sum_{n\le x}\frac{1}{n}\sim\log x+\gamma+O\left(\frac{1}{x}\right)$$

where one again would not expect the constant γ to be correct. Indeed, conjecturing *lower* order terms cannot be approached in this way.

Sums and products in p

We begin with

$$\sum_{p\le x}p^\alpha=\frac{x^{\alpha+1}}{(\alpha+1)\log x}+O\left(\frac{x^{\alpha+1}}{\log^2 x}\right),\quad\text{for }\alpha>-1,$$

(E.3) $$\sum_{p\le x}\frac{1}{p^\alpha}=P(\alpha)-\frac{1}{(\alpha-1)x^{\alpha-1}\log x}+O\left(\frac{1}{x^{\alpha-1}\log^2 x}\right),\quad\text{for }\alpha>1,$$

where $P(\alpha)$ is the value of the "prime zeta function" defined in equation (3.21). In particular, as $\alpha\downarrow 1$,

(E.4) $$\lim_{x\to\infty}\sum_{p\le x}\frac{1}{p^\alpha}=P(\alpha)\sim\log\left(\frac{1}{\alpha-1}\right).$$

As with all convergent sums, equation (E.3) can be recast as a formula for the tail of the sum:

$$\sum_{p>x}\frac{1}{p^\alpha}=\frac{1}{(\alpha-1)x^{\alpha-1}\log x}+O\left(\frac{1}{x^{\alpha-1}\log^2 x}\right),\quad\text{for }\alpha>1.$$

Some special cases are

$$\sum_{p\le x} p = \frac{x^2}{2\log x} + O\left(\frac{x^2}{\log^2 x}\right), \quad \text{and} \sum_{p\le x} p^2 = \frac{x^3}{3\log x} + O\left(\frac{x^3}{\log^2 x}\right).$$

We also have

$$\sum_{p>x} \frac{1}{p^2} = \frac{1}{x\log x} + O\left(\frac{1}{x\log^2 x}\right), \quad \text{and} \sum_{p\le x} \frac{1}{p} \sim \sum_{n\le x} \frac{1}{n\log n} \sim \log\log x.$$

More precise versions for the last of these are

$$\sum_{p\le x} \frac{1}{p} = \log\log x + B + O\left(\frac{1}{\log x}\right), \quad \text{where } B = \gamma - \sum_p \sum_{k=2}^{\infty} \frac{1}{kp^k} \approx 0.261497\ldots$$

$$\sum_{\sqrt{x}<p\le x} \frac{1}{p} = \log 2 + O\left(\frac{1}{\log x}\right), \quad \text{as } x \to \infty.$$

We include next a result which follows from that of $\sum \frac{1}{p}$:

$$\sum_{n\le x} \frac{\Lambda^2(n)}{n\log^2 n} = \log\log x + B' + O\left(\frac{1}{\log x}\right), \quad \text{where } B' = \gamma + \sum_p \sum_{k=2}^{\infty} \frac{1}{p^k}\left(\frac{1}{k^2} - \frac{1}{k}\right).$$

Formulas for the four cases of

$$\prod_{p\le x} \left(1 \pm \frac{1}{p}\right)^{\pm 1}$$

may be found in Sections 2.1 and 2.11 and are not repeated here. By multiplying the results for $\prod(1 - 1/p)$ and $\prod(1 + 1/p)$ one obtains

$$\prod_{p\le x} \left(1 - \frac{1}{p^2}\right) = \frac{6}{\pi^2}\left(1 + O\left(\frac{1}{\log x}\right)\right) = \frac{6}{\pi^2} + O\left(\frac{1}{\log x}\right).$$

We also have

$$\prod_{p\le x} \left(1 + \frac{\kappa}{p}\right) = (C\log^{\kappa} x)\left(1 + O\left(\frac{1}{\log x}\right)\right), \quad \kappa \in \mathbb{R}$$

as long as any term with $(1 + \kappa/p) = 0$ is omitted from the product.

Finally we have

$$\sum_{pq\le x} \frac{1}{pq} = (\log\log x)^2 + O(\log\log x),$$

where p and q both range over primes.

Sums in $\log p$

$$\sum_{p\le x} \log^{\alpha} p = x\log^{\alpha-1} x + O\left(x\log^{\alpha-2} x\right), \quad \alpha \in \mathbb{R}$$

with special cases

$$\sum_{p\le x} \log p = x + O\left(xe^{-c\lambda(x)}\right), \qquad \sum_{p\le x} \log^2 p = x\log x + O(x).$$

Under RH the error term in the first of these would be $O\left(\sqrt{x}\log^2 x\right)$. A consequence of this first formula is

$$\prod_{p\le x} p \;=\; \exp(x(1+o(1)).$$

With p and q both ranging over primes, we also have

$$\sum_{p\le x}\log^2 p + \sum_{pq\le x}\log p\log q = 2x\log x + O(x),$$

$$\sum_{p\le x}\log p + \sum_{pq\le x}\frac{\log p\log q}{\log pq} = 2x + O\left(\frac{x}{\log x}\right).$$

Sums in p and $\log p$

We first consider sums of the form $\sum p^\alpha \log^\beta p$. The cases $\alpha = 0$ and $\beta = 0$ appeared above.

The divergent case $\alpha > -1$: We have in general

$$\sum_{p\le x} p^\alpha \log^\beta p = \frac{x^{\alpha+1}}{\alpha+1}\log^{\beta-1} x + O(x^{\alpha+1}\log^{\beta-2} x), \quad \alpha > -1.$$

When β is a positive integer one can do better:

$$\begin{aligned}\sum_{p\le x} p^\alpha \log^\beta p \;=\; & \frac{x^{\alpha+1}}{\alpha+1}\log^{\beta-1} x - (\beta-1)x^{\alpha+1}\sum_{k=0}^{\beta-2}\frac{(-1)^{\beta-k}}{(\alpha+1)^{\beta-k}}\frac{(\beta-2)!}{k!}\log^k x\\ & + \; O(x^{\alpha+1}e^{-c\lambda(x)}), \quad \alpha > -1,\ \beta\in\mathbb{N}.\end{aligned}$$

The convergent case $\alpha < -1$: We have in general

$$\sum_{p\le x} p^\alpha \log^\beta p \;=\; C - \frac{x^{\alpha+1}}{\alpha+1}\log^{\beta-1} x + O\left(x^{\alpha+1}\log^{\beta-2} x\right), \quad \alpha < -1.$$

When $\alpha = -1$ there are three cases according to whether $\beta > 0$, $\beta < 0$, or $\beta = 0$.

The divergent case $\alpha = -1$ and $\beta > 0$: We have in general

$$\sum_{p\le x}\frac{\log^\beta p}{p} = \frac{\log^\beta x}{\beta} + C + O\left(e^{-c\lambda(x)}\right), \quad \beta > 0.$$

In the special case $\beta = 1$, one can identify the constant:

$$\sum_{p\le x}\frac{\log p}{p} = \log x - \left(\gamma + \sum_p\sum_{k=2}^{\infty}\frac{\log p}{p^k}\right) + O\left(e^{-c\lambda(x)}\right).$$

The constant appearing within the brackets here is $\left(\gamma + \sum\sum\right) \approx 1.33258$.

The divergent case $\alpha = -1$ and $\beta = 0$: This is the special case (noted earlier)

$$\sum_{p\le x}\frac{1}{p} = \log\log x + \left(\gamma + \sum_p\sum_{k=2}^{\infty}\frac{1}{kp^k}\right) + O\left(e^{-c\lambda(x)}\right).$$

The convergent case $\alpha = -1$ and $\beta < 0$: In this case we have

$$\sum_{p \le x} \frac{\log^\beta p}{p} = C - \frac{\log^\beta x}{\beta} + O\left(e^{-c\lambda(x)}\right), \quad \beta < 0.$$

The following sum diverges when $\delta \le 0$, but for $\delta \downarrow 0$ it satisfies

$$\sum_{p \le x} \frac{1}{p \log^\delta p} = \frac{1}{\delta} + C + o(1), \text{ where } C = 0.261497\ldots.$$

Finally, we state a few results involving p and $\log p$ not of the form $\sum p^\alpha \log^\beta p$:

$$\sum_{p \le x/2} \frac{1}{p \log(x/p)} = O\left(\frac{\log\log x}{\log x}\right), \quad \sum_{p \le x} \frac{\log p}{p \log \frac{2x}{p}} = O(\log\log x),$$

$$\sum_p \frac{1}{p(\log\log p)^k} \text{ converges if } k > 1, \text{ and diverges if } k \le 1.$$

In particular,

$$\sum_{p \le x} \frac{1}{p \log\log p} = \log\log\log x + C + O\left(\frac{1}{\log\log x}\right).$$

With p and q both ranging over the primes we also have

$$\sum_{pq \le x} \frac{\log p \log q}{pq} = \frac{1}{2}\log^2 x + O(\log x), \quad \sum_{pq \le x} \frac{\log p \log q}{pq \log pq} = \log x + O(\log\log x).$$

Remarks concerning infinite products

Infinite products $\prod_{n=1}^\infty (1 + c_n) \equiv \lim_{k \to \infty} \prod_{n=1}^k (1 + c_n)$ can be studied by relating them to their logarithmic form $\lim_{k \to \infty} \sum_{n=1}^k \log(1 + c_n)$. However two subtleties arise. The first is that the logarithm tends to $-\infty$ when the product tends to 0. Some books adopt the convention of requiring that the limit not be 0 before a product can be said to converge.[3] We shall not do so; *convergence* of a product will mean convergence to any finite value. If 0 is to be excluded we will refer to this explicitly as *convergence to a nonzero value*. (On the other hand, it is not impolite to refer to a product which tends to zero as '*diverging to zero*'.) The second subtlety is that the range of the logarithm is determined only to within multiples of $2\pi i$; this compromises the link between convergence of the product and convergence of the logarithm. On the other hand, if the terms $(1 + c_n)$ of the product are all required to have positive real parts, and the argument of the logarithm is restricted to $(-\frac{\pi}{2}, \frac{\pi}{2})$ then the two convergences will be comparable. And one last point: To avoid trivialities, when formulating results concerning infinite products, factors of 0 are prohibited. The fundamental result may now be stated as follows.

LEMMA E.1. *Assume that no term in any of the products below equals 0. Then:*
(i) $\prod_{n=1}^\infty (1 + c_n)$ *converges to a finite nonzero limit if and only if we have* $\lim_{k \to \infty} \prod_{n=k}^\infty (1 + c_n) = 1$, *and in this case the individual factors* $1 + c_n$ *converge to 1, so the* c_n *converge to 0.*

[3]The terse treatment in Whittaker & Watson (1927, [**1100**, §2.7]) shows how this convention can be confusing.

(ii) *If* $\sum_{n=1}^{\infty} |c_n| < \infty$ *then* $\prod_{n=1}^{\infty}(1+c_n)$ *converges to a nonzero limit and it is unaffected by arbitrary rearrangements of its factors.*

(iii) *If* $\prod_{n=1}^{\infty}(1+|c_n|)$ *converges to a nonzero limit then* $\prod_{n=1}^{\infty}(1+c_n)$ *does as well. (The converse is false.)*

(iv) *If* $\sum_{n=1}^{\infty} |c_n|^2 < \infty$, *the convergence of* $\prod_{n=1}^{\infty}(1+c_n)$ *to a nonzero limit and the convergence of* $\sum_{n=1}^{\infty} c_n$ *are equivalent.*

(v) *If* $\Re(1+c_n) > 0$ *for all* n, *and the arguments of the logarithms are restricted to* $(-\frac{\pi}{2}, \frac{\pi}{2})$, *then* $\prod_{n=1}^{\infty}(1+c_n)$ *converges to a nonzero limit if and only if* $\sum_{n=1}^{\infty} \log(1+c_n)$ *converges.*

(vi) *If the* a_n *are real valued with* $a_n \geq 0$, *then* $\prod_{n=1}^{\infty}(1+a_n)$ *converges to a finite (nonzero) limit if and only if* $\sum_{n=1}^{\infty} a_n$ *converges. If* $0 \leq a_n < 1$, *then* $\prod_{n=1}^{\infty}(1-a_n)$ *converges to a nonzero limit if and only if* $\sum_{n=1}^{\infty} a_n$ *converges, and otherwise the product diverges to* 0.

REMARKS E.2.

(1) The result in (i) explains why one speaks of the form $\prod_{n=1}^{\infty}(1+c_n)$ rather than $\prod_{n=1}^{\infty} c_n$.

(2) In connection with (ii) note that the convergence of $\sum_{n=1}^{\infty} c_n$ is neither necessary nor sufficient for $\prod_{n=1}^{\infty}(1+c_n)$ to converge to a nonzero limit. For on the one hand consider the case $c_n = (-1)^n/\sqrt{n}$, for which the sum converges (to $(\sqrt{2}-1)\zeta(\frac{1}{2})$) but the product diverges to ∞; and on the other hand consider the case $c_{2k-1} = 1/\sqrt{k}$ and $c_{2k} = -1/(1+\sqrt{k})$, for which the product converges (to 1) but the sum diverges to ∞.

(3) In further connection with (ii), it is not hard to show that $\sum_{n=1}^{\infty} |c_n| < \infty$ if and only if $\sum_{n=1}^{\infty} |\log(1+c_n)| < \infty$. This gives an alternate necessary and sufficient condition for $\prod_{n=1}^{\infty}(1+c_n)$ to converge to a nonzero limit.

Proofs for the various components of this lemma appear in numerous sources. Titchmarsh (1939, [**1058**, §1.4–1.43]) gives a lucid treatment. Other references include Bromwich (1926, [**155**]), Knopp (1990, [**640**]), Conway (1978, [**232**, Chapter VII, § 5]), and Trench (1999, [**1065**]). Theorems 1 & 2, of §2 in the appendix of Karatsuba & Voronin (1992, [**622**, p. 327]), for example, deal with products of analytic functions.

EXAMPLES E.3.

(a) Using part (vi) of the Lemma, because the sum $\sum_p p^{-1}$ diverges, we see that the product $\prod_p(1-p^{-1})$ diverges to 0.

(b) Using part (ii) of Lemma E.1 with $c_n = p_n^{-s}$ for some complex s with $\sigma > 1$, where p_n is the nth prime, we see that $\zeta(s)^{-1} = \prod_p (1-p^{-s})$ converges to a nonzero limit, and therefore the same is true for $\zeta(s) = \prod_p (1-p^{-s})^{-1}$.

APPENDIX F

Some Mathematical Points

In this appendix we provide reference material on the topics of Riemann–Stieltjes integration, partial summation (summation by parts), and Euler–Maclaurin summation, as well as a synopsis of Fourier analysis.

Riemann–Stieltjes integration. The Riemann–Stieltjes integral, written as $\int g(x)\,dF(x)$ is defined as the limit (whenever it exists)

$$\int_a^b g(x)\,dF(x) \equiv \lim \sum_{i=1}^{n} g(x_i^*)[F(x_i) - F(x_{i-1})], \quad a = x_0 < x_1 < \cdots < x_n = b,$$

where $x_{i-1} \le x_i^* \le x_i$, $\max_{1\le i\le n}(x_i - x_{i-1}) \to 0$, and $n \to \infty$. It is analogous to the ordinary Riemann integral defined via limits such as

$$\int_a^b g(x)\,dx \equiv \lim \sum_{i=1}^{n} g(x_i^*)[x_i - x_{i-1}], \quad a = x_0 < x_1 < \cdots < x_n = b.$$

(It may be helpful, at first, to think of $F(x)$ as a cumulative distribution function.) If $f(x) = dF(x)/dx$, then typically $\int g(x)\,dF(x)$ will equal $\int g(x)f(x)\,dx$, while if $F(x)$ has jumps of heights w_i at positions x_i but is otherwise constant, and $g(x)$ is continuous, then $\int g(x)\,dF(x)$ will equal $\sum_i w_i g(x_i)$. The properties of the Riemann–Stieltjes integral closely mirror those of the Riemann integral, while the formula for integration by parts becomes

$$\int_a^b G(x)\,dF(x) = G(x)F(x)\Big|_a^b - \int_a^b F(x)\,dG(x).$$

Further details on the Riemann–Stieltjes integral can be found in many textbooks.

The partial summation formulas. This discrete version of integration by parts comes in assorted flavours. Let $\{a_n\}_{n=1}^{\infty}$ and $\{b_n\}_{n=1}^{\infty}$ be sequences of complex numbers, and let $A_n = a_1 + a_2 + \cdots + a_n$. Two equivalent versions of the formula are

$$\text{(F.1)} \qquad \sum_{n=M+1}^{N} a_n b_n = A_N b_{N+1} - A_M b_{M+1} - \sum_{n=M+1}^{N} A_n\,(b_{n+1} - b_n),$$

and

$$\text{(F.2)} \qquad \sum_{n=M}^{N} (a_n - a_{n-1})\,b_n = \sum_{n=M}^{N} a_n\,(b_n - b_{n+1}) - a_{M-1}b_M + a_N b_{N+1},$$

for integers $M \le N$. Such formulas are proved by means of telescopic reduction.

Abel's summation formula or summation by parts. If $f(x)$ has a continuous derivative and $\{a_n\}_{n=1}^{\infty}$ is a sequence of complex numbers, then with $A(x) = \sum_{n\le x} a(n)$ we have

$$\sum_{n\le x} a(n)f(n) = A(x)f(x) - \int_1^x A(y)f'(y)\,dy. \tag{F.3}$$

PROOF. For simplicity assume $x = N$ is an integer. Rearrange the sum as

$$\sum_{n=1}^{N} a(n)f(n) = \sum_{n=1}^{N}\{A(n)-A(n-1)\}f(n) = \sum_{n=1}^{N-1} A(n)\{f(n)-f(n+1)\}+A(N)f(N),$$

make the substitution

$$f(n+1) - f(n) = \int_n^{n+1} f'(y)\,dy,$$

and assemble the components. □

REMARK F.1. Using $\sum_{a<n\le b} = \sum_{x\le b} - \sum_{x\le a}$ Abel's formula can be generalized to

$$\sum_{a<n\le b} a(n)f(n) = A(b)f(b) - A(a)f(a) - \int_a^b A(y)f'(y)\,dy.$$

EXAMPLES F.2.

(1) Set $a_n \equiv 1$ and $b_n = 1/n^s$ in (F.1) to obtain

$$\sum_{n=1}^{N} \frac{1}{n^s} = \frac{N}{(N+1)^s} + \sum_{n=1}^{N} n\left(\frac{1}{n^s} - \frac{1}{(n+1)^s}\right).$$

Taking $N \to \infty$ limits gives, provided $\Re s > 1$,

$$\zeta(s) = \sum_{n=1}^{\infty} \frac{1}{n^s} = \sum_{n=1}^{\infty} n\left(\frac{1}{n^s} - \frac{1}{(n+1)^s}\right).$$

(2) With $a(n) \equiv 1$ and $f(x) = x^{-s}$, we have $A(x) \equiv \lfloor x \rfloor$, and so

$$\sum_{n=1}^{\lfloor x \rfloor} \frac{1}{n^s} = \frac{\lfloor x \rfloor}{x^s} + s\int_1^x \frac{\lfloor y \rfloor}{y^{1+s}}\,dy.$$

Letting $x \to \infty$ then gives $\zeta(s) = s\int_1^{\infty} \lfloor y \rfloor / y^{1+s}\,dy$ for $\Re s > 1$.

Euler's summation formula. If $f(t)$ is a complex-valued function defined on the real interval $[a, b]$ with a Riemann integrable derivative $f'(t)$, then

$$\sum_{a<n\le b} f(n) = \int_a^b f(t)\,dt + \int_a^b \{t\}f'(t)\,dt + \{a\}f(a) - \{b\}f(b), \tag{F.4}$$

where $0 \le \{t\} \equiv t - \lfloor t \rfloor < 1$ is the "fractional part" function. Note that the $\{a\}f(a)$ and $\{b\}f(b)$ terms will be 0 when a and b are integers.

PROOF. For simplicity, assume that a and b are integers. Then

$$\int_a^b \{t\}f'(t)\,dt = \int_a^b tf'(t)\,dt + \int_a^b \lfloor t \rfloor f'(t)\,dt.$$

Evaluating the first integral on the right using integration by parts gives

$$\int_a^b tf'(t)\,dt = tf(t)\Big|_a^b - \int_a^b f(t)dt = bf(b) - af(a) - \int_a^b f(t)dt,$$

while splitting up the second integral on the right gives

$$\int_a^b \lfloor t\rfloor f'(t)\,dt = \sum_{n=a+1}^{b} \int_{n-1}^{n} (n-1)f'(t)\,dt = \sum_{n=a+1}^{b} (n-1)\{f(n) - f(n-1)\}.$$

The result follows upon collapsing the sum. □

EXAMPLES F.3.

(1) Setting $a = 1$, $b = N$ and $f(t) = \frac{1}{t}$ in (F.4) gives

$$\sum_{1<n\le N} \frac{1}{n} = \int_1^N \frac{dt}{t} - \int_1^N \frac{\{t\}}{t^2}dt - 0\cdot 1 + 0\cdot\frac{1}{N} = \log N - \int_1^\infty \frac{\{t\}}{t^2}dt + \int_N^\infty \frac{\{t\}}{t^2}dt.$$

The last integral is bounded by $\int_N^\infty \frac{1}{t^2}\,dt = O(\frac{1}{N})$ so that altogether we have

$$\sum_{n=1}^{N} \frac{1}{n} = \log N + \gamma + O\left(\frac{1}{N}\right), \quad \text{where } \gamma = 1 - \int_1^\infty \frac{\{t\}}{t^2}dt.$$

The constant γ here is the same as the Euler–Mascheroni constant; see (B.6). (Remark: The $O(\frac{1}{N})$ here cannot be replaced by $o(\frac{1}{N})$.) □

(2) Setting $a = 1$, $b = N$ and $f(t) = \log t$ in (F.4) gives

$$\log N! = \sum_{1<n\le N} \log n = \int_1^N \log t\,dt + \int_1^N \frac{\{t\}}{t}\,dt - 0\cdot\log N + 0\cdot\log 1$$

$$= (t\log t - t)\Big|_1^N + O(\log N) = N\log N - N + O(\log N).$$

Stirling's formula can also be established in this way but needs a more involved argument. Note that the $O(\log N)$ here cannot be replaced by $o(\log N)$. □

(3) (Adapted from Montgomery & Vaughan, (2006, [**774**, Theorem 1.12]).) We will show that

$$\sum_{n\le x} \frac{1}{n^\alpha} = \zeta(\alpha) + \frac{x^{1-\alpha}}{1-\alpha} + O\left(\frac{1}{x^\alpha}\right), \quad \text{as } x\to\infty,$$

for $\alpha > 0$, $\alpha \ne 1$, with the implied constant not depending on α. At first, take s to be a complex number with $\Re s > 1$. For any positive real numbers $x < y$, put $f(n) = \frac{1}{n^s}$ in equation (F.4), to get

$$\sum_{x<n\le y} \frac{1}{n^s} = \int_x^y \frac{1}{t^s}\,dt + \int_x^y \{t\}\frac{-s}{t^{s+1}}\,dt + \frac{\{x\}}{x^s} - \frac{\{y\}}{y^s}.$$

Taking the limit of both sides as $y\to\infty$ then gives

$$\sum_{n>x} \frac{1}{n^s} = \int_x^\infty \frac{1}{t^s}\,dt - s\int_x^\infty \frac{\{t\}}{t^{s+1}}\,dt + \frac{\{x\}}{x^s} - 0 = \frac{x^{1-s}}{s-1} - s\int_x^\infty \frac{\{t\}}{t^{s+1}}\,dt + \frac{\{x\}}{x^s},$$

from which it follows that

$$\zeta(s) = \sum_{n\le x} \frac{1}{n^s} + \sum_{n>x} \frac{1}{n^s} = \sum_{n\le x} \frac{1}{n^s} + \frac{x^{1-s}}{s-1} + \frac{\{x\}}{x^s} - s\int_x^\infty \frac{\{t\}}{t^{s+1}}\,dt. \tag{F.5}$$

This was established on the half-plane $\Re s > 1$. However, both sides are analytic functions of s for the larger half-plane[1] $\Re s > 0$, except for their poles at $s = 1$. By the principle of analytic continuation, the identity continues to hold for this larger region, and in particular,

$$\begin{aligned}\zeta(s) &= \sum_{n\le x}\frac{1}{n^s} + \frac{x^{1-s}}{s-1} + O\Big(\frac{1}{x^{\Re s}} + |s|\int_x^\infty \frac{1}{t^{\Re s+1}}\,dt\Big)\\ &= \sum_{n\le x}\frac{1}{n^s} + \frac{x^{1-s}}{s-1} + O\Big(\frac{1}{x^{\Re s}} + \frac{|s|/\Re s}{x^{\Re s}}\Big).\end{aligned}$$

When $s = \alpha$ is a positive real number with $\alpha \neq 1$, this asympotic formula can be written as

$$\sum_{n\le x}\frac{1}{n^\alpha} = \zeta(\alpha) + \frac{x^{1-\alpha}}{1-\alpha} + O\Big(\frac{1}{x^\alpha}\Big), \qquad x\to\infty,$$

where the implicit constant is absolute. (Note that the case $\alpha = 1$ really is an exception, as we saw in the first example above.) □

The Euler–Maclaurin summation formulas. These are a series of approximations (which are typically successively better) for the difference between $\sum_{a\le n\le b} f(n)$ and $\int_a^b f(x)\,dx$. They involve the derivatives of $f(x)$ and they have exact, known remainder terms. When $a < b$ are *integers*, with f continuous and f' piecewise continuous on $[a,b]$, Euler's formula (F.4) can be written as

$$\sum_{a<n\le b} f(n) = \int_a^b f(x)\,dx + \frac{f(b)-f(a)}{2} + \int_a^b S(x)f'(x)\,dx,$$

where $S(x) = x - \lfloor x\rfloor - \frac12$ is a sawtooth function. Note that if the sum on the left were to be taken as $\sum_{a\le n\le b}$ then the right side must be increased by $f(a)$ so that the term $(f(b)-f(a))/2$ would become $(f(b)+f(a))/2$.

When more derivatives of f exist, integration by parts can be applied repeatedly to the second of the two integrals above. Then using the relation $B_k'(x) = kB_{k-1}(x)$ for the Bernoulli polynomials (and with $a < b$ still integers) one may obtain the general formulas[2]

$$\sum_{a<n\le b} f(n) = \int_a^b f(x)\,dx + \frac{f(b)-f(a)}{2} + \sum_{r=1}^{k}\frac{B_{2r}}{(2r)!}\Big(f^{(2r-1)}(b) - f^{(2r-1)}(a)\Big) + R_k,$$

where

$$R_k = -\frac{1}{(2k)!}\int_a^b f^{(2k)}(x)P_{2k}(x)\,dx, \quad \text{with } |R_k| \le \frac{2\zeta(2k)}{(2\pi)^{2k}}\int_a^b \Big|f^{(2k)}(x)\Big|\,dx,$$

and $P_k(x) \equiv B_k(\{x\})$ is the period-1 extension of the Bernoulli polynomial $B_k(x)$. The fact that only every other derivative of f enters into these formulas corresponds to the fact that (other than $B_1 = -\frac12$) the odd Bernoulli numbers are all zero. H. Cohen (2007, [**211**, vol. II, Chapter 9]) gives a detailed treatment of the various Euler–McLaurin formulas, together with corollaries, equivalent forms, and far-reaching applications.

---

[1]The final integral converges when $\Re s > 0$ by comparison with $\int_x^\infty 1/t^{\Re s+1}\,dt$, and therefore represents an analytic function of s on that region since the integrand is analytic.

[2]The Bernoulli numbers and Bernoulli polynomials are defined and discussed in Chapter 5.

EXAMPLE F.4 (Adapted from Montgomery & Vaughan, (2006, [**774**, Appendix B])). We will show that the analytic continuation of $\zeta(s)$ can be carried out via an Euler–Maclaurin summation.

If we take $f(n) = \frac{1}{n^s}$ in the above equation, and take the limits as $a \to 1$ and $b \to \infty$, after some work we obtain

$$\zeta(s) = \frac{1}{2} + \frac{1}{s-1} - \sum_{r=1}^{k} \binom{-s}{2r-1} \frac{B_{2r}}{2r} - \binom{-s}{2k} \int_1^\infty P_{2k}(x) x^{-s-2k}\, dx. \tag{F.6}$$

(Here the binomial coefficient $\binom{-s}{m}$ is defined to be the polynomial

$$-s(-s-1)\cdots(-s-(m-1))/m!,$$

agreeing with its value when $-s$ is replaced by an integer $n \geq m$.) The identity (F.6) was derived under the assumption $\Re s > 1$, but the right-hand side is analytic in the larger half-plane[3] $\Re s > 1 - 2k$ (except when $s = 1$), and therefore this formula establishes the analytic continuation of $\zeta(s)$ to that larger punctured half-plane. Since k can be taken arbitrarily large, this Euler–Maclaurin summation method shows that $\zeta(s)$ can be continued to an analytic function on the entire complex plane, except for a simple pole at $s = 1$.

There is an interesting byproduct of this identity: if we take $s = -n$ to be a nonpositive integer, any binomial coefficient $\binom{-s}{m} = \binom{n}{m}$ will vanish as soon as $m \geq n + 1$. Therefore when k is sufficiently large, equation (F.6) becomes

$$\zeta(-n) = \frac{1}{2} - \frac{1}{n+1} - \sum_{r=1}^{\lfloor (n+1)/2 \rfloor} \binom{n}{2r-1} \frac{B_{2r}}{2r}.$$

In particular, this formula implies the perhaps surprising fact that $\zeta(0), \zeta(-1), \zeta(-2), \ldots$, are all rational numbers! Indeed, some additional simplification involving recursive formulas for the Bernoulli numbers allows one to simplify this formula even further to the tidy

$$\zeta(-n) = (-1)^n \frac{B_{n+1}}{n+1}, \quad n = 0, 1, 2, \ldots,$$

which in particular shows that $\zeta(-2) = \zeta(-4) = \zeta(-6) = \cdots = 0$, these being the "trivial zeros" of $\zeta(s)$.

Some references that deal with Euler–Maclaurin formulas include Apostol (1976, [**25**] and 1999, [**28**]), Borwein, Borwein & Dilcher (1989, [**136**]), Hardy (1949, [**490**, Chapter 13]), Havil (2003, [**517**, §10.2]), Iwaniec & Kowalski (2004, [**594**, Chapter 4]), Knopp (1928, [**640**, §64]), Mongomery & Vaughan (2006, [**774**, Appendix B, pp. 495–519]), Vardi (1991, [**1081**]), and Whittaker & Watson (1927, [**1100**, §7.21]).

Fourier analysis, a synopsis. Jean-Baptiste Joseph Fourier (1768–1830)—author of the seminal 1822 treatise *Théorie Analytique de la Chaleur*—was a physicist and a devotee of Napoleon. Relying on intuition, he made a number of assertions concerning the representation of functions by infinite sums of trigonometric terms, together with an important inversion formula. These turned out to, by and large,

[3] As in equation (F.5), the integral converges when $\Re s > 1 - 2k$ by comparison with $\int_x^\infty 1/x^{\Re s + 2k}\, dx$, and therefore represents an analytic function of s on that region since the integrand is analytic.

be true and to have a wide range of applications. Prestini (2016, [**868**]) provides a nice biography.

If $f(x)$ is a complex-valued function of a real variable defined on $-\pi \leq x \leq \pi$, or is a 2π-periodic version of such a function, one can often justify a representation such as

$$f(x) = \frac{1}{2}a_0 + \sum_{n=1}^{\infty} \big(a_n \cos(nx) + b_n \sin(nx)\big), \tag{F.7}$$

where

$$a_0 = \frac{1}{\pi}\int_{-\pi}^{\pi} f(x)\, dx, \ a_n = \frac{1}{\pi}\int_{-\pi}^{\pi} f(x)\cos(nx)\, dx, \ b_n = \frac{1}{\pi}\int_{-\pi}^{\pi} f(x)\sin(nx)\, dx. \tag{F.8}$$

Behind this representation lies the fact that the indicated sine and cosine functions are all 'orthogonal' to each other (so that $\int_{-\pi}^{\pi}$ of the product of any distinct pair of them is 0) and that, as a set, they are *'complete'* enough to 'span' a significant portion of the space of functions defined on $[-\pi, \pi]$. An alternate version of the *Fourier series* (F.7) and its inverse (F.8) can be stated elegantly as

$$f(x) = \sum_{n=-\infty}^{\infty} c_n e^{inx}, \text{ where } c_n = \frac{1}{2\pi}\int_{-\pi}^{\pi} f(x)e^{-inx}\, dx, \ n = 0, \pm 1, \pm 2, \dots. \tag{F.9}$$

The representations (F.7)/(F.8) and (F.9) are linked through $c_n = \frac{1}{2}(a_n - ib_n)$ for $n \geq 0$ and $c_{-n} = \frac{1}{2}(a_n + ib_n)$ for $n \geq 1$, or equivalently as $a_0 = 2c_0$, $a_n = c_n + c_{-n}$ and $b_n = i(c_n - c_{-n})$ for $n \geq 1$. The function $f(x)$ and its Fourier coefficients c_n satisfy the *Parseval formula*

$$\frac{1}{2\pi}\int_{-\pi}^{\pi} |f(x)|^2\, dx = \sum_{n=-\infty}^{\infty} |c_n|^2, \tag{F.10}$$

a result that stems from the deeper fact that the 'Hilbert spaces' $L^2(-\pi, \pi)$ of functions that are 'square-integrable' on $[-\pi, \pi)$, and ℓ^2 of sequences that are 'square-summable', are 'isomorphic' to each other—meaning, fundamentally, that they are mathematically identical.

EXERCISES F.5.

(1) Show that if $f(x)$ satisfies (F.7) then its coefficients satisfy (F.8). Suggest any reasonable set of sufficient conditions for this result to hold.
(2) Express the right-hand side of (F.10) in terms of the a_n and b_n.
(3) Show that if $f(x) = \sum_n \alpha_n \exp(i\beta_n x)$, where the β_n are real-valued and are not commensurate (meaning that they satisfy no linear relation with integer coefficients) then (under reasonable conditions) the coefficients α_n may be recovered through

$$\alpha_n = \lim_{T\to\infty} \frac{1}{2T}\int_{-T}^{T} f(x)e^{-i\beta_n x}\, dx.$$

(This is relevant for recovering the discrete components of an almost periodic function.)

Next, suppose that $f(x)$ is a complex-valued function defined for $x \in \mathbb{R}$. Then under certain conditions one may define its *Fourier transform* $\widehat{f}(t)$, $t \in \mathbb{R}$, and also invert it to recover $f(x)$, according to

$$\widehat{f}(t) \equiv \int_{-\infty}^{\infty} f(x)e^{-2\pi itx}\,dx, \quad f(x) = \int_{-\infty}^{\infty} \widehat{f}(t)e^{2\pi itx}\,dt. \tag{F.11}$$

(It suffices for $|f(x)|$ and $|\hat{f}(t)|$ to both be integrable for this to hold.) Here one may think of $f(x)$ as representing the function in the *time* or *spatial domain*, and $\widehat{f}(t)$ as representing the same function but in the *frequency domain.* When the integrals involved both converge $f(x)$ and $\widehat{f}(t)$ will satisfy the *Plancherel formula*

$$\int_{-\infty}^{\infty} \left|\widehat{f}(t)\right|^2 dt = \int_{-\infty}^{\infty} |f(x)|^2\,dx, \tag{F.12}$$

a relation that reflects an isomorphism between $L^2(\mathbb{R})$ with itself. Using the Fourier transform pairs $f, \widehat{f}$ and $g, \widehat{g}$ together with the *polarization identities*,[4] the relation (F.12) can be recast in the form

$$\int_{-\infty}^{\infty} f(x)\overline{g(x)}\,dx = \int_{-\infty}^{\infty} \widehat{f}(t)\overline{\widehat{g}(t)}\,dt, \tag{F.13}$$

or alternately as

$$\int_{-\infty}^{\infty} f(x)\overline{\widehat{g}(x)}\,dx = \int_{-\infty}^{\infty} \widehat{f}(x)\overline{g(x)}\,dx. \tag{F.14}$$

Under certain conditions a Fourier transform pair satisfies the *Poisson summation formula*

$$\sum_{n=-\infty}^{\infty} f(n) = \sum_{k=-\infty}^{\infty} \widehat{f}(k), \tag{F.15}$$

or its more general version

$$\sum_{n=-\infty}^{\infty} f(x+n) = \sum_{k=-\infty}^{\infty} \widehat{f}(k)e^{2\pi ikx}. \tag{F.16}$$

A further important result satisfied by Fourier transformation is the 'convolution' property:

$$h(x) = \int_{-\infty}^{\infty} f(y)g(x-y)\,dy \iff \widehat{h}(t) = \widehat{f}(t)\cdot\widehat{g}(t). \tag{F.17}$$

Sufficient conditions for these various results to hold can be found in many sources.

---

[4] If a vector space over the real numbers has both a norm $\|\cdot\|$ and an inner product $(\cdot,\cdot)$, related according to $\|u\| = \sqrt{(u,u)}$, then the norm will satisfy the *parallelogram law*

$$2\|u\|^2 + 2\|v\|^2 = \|u+v\|^2 + \|u-v\|^2.$$

The *polarization identities* are a set of identities each equivalent to the parallelogram law, namely

$$\begin{aligned}(u,v) &= \frac{1}{4}\left(\|u+v\|^2 - \|u-v\|^2\right)\\ &= \frac{1}{2}\left(\|u+v\|^2 - \|u\|^2 - \|v\|^2\right)\\ &= \frac{1}{2}\left(\|u\|^2 + \|v\|^2 - \|u-v\|^2\right).\end{aligned}$$

For complex vector spaces these formulas must be adjusted. See for example Rudin (1991, [**926**]).

EXAMPLE F.6. The function $f(x) = e^{-ax^2}$, $a > 0$ has the Fourier transform $\widehat{f}(t) = \sqrt{\frac{\pi}{a}}e^{-\pi^2t^2/a}$. Applying (F.15) gives the (nontrivial) *theta function* identity

$$\sum_{n=-\infty}^{\infty} e^{-an^2} = \sqrt{\frac{\pi}{a}} \sum_{n=-\infty}^{\infty} e^{-\pi^2n^2/a}.$$

It is valid whenever $\Re a > 0$.

Among the more interesting properties of Fourier transforms is the quantum-mechanical *uncertainty principle* which states that a function and its Fourier transform cannot be simultaneously 'localized'. If we normalize the functions, so that $\int |f|^2 = \int |\widehat{f}|^2 = 1$ (in which case $|f(x)|^2$ and $|\widehat{f}(t)|^2$ become probability densities) this property can be expressed in terms of the product of their 'variances':

$$\left(\int_{-\infty}^{\infty} (x-x_0)^2 |f(x)|^2\, dx\right)\left(\int_{-\infty}^{\infty} (t-t_0)^2 |\widehat{f}(t)|^2\, dt\right) \geq \frac{1}{16\pi^2}$$

for all $x_0, t_0 \in \mathbb{R}$. The 'centring' choices $x_0 = \int x\,|f(x)|^2\,dx$ and $t_0 = \int t\,|\widehat{f}(t)|^2\,dt$ minimizes the left-hand side, and equality is then achieved if and only if f (and therefore $\widehat{f}$ as well) is proportional to the Gaussian form e^{-ax^2}, where $a > 0$ is a constant.

The definition in (F.11) of the Fourier transform, and its corresponding inverse relation, is not the only one in use. According to one alternate version, (F.11) is replaced by

$$\text{(F.18)} \qquad \widehat{f}(t) \equiv \int_{-\infty}^{\infty} f(x)e^{itx}\,dx, \quad f(x) = \frac{1}{2\pi}\int_{-\infty}^{\infty} \widehat{f}(t)e^{-itx}\,dt.$$

The various factors of π which appear in all of the formulas above—from (F.11) down to (F.18)—cannot be eliminated without consequence; factors of π must surface somewhere. All versions of the Fourier transform continue to satisfy the symmetrical relation (F.14); however when using the definition in (F.18) for example, the formulas (F.12), (F.13), (F.15) and (F.16) must respectively be replaced by

$$\int_{-\infty}^{\infty} \left|\widehat{f}(t)\right|^2 dt = 2\pi \int_{-\infty}^{\infty} |f(x)|^2\,dx,$$

$$\int_{-\infty}^{\infty} \widehat{f}(t)\overline{\widehat{g}(t)}\,dt = 2\pi \int_{-\infty}^{\infty} f(x)\overline{g(x)}\,dx,$$

$$\sum_{n=-\infty}^{\infty} f(n) = \sum_{k=-\infty}^{\infty} \widehat{f}(2\pi k),$$

$$\sum_{n=-\infty}^{\infty} f(x+n) = \sum_{k=-\infty}^{\infty} \widehat{f}(2\pi k)e^{ikx}.$$

Fourier methods have an enormous range of applications and many books have been written on this subject. The treatments by Korner (1988, [**653**]), Stein & Shakarchi (2003, [**1022**]), and Goldberg (1961, [**418**]) are particularly nice. Other references include Bracewell (1986, [**140**]), Dym & McKean (1972, [**306**]), Gray & Goodman (1995, [**463**]), and Papoulis (1960, [**832**]). At a much higher level is the elegant book by Katznelson (2004, [**625**]). Of material available on the Web, the 2007 Stanford EE261 lecture notes by Brad Osgood are especially nice. A riveting three-page history of the subject—starting from Babylonian cuneiform tablets which predate Pythagoras (560–480 BCE) by well over a millennium, and ending at the era of modern harmonic analysis—is contained in the review by Roger Cooke (1996, [**233**]) of *Harmonic Analysis* by Henry Helson. For a much longer but nonetheless extremely interesting historical account, see the article by MacKey (1980, [**719**]).

Acknowledgments

On revient toujours à ses premières amours

The authors gratefully acknowledge support by the Natural Sciences and Engineering Research Council of Canada, as well as facilities provided by the Universities of Toronto and British Columbia, the Fields Institute, and the Simons-Laufer Mathematical Sciences Research Institute. This work of AF was further supported by a gratefully acknowledged unrestricted grant from Google LLC. We are also indebted to many individuals who have helped us in ways both large and small; we thank them all. Unfortunately, any errors or omissions are our faults alone.

For reading segments of our manuscript and answering many questions we wish to thank Yan Fyodorov (FHK conjecture), John Harnad (RMT), Adam Harper (chaos models), and Maksym Radziwiłł (Chowla conjecture).

We also thank Daryl Pregibon for encouragement as our liaison at Google, Andrew Odlyzko for data on zeros at an early stage of this project, as well as the late Paul Chernoff, Lester Dubins, Jack Feldman, and Marina Ratner (all of UC Berkeley) for lively conversations on mathematical topics over many years.

For replying to emails, for personal communications, and for many other forms of assistance—both large and small—we also thank David Andrews, Juan Arias de Renya, Zhidong Bai, Rudy Beran, Enrico Bombieri, David Brillinger, Alexandra Chouldechova, Brian Conrey, Zohar Freeman (library assistant), Erika Giddens (Folger Shakespeare Librarian), Daniel Goldston, Ofir Gorodetsky, Alia Hamieh, Tom Hunt (computer consultant), Alan Izenman, Habiba Kadiri, Marek Kanter, Peter Kim, Keith Knight, Hugh Montgomery, Nathan Ng, Alexandru Pintilie, Jim Pitman, Linda Rewe (SLMSRI librarian), Nancy Reid, Jeff Rosenthal, David Sinclair, Trevor Wooley, Vova Vinogradov, Dan-Virgil Voiculescu, Aled Walker, Bohan Wang (indexing assistant), Eli Yablonovitch, Zhou Zhou, and Tamara Ziegler.

Very special thanks to Ina Mette of the American Mathematical Society who believed in our concept from the get-go, and to the anonymous reviewers of our manuscript who so generously offered suggestions and corrections; we apologize to any who may have preferred for us to write a different kind of book.

The plots appearing in this book were prepared using the R programming language. The quote appearing above (One always returns to their first love) originates with Charles-Guillaume Étienne (1778–1845).

Bibliography

This is a partial bibliography. With minor exceptions, it includes only references cited in the text. A more complete bibliography may be downloaded here: `https://www.ams.org/publications/authors/books/postpub/mbk-156`.

Surname particles such as de, de la, el, 't, te, van der, and von, at the beginning of surnames are ignored in the alphabetization.

Although major contributions have been made in French, German, Hungarian, Russian, and other languages, our reference list is unavoidably (and with apologies) somewhat biased towards publications in English.

Other notes:

(i) A particularly valuable list of references appears in Narkiewicz (2000, [**797**]), but many other such lists may be found.

(ii) We use simply 'Springer' as an abbreviation for the frequently cited publisher Springer-Verlag; it has offices in many cities.

(iii) All arXiv references are from the 'math' subject groupings unless otherwise noted.

[1] Ablowitz, M. J. and Fokas, S. (2003). Complex Variables: Introduction and Applications, 2nd ed. Cambridge Univ. Press.

[2] Abramowitz, M. and Stegun, I. A., eds. (1964). Handbook of Mathematical Functions with Formulas, Graphs, and Mathematical Tables. National Bureau of Standards, USA. (Reprinted by Dover, NY.)

[3] Adamchik, V. S. and Srivastava, H. M. (1998). Some series of the zeta and related functions. Analysis, 18, 131–144.

[4] Adleman, L. M., Pomerance, C., and Rumely, R. S. (1983). On distinguishing prime numbers from composite numbers. Ann. Math., 117, 173–206.

[5] Adleman, L. M., Rivest, R. L., and Shamir, A. (1978). A method for obtaining digital signatures and public-key cryptosystems. Comm. of the ACM, 21, 121–126.

[6] Advances in Cryptology – CRYPTO 2010 by Springer.

[7] Agrawal, M., Kayal, N., and Saxena, N. (2004). Primes is in P. Ann. Math., 160, 781–793.

[8] Ahlfors, L. V. (1979). Complex Analysis, 3rd ed. McGraw-Hill, NY.

[9] Aigner, M. and Ziegler, G. M. (2014). Proofs from the Book, 5th ed. Springer.

[10] Aistleitner, C. (2016). Lower bounds for the maximum of the Riemann zeta function along vertical lines. Mathematische Annalen, 365, 473–496.

[11] Akemann, G., Baik, J., and Di Francesco, P., eds. (2011). The Oxford Handbook of Random Matrix theory. Oxford Univ. Press.

[12] Albers, D. J., Alexanderson, D. L., and Dunham, W., eds. (2015). The G. H. Hardy Reader. Cambridge Univ. Press.

[13] Alcantara-Bode, J. (1993). An integral equation formulation of the Riemann Hypothesis. Integral Equations Operator Theory, 17, 151–168.

[14] Alfors, L. C. (1978). Complex Analysis: An Introduction to the Theory of Analytic Functions of One Complex Variable, 3rd ed. McGraw-Hill.

[15] Alkan, E. (2021). Variations on criteria of Pólya and Turán for the Riemann hypothesis. J. Number Theory, 225, 90–124.

[16] Alladi, K. (1982). A Turán-Kubilius inequality for integers without large prime factors. J. Reine Angew. Math., 335, 180–196.
[17] Alladi, K. (1987). An Erdős–Kac theorem for integers without large prime factors. Acta Arith., 49, 81–105.
[18] Alladi, K. and Erdős, P. (1977). On an additive arithmetic function. Pacific J. Math., 71, 275–294.
[19] Alladi, K. and Erdős, P. (1979). On the aymptotic behavior of large prime factors of integers. Pacific J. Math., 82, 295–315.
[20] Alt, F. L. (1972). Archaeology of computers: Reminiscences, 1945–1947. Communications of the ACM, 15, 693–694.
[21] Anderson, G. W., Guionnet, A., and Zeitouni, O. (2010). An Introduction to Random Matrices. Cambridge Univ. Press.
[22] Andrica, D. (1986). Note on a conjecture in prime number theory. Studia Univ. Babeş-Bolyai Math., 31, 44–48.
[23] Apéry, R. (1979). Irrationalité de $\zeta(2)$ et $\zeta(3)$. Astérisque, 61, 11–13.
[24] Apostol, T. M. (1970). Möbius functions of order k. Pacific J. Math., 32, 21–27.
[25] Apostol, T. M. (1976). Intoduction to Analytic Number Theory. Springer.
[26] Apostol, T. M. (1985). Formulas for higher derivatives of the Riemann zeta function. Math. Comp., 44, 223–232.
[27] Apostol, T. M. (1990). Modular Functions and Dirichlet Series in Number Theory, 2nd ed. Springer, NY.
[28] Apostol, T. M. (1999). An elementary view of Euler's summation formula. Amer. Math. Monthly, 106, 409–418.
[29] Arguin, L.-P. (2017). Extrema of log-correlated random variables, principles and examples. In: Advances in Disordered Systems, Random Processes and Some Applications. P. Contucci & C. Giardiná, eds. Cambridge Univ. Press, 166–204.
[30] Arguin, L.-P., Belius, D., and Bourgade, P. (2017). Maximum of the characteristic polynomial of random unitary matrices. Comm. Math. Phys., 349, 703–751.
[31] Arguin, L.-P., Belius, D., Bourgade, P., Radziwiłł, M., and Soundararajan, K. (2019). Maximum of the Riemann zeta function on short intervals of the critical line. Comm. Pure Appl. Math., 72, 500–535.
[32] Arguin, L.-P., Belius, D., and Harper, A. J. (2017). Maxima of a randomized Riemann zeta function, and branching random walks. Ann. Appl. Probab., 27, 178–215.
[33] Arguin, L.-P., Bourgade, P., and Radziwiłł, M. (2020). The Fyodorov-Hiary-Keating conjecture, I. pp1–49. `arXiv:2007.00988`.
[34] Arguin, L.-P., Bourgade, P., Radziwiłł, M., and Soundararjan, K. (2019). Maximum of the Riemann zeta function on a short interval of the critical line. Comm. Pure Appl. Math., 72, 500–535.
[35] Arguin, L.-P., Dubach, G., and Hartung, L. (2022). Maxima of a random model of the Riemann zeta function over intervals of varying length. `arXiv:2103.04817v2`.
[36] Arguin, L.-P., Hartung, L., and Kistler, N. (2022). High points of a random model of the Riemann-zeta function and Gaussian multiplicative chaos. Stochastic Processes and Applics., 151, 174–190.
[37] Arguin, L.-P., Ouimet, F., and Radziwiłł, M. (2021). Moments of the Riemann zeta function on short intervals of the critical line. Ann. Probab., 49, 3106–3141.
[38] Arias de Reyna, J. (2008). A test for the Riemann hypothesis. Funct. Approx. Comment. Math., 38, 159–170.
[39] Arias de Reyna, J. and Toulisse, J. (2013). The n-th prime asymptotically. J. Theor. Nombres Bordeaux, 25, 521–555.
[40] Arratia, R., Barbour, A. D., and Tavaré, S. (1997). Random combinatorial structures and prime factorizations. Notices Amer. Math. Soc., 44, 903–910.
[41] Arratia, R., Barbour, A. D., and Tavaré, S. (1999). The Poisson Dirichlet distribution and the scale invariant Poisson process. Combin., Probab. and Comput., 8, 407–416.
[42] Artin, E. (1964). The Gamma Function. Holt Rinehart and Winston, NY. Translation by M. Butler of the 1931 Einführung in die Theorie der Gamma-Funktion, Hamburger Math. Einzelschriften, 11, Teubner, Leipzig.
[43] Atkinson, F. V. (1939). The mean value of the zeta-function on the critical line. Quart. J. Math., Oxford, 10, 122–128.

[44] Aumasson, J.-P. (2018). Serious Cryptography: A Practical Introduction to Modern Encryption. No Starch Press, San Francisco (free download).
[45] Axler, C. (2017). A new upper bound for the sum of divisors function. Bull. Australian Math. Soc., 96, 374–379.
[46] Ayoub, R. G. (1963). An Introduction to the Analytic Theory of Numbers. Mathematical Surveys, No. 10. Amer. Math. Soc., Providence, RI.
[47] Ayoub, R. G. (1974). Euler and the zeta function. Amer. Math. Monthly, 81, 1067–1086.
[48] Ayoub, R. G., Huard, J. G., and Williams, K. S. (1998). Sarvadaman Chowla (1907–1995). Notices Amer. Math. Soc. 45, 594–598.
[49] Ayyad, A., Cochrane, T., and Zheng, Z. (1996). The congruence $x_1x_2 = x_3x_4(\mod p)$, the equation $x_1x_2 = x_3x_4$, and mean values of character sums. J. Number Theory 59, 398–413.
[50] Baas, N. A. and Skau, C. F., eds. (1989). The Collected Ppaers of Atle Selberg, Vol. I, Math. Assoc. America.
[51] Baas, N. A. and Skau, C. F. (2008). The lord of the numbers, Atle Selberg: On his life and mathematics. Bull. Amer. Math. Soc., 45, 617–649.
[52] Babai, L., Pomerance, C., and Vértesi, P. (1998). The mathematics of Paul Erdős. Notices Amer. Math. Soc., 45, 19–31.
[53] Babu, G. J. (1973). Some results on the distribution of additive arithmetical functions, II, III. Acta Arith., 23, 315–328; 25, 39–49.
[54] Bach, E. and Shallit, J. (1996). Algorithmic Number Theory, Vol. I: Efficient Algorithms. MIT Press.
[55] Bäcklund, R. J. (1918/1919). Über die beziechung zwischen Anwachsen und Nullstellen der Zetafunktion. Öfversigt Finska Vetensk. Soc., 61, No. 9.
[56] Baez, J. C. (2020). The tenfold way. Notices Amer. Math. Soc., November, 1599–1601.
[57] Baez-Duarte, L. (2003). A strengthening of the Nyman-Beurling criterion for the Riemann hypothesis. Rend. Lincei Mat. Appl., (9) 14, 5–11.
[58] Bagchi, B. (1981). Statistical behaviour and universality properties of the Riemann zeta function and other allied Dirichlet series. PhD Thesis, Indian Statistical Institute, Calcutta.
[59] Bagchi, B. (2006). On Nyman, Beurling and Baez-Duarte's Hilbert space reformulation of the Riemann hypothesis. Proc. Indian Acad. Sci. Math. Sci., 116, 137–146.
[60] Bai, Z. (1997). Circular law. Ann. Probab., 25, 494–529.
[61] Bai, Z. (1999). Methodologies in spectral analysis of large dimensional random matrices, a review. Statistia Sinica, 9, 611–677.
[62] Baik, J., Deift, P., and Johansson, K. (1999). On the distribution of the length of the longest increasing subsequence of random permutations. J. Amer. Math. Soc., 12, 1119–1178.
[63] Baik, J., Deift, P., and Suidan, T. (2016). Combinatorics and Random Matrix Theory. Amer. Math. Soc., Providence, RI.
[64] Baillaud, B. and Bourget, H. (1905). Correspondence d'Hermite et de Stieltjes, Tome I. Gauthier-Villars, Paris.
[65] Bak, J. and Newman, D. J. (1997). Complex Analysis, 2nd ed. Springer.
[66] Baker, A. (1984). A Concise Introduction to the Theory of Numbers. Cambridge Univ. Press.
[67] Baker, A. (2012). A Comprehensive Course in Number Theory. Cambridge Univ. Press.
[68] Baker, A., Bollobas, B., and Hajnal, A., eds. (1990). A Tribute to Paul Erdős. Cambridge Univ. Press.
[69] Baker, R. C. and Harman, G. (1991). Exponential sums formed with the Möbius function. J. London Math. Soc., 43, 193–198.
[70] Baker, R. C., Harman, G., and Pintz, J. (2001). The difference between consecutive primes, II. Proc. London Math. Soc., 83, 532–562.
[71] Balazard, M. (2003). Completeness problems and the Riemann hypothesis: An annotated bibliography. In: Surveys in Number Theory, Papers from the Millennial Conference on Number Theory, A. K. Peters.
[72] Balazard, M. and de Roton, A. (2008). Notes de lecture de l'article "Partial sums of the M obius function" de Kannan Soundararajan. `arXiv:0810.3587` [math.NT].
[73] Balazard, M., Saias, E., and Yor, M. (1999). Notes sur la fonction ζ de Riemann, 2. Advances Math, 143, 284–287.

[74] Banks, W. D., Ford, K., and Tao, T. (2023). Large prime gaps and probabilistic models. Inventiones Mathematicae, 233, 1471–1518.

[75] Banks, W. D. and Kang, S. (2012). On repeated values of the Riemann zeta function on the critical line. Exp. Math., 21, 132–140.

[76] Barak, B. (2021). An Intensive Introduction to Cryptography. `https://files.boazbarak.org/crypto/lnotes_book.pdf`

[77] Barnabei, M., Brini, A., and Rota, G.-C. (1986). Theory of Möbius functions. Uspekhi Mat. Nank, 41, 113–157; Russian Math. Surveys, 41, 135–188. Also: An introduction to the theory of Möbius functions (Italian). Matroid theory and its applications, 5–109, C.I.M.E. Summer Sch., 83, Springer, 2010. Also: An introduction to the theory of Möbius functions. Liguori Editore, Naples, 1982, 7–109.

[78] Barndorff-Nielsen, O. E. (1961). On the rate of growth of the partial maxima of a sequence of independent identically distributed random variables. Math. Scand., 9, 383–394.

[79] Barndorff-Nielsen, O. E. (1991). Modified signed log likelihood ratio. Biometrika, 78, 557–563.

[80] Basiuk, Y. V. and Tarasyuk, S. I. (2016). Fourier coefficients associated with the Riemann zeta-function. Carpathian Math. Publ., 8, 16–20.

[81] Bateman, P. T. and Diamond, H. G. (1996). A hundred years of prime numbers. Amer. Math. Monthly, 103, 729–741.

[82] Bateman, P. T. and Diamond, H. G. (2004). Analytic Number Theory: An Introductory Course. World Scientific, Singapore.

[83] Bauer, H. (1996, 2011). Probability Theory. de Gruyter, Berlin.

[84] Bays, C. and Hudson, R. (2000). A new bound for the smallest x with $\pi(x) > \mathrm{li}(x)$. Math. Comp., 69, 1285–1296.

[85] Behrend, F. (1946). On sets of integers which contain no three terms in arithmetical progression. Proc. Nat. Acad. Sci. USA, 32, 331–332.

[86] Ben Arous, G. and Bourgade, P. (2013). Extreme gaps between eigenvalues of random matrices. Ann. Probab., 41, 2648–2681.

[87] Benjamin, A. T. and Brown, E., eds. (2009). Biscuits of Number Theory. Math. Assoc. America.

[88] Berline, N. and Sabbah, C. (2003). La Fonction Zêta. Les Éditions de L'École Polytechunique.

[89] Berndt, B. C. (1970). On the average order of some arithmetical functions. Bull. Amer. Math. Soc., 76, 856–859.

[90] Berry, M. V. (1988). Semiclassical formula for the number variance of the Riemann zeros. Nonlinearity, 1, 399–407.

[91] Berry, M. V. (2010). Functions of Number Theory. Cambridge Univ. Press, 637–641.

[92] Berry, M. V. (2012). Hearing the music of the primes: auditory complementarity and the siren song of zeta. (With supplementary audio.) J. Phys., A45, 7pp.

[93] Berry, M. V. and Keating, J. P. (1992). A new asymptotic representation for $\zeta(\frac{1}{2} + it)$ and quantum spectral determinants. Proc. Royal Soc. London, 437, 151–173.

[94] Berry, M. V. and Keating, J. P. (1999). The Riemann zeros and eigenvalue asymptotics. SIAM Review, 41, 236–266.

[95] Bertrand, J., Bertrand, P., and Ovarlez, J. (2000). The Mellin transform. In: The Transforms and Applications Handbook, CRC Press.

[96] Besicovitch, A. S. (1932). Almost periodic functions. Cambridge Univ. Press. Reprinted 1954 by Dover Publications.

[97] Best, D. G. and Trudgian, T. S. (2015). Linear relations of zeroes of the zeta-function. Math. Comp., 84, 2047–2058.

[98] Beurling, A. (1937). Analyse de la loi asymptotique de la distribution des nombres premiers généralisés. Acta Math., 68, 255–291.

[99] Bhattacharya, A. (2022). The Man From the Furure: The Visionary Life of John von-Neumann. W.W. Norton & Company, NY.

[100] Biane, P. (2003). La fonction zêta de Riemann et les probabilités. In: La Fonction Zêta, École Polytech., Palaiseau, 165–193. (A 2015 English translation by Biane is available.)

[101] Biane, P., Pitman, J., and Yor, M. (2001). Probability laws related to the Jacobi theta and Riemann zeta functions, and Brownian excursions. Bull. Amer. Math. Soc., 38, 435–465.

[102] Billingsley, P. (1969). On the central limit theorem for the prime divisor function. Amer. Math. Monthly, 76, 132–139.
[103] Billingsley, P. (1972). On the distribution of large prime divisors. Periodica Mathematica Hungarica, 2, 283–289.
[104] Billingsley, P. (1974). The probability theory of additive arithmetic functions. Ann. Probab., 2, 749–791.
[105] Binder, T., Pauli, S., and Saidak, F, (2015). Zeros of high derivatives of the Riemann zeta function. Rocky Mountain J. Math., 45, 903–926.
[106] Bingham, N. H. (1986). Variants on the law of the iterated logarithm. Bull. London Math. Soc., 18, 433–467.
[107] Binyon, L. (1914). For the Fallen. Poetry Foundation. Stanza 4, lines 3–4. https://www.poetryfoundation.org/poems/57322/for-the-fallen
[108] Birch, B. J. (1967). Multiplicative functions with non-decreasing normal order. J. London Math. Soc., 42, 149–151.
[109] Blake, I. F., Seroussi, G., and Smart, N. P. (1999). Elliptic Curves in Cryptography. Cambridge Univ. Press.
[110] Blei, R. and Janson, S. (2004). Rademacher chaos: tail estimates vs limit theorems. Ark. Mat., 42, 13–29.
[111] Blomer, V. (2008). The theorem of Green-Tao. European Math. Soc. Newsletter, March 2008, 13–18.
[112] Bloom, T. F. (2016). A quantitative improvement for Roth's theorem on arithmetic progressions. J. London Math. Soc., 93, 643–663.
[113] Bloom, T. F. and Sisask, O. (2019). Logarithmic bounds for Roth's theorem via almost-periodiciry. Dicrete Analysis, 4, 20 pp.
[114] Bloom, T. F. and Sisask, O. (2020). Breaking the logarithmic barrier in Roth's theorem on arithmetic progressions. arXiv:2007.03528.
[115] Bloom, T. F. and Sisask, O. (2023). An improvement to the Kelley-Meka bounds on three-term arithmetic progressions. arXiv:2309.02353.
[116] Boas, R. P. (2010). Invitation to Complex Analysis, 2nd ed. Revised by H. P. Boas. Math. Assoc. America, Washington, DC.
[117] Bohman, J. and Froberg, C.-E. (1988). The Stieltjes function – definition and properties. Math. Comp., 51, 281–289.
[118] Bohr, H. (1947). Almost Periodic Functions. Chelsea, NY.
[119] Bohr, H. and Jessen, B. (1930, 1932). Über die wertverteilung der Riemannschen Zetafunktion, I and II. Acta Math., 54, 1–35; and 58, 1–55.
[120] Bohr, H. and Landau, E. (1914). Sur les zéros de la fonction $\zeta(s)$ de Riemann. Comptes Rendus de l'Acad. des Sciences, Paris, 158, 106–110.
[121] Bombieri, E. (2006). Riemann Hypothesis. The Millenium Prize Problems. Clay Math. Inst., Cambridge, Mass., 107–124.
[122] Bombieri, E. and Davenport, H. (1966). Small differences between prime numbers. Proc. Royal Soc. London, Ser. A, 293, 1–18.
[123] Bombieri, E. and Iwaniec, H. (1986). On the order of $\zeta(1/2+it)$. Ann. Scuola Norm. Sup., Pisa, Cl. Sci., 13, 449–472.
[124] Bombieri, E. and Lagarias, J. C. (1999). Complements to Li's criterion for the Riemann hypothesis. J. Number Theory, 77, 274–287.
[125] Bondarenko, A., Brevig, O. F., Saksman, E., Seip, K., and Zhao, J. (2018). Pseudomoments of the Riemann zeta function. Bull. London Math. Soc., 50, 709–724.
[126] Bondarenko, A., Heap, W., and Seip, K. (2015). An inequality of Hardy–Littlewood type for Dirichlet polynomials. J. Number Theory, 150, 191–205.
[127] Bondarenko, A. and Seip, K. (2016). Helson's problem for sums of a random multiplicative function. Mathematika, 62, 101–110.
[128] Bondarenko, A. and Seip, K. (2017). Large greatest common divisor sums and extreme values of the Riemann zeta function. Duke Math. J., 166, 1685–1701.
[129] Bondarenko, A. and Seip, K. (2018). Extreme values of the Riemann zeta function and its argument. Math. Ann., 372, 999–1015.
[130] Booker, A. R. (2006). Turing and the Riemann hypothesis. Notices Amer. Math. Soc., 53, 1208–1211.

[131] Booker, A. R. (2016). Turing and the primes. In: The Once and Future Turing: Computing the World, S. S. Cooper & A. Hodges, eds., Cambridge Univ. Press, 34–52.

[132] Borwein, J. M., Bradley, D. M., and Crandall, R. E. (2000). Computational strategies for the Riemann zeta function. J. Comput. Appl. Math., 121, 247–296.

[133] Borwein, P., Choi, S., Rooney, B., and Weirathmueller, A. (2008). The Riemann Hypothesis: A Resource for the Afficionado and Virtuoso Alike. Canadian Math. Soc., Springer.

[134] Borwein, P., Fee, G., Ferguson, R., and van der Waall, A. (2007). Zeros of partial sums of the Riemann zeta-function. Experimental Math., 16, 21–39.

[135] Borwein, P., Ferguson, R., and Mossinghoff, M. J. (2008). Sign changes in sums of the Liouville function. Math. Comp. 77, 1681–1694.

[136] Borwein, J. M., Borwein, P., and Dilcher, K. (1989). Pi, Euler numbers, and asymptotic expansions. Amer. Math. Monthly 96, 681–687.

[137] Bose, A., Chatterjee, S., and Gangopadhyay, S. (2003). Limiting spectral distributions of large dimensional random matrices. J. Indian Statist. Assoc., 41, 221–259.

[138] Bourgain, J. (2000). On large values estimates for Dirichlet polynomials and the density hypothesis for the Riemann zeta function. Internat. Math. Res. Notices, 3, 133–146.

[139] Bourgain, J. (2017). Decoupling, exponential sums and the Riemann zeta function. J. Amer. Math. Soc., 30, 205–224.

[140] Bracewell, R. N. (1986). The Fourier Transform and its Applications. McGraw-Hill.

[141] Bredberg, J. (2011). Large gaps between consecutive zeros, on the critical line, of the Riemann zeta-function. Unpublished manuscript. `arXiv:1101.3197v3`.

[142] Breiman, L. (1992). Probability. S.I.A.M. (Corrected republication of the 1968 Addison-Wesley edition.)

[143] Breitzman, A. (2016). Major milestones in the twin prime conjecture. Math. Scientist, 41, 3–15.

[144] Brent, R. P. (1979). On the zeros of the Riemann zeta function in the critical strip. Math. of Computation, 33, 1361–1372.

[145] Brent, R. P. and van de Lune, J. (2011). A note on Pólya's observation concerning Liouville's function. Manuscript, `arXiv:1112.4911`.

[146] Brent, R. P., van de Lune, J., te Riele, H. J. J., and Winter, D. T. (1982). The first 200,000,001 zeros of Riemann's zeta function. Computational Methods in Number Theory, Part II, 389–403, Math. Centre Tracts, 155, Amsterdam.

[147] Bressoud, D. (1989). Factorization and Primality Testing. Springer.

[148] de la Bretèche, R. (2014). Petit écartes entre les nombres premiers et polymath; une nouvelle manière de faire de la recherche en mathématiques? SMF Gazette des Mathématiciens, 140, 19–31.

[149] de la Bretèche, R. and Tenenbaum, G. (2002). Sur les lois locales de la répartition du k-ième diviseur d'un entier. [Local distribution of the kth divisor of an integer.] Proc. London Math. Soc., 84, 289–323.

[150] de la Bretèche, R. and Tenenbaum, G. (2019). Sommes de Gál et applications. Proc. London Math. Soc., 119, 104–134.

[151] Brevig, O. F. and Heap, W. (2019). High pseudomoments of the Riemann zeta function J. Number Theory, 197, 383–410.

[152] Briggs, W. E. (1955). Some constants associated with the Riemann zeta function. Michigan Math. J., 3, 117–121.

[153] Briggs, W. E. and Chowla, S. (1955). The power series coefficients of $\zeta(s)$. Amer. Math. Monthly, 62, 323–325.

[154] Brody, T. A., Flores, J., French, J. B., Mello, P. A., Pandey, A., and Wong, S. S. M. (1981). Random-matrix physics: spectrum and strength fluctuations. Rev. Modern Physics, 53, 385–479.

[155] Bromwich, T. J. (1926). An Introduction to the Theory of Infinite Series, 2nd ed. Macmillan, London, NY.

[156] Brouder, F., ed. (1976). Mathematical Developments Arising from Hilbert Problems. Amer. Math. Soc., Providence, RI.

[157] Broughan, K. (2017). Equivalents of the Riemann hypothesis: Volume One: Arithmetic Equivalents, and Volume Two: Analytic Equivalents. Cambridge Univ. Press.

[158] de Bruijn, N. G. (1951). The asymptotic behaviour of a function occurring in the theory of primes. J. Indian Math. Soc., 15, 25–32.

[159] Brun, V. (1915). Über das Goldbachsche Gesetz und die Anzahl der Primzahlpaare. Archiv for Mathematik og Naturvidenskab, B34, 3–19.

[160] Brun, V. (1919). La crible d'Eratosthène et la théorème de Goldbach. Comptes Rendus de l'Acad. Sciences, Paris, 168, 544–546.

[161] Brun, V. (1919). La série $1/5 + 1/7 + 1/11 + 1/13 + \ldots,$ ou les dénominateurs sont nombres premiers jumeaux est convergente ou finie. Bull. des Sciences Mathématiques, 43, 100–104, 124–128.

[162] Buchstab, A. A. (1951). On an asymptotic estimate of the number of numbers of an arithmetic progression which are not divisible by "relatively" small prime numbers. Sbornik N. S., 28, 165–184.

[163] Bui, H. M. (2011). Large gaps between consecutive zeros of the Riemann zeta-function. J. Number Th., 131, 67–95.

[164] Bui, H. M. (2014). Large gaps between consecutive zeros of the Riemann zeta-function, II. Acta Arith., 165, 101–122.

[165] Bui, H. M. and Heath-Brown, D. R. (2013). On simple zeros of the Riemann zeta-function. Bull. London Math. Soc., 45, 953–961.

[166] Bui, H. M. and Keating, J. P. (2006). On twin primes associated with the Hawkins random sieve. J. Number Theory, 119, 284–296.

[167] Bui, H. M., Lester, S. J., and Milinovich, M. B. (2014). On Balazard, Saias, and Yor's equivalence to the Riemann hypothesis. J. Math. Anal. Appl., 409, 244–253.

[168] Bui, H. M. and Milinovich, M. B. (2018). Gaps between zeros of the Riemann zeta function. Quart. J. Math., 69, 403–423.

[169] Bui, H. M., Milinovich, M. B., and Ng, N. C. (2010). A note on the gaps between consecutive zeros of the Riemann zeta-function. Proc. Amer. Math. Soc., 138, 4167–4175.
See also: `https://www.cs.uleth.ca/~nathanng/RESEARCH/gaps_submitted.pdf`

[170] Büthe, J. (2015). On the first sign change in Mertens' theorem. Acta Arith., 171, 183–195.

[171] Büthe, J. (2016). Estimating $\pi(x)$ and related functions under partial RH assumptions. Math. Comp., 85, 2483–2498.

[172] Butler, R. W. (2007). Saddlepoint Approximations with Applications. Cambridge Univ. Press.

[173] Buttkewitz, Y. and Elsholtz, C. (2011). Patterns and complexity of multiplicative functions. J. London Math. Soc., 84, 578–594.

[174] Butzer, P. L. and Jansche, S. (1997). A direct approach to the Mellin transform. J. Fourier Analysis Applics., 3, 325–376.

[175] Butzer, P. L. and Jongmans, F. (1999). P. L. Chebyshev (1821–1894): A guide to his life and work. J. Approximaton Theory, 96, 111–138.

[176] Cai, Y. C. (2002). A remark on Chen's theorem. Acta Arith., 339–352.

[177] Cai, Y. C. and Lu, M. G. (2003). On the upper bound for $\pi_2(x)$. Acta Arith., 110, 275–298.

[178] Caich, R. (2024). Almost sure upper bound for random multiplicative functions. `arXiv:2304.00943`.

[179] Canfield, E., Erdős, P., and Pomerance, C. (1983). On a problem of Oppenheim concerning "factorisatio numerorum." J. Number Theory, 17, 1–28.

[180] Cardinal, J.-P. (2010). Symmetric matrices related to the Mertens function. Linear Alg. and its Applics., 432, 161–172.

[181] Carneiro, E., Chandee, V., and Milinovich, M. B. (2013). Bounding $S(t)$ and $S_1(t)$ on the Riemann hypothesis. Math. Ann., 356, 939–968.

[182] Carneiro, E., Milinovich, M. B., and Soundararajan, K. (2019). Fourier optimization and prime gaps. Comentarii Math. Helvetici, 94, 533–568.

[183] Cassaigne, J., Ferenczi, S., Mauduit, C., Rivat, J., and Sárkőzy, A. (1999). On finite pseudorandom binary sequences III: The Liouville function, I. Acta Arith., 87, 367–390.

[184] Cassaigne, J., Ferenczi, S., Mauduit, C., Rivat, J., and Sárkőzy, A. (2000). On finite pseudorandom binary sequences IV: The Liouville function II. Acta Arith. XCV, 343–359.

[185] Cellarosi, F. and Sinai, Ya. G. (2011). The Möbius function and statistical mechanics. Bull. Math. Sciences, 1, 245–275.

[186] Cellarosi, F. and Sinai, Ya. G. (2013). Non-standard limit theorems in number theory. In: A. N. Shiryaev et al. (eds.), Prokhorov and Contemporary Probability Theory, Springer, 197–213.

[187] Chandee, V. and Soundararajan, K. (2011). Bounding $|\zeta(\frac{1}{2}+it)|$ on the Riemann Hypothesis. Bull. London Math. Soc., 243–250.

[188] Chandrasekharan, K. (1968). Introduction to Analytic Number Theory. Springer.

[189] Chandrasekharan, K. (1970). Arithmetical Functions. Springer.

[190] Chao, H. (1987). A generalization of a theorem on prime numbers. (Chinese, with English summary). Adv. in Math. (Beijing), 16, 419–426.

[191] Chatterjee, S. and Soundararajan, K. (2012). Random multiplicative functions in short intervals. Int. Math. Res. Notes, 3, 479–492.

[192] Chebyshev, P. L. (1852). Sur la totalité des nombres premiers inférieurs à une limite donnée. J. de Mathématiques Pures et Appliqués, 17, 341–365.

[193] Chebyshev, P. L. (1852). Memoire sur les nombres premiers. J. de Mathématiques Pures et Appliqués, 17, 366–390.

[194] Chen, J. (1973). On the representation of a large even integer as the sum of a prime and the product of at most two primes. Scientia Sinica, 16, 157–176. (Originally announced in Kexue Tongbao (Chinese), 1966, 17, 385–386.)

[195] Chen, J. (1975). On the representation of a large even integer as the sum of a prime and the product of at most two primes. Scientia Sinica, 18, 702–734.

[196] Chen, J. (1978). On the representation of a large even integer as the sum of a prime and a product of at most two primes. Scientia Sinica, 21, 693–726.

[197] Chen, N. (2010). Möbius Inversion in Physics. World Scientific, Singapore.

[198] Cheng, Y. and Graham, S. W. (2004). Explicit estimates for the Riemann zeta function. Rocky Mountain J. Math. 34 (2004), no. 4, 1261–1280.

[199] Chernoff, P. R. (2000). A pseudo zeta function and the distribution of primes. Proc. Nat. Acad. Sci. USA, 97, 7697–7699.

[200] Cherwell, L. and Wright, E. M. (1960). The frequency of prime-patterns. Quart. J. Math., Oxford, 11, 60–63.

[201] Chih, T. (1950). A divisor problem. Acad. Sinica Science Record, 3, 177–182.

[202] Choudhury, B. K. (1995). The Riemann zeta function and its derivatives. Proc. Royal Soc. London, A, 450, 477–499.

[203] Chow, Y. S. and Teicher, H. (1997). Probability Theory, 3rd ed. Springer.

[204] Chowla, S. (1965). The Riemann Hypothesis and Hilbert's Tenth Problem. Gordon and Breach Science Publishers, NY.

[205] Chudakov, N. G. (1936). On zeros of the function $\zeta(s)$. Dokl. Akad. Nauk SSSR 1, 201–204.

[206] Chung, K. L. (1948). On the maximum partial sums of sequences of independent random variables. Trans. Amer. Math. Soc. 64, 205–233.

[207] Chung, K. L. (2001). A Course in Probability Theory, 3rd ed. Academic Press, NY.

[208] Chung, K. L. and AitSahlia, F. (2003). Elementary Probability Theory. Springer.

[209] Churchill, R. V. and Brown, J. W. (2009). Complex Variables and Applications, 8th ed. McGraw-Hill.

[210] Cipolla, M. (1902). La determinazione assintotica dell' n^{imo} numero primo. Rend. Accad. Sci. Fis.-Mat. Napoli, 8, 132–166.

[211] Cohen, H. (2007). Number Theory, vol. I: Tools and Diophantane Equations, and Number Theory, vol. II: Analytic and Modern Tools. Springer.

[212] Cohen, J. E. (2022). Multiplicative inequalities for primes and the prime counting function. Integers, 22, A106.

[213] Coil, A. L. (2013). The large-scale structure of the universe. In: Planets, Stars and Stellar Systems, vol. 6, Springer, 387–421.

[214] Cojocaru, A. C. and Murty, R. (2006). An Introduction to Sieve Methods and their Applications. Cambridge Univ. Press.

[215] Conlon, D., Fox, J., and Zhao, Y. (2014). The Green–Tao theorem: An exposition. EMS Surveys in Mathematical Sciences. 1, 249–282. Also in `arXiv:1403.2957`.

[216] Connes, A. (1999). Trace formula in noncommutative geometry and the zeros of the Riemann zeta function. Selecta Mathematica, 5, 29–106.

[217] Connes, A. (2016). An essay on the Riemann hypothesis. In: Open Problems in Mathematics, Springer, 225–257.

[218] Conrey, J. B. (1989). At least two fifths of the zeros of the Riemann zeta function are on the critical line. Bull. Amer. Math. Soc., 20, 79–81

[219] Conrey, J. B. (1989). More than two fifths of the zeros of the Riemann zeta function are on the critical line. J. Reine Angew. Math., 399, 1–26.
[220] Conrey, J. B. and Farmer, D. W. (2000). Mean values of L-functions and symmetry. Internat. Math. Res. Notices, 17, 883–908.
[221] Conrey, J. B. and Farmer, D. W. (2012). Equivalences to the Riemann hypothesis. Manuscript, 9 pp. Available at:
`https://web.archive.org/web/20120731034246/http://aimath.org/pl/rhequivalences`
[222] Conrey, J. B., Farmer, D. W., Keating, J. P., Rubenstein, M. O., and Snaith, N. C. (2005). Integral moments of L-functions. Proc. London Math. Soc., 91, 33–104.
[223] Conrey, J. B. and Gamburd, A. (2006). Pseudomoments of the Riemann zeta function and pseudomagic squares. J. Number Theory, 117, 263–278.
[224] Conrey, J. B. and Ghosh, A. (1984). On mean values of the zeta-function. Mathematika, 31, 159–161.
[225] Conrey, J. B. and Ghosh, A. (1998). A conjecture for the sixth power moment of the Riemann zeta-function. Internat. Math. Res. Notices, 15, 775–780.
[226] Conrey, J. B., Ghosh, A., and Gonek, S. M. (1984). A note on gaps between zeros of the zeta function. Bull. London Math. Soc., 16, 421–424.
[227] Conrey, J. B., Ghosh, A., and Gonek, S. M. (1986). Large gaps between zeros of the zeta-function. Mathematica, 33, 212–238.
[228] Conrey, J. B., Ghosh, A., and Gonek, S. M. (1998). Simple zeros of the Riemann zeta-function. Proc. London Math. Soc., 76, 497–522.
[229] Conrey, J. B. and Gonek, S. M. (2001). High moments of the Riemann zeta-function. Duke Math. J., 107, 577–604.
[230] Conrey, J. B. and Snaith, N. C. (2008). Correlations of eigenvalues and Riemann zeros. Commun. Number Theory Physics, 2, 477–453.
[231] Conrey, J. B. and Snaith, N. C. (2014). In support of n-correlation, Comm. Math. Physics, 330, 639–653.
[232] Conway, J. B. (1978). Functions of One Complex Variable, 2nd ed. Springer.
[233] Cooke, R. (1996). Review of Harmonic Analysis, 2nd ed. by Henry Helson. Bull. Amer. Math. Soc., 33, 505–508.
[234] Coram, M. and Diaconis, P. (2003). New tests of the correspondence between unitary eigenvalues and the zeros of Riemann's zeta function. Random matrix theory. J. Phys., A36, 2883–2906.
[235] van der Corput, J. G. (1928). Zum Teilerproblem. Math. Ann., 98, 697–716.
[236] van der Corput, J. G. (1937). Sur l'hypothèse de Goldbach pour presque tous es nombres premieres. Acta Arith., 2, 266–290.
[237] Costa, E. and Harvey, D. (2014). Faster deterministic integer factorization. Math. Comp., 83, 339–345.
[238] Costa Pereira, N. (1985). Estimates for the Chebyshev Function $\psi(x) - \theta(x)$. Math. Comput., 44, 211–221. Corrigendum, 1987, Math. Comput., 48, 447, 1987.
[239] Cox, D. R. and Isham, V. (1980). Point Processes. C.R.C. Press, Chapman and Hall, London.
[240] Cramér, H. (1918). Über die Nullstellen der Zetafunktion. Math. Zeitschrift, 2, 237–241.
[241] Cramér, H. (1919). Studien über die Nullstellen der Riemannschen Zetafunktion. Math. Zeitschrift, 4, 104–130.
[242] Cramér, H. (1920). On the distribution of primes. Proc. Camb. Phil. Soc., 20, 272–280.
[243] Cramér, H. (1921). Some theorems concerning prime numbers. Arkiv. for Mat. Astronom. o. Fys., 15, 1–32. Also: Acta Arith., 2, 23–46, 1936.
[244] Cramér, H. (1935). Prime numbers and probability. Proc. 8th Scand. Math. Congress, Stockholm, Skand. Mat.-Kongr., 8, 107–115.
[245] Cramér, H. (1936). On the order of magnitude of the difference between consecutive prime numbers. Acta Arith., 2, 23–46.
[246] Cramér, H. (1937, 1970). Random variables and Probability Distributions, 3rd ed. Cambridge Univ. Press.
[247] Cramér, H. (1946). Mathematical Methods of Statistics.
[248] Crandall, R. and Pomerance, C. (2005). Prime Numbers: A Computational Perspective, 2nd ed. Springer.

[249] Cranston, M. and Mountford, T. (2924). A new proof of the Erdős–Kac Central Limit Theorem. Trans. Amer. Math. Soc., 377, 1475–1503.

[250] Cranston, M. and Peltzer, A. (2022). On properties of the Riemann zeta distribution. Rocky Mountain J. Math., 52, 843–875.

[251] Croot, E. S. (2003). On the oscillations of multiplicative functions taking values ± 1. J. Number Theory, 98, 184–194.

[252] Daboussi, H. (1981). On the limiting distribution of nonnegative additive functions. Compositio Math., 43, 101–105.

[253] Daboussi, H. (1981). Sur les fonctions multiplicatives ayant une valeur moyen non nulle. Bull. Soc. Math. France, 109, 183–205.

[254] Daboussi, H. (1984). Sur le théoreme des nombres premiers. C.R. Acad. ScI., Paris, 298, Ser. I, 161–164.

[255] Daboussi, H. and Delange, H. (1982). On multiplicative arithmetical functions whose modulus does not exceed one. J. London Math. Soc., 26, 245–264.

[256] Dabrowski, A. (1996). A note on values of the Riemann zeta function at positive odd integers. Nieuw Archief voor Wiskunde, 14, 199–207.

[257] Daley, D. J. and Vere-Joners, D. (2003). An Introduction to the Theory of Point Processes, vol I: Elementary Theory and Methods. Springer.

[258] Daniels, H. E. (1954). Saddlepoint approximations in statistics. Ann. Math. Statistics, 25, 631–650.

[259] Dartyge, C. (2019). Friable integers: An overview. Eur. Math. Soc. Newsl., 111, 6–13.

[260] Davenport, H. (1933). Über numeri abundantes. Preuss. Akad. Wiss. Sitzungsber., 830–837.

[261] Davenport, H. (1937). On some infinite series involving arithmetic function, i and ii. Quart. J. Math., 8, 8–13 and 313–320.

[262] Davenport, H. (2000). Multiplicative Number Theory, 3rd ed. (Revised by H. L. Montgomery.) Springer.

[263] Davenport, H. (2008). The Higher Arithmetic, 8th ed. Cambridge Univ. Press.

[264] Davis, P. J. (1959). Leonhard Euler's integral: A historical profile of the gamma function. Amer. Math. Monthly, 66, 849–869.

[265] Deheuvels, P. (1985). Asymptotic results for the pseudo-prime sequence generated by Hawkins' random sieve: Twin primes and Riemann's hypothesis. Proc. Seventh Conf. on Probab. Theory (Brasov, Romania, 1982). M. Iosifescu, T. Ostelnicu and S. Grigorescu, eds. VNU Sci. Press, Utrecht, 109–115.

[266] Delange, H. (1953). Sur le nombre des diviseurs premiers de n. C.R. Acad. Sci., Paris, 237, 542–544.

[267] Delange, H. (1956). Sur un théoreme d'Erdős et Kac. Acad. Roy. Belg. Bull. Cl. Sci., 42, 130–144.

[268] Delange, H. (1957). Sur les fonctions arithmétiques fortement additives. C.R. Acad. Sci. Paris. 244, 1307–1309.

[269] Delange, H. (1961). Sur les fonctions arithmétiques multiplicatives. Ann. Sci. École Norm. Sup., 78, 273–304.

[270] Delange, H. (1971). Sur des formules de Atle Selberg. Acta Arith., 19, 105–146.

[271] Delange, H. and Halberstam, H. (1957). A note on additive functions. Pacific J. Math., 7, 1551–1556.

[272] Deleglise, M. and Rivat, J. (1996). Computing $\pi(x)$: The Meissel, Lehmer, Lagarias, Miller, Odlyzko method. Math. Comp., 65, 235–245.

[273] Delone, B. N. (2005). The St. Petersburg School of Number Theory. Amer. Math. Soc. and London Math. Soc.

[274] Denjoy, A. (1931). L'hypothèse de Riemann sur la distribution des zéros de $\zeta(s)$, reliee a la théorie des probabilités. Comptes Rendus, 192, 656–658.

[275] Denjoy, A. (1964). Probabilités confirmant l'hypothèse de Riemann sur les zéros de $\zeta(s)$. Comptes Rendus Acad. Sci. Paris, 259, 3143–3145.

[276] Deshouillers, J.-M., Dress, F., and Tenenbaum, G. (1979). Lois de répartition des diviseurs I. Acta Arith., 34, 273–285.

[277] Deshouillers, J.-M., Effinger, G., te Riele, H., and Zinoviev, D. (1997). A complete Vinogradov 3-primes theorem under the Riemann hypothesis. Electron. Res. Announc. Amer. Math. Soc., 3. 99–104.

[278] Diaconis, P. (2002). G. H. Hardy and probability??? Bull. London Math. Soc., 34, 385–402.

[279] Diaconis, P. (2003). Patterns in eigenvalues: The 70th Josiah Willard Gibbs lecture. Bull. Amer. Math. Soc., 40, 155–178.

[280] Diaconis, P. and Skyrms, B. (2018). Ten Great Ideas About Chance. Princeton Univ. Press.

[281] Diaconu, A., Goldfeld, D., and Hoffstein, J. (2003). Multiple Dirichlet series and moments of zeta and L-functions. Compositio Mathematica., 139, 297–360.

[282] Diamond, H. G. (1975). Changes of sign of $\pi(x) - \mathrm{li}(x)$. l'Enseignement Math., 21, 1–14.

[283] Diamond, H. G. (1982). Elementary methods in the study of the distribution of prime numbers. Bull. Amer. Math. Soc., 7, 553–589.

[284] Diamond, H. G. and Erdős, P. (1980). On sharp elementary prime number estimates. l'Enseignement Math., 26, 313–321.

[285] Diamond, H. G. and Steinig, J. (1970). An elementary proof of the prime number theorem with a remainder term. Inventiones Mathematicae, 11, 199–258.

[286] Dickman, K. (1930). On the frequency of numbers containing prime factors of a certain relative magnitude. Arkiv for Mat. Astron. och Fysik, 22, 1–14.

[287] Dickson, L. E. (1904). A new extension of Dirichlet's theorem on prime numbers. Messenger of Math., 33, 155–161.

[288] Dieng, M. and Tracy, C. A. (2011). Application of random matrix theory to multivariate statistics. In: Random matrices, random processes and integrable systems, CRM Ser. Math. Phys., Springer, 443–507.

[289] Dilcher, K. and Slavutskii, I. Sh. (2007). A bibliography of Bernoulli numbers. Website: `https://www.mscs.dal.ca/~dilcher/bernoulli.html`

[290] Dirichlet, P. G. L. (1837). Beweis des Satzes, dass jede unbegrenzte arithmetische Progression, deren erstes Glied und Differenz ganze Zahlen ohne gemeinschaftlichen Factor sind, unendlich viele Primzahlen enhält. [Proof of the theorem that every unbounded arithmetic progression, whose first term and common difference are integers without common factors, contains infinitely many prime numbers.] Abhandlungen der Königlichen Preussischen Akademie der Wissenschaften zu Berlin, 48: 45–71.

[291] Donnelly, P. and Grimmett, G. (1993). On the asymptotic distribution of large prime factors. J. London Math. Soc., 47, 395–404.

[292] Doray, L. G. and Arsenault, M. (2002). Estimators of the regression parameters of the zeta distribution. Insurance: Mathematics and Economics, 30, 439–450.

[293] Doray, L. G. and Luong, A. (1995). Quadratic estimators for the zeta family. Insurance: Mathematics and Economics, 16, 255–260.

[294] Douglas, A. A. (2016). 4044 William Shakespeare Quotes. UB Tech.

[295] Doyle, A. C. (1893). The Adventure of the Cardboard Box, Strand Magazine, UK.

[296] Ducray, S. (1964). Probability and prime numbers. Proc. Indian Acad. Sci. A, 60, 159–164.

[297] Dudek, A. W. (2016). An explicit result for primes between cubes. Funct. Approx. Comment. Math., 55, 177–197.

[298] Dudek, A. W., Grenié, L., and Molteni, G. (2016). Primes in explicit short intervals on RH. Int. J. Number Theory, 12, 1391–1407.

[299] Dumitriu, I. and Edelman, A. (2002). Matrix models for beta ensembles. J. Math. Phys., 43, 5830–5847.

[300] Durrett, R. (1999/2012/2016). Essentials of stochastic processes. Springer.

[301] Dusart, P. (1998S). Sharper bounds for $\psi, \theta, \pi p_k$. Rapport de recherche #1998-06. Laboratoire d'Arithmétique de Calcul formel et d'Optimisation

[302] Dusart, P. (1999). The k-th prime is greater than $k(\log k + \log\log k - 1)$ for $k \geq 2$. Math. Comp., 68, 411–415.

[303] Dusart, P. (1999). Inégalités explicites pour $\psi(X)$, $\theta(X)$, $\pi(X)$ et les nombres premiers. C.R. Math. Acad. Sci., Canada, 21, 53–59.

[304] Dusart, P. (2018). Explicit estimates of some functions over primes. Ramanujan J., 45, 227–251.

[305] Dusart, P. (2018). Estimates of the kth prime under the Riemann hypothesis. Ramanujan J., 47, 141–154.

[306] Dym, H. and McKean, H. P. (1972). Fourier Series and Integrals. Academic press.

[307] Dyson, F. J. (1962a,b,c, and 1963). Statistical theory of energy levels of complex systems, I, II, III and IV. J. Mathematical Phys., 3, 140–156, 157–165, 166–175; and 4, 701–712. (Publication IV is joint with M. L. Mehta.)

[308] Dyson, F. J. (1979). Disturbing the Universe. Harper and Rown, NY.

[309] Edelman, A. (1989). Eigenvalues and condition numbers of random matrices. PhD thesis, Math. Dept., Mass. Inst. Tech.

[310] Edwards, H. (1974). Riemann's Zeta Function. Academic Press, NY. (Reprinted 2001 by Dover, Mineola, NY.)

[311] Efron, B. (2023). Exponential Families in Theory and Practice. Cambridge Univ. Press.

[312] Ehm, W. (2001). A family of probability densities related to the Riemann zeta function. Contemp. Math., 287, Amer. Math. Soc., 63–74.

[313] Ehm, W. (2007). A Riemann zeta stochastic process. C.R. Math. Acad. Sci. Paris, 345, 279–282.

[314] Eisen, M. (1969). Introduction to Mathematical Probability Theory. Prentice-Hall, Englewood Cliffs, NJ, 1969.

[315] Elliott, P. D. T. A. (1972). The Riemann zeta function and coin tossing. J. Reine Angew. Math., 254, 100–109.

[316] Elliott, P. D. T. A. (1973). On additive functions whose limiting distributions possess a finite mean and variance. Pacific J. Math., 48, 47–55.

[317] Elliott, P. D. T. A. (1975). The law of large numbers for additive arithmetic functions. Math. Proc. Cambridge Phil. Soc., 78, 33–71.

[318] Elliott, P. D. T. A. (1979). Probabilistic Number Theory I. Mean-Value Theorems. Springer.

[319] Elliott, P. D. T. A. (1980). Probabilistic Number Theory II. Central Limit Theorems. Springer.

[320] Elliott, P. D. T. A. (1987). Functional analysis and additive arithmetic functions. Bull. Amer. Math. Soc., 16, 179–223.

[321] Elliott, P. D. T. A. (1992). On the correlation of multiplicative functions. Notas de la Sociedad de Matematica de Chile, 11, 1–11.

[322] Elliott, P. D. T. A. (1997). Duality in Analytic Number Theory. Cambridge Univ. Press.

[323] Elliott, P. D. T. A. and Halberstam, A. (1970). A conjecture in prime number theory. In: Symposia Mathematica, Vol. IV, (INDAM, Rome, 1968/69), Academic Press, London, pp 59–72.

[324] Ellison, F. and Ellison, W. J. (1985). Prime Numbers. Wiley, NY. (English translation of Ellison & Mendès-France 1975.)

[325] Ellison, W. J. and Mendès-France, M. (1975). Les Nombres Premiers. Publications de l'Institut de Mathématique de l'Université de Nancago, IX. Actualités Scientifiques et Industrielles, 1366, Herman, Paris.

[326] Elsholtz, C. (2001). The inverse Goldbach problem. Mathematika, 48, 151–158.

[327] Elsholtz, C. (2006). Additive decompositions of multiplicatively defined sets. Functiones et Approximatio, XXXV, 61–77.

[328] Elsholtz, C. (2017). Golomb's conjecture on prime gaps. Amer. Math. Monthly, 124, 365–368.

[329] Elsholtz, C. (2021). Fermat's last theorem implies Euclid's infinitude of primes. Amer. Math. Monthly, 128, 250–257.

[330] Elsholtz, C. and Harper, A. J. (2015). Additive decompositions of sets with restricted prime factors. Trans. Amer. Math. Soc., 367, 7403–7427.

[331] Erdős, L., Schlein, B., and Yau, H.-T. (2009). Local semicircle law and complete delocalization for Wigner random matrices. Comm. Math. Phys., 287, 641–655.

[332] Erdős, P. (1932). Beweis eines Satzes von Tschebyschef. Acta Sci. Math. (Szeged), 5, 194–198.

[333] Erdős, P. (1934). On the density of abundant numbers. J. London Math. Soc., 9, 278–282.

[334] Erdős, P. (1935). On the difference of consecutive primes. Quart. J. Math., Oxford, 6, 124–128.

[335] Erdős, P. (1935). On the normal order of prime factors of $p-1$ and some related problems concerning Euler's φ-functions. Quart. J. Math., Oxford, 6, 205–213.

[336] Erdős, P. (1935, 1937, 1938). On the density of some sequences of numbers, I, II, and III. J. London Math. Soc., 10, 120–125; 12, 7–11; 13, 119–127.

[337] Erdős, P. (1939). On the smoothness of the asymptotic distribution of additive arithmetical functions. Amer. J. Math., 61, 722–725.

[338] Erdős, P. (1940). The difference of consecutive primes. Duke Math. J., 6, 438–441.

[339] Erdős, P. (1946). Some remarks about additive and multiplicative functions. Bull. Amer. Math. Soc., 52, 527–537.

[340] Erdős, P. (1949). On a new method in elementary number theory which leads to an elementary proof of the prime number theorem. Proc. Natl. Acad. Sci. USA, 35, 374–384.
[341] Erdős, P. (1950). On integers of the form $2^k + p$ and some related problems. Summa Brasil. Math., 2, 113–123.
[342] Erdős, P. (1951). Some problems and results in elementary number theory. Publ. Math. Debrecen, 2, 103–109.
[343] Erdős, P. (1955). Problems and results in additive number theory. Colloque sur la theorie des nombres, 127–137, Brussels.
[344] Erdős, P. (1957). Some unsolved problems. Michigan Math. J., 291–300.
[345] Erdős, P. (1961). Some unsolved problems. Magyar Tu. Acad. Mat. Kut. Int. Kozl., 6, 221–254.
[346] Erdős, P. (1963). Quelques problemes de théorie des nombres. Monographies de l'Enseignement Math., 6, 81–135, Université Genève.
[347] Erdős, P. (1976). Problems and results in number theory. In: Number Theory Day, NY, Springer, 626, 43–72.
[348] Erdős, P. (1977). Some unsolved problems in number theory. 405–414.
[349] Erdős, P. (1979). Some unconventional problems in number theory. Acta Math. Acad. Sci. Hungary, 33, (1979), 1–2, 71–80. Also: Math. Magazine, 52, 67–70. Also: Astérisque, 61, 73–82.
[350] Erdős, P. (1985). Some solved and unsolved problems of mine in number theory. In: S. W. Graham and J. D. Vaaler, Topics in Analytic Number Theory, Univ. of Texas Press, Austin.
[351] Erdős, P. (1985). On some of my problems in number theory I would most like to see solved. In: Number theory (Ootacamund, 1984), Springer, 74–84.
[352] Erdős, P. (1985). Some applications of probability methods to number theory. In: Mathematical Statistics and Applications, Vol. B, Proc. Fourth Pannonian Symp. on Math. Stat., (Bad Tatzmannsdorf, 1983), Reidel, Dordrecht, 1–18.
[353] Erdős, P. (1986). Some problems on number theory. In: Analytic and Elementary Number Theory (Marseille, 1983), Publ. Math., Univ. Paris XI, Orsay, 86, 53–67.
[354] Erdős, P. (1995). Some of my favourite problems in number theory, combinatorics, and geometry. Resenhas, 2, 165–186.
[355] Erdős, P. and Jabotinsky, E. (1958). On sequences of integers generated by a sieving process, I, II. Nederl. Akad. Wetensch. Proc. Ser. A6, Indag. Math., 20, 115–128.
[356] Erdős, P. and van Lint, J. H. (1982). On the average ratio of the smallest and largest prime divisor of n. Nederl. Akad. Wetensch. Indag. Math., 44, 127–132.
[357] Erdős, P. and Kac, M. (1939). On the Gaussian law of errors in the theory of additive functions. Proc. Nat. Acad. Sci., USA, 25, 206–207.
[358] Erdős, P. and Kac, M. (1940). The Gaussian law of errors in the theory of additive number-theoretic functions. Amer. J. Math., 62, 738–742.
[359] Erdős, P. and Rényi, A. (1959). On Cantor's series with convergent $\sum 1/q_n$. Ann. Univ. Sci. Budapest. Sect. Math., 2, 93–109.
[360] Erdős, P. and Rényi, A. (1970). On a new law of large numbers. J. Analyse Math., 23, 103–111.
[361] Erdős, P. and Shapiro, H. N. (1951). On the changes of signs of a certain error function. Canadian J. Math., 3, 375–385.
[362] Erdős, P. and Tenenbaum, G. (1989). Sur les fonctions arithmétiques liées aux diviseurs consécutifs [On arithmetical functions associated with consecutive divisors.] J. Number Theory, 31, 285–311.
[363] Erdős, P. and Turán, P. (1936). On some sequences of integers. J. London Math. Soc., 11, 261–264.
[364] Erdős, P. and Wintner, A. (1939). Additive arithmetical functions and statistical independence. Amer. J. Math., 61, 713–721.
[365] Estermann, T. (1928). On certain functions represented by Dirichlet series. Proc. London Math. Soc., 27, 435–448.
[366] Estermann, T. (1928). On a problem of analytic continuation. Proc. London Math. Soc., 27, 471–482.
[367] Estermann, T. (1952). Introduction to Modern Prime Number Theory. Cambridge Univ. Press.

[368] Everest, G. and Ward, T. (2005). An Introduction to Number Theory. Springer.

[369] Farmer, D. W., Gonek, S. M., and Hughes, C. P. (2007). The maximum size of L-functions. J. Reine Angew. Math., 609, 215–236.

[370] Farr, R. E., Pauli, S., and Saidak, F. (2018). On fractional Stieltjes constants. Indag. Math., 29, 1425–1431.

[371] Fazzari, A. (2021). Weighted value distributions of the Riemann zeta function on the critical line. Forum Math., 33. 579–592.

[372] Fazzari, A. (2021). A weighted central limit theorem for $\log|\zeta(1/2+it)|$. Mathematika, 67, 324–341.

[373] Feller, W. (1970, 1971). An Introduction to Probability Theory and its Applications, Vols. I and II. Wiley, NY.

[374] Feng, S. (2012). Zeros of the Riemann zeta-function on the critical line. J. Number Theory, 132, 511–542.

[375] Feng, S. and Wu, X. (2012). On gaps between zeros of the Riemann zeta-function. J. Number Theory, 132, 1385–1397.

[376] Feng, S. and Wu, X. (2013). On large spacing of the zeros of the Riemann zeta-function. J. Number Theory 133, 2538–2566.

[377] Ferenczi, S., Kułaga-Przymus, J., and Lemanćzyk, M. (2018). Sarnak's conjecture - what's new. In: Ergodic Theory and Dynamical Systems in their Interactions with Arithmetics and Combinatorics, Springer, 163–235.

[378] Feuerverger, A. and Martin, G. (2000). Biases in the Shanks-Rényi prime numbers race. J. Experimental Math., 9, 535–570.

[379] Filaseta, M. (1992). Squarefree values of polynomials. Acta Arith., 60, 213–231.

[380] Finch, S. R. (2003). Mathematical Constants. Cambridge Univ. Press.

[381] Fine, B. and Rosenberger, G. (2016). Number theory: An introduction via the Density of Primes, 2nd ed. Birkhäuser/Springer.

[382] Fisher, S. D. (1990). Complex Variables, 2nd ed. Reprinted by Dover, 1999.

[383] Ford, K. (1998). The distribution of totients. Ramanujan J., 2, 67–151.

[384] Ford, K., Green, B., Konyagin, S., Maynard, J., and Tao, T. (2018). Long gaps between primes. J. Amer. Math. Soc., 31, 65–105.

[385] Ford, K., Green, B., Konyagin, S., Maynard, J., and Tao, T. (2018). Long gaps between primes. J. Amer. Math. Soc., 31, 65–105.

[386] Fornaess, J. E., Forstnerič, F., and Wold, E. F. (2020) Holomorphic approximation: The legacy of Weierstrass, Runge, Oka-Weil, and Mergelyan. Springer, 133–192.

[387] Fornberg, B. and Piret, C. (2020). Compex Analysis and Analytic Functions, an Illustrated Introduction. S.I.A.M.

[388] Forrester, P. J. (2010). Log-gases and Random Matrices. Princeton Univ. Press.

[389] Forrester, P. J., Snaith, N. C., and Verbaarschot, J. J. M. (2003). Developments in random matrix theory. J. of Physics, Mathematical and General. Special Issue on Random Matrix Theory, A 36, 2859–3646 and R1–R10.

[390] Foster, D. and Williams, D. (1978). The Hawkins sieve and Brownian motion. Composito Mathematica, 37, 277–289.

[391] Franel, J. (1924). Les suites de Farey et le probleme des nombres premiers. Gott. Nachr., 1924, 198–201.

[392] Freiberg (2018). A note on the distribution of primes in intervals. Springer, 23–44.

[393] Friedlander, J. B. (2015). Prime numbers: A much needed gap is finally found. Notices Amer. Math. Soc., 62, 660–664.

[394] Frisch, O. R. (1974). Somebody turned the sun on with a switch. Science & Public Affairs, Bulletin of the Atomic Scientists, vol. XXX, April, 13–18, p. 17.

[395] Fröberg, C.-E. (1968). On the prime zeta function. Nordisk Tidskr. Informationsbehandling (BIT), 8, 187–202.

[396] Fujii, A. (1975). On the distribution of the zeros of the Riemann zeta-function in short intervals. Bull. Amer. Math. Soc., 81, 139–142.

[397] Fujii, A. (1981). On the zeros of Dirichlet L-functions, II. Trans. Amer. Math. Soc., 267, 33–40.

[398] Fujii, A. (1990). On the gaps between consecutive zeros of the Riemann zeta function. Proc. Japan Acad., 66, Ser.Ã, 97–100.

[399] Fujii, A. (2004). An explicit estimate in the theory of the distribution of the zeros of the Riemann zeta function. Comment. Math. Univ. St. Pauli 53, 85–114.

[400] Fyodorov, Y. V., Hiary, G. A., and Keating, J. P. (2012). Freezing transitions, characteristic polynomials of random matrices, and the Riemann zeta-function. Phys. Rev. Lett., 108,1–4.

[401] Fyodorov, Y. V. and Keating, J. P. (2014). Freezing transitions and extreme values: Random matrix theory, $\zeta(1/2+it)$, and disordered landscapes. Philos. Trans. Royal Soc. London, A. 372, 1–32.

[402] Galambos, J. (1970). Distribution of arithmetical functions. A survey. Ann. de l'Inst. H. Poincaré Sect. B, 6, 281–305.

[403] Galambos, J. (1976). The sequence of prime divisors of integers. Acta Arith., 31, 213–218.

[404] Galambos, J. and Szüsz, P. (1986). On the distribution of multiplicative arithmetical functions. Acta Arith., 47, 57–62.

[405] Gallagher, P. X. (1976). On the distribution of primes in short intervals. Mathematika, 23, 4–9. Corrigendum 1981, 28, 86.

[406] Galton, F. (1889). Natural Inheritance. Macmillan & Co., NY, p. 66.

[407] Gamelin, T. W. (2001). Complex Analysis. Springer.

[408] Ganguli, H. S. (2013). On the correlation of completely multiplicative functions. PhD Thesis, Simon Fraser Univ. (Canada).

[409] Garcia-Pérez, G., Serrano, A. M., and Boguñá, M. (2014). The complex architecture of primes and natural numbers. Phys. Rev., E 90.

[410] Gardiner, V., Lazarus, R., Metropolis, N., and Ulam, S. (1956). On certain sequences of integers defined by sieves. Math. Magazine, 29, 117–122.

[411] Gaudin, M. L. (1961). Sur la loi limite de l'espacement des valeurs propres d'une matrice aléatoire. Nulclear Physics, 25, 447–458.

[412] Germain, S. (1819). S. Germain to Gauus. May 12, 1819.

[413] Gerspach, M. and Lamzouri, Y. (2022). Low pseudomoments of Euler products. Quart. J. Math., 73, 517–537

[414] Ghosh, A. (1983). On the Riemann zeta function – Mean value theorems and the distribution of $|S(T)|$. J. Number Th., 17, 93–102.

[415] Ginibre, J. (1965). Statistical ensembles of complex, quaternion, and real matrices. J. Mathematical Phys., 6, 440–449.

[416] Girko, V. L. (1984). On the circle law. Theory Probab. Math. Statist., 28, 15–23.

[417] Gnedenko, B. V. and Kolmogorov, A. N. (1968). Limit Distributions for Sums of Independent Random Variables. Addison-Wesley, Reading, MA. Translated from the 1949 Russian and revised by K. L. Chung.

[418] Goldberg, R. R. (1961). Fourier Transforms. Cambridge Univ. Press.

[419] Goldfeld, D. (2003). The elementary proof of the prime number theorem: An historical perspective. In: D. Chudnovsky, G. Chudnovsky, and M. B. Nathanson (eds.), Number Theory (NY, 2003). Springer.

[420] Goldstein, L. J. (1973). A history of the prime number theorem. Amer. Math. Monthly, 80, 599–615.

[421] Goldston, D. A. (1987). On the function $S(T)$ in the theory of the Riemann zeta-function. J. Number Theory, 27, 149–177.

[422] Goldston, D. A. (1988). On the pair correlation conjecture for the zeros of the Riemann zeta-function. J. Reine Angew. Math., 385, 24–40.

[423] Goldston, D. A. (2005). Notes on the pair correlation of zeros and prime numbers. In: Recent Perspectives in Random Matrix Theory and Number Theory. F. Mezzadri and N. Snaith, eds., Cambridge Univ. Press, 79–110.

[424] Goldston, D. A. and Gonek, S. M. (2007). A note on $S(t)$ and the zeros of the Riemann zeta-function. Bull. London Math. Soc., 39, 482–486.

[425] Goldston, D. A., Gonek, S. M., Ozluk, A. E., and Snyder, C. (2000). On the pair correlation of zeros of the Riemann Zeta-function. Proc. London Math. Soc., 80, 31–49.

[426] Goldston, D. A. and Heath-Brown, D. R. (1984). A note on the difference between consecutive primes. Math. Ann., 266, 317–320.

[427] Goldston, D. A., Motohashi, Y., Pintz, J., and Yildirim, C. Y. (2006). Small gaps between primes exist. Proc. Japan Acad., Series A, Math. Sci., 82, 61–65.

[428] Goldston, D. A., Pintz, J., and Yildirim, C. Y. (2007). The path to recent progress on small gaps between primes. Analytic Number Theory, Clay Math. Proceedings, 7, 129–139. Amer. Math. Soc., Providence, RI.

[429] Goldston, D. A., Pintz, J., and Yildirim, C. Y. (2009). Primes in tuples I. Annals Math., 170, 819–862.

[430] Goldston, D. A., Pintz, J., and Yildirim, C. Y. (2010). Primes in tuples II. Acta Math., 204, 1–47.

[431] Goldston, D. A., Pintz, J., and Yildirim, C. Y. (2013). Primes in tuples IV: Density of small gaps between consecutive primes. Acta Arith., 160, 37–53.

[432] Goldston, D. A., Pintz, J., and Yildirim, C. Y. (2014). Small gaps between primes. Proc. Int. Congress of Mathematicians – Seoul 2014, Vol. II, Kyung Moon Sa, Seoul, 419–441.

[433] Goldston, D. A., Trudgian, T. S., and Turnage-Butterbaugh, C. L. (2023). On the Montgomery-Odlyzko method regarding gaps between zeros of the zeta-function. J. Math. Anal. Appl., 527, Paper No. 127548, 7 pp.

[434] Goldston, D. A. and Turnage-Butterbaugh, C. L. (2019). A note on small gaps between zeros of the Riemann zeta-function. Preprint: `arXiv:1904.06001`.

[435] Golomb, S. W. (1962). On the ratio of N to $\pi(N)$. Amer. Math. Monthly, 69, 36–37.

[436] Golomb, S. W. (1970). A class of probability distributions on the integers. J. Number Theory, 2, 189–192.

[437] Gomilko, A., Kwietniak, D., and Lemańczyk, M. (2018). Sarnak's conjecture implies the Chowla conjecture along a subsequence. In: Ergodic Theory and Dynamical Systems in their Interactions with Arithmetics and Combinatorics, S. Ferenczi et al. eds., Springer, 237–247.

[438] Gonek, S. M. (1984). Mean values of the Riemann zeta-function and its derivatives. Inventiones Math, 75, 123–141.

[439] Gonek, S. M. (2012). Finite Euler products and the Riemann hypothesis. Trans. Amer. Math. Soc., 364, 2157–2191.

[440] Gonek, S. M., Graham, S., and Lee, Y. (2020). The Lindelöf hypothesis for primes is equivalent to the Riemann hypothesis. Proc. Amer. Math. Soc., 148, 2863–2875.

[441] Gonek, S. M., Hughes, C. R., and Keating, J. P. (2007). A hybrid Euler–Hadamard product for the Riemann zeta function. Duke Math. J., 136, 507–549.

[442] Good, I. J. and Churchhouse, R. F. (1968). The Riemann hypothesis and pseudorandom features of the Möbius sequence. Math. Comp., 22, 857–861.

[443] Gorodetsky, O. and Wong, M. D. (2024). Martingale central limit theorem for random multiplicative functions. `arXiv:2405.20311`.

[444] Gorodetsky, O. and Wong, M. D. (2025). On the limiting distribution of sums of random multiplicative functions. `arXiv:2508.12956`.

[445] Gourdon, X. and Demichel, P. (2004). Computation of zeros of the zeta function. `http://numbers.computation.free.fr/Constants/Miscellaneous/zetazeroscompute.html`

[446] Gourdon, X. and Sebah, P. (2003). The pHxL project. `http://numbers.computation.free.fr/Constants/constants.html`

[447] Gradshteyn, I. S. and Ryzhik, I. M. (2014). Tables of Integrals, Series, and Products, 28th ed. Elsevier, Amsterdam.

[448] Graham, S. W. and Hensley, D. (1986 Solution to problem E3025. Amer. Math. Monthly, 93, 655.

[449] Gram, J. P. (1903). Sur les zéros de la fonction $\zeta(s)$ de Riemann. Acta Math., 27, 289–304.

[450] Granville, A. J. (1995). Harold Cramér and the distribution of prime numbers. (Harold Cramér Symposium, Stockholm, 1993). Scand. Actuarial J., 1, 12–28.

[451] Granville, A. J. (1995). Unexpected irregularities in the distribution of prime numbers. Proc. Int. Congress of Mathematicians (Zürich 1994), Vol. 1, 2, 388–399. Birkhäuser, Basel.

[452] Granville, A. J. (2008). Smooth numbers: Computational number theory and beyond. In: Algorithmic Number Theory: Lattices, Number Fields, Curves and Cryptography. MSRI Publications, vol. 44, Cambridge Univ. Press, 267–323.

[453] Granville, A. J. (2008). Prime number patterns. Amer. Math. Monthly, 115, 279–296.

[454] Granville, A. J. (2009). Pretentiousness in analytic number theory. J. Théor. Nombres Bordeaux, 21, 159–173.

[455] Granville, A. J. (2015). Primes in intervals of bounded length. Bull. Amer. Math. Soc., 52, 171–222.

[456] Granville, A. J. (2017). A Panoply of proofs that there are infinitely many primes London Math. Soc. Newsletter, 472, 23–27.
[457] Granville, A. J. (2025). Analytic Number Theory Revealed: A First Guide to Prime Numbers. Amer. Math. Soc., Providence, RI.
[458] Granville, A. J. and Martin, G. (2006). Prime number races. Amer. Math. Monthly, 113, 1–33.
[459] Granville, A. J. and Soundararajan, K. (2006). Extreme values of $|\zeta(1+it)|$. In: The Riemann Zeta Function and Related Themes (in honour of K. Ramachandra). Ramanujan Math. Soc. Lect. Notes Ser., 2, 65–80.
[460] Granville, A. J. and Soundararajan, K. (2007). Sieving and the Erdős-Kac theorem. In: A. Granville and Z. Rudnick, eds., Equidistribution in Number Theory, An Introduction (Montréal 2005). Proc. NATO Advanced Study Inst. on Equidistribution in Number Theory, Montréal, 2005. Springer, 15–27.
[461] Granville, A. J. and Soundararajan, K. (2007). An uncertainty principle for arithmetic sequences. Ann. Math., 165, 593–635.
[462] Granville, A. J. and Tucker, T. (2002). It's as easy as *abc*. Notices Amer. Math. Soc., 49, 1224–1231.
[463] Gray, R. M. and Goodman, J. W. (1995). Fourier Transforms. Kluwer.
[464] Green, B. and Tao, T. (2008). The primes contain arbitrarily long arithmetic progressions. Ann. Math., 167, 481–547.
[465] Green, B. and Tao, T. (2010). Linear equations in primes. Ann. Math., 171, 1753–1850.
[466] Gritsenko, S. A. (1997). On zeros of linear combinations of analogues of the Riemann function. Tr. Mat. Inst. Steklova 218, 134–150 (Russian); translation in Proc. Steklov Inst. Math., 218, 129–145.
[467] Gumbel, E. J. (1954). Statistical Theory of Extreme Values and Some Practical Applications. Applied Math. Series, Vol. 33, US Dept. Commerce, National Bureau of Standards.
[468] Gut, A. (2006). Some remarks on the Riemann zeta distribution. Revue Roumaine Math. Pures Appl., 51, 205–217.
[469] Gut, A. (2009). An Intermediate Course in Probability. 2nd ed. Springer.
[470] Gut, A. (2013). Probability: A Graduate Course, 2nd ed. Springer.
[471] Gut, A. and Spătaru, A. (2000). Precise asymptotics in the law of the iterated logarithm. Ann. Probab., 28, 1870–1883.
[472] Guth, L. and Maynard, J. (2024). New large value estimates for Dirichlet polynomials. `arXiv:2405.20552`.
[473] Guy, R. K. (1975). How to factor a number. Proc. Fifth Manitoba Conf. Numerical Math., 49–89.
[474] Guy, R. K. (2004). Unsolved Problems in Number Theory, 3rd ed. Springer.
[475] Hadamard, J. (1896). Sur la distribution des zéros de la fonction $\zeta(s)$ et ses consequences arithmétiques. Bull. Soc. Math. France, 14, 199–220.
[476] Halász, G. (1968). Über die Mittelwerte multiplikativer zahlentheoretischer Funktionen. Acta Math. Acad. Sci. Hungarica., 19, 365–403.
[477] Halász, G. (1971). On the distribution of additive and the mean values of multiplicative arithmetic functions. Studia Sci. Math. Hungarica, 211–233.
[478] Halász, G. (1983). On random multiplicative functions. In: Proc. Hubert Delange Colloq. (Orsay 1982), Publ. Math. d'Orsay, 83, Univ. Paris XI, Orsey, 74–96.
[479] Halberstam, H. (1955–1956). On the distribution of additive number-theoretic functions, I, II, III. J. London Math. Soc., 30, 43–53; 31, 1–14 and 14–27.
[480] Halberstam, H. and Richert, H. E. (1974). Sieve Methods. Academic Press, London. Dover Publications has a 2010 edition.
[481] Hall, P. G. and Heyde, C. C. (1980). Martingale Limit Theory and its Application. Academic Press, NY.
[482] Hall, R. R. (1999). The behaviour of the Riemann zeta-function on the critical line. Mathematika, 46, 281–313.
[483] Hall, R. R. (2008). Large spaces between the zeros of the Riemann zeta-function and random matrix theory. II. J. Number Theory, 128, 2836–2851.
[484] Hanson, D. L. (1972). On the product of primes. Canadian Math. Bull., 15, 33–37.
[485] Hardy, G. H. (1914). Sur les zéros de la fonction $\zeta(s)$ de Riemann. C.R. Acad. Sci. Paris, 158, 1012–1014.

[486] Hardy, G. H. (1920). Some famous problems in the theory of numbers. Inaugural Lecture, Oxford Univ. Press.

[487] Hardy, G. H. (1929). An introduction to the theory of numbers. Bull. Amer. Math. Soc., 35, 778–818. (Also in Albers et al.)

[488] Hardy, G. H. (1937). The Indian Mathematician Ramanujan. Amer. Math. Monthly, 44,137–155. (Also in Albers et al.)

[489] Hardy, G. H. (1940). Ramanujan: Twelve lectures on Subjects Suggested by His Life and Work. Cambridge Univ. Press.

[490] Hardy, G. H. (1949). Divergent Series. Oxford Univ. Press.

[491] Hardy, G. H. (1967). A Mathematicians Apology, with a foreword by C. P. Snow. Cambridge Univ. Press.

[492] Hardy, G. H. and Littlewood, J. E. (1916). Contributions to the theory of the Riemann zeta-function and the theory of the distribution of primes. Acta Math., 41, 119–196.

[493] Hardy, G. H. and Littlewood, J. E. (1921). The zeros of Riemann's zeta function on the critical line. Math. Zeitschrift, 10, 283–317.

[494] Hardy, G. H. and Littlewood, J. E. (1923). On Lindelöf's hypothesis concerning the Riemann zeta-function. Proc. Royal Soc. England, A103, 403–412.

[495] Hardy, G. H. and Littlewood, J. E. (1923). Some problems of 'Partitio Numerorum', III: On the expression of a number as a sum of primes. Acta Math., 44, 1–70.

[496] Hardy, G. H. and Littlewood, J. E. (1924). Some Problems of 'Partitio Numerorum' (V): A Further Contribution to the Study of Goldbach's Problem. Proc. London Math. Soc., 22, 46–56.

[497] Hardy, G. H. and Ramanujan, S. (1917). The normal order of prime factors of a number n. Quart. J. Math., 48, 76–92.

[498] Hardy, G. H. and Riesz, M. (1915). The General Theory of Dirichlet's Series. Cambridge Univ. Press. (Reprinted 1964 by Stechert-Hafner, NY.)

[499] Hardy, G. H. and Wright, E. M. (2008). An Introduction to the Theory of Numbers, 6th ed. Updated by D. R. Heath-Brown & J. H. Silverman. Oxford Univ. Press. (Previous editions were in 1938, 1945, 1954, 1960, and 1969.)

[500] Hardy, S. (2025). The distribution of partial sums of random multiplicative functions with a large prime factor. ArXiv.

[501] Hargittai, M. and Hargittai, I. (2020). Memorials of mathematicians in Moscow. Math. Intelligencer, 42, 34–40.

[502] Harman, G., Pintz, J., and Wolke, D. (1985). A note on the Möbius and Liouville functions. Studia Scientiarum Mathematicarum Hungarica, 20, 295–299.

[503] Harper, A. (2009). Two new proofs of the Erdős-Kac Theorem, with bound on the rate of convergence, by Stein's method for distributional approximations. Math. Proc. Cambridge Phil. Soc., 147, 95–114.

[504] Harper, A. J. (2013). Bounds on the suprema of Gaussian processes, and omega results for the sum of a random multiplicative function. Ann. Appl. Probab., 23, 584–616.

[505] Harper, A. J. (2013). On the limit distributions of some sums of a random multiplicative function. J. Reine Angew. Math., 678, 95–124.

[506] Harper, A. J. (2013). Sharp conditional bounds for moments of the zeta function. `arXiv:1305.4618`.

[507] Harper, A. J. (2013). A note on the maximum of the Riemann zeta function, and log-correlated random variables. Preprint, pp. 1–26. `arXiv:1304.0677`.

[508] Harper, A. J. (2019). Moments of random multiplicative functions, II: High moments. Algebra Number Theory, 13, 2277–2321.

[509] Harper, A. J. (2019). On the partition function of the Riemann zeta function, and the Fyodorov–Hiary–Keating conjecture. `arXiv:1906.05783`.

[510] Harper, A. J. (2019). The Riemann zeta function in short intervals [after Najnudel, and Arguin, Belius, Bourgade, Radziwiłł and Soundararajan]. Astérisque, 422, Séminaire Bourbaki, Vol. 2018/2019. Exposes 1151–1165, 391–414. Alternately: Séminaire Bourbaki, 71e annee, no 1159.

[511] Harper, A. J. (2020). Moments of random multiplicative functions, I: Low moments, better than squareroot cancellation, and critical multiplicative chaos. Forum Math., Pi 8, e1, 95 pp.

[512] Harper, A. J. (2023). Almost sure large fluctuations of random multiplicative functions. Int. Math. Res. Not., no. 3, 2095–2138.

[513] Harper, A. J. (2024). Moments of random multiplicative functions, III: A short review. arXiv:2410.11523.

[514] Harper, A. J., Nikeghbali, A., and Radziwiłł, M. (2015). A note on Helson's conjecture on moments of random multiplicative functions. In: Analytic Number Theory: In Honor of Helmut Maier's 60th Birthday. Springer, 145–169.

[515] Hartman, P. and Wintner, A. (1941). On the law of the iterated logarithm. Amer. J. Math., 63, 169–176.

[516] Haselgrove, C. B. (1958). A disproof of a conjecture of Pólya. Mathematika, 5, 141–145.

[517] Havil, J. (2003). Gamma: Exploring Euler's Constant. Foreword by Freeman Dyson. Princeton Univ. Press.

[518] Hawkins, D. (1957). The random sieve. Math. Magazine, 31, 1–3.

[519] Hawkins, D. (1974). Random sieves, II. J. Number Theory, 6, 192–200.

[520] Hawkins, D. (1987). The spirit of play: A memoir for Stan Ulam. Los Almos Science. Special Issue, 39–51.

[521] Hayes, B. (2003). The spectrum of Riemannium. American Scientist, 91, 296–300.

[522] Heap, W. and Lindqvist, S. (2016). Moments of random multiplicative functions and truncated characteristic polynomials. Quart, J. Math., 67, 683–714.

[523] Heap, W., Radziwiłł, M., and Soundararajan, K. (2019). Sharp upper bounds for fractional moments of the Riemann zeta function. Quart. J. Math., 70, 1387–1396.

[524] Heap, W. and Soundararajan, K. (2022). Lower bounds for moments of zeta and L-functions revisited. Mathematika, 68, 1–14.

[525] Heath-Brown, D. R. (1978). The twelfth power moment of the Riemann zeta function. Quart. J. Math. Oxford, 29, 443–462.

[526] Heath-Brown, D. R. (1979). The fourth power moment of the Riemann zeta function. Proc. London Math. Soc., 38, 385–422.

[527] Heath-Brown, D. R. (1980). Differences between consecutive primes. Univ. Bordeaux I, Talence, Exp. No. 14, 7 pp.

[528] Heath-Brown, D. R. (1981). Fractional moment of the Riemann zeta-function. J. London Math. Soc., 24, 65–78.

[529] Heath-Brown, D. R. (1982). Prime numbers in short intervals and a generalized Vaughan identity. Canadian J. Math., 34, 1365–1377.

[530] Heath-Brown, D. R. (1988). Differences between consecutive primes. Jahresber. Deutsch. Math.-Verein., 90, 71–89.

[531] Heath-Brown, D. R. and Goldston, D. A. (1984). A note on the difference between consecutive primes. Math. Ann., 266, 317–320.

[532] Hejhal, D. A. (1989). On the distribution of $\log \zeta'(1 + it)$. In: Number Theory, Trace Formulas and Discrete Groups: Symposium in Honor of Atle Selberg, July 14–21, 1987. K. E. Aubert, E. Bombieri, and D. Goldfeld, eds. Academic Press, Boston, 343–370.

[533] Hejhal, D. A. (1990). On a result of G. Pólya concerning the Riemann ζ-function. J. Analyse Math., 55, 59–95.

[534] Hejhal, D. A. and Odlyzko, A. M. (2013). Alan Turing and the Riemann zeta function. In: Alan Turing – His Work and Impact, S. Barry Cooper and J. van Leeuwen, eds., Elsevier, 265–279.

[535] Helfgott, H. A. (2014). The ternary Goldbach problem. Proc. Int. Congress Mathematicians, Vol. II, Kyung Moon Sa, Seoul, 2014, 391–418.

[536] Helfgott, H. A. and Platt, D. J. (2013). Numerical verification of the ternary Goldbach conjecture up to $8.875 \cdot 10^{30}$. Exp. Math., 22, 406–409.

[537] Helfgott, H. A. and Radziwiłł, M. (2021). Expansion, divisibility and parity. arXiv:2103.06853 [math.NT].

[538] Helson, H. B. (2006). Hankel forms and sums of random variables. Studia Mathematica, 176, 85–92.

[539] Helson, H. B. (2010). Hankel forms. Studia Mathematica, 198, 79–84.

[540] Henrici, P. (1977). Applied and Computational Complex Analysis, Vols. 1 and 2. Wiley, NY.

[541] Hensley, D. and Richards, I. (1973). On the incompatibility of two conjectures concerning primes. In: Analytic Number Theory, Proc. Sympos. Pure Math., 24, Amer. Math. Soc., Providence, RI, 123–127.
[542] Hewitt, E. and Williamson, H. (1957). Note on absolutely convergent Dirichlet series. Proc. Amer. Math. Soc., 8, 863–868.
[543] Heyde, C. C. (1969). A note concerning behaviour of iterated logarithm type. Proc. Amer. Math. Soc., 23, 85–90.
[544] Heyde, C. C. (1976). On asymptotic behaviour for the Hawkins random sieve. Proc. Amer. Math. Soc., 56, 277–280.
[545] Heyde, C. C. (1978). A log log improvement to the Riemann Hypothesis for the Hawkins random sieve. Ann. Probab., 6, 870–875.
[546] Hiary, G. A., Patel, D., and Yang, A. (2022). An improved explicit estimate for $\zeta(1/2+it)$. arXiv:2207.02366.
[547] Hilbert, D. (1900). Mathematische Probleme. In: Nachrichten von der Gesellschaft der Wissenschaften zu Göttingen, Mathematisch-Physikalishce Klasse, vol. 1900, 253–297. https://gdz.sub.uni-goettingen.de/id/PPN252457811_1900
[548] Hildebrand, A. J. (1983). An asymptotic formula for the variance of an additive function. Math.Z̃., 183, 145–170.
[549] Hildebrand, A. J. (1984). Integers free of large prime factors and the Riemann hypothesis. Mathematika, 31, 258–271.
[550] Hildebrand, A. J. (1986). On consecutive values of the Liouville function. l'Enseignement Math., 32, 219–226.
[551] Hildebrand, A. J. (1986). Multiplicative functions at consecutive integers. Math. Proc. Cambridge Philos. Soc., 100, 229–236.
[552] Hildebrand, A. J. (1986). The prime number theorem via the large sieve. Mathematika, 33, 23–30.
[553] Hildebrand, A. J. (1997). Multiplicative properties of consecutive Integers. In: Analytic number theory (Kyoto, 1966), Y. Motohashi, ed. London Math. Soc. Lecture Note Series (No. 247), Cambridge Univ. Press, 103–118.
[554] Hildebrand, A. and Maier, H. (1989). Irregularities in the distribution of primes in short intervals. J. Reine Angew. Math., 397, 162–193.
[555] Hildebrand, A. and Tenenbaum, G. (1986). On integers free of large prime factors. Trans. Amer. Math. Soc., 296, 265–290.
[556] Hildebrand, A. and Tenenbaum, G. (1993). Integers without large prime factors. J. de Théorie des Nombres de Bordeaux, 5, 411–484.
[557] Hodges, A. (2014) Alan Turing: The Enigma. Vantage (Random House).
[558] Hoel, P. G., Port, S. C., and Stone, C. J. (1971). Introduction to Probability Theory. Houghton Mifflin, Boston.
[559] Hoffman, P. (1998), The Man Who Loved Only Numbers. Hyperion, NY.
[560] Holst, L. (2001). The Poisson-Dirichlet distribution and its relatives revisited. Tech. report, Dept. Math., Royal Inst. Tech., Stockholm.
[561] Hough, R. (2011). Summation of a random multiplicative function on numbers having few prime factors. Math. Proc. Cambridge Philos. Soc., 150, 193–214.
[562] Hough, J. B., Krishnapur, M., Peres, Y., and Virag, B. (2006). Determinantal processes and independence. Probability Surveys, 3, 206–229.
[563] Hough, J. B., Krishnapur, M., Peres, Y., and Virag, B. (2009). Zeros of Gaussian Analytic Functions and Determinantal Point Processes. Amer. Math. Soc.
[564] Hu, C., Iksanov, A. M., Lin, G. D., and Zakusylo, O. K. (2006). The Hurwitz zeta distribution. Aust. N.Z. J. Stat., 48, 1–6.
[565] Hughes, B. D. (1996). Random Walks and Random Environments, vols. 1 & 2. Clarendon Press, Oxford.
[566] Hughes, C. P. and Pearce-Crump, A. (2022). A discrete mean-value theorem for the higher derivatives of the Riemann zeta function. J. Number Theory, 241, 142–164.
[567] Humphries, P. (2012). The distribution of weighted sums of the Liouville function and Pólya's conjecture. J. Number Theory 133, 545–582.
[568] Hurst, G. (2018). Computations of the Mertens function and improved bounds on the Mertens conjecture. Math. Comp., 87, 1013–1028.

[569] Huxley, M. N. (1972). On the difference between consecutive primes. Inventiones Mathematicae, 15, 164–170.
[570] Huxley, M. N. (1972). The Distribution of Prime Numbers: Large Sieves and Zero Density Theorems. Clarendon Press, Oxford.
[571] Huxley, M. N. (1977). Small differences between consecutive primes, II. Mathematika, 24, 142–152.
[572] Huxley, M. N. (2003). Exponential sums and lattice points, III. Proc. London Math. Soc., 87, 591–609.
[573] Huxley, M. N. (2005). Exponential sums and the Riemann zeta function, V. Proc. London Math. Soc., 90, 1–41.
[574] Huxley, M. N. and Ivić, A. (2007). Subconvexity for the Riemann zeta-function and the divisor problem. Bulletin CXXXIV de l'Académie Serbe des Sciences et des Arts, Classe des Sciences Mathématiques et Naturelles, Sciences Mathématiques, 32, 13–32. Available at: `arXiv:math/0611809`
[575] Ibragimov, I. and Linnik, Y. (1971). Independent and Stationary Sequences of Random Variables. Wolters-Noordhoff, Groningen. (Addison-Wesley, Reading, MA, 1968; 1965 Nauka.)
[576] Indlekofer, K.-H. (1981). Limiting distributions and mean-values of multiplicative arithmetical functions. J. Reine Angew. Math., 328, 116–127.
[577] Indlekofer, K.-H. (1984). On multiplicative arithmetical functions. iN: Topics in Classical Number Theory, Vols. I, II. Coll. Math. Soc. János Bolyai, 34, Budapest, (G. Halász, ed.), 731–748.
[578] Indlekofer, K.-H. (2002). Number Theory – Probabilistic, Heuristic, and Computational Approaches. Comput. Math. Appl., 43, 1035–1061.
[579] Ingham, A. E. (1926). Mean-value theorems in the theory of the Riemann zeta-function. Proc. London Math. Soc., 27, 273–300.
[580] Ingham, A. E. (1932). The Distribution of Prime Numbers. Reissued 1964 by Stechert-Hafner, NY, and 1990 by Cambridge Univ. Press.
[581] Ingham, A. E. (1937). On the difference between consecutive primes. Quart. J. Math. Oxford, 8, 255–266.
[582] Ingham, A. E. (1940). On the estimation of $N(\sigma, T)$. Quart. J. Math., 11, 291–292.
[583] Ingham, A. E. (1942). On two conjectures in the theory of numbers. Amer. J. Math., 64, 313–319.
[584] Inoue, S. (2019). On the logarithm of the Riemann zeta-function and its iterated integrals. `arXiv:1909.03643`.
[585] Ishikawa, H. (1934). Über die verteilung der primzahlen. Sci. Rep. Tokyo Univ., A2, 27–40.
[586] Ivić, A. (1985). The Riemann Zeta-Function: The Theory of the Riemann Zeta-Function with Applications. Wiley, NY. Reprinted by Dover, 2003.
[587] Ivić, A. (1991). Lectures on Mean Values of the Riemann Zeta-Function. Lectures on Mathematics and Physics, Vol. 82, Tata Inst. of Fundamental Research, Bombay. Springer.
[588] Ivić, A. (1995). On sums of gaps between the zeros of $\zeta(s)$ on the critical line. Univ. Beograd. Publ., Elektrotehn. Fak. Ser. Mat., 6, 55–62.
[589] Ivić, A. (2003). On some reasons for doubting the Riemann hypothesis. `arXiv:math/0311162`.
[590] Ivić, A. (2013). The Theory of Hardy's Z-Function. Cambridge Univ. Press.
[591] Ivić, A. (2018). Some remarks on the differences between ordinates of consecutive zeta zeros. Funct. Approx. Comment. Math., 58, 23–35.
[592] Iwaniec, H. (1978). Almost-primes represented by quadratic polynomials. Inventiones Mathematicae, 47, 171–188.
[593] Iwaniec, H. (2014). Lectures on the Riemann Zeta Function. Amer. Math. Soc., Providence, RI.
[594] Iwaniec, H. and Kowalski, E. (2004). Analytic Number Theory. Amer. Math. Soc., Providence, RI.
[595] Iwaniec, H. and Mozzochi, C. J. (1988). On the Divisor and Circle Problem. J. Numb. Th., 29, 60–93.
[596] Izenman, A. J. (2021). Random matrix theory and its applications. Statistical Science, 36, 421–442.

[597] Jabbarov, I. Sh. and Hasanova, G. K. (2023). On asymptotic formula for the first moment of the Riemann zeta-function on the critical line (in Russian). arXiv:1804.08448.

[598] Jain, D. V. S., Mehta, M. L., and des Cloizeaux, J. (1972). The probabilities for several consecutive eigenvalues of a random matrix. Indian J. Pure Appl. Math., 3, 329–351.

[599] Jensen, J. L. W. V. (1916). An elementary exposition of the theory of the Gamma function. Annals Math., 17, 124–166.

[600] Jessen, B. and Winter, A. (1935). Distribution functions and the Riemann zeta function. Trans. Amer. Math. Soc., 38, 48–88.

[601] Jia, C. (1993). The distribution of square-free numbers. Sci. China, A36, 154–169.

[602] Johansson, K. (2004). Determinantal processes with number variance saturation. Commun. Math. Phys., 252, 111–148.

[603] Johansson, K. (2006). Random matrices and determinantal processes. Mathematical Statistical Physics, Elsevier, Amsterdam, 1–55.

[604] Johnson, N. L., Kotz, S., and Kemp, A. W. (1992). Univariate Discrete Distributions, 2nd ed. Wiley, NY.

[605] Johnstone, I. M. (2007). High dimensional statistical inference and random matrices. European Math. Soc., Zürich, 307–333.

[606] Jörgensen, B. (1986). Some properties of exponential dispersion models. Scand. J. Statistics., 13, 187–197.

[607] Jörgensen, B. (1997). The Theory of Dispersion Models. Chapman and Hall, London.

[608] Joyner, D. (1992). On the Dyson-Montgomery hypothesis. Proc. Amalfi Conf. Analytic Number Theory, 1989, Univ. Salerno, 257–261.

[609] Jurkat, W. and Peyerimhoff, A. (1976). A constructive approach to Kronecker approximation and its application to the Mertens conjecture. J. Riene Angew. Math., 286/287, 322–340.

[610] Jutila, M. (1983). Zeros of the zeta-function near the critical line. In: P. Erdős et al., eds., Studies in Pure Mathematics, to the Memory of Paul Turán. Birkhäuser, 385–394.

[611] Jutila, M. (1983). On the value distribution of the zeta-function on the critical line. Bull. London Math. Soc., 15, 513–518.

[612] Kac, M. (1949). Probability methods in some problems of analysis and number theory. Bull. Amer. Math. Soc., 55, 641–665.

[613] Kac, M. (1959). Statistical Independence in the Theory of Numbers. Wiley, NY.

[614] Kac, M. (1979). Mark Kac: Probability, Number Theory, and Statistical Physics, Selected Papers. K. Baclawski & M. D. Donsker, eds., M.I.T. Press.

[615] Kac, M. (1985). Enigmas of Chance: An Autobiography. Harper and Row, NY. Also Univ. of California Press, 1987.

[616] Kaczorowski, J. (1984, 1985). On sign-changes in the remainder-term of the prime-number formula, I & II. Acta Arith., 44, 365–377, and 45, 65–74.

[617] Kadiri, H. (2005). Une région explicite sans zéros pour la fonction ζ de Riemann (in French). Acta Arith., 117, 303–339.

[618] Kahane, J. P. (1997). A Fourier formula for prime numbers. Canadian Math. Soc. Conf. Proc., 21, 89–102.

[619] Kanigel, R. (1991). The Man Who Knew Infinity: The Life of the Genius Ramanujan. Scribner's, NY.

[620] Kaplan, W. (1966). Introduction to Analytic Functions. Addison-Wesley.

[621] Karatsuba, A. A. and Korolev, M. A. (2005). The argument of the Riemann zeta function. Russian Math. Surveys, 60, 433–488.

[622] Karatsuba, A. A. and Voronin, S. M. (1992). The Riemann Zeta-Function. Transl. from the Russion by N. Koblitz. de Gruyter, Berlin. Reprinted 2011.

[623] Kátai, I. (1969). On the distribution of arithmetical functions. Acta. Math. Acad. Sci. Hungarica., 20, 69–87.

[624] Kátai, I. (1996). Research Problems in Number Theory II. Ann. Univ. Sci. Budapest, Sect. Comput., 223–251.

[625] Katznelson, Y. (2004). An Introduction to Harmonic Analysis, 3rd ed. Cambridge Univ. Press.

[626] Keating, J. P., Rodgers, B. W., Roditty-Gershon, E., and Rudnick, Z. (2018). Sums of divisor functions in $F_q(t)$ and matrix integrals. Math. Zeitschrift, 288, 167–198.

[627] Keating, J. P. and Snaith, N. C. (2000). Random matrix theory and $\zeta(1/2+it)$. Comm. Math. Phys., 214, 57–89.

[628] Keating, J. P. and Snaith, N. C. (2000). Random matrix theory and L-functions at $s = 1/2$. Commun. Math. Phys., 214, 91–110.

[629] Keating, J. P. and Snaith, N. C. (2011). Number theory. In: The Oxford Handbook of Handom Matrix Theory, Oxford Univ. Press, 491–509.

[630] Keiper, J. B. (1992). Power series expansions of Riemann's ζ function. Math. Comp., 58, 765–773.

[631] Kelly, Z. and Meka, R. (2023). Strong bounds for 3-progressions. `arXiv:2302.05537`.

[632] Kesten, H. (1972). Sums of independent random variables–without moment conditions. Ann. Math. Statist., 43, 701–732.

[633] Khan, R. (2010). Spacings between integers having typically many prime factors. Canadian Math. Bull., 53, 102–117.

[634] Khinchin, A. Ya. (1927). Über das gesetz der grossen zahlen. Math. Ann., 96, 152–168.

[635] Khinchin, A. Ya. (1938). Limit Theorems for Sums of Indepenent Random Variables (in Russian). GONTI, Moscow and Leningrad. (A translation by S. Rogosin into English is available.)

[636] Kingman, J. F. C. (1993). Poisson Processes. Oxford Univ. Press.

[637] Klyve, D. (2007). Explicit bounds on twin primes and Brun's constant. PhD Thesis, Dartmough College, Hanover, NH.

[638] Knessl, C. and Coffey, M. W. (2011). An effective asymptotic formula for the Stieltjes constants. Math. Comp., 80, 379–386.

[639] Knight, K. (2000). Mathematical Statistics. Chapman & Hall/CRC, Boca Raton, FL.

[640] Knopp, K. (1928–1990). Theory and Application of Infinite Series. Reprinted: Hafner, 1947; Blackie, 1954; Olver, 1974; Dover, NY, 1990.

[641] Knuth, D. E. and Trabb Pardo, L. (1976). Analysis of a simple factorization algorithm. Theoretical Computer Science, 3, 321–348.

[642] von Koch, N. F. H. (1901). Sur la distribution des nombres premiers. Acta Math., 24, 159–182.

[643] Kochen, S. P. and Stone, C. J. (1964). A note on the Borel–Cantelli lemma. Illinois J. Math., 8, 248–251.

[644] Kolassa, J. E. (2006). Series Approximation Methods in Statistics, 3rd ed. Springer.

[645] Kolesnik, G. A. (1969). An improvement of the remainder term in the divisor problem. Mat. Zametki, 6, 545–554.

[646] Kolesnik, G. A. (1985). On the method of exponent pairs. Acta Arith., 45, 115–143.

[647] Kolmogorov, A. N. (1929). Über das Gesetz des iterierten Logarithmus. Math. Ann., 101, 126–135.

[648] Kondratyuk, A. and Yatsulka, P. (2008). Summation of the Riemann zeta-function logarithm on the critical line. Proc. Fourth Int. Conf. Analytic Number Theory and Spatial Tessellations "Voronoy's Impact on Modern Science", Kiev, Ukraine, 59–63.

[649] de Koninck, J.-M. and Ivić, A. (1980). Topics in Arithmetical Functions. Notas de Matematica, 72, North-Holland, Amsterdam.

[650] de Koninck, J.-M. and Luca, F. (2012). Analytic Number Theory: Exploring the Anatomy of Integers. Amer. Math. Soc, Providence, RI.

[651] de Koninck, J.-M. and Tenenbaum, G. (2002). Sur la loi de repartition du k-ieme facteur premier d'un entier. [On the distribution law of the kth prime factor of an integer.] Math. Proc. Cambridge Philos. Soc., 133, 191–204.

[652] Korevaar, J. (1982). On Newman's quick way to the Prime number Theorem. Math. Intelligencer, 4, 108–115.

[653] Korner, T. W. (1988). Fourier Analysis, 2nd ed. Cambridge Univ. Press.

[654] Korobov, M. N. (1958). Estimates of trigonometric sums and their applications (Russian). Uspekhi Matematicheskikh Nauk., 13, 185–192.

[655] Kosambi, D. D. (1963). The sampling distribution of primes. Proc. Nat. Acad. Sci. USA, 49, 20–23.

[656] Kotnik, T. and te Riele, H. J. J. (2006). The Mertens conjecture revisited. In: Algorithmic Number Theory, 156–167, Lecture Notes in Comput. Sci., 4076, Springer.

[657] Kotnik, T. and van de Lune, J. (2004). On the order of the Mertens function. Experimental Math., 13, 473–481.

[658] Koukoulopoulos, D. (2019). The Distribution of Prime Numbers. Amer. Math. Soc.

[659] Kovaleskaya, S. (1951/1961). Memories and Letters (in Russian). Shtraikh, S. Y., ed., USSR Academy of Sciences, Moscow.

[660] Kowalski, E. (2015). Gaps between prime numbers and primes in arithmetic progressions (after Y. Zhang and J. Maynard). Astérisque, 367–368, Exp. No. 1084, 327–366.

[661] Kowalski, E. (2021). An Introduction to Probabilistic Number Theory. Cambridge Univ. Press.

[662] Kubilius, J. (1955). On the distribution of values of additive arithmetic functions. Dokl. Akad. Nauk SSSR, 100, 623–626.

[663] Kubilius, J. (1964). Probabilistic Methods in the Theory of Numbers. Translation of 2nd Russian edition of 1962. Translations of mathematical monographs, 11. Amer. Math. Soc. Transl., Providence, RI. (5th printing 1997, Amer. Math. Soc.) 47–85.

[664] Kubilius, J. (1983). Estimate of the second central moment for any additive arithmetic functions. Litovsk. Mat. Sb., 23, 110–117.

[665] Kulesza, J. A. and Taskar, B. (2012). Learning determinantal point processes. `arXiv:1202.3738`.

[666] Kunik, M. and Lucht, L. G. (2012). Power series with the von Mangodt function. Functiones et Approximatio, 47, 15–33.

[667] Lagarias, J. C. (2002). An elementary problem equivalent to the Riemann hypothesis. Amer. Math. Monthly, 109, 534–543.

[668] Lagarias, J. C. (2013). Euler's constant: Euler's work and modern developments. Bull. Amer. Math. Soc., 50, 527–628.

[669] Lagarias, J. C. and Li, W.-C. W. (2012). The Lerch zeta function, I: Zeta integrals. Forum Math., 24, 1–48. This is the first in a series of four articles.

[670] Lagarias, J. C. and Suzuki, M. (2006). The Riemann hypothesis for certain integrals of Eisenstein series. J. Number Theory, 118, 98–122.

[671] Lai, T. L. (1974). Limit theorems for delayed sums. Ann. Probab., 2, 432–440.

[672] Lamperti, J. W. (1977). Stochastic Processes: A Survey of the Mathematical Theory. Springer.

[673] Lamperti, J. W. (1996). Probability: A Survey of the Mathematical Theory, 2nd ed. Wiley.

[674] Lamzouri, Y. (2008). The two dimensional distribution of values of $\zeta(1+it)$. Int. Math. Res. Notices, Art. ID rnn 106, 48 pp.

[675] Lamzouri, Y. (2011). On the distribution of extreme values of zeta and L-functions in the strip $\frac{1}{2} < \sigma < 1$. Int. Math. Res. Notices, 23, 5449–5503.

[676] Landau, E. (1909). Handbuch der Lehre von der Verteilung der Primzahlen. Teubner, Leipzig, Berlin. 2nd ed. with appendix by P. T. Bateman (1953); 3rd ed. with appendix by P. T. Bateman (1974). Chelsea, NY.

[677] Landau, E. (1920). Über die Nullstellen der Zetafunktion. Math. Zeitschrift, 6, 151–154.

[678] Landau, E. (1927). Vorlesungen über Zahlentheorie. Three vols. Hirzel, Leipzig. Reprinted 1947 by Chelsea, NY.

[679] Lau, Y.-K., Tenenbaum, G., and Wu, J. (2013). On mean values of random multiplicative functions. Proc. Amer. Math. Soc., 141, 409–420.

[680] Laubenbacher, R. and Pengelley, D. (2010). Voici ce que j'ai trouveé: Sophie Germain's grand plan to prove Fermat's Last Theorem. Historia Mathematica, 37, 641–692.

[681] Laurinčikas, A. (2003). The universality of zeta functions. Acta Appl. Math., 78, 251–271.

[682] Laurinčikas, A., Manstavičius, E., and Stakėnas, V., eds. (1977). Analytic and Probabilistic Methods in Number Theory, Vol. 4. New Trends in Probability and Statistics. VSP Science.

[683] Lee, J. (1989). On the constant in the Turán-Kubilius inequality. PhD Thesis, Univ. of Michigan. 83 pp.

[684] Lee, J. (1992). The second central moment of additive functions. Proc. Amer. Math. Soc., 114, 887–895.

[685] Lehman, R. S. (1960). On Liouville's function. Math. Comp., 14, 311–320.

[686] Lehman, R. S. (1966). On the difference $\pi(x) - \mathrm{li}(x)$. Acta Arith., 11, 397–410.

[687] Lehmer, D. H. (1956). On the roots of the Riemann zeta function. Acta Math., 95, 291–298.

[688] Lenstra, H. W. and Pomerance, C. (2019). Primality testing with Gaussian periods. J. Eur. Math. Soc., 21, 1229–1269.

[689] Letac, G. (1992). Lectures on natural exponential families and their variance functions. Monografas de Matematica [Mathematical Monographs], 50. Instituto de Matematica Pura e Aplicada (IMPA), Rio de Janeiro.

[690] Letac, G. and Mora, M. (1986). Sur les fonctions-variance des familles exponentielles naturelles sur R. (French.) [On the variance functions of natural exponential families on R.] C.R. Acad. Sci. Paris Sr. I, Math. 302, no. 15.

[691] LeVeque, W. J. (1949). On the size of certain number–theoretic functions. Trans. Amer. Math. Soc., 66, 440–463.

[692] LeVeque, W. J. (1996). Fundamentals of Number Theory. Dover, NY. (Originally 1977, Addison-Wesley.)

[693] Levin, B. Ja. and Fainleib, A. S. (1966). Distribution of values of additive arithmetical functions (in Russian). Dokl. Acad. Nauk SSSR, 171, 281–284.

[694] Levinson, N. (1969). A motivated account of an elementary proof of the Prime Number Theorem. Amer. Math. Monthly, 76, 225–245.

[695] Levinson, N. (1974). More than one-third of zeros of Riemann's zeta function are on $\sigma = 1/2$. Advances in Math., 13, 383–436. See also 1975, 18, 239–242.

[696] Lévy, P. (1931). Sur les séries dont les termes sont des variables éventuelles indépendantes. Studia Mathematica, 3, 119–155.

[697] Li, X. (1997). The positivity of a sequence of numbers and the Riemann Hypothesis. J. Number Theory, 65, 325–333.

[698] Lichtman, J. D. (2023). Primes in arithmetic progressions to large moduli, and Goldbach beyond the square-root barrier. `arXiv:2309.08522`.

[699] Lichtman, J. D. and Pomerance, C. (2018). Explicit estimates for the distribution of numbers free of large prime factors. J. Number Theory, 183, 1–23.

[700] Lin, G. and Hu, C. (2001). The Riemann zeta distribution. Bernoulli, 7, 817–828.

[701] Littlewood, J. E. (1912). Quelques conséquences de l'hypothèse que la fonction $\zeta(s)$ de Riemann n'a pas de zéros dans le demi-plan $R(s) > 1/2$. Comptes Rendus de l'Acad. des Sciences (Paris), 154, 263–266.

[702] Littlewood, J. E. (1914). Sur la distribution des nombres premieres. Comptes Rendus de l'Acad. Sciences, Paris, 158, 1869–1872.

[703] Littlewood, J. E. (1922). Researches in the theory of the Riemann ζ-function. Proc. London Math. Soc., 20, xxii–xxviii.

[704] Littlewood, J. E. (1924). Two notes on the Riemann zeta-function. Proc. Cambridge Philos. Soc., 22, 263–266.

[705] Littlewood, J. E. (1924). On the zeros of the Riemann zeta-function. Proc. Cambridge Phil. Soc., 22, 295–318.

[706] Littlewood, J. E. (1926). On the Riemann zeta-function. Proc. London Math. Soc., 24, 175–201.

[707] Littlewood, J. E. (1986). Littlewood's Miscellany. Edited with foreword by Bela Bollobás, Cambridge Univ. Press. Formerly: A Mathematicians Miscellany, 1957, Methuen, London.

[708] Liu, M.-C. and Wang, T. (2002). On the Vinogradov bound in the three primes Goldbach conjecture. Acta Arith., 105, 133–175.

[709] Lloyd, S. P. (1984). Ordered prime divisors of a random integer. Ann. Probab., 12, 1205–1212.

[710] Lorch, J. (2007). A generalized Hawkins sieve and prime k-tuplets. Rocky Mountain J. of Math., 37, 533–550.

[711] Lorch, J. and Okten, G. (2007). Primes and probability: The Hawkins random sieve. Math. Magazine, 80, 112–119.

[712] Lugannani, R. and Rice, S. (1980). Saddle point approximation for the distribution of the sum of independent random variables Adv. Appl. Probab., 12, 475–490.

[713] Lugar, M. M., Milinovich, M. B., and Quesada-Herrera, E. (2023). On the number variance of zeta zeros and conjecture of Berry. Mathematika, 203, 303–348.

[714] Lukacs, E. (1970). Characteristic Functions, 2nd ed. Hafner, NY.

[715] van de Lune, J. and te Riele, H. J. J. (1983). On the zeros of the Riemann zeta function in the critical strip, III. Math. Comp., 41, 759–767.

[716] van de Lune, J., te Riele, H. J. J., and Winter, D. T. (1986). On the zeros of the Riemann zeta function in the critical strip, IV. Math. Comp., 46, 667–681.

[717] Luong, A. and Doray, L. G. (1996). Goodness of fit test statistics for the zeta family. Insurance: Mathematics and Economics, 19, 45–53.

[718] Mačiulis, A. (1983). A lemma on large deviations of arithmetic functions. Lithuanian Math. J., 23, 141–161.

[719] Mackey, G. W. (1980). Harmonic analysis as the exploitation of symmetry – a historical survey. Bull. Amer. Math. Soc., 3, 543–698.

[720] Mahajan, S. (1998). Order of Magnitude Physics: A Textbook with Applications to the Retinal Rod and to the Density of the Prime Numbers. PhD Thesis, California Inst. Technology.

[721] Maier, H. (1981). Chains of large gaps between consecutive primes. Adv. in Math. 39, 257–269.

[722] Maier, H. (1985). Primes in short intervals. Michigan Math. J., 32, 221–225.

[723] Maier, H. (1988). Small differences between prime numbers. Michigan Math. J., 35, 323–344.

[724] Maier, H. and Pomerance, C. (1990). Unusually large gaps between consecutive primes. Theorie des Nombres (Quebec, PQ, 1987), de Gruyter, Berlin. Trans. Amer. Math. Soc., 322, 201–237.

[725] Maier, H. and Stewart, C. L. (2007). On intervals with few prime numbers. J. Reine Angew. Math., 608, 183–199.

[726] Maller, R. A. (1980). An extension of Kesten's generalised law of the iterated logarithm Bull. Australian Math. Soc., 21, 393–406.

[727] Mandelbrojt, S. (1972) Dirichlet Series: Principles and Methods. Reidel, Dordrecht-Holland.

[728] von Mangoldt, H. (1895). Zu Riemann's Abhandlung "Über die Anzahl der Primzahlen unter einer gegebenen Grösse." J. Reine Angew. Math., 114, 225–305.

[729] Marčenko, V. A. and Pastur, L. A. (1967). Distribution of eigenvalues for some sets of random matrices. Math. USSR Sbornik, 1, 457–483.

[730] Mardia, K. V. and Jupp, P. E. (2000). Directional Statistics. John Wiley & Sons, Chichester, NY.

[731] Marsden, J. E. and Hoffman, M. J. (1999). Basic Complex Analysis, 3rd ed. W. H. Freeman, NY.

[732] Martin, G., and Miao, W. (2016). *abc* triples. Funct. Approx. Comment. Math., 55, 145–176.

[733] Martin, G., Mossinghoff, M. J., and Trudgian, T. S. (2023). Fake mu's. Proc. Amer. Math. Soc., 151, 3229–3244.

[734] Martikainen, A. I. and Petrov, V. V. (1990). On the Borel-Cantelli lemma. Zap. Nauchn. Sem. Leningrad, Otdel. Mat. Inst. Steklov. (LOMI), 184, 200–207, 324. J. Math. Sci. 68, (1994), 540–544.

[735] Masser, D. W. (1985). Open problems. Proc. Symp. Analytic Number Theory (W.W.L. Chen, ed.), Imperial College, London.

[736] Massias, J.-P. and Robin, G. (1996). Effective bounds for certain functions involving primes (French). J. Théor. Nombres Bordeaux, 8, 215–242.

[737] Matomäki, K. and Radziwiłł, M. (2016). Multiplicative functions in short intervals. Annals Math., 183, 1015–1056.

[738] Matomäki, K. and Radziwiłł, M. (2020). Multiplicative functions in short intervals, II. `arXiv:2007.04290`.

[739] Matomäki, K., Radziwiłł, M., and Tao, T. (2015). An averaged form of Chowla's conjecture. Algebra and Number Theory, 9, 2167–2196.

[740] Matomäki, K., Radziwiłł, M., and Tao, T. (2016). Sign patterns of the Liouville and Möbius functions. Forum Math., Sigma, 4, e14, 44 pp.

[741] Matomäki, K., Radziwiłł, M., Tao, T., Teräväinen, J., and Zieger, T. (2023). Higher uniformity of bounded multiplicative functions in short intervals on average. Ann. Math., 197, 739–857.

[742] Matomäki, K. and Teräväinen, J. (2023). On the Möbius function in all short intervals. J. Eur. Math. Soc., 25, 1207–1225.

[743] Maynard, J. (2015). Small gaps between primes. Annals Math., 181, 383–413.

[744] Maynard, J. (2016). Large gaps between primes. Ann. Math., 183, 915–933.

[745] Mazur, B. and Stein, W. A. (2016). Prime Numbers and the Riemann Hypothesis. Cambridge Univ. Press.

[746] McCarthy, P. J. (1986). Introduction to Arithmetical Functions. Springer.
[747] McCullagh, P. and Nelder, J. A. (1989). Generalized Linear Models, 2nd ed. Chapman & Hall, London.
[748] McLeish, D. L. (1974). Dependent central limit theorems and invariance principles. Ann. Probab. 2, 620–628.
[749] Mehta, M. L. (2004). Random Matrices. 3rd ed. Elsevier/Academic, Amsterdam.
[750] Mehta, M. L. and Gaudin, M. (1960). On the density of the eigenvalues of a random matrix. Nuclear Physics, 18, 420–427.
[751] Mertens, F. (1874). Über einige asymptotische gesetze der zahlentheorie. J. Reine Angew. Math., 77, 289–338.
[752] Mertens, F. (1874). Ein Beitrag zur analytischen Zahlentheorie. J. Reine Angew. Math., 78, 46–62.
[753] Mertens, F. (1897). Über eine zahlentheoretische Funktion. Sitzungsberichte Kais. Akad. Wiss. Wien, IIa, 106, 761–830.
[754] Meštrović, R. (2018). Euclid's theorem on the infinitude of primes: A historiclal survey of its proofs (300 B.C. – 2017). arXiv:1202.3670.
[755] Mezzadri, F. and Snaith, N. C., eds. (2005). Recent Perspectives in Random Matrix Theory and Number Theory. London Math. Soc., Cambridge Univ. Press.
[756] Miller, S. J. and Takloo-Bighash, R. (2006). An Invitation to Modern Number Theory. Princeton Univ. Press.
[757] Mincu, G. and Panaitopol, L. (2008). Properties of some functions connected to prime numbers. J. Inequal. Pure Appl. Math., 9, Article 12.
[758] Mitrinović, D. S. and Popadić, M. S. (1978). Inequalities in Number Theory. Naucni Podmladak, Univ. of Nis.
[759] Mitrinović, D. S., Sándor, J., and Crstici, B. (1995). Handbook of Number Theory. Kluwer Acad. Publ.
[760] Mitrinović, S. J., Dragoslav S., and Crstici, B. (2006). Handbook of Number Theory, I. Springer.
[761] Molinari, G. (2015). Elementary prime identities. Amer. Math. Monthly, 122, 151–154.
[762] Monastyrsky, M. I. (2008). Grisha Margulis: Recollections of an old friend. J. Modern Dynamics, 2, 7–13.
[763] Montgomery, H. L. (1973). The pair correlation of zeros of the zeta-function. In: Symp. on Analytic Number Theory, H. G. Diamond, ed., Proc. Symposia Pure Math., 24, St. Louis, MO, Amer. Math. Soc., Providence, RI, 1972, pp. 181–193.
[764] Montgomery, H. L. (1977). Extreme values of the Riemann zeta function. Commentarii Mathematici Helvetici., 52, 511–518.
[765] Montgomery, H. L. (1980). The zeta function and prime numbers. Proc. Queen's Number Theory Conf., (Kingston, Ont., 1979), Queen's Papers in Pure and Appl. Math., 54, Queen's Univ., Kingston, 1–31.
[766] Montgomery, H. L. (1983). Zeros of approximations to the zeta function. In: Studies in Pure Mathematics: To the Memory of Paul Turán, Birkhäuser, Basel, 497–506.
[767] Montgomery, H. L. (1987). Fluctuations in the mean of Euler's phi function. Proc. Indian Acad. Sci. (Math. Sci.), 97, 239–245.
[768] Montgomery, H. L., Nikeghbali, A., and Rassias, M. Th., eds. (2017). Exploring the Riemann Zeta Function: 190 years from Riemann's Birth. Springer.
[769] Montgomery, H. L. and Odlyzko, A. M. (1984). Gaps between zeros of the zeta-function. In: Topics in Classical Number Theory, Colloquia Mathematica Societatis János Bolyai, Budapest 1981, 34, 1079–1106. North Holland, Amsterdam.
[770] Montgomery, H. L. and Odlyzko, A. M. (1988). Large deviations of sums of independent random variables. Acta Arith., 49, 427–434.
[771] Montgomery, H. L. and Vaughan, R. C. (1973). The large sieve. Mathematika, 20, 119–134.
[772] Montgomery, H. L. and Vaughan, R. C. (1975). The exceptional set in Goldbach's problem. Acta Arith., 27, 353–370.
[773] Montgomery, H. L. and Vaughan, R. C. (1981). The distribution of squarefree numbers. In: Recent progress in Analytic Number Theory, vol. 1, London, Academic Press, 247–256.
[774] Montgomery, H. L. and Vaughan, R. C. (2006). Multiplicative Number Theory I. Classical Theory. Cambridge Univ. Press.

[775] Montgomery, H. L. and Wagon, S. (2006). A heuristic for the Prime Number Theorem. Math. Intelligencer, 28, 6–9.

[776] Moore, T. O. and Hadlock, E. H. (1991). Complex Analysis. World Scientific, Singapore.

[777] Morris, D. W. (2005). Ratner's theorems on unipotent flows. Univ. Chicago Press.

[778] Mossinghoff, M. J. and Trudgian, T. S. (2012). Between the problems of Pólya and Turán. J. Australian Math. Soc., 93, 157–171.

[779] Mossinghoff, M. J. and Trudgian, T. S. (2017). The Liouville function and the Riemann hypothesis. In: H. Montgomery et al., eds., Exploring the Riemann Zeta Function 190 Years From Riemann's Birth. Springer, 201–221.

[780] Mosteller, F. (1972). An empirical study of the distribution of primes and litters of primes. In: Statistical Papers in Honour of G. W. Snedecor, Iowa State Univ. Press.

[781] Mosteller, F. (1972). A data analytic look at Goldbach counts. Statistsica Neerlandica 26, 227–242.

[782] Motohashi, Y. (1997). Spectral Theory of the Riemann zeta-function. Cambridge Univ. Press.

[783] Muckrock. (2014). Paul Erdos FBI file. `https://www.muckrock.com/foi/united-states-of-america-10/paul-erdos-fbi-file-6317/#file-24580`.

[784] Mueller, J. (1981). Gaps between consecutive zeta zeros. In: Analytic Number Theory (Philadelphia, Pa., 1980), Springer, 141–147.

[785] Murty, M. R. (2013). The twin prime problem and generalizations (après Yitang Zhang). Resonance, 712–731.

[786] Musielak, D. (2020). Sophie Germain: Revolutionary Mathematician, 2nd ed., Springer.

[787] Nagura, J. (1952). On the interval containing at least one prime number. Proc. Japan Acad. Sci., 28, 177–181.

[788] Nahin, P. J. (1998). An Imaginary Tale: The Story of $\sqrt{-1}$. Princeton Univ. Press.

[789] Nahin, P. J. (2021). In Pursuit of Zeta-3: The World's Most Mysterious Unsolved Math Problem. Princeton Univ. Press.

[790] Najnudel, J. (2018). On the extreme values of the Riemann zeta function on random intervals of the critical line. Prob. Theory & Related Fields, 172, 387–452.

[791] Najnudel, J. (2020). Exponential moments of the argument of the Riemann zeta function on the critical line. Mathematika, 66, 612–621.

[792] Najnudel, J. (2020). On consecutive values of random completely multiplicative functions. Electronic J. Probab., 25, Paper 59, 28 pp.

[793] Nakamura, T. (2015). A modified Riemann zeta distribution in the critical strip. Proc. Amer. Math. Soc., 143, 897–905.

[794] Nakamura, T. (2015). A complete Riemann zeta distribution and the Riemann hypothesis. Bernoulli, 21, 604–617.

[795] Nanopoulos, P. (1975). Loi de Dirichlet sur N^* et pseudo-probabilités. Comptes Rendus Acad. Sci. Paris, A 280, 1543–1546.

[796] Narkiewicz, W. (1983). Number Theory. (Translated from the Polish by S. Kanemitsu.) World Scientific, Singapore.

[797] Narkiewicz, W. (2000). The Development of Prime Number Theory: From Euclid to Hardy and Littlewood. Springer.

[798] Neale, V. (2017). Closing the Gap: The Quest to Understand Prime Numbers. Oxford Univ. Press.

[799] Neudecker, W. (1975). On twin "primes" and gaps between successive "primes" for the Hawkins random sieve. Math. Proc. Cambridge Phil. Soc., 77, 365–367.

[800] Neudecker, W. and Williams, D. (1974). The "Riemann hypothesis" for the Hawkins random sieve. Compositio Math., 29, 197–200.

[801] Nevanlinna, V. (1967). A refinement of Selberg's asymptotic equation. Pacific J. Math., 21, 537–540.

[802] Newman, D. J. (1980). Simple analytic proof of the prime number theorem. Amer. Math. Monthly, 87, 693–696.

[803] Ng, N. (2004). The distribution of the summatory function of the Möbius function. Proc. London Math. Soc., 89, 361–389.

[804] Ng, N. (2008). Large gaps between the zeros of the Riemann zeta function. J. Number Theory, 128, 509–556.

[805] Ng, N. (2008). The Möbius function in short intervals. Anatomy of Integers, CRM Proc. Lecture Notes, 46, Amer. Math. Soc., 247–257.

[806] Nicely, T. R. (1996). Enumeration to 10^{14} of the twin primes and Brun's constant. Virginia J. of Science,, 46, 195–204.

[807] Nicely, T. R. (2001). A new error analysis for Brun's constant. Virginia J. of Science, 52, 45–55.

[808] Nicolas, J.-L. (1983). Petites valeurs de la fonction d'Euler et hypothèse de Riemann. [Small values of the Euler function and the Riemann hypothesis.] Progr. Math., 38, Birkhäuser, Boston, 207–218.

[809] Nicolas, J.-L. (2012). Small values of the Euler function and the Riemann hypothesis. Acta Arith., 55, 311–321.

[810] Nicolas, J.-L. (2022). The sum of divisors function and the Riemann hypothesis. Ramanujan J., 58, 1113–1157.

[811] Nielsen, N. (1906). Handbuch der Theorie der Gammafunktion. Teubner, Leipzig. Reprinted 1965, Chelsea Publ. Co.

[812] NIST. (2010). Handbook of Mathematical Functions. F. W. Olver, D. W. Lozie, R. F. Boisvert and C. W. Clark, eds. Natl. Inst. Standards and Technology, Cambridge Univ. Press.

[813] Niven, I. (1951). Functions which represent prime numbers. Proc. Amer. Math. Soc., 2, 753–755.

[814] Niven, I., Zuckerman, H. S., and Montgomery, H. L. (1991). An Introduction to the Theory of Numbers, 5th ed. Wiley, NY.

[815] Nourdin, I., Pecatti, G., and Reinert, G. (2010). Stein's method and stochastic analysis of Rademacher functionals. Electron. J. Probab., 15, 1703–1742.

[816] Novoselov, E. V. (1964). A new method in probabilistic number theory (in Russian). Izv. Akad. Nauk. SSSR Ser. Mat., 28, 307–364.

[817] Nyman, B. (1950). On the one-dimensional translation group and semi-group in certain function spaces. PhD Thesis, Univ. of Uppsala.

[818] Odlyzko, A. M. (1987). On the distribution of spacings between zeros of the zeta function. Math. Comp., 48, 273–308.

[819] Odlyzko, A. M. (1989). The 10^{20}-th zero of the Riemann zeta-function and 70 million of its neighbours. `http://www.research.att.com/~amo/unpublished/index.html`

[820] Odlyzko, A. M. (1992). The 10^{20}-th zero of the Riemann zeta function and 175 million of its neighbors. 163 pages. `https://www-users.cse.umn.edu/~odlyzko/unpublished/zeta.10to20.1992.pdf`.

[821] Odlyzko, A. M. (1998). Tables of zeros of the Riemann zeta function. `https://www-users.cse.umn.edu/~odlyzko/zeta_tables/`

[822] Odlyzko, A. M. and te Riele, H. J. J. (1985). Disproof of the Mertens conjecture. J. Reine Angew. Math., 357, 138–160.

[823] Odlyzko, A. M. and Schonhage, A. (1988). Fast algorithms for multiple evaluations of the Riemann zeta function. Trans. Amer. Math. Soc., 309, 797–809.

[824] Oesterlé, J. (1988). Nouvelles approches du "théoreme" de Fermat. Astérisque, no. 161–162, Exp. No. 694, 4, 165–186.

[825] Oliveira e Silva, T., Herzog, S., and Pardi, S. (2014). Empirical verification of the even Goldbach conjecture and computation of prime gaps up to $4 \cdot 10^{18}$. Math. Comp., 83, 2033–2060.

[826] Oliveira e Silva, R. (2011). Properties of the Beurling generalized primes. J. Number Theory, 131, 45–58.

[827] Ostmann, H.-H. (1956) Additive Zahlentheorie, 2 vols. Springer. Reprinted 1968.

[828] Overholt, M. (2015). A Course in Analytic Number Theory. Amer. Math. Soc., Providence, RI.

[829] Painlevé, P. (1900). Analyse des Travaux Scientifiques (in French), Gauthier-Villars, Paris. (Remark: A version of the last sentence is commonly attributed to Jacques Hadamard.)

[830] Panaitopol, L. (1998). On the inequality $p_a p_b > p_{ab}$. Bull. Math. Soc. Sc. Math., Roumainie, 41, 135–139.

[831] Panaitopol, L. (2001). Inequalities involving prime numbers. Math. Reports Bucureşti, 3, 251–256.

[832] Papoulis, A. (1960). The Fourier Transform and its Applications. McGraw-Hill.

[833] Paquette, E. and Zeitouni, O. (2018). The maximum of the CUE field. Int. Math. Res. Not., 16, 5028–5119.

[834] Pastur, L. and Shcherbina, M. (2011). Eigenvalue distribution of large random matrices. Math. Surveys Monogr., 171, Amer. Math. Soc., Providence, RI.

[835] Patel, D. and Yang, A. (2023). An explicit sub-Weyl bound for $\zeta(1/2 + it)$. arXiv:2302.13444.

[836] Patterson, S. J. (1995). An Introduction to the Theory of the Riemann Zeta Function. Cambridge Univ. Press.

[837] Paul, D. and Aue, A. (2014). Random matrix theory in statistics: A review. J. Statistical Planning and Inference, 150, 1–29.

[838] Peebles, P. J. E. (1980). The Large-Scale Structure of the Universe. Princeton Univ. Press.

[839] Peres, Y. and Virág, B. (2005). Zeros of the i.i.d. Gaussian power series: A conformally invariant determinantal process. Acta Math., 194 1–35.

[840] Perret–Gentil, C. (2019). Some recent interactions of probability and number theory. European Math. Soc. Newsletter, March 2019.

[841] Petrov, V. V. (1966). On the law of iterated logarithm. Vestnik Leningrad. Univ., 21, 63–67.

[842] Petrov, V. V. (1975). Sums of Independent Random Variables. Springer.

[843] Petrov, V. V. (2002). A note on the Borel-Cantelli lemma. Statistics and Probability Letters, 58, 283–286.

[844] Petrov, V. V. (2004). A generalization of the Borel-Cantelli lemma. Statistics and Probability Letters, 67, 233–239.

[845] Pilatte, C. (2023). Improved bounds for the two-point logarithmic Chowla conjecture. arXiv:2310.19357 [math.NT].

[846] Pintz, J. (1980). On the remainder term of the prime number formula, II: On a theorem of Ingham. Acta Arith., 37, 209–220.

[847] Pintz, J. (1984). On the remainder term of the prime number formula and the zeros of Riemann's zeta-function. In: Number Theory, Noordwijkerhout, 1983, 186–197. LNM., 1068, Springer.

[848] Pintz, J. (1997). Very large gaps between consecutive primes. J. Number Theory, 63, 286–301.

[849] Pintz, J. (2007). Cramér vs. Cramér. On Cramér's probabilistic model for primes. Functiones et Approximatio Commentarii Mathematici, 37, 361–376.

[850] Pintz, J. (2007). Approximations to the Goldbach and twin prime problem and gaps between consecutive primes. In: Probability and Number Theory – Kanazawa 2005, T. Kohno, ed., Advanced Studies in Pure Math., 49, Math. Soc. of Japan, 323–365.

[851] Pintz, J. (2013). Paul Erdős and the difference of primes. Erdős Centennial, 485–513, Bolyai Soc. Math. Stud., 25, János Bolyaii Math. Soc., Budapest.

[852] Pintz, J. (2014). Some new results on gaps between consecutive primes. In: Number Theory, Analysis, and Combinatorics, 261–278, de Gruyter Proc. Math., de Gruyter, Berlin.

[853] Platt, D. J. (2016). Numerical computations concerning the GRH. Math. Comp., 85, 3009–3027.

[854] Platt, D. J. and Trudgian, T. S. (2015). An improved explicit bound on $|\zeta(\frac{1}{2}+it)|$. J. Number Theory, 147, 842–851.

[855] de Polignac, A. (1849). Recherches sur les nombres premiers. Comptes Rendus Acad. Sci. Paris, 45, 406–410, 431–434, 575–580, 882–886.

[856] de Polignac, A. (1849). Six propositions arithmologiques déduites du crible d'Ératosthene. Nouvelles Annales de Math., 8, 423–429.

[857] de Polignac, A. (1849). Recherches nouvelles sur les nombres premiers. Comptes Rendus Acad. Sci. Paris, 29, 397–401, 738–739.

[858] de Polignac, A. (1851). Recherches Nouvelles sur les Nombres Premiers. Bachelier, Paris.

[859] Pollack, P. (2013). On Mertens' theorem for Beurling primes. Canadian Math. Bull., 56, 829–843.

[860] Polvani, G. (1933). Sulla frequenza dei numeri primi. Tech. report, Physics Dept., Univ. of Milan.

[861] Pólya, G. (1919). Verschiedene bemerkungen zur zahlentheorie. Jahresbericht der Deutschen Mathematiker-Vereinigung, 28, 31–40.

[862] Pólya, G. (1959). Heuristic reasoning in the theory of numbers. Amer. Math. Monthly. 66, 375–384.

[863] Pólya, G. (1962). Mathematical discovery. On understanding, learning, and teaching problem solving. Vol. I, Wiley, New York-London.

[864] van der Poorten, A. (1978). A proof that Euler missed ... Apery's proof of the irrationality of $\zeta(3)$, An informal report. Math. Intelligencer, 1, 195–203.

[865] Postnikov, A. G. (1988). Introduction to Analytic Number Theory. Amer. Math. Soc. Translations, Providence, RI.

[866] Prachar, K. (1957). Primzahlverteilung. Reprinted 1978 by Springer.

[867] Pratt, K., Robles, N., Zaharescu, A., and Zeindler, D. (2020). More than five-twelfths of the zeros of ζ are on the critical line. Research Math. Sciences, 7, 2.

[868] Prestini, E. (2016). Joseph Fourier: The man and the mathematician. In: The Evolution of Applied Harmonic Analysis, Birkhäuser, 1–24.

[869] Priestley, H. A. (2003). Introduction to Complex Analysis, 2nd ed. Oxford Univ. Press.

[870] Purdon, P. and Williams, J. H. (1968). Cycle length in a random function. Trans. Amer. Math. Soc., 133, 547–551.

[871] Pustylnikov, L. D. (1999). On a property of the classical zeta-function associated with the Riemann conjecture on zeros. Russian Math. Surveys 54, 262–263.

[872] Rabin, M. O. (1976). Probabilistic algorithms. In: Algorithms and Complexity, Proc. Symp., Carnegie-Mellon Univ., Academic Press, NY, 21–39.

[873] Rabin, M. O. (1980). Probabilistic algorithm for testing primality. J. Number Theory, 12, 128–138.

[874] Rademacher, H. (1964). Lectures on Elementary Number Theory. Blaisdell, NY. Reprinted 1977, R. E. Krieger Publishing Co., Huntington, NY.

[875] Rademacher, H. (1973). Topics in Analytic Number Theory. Edited by E. Grosswald, J. Lehner and M. Newman. Springer.

[876] Radziwiłł, M. (2011). Large deviations in Selberg's central limit theorem. `arXiv:1108.5092`.

[877] Radziwiłł, M. (2011). A structure theorem in probabilistic number theory `arXiv:1109.0033`.

[878] Radziwiłł, M. and Soundararajan, K. (2013). Continuous lower bounds for moments of zeta and L-functions. Mathematika, 59, 119–128.

[879] Radziwiłł, M. and Soundararajan, K. (2017). Selberg's central limit theorem for $\log|\zeta(1/2+it)|$. l'enseignement Math., 63, 1–19. Also 2015, `arXiv:1509.06827`.

[880] Ramachandra, K. (2007). Theory of Numbers: A Textbook. Alpha Science International, Oxford.

[881] Ramachandra, K. and Sankaranarayanan, A. (1993). On some theorems of Littlewood and Selberg. I. J. Number Theory 44, 281–291.

[882] Ramaré, O. (1995). On Schnirelmann's constant. Ann. Scuola Norm. Sup. Pisa Cl. Sci., 22, 645–706.

[883] Ramaré, O. (2013). Explicit estimates for the summatory function of $\Lambda(n)/n$ from the one of $\Lambda(n)$. Acta Arith., 159, 113–122.

[884] Ramaré, O. (2018). Chowla's conjecture: from the Liouville function to the Moebius function. LNM, 2213, Springer, 317–323.

[885] Rankin, R. A. (1938). The difference between consecutive prime numbers. J. London Math. Soc., 13, 242–247.

[886] Rao, K. S. (2021). Srinivasa Ramanujan: Life and Work of a Natural Mathematical Genius, Swayambhu. Springer.

[887] Redheffer, R. M. (1977). Eine explizit losbare optimierungsaufgabe. (Tagung, Math. Forschungsinst., Oberwolfach, 1976) Internat. Schriftenreihe Numer. Math., 36, 213–216.

[888] Reisel, H. and Gohl, G. (1970). Some calculations related to Riemann's prime number formula. Math. Comp., 24, 969–983.

[889] Reisel, H. and Vaughan, R. C. (1983). On sums of primes. Arkiv Mat., 21, 45–74.

[890] Remmert, R. (1991). Theory of Complex Functions. Springer, NY. (English translation of the 1989 German 2nd ed.)

[891] Remmert, R. (1998). Classical Topics in Complex Function Theory. Springer.

[892] Remy, G. (2020). The Fyodoro–Bouchaud formula and Liouville conformal field theory. Duke Math. J., 169, 177–211.

[893] Rényi, A. (1960). Probabilistic Methods in Number Theory. Proc. Internat. Congress of Mathematicians, 1958, 529–539. Cambridge Univ. Press.

[894] Rényi, A. (1963). On the distribution of values of additive number-theoretical functions. Publ. Math. Debrecen, 10, 264–273.

[895] Rényi, A. (1965). A new proof of a theorem of Delange. Publ. Math. Debrecen, 12, 323–329.

[896] Rényi, A. (1965). Additive and multiplicative number theoretic functions. Lecture notes, Univ. Michigan, Ann Arbor, 1–123.

[897] Rényi, A. and Turán, P. (1958). On a theorem of Erdős and Kac. Acta Arith., 4, 71–84.

[898] Resnick, S. I. (1992). Adventures in stochastic processes. Birkhäuser, Boston.

[899] Révész, P. (2013). Random Walk in Random and Non-Random Environments, 3rd ed. World Scientific, Singapore.

[900] Rezgui, H. (2017). Conjecture of twin primes (still unsolved problem in number theory). An expository essay. Surv. Math. Appl., 12, 229–252.

[901] Ribenboim. P. (1996). The New Book of Prime Number Records, 3rd ed. Springer.

[902] Ricardo, H. J. (2005). Goldbach's conjecture implies Bertrand's postulate. Amer. Math. Monthly, 112, 492.

[903] Richards, I. (1974). On the incompatibility of two conjectures concerning primes. Bull. Amer. Math. Soc., 80, 419–438.

[904] te Riele, H. J. J. (1985). Some historical and other notes about the Mertens conjecture and its recent disproof. Nieuw Archief voor Wiskunde, 3, 237–243.

[905] te Riele, H. J. J. (1987). On the sign of the difference $\pi(x) - \mathrm{li}(x)$. Math. Comput., 48, 323–328.

[906] Riemann, B. (1859). Über die anzahl der primzahlen unter einer gegebenen grösse. Monatsber. Akad., Berlin, 671–680. (A transcription by D. R. Williams is available.)

[907] Riesel, H. (1994). Prime Numbers and Computer Methods for Factorization, 2nd ed. Birkhäuser, Boston.

[908] Riesz, M. (1916). Ein Konvergenzsatz fur Dirichletsche Reihen. Acta Math., 40, 349–361.

[909] Rivest, R. L., Shamir, A., and Adleman, L. M. (1978). A method for obtaining digital signatures and public-key cryptosystems. Comm. ACM, 21, 120–126.

[910] Rivoal, T. (2008). On the distribution of the Hawkins' random "primes." J. Théor. Nombres Bordeaux 20, 799–809.

[911] Robin, G. (1983). Estimation de la fonction de Tchebychef ϑ sur le k-ième nombre premier et grandes valeurs de la fonction $\omega(n)$ nombre de diviseurs premiers de n. Acta Arith., 42, 367–389.

[912] Robin, G. (1984). Grandes valeurs de la fonction somme des diviseurs et hypothèse de Riemann. J. Math. Pures Appl., 63, 187–213.

[913] Roesler, F. (1986, 1987, and 1990). Riemann's hypothesis as an eigenvalue problem, I, II and III. Linear Algebra Appl., 81, 153–198; 92, 45–73; 141, 1–46.

[914] Rose, H. E. (1994). A Course in Number Theory. 2nd ed. Clarendon Press, Oxford.

[915] Ross, P. M. (1975). On Chen's Theorem that each large even number has the form $p_1 + p_2$ or $p_1 + p_2p_3$. J. London Math. Math. Soc., 10, 500–506.

[916] Ross, S. M. (1996). Stochatic Processes. Wiley, NY.

[917] Rosser, J. B. (1939). The n-th prime is greater than $n \log n$. Proc. London Math. Soc., 45, 21–44.

[918] Rosser, J. B. (1941). Explicit bounds for some functions of prime numbers. Amer. J. Math., 63, 211–232.

[919] Rosser, J. B. and Schoenfeld, L. (1962). Approximate formulas for some functions of prime numbers. Illinois J. Math., 6, 64–94.

[920] Rosser, J. B. and Schoenfeld, L. (1975). Sharper Bounds for the Chebyshev Functions $\theta(x)$ and $\psi(x)$. Math. Comput., 29, 243–269.

[921] Rosser, J. B., Yohe, J. M., and Schoenfeld, L. (1969). Rigorous computation of the zeros of the Riemann zeta-function (with discussion). Information Processing 68, Proc. IFIP Congress, Edinburgh, 1968, Vol. 1, Math., Software, North-Holland, Amsterdam, 70–76.

[922] Rota, G.-C. (1964). On the foundations of combinatorial theory, I: Theory of Möbius functions. Z. Wahrsch. Verw. Gebiete, 2, 340–368.

[923] Roth, K. F. (1953). On certain sets of integers. J. London Math. Soc., 28, 104–109.

[924] Rudin, W. (1976). Principles of Mathematical Analysis, 3rd ed. McGraw-Hill, NY

[925] Rudin, W. (1987). Real and Complex Analysis, 3rd ed. McGraw-Hill, NY.

[926] Rudin, W. (1991). Functional Analysis, 2nd ed. McGraw-Hill, NY.

[927] Rupert, M. (2013). Extending Erdős-Kac and Selberg-Sathe to Beurling Primes with Controlled Integer Counting Functions. MSc Thesis, Univ. British Columbia.

[928] Rusza, I. Z. (1996). Paul Erdős – from an epsilon's-eye view. Periodica Mathernatica Hungarica, 33, 73–81.
[929] Ruzsa, I. Z. (1999). Erdős and the integers. J. Number Theory, 79, 115–163.
[930] Saidak, F. (2002). An elementary proof of a theorem of Delange. C. R. Math. Acad. Sci. Soc., Canada, 24, 144–151.
[931] Saidak, F. (2008). On the prime number lemma of Selberg. Math. Scand., 103, 5–10.
[932] Saito, S. and Tanaka, T. (2012). A note on infinite divisibility of zeta distributions. Applied Mathematical Sciences, 6, 1455–1461.
[933] Saksman, E. and Seip, K. (2016). Some open questions in analysis for Dirichlet series. In: Recent Progress on Operator Theory and Approximation in Spaces of Analytic Functions, Contemp. Math., vol. 679, Amer. Math. Soc., Providence, RI, 179–191.
[934] Saksman, E. and Webb, C. (2016). Multiplicative chaos measures for a random model of the Riemann zeta function. arXiv:1604.08378.
[935] Saksman, E. and Webb, C. (2020). On the Riemann zeta function and Gaussian multiplicative chaos. In: D. Breaz & M. Rassias, eds., Advancements in Complex Analysis, Springer., 473–496.
[936] Saksman, E. and Webb, C. (2020). The Riemann zeta function and Gaussian multiplicative chaos: statistics on the critical line. Ann. Probab., 48, 2680–2754.
[937] Salem, R. (1953). Sur une proposition equivalente a l'hypothese de Riemann (in French). C.R. Acad. Sci. Paris 236, 1127–1128.
[938] Sandifer, E. (2007). Euler's solution of the Basel problem – the longer story. In: Euler at 300, MAA Spectrum, Math. Assoc. Amer., Washington DC, 105–117.
[939] Sándor, J. and Crstici, B., eds. (2004). Handbook of number theory II. Dordrecht, Kluwer Academic.
[940] Sándor, J., Mitrinović, D. S., and Crstici, B., eds. (2006). Handbook of Number Theory I. Dordrecht, Springer.
[941] Saouter, Y., Trudgian, T., and Demichel, P. (2015). A still sharper region where $\pi(x) - \mathrm{li}(x)$ is positive. Math. Comput., 84, 2433–2446,
[942] Sárközy, A. (2001). Unsolved Problems in Number Theory. Periodica Mathematica Hungarica, 42, 17–35.
[943] Sarnak, P. (2000). The Riemann Hypothesis. Clay Problems of the Millenium.
[944] Sarnak, P. (2011). Three lectures on the Möbius function, randomness and dynamics. Tech. report, Inst. Adv. Studies. http://publications.ias.edu/sarnak/.
[945] Sathe, L. G. (1953–1954). On a problem of Hardy on the distribution of integers having a given number of prime factors, I, II, III, IV. J. Indian Math. Soc, 17, 62–83, 83–141; 18, 27–42, 43–81.
[946] Sato, K. (1999). Lévy Processes and Infinitely Divisible Distributions. Cambridge Univ. Press.
[947] du Sautoy, M. (2003). The Music of the Primes. Harper Collins.
[948] Saxena, R. K., Pogany, T. K., Saxena, R., and Jankov, D. (2011). On generalized Hurwitz-Lerch zeta distributions occuring in statistical inference. Acta Univ. Sapientiae, Mathematica, 3, 43–59.
[949] Schechter, B. (1998). My Brain is Open; The Mathematical Journeys of Paul Erdős. Simon & Schuster, NY.
[950] Schmidt, E. (1903). Über die anzahl der primzahlen unter gegebener grenze. Mathematische Annalen, 57, 195–204.
[951] Schnirelmann, L. G. (1930/1933/1939). On the additive properties of numbers (in Russian). Proc. Don Polytechnic Inst. Novocherkassk, vol. XIV (1930). Reprinted: Über additive Eigenschaften von Zahlen (in German). Math. Ann., 107, (1933), 649–690. Reprinted, Uspekhi Matematicheskikh Nauk (in Russian), 1939, 6, 9–25.
[952] Schoenberg, I. J. (1928). Über die asymptotische Verteilung reeller Zahlen mod 1. Math. Zeitschrift, 28, 171–199.
[953] Schoenberg, I. J. (1936). On asymptotic distributions of arithmetical functions, Trans. Amer. Math. Soc., 39, 315–330.
[954] Schoenfeld, L. (1976). Sharper bounds for Chebyshev functions $\theta(x)$ and $\psi(x)$, II. Math. Comput., 30, 337–360.
[955] Schröeder, M. R. (2006). Number Theory in Science and Communication, 4th ed. Springer.

[956] Schur, I. and Schur, G. (1993). Multiplikativ signierte Folgen positiver ganzer Zahlen. Gesammelte Abhandlungen von Isaac Schur, vol. 3, Springer, 392–399.

[957] Schwarz, W. (1994). Some remarks on the history of the Prime Number Theorem from 1896 to 1960. In: Development of Mathematics 1900-1950 (Luxembourg 1992). Birkhäuser, Basel.

[958] Schwarz, W. (1994). Some aspects of the development of probabilistic number theory. Prob. Theory and Math. Stat. (Vilnius, 1993) VSP, (B. Grigelionis et al., eds.), 661–701.

[959] Schwarz, W. (2007). Some highlights from the history of probabilistic number theory. In: Probability and Number Theory – Kanazawa 2005, T. Kohno, ed., Math. Soc. of Japan, 367–419.

[960] Schwarz, W. and Spilker, J. (1994). Arithmetical Functions: An Introduction to Elementary and Analytic Properties of Arithmetic Functions and to Some of their Almost-Periodic Properties. London Math. Soc. Lecture Note Series, 184. Cambridge Univ. Press.

[961] Schweiger, F. and Manstavičius, E., eds. (1992). New Trends in Probability and Statistics, vol. 2: Analytic and Probabilistic Methods in Number Theory. Proc. Int. Conf. in Honour of J. Kubilius, Palanga, Lithuania, 1991. V.S.P. BV Vilnius Univ.

[962] Sekatskii, S. K. (2016). Analysis of Voros criterion equivalent to the Riemann hypothesis. Anal. Geom. Number Theory, 95–102.

[963] Sekatskii, S. K., Beltraminelli, S., and Merlini, D. (2012). On equalities involving integrals of the logarithm of the Riemann ζ-function and equivalent to the Riemann hypothesis Ukrainian Math. J. 64, 247–261.

[964] Sekatskii, S. K. (2014). Generalized Bombieri-Lagarias' theorem and generalized Li's criterion with its arithmetic interpretation Ukrainian Math. J. 66, 415–431.

[965] Sekatskii, S. K. (2021). On the generalized Li's criterion equivalent to the Riemann hypothesis and its first applications. Proc. Math. Stat., 348, Springer, 241–254.

[966] Sekatskii, S. K. and Beltraminelli, S. (2023). Some simplest integral equalities equivalent to the Riemann hypothesis. Ukrainian Math. J. 74, 1433–1440.

[967] Sekatskii, S. K., Beltraminelli, S., and Merlini, D. (2015). On equalities involving integrals of the logarithm of the Riemann ζ-function with exponential weight which are equivalent to the Riemann hypothesis. Int. J. Anal., Art. ID 980728, 8 pp.

[968] Selberg, A. (1942). On the zeros of Riemann's zeta-function Der Kong. Skr. Norske Vidensk. Selsk. Forhand., Oslo, 10, 59 pp.

[969] Selberg, A. (1943). On the normal density of primes in small intervals and the difference between consecutive primes. Arch. Math. Naturvid. B47, 87–105. (Also in: Collected papers, 1989 Vol. I. Springer, 160–178.)

[970] Selberg, A. (1944). On the remainder in the formula for $N(T)$, the number of zeros of $\zeta(s)$ in the strip $0 < t < T$. Avhandlinger Norske Vid. Akad. Oslo, 1, 27 pp.

[971] Selberg, A. (1946). The zeta-function and the Riemann Hypothesis. Skandinaviske Matematikerkongres, 10, 187–200.

[972] Selberg, A. (1946). Contributions to the theory of the Riemann zeta-function. Archiv for Mathematik og Naturvidenskab B, 48, 89–155.

[973] Selberg, A. (1946). The zeta-function and the Riemann hypothesis. C. R. Dixieme Congres Math., Scandinaves, 187–200.

[974] Selberg, A. (1947). On an elementary method in the theory of primes. Det Kong. Norske Vid. Selsk. Forh., Trondheim, B19, 64–67.

[975] Selberg, A. (1949). An elementary proof of the prime number theorem. Ann. Math., 50, 305–313.

[976] Selberg, A. (1949). On elementary methods in primenumber-theory and their limitations. Proc. 11th Scand. Math. Cong., Trondheim, 1949. In: Collected Works, 1989, Vol. I, Springer, 388–397.

[977] Selberg, A. (1954). Note on a paper by L.G. Sathe. J. Indian Math. Soc., 18, 83–87.

[978] Selberg, A. (1989). Collected Papers, vol. I. Springer.

[979] Selberg, A. (1992). Old and new conjectures and results about a class of Dirichlet series. In: Proc. Amalfi Conf. Analytic Number Th., E. Bombieri et al., eds., Universita di Salerno, Salerno, 367–385.

[980] Shakespeare, W. (1594). The Taming of the Shrew. 1.1.150–151.

[981] Shanks, D. (1985). Solved and Unsolved Problems in Number Theory, 3rd ed. Chelsea, NY.

[982] Shapiro, H. N. (1956). Distribution functions of additive arithmetic functions. Proc. Nat. Acad. Sci. USA, 42, 426–430.
[983] Shapiro, H. N. (1983). On the Erdős-Wintner theorem of probabilistic number theory. Comm. Pure Appl. Math., 36, 665–678.
[984] Shen, Jianhong (2001). On the singular values of Gaussian random matrices. Linear Alg. Applics., 326, 1–14.
[985] Shepp, L. A. and Lloyd, S. P. (1966). Ordered cycle length in a random permutation. Trans. Amer. Math. Soc., 121, 340–357.
[986] Shor, P. W. (1994). Algorithms for quantum computation: discrete logarithms and factoring. Proc. 35th Symp. Foundations of Computer Science, IEEE Comput. Soc. Press, 124–134.
[987] Shor, P. W. (1997). Polynomial-time algorithms for prime factorization and discrete logarithms on a quantum computer. SIAM J. Comput., 26, 1484–1509.
[988] Shuster, J. (1970). On the Borel-Cantelli problem. Canadian Math. Bull., 13, 273–275.
[989] Sierpiński, W. (1964). A Selection of Problems in the Theory of Numbers. Pergamon Press, Oxford.
[990] Simonič, A. (2020). Explicit zero density estimate for the Riemann zeta-function near the critical line. J. Math. Anal. Appl., 491, no. 1, 124303, 41 pp.
[991] Simonič, A. (2022). On explicit estimates for $S(t)$, $S_1(t)$, and $\zeta(1/2+it)$ under the Riemann Hypothesis. J. Number Theory, 231, 464–491.
[992] Simonič, A., Trudgian, T. S., and Turnage-Butterbaugh, C. L. (2022). Some explicit and unconditional results on gaps between zeroes of the Riemann zeta-function. Trans. Amer. Math. Soc., 375, no. 5, 3239–3265.
[993] Sivaramakrishnan, R. (1989). Classical Theory of Arithmetic Functions. Marcel Dekker, NY.
[994] Skewes, S. (1933). On the difference $\pi(x) - \mathrm{li}(x)$. J. London Math. Soc., 8, 277–283.
[995] Skewes, S. (1955). On the difference $\pi(x) - \mathrm{li}(x)$, II J. London Math. Soc., 5, 48–70.
[996] Snaith, N. C. (2000). Random Matrix Theory and Zeta Functions. PhD Thesis, Univ. of Bristol.
[997] Snaith, N. C. (2010). Riemann zeros and random matrix theory. Milan J. Math. 78, 135–152.
[998] Snaith, N. C., Forrester, P. J., and Verbaarschot, J. J. M. (2003). Developments in Random Matrix Theory. J. Phys. A36, R1–R12.
[999] Sós, V. T. (2002). Turbulent years: Erdős in his correspondence with Turán from 1934 to 1940. In Paul Erdős and his mathematics, I . (Budapest, 1999), vol. 11 of Bolyai Soc. Math. Stud., J. Bolyai Math. Soc., Budapest, 85–146.
[1000] Sossinsky, A. B. (2007). In the other direction. In: Golden Years of Moscow Mathematics, 2nd ed., S. Zdravkovska and P.L. Duren, eds. Amer. Math. Assoc., Providence, RI, 237–258.
[1001] Soundararajan, K. (1995). Mean-values of the Riemann zeta-function. Mathematika, 42, 158–174.
[1002] Soundararajan, K. (1996). On the distribution of gaps between zeros of the Riemann zeta-function. Quart. J. Math., Oxford, 47, 383–387.
[1003] Soundararajan, K. (2007)/ Small gaps between prime numbers: The work of Goldston-Pintz-Yilidrim. Bull. Amer. Math. Soc., 44, 1–18.
[1004] Soundararajan, K. (2007). The distribution of prime numbers. In A. Granville and Z. Rudnick, eds., Equidistribution in Number Theory, An Introduction. Proc. NATO Advanced Study Inst. on Equidistribution in Number Theory, Montréal, 2005. NATO Sci. Ser. II Math. Phys. Chem., 237, Springer. 59–83.
[1005] Soundararajan, K. (2008). Extreme values of zeta and L-functions. Mathematische Annalen, 342, 467–486.
[1006] Soundararajan, K. (2009). Moments of the Riemann zeta-function. Ann. Math., 170, 981–993.
[1007] Soundararajan, K. (2009). Partial sums of the Möbius function. J. Reine Angew. Math., 631, 141–152.
[1008] Soundararajan, K. (2017). The Liouville function in short intervals. Séminaire Bourbaki, vol. 2015/2016. Exposés 1104–1119. Astérisque No. 390, Exp. No. 1119, 453–479.
[1009] Soundararajan, K. and Xu, M. A. (2023). Central limit theorems for random multiplicative functions. J. d'Analyse Mathématique, 151, 343–374.

[1010] Soundararajan, K. and Zaman, A. (2022). A model problem for multiplicative chaos in number theory. l'Enseignement Math., 68, 307–340.

[1011] Speiser, A. (1934). Geometrisches zur Riemannschen Zetafunktion. Math. Ann., 110, 514–521.

[1012] Spencer, J. and Graham, R. (2009). The elementary proof of the Prime Number Theorem Math. Intelligencer, 31, 18–23.

[1013] Spira, R. (1968). Zeros of sections of the zeta function, II. Math. Comp. 22, 163–173.

[1014] Spitzer, F. (1964). Principles of Random Walk. Van Nostrand, Princeton, NJ.

[1015] Springel, V., White, S. D. M., Jenkins, A., Frenk, C. S., Yoshida, N., Gao, L., Navarro, J., Thacker, R., Croton, D., Helly, J., Peacock, J. A., Cole, S., Thomas, P., Couchman, H., Evrard, A., Colberg, J., and Pearce, F. (2005). Simulating the joint evolution of quasars, galaxies and their large-dcale distribution. arXiv: astro-ph 0504097v2.

[1016] Srinivasan, A. (2022). Infinitely many sign changes of the Liouville function on x^2+d Proc. Amer. Math. Soc., 150, 3799–3809.

[1017] Srivastava, H. M. and Choi, J. (2001). Series Associated with the Zeta and Related Functions. Kluwer Academic, Dordrecht, Netherlands.

[1018] Stakėnas, V. (2015). Jonas Kubilius and genesis of probabilistic number theory. Lithuanian Math. J., 55, 25–47.

[1019] Stankova-Frenkel, Z. (1998-99). Möbius inversion formula; Multiplicative functions. Berkeley Math Circle notes. `https://math.berkeley.edu/~stankova/MathCircle/Multiplicative.pdf`

[1020] Stein, Ch. M. (1984). On the Turán-Kubilius inequality. Tech. report 220, Dept. of Statistics, Stanford Univ.

[1021] Stein, E. M. and Shakarchi, R. (2003). Complex Analysis. Princeton Univ. Press.

[1022] Stein, E. M. and Shakarchi, R. (2003). Fourier Analysis. Princeton Univ. Press.

[1023] Stein, E. M. and Shakarchi, R. (2005). Real Analysis. Princeton Univ. Press.

[1024] Stein, E. M. and Shakarchi, R. (2011). Functional Analysis. Princeton Univ. Press.

[1025] Stein, M. L. and Ulam, S. M. (1967). An observation on the distribution of primes. Amer. Math. Monthly 74, 43–44.

[1026] Steuding, J. (2002). Probabilistic Number Theory. Course notes, Universitat Würzburg.

[1027] Steuding, J. (2007). Value Distribution of L-functions. Springer.

[1028] Steuding, J. (2013). Aspects of analytic number theory: The universality of the Riemann zeta-function. Topics in Modern Function Theory, Ramanujan Math. Soc., Mysore, 84–165.

[1029] Steutel. F. W. and van Harn, K. (2004). Infinite Divisibility of Probability Distributions on the Real Line. Marcel Dekker, London.

[1030] Stewart, A. and Tall, D. (2018). Complex Analysis (The Hitch Hiker's Guide to the Plane). Cambridge Univ. Press.

[1031] Stewart, C. L. and Tijdeman, R. (1986). On the Oesterle-Masser conjecture. Monatsh. Math., 102, 251–257.

[1032] Stewart, C. L. and and Yu, K. (1996). On the abc conjecture, II. Duke Math. J., 108, 169–181.

[1033] Stieltjes, T. J. (1905). Letters to C. Hermite (July–August 1885). In: Correspondance d'Hermite et de Stieltjes, Vol. II, Baillaud and Bourget, Eds., Gauthier-Villars, Paris.

[1034] Stopple, J. (2003). A Primer of Analytic Number Theory: From Pythagoras to Riemann. Cambridge Univ. Press.

[1035] Stout, W. F. (1974). Almost sure convergence. Academic Press, NY.

[1036] Strömbergsson, A. (2021). Analytic number theory – lecture notes. Univ. of Uppsala. Also: `http://www2.math.uu.se/~astrombe/analtalt08/www_notes.pdf`

[1037] Sudo, M. (1984). On totally multiplicative signatures of natural numbers. Proc. Japan Acad., Ser. A Math. Sci., 60, no 7, 273–275.

[1038] Szanton, A. (1992). The Recollections of Eugene P. Wigner (as told to Andrew Szanton). Plenum Press. Reprinted by Springer.

[1039] Tanaka, M. (1980). On the Möbius and allied functions. Tokyo J. Math., 3, 215–218.

[1040] Tagore, R. (1915). Let this be my parting word. In: Gitanjali. Macmillan, New York, 96, p. 88. (Translated from the Bengali by the author.)

[1041] Tao, T. C. (2012). Topics in Random Matrix Theory. Amer. Math. Soc., Providence, RI.

[1042] Tao, T. C. (2014). Every odd number greater than 1 is the sum of at most five primes. Math. Comp., 83, 997–1038.

[1043] Tao, T. C. (2016). The logarithmically averaged Chowla and Elliott conjectures for two-point correlations. Forum of Math., Pi4, e8, 36 pp.

[1044] Tao, T. C. (2017). Equivalence of the logarithmically averaged Chowla and Sarnak conjectures. In: Number Theory—Diophantine Problems, Uniform Distribution and Applications: Festschrift in Honour of Robert F. Tichy's 60th Birthday, C. Elsholtz and P. Grabner, eds., Springer, 391–421.

[1045] Tao, T. C. and Teräväinen, J. (2018). Odd order cases of the logarithmically averaged Chowla conjecture. J. Théor. Nombres Bordeaux 30, 997–1015.

[1046] Tao, T. C. and Teräväinen, J. (2019). The structure of correlations of multiplicative functions at almost all scales, with applications to the Chowla and Elliott conjectures. Algebra Number Theory 13, 2103–2150.

[1047] Tao, T. C. and Teräväinen, J. (2019). The structure of logarithmically averaged correlations of multiplicative functions, with applications to the Chowla and Elliott conjectures. Duke Math. J., 168, 1977–2027.

[1048] Tao, T. C. and Teräväinen, J. (2019). Value patterns of multiplicative functions and related sequences. Forum Math. Sigma 7, Paper No. e33, 55 pp.

[1049] Tao, T. C. and Vu, V. (2006). Additive combinatorics. Cambridge Univ. Press.

[1050] Tao, T. C., Vu, V., and Krishanpur, M. (2010). Random matrices: Universality of ESDs and the circular law. Annals Probab., 38, 2023–2206.

[1051] Tao, T. C. and Ziegler, T. (2023). Infnite partial sumsets in the primes. J. Anal. Math., 151, 375–389.

[1052] Taylor, P. R. (1945). On the Riemann zeta function. Quart. J. Math., Oxford, 16, 1–21.

[1053] Tenenbaum, G. (2015). Introduction to Analytic and Probabilistic Number Theory, 3rd ed. Amer. Math. Soc., Providence, RI. Complete solutions to all exercises by Jie Wu, published by Belin, Paris.

[1054] Tenenbaum, G. and Mendès-France, M. (1997). Les Nombres Premiers [Prime Numbers]. Que Sais-Je? Presses Universitaires de France, Paris.

[1055] Tenenbaum, G. and Mendès-France, M. (2000). The Prime Numbers and their Distribution. Amer. Math. Soc., Providence, RI. Translated from the 1997 French by P. G. Spain.

[1056] Tenenbaum, G. and Verwee, J. (2021). Effective Erdős-Wintner theorems. Proc. Steklov Inst. Math., 314, 264–278.

[1057] Teräväinen, J. (2023). On the Liouville function at polynomial arguments. `arXiv:2010.07924` [math.NT].

[1058] Titchmarsh, E. C. (1939). The Theory of Functions, 2nd ed. Clarendon Press, Oxford Univ. Press. (Subsequently reprinted with corrections.)

[1059] Titchmarsh, E. C. (1986). The Theory of the Riemann Zeta-Function, 2nd ed. Updated version by D. R. Heath-Brown of the 1951 text. Clarendon Press, Oxford.

[1060] Tomkins, R. J. (1971). Some iterated logarithm results related to the central limit theorem. Trans. Amer. Math. Soc., 156, 185–192.

[1061] Tomkins, R. J. (1992). Refinements of Kolmogorov's law of the iterated logarithm. Statist. Probab. Letters, 14, 321–325.

[1062] Totsuji, H. and Kihara, T. (1969). The correlation function for the distribution of galaxies. Publications Astonomical Soc. Japan, 21, 221–229.

[1063] Toulmonde, V. (2006). Module de continuité de la fonction de répartition de $\varphi(n)/n$. Acta Arith., 121, 367–402.

[1064] Tracy, C. A. and Widom, H. (2000). The distribution of the largest eigenvalue in the Gaussian ensembles: $\beta = 1, 2, 4$. CRM Ser. Math. Phys., Springer, 461–472.

[1065] Trench, W. F. (1999). Conditional convergence of infinite products. Amer. Math. Monthly, 106, 646–651.

[1066] Trost, E. (1953). Primzahlen. Birkhäuser, Basel & Stuttgart.

[1067] Tsang, K.-M. (1986). Some Ω-theorems for the Riemann zeta-function. Acta Arith., 46, 369–395.

[1068] Tucker, H. G. (1967) A Graduate Course in Probability. Academic Press, NY.

[1069] Turán, P. (1934). On a theorem of Hardy and Ramanujan. J. London Math. Soc., 9, 274–276.

[1070] Turán, P. (1936). Über einige Verallgemeinerungen eines Zatses von Hardy und Ramanujan. J. London Math. Soc., 11, 125–133.

[1071] Turán, P. (1948). On some approximative Dirichlet-polynomials in the theory of the zeta-function of Riemann. Danske Vid. Selsk. Mat.-Fys. Medd., 24, 1–36.

[1072] Turán, P. (1950). On the remainder-term in the prime-number formula, I, II. Acta. Math. Acad. Sci. Hungarica, 1, 48–63 and 155–166.

[1073] Turán, P. (1959). Nachtrag zu meiner Abhandlung "On some approximative Dirichlet polynomials in the theory of zeta function of Riemann." Acta Math. Acad. Sci. Hungarica, 10, 277–298.

[1074] Turing, A. (1943). A Method for the calculation of the Zeta-Function. Proc. London Math. Soc., ser. 2, 48, 180–197.

[1075] Turing, A. (1953). Some calculations of the Riemann Zeta-Function. Proc. London Math. Soc., ser. 3, 3, 99–117.

[1076] Udrescu, V. S. (1975). Some remarks concerning the conjecture $\pi(x + y) \leq \pi(x) + \pi(y)$. Revue Roumaine de Math. Pures et Appliquées, 20, 12-1-1208.

[1077] Ulam, S. (1983). Adventures of a Mathematician. Charles Scribner's Sons, NY.

[1078] Vaidyanathaswamy, R. (1931). The theory of multiplicative arithmetic functions. Trans. Amer. Math. Soc., 33, 579–662.

[1079] de la Vallée-Poussin, Ch. J. (1896). Récherches analytiques sur la thëorie des nombres premiers. Premiere partie. La fonction $\zeta(s)$ de Riemann et les nombre premiers en general. Annales de la Societe Scientifique de Bruxelles, 20, 183–256. Parts II and III on pages 281–362, 363–397.

[1080] Vandiver, H. S. (1961). On developments in an arithmetic theory of the Bernoulli and allied numbers. Scripta Math., 25, 273–303.

[1081] Vardi, I. (1991). The Euler–Maclaurin formula. In: Computational Recreations in Mathematica, §8.3 Addison-Wesley, Reading, MA, 159–163.

[1082] Vassilev-Missana, M. (2016). A note on prime zeta function and Riemann zeta function. Notes on Number Theory and Discrete Math., 22, 12–15.

[1083] Vaughan, R. C. (1993). On the eigenvalues of Redheffer's matrix, I. In: Number Theory with an Emphasis on the Markoff Spectrum (Provo, UT, 1991). M. Dekker, NY, 283–296.

[1084] Vaughan, R. C. (1996). On the eigenvalues of Redheffer's matrix, II. J. Australian Math. Soc., A 60, 260–273.

[1085] Vehka, T. (1979). Explicit construction of an admissible set for the conjecture that sometimes $\pi(x + y) > \pi(y) + \pi(x)$. Notices Amer. Math. Soc., 26, A453.

[1086] Vilaplana, J. P. (1987). The Hurwitz distribution. In: New Perspectives in Theoretical and Applied Statistics, M. L. Puri et al., eds., (Bilbao, 1986). Wiley, NY.

[1087] Vilkas, E. (1957). On the normal distribution of additive arithmetical functions. Uc. Trudy Viln. Gosud. Inst., 16, 23–31.

[1088] Villani, C. (2016). Birth of a Theorem: A Mathematical Adventure. Farrar, Straus and Giroux, NY.

[1089] Vinogradov, I. M. (1937). Representation of an odd number as the sum of three primes (in Russian). Dokllady Akad. Nauk., SSSR, 15, 291–294. OR IS IT: 16, 139–142.

[1090] Vinogradov, I. M. (1958). A new estimate for $\zeta(1 + it)$. Izvestiya Akademii Nauk SSSR, Ser. Mat., 22, 161–164.

[1091] Volchkov, V. V. (1995). On an equality equivalent to the Riemann hypothesis. Ukrainian Math. J., 47, 491–493.

[1092] Voronin, S. M. (1975). Theorem on the 'universality' of the Riemann zeta-function (Russian). Izv. Akad. Nauk SSSR Ser. Mat. 39, 475–486. English translation in Math. USSR Isvestija 9 (1975), 443–445.

[1093] Voronoi, G. (1903). Sur un problème du calcul des fonctions asymptotiques. J. Reine Angew. Math., 126, 241–282.

[1094] Voros, A. (2006). Sharpenings of Li's criteion of the Riemann hypothesis. Mathematical Physics, Analysis, and Geometry, 9, 53–63.

[1095] Wagstaff, S. S. (2013). The Joy of Factoring. Amer. Math. Soc., Providence, RI.

[1096] Walfisz, A. (1963). Weylsche Exponentialsummen in der neueren Zahlentheorie. Mathematische Forschungsberichte, VEB Deutscher Verlag der Wissenschaften, Berlin.

[1097] Walsh, M. N. (2025). Stability under scaling in the local phases of multiplicative functions. Inventiones Mathematicase, 241, 325–362.

[1098] Weingartner, A. (2007). The distribution functions of $\sigma(n)/n$ and $n/\varphi(n)$. Proc. Amer. Math. Soc., 135, 2677–2681.

[1099] Westzynthius, E. (1931). Über die Verteilung der Zahlen, die zu der n ersten Primzahlen teilerfremd sind. Comment. Phys.-Math. Soc. Sci. Fennica, Helsingfors, 25, 1–37.

[1100] Whittaker, E. T. and Watson, G. N. (1927). A course of Modern Analysis, 4th ed. Cambridge Univ. Press. (Reprinted, 1962.)

[1101] Widder, D. V. (1946). The Laplace Transform. Princeton Univ. Press.

[1102] Widder, D. V. (1971). An Introduction to Transform Theory. Academic Press, NY.

[1103] Wieand, K. (2000). Eigenvalue distributions of random permutation matrices. Ann. Probab., 28, 1563–1587.

[1104] Wiener, N. (1932). Tauberian theorems. Ann. Math., 33, 1–100 & 787.

[1105] Wiener, N. and Wintner, A. (1957). Notes on Pólya's and Turán's hypotheses concerning Liouville's factor. Rend. Circ. Mat. Palermo, 6, 240–248.

[1106] Wigner, E. (1958). On the distribution of the roots of certain symmetric matrices, Ann. Math., 67, 325–327.

[1107] Wilf, H. S. (1987). A greeting: And a view of Riemann's hypothesis. Amer. Math. Monthly, 94, 3–6.

[1108] Wilkins, D. R. (2008). Translation of Riemann's paper. In: P. Borwiein et al., The Riemann Hypothesis: A Resource for the Officionado and Virtuoso Alike, Springer, 190–198.

[1109] Wilkinson, A. (2015). The pursuit of beauty: Yitang Zhang solves a pure-math mystery. NYer Magazine, Profiles, February 2, 2015.

[1110] Williams, D. (1974). Brownian motions and diffusions as Markov processes. Bull. London Math. Soc., 6. 257–303.

[1111] Williams, D. (1990). Brownian motion and the Riemann zeta function. In: Disorder in Physical Systems (in Honour of J. M. Hammersley), G. R. Grimmett & D. J. A. Welsh, eds., Oxford Univ. Press, 361–372.

[1112] Williams, H. C. (1984). An overview of factoring. In: Advances in Cryptology. D. Chaum, ed. 71–80. Plenum, NY.

[1113] Williams, H. C. (1984). Factoring on a computer. Math. Intelligencer, 6, 29–36.

[1114] Wintner, A. (1935) On the asymptotic distribution of the remainder term of the prime number theorem. Amer. J. Math., 57, 534–538.

[1115] Wintner, A. (1944). Random factorizations and Riemann's hypothesis. Duke Math. J., 11, 267–275.

[1116] Wirsing, E. A. (1967). Das asymptotische verhalten von Summen über multiplikative Funktionen, II. (In German) Acta Math. Acad. Sci. Hungarica, 18, 411–467.

[1117] Wu, J. (1990). Sur la suite des nombres premiers jumeaux. Acta Arith., 55, 365–394.

[1118] Wu, J. (2004). Chen's double sieve, Goldbach's conjecture and the twin prime problem. Acta Arith., 114, 215–273.

[1119] Wunderlich, M. (1974). A probabilistic setting for prime number theory. Acta Arith., 26, 59–81.

[1120] Wunderlich, M. (1976). The prime number theorem for random sequences. J. Number Theory, 8, 369–371.

[1121] Yakubovich, S. (2014). Integral and series transformations via Ramanujan's identities and Salem's type equivalences to the Riemann hypothesis. Integral Transforms and Special Functions, 25, 255–271.

[1122] Zagier, D. (1977). The first 50 million prime numbers. Math. Intelligencer, 1, 7–19.

[1123] Zagier, D. (1997). Newman's short proof of the prime number theorem. Amer. Math. Monthly, 104, 705–708.

[1124] Zhang, Y. (2014). Bounded gaps between primes. Ann. Math., 179, 1121–1174.

[1125] Zipf, G. K. (1949). Human Behaviour and the Principle of Least Effort. Addison-Wesley, Cambridge, MA.

Index

Kolibri 360 GmbH
Ferdinand-Jühlke-Straße 7
99095 Erfurt, Deutschland
produktsicherheit@kolibri360.de